AF775824

Classical Field Theory and the Stress–Energy Tensor (Second Edition)

Classical Field Theory and the Stress–Energy Tensor (Second Edition)

Mark S Swanson
Emeritus Professor of Physics, University of Connecticut, USA

IOP Publishing, Bristol, UK

© IOP Publishing Ltd 2022

All rights reserved. No part of this publication may be reproduced, stored in a retrieval system or transmitted in any form or by any means, electronic, mechanical, photocopying, recording or otherwise, without the prior permission of the publisher, or as expressly permitted by law or under terms agreed with the appropriate rights organization. Multiple copying is permitted in accordance with the terms of licences issued by the Copyright Licensing Agency, the Copyright Clearance Centre and other reproduction rights organizations.

Permission to make use of IOP Publishing content other than as set out above may be sought at permissions@ioppublishing.org.

Mark S Swanson has asserted his right to be identified as the author of this work in accordance with sections 77 and 78 of the Copyright, Designs and Patents Act 1988.

ISBN 978-0-7503-3455-6 (ebook)
ISBN 978-0-7503-3453-2 (print)
ISBN 978-0-7503-3456-3 (myPrint)
ISBN 978-0-7503-3454-9 (mobi)

DOI 10.1088/978-0-7503-3455-6

Version: 20220401

IOP ebooks

British Library Cataloguing-in-Publication Data: A catalogue record for this book is available from the British Library.

Published by IOP Publishing, wholly owned by The Institute of Physics, London

IOP Publishing, Temple Circus, Temple Way, Bristol, BS1 6HG, UK

US Office: IOP Publishing, Inc., 190 North Independence Mall West, Suite 601, Philadelphia, PA 19106, USA

This book is dedicated to all my family members, friends, colleagues, and mentors who made it possible. Your encouragement and patience were a gift beyond measure.

Contents

Preface		**xiii**
Author biography		**xvi**
Introduction		**xvii**
1	**Geometry and physics**	**1-1**
1.1	Manifolds	1-1
1.2	Coordinate systems	1-3
1.3	The Jacobian	1-5
1.4	Contravariant and covariant quantities	1-8
1.5	The summation convention	1-9
1.6	Vectors and direction vectors	1-11
1.7	Vector addition and the scalar product	1-18
1.8	The metric tensor and distance in manifolds	1-21
1.9	The metric tensor and raising and lowering indices	1-22
1.10	General tensors and tensor densities	1-27
1.11	Trajectories and tangent spaces	1-29
1.12	The vector product	1-32
1.13	The gradient	1-34
1.14	The divergence, the Laplacian, and the curl	1-38
1.15	Differential forms and the wedge product	1-41
1.16	Differential forms and Stokes' theorem	1-48
1.17	The Lie derivative	1-50
2	**Newtonian mechanics and functional methods**	**2-1**
2.1	Newton's second law	2-1
2.2	Newtonian trajectories and tangent vectors	2-3
2.3	Newton's first law and Galilean relativity	2-4
2.4	Functionals and the calculus of variations	2-10
	2.4.1 The Dirac delta	2-12
	2.4.2 The functional derivative and differential	2-15
	2.4.3 Complex vectors, Fourier transforms, and the Dirac delta	2-18
2.5	The action approach to Newtonian mechanics	2-22

3 Basic field theory **3-1**

3.1 The mechanical properties of a stretched string 3-1
3.2 The stretched string as a field theory 3-3
3.3 The Euler–Lagrange equation for the stretched string 3-5
3.4 Solving the Euler–Lagrange equation 3-7
3.5 Galilean relativity and wave solutions 3-13
3.6 Momentum and energy in field theories 3-15
3.7 The stress–energy tensor 3-20
3.8 Static sources and Green's function techniques 3-24
3.9 The catenary, the Beltrami identity, and constraints 3-28
3.10 Functional derivatives and Poisson brackets 3-33

4 Newtonian fluid dynamics **4-1**

4.1 Fluid flow from Newtonian physics 4-2
4.2 The equation of continuity 4-7
4.3 Viscosity 4-10
4.4 The Navier–Stokes equation and the stress–energy tensor 4-15
4.5 Basic solutions to the Navier–Stokes equation 4-17
4.5.1 The pressure–depth relation for a static fluid 4-18
4.5.2 The Bernoulli relation 4-19
4.5.3 The velocity vortex 4-21
4.5.4 Vorticity transport and stability 4-24
4.5.5 Viscous flow rate 4-26
4.5.6 The Reynolds number 4-28
4.6 Homentropic flow 4-30
4.7 The action formulation for homentropic flow 4-33
4.8 The homentropic stress–energy tensor 4-36
4.9 The symmetric fluid stress–energy tensor 4-39
4.10 Fluctuations around solutions and stability 4-41
4.11 Spherical sound waves, power, and the Doppler effect 4-47

5 Galilean covariant complex fields **5-1**

5.1 The complex classical nonrelativistic field 5-2
5.2 The Euler–Lagrange equation and its solutions 5-4
5.3 Symmetries of the Lagrangian 5-7
5.4 Galilean covariance 5-10

5.5 Complex analysis and Cauchy's theorem 5-13
5.6 Scattering and the Dirac delta potential 5-17
5.7 Bose–Einstein condensation 5-20
5.8 Condensate fluctuations 5-23
5.9 Vortices and the healing length 5-26

6 Basic special relativity 6-1

6.1 Maxwell's equations 6-1
6.2 The problem with electromagnetic waves 6-3
6.3 Lorentz transformations 6-7
6.4 Observational effects of special relativity 6-11
6.4.1 Simultaneity 6-11
6.4.2 Length contraction 6-11
6.4.3 Time dilation 6-12
6.4.4 The relativistic Doppler effect 6-13
6.4.5 Velocity addition and the concept of space-time 6-15
6.5 The Minkowski metric and space-time 6-17
6.6 Relativistic energy and momentum 6-24
6.7 Proper velocity and accelerated motion 6-28
6.8 Relativistic action in the presence of force 6-36
6.8.1 Conservation of relativistic energy 6-43
6.8.2 One-dimensional constant-force motion 6-44
6.8.3 Relativistic gravitational infall 6-47
6.8.4 Relativistic centripetal force and circular orbits 6-49
6.9 Relativistic quantities 6-50

7 Linear algebra and group theory 7-1

7.1 Linear algebra and matrices 7-1
7.2 Basic group theory 7-14
7.2.1 Spatial rotations and continuous groups 7-14
7.2.2 Generators and the exponential map 7-17
7.2.3 Lie algebras and structure constants 7-20
7.2.4 **SU**(n) and spinor rotations 7-23
7.3 SO (3,1) and the Lorentz group 7-29
7.4 Spinor representations of the Lorentz group 7-36
7.4.1 The Weyl spinor 7-36
7.4.2 The Dirac spinor 7-42

8 Scalar and spinor field theories **8-1**

8.1 Classical point particles 8-1
8.2 Lorentz invariant actions 8-4
8.3 Relativistic scalar field theory 8-6
8.3.1 The real Klein–Gordon field 8-6
8.3.2 Solving the Klein–Gordon equation 8-8
8.3.3 Energy and momentum in the Klein–Gordon field 8-13
8.3.4 The Klein–Gordon field and point particle interactions 8-16
8.4 Classical scalar solutions and broken symmetry 8-23
8.5 Relativistic spinor fields and quadratic actions 8-27
8.5.1 The Weyl spinor and its solutions 8-27
8.5.2 The Dirac bispinor and its solutions 8-33
8.6 Symmetry and conservation laws 8-37
8.6.1 Noether's theorem 8-38
8.6.2 Matter currents and the Weyl spinor 8-42
8.6.3 Poincaré invariance and the stress–energy tensor 8-43
8.6.4 Angular momentum 8-45

9 Classical relativistic electrodynamics **9-1**

9.1 Aspects of Maxwell's equations 9-1
9.2 The Helmholtz decomposition and the Coulomb potential 9-2
9.3 The field strength tensor 9-5
9.4 Electromagnetic fields and the gauge field 9-9
9.5 Gauge transformations and gauge conditions 9-11
9.6 Natural units 9-13
9.7 The gauge field action and minimal coupling 9-15
9.8 Relativistic point charges and electromagnetic interactions 9-19
9.9 The stress–energy tensor and electrodynamics 9-23
9.10 Angular momentum for gauge and spinor fields 9-29
9.10.1 Gauge field angular momentum 9-30
9.10.2 Spinor field angular momentum 9-33
9.11 Electromagnetic waves and spin 9-36
9.12 The Proca field 9-45
9.13 Green's functions and electromagnetic radiation 9-48
9.14 The gauge field as a differential form 9-62
9.15 Magnetic monopoles 9-64

10 General relativity and gravitation **10-1**

10.1 The metric tensor and Einstein's principle of equivalence 10-3
10.2 The affine connection and the covariant derivative 10-14
10.2.1 Parallel transport of vectors and the covariant derivative 10-15
10.2.2 Torsion 10-22
10.2.3 Metric compatibility and the Levi-Civita connection 10-24
10.2.4 The affine connection and the tetrad formalism 10-26
10.2.5 Examples: the Euclidean plane and the sphere 10-30
10.3 The curvature tensor 10-36
10.3.1 Defining the curvature tensor 10-36
10.3.2 Properties of the curvature tensor 10-38
10.3.3 Two-dimensional examples of curvature 10-40
10.3.4 Tensors derived from the curvature 10-41
10.4 The connection and curvature in differential geometry 10-43
10.5 Variational techniques in general relativity 10-48
10.5.1 General variational properties 10-49
10.5.2 Point-mass motion in curved space-time 10-52
10.5.3 Diffeomorphic invariant field theory 10-54
10.6 The generalized stress–energy tensor 10-55
10.7 Einstein's field equation 10-60
10.7.1 The variational approach 10-60
10.7.2 The Newtonian limit 10-63
10.7.3 Gravitational time dilation and Doppler shift 10-67
10.8 Vacuum solutions to Einstein's equation 10-69
10.8.1 Gravitational waves 10-69
10.8.2 The Schwarzschild metric 10-73
10.8.3 Killing vectors and motion 10-75
10.8.4 Gravitational infall for the Schwarzschild metric 10-79
10.9 Kaluza–Klein theory 10-81
10.10 Basic cosmology 10-85
10.10.1 The cosmological principle 10-85
10.10.2 The Friedmann–Robertson–Walker metric and Hubble's law 10-89
10.10.3 The Big Bang model 10-92

11 Yang–Mills fields and connections **11-1**

11.1 Unitary symmetry and isospin 11-1
11.2 Nonabelian gauge fields 11-4

11.3 The Yang–Mills stress–energy tensor and force equation 11-11
11.4 Spontaneous breakdown of symmetry 11-12
11.5 Aspects of classical solutions for Yang–Mills fields 11-17
11.5.1 The Gribov ambiguity 11-19
11.5.2 The Wu–Yang monopole 11-22
11.6 Yang–Mills fields, forms, and connections 11-23
11.7 Spinor fields in general relativity 11-29
11.8 Yang–Mills fields and the Gribov instability 11-33
11.9 Classical string theory 11-44
11.9.1 The string action 11-44
11.9.2 The string equation of motion 11-47
11.9.3 D-branes 11-51

Preface

While it is fair to say that the last century of physics has been dominated both experimentally and theoretically by the study of quantum processes, the framework for the understanding of these phenomena consists of concepts developed in classical physics and extended to quantum mechanical systems. In that regard, the necessity of generalizing the concepts of energy, momentum, and angular momentum to the case of *classical fields* provided many of the tools used in the wave-particle duality of quantum mechanics. Classical electrodynamics and nonabelian gauge theories have become successful quantum field theories, yielding deep insights into quantum processes. At the time of this writing, developing a quantum mechanically consistent version of gravitation is a vigorous and unsettled area of research, yet the insights provided by Einstein's classical theory of general relativity remain the basis for much of the effort to create such a theory.

The purpose of this monograph is to present a reasonably concise introduction to the key concepts of classical field theory. Its intended audiences are beginning graduate students and advanced undergraduate students who wish to study the unifying structures and physical insights provided by classical field theory without dealing with the additional complication of quantization. In that regard, there are many important aspects of field theory that can be understood without quantizing the fields. These include the action formulation, Galilean and relativistic invariance, traveling and standing waves, spin angular momentum, gauge invariance, subsidiary conditions, fluctuations, spinor and vector fields, conservation laws and symmetries, and the Higgs mechanism, all of which are often treated briefly in a course on quantum field theory. In writing such a monograph, there are many decisions to be made about the level of presentation and the choice of topics and examples. I have chosen to assume the reader has a good working knowledge of undergraduate Newtonian mechanics and electricity and magnetism. This includes the Newtonian formulation of mechanics, the formulation of Maxwell's equations in terms of a vector and scalar potential, and a brush with special relativity and its relationship to these two areas. It assumes no background in continuum mechanics and fluid dynamics, spinor fields, general relativity, nonabelian gauge theories, differential geometry, or group theory. The intent is to connect these diverse areas together in terms of variational principles, conservation laws and symmetries, coordinate transformations, and the insights given by the stress–energy tensor. In so doing, pedagogical goals have superseded a rigorous mathematical presentation. This stems from my belief that presenting these connections succinctly with clear and reasonably basic examples will extend the reader's understanding much more quickly and effectively than pointing out the sometimes bewildering subtleties that require a more rigorous approach. However, one topic that is present throughout the monograph is the nature of wave disturbances in the classical fields, and it serves to both unite and differentiate the various field theories treated.

It is the intention of this monograph to assist the reader in gaining a basic foundation in the structures and concepts commonly used in classical field theory.

The references and recommended reading listed at the end of each chapter have been chosen from books and research papers I have found useful for further study of the details and applications presented in the chapter. It is my sincere hope that the reader will use the recommended texts for further study of the myriad details and applications that classical fields possess, as well as eventually proceeding to the study of quantum field theory. Because this monograph is primarily pedagogical in nature and designed to cover a wide variety of topics, most of the original research papers in classical field theory are, with a few exceptions, not listed. I sincerely apologize to the giants of physics, both living and dead, whose research, insights, and hard work laid the foundations of field theory, but a proper list of their papers would be extremely long and largely irrelevant to the intended goal of the monograph. In that regard, some concepts presented in this text have become so widely used that they are known simply by the name of their originator, and this form of attribution is followed whenever possible.

The second edition of this monograph has given me the opportunity to more clearly develop and integrate the mathematical tools useful in analyzing classical field theory and to weed out the errors of the first edition. It is assumed the reader has a working knowledge of multivariable calculus and some experience solving simple differential equations, along with some acquaintance with matrix algebra. The book presents an overview of those mathematical ideas and techniques used in field theory that may be unfamiliar to the intended reader, when and where they are essential to the development of the physical concepts. The revised version begins with an introduction to the concepts of differential geometry, since it undergirds classical field theory and especially general relativity. It then develops variational techniques and their application to classical physics, along with a brief overview of Fourier series and transforms, the Dirac delta, and their relationship to solving differential equations. Field theories consistent with Newtonian mechanics and Galilean relativity are derived and used to develop basic concepts present in all field theories. These include the continuum mechanics of the stretched string and fluid motion. Complex variable techniques are introduced in the context of complex fields that can be used to describe classical aspects of the Bose–Einstein condensate, as well as serving as the basis of quantum wave mechanics. A review of linear algebra and an introduction to Lie group theory present the mathematical tools to develop classical relativistic field theory. An advanced reader may wish to skip the review sections covering these topics and use the monograph as a reference for these ideas, while the introductory reader can obtain an overview of these essential areas of mathematics, identifying those topics that may require further study. These concepts are used to develop the representations of the Lorentz group that provide the types of relativistic fields. These include scalar fields, spinor fields, the vector field of electrodynamics, and the metric tensor version of gravitation, including both the theory and selected applications. A final chapter presents the classical aspects of nonabelian gauge theories and the briefest of introductions to string theory. All of these are presented without mathematical rigor and aimed at the level of understanding appropriate to the intended audience.

There are several standard notational conventions used throughout this text. Wherever possible, the summation convention is employed, so that repeated indices indicate a sum over those indices that is appropriate to the range of indices. When applied to spatial coordinates, capital Latin letters indicate a range of two, lower-case Latin letters indicate a range of three, and lower-case Greek letters indicate a range of four, so that

$$x^j x_j = \sum_{j=1}^{3} x^j x_j, \tag{0.1}$$

while

$$x^\mu x_\mu = \sum_{\mu=0}^{3} x^\mu x_\mu. \tag{0.2}$$

When a repeated index is *not* to be summed, it is indicated with parentheses, so that $\lambda^{(a)}\mathbf{e}^a$ is *not* summed over the index a; instead, $\lambda^{(a)}$ is simply a multiplicative factor associated with the index a. Spatial vectors are indicated in bold font rather than with a vector sign, so that $\boldsymbol{x}$ typically indicates a three-dimensional vector, while x or $|\boldsymbol{x}|$ represents the magnitude of the vector $\boldsymbol{x}$, and x^j is the jth Cartesian component of the vector $\boldsymbol{x}$. In a relativistic context, this book uses the signature of $\eta^\nu{}_\nu = -2$ for the Minkowski metric, so that $\eta^{00} = 1$, while $\eta^{ij} = -1$ if $i = j$ and is zero otherwise. Throughout the text, several symbols are used as shorthand for a phrase. These include '$\equiv$' for 'is defined as,' '$\Longrightarrow$' for 'implies that,' '$\Longleftrightarrow$' for 'are equivalent to each other,' '$\in$' for 'is contained in,' '$\longrightarrow$' for 'becomes,' '$\cdots$' for 'all the product terms in between,' and '$\ldots$' for 'all the subscripts in between.'

I would like to thank John Navas and the staff at IOP for the opportunity to develop this second edition.

Author biography

Mark S Swanson

Photo courtesy of Nancy Ruth Levine.

Mark Swanson received his PhD in physics from the University of Missouri at Columbia in 1976 under the supervision of Justin Huang. After postdoctoral appointments at the University of Alberta and the University of Connecticut, he held a faculty appointment at the University of Connecticut at Stamford from 1983 to 2014, during which time he spent six years in university administrative roles. He is the author of 25 research articles and three other monographs with an emphasis on field theory and path integral techniques. He is currently Emeritus Professor of Physics at the University of Connecticut.

Introduction

The term *field* has its origins in characterizing the action of forces between spatially separated material objects. It was first used by Faraday in the context of electricity and magnetism, although the earlier work of Euler describing fluid flow as well as the description of the propagation of waves through a medium used similar concepts. In order to be the agent of *action at a distance* between material objects, a field must exist throughout a volume of space and evolve over time. Because a field can be present throughout a volume of space, it is very different from the Newtonian concept of a material *point particle*, which always has a definite position at a given time. Newton's laws of motion describe how a particle responds to forces and how it reacts back on the agents of those forces. Newton's laws and their subsequent formulation in terms of a variational principle give rise to the generalized concepts of energy and momentum as well as conservation laws. However, Newton's laws of motion do not predict or explain what forces exist in nature, how they are transmitted, or if they are fundamental or the secondary result of other more fundamental forces. Answering these questions requires the interaction of experiment and theory. In that process, it became apparent to 19th-century physicists that the field concept was central to the mathematical description of electric and magnetic phenomena. This culminated in the development of Maxwell's equations, a cornerstone of modern physics as well as a guidepost to understanding the nature of fundamental forces.

Because physical phenomena do not come with a coordinate system or clock, observers must choose a coordinate system to describe the location and a clock to record the time at which their observations occur. Once a coordinate system for spatial position $\boldsymbol{r}$ and time t is chosen, a field is represented by a mathematical function of $\boldsymbol{r}$ and t. At the foundation of physics is the axiom that the choice of coordinate system cannot affect the physical laws of nature, although the choice can clearly affect how observations of those physical laws are reported. At the inception of modern physics it was also recognized that relative motion plays a profound role in the observation of physical phenomena. The mathematical quantity that represents the field must therefore have a well-defined set of rules for how it changes or *transforms* when a different coordinate system is chosen *or* when relative motion between different observers is present. These rules allow observations quantified by different observers to be compared or *related to each other*. The set of rules for comparing observations in different coordinate systems or coordinate systems in relative motion therefore goes by the generic term *relativity*. These requirements need considerable further explanation, but eventually lead to the identification of the possible types of mathematical quantities that can be used to represent fields, as well as the imposition of restrictions on the types of field theories that can be formulated.

In the process of developing *field theory* it will be seen that a field contains energy, momentum, and angular momentum that is distributed throughout a volume of space. In addition, a field may serve as an agent that transmits force by interacting with material objects, and in so doing may transfer energy, momentum, and angular momentum to the object. It is therefore necessary to generalize the Newtonian concepts of energy and momentum to include the contributions from both fields and particulate matter. It will be seen that each field is associated with a contribution to what is commonly called the *stress–energy tensor*, and it is the stress–energy tensor which quantifies the energy and momentum present in the field. It is often the case that a field theory predicts the existence and nature of *waves* associated with the field. Waves are the manifestation of energy and momentum moving through a volume of space in the form of the field. In the case of electricity and magnetism, the wave was identified as light, giving rise to physical optics. It was the experimental study of light propagation that revealed a flaw in Newtonian physics. In order to remedy this flaw, it was necessary to allow the relative motion between two observers to affect the way in which those two observers measure distances and time intervals, leading to the *special theory of relativity*. The subsequent generalization to observers in arbitrary relative motion led to the *general theory of relativity*, in which gravitation is the result of the interaction of space and time with mass and energy, with the latter quantified by the stress–energy tensor.

It is often said that classical physics would govern a world where Planck's constant $\hbar$ is zero. Strictly speaking, this is true, since the quantum mechanical deviations from Newtonian physics and its relativistic version can be characterized in terms of powers of $\hbar$. Maxwell's equations and Einstein's general theory of relativity have no explicit reference to $\hbar$ and can therefore be analyzed with no reference to $\hbar$. However, ignoring field theories with an explicit dependence on $\hbar$ in their constants overlooks the contribution that the techniques of classical field theory can provide for such theories. Classical field techniques give an important perspective on the energy and momentum content of such theories, as well as the nature and meaning of the conservation laws these theories obey. As a result, the techniques and results of classical field theory can often shed light on a field theory that requires $\hbar$ in its definition. An example of this will be found in chapter 5, where a classical version of Bose–Einstein condensation is examined by treating the nonlinear Schrödinger equation as a classical field theory.

The first step in formulating a field theory is to choose the mathematical quantities that will represent the field. These are chosen from the set of possible mathematical functions whose transformation properties allow different observers to relate their observations. The second step is to determine how these quantities can reflect the physical phenomena and interactions being observed. Because fields are functions of space and time, the process of cataloging the possible types of fields first requires an understanding of the geometric properties of coordinate systems and their relation to the space the coordinate system is describing. Subsequently, the relationship between different coordinate systems will determine the mathematical quantities that can represent the fields present in the space. It is also critical to understand how physical laws affect the relationship between different observers, in particular observers in relative motion, and how this also determines the types of fields that are consistent with physical laws. In addition, classical field theory must have a clear relation to Newtonian physics and its modifications for consistency with special relativity and general relativity. In that regard, the mathematical foundation for field theory is multivariable calculus and its extensions into differential geometry, partial differential equations, and group theory. In an effort to be complete, this text will attempt to provide the reader with a physics perspective on these relevant mathematical topics. In the process it is hoped that the reader will gain an appreciation of the remarkable intellectual achievement that classical field theory represents, as well as understanding how it forms the foundation of quantum physics.

IOP Publishing

Classical Field Theory and the Stress–Energy Tensor (Second Edition)

Mark S Swanson

Chapter 1

Geometry and physics

Properly formulating classical field theory requires some basic concepts from differential geometry. This book will develop these ideas and techniques specifically for application to physical systems, and so the presentation is neither complete nor rigorous. Instead, the intent is to place physical concepts familiar from basic physics into a more unified context. In particular, coordinate systems, coordinate transformations, vectors, the metric tensor, the gradient, the Laplacian, and differential forms are critical ingredients in formulating classical field theory and also form the foundation of basic differential geometry. These ideas and constructs will be used throughout this book and so familiarity with these tools is important.

1.1 Manifolds

Newtonian physics was formulated to describe the phenomenon of motion or *dynamics*. The *classical dynamics* of Newton makes a number of assumptions, often taken for granted in the presentation of elementary physics, regarding the properties of the space in which motion occurs. The most basic assumption of Newtonian physics is that the space available for motion is a *smooth* collection of *points* and that, as a point particle moves in that space, its position is a continuous or smooth *trajectory* through those points. It is therefore possible to install a coordinate system, at least *locally*, that allows the definition of the calculus operations of differentiation and integration. Loosely speaking, the term *local* means that there is a neighborhood of each point where this is possible. The installation of a coordinate system is understood to be a local mapping of the points in the space into the coordinates of the point or a function that meets the same objective. There may be overlapping neighborhoods with different coordinate systems, and where they overlap these coordinate mappings must be invertible and one-to-one, which requires their differentiability. A collection of points is referred to mathematically as a *manifold*,

doi:10.1088/978-0-7503-3455-6ch1

© IOP Publishing Ltd 2022

generically designated $\mathcal{M}$, and so such a smooth collection of points is known as a *differentiable manifold.* This brief description leaves out many subtleties, and the reader interested in a more mathematically rigorous development of manifolds is referred to one of the many excellent books on modern analysis.

Classical physics typically assumes that the spatial manifold where motion occurs possesses three independent dimensions or *directions.* Although motion may be restricted to a two-dimensional plane or a one-dimensional line, in general, the motion of a Newtonian particle is a three-dimensional phenomenon. Modern physics has relaxed this assumption to allow the consideration of physical manifolds that are n-dimensional. Stated simply, this means there are n independent directions available for motion. This step was required to accommodate the concept of four-dimensional space-time, but has also emerged in the context of higher-dimensional attempts to unify forces, such as Kaluza–Klein theory and string theory. The physics of Newton and contemporaries further assumed that both space and time, equivalently distance and duration, were absolute properties of the world that all observers could agree upon. A significant offshoot of that worldview was the assumption that three-dimensional Euclidean geometry correctly describes the relationships between points in the spatial manifold of motion. A manifold where Euclid's axioms and theorems hold will be referred to as *Euclidean.* It was understood that Euclidean geometry did not hold on a curved surface, such as the surface of the Earth, and so Euclidean manifolds are often referred to as *flat manifolds.* The n-dimensional extension of the Euclidean plane is known mathematically as $\mathbb{R}^n$, where n copies of the real line $\mathbb{R}$ serve as the Euclidean degrees of freedom.

However, experimental observations and the subsequent development of the special and general theories of relativity forced a revision of both these assumptions. In particular, the special theory of relativity showed that the measurements of distance and time intervals by different observers are affected by their relative motion, and that correctly describing motion required merging space and time into the four-dimensional manifold referred to as *space-time.* Further, the principle of general relativity incorporates a description of gravitation in terms of a space-time that is *dynamic and non-Euclidean,* so that distances as well as durations are no longer governed by Euclidean geometry. These changes required a significant revision of Newtonian physics as well as a fundamental overhaul of how physical theories are formulated.

An effective presentation of classical field theory must therefore take into account the changes in assumptions regarding the manifold for motion and fields that have occurred over the last century. Although space, together with time, are still assumed to define a differentiable manifold, understanding both the arbitrary dimensionality and the deviation from Euclidean behavior benefit greatly from adopting the more flexible techniques of modern differential geometry. Some of these techniques may be unfamiliar, and so this chapter will present a simplified introduction to the more unified approach to manifolds, coordinates, and vector spaces that basic differential geometry provides. The standard Euclidean treatment of manifolds and vectors is still contained in this approach, but it allows the generalizations required by modern classical field theory.

1.2 Coordinate systems

Quantifying position and motion in a differentiable manifold requires a choice of coordinate system. This consists of choosing an *origin point* and mapping the points of the manifold onto a *coordinate grid* around the origin. Figure 1.1 depicts the same point in two commonly used coordinate systems for the two-dimensional Euclidean plane. The choice of coordinate system and origin is arbitrary, and it is stressed that coordinate systems themselves have no physical meaning. However, once a coordinate system is chosen, the coordinates of a point in the space around the origin are determined by its place on the coordinate grid in relation to the origin.

In two dimensions *Cartesian coordinates* consist of two straight perpendicular lines or *coordinate axes* through the origin and a coordinate grid consisting of straight lines parallel to these two axes. The Cartesian coordinates (x, y) of a point are defined by *parallel projection*, and the two gridlines that determine the position of a point form a *parallelogram* with the coordinate axes. The point of intersection of the line with the axis gives the coordinate of the point. For the choice of two-dimensional *polar coordinates* the grids consist of circles of radius r around the origin coupled with radial lines at each angle θ. The polar coordinates of the point are then given by (r, θ). In order for the two coordinate systems to share the same neighborhood of the manifold, they must be related by differentiable and invertible functions. If the plane is Euclidean, these relations are given by the familiar Euclidean expressions,

$$x = r\cos\theta, \quad y = r\sin\theta \iff r = \sqrt{x^2 + y^2}, \quad \theta = \tan^{-1}(y/x). \tag{1.1}$$

However, depending on the manifold or the choice of coordinate system, the process of installing a coordinate system can become subtle. For example, it is possible to choose coordinate axes that are not perpendicular as well as defining coordinates by *perpendicular projection*. In addition, if the manifold is non-Euclidean, the expressions of (1.1) require modification.

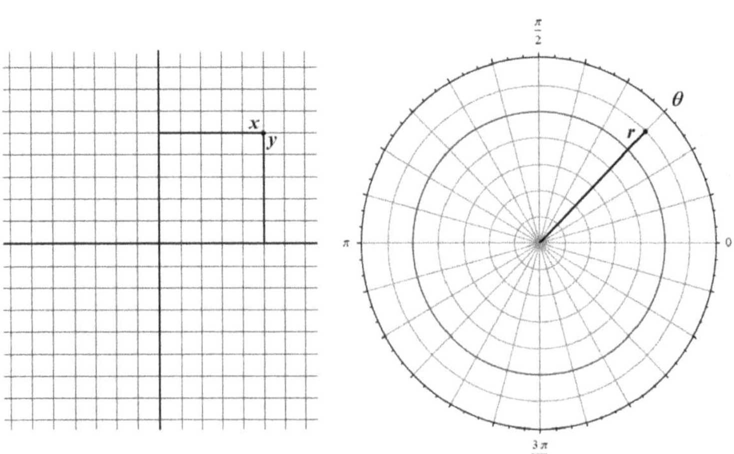

Figure 1.1. Cartesian and polar coordinate grids.

The upshot of choosing a coordinate system in an n-dimensional manifold is to assign or *map* each point p into a set of n coordinates, denoted x^j, where the superscript j can take on any value from 1 to n. The coordinates of each point are assumed to be unique and unambiguous. Two important quantities are derived from the coordinates of a point, regardless of the choice of coordinate system or whether the manifold is Euclidean or not. The first is the *differential* $\mathrm{d}x^j$, which is an infinitesimal change in the jth coordinate of a point. The second is the *partial derivative* with respect to x^j, denoted $\partial_j \equiv \partial/\partial x^j$, which quantifies the rate of change along the jth coordinate axis for a multivariable function of the coordinates of the point, $f(x^1, \ldots, x^n)$. These two quantities allow the operations of calculus to be defined and are clearly dependent upon the specific choice of coordinate system for their definition.

Since coordinate systems are chosen by an observer, another observer may choose a different coordinate system, and this is referred to as a *coordinate transformation.* Doing so changes the coordinates of a point to a different set of values, denoted y^j, and changes the differentials to $\mathrm{d}y^j$ and the partial derivatives to $\partial/\partial y^j$. However, the change of coordinates does *not* change the relationship of each point to the points around it, and so it is referred to as a *passive transformation.* The ratio of distances remains the same in both coordinate systems, as do the angles in geometric shapes such as triangles. In order to compare the results of an observation or evaluate a mathematical expression, it is critical to be able to relate the quantities measured or defined at a point in either coordinate system. It is therefore assumed that the y-coordinates can be expressed as *invertible and differentiable functions* of the x-coordinates, so that $y^j = y^j(x^1, \ldots, x^n) \equiv y^j(x)$ gives the y-coordinates of the point p whose x-coordinates are x^j. Since these functions are invertible, there is also a set of differentiable functions that yields the x-coordinates $x^j = x^j(y^1, \ldots, y^n) \equiv x^j(y)$ given the y-coordinates. It is assumed that the transformation between the two coordinate systems is both unique and one-to-one, which is referred to as an *isomorphism*, for all the points in the manifold where the two coordinate systems are both defined.

For the moment, it will be assumed that the two coordinate systems share the same point as an origin, so that $y^j(0, \ldots, 0) = 0$. The coordinates must be independent degrees of freedom in both coordinate systems. Stated mathematically, this means that both coordinate systems satisfy

$$\frac{\partial y^j}{\partial y^k} = \frac{\partial x^j}{\partial x^k} = \delta_k^j = \begin{cases} 1 & \text{if } j = k \\ 0 & \text{if } j \neq k \end{cases}, \tag{1.2}$$

where $\delta^j{}_k$ is known as the *Kronecker delta.* Requirement (1.2) simply means that varying one independent degree of freedom creates no variation in the other independent degrees of freedom. The n-dimensional Kronecker delta has the properties that

$$\sum_{k=1}^{n} \delta^j{}_k \, \mathrm{d}x^k = \mathrm{d}x^j, \quad \sum_{j=1}^{n} \delta^j{}_k \frac{\partial}{\partial x^j} = \frac{\partial}{\partial x^k}, \quad \sum_{k=1}^{n} \delta^j{}_k \delta^k{}_\ell = \delta^j{}_\ell. \tag{1.3}$$

These two sets of coordinates must also obey the chain rule, which states that

$$\sum_{k=1}^{n} \frac{\partial y^j(x)}{\partial x^k}\frac{\partial x^k(y)}{\partial y^\ell} = \frac{\partial y^j}{\partial y^\ell} = \delta^j{}_\ell, \quad \sum_{k=1}^{n} \frac{\partial x^j(y)}{\partial y^k}\frac{\partial y^k(x)}{\partial x^\ell} = \frac{\partial x^j}{\partial x^\ell} = \delta^j{}_\ell. \tag{1.4}$$

The proof of (1.4) follows from the definition of differentiable mappings in a manifold and is available in the references for multivariable calculus. Two isomorphic coordinate systems satisfying (1.2) and (1.4) are said to be *diffeomorphic*.

Exercise 1.1. Show explicitly that the Cartesian to polar coordinate transformation (1.1) satisfies

$$\frac{\partial r}{\partial x}\frac{\partial x}{\partial \theta} + \frac{\partial r}{\partial y}\frac{\partial y}{\partial \theta} = 0, \qquad \frac{\partial r}{\partial x}\frac{\partial x}{\partial r} + \frac{\partial r}{\partial y}\frac{\partial y}{\partial r} = 1.$$

1.3 The Jacobian

A coordinate system may result in ambiguous or undefined coordinates for some points. An example occurs at the origin in polar coordinates, where the polar angle is arbitrary and therefore undefined. For the sake of brevity this book will only peripherally touch on such subtleties and instead assume that the coordinate system provides unambiguous and unique coordinates for all but a few points. In that regard, the coordinate transformation is invertible at all points where the *Jacobian of the coordinate transformation* is nonzero.

Defining the Jacobian, as well as many other important quantities in matrix theory, group theory, and differential geometry, requires the *Levi-Civita symbol*, which is denoted $\varepsilon^{abc\ldots}$. The Levi-Civita symbol can have an arbitrary number of superscripts, but once the number of subscripts is specified, the range of the subscripts must match that number. For example, the Levi-Civita symbol ε^{ab} has two superscripts, each of which can take the values of one or two. The Levi-Civita symbol ε^{ijk} has three subscripts, each of which can take the values of one, two, or three. The Levi-Civita symbol $\varepsilon^{\mu\nu\rho\sigma}$ has four subscripts, each of which can take the values of zero, one, two, or three when used in space-time physics applications. The Levi-Civita symbol has one of three values, given by

$$\varepsilon^{abc\ldots} = \begin{cases} +1 & \text{if } abc\ldots \text{ is an even permutation of } 1, 2, 3, \ldots \\ -1 & \text{if } abc\ldots \text{ is an odd permutation of } 1, 2, 3, \ldots \\ 0 & \text{if any of the abc} \ldots \text{ are the same} \end{cases}. \tag{1.5}$$

For the case of two superscripts the Levi-Civita symbol ε^{ab} has the values $\varepsilon^{12} = -\varepsilon^{21} = 1$, while $\varepsilon^{11} = \varepsilon^{22} = 0$. For three superscripts, $\varepsilon^{123} = -\varepsilon^{132} = 1$ and $\varepsilon^{112} = \varepsilon^{223} = 0$, and so on. It is worth noting that $\varepsilon^{3ab} = \varepsilon^{ab}$, so that the lower-order versions of the Levi-Civita symbol descend from the higher-order versions. For that reason, the Levi-Civita symbol obeys many identities.

If the Levi-Civita symbol is defined in a Euclidean manifold using Cartesian coordinates, it is possible to define the Levi-Civita symbol with subscripts $\varepsilon_{abc\ldots}$ by simply equating it:

$$\varepsilon_{abc\ldots} = \varepsilon^{abc\ldots}. \tag{1.6}$$

However, if the manifold is non-Euclidean, care must be taken, since the two forms are not simply related. In particular, the space-time manifold that forms the basis of modern field theory requires a more careful analysis to find the correct relationship. This remark will be clarified as further aspects of manifolds are developed.

Exercise 1.2. Using (1.6), demonstrate the two very useful properties of the three-dimensional Levi-Civita symbol in a Euclidean space:

$$\sum_{k=1}^{3} \varepsilon^{k\ell m}\varepsilon_{kij} = \delta^{\ell}{}_{i}\delta^{m}{}_{j} - \delta^{\ell}{}_{j}\delta^{m}{}_{i}, \tag{1.7}$$

$$\sum_{j,k=1}^{3} \varepsilon^{\ell jk}\varepsilon_{ijk} = 2\,\delta^{\ell}{}_{i}. \tag{1.8}$$

It is tedious but easy to show both sides are the same by writing out all possible index combinations.

It is easy to see that

$$\sum_{a,b=1}^{2} \varepsilon^{ab}\varepsilon_{ab} = \varepsilon^{11}\varepsilon_{11} + \varepsilon^{12}\varepsilon_{12} + \varepsilon^{22}\varepsilon_{22} + \varepsilon^{21}\varepsilon_{21} = 2 = 2!, \tag{1.9}$$

while (1.8) gives

$$\sum_{i,j,k=1}^{3} \varepsilon^{ijk}\varepsilon_{ijk} = 6 = 3!. \tag{1.10}$$

Using induction shows that

$$\sum_{i,j,k,\ldots=1}^{n} \varepsilon^{ijk\ldots}\varepsilon_{ijk\ldots} = n!, \tag{1.11}$$

where n is the number of indices. This follows by noting that the n-dimensional case of (1.11) is simply n copies of the $n - 1$- dimensional case, giving $n \cdot (n - 1)! = n!$, completing the inductive proof.

For the case of a coordinate transformation from x-coordinates to y-coordinates in n dimensions, given by the functions $x^j = x^j(y)$ and $y^j = y^j(x)$, the Jacobian J is given by

$$J(y) = \sum_{j_1, j_2, \ldots, j_n = 1}^{n} \varepsilon_{j_1 j_2 \ldots j_n} \frac{\partial x^{j_1}(y)}{\partial y^1} \frac{\partial x^{j_2}(y)}{\partial y^2} \cdots \frac{\partial x^{j_n}(y)}{\partial y^n}. \tag{1.12}$$

The reader may recognize (1.12) as the determinant of the matrix whose elements are $\partial x^j / \partial y^k$. This is discussed in more detail in chapter 7, where linear algebra is reviewed. The properties of the Jacobian are discussed later in this chapter, in particular, where differential forms are presented. Wherever the Jacobian is nonzero, the relationship between the y-coordinates and the x-coordinates is invertible. For example, (1.1) relates the Cartesian coordinates $x^1 = x$ and $x^2 = y$ for the two-dimensional Euclidean plane to the polar coordinates $y^1 = r$ and $y^2 = \theta$ for the same two-dimensional Euclidean plane. The Jacobian for transforming from Cartesian to polar coordinates is given by

$$J(r, \theta) = \sum_{a,b=1}^{2} \varepsilon_{ab} \frac{\partial x^a}{\partial y^1} \frac{\partial x^b}{\partial y^2} = \varepsilon_{12} \frac{\partial x}{\partial r} \frac{\partial y}{\partial \theta} + \varepsilon_{21} \frac{\partial y}{\partial r} \frac{\partial x}{\partial \theta} = r \cos^2 \theta + r \sin^2 \theta = r, \tag{1.13}$$

which is nonzero everywhere except at the origin, $r = 0$. As mentioned earlier, polar coordinates are not invertible at the origin, since $\theta = \tan^{-1}(y/x)$ is undefined there.

There is a second form of the Jacobian expression (1.12) that will be useful, given by

$$\sum_{j_1, j_2, \ldots, j_n = 1}^{n} \varepsilon_{j_1 j_2 \ldots j_n} \frac{\partial x^{j_1}(y)}{\partial y^{k_1}} \frac{\partial x^{j_2}(y)}{\partial y^{k_2}} \cdots \frac{\partial x^{j_n}(y)}{\partial y^{k_n}} = \varepsilon_{k_1 k_2 \ldots k_n} J(y). \tag{1.14}$$

The proof that (1.14) is equivalent to (1.12) follows from relabeling any two of the summed dummy indices, say j_1 and j_2, and noting that rearranging the first two factors gives

$$\begin{aligned} &\sum_{j_1, \ldots, j_n = 1}^{n} \varepsilon_{j_2 j_1 \ldots j_n} \frac{\partial x^{j_2}(y)}{\partial y^{k_1}} \frac{\partial x^{j_1}(y)}{\partial y^{k_2}} \cdots \frac{\partial x^{j_n}(y)}{\partial y^{k_n}} \\ &= - \sum_{j_1, \ldots, j_n = 1}^{n} \varepsilon_{j_1 j_2 \ldots j_n} \frac{\partial x^{j_1}(y)}{\partial y^{k_2}} \frac{\partial x^{j_2}(y)}{\partial y^{k_1}} \cdots \frac{\partial x^{j_n}(y)}{\partial y^{k_n}}. \end{aligned} \tag{1.15}$$

This shows that the left-hand side of (1.14) changes sign under a permutation of the k indices and therefore vanishes if any two of the k indices are the same. This is precisely the property of the Levi-Civita symbol on the right-hand side of (1.14). Examining the expression for $n = 2$, where $\varepsilon_{11} = \varepsilon_{22} = 0$, shows that

$$\sum_{a,b=1}^{2} \varepsilon_{ab} \frac{\partial x^a(y)}{\partial y^c} \frac{\partial x^b(y)}{\partial y^d} = \frac{\partial x^1(y)}{\partial y^c} \frac{\partial x^2(y)}{\partial y^d} - \frac{\partial x^2(y)}{\partial y^c} \frac{\partial x^1(y)}{\partial y^d}, \tag{1.16}$$

which vanishes if $c = d$ and is given by $\varepsilon_{cd} J(y)$ for $c \neq d$, inductively corroborating (1.14).

The Jacobian also appears in volume integrations for an n-dimensional manifold. The infinitesimal n-dimensional volume elements in the two coordinate systems, denoted $\mathrm{d}^n x$ and $\mathrm{d}^n y$, are related by

$$d^n x = |J(y)|\, d^n y, \tag{1.17}$$

where the absolute value preserves the positivity of the volume element, allowing the limits on the integration to remain the same. This result is derived when differential forms are discussed later in this chapter. Using the Jacobian (1.13) gives the well-known relation between the Cartesian surface element dS_c and the polar surface element dS_p in two dimensions,

$$dS_c = d^2x = dx\, dy = \left|J(r, \theta)\right| dr\, d\theta = r\, dr\, d\theta = d^2r = dS_p. \tag{1.18}$$

The Leibniz property of the differential and (1.1) combine to give

$$\begin{aligned} dx = d(r\cos\theta) &= dr\frac{\partial}{\partial r}(r\cos\theta) + d\theta\frac{\partial}{\partial\theta}(r\cos\theta) \\ &= dr\cos\theta - r\sin\theta\, d\theta, \\ dy = d(r\sin\theta) &= dr\frac{\partial}{\partial r}(r\sin\theta) + d\theta\frac{\partial}{\partial\theta}(r\sin\theta) \\ &= dr\sin\theta + r\cos\theta\, d\theta. \end{aligned} \tag{1.19}$$

However, it is important to note that the surface element in polar coordinates is *not* given by simply multiplying together the dx and dy given by (1.19). Instead, the coordinate differentials of (1.19) will be used later in this chapter to motivate the *wedge product* that gives infinitesimal surface and volume elements.

Exercise 1.3. If attention is restricted to the first quadrant, where $(x, y) > 0$ in Cartesian coordinates, *hyperbolic coordinates* (u, v) are defined by

$$u = \frac{1}{2}\ln(x/y), \quad v = \sqrt{xy} \iff x = v\exp(u), \quad y = v\exp(-u). \tag{1.20}$$

Show that the Jacobian for this coordinate transformation is $J(u, v) = 2v$, so that hyperbolic coordinates are uniquely invertible for $x, y > 0$.

1.4 Contravariant and covariant quantities

Result (1.19) is a specific case of how a coordinate transformation relates the infinitesimal differentials of the two coordinate systems, dx^k and dy^j, through the Leibniz property,

$$dy^j = \sum_{k=1}^{n} \frac{\partial y^j(x)}{\partial x^k}\, dx^k. \tag{1.21}$$

The mapping (1.21) between dy^j and the original dx^j is referred to as a *contravariant transformation*, and it identifies the differentials dx^j as *contravariant quantities.*

These differentials are not the only contravariant quantities of importance in field theory. Such contravariant quantities are notationally identified using a *superscript*, so that V^j transforms identically to (1.21) in another coordinate system.

Similarly, the relationship of the partial derivatives in the two coordinate systems $\partial/\partial x^k$ and $\partial/\partial y^j$ is given by the inverse functions from the chain rule (1.4),

$$\frac{\partial}{\partial y^j} = \sum_{k=1}^{n} \frac{\partial x^k(y)}{\partial y^j} \frac{\partial}{\partial x^k}. \tag{1.22}$$

The mapping (1.22) between $\partial/\partial y^j$ and the original $\partial/\partial x^j$ is referred to as a *covariant transformation*, and it identifies the partial derivatives as *covariant quantities*. The partial derivative is not the only covariant quantity of importance in field theory. Such covariant quantities are notationally identified using a *subscript*, so that $\partial/\partial x^j \equiv \partial_j$ indicates the covariant transformation property of the partial derivative with respect to x^j. All other covariant quantities V_j transform according to (1.22) when a coordinate transformation occurs.

The two types of transformations, (1.21) and (1.22), provide observers with a well-defined mathematical recipe to compare covariant and contravariant quantities measured in different coordinate systems. Therefore, it is a basic assumption of field theory that covariant and contravariant quantities, and their extension to *spinor quantities* (to be discussed), provide the building blocks of all physically viable theories.

As an example, using (1.21), (1.22), and the chain rule (1.4) shows that both coordinate systems result in the *same form* for the *differential operator* d,

$$\begin{aligned} \mathrm{d} &= \sum_{j=1}^{n} \mathrm{d}y^j \frac{\partial}{\partial y^j} = \sum_{j=1}^{n}\left(\sum_{\ell=1}^{n} \frac{\partial y^j}{\partial x^\ell}\mathrm{d}x^\ell\right)\left(\sum_{k=1}^{n} \frac{\partial x^k}{\partial y^j}\frac{\partial}{\partial x^k}\right) \\ &= \sum_{k,\ell=1}^{n} \mathrm{d}x^\ell \left(\sum_{j=1}^{n} \frac{\partial x^k}{\partial y^j}\frac{\partial y^j}{\partial x^\ell}\right)\frac{\partial}{\partial x^k} = \sum_{k,\ell=1}^{n} \mathrm{d}x^\ell\, \delta^k{}_\ell \frac{\partial}{\partial x^k} = \sum_{k=1}^{n} \mathrm{d}x^k \frac{\partial}{\partial x^k}. \end{aligned} \tag{1.23}$$

Applying (1.23) to $y^\ell(x)$ correctly reproduces the contravariant transformation law (1.21) for differentials. Any quantity that takes the same form in different coordinate systems is referred to as a *scalar quantity*. Scalar quantities are central to field theory, since they are *independent* of the choice of coordinate system and so all observers agree upon their value.

1.5 The summation convention

Because summing over all n degrees of freedom is so common and the summation is typically over a pair of contravariant and covariant *indices*, equations such as (1.23) are greatly streamlined by employing the *summation convention*. This means that a pair of identical indices, one contravariant and one covariant, implies a summation over the entire range of that index. From this point forward, the summation convention will be in effect unless the indices are placed in parentheses. Using this convention, it follows that

$$\alpha^j\beta_j = \sum_{j=1}^{n} \alpha^j\beta_j, \quad \alpha^{(j)}\beta_j \neq \sum_{j=1}^{n} \alpha^{(j)}\beta_j, \quad \delta_{(jj)} \neq \sum_{j=1}^{n} \delta_{jj}\,, \tag{1.24}$$

where the last two expressions stand for the product of $\alpha^{(j)}$ with β_j and the element $\delta_{(jj)}$, respectively, each with no sum over the index j. The demonstration of (1.23) is repeated, step for step, using the summation convention:

$$\begin{aligned} \mathrm{d} = \mathrm{d}y^j \frac{\partial}{\partial y^j} &= \left(\frac{\partial y^j}{\partial x^\ell}\,\mathrm{d}x^\ell\right)\left(\frac{\partial x^k}{\partial y^j}\frac{\partial}{\partial x^k}\right) \\ &= \mathrm{d}x^\ell\left(\frac{\partial x^k}{\partial y^j}\frac{\partial y^j}{\partial x^\ell}\right)\frac{\partial}{\partial x^k} = \mathrm{d}x^\ell\,\delta^k_\ell\,\frac{\partial}{\partial x^k} = \mathrm{d}x^k\frac{\partial}{\partial x^k}. \end{aligned} \tag{1.25}$$

The Kronecker delta gives

$$\delta^k_k = \sum_{k=1}^{n} \delta^k{}_k = \sum_{k=1}^{n} 1 = n, \quad \delta^{(k)}_{(k)} = 1. \tag{1.26}$$

The value of n is determined *contextually* from the *dimension* of the quantity, which requires the reader to be aware of the dimension of the manifold under analysis. The repeated index is a *dummy index*, since the choice of repeated symbol is irrelevant, so that $\delta^j_j = \delta^k_k = n$. As with the Levi-Civita symbol, it is common practice in physics to use the lower-case Latin letters, such as j and k, to indicate two or three dimensions and lower-case Greek letters, such as μ and ν, to indicate four and higher dimensions.

Given a contravariant quantity $A^j(x)$ and a covariant quantity $B_k(x)$, it is possible to form a scalar quantity $S(x)$ through *contraction*, which is the sum over the product of the components,

$$S(x) = A^j(x)B_j(x). \tag{1.27}$$

Demonstrating that $S(x)$ is a scalar is identical to the demonstration (1.25),

$$\begin{aligned} S'(y) = A'^j(y)B'_j(y) &= \left(\frac{\partial y^j}{\partial x^\ell}\,A^\ell(x(y))\right)\left(\frac{\partial x^k}{\partial y^j}\,B_k(x(y))\right) \\ &= A^\ell(x)\left(\frac{\partial x^k(y)}{\partial y^j}\frac{\partial y^j(x)}{\partial x^\ell}\right)B_k(x) = A^\ell(x)\,\delta^k_\ell\,B_k(x) \\ &= A^k(x)B_k(x) = S(x). \end{aligned} \tag{1.28}$$

The Kronecker delta δ^j_k defined by (1.2) is a purely mathematical construct, and so it must be invariant under coordinate transformations. This will be true as long as the j index transforms contravariantly, like (1.21), and the k index transforms covariantly, like (1.22). Denoting the Kronecker delta in the y-coordinate system as δ'^i_j, it follows from the properties of the Kronecker delta and the chain rule that

$$\delta'^i_j = \frac{\partial y^i(x)}{\partial x^k}\frac{\partial x^\ell(y)}{\partial y^j}\,\delta^k_\ell = \frac{\partial y^i(x)}{\partial x^k}\frac{\partial x^k(y)}{\partial y^j} = \frac{\partial y^i}{\partial y^j} = \delta^i_j, \tag{1.29}$$

demonstrating that the Kronecker delta is the same in all coordinate systems, as it must be.

So far, there have been no restrictions on the specific form of the coordinate transformation other than giving a nonzero value for the Jacobian (1.12). In physical applications, there is a significant possibility that a coordinate transformation may involve two coordinate systems whose origins are in relative motion. Understanding the effects of relative motion for different observers requires a physical principle for resolution. This is because relative motion involves the interplay of time and distance measurements, and may result in effects similar to those of forces. This will be discussed in detail later.

1.6 Vectors and direction vectors

The spatial relationship of two points in a manifold allows the definition of a *vector*, a quantity with both *direction* and *magnitude*. While vectors can be defined between any two points in the manifold, initial attention will be given to the *position vector* of a point p. In the case of a Euclidean manifold, it is defined as an *oriented straight line* or *arrow* from the *origin* to the point p, with the latter point referred to as the *vector tip*. The position vector therefore has both a direction as well as a magnitude given by its length. It is denoted in boldface as $\boldsymbol{r}_p$, but the p is often dropped and $\boldsymbol{r}$ is then understood to be the position vector of an arbitrary point in the manifold. In the x-coordinate system the position vector will be a function of the coordinates of the point p, and so it will often be written $\boldsymbol{r}(x_p)$ or $\boldsymbol{r}(x)$ for an arbitrary point.

It is important to note that the position vector lies in the manifold, and if the manifold is non-Euclidean, the position vector is not a straight line in the traditional Euclidean sense. As an example, the position vector on a sphere can be visualized as an arrow emanating from the north pole and coinciding with a segment of a longitude line that intersects the point of interest. The longitude gridlines are great circles on the sphere and correspond to the spherical version of a straight line. In a general non-Euclidean manifold the space between the origin and the point of interest is not flat and, unlike the sphere, may *not* even be uniformly curved. Determining the length between the origin and the point is therefore more difficult and renders the *finite* position vector a far less useful concept for a non-Euclidean manifold. However, in the Euclidean case, the straight-line position vector is uniquely determined by the choice of origin and the point defining the vector tip. It is important to note that using different coordinate systems does not alter the Euclidean vector $\boldsymbol{r}_p$ *as long as the two coordinate systems share the same origin.*

In order to give a mathematical representation to $\boldsymbol{r}_p$, a set of *n direction vectors* $\{\mathbf{e}_j\}$ is introduced *at every point in the manifold.* The direction vectors are a complete set of *independent directions* at each point, referred to as a *local basis*, and are defined in a manner consistent with the chosen coordinate system. The definition of the set $\{\mathbf{e}_j\}$ is obtained from the relationship between partial derivatives and directions. For a particular choice of coordinate system, the position vector of a point p will be a function of the coordinates x of the point p, so that $\boldsymbol{r} = \boldsymbol{r}(x)$. The direction vector $\mathbf{e}_j(x)$ at the point p with coordinates x is defined using the position vector $\boldsymbol{r}(x)$ and the partial derivative $\partial/\partial x^j$,

$$\mathbf{e}_j(x) = \frac{\partial \boldsymbol{r}(x)}{\partial x^j} = \lim_{\Delta x^j \to 0} \frac{\Delta \boldsymbol{r}}{\Delta x^j}$$
$$= \lim_{\Delta x^j \to 0} \frac{\boldsymbol{r}(x^1, \ldots, x^j + \Delta x^j, \ldots, x^n) - \boldsymbol{r}(x^1, \ldots, x^j, \ldots, x^n)}{\Delta x^j}. \quad (1.30)$$

Definition (1.30) shows that there are n basis vectors $\mathbf{e}_j$, since there are n coordinates.

Understanding definition (1.30) begins by noting that the numerator is the difference in the position vector at infinitesimally separated points,

$$\Delta \boldsymbol{r} = \boldsymbol{r}(x^1, \ldots, x^j + \Delta x^j, \ldots, x^j) - \boldsymbol{r}(x^1, \ldots, x^j, \ldots, x^n). \quad (1.31)$$

Because the partial derivative $\partial/\partial x^j$ in definition (1.30) holds all coordinates except x^j constant, the difference $\Delta \boldsymbol{r}$ can be visualized in the following way. As Δx^j changes, the tip of the position vector directed to the point p moves along the line of constant value for all other coordinates, which is the jth coordinate gridline, and this defines the vector $\Delta \boldsymbol{r}$. The direction of $\Delta \boldsymbol{r}$ is therefore tangent to the jth gridline, and this is the direction of $\mathbf{e}_j$.

The case of the Euclidean plane is depicted in figure 1.2 for both Cartesian and polar coordinates with exaggerated values of Δx^j. For the case of Cartesian coordinates (x, y) in two dimensions, the direction vector $\mathbf{e}_x \sim \Delta \boldsymbol{r}/\Delta x$ is tangent to gridlines of constant y, so that $\mathbf{e}_x$ is *everywhere* parallel to the x-axis. Although not depicted, $\mathbf{e}_y$ is tangent to the gridlines of constant x and is *everywhere* parallel to the y-axis. The Cartesian direction vectors therefore maintain their orientations *throughout a Euclidean manifold.* However, choosing polar coordinates (r, θ) gives a direction vector $\mathbf{e}_\theta \sim \Delta \boldsymbol{r}/\Delta\theta$ that is tangent in the limit $\Delta\theta \to 0$ to the circular gridlines of constant r. Although not depicted, $\mathbf{e}_r$ is tangent to gridlines of constant θ, which are straight lines radiating from the origin. As a result, $\mathbf{e}_r$ is radial at every point around the origin. It should be clear that both $\mathbf{e}_r$ and $\mathbf{e}_\theta$ change orientation at different points in the plane, and that the local directions of both are dependent on the angular coordinate, θ. Although not always displayed, the polar coordinate direction vectors are therefore functions of θ and should be written $\mathbf{e}_r(\theta)$ and $\mathbf{e}_\theta(\theta)$. As a result, a general direction vector $\mathbf{e}_j$ may depend on the point where it is defined,

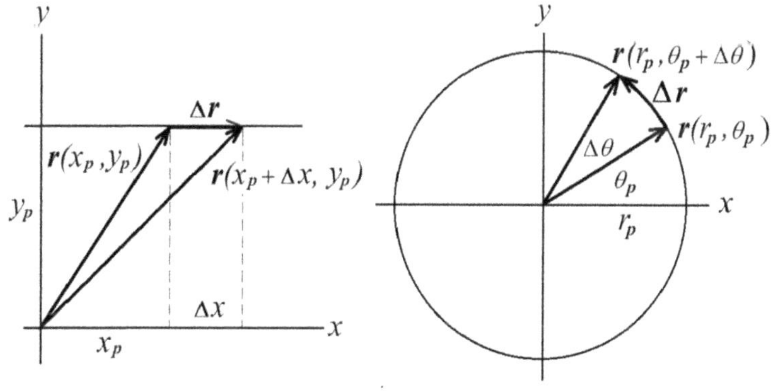

Figure 1.2. Euclidean direction vectors in Cartesian and polar coordinates.

and so it will be written $\mathbf{e}_j(x)$ when it is important to remind the reader of this dependence.

It is important to note that the direction vectors are *tangent* to the gridlines at every point, even if the gridlines are not straight, as is the case with polar coordinates. If Δx^j has units of length, then $\mathbf{e}_j$ has no physical dimensions and solely represents the jth direction, an intrinsic property of space. This is the case with Cartesian coordinates in a flat Euclidean space, where figure 1.2 shows that the length of the arrow $\Delta \boldsymbol{r}$ is congruent to the length of Δx by parallel projection. The ratio of the lengths in $\Delta \boldsymbol{r}/\Delta x \sim \mathbf{e}_x$ is therefore one in Cartesian coordinates, and the Cartesian direction vectors are said to be *normalized.* Because the gridlines are perpendicular to each other in Cartesian coordinates, the associated direction vectors are also perpendicular or *orthogonal.* The Cartesian direction vectors are therefore *orthonormal.* However, in the case of polar coordinates, the arc length of $\Delta \boldsymbol{r}$ is given by $r\ \Delta\theta$, and so the ratio of the lengths in $\Delta \boldsymbol{r}/\Delta\theta \sim \mathbf{e}_\theta$ is given by r. While $\mathbf{e}_r$ is normalized, $\mathbf{e}_\theta$ is not. This will be remedied later when the gradient is introduced. The definition (1.30) for direction vectors generalizes to non-Euclidean manifolds, where all members of the set of direction vectors $\{\mathbf{e}_j(x)\}$ are *tangent* to the manifold at the point described by the coordinates x.

In a non-Euclidean manifold, where finite position vectors are not well defined, these definitions can still be made in the neighborhood of a point p. The differential operator d can be applied to the position vector $\boldsymbol{r}(x)$ to find the infinitesimal change $\mathrm{d}\boldsymbol{r}(x)$ at the point p for arbitrary infinitesimal coordinate changes $\mathrm{d}x^j$. Combining its Leibniz property with the definition (1.30) and the summation convention yields

$$\mathrm{d}\boldsymbol{r}(x) = \mathrm{d}x^j\ \frac{\partial \boldsymbol{r}(x)}{\partial x^j} = \mathrm{d}x^j\ \mathbf{e}_j(x), \tag{1.32}$$

where $\{\mathbf{e}_j(x)\}$ is the same set of direction vectors that were defined by (1.30). The differential $\mathrm{d}\boldsymbol{r}(x)$ is an infinitesimal vector that is a *linear sum* of *infinitesimal component vectors* $\mathrm{d}x^{(j)}\ \mathbf{e}_j(x)$. The quantity $\mathrm{d}x^j$ is referred to as the jth *component* of the vector $\mathrm{d}\boldsymbol{r}(x)$ and determines the magnitude of the contribution from $\mathbf{e}_j$. The method used to add these component vectors so that they sum to the original vector $\mathrm{d}\boldsymbol{r}$ will be discussed in detail in the next section. However, the form of (1.32) shows that the set of direction vectors $\{\mathbf{e}_j(x)\}$ provides a *basis set* to expand a vector into a linear sum, *at least locally* in the neighborhood of the point with the coordinates x. It is important to note that the form of (1.32) allows the position vector to change direction at a point through the choice of $\mathrm{d}x^j$ consistent with the curvature of the manifold.

In a Euclidean manifold the result (1.32) can be extended to a *finite* position vector $\boldsymbol{r}(x)$ by integration. For example, for the Euclidean plane the finite position vector for the point (x, y) in Cartesian coordinates is given by

$$\begin{aligned}\boldsymbol{r}(x) &= \int_0^{\boldsymbol{r}(x)} \mathrm{d}\boldsymbol{r} = \int_0^{\boldsymbol{r}(x)} \left(\mathrm{d}x\ \mathbf{e}_x + \mathrm{d}y\ \mathbf{e}_y\right) = \mathbf{e}_x \int_0^x \mathrm{d}x + \mathbf{e}_y \int_0^y \mathrm{d}y \\ &= x\ \mathbf{e}_x + y\ \mathbf{e}_y.\end{aligned} \tag{1.33}$$

Result (1.33) breaks into straight-line integrations from the origin to the vector tip along the component vectors of d$\boldsymbol{r}$, reflecting the independence of the directions. Because the Cartesian unit vectors are *constant* throughout a Euclidean manifold, they can be factored from the integrand. Result (1.33) shows that the use of Cartesian coordinates in a Euclidean space results in a position vector with the Cartesian coordinates of the tip as its components, so that $r^k(x) = x^k$. Using the Cartesian result $\partial \mathbf{e}_k/\partial x^j = 0$ shows that (1.33) satisfies

$$\frac{\partial \boldsymbol{r}(x)}{\partial x^k} = \frac{\partial}{\partial x^k}(x^j \mathbf{e}_j) = \delta_k^j \, \mathbf{e}_j + x^j \frac{\partial \mathbf{e}_j}{\partial x^k} = \mathbf{e}_k \tag{1.34}$$

for all k, and so it is consistent with definition (1.30).

By extension of (1.33), a finite Euclidean position vector in an arbitrary coordinate system will take the general form

$$\boldsymbol{r}(x) = r^1(x)\,\mathbf{e}_1(x) + r^2(x)\,\mathbf{e}_2(x) + \cdots + r^n(x)\,\mathbf{e}_n(x) = r^j(x)\,\mathbf{e}_j(x), \tag{1.35}$$

where $\{\mathbf{e}_j(x)\}$ is the set of direction vectors *at the tip of the vector*. In expression (1.35) $r^j(x)$ are the *components* of the vector $\boldsymbol{r}(x)$. In a general coordinate system, $r^j(x)$ are functions of the coordinates x of the point p that defines the vector. The form of the components will also depend on the direction vectors, which are determined by the choice of coordinate system. However, in an arbitrary coordinate system the components r^j in general *do not* coincide with the coordinates of the vector tip.

Definitions (1.30) and (1.32) show that the direction vectors transform identically to the partial derivative under a coordinate transformation. Denoting the direction vectors in the y-coordinate system as $\{\mathbf{e}'_j(y)\}$, it follows that

$$\mathbf{e}'_j(y) = \frac{\partial \boldsymbol{r}(y)}{\partial y^j} = \frac{\partial x^k(y)}{\partial y^j}\frac{\partial \boldsymbol{r}(y(x))}{\partial x^k} = \frac{\partial x^k(y)}{\partial y^j}\,\mathbf{e}_k(x(y)). \tag{1.36}$$

Combining the transformation (1.36) with the chain rule shows that the differential vector d$\boldsymbol{r}(x)$, like the differential d itself, is invariant under coordinate transformations,

$$\mathrm{d}\boldsymbol{r}(y) = \mathrm{d}y^j \frac{\partial \boldsymbol{r}(y)}{\partial y^j} = \mathrm{d}x^k \frac{\partial y^j}{\partial x^k}\frac{\partial \boldsymbol{r}(y)}{\partial y^j} = \mathrm{d}x^k \frac{\partial \boldsymbol{r}(y(x))}{\partial x^k} = \mathrm{d}x^k \mathbf{e}_k(x) = \mathrm{d}\boldsymbol{r}(x). \tag{1.37}$$

An example of using (1.36) is the coordinate transformation from Cartesian to polar coordinates given by (1.1). The polar direction vectors are given by

$$\mathbf{e}_r = \frac{\partial \boldsymbol{r}}{\partial r} = \frac{\partial x}{\partial r}\frac{\partial \boldsymbol{r}}{\partial x} + \frac{\partial y}{\partial r}\frac{\partial \boldsymbol{r}}{\partial y} = \cos\theta\,\mathbf{e}_x + \sin\theta\,\mathbf{e}_y, \tag{1.38}$$

$$\mathbf{e}_\theta = \frac{\partial \boldsymbol{r}}{\partial \theta} = \frac{\partial x}{\partial \theta}\frac{\partial \boldsymbol{r}}{\partial x} + \frac{\partial y}{\partial \theta}\frac{\partial \boldsymbol{r}}{\partial y} = -r\sin\theta\,\mathbf{e}_x + r\cos\theta\,\mathbf{e}_y. \tag{1.39}$$

These two results can be found using Euclidean geometry and trigonometry, but are much more easily obtained from the covariant transformation formula (1.36).

Result (1.39) shows that $\mathbf{e}_\theta$ is indeed proportional to r. It will be useful to note that (1.38) allows the two-dimensional finite Cartesian position vector (1.33) to be expressed in polar coordinates as

$$\boldsymbol{r} = x\,\mathbf{e}_x + y\,\mathbf{e}_y = r\cos\theta\,\mathbf{e}_x + r\sin\theta\,\mathbf{e}_y = r(\cos\theta\,\mathbf{e}_x + \sin\theta\,\mathbf{e}_y) = r\,\mathbf{e}_r. \quad (1.40)$$

Using the chain rule, the transformation (1.36) can be inverted to find

$$\begin{aligned}\mathbf{e}_k(x) = \delta^\ell_k\,\mathbf{e}_\ell(x) &= \left(\frac{\partial x^\ell(y)}{\partial y^j}\frac{\partial y^j(x)}{\partial x^k}\right)\mathbf{e}_\ell(y(x))\\ &= \frac{\partial y^j(x)}{\partial x^k}\left(\frac{\partial x^\ell(y)}{\partial y^j}\,\mathbf{e}_\ell(y(x))\right) = \frac{\partial y^j(x)}{\partial x^k}\,\mathbf{e}'_j(y(x)),\end{aligned} \quad (1.41)$$

which will be useful in understanding the behavior of the vector components.

Exercise 1.4. The relationship between Cartesian coordinates (x, y, z) and spherical coordinates (r, θ, ϕ) in three dimensions is given by

$$\begin{aligned}&x = r\sin\theta\cos\phi,\ \ y = r\sin\theta\sin\phi,\ \ z = r\cos\theta \Longleftrightarrow\\ &r^2 = x^2 + y^2 + z^2,\ \ \theta = \cos^{-1}\frac{z}{r},\ \ \phi = \tan^{-1}\frac{y}{x}.\end{aligned} \quad (1.42)$$

Show that the direction vectors in spherical coordinates are given in terms of the Cartesian direction vectors by

$$\begin{aligned}\mathbf{e}_r &= \sin\theta\cos\phi\,\mathbf{e}_x + \sin\theta\sin\phi\,\mathbf{e}_y + \cos\theta\,\mathbf{e}_z,\\ \mathbf{e}_\theta &= r\cos\theta\cos\phi\,\mathbf{e}_x + r\cos\theta\sin\phi\,\mathbf{e}_y - r\sin\theta\,\mathbf{e}_z,\\ \mathbf{e}_\phi &= -r\sin\theta\sin\phi\,\mathbf{e}_x + r\sin\theta\cos\phi\,\mathbf{e}_y,\end{aligned} \quad (1.43)$$

and that the Cartesian position vector (1.33) in three dimensions, $\boldsymbol{r} = x\,\mathbf{e}_x + y\,\mathbf{e}_y + z\,\mathbf{e}_z$, is given by

$$\boldsymbol{r} = r\,\mathbf{e}_r. \quad (1.44)$$

Because both $\mathbf{e}_j$ and ∂_j play the same directional role, one for vectors and the other for functions, there is a directional equivalence between the two sets, denoted by $\{\mathbf{e}_j\} \leftrightarrow \{\partial_j\}$. This directional equivalence is at the foundation of differential geometry, and since the work of Cartan the two sets, $\{\mathbf{e}_j\}$ and $\{\partial_j\}$, have been used interchangeably as a direction basis. In Cartan's notation the vector $\boldsymbol{v}$ is written

$$\boldsymbol{v}(\boldsymbol{x}) = v^j(\boldsymbol{x})\frac{\partial}{\partial x^j}, \quad (1.45)$$

and is made concrete by examining its effect on a scalar function $\varphi(\boldsymbol{x})$, which is given by $\boldsymbol{v}(\boldsymbol{x})\varphi(\boldsymbol{x})$. Because $\varphi(\boldsymbol{x})$ is an arbitrary function, this creates a vector space at point $\boldsymbol{x}$. This is discussed again in the sections on trajectories and tangent spaces.

It is the covariant property of $\{\mathbf{e}_j\}$ that renders the combination (1.37) invariant under coordinate changes. The basis vectors $\{\mathbf{e}_j\}$ are referred to as the *legs* of a vector, and so they are often referred to as zweibein, dreibein, vierbein, and vielbein, depending on whether there are two, three, four, or many legs.

It is possible to consider a transformation between two coordinate systems that changes the origin. An example of such a transformation is depicted for the two-dimensional case in figure 1.3. For simplicity, the x-coordinate system will be chosen to be Cartesian. The inverse coordinate transformation $x^k(y)$ has the property that $x^k(0) = a^k$, which are the x-coordinates of the origin in the y-coordinate system. For such a case, the *finite position vector* can be written

$$\begin{aligned} \boldsymbol{r}(x) = x^k(x)\,\mathbf{e}_k(x) &= (x^k(x) - a^k)\mathbf{e}_k(x) + a^k\,\mathbf{e}_k(x) \\ &\equiv y^k(x)\,\mathbf{e}_k(x) + a^k\,\mathbf{e}_k(x) \\ &= \tilde{\boldsymbol{r}}(x) + \boldsymbol{a}(x). \end{aligned} \tag{1.46}$$

The original vector $\boldsymbol{r}(x)$ is now expressed as the sum of two vectors, $\tilde{\boldsymbol{r}}(x)$ and $\boldsymbol{a}(x)$, where $\boldsymbol{a}$ is the position vector of the y-coordinate origin in the x-coordinate system, while $\tilde{\boldsymbol{r}}(x)$ is the position vector of the point p *in the new coordinate system*, where $\tilde{\boldsymbol{r}}$ has the components $y^k = x^k - a^k$. The latter follows from the fact that the components of $\tilde{\boldsymbol{r}}(x)$ are zero when $y^j = 0$, which occurs if $x^k = a^k$. The direction vectors are defined at point p, and so $\tilde{\boldsymbol{r}}$ is the position vector of point p defined using the new origin. In an arbitrary second coordinate system, (1.41) gives

$$\tilde{\boldsymbol{r}}(x) = \tilde{r}^k(x)\,\mathbf{e}_k(x) = \mathbf{e}'_j(y)\left(\tilde{r}^k(x)\,\frac{\partial y^j(x)}{\partial x^k}\right)\Bigg|_{x(y)} = \tilde{r}'^j(y)\,\mathbf{e}'_j(y) = \tilde{\boldsymbol{r}}(y). \tag{1.47}$$

The two vectors, $\tilde{\boldsymbol{r}}(x)$ and $\tilde{\boldsymbol{r}}(y)$, are the *same vector* since they share the same origin and the same tip. Result (1.47) gives

$$\tilde{r}'^j(y) = \tilde{r}^k(x)\,\frac{\partial y^j(x)}{\partial x^k}\Bigg|_{x(y)}. \tag{1.48}$$

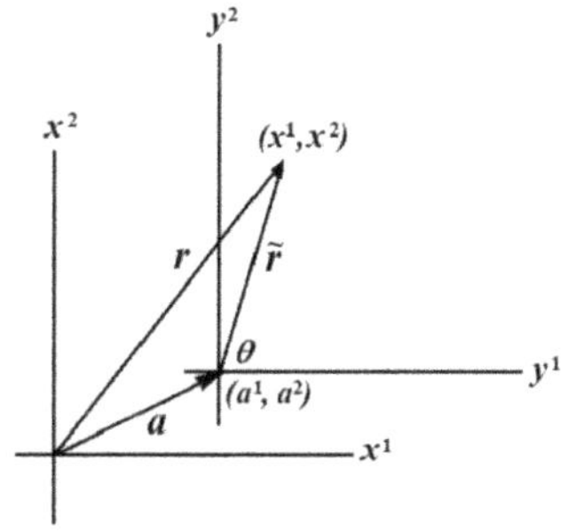

Figure 1.3. A coordinate transformation involving a translation of the origin.

This shows that the finite position vector components transform as contravariant quantities, but only for the components of the vector emanating from the origin of the new coordinate system. This is consistent with the contravariant nature of the infinitesimal components $\mathrm{d}x^j$ of $\mathrm{d}\boldsymbol{r}$. Because the origin has been moved or translated, such a coordinate transformation is said to include a *translation*. If there is no translation of the origin, the $\tilde{r}^k(x)$ components in (1.48) are replaced with $r^k(x)$.

Result (1.48) can be clarified by considering the transformation in the Euclidean plane from two-dimensional Cartesian coordinates (x^1, x^2) to two-dimensional polar coordinates $(\tilde{r}, \theta)$ for the case depicted in figure 1.3. Such a transformation is expressed mathematically by

$$\tilde{r} = \sqrt{(x^1 - a^1)^2 + (x^2 - a^2)^2}, \quad \theta = \tan^{-1}\left(\frac{x^2 - a^2}{x^1 - a^1}\right) \Longleftrightarrow \\ x^1 = a^1 + \tilde{r}\cos\theta, \quad x^2 = a^2 + \tilde{r}\sin\theta. \tag{1.49}$$

The vector $\boldsymbol{a} = a^j\mathbf{e}_j$ is a position vector from the original Cartesian coordinate origin to the point (a^1, a^2) in Cartesian coordinates. Applying formula (1.48) requires the identification of $\tilde{x}^1 = x^1 - a^1$ and $\tilde{x}^2 = x^2 - a^2$ as the Cartesian coordinates of the point measured with respect to the *new* origin. Using these in (1.48) in combination with (1.49) gives the position vector components in the translated polar coordinate system,

$$\begin{aligned} \tilde{r}'^1 &= \left(\tilde{x}^1\frac{\partial\tilde{r}}{\partial x^1} + \tilde{x}^2\frac{\partial\tilde{r}}{\partial x^2}\right)\Bigg|_{x(\tilde{r})} = \left(\frac{(x^1 - a^1)^2 + (x^2 - a^2)^2}{\sqrt{(x^1 - a^1)^2 + (x^2 - a^2)^2}}\right)\Bigg|_{x(\tilde{r})} = \tilde{r}, \\ \tilde{r}'^2 &= \left(\tilde{x}^1\frac{\partial\theta}{\partial x^1} + \tilde{x}^2\frac{\partial\theta}{\partial x^2}\right)\Bigg|_{x(\tilde{r})} \\ &= \left(\frac{-(x^1 - a^1)(x^2 - a^2) + (x^2 - a^2)(x^1 - a^1)}{(x^1 - a^1)^2 + (x^2 - a^2)^2}\right)\Bigg|_{x(\tilde{r})} = 0. \end{aligned} \tag{1.50}$$

> **Exercise 1.5.** Show that the polar coordinate direction vectors $\mathbf{e}_j(y)$ in the translated coordinate system take the same form as (1.38) and (1.39), although r is replaced by $\tilde{r}$, while the trigonometric functions stand for $\cos\theta = (x^1 - a^1)/\tilde{r}$ and $\sin\theta = (x^2 - a^2)/\tilde{r}$.

Using (1.50) shows that the finite position vector $\boldsymbol{r}$ in the original system is therefore given as the sum

$$\boldsymbol{r}(x) = \tilde{\boldsymbol{r}}(y) + \boldsymbol{a} = \tilde{r}\,\mathbf{e}_r(y) + \boldsymbol{a}. \tag{1.51}$$

The vector $\tilde{r}(x)$ does not involve $\mathbf{e}_\theta$, since the position vector $\tilde{r}(x)$ points radially from the new origin at $\boldsymbol{a}$.

It is worth repeating that a finite position vector $\boldsymbol{r}$ can be expressed in the general form (1.35) or the specific Cartesian form (1.33) *only in a Euclidean manifold.* This is possible because a Euclidean manifold allows unique straight lines everywhere throughout the manifold. As a result, specifying the origin and the point of interest in a Euclidean manifold is sufficient to determine the unique straight line between the two, in turn defining the finite position vector. However, in a general non-Euclidean manifold, only the infinitesimal change in the position vector $\mathrm{d}\boldsymbol{r}_p$, which is defined at each point p by (1.32), is available. This is because the orientation of the unit vectors at each point will depend on the curvature of the manifold *at that point.* As a result, the components of the vector will depend upon the nature of the manifold between the origin and the final position as well as the path between the two points. In a non-Euclidean manifold, finite position vectors become cumbersome and so are seldom, if ever, used. Alternate methods for determining distances and angles will be developed.

1.7 Vector addition and the scalar product

For position vectors in a Euclidean manifold, the method used to add vectors is deduced from relation (1.31), which gives

$$\boldsymbol{r}(x^1, \ldots, x_j + \Delta x^j, \ldots, x^j) = \boldsymbol{r}(x^1, \ldots, x^j, \ldots, x^n) + \Delta \boldsymbol{r}. \qquad (1.52)$$

Examining figure 1.2 for the Cartesian case shows that the two vectors on the right-hand side of (1.52) form two sides of a parallelogram. Their sum, the vector on the left-hand side of (1.52), is given by the diagonal of the parallelogram. The same graphical relationship holds in figure 1.3, where the vector $\boldsymbol{r} = \tilde{\boldsymbol{r}} + \boldsymbol{a}$ is the diagonal of the parallelogram formed by $\boldsymbol{a}$ and $\tilde{\boldsymbol{r}}$. This gives the well-known *parallelogram law of vector addition in a Euclidean manifold.* The two vectors being added are treated as the adjacent sides of a parallelogram, and the *resultant vector* is the diagonal of the parallelogram. The parallelogram law can be applied to the component vectors in (1.35) by adding them sequentially. In the case of Cartesian coordinates, the finite result (1.33) shows that the two perpendicular components $x\ \mathbf{e}_x$ and $y\ \mathbf{e}_y$ sum to the vector directed to the point (x, y). This is consistent with the parallelogram law, since the two component vectors define a rectangle whose diagonal is their sum. While this process has been introduced graphically, it can now be expressed mathematically.

The mathematical method used to implement the parallelogram law requires the *scalar product* for vectors, also known as the *inner product* or *dot product.* In Euclidean $\mathbb{R}^n$ the *scalar product* of two finite position vectors $\boldsymbol{r}_1$ and $\boldsymbol{r}_2$ is defined as

$$\boldsymbol{r}_1 \cdot \boldsymbol{r}_2 = \boldsymbol{r}_2 \cdot \boldsymbol{r}_1 = |\boldsymbol{r}_1|\, |\boldsymbol{r}_2| \cos \phi_{12}, \qquad (1.53)$$

where $|\boldsymbol{r}_1|$ and $|\boldsymbol{r}_2|$ are the lengths or *magnitudes* of the position vectors $\boldsymbol{r}_1$ and $\boldsymbol{r}_2$, respectively, while ϕ_{12} is the angle between the two vectors measured in the plane they define. When there is no chance of confusion, the magnitude of the vector $\boldsymbol{r}$ will be written $|\boldsymbol{r}| = r$. The scalar product of two finite vectors creates a directionless *scalar quantity* from them that depends on both their magnitudes and their relative directions, and in a Euclidean manifold is the *perpendicular projection* of $\boldsymbol{r}_1$ onto $\boldsymbol{r}_2$. If the scalar product of two nonzero vectors is zero, then the two vectors are perpendicular or *orthogonal*, since $\cos\phi = 0$ is satisfied only if $\phi = \pm\pi/2$. Since the angle between a vector $\boldsymbol{r}$ and itself is zero, the magnitude r of the position vector $\boldsymbol{r}$ in $\mathbb{R}^n$ is given by

$$\boldsymbol{r} \cdot \boldsymbol{r} = r \cdot r \cdot \cos 0 = r^2. \tag{1.54}$$

The scalar product therefore yields the magnitude or length of a vector in $\mathbb{R}^n$. The length of the position vector $\boldsymbol{r}$ in a Euclidean manifold is understood as the distance to the origin for the point defining the vector. One of the immediate ramifications of (1.54) is that the vector $a\boldsymbol{r}$, where a is an arbitrary real number, has the magnitude ar.

The parallelogram law for adding two vectors in a Euclidean manifold can now be formalized. If $\boldsymbol{r}_1$ and $\boldsymbol{r}_2$ are two position vectors emanating from a common origin, it follows that the magnitude of the vector $\boldsymbol{r} = \boldsymbol{r}_1 + \boldsymbol{r}_2$ that results from adding them is given by

$$\begin{aligned} r^2 &= \boldsymbol{r} \cdot \boldsymbol{r} = (\boldsymbol{r}_1 + \boldsymbol{r}_2) \cdot (\boldsymbol{r}_1 + \boldsymbol{r}_2) = \boldsymbol{r}_1 \cdot \boldsymbol{r}_1 + \boldsymbol{r}_2 \cdot \boldsymbol{r}_2 + 2\, \boldsymbol{r}_1 \cdot \boldsymbol{r}_2 \\ &= r_1^2 + r_2^2 + 2\, r_1 r_2 \cos\phi_{12}. \end{aligned} \tag{1.55}$$

This is the familiar law of cosines for finding the third leg of an oblique triangle, given two sides that have the angle $\pi - \phi_{12}$ between them. This shows that the zero vector, defined by $0 = \boldsymbol{r} - \boldsymbol{r}$, has zero magnitude and is therefore a point. Once r is known, the direction of the resultant vector can obtained from the angles between one of the original vectors and the resultant vector. For example, the angle $\phi_{r\,r_1}$ between the resultant vector $\boldsymbol{r}$ and $\boldsymbol{r}_1$ is given by

$$\cos\phi_{r\,r_1} = \frac{\boldsymbol{r} \cdot \boldsymbol{r}_1}{r\, r_1} = \frac{(\boldsymbol{r}_1 + \boldsymbol{r}_2) \cdot \boldsymbol{r}_1}{r\, r_1} = \frac{r_1}{r} + \frac{r_2}{r} \cos\phi_{12}. \tag{1.56}$$

Because of the similarity between (1.55) and the law of cosines, vector addition is often depicted as placing one vector on the tip of the other vector and drawing the third leg of the triangle that is formed. This is consistent with the intuitive idea that adding the two position vectors is identical to performing sequential displacements and determining the final position from vector addition. However, both vectors in (1.55) and (1.56) are defined with respect to the *same origin*. As a result, forming a triangle, or equivalently a parallelogram, requires the second position vector to be moved or *transported* to the tip of the other position vector so that it remains parallel to itself, a process known as *parallel transport*. Parallel transport in a Euclidean manifold is straightforward, while the parallel transport of vectors in a non-Euclidean manifold will be developed in chapter 10.

For a Euclidean manifold, result (1.54) can be combined with (1.35) to obtain the magnitude of a vector from its components,

$$r^2 = \boldsymbol{r} \cdot \boldsymbol{r} = \left(r^j \mathbf{e}_j\right) \cdot \left(r^k \mathbf{e}_k\right) = r^j r^k\, \mathbf{e}_j \cdot \mathbf{e}_k. \tag{1.57}$$

Result (1.57) can be evaluated for the case in which the vector is in a Cartesian coordinate system. It was established earlier that Cartesian direction vectors have a unit length and are tangent to orthogonal coordinate gridlines, which gives

$$\mathbf{e}_j \cdot \mathbf{e}_k = 1 \cdot 1 \cdot \cos \phi_{jk} = \begin{cases} 1 & \text{if } j = k \\ 0 & \text{if } j \neq k \end{cases}, \tag{1.58}$$

so that the Cartesian direction vectors form an *orthonormal set* and maintain their orientation everywhere in a Euclidean manifold. It was shown in (1.33) that a finite Cartesian vector has the Cartesian coordinates of its tip as the components $r^j = x^j$, with the result that (1.57) becomes

$$r^2 = \sum_{j=1}^{n} (x^j)^2. \tag{1.59}$$

Result (1.59) is simply the Pythagorean theorem extended to n dimensions.

There is a second useful form of the Euclidean scalar product of two vectors in Cartesian coordinates, $\boldsymbol{r}_1 = x_1\, \mathbf{e}_x + y_1\, \mathbf{e}_y$ and $\boldsymbol{r}_2 = x_2\, \mathbf{e}_x + y_2\, \mathbf{e}_y$. Using (1.58) gives

$$\boldsymbol{r}_1 \cdot \boldsymbol{r}_2 = x_1 x_2\, \mathbf{e}_x \cdot \mathbf{e}_x + x_1 y_2\, \mathbf{e}_x \cdot \mathbf{e}_y + y_1 x_2\, \mathbf{e}_y \cdot \mathbf{e}_x + y_1 y_2\, \mathbf{e}_y \cdot \mathbf{e}_y = x_1 x_2 + y_1 y_2. \tag{1.60}$$

Form (1.60) is equivalent to (1.53), which is seen by switching to polar coordinates and using $x_1 = r_1 \cos \theta_1$, $y_1 = r_1 \sin \theta_1$, $x_2 = r_2 \cos \theta_2$, and $y_2 = r_2 \sin \theta_2$:

$$\boldsymbol{r}_1 \cdot \boldsymbol{r}_2 = x_1 x_2 + y_1 y_2 = r_1 r_2 (\cos \theta_1 \cos \theta_2 + \sin \theta_1 \sin \theta_2) = r_1 r_2 \cos(\theta_1 - \theta_2). \tag{1.61}$$

Since the angle between the two vectors is $\phi_{12} = \theta_1 - \theta_2$, result (1.61) is identical to (1.53).

Even in a Euclidean manifold, a finite position vector in another coordinate system can take a very different form. For example, it was found in (1.51) that, after a translation and transformation from two-dimensional Cartesian coordinates into polar coordinates, the vector $\boldsymbol{r}$ was given by $\boldsymbol{r} = \tilde{\boldsymbol{r}} + \boldsymbol{a}$, where $\tilde{\boldsymbol{r}} = \tilde{r}\, \mathbf{e}_r$ and $\boldsymbol{a} = a^j \mathbf{e}_j = a^1 \mathbf{e}_x + a^2 \mathbf{e}_y$. Result (1.38) gives $\mathbf{e}_r = \cos \theta\, \mathbf{e}_x + \sin \theta\, \mathbf{e}_y$ in terms of the original Cartesian direction vectors, where θ is the polar angle in the new coordinate system. The direction vectors give $\mathbf{e}_r \cdot \mathbf{e}_r = 1$, $\mathbf{e}_r \cdot \mathbf{e}_x = \cos \theta$, and $\mathbf{e}_r \cdot \mathbf{e}_y = \sin \theta$. This gives

$$\begin{aligned} r^2 = \boldsymbol{r} \cdot \boldsymbol{r} &= \tilde{r}^2 \mathbf{e}_r \cdot \mathbf{e}_r + a^2 + 2\tilde{r}\, \mathbf{e}_r \cdot \boldsymbol{a} = \tilde{r}^2 + a^2 + 2\tilde{r} a^j \mathbf{e}_r \cdot \mathbf{e}_j \\ &= \tilde{r}^2 + a^2 + 2\tilde{r} \cos \theta\, a^1 + 2\tilde{r} \sin \theta\, a^2. \end{aligned} \tag{1.62}$$

The last term in (1.62) is identical to $2\tilde{\boldsymbol{r}} \cdot \boldsymbol{a}$ in the Cartesian form (1.60), since

$$\tilde{\boldsymbol{r}} \cdot \boldsymbol{a} = \tilde{r}^1 a^1 + \tilde{r}^2 a^2 = \tilde{r} \cos \theta\, a^1 + \tilde{r} \sin \theta\, a^2. \tag{1.63}$$

Result (1.62) is therefore identical in form to (1.55).

1.8 The metric tensor and distance in manifolds

If a manifold is non-Euclidean, the finite component formula (1.57) and its Cartesian version (1.59) are no longer valid. However, it is still possible to use the *local length* $\mathrm{d}s$ of the infinitesimal position vector element $\mathrm{d}\boldsymbol{r}$. This is obtained by applying (1.54) to (1.32), which gives

$$\begin{aligned} \mathrm{d}s^2 &\equiv \mathrm{d}\boldsymbol{r} \cdot \mathrm{d}\boldsymbol{r} = \left(\mathrm{d}x^j \mathbf{e}_j(x)\right) \cdot \left(\mathrm{d}x^k \mathbf{e}_k(x)\right) = \mathbf{e}_j(x) \cdot \mathbf{e}_k(x)\, \mathrm{d}x^j \mathrm{d}x^k \\ &\equiv g_{jk}(x)\, \mathrm{d}x^j \mathrm{d}x^k. \end{aligned} \tag{1.64}$$

Because of their association with length, the quantities given by

$$g_{jk}(x) = \mathbf{e}_j(x) \cdot \mathbf{e}_k(x) \tag{1.65}$$

are referred to as the components of the *metric tensor* $\boldsymbol{g}$. The metric tensor and its relation to distance through (1.64) is the key tool used to develop non-Euclidean geometry and general relativity. From their definition, it follows that the components are symmetric, since the scalar product *commutes*:

$$g_{jk}(x) = \mathbf{e}_j(x) \cdot \mathbf{e}_k(x) = \mathbf{e}_k(x) \cdot \mathbf{e}_j(x) = g_{kj}(x). \tag{1.66}$$

As a result, in n dimensions, the metric tensor has $1 + 2 + \cdots + n = \frac{1}{2}n(n+1)$ independent elements.

The metric tensor inherits the transformation property (1.36) of the two direction vectors that define it,

$$\begin{aligned} g'_{ij}(y) = \mathbf{e}'_i(y) \cdot \mathbf{e}'_j(y) &= \frac{\partial x^k(y)}{\partial y^i}\frac{\partial x^\ell(y)}{\partial y^j}\, \mathbf{e}_k(x(y))\cdot \\ \mathbf{e}_\ell(x(y)) &= \frac{\partial x^k(y)}{\partial y^i}\frac{\partial x^\ell(y)}{\partial y^j}\, g_{k\ell}(x(y)). \end{aligned} \tag{1.67}$$

Because two occurrences of the covariant transformation factor $\partial x^k/\partial y^j$ appear in (1.67), the metric tensor g_{jk} is referred to as a *second-rank covariant tensor*. It is important to note that the transformation property (1.67) combines with the transformation property (1.21) of $\mathrm{d}x^j$ to render the infinitesimal length (1.64) *invariant under coordinate transformations*. Expression (1.64) is, in effect, a double contraction, and is therefore invariant for the same reason that the single contraction of (1.27) is invariant.

The distance along any continuous path between two arbitrary points in a manifold can, in principle, be computed by integrating the form for $\mathrm{d}s$ obtained from (1.64) along that path. It is important to note that (1.64) gives the infinitesimal length along the path element associated with *arbitrary* values for $\mathrm{d}x^j$, and so it is not restricted to the Euclidean straight-line path between two points. It includes the effect of curvature on length through the changing orientation of the local direction vectors along the path. A simple application of (1.64) shows that it yields the Pythagorean expression for distance in the Euclidean plane. In Cartesian coordinates, the use of (1.58) shows that the metric tensor for the Euclidean plane is given by $g_{11} = \mathbf{e}_x \cdot \mathbf{e}_x = 1$, $g_{22} = \mathbf{e}_y \cdot \mathbf{e}_y = 1$, and $g_{12} = g_{21} = \mathbf{e}_x \cdot \mathbf{e}_y = 0$, so that (1.64) reduces to

$$ds = \sqrt{g_{jk} dx^j dx^k} = \sqrt{(dx)^2 + (dy)^2}. \tag{1.68}$$

The length of a straight-line vector from the origin to the point (X, Y) can be found using the relation $y = Yx/X$ for the straight line, so that $dy = Y\,dx/X$. This gives the distance from the origin to the point

$$\begin{aligned} r &= \int_0^r ds = \int_0^X \sqrt{(dx)^2 + Y^2(dx)^2/X^2} = \int_0^X dx\, \sqrt{1 + Y^2/X^2} \\ &= X\sqrt{1 + Y^2/X^2} = \sqrt{X^2 + Y^2}, \end{aligned} \tag{1.69}$$

which is the Pythagorean result (1.59) in two dimensions. It should be stressed that integrating (1.64) requires choosing a path, and that the choice of path is arbitrary. In the case just treated, a straight-line path was used, since that is how distance is defined in a Euclidean manifold. However, non-Euclidean manifolds require redefining the distance between two points as the length of the shortest path between the two points, referred to as a *geodesic*. This will be revisited in the section on the calculus of variations, where the differential equation for the geodesic will be derived using (1.64).

Exercise 1.6. Use (1.43) to show that the metric tensor in spherical coordinates has the components

$$g_{rr} = 1, \quad g_{\theta\theta} = r^2, \quad g_{\phi\phi} = r^2 \sin^2\theta, \tag{1.70}$$

and that all other components vanish.

1.9 The metric tensor and raising and lowering indices

In addition to providing the infinitesimal element of length, the metric tensor enables the raising and lowering of indices, in effect converting contravariant and covariant indices into one another. This can be demonstrated by considering the scalar product $\mathbf{e}_j(x) \cdot d\boldsymbol{r}$,

$$\mathbf{e}_j(x) \cdot d\boldsymbol{r} = \mathbf{e}_j(x) \cdot dx^k \mathbf{e}_k(x) = \big(\mathbf{e}_j(x) \cdot \mathbf{e}_k(x)\big) dx^k = g_{jk}(x)\, dx^k \equiv dx_j. \tag{1.71}$$

The definition of the scalar product (1.53) shows that dx_j is the *perpendicular projection* of $d\boldsymbol{r}$ onto $\mathbf{e}_j$. This follows from $\mathbf{e}_j \cdot d\boldsymbol{r} = e_j\, dr \cos\theta$, where θ is by definition the angle between $\mathbf{e}_j$ and $d\boldsymbol{r}$. The quantity dx_j transforms covariantly, since (1.67) and (1.21) combine to give the relationship between dy_j and dx_k,

$$
\begin{aligned}
\mathrm{d}y_j = g_{jk}^{'}(y)\,\mathrm{d}y^k &= \frac{\partial x^i(y)}{\partial y^j}\frac{\partial x^\ell(y)}{\partial y^k}\,g_{i\ell}(x(y))\,\frac{\partial y^k(x)}{\partial x^m}\,\mathrm{d}x^m \\
&= \frac{\partial x^i(y)}{\partial y^j}\,\delta^\ell_m\,g_{i\ell}(x)\,\mathrm{d}x^m = \frac{\partial x^i(y)}{\partial y^j}\,g_{i\ell}(x)\,\mathrm{d}x^\ell = \frac{\partial x^i(y)}{\partial y^j}\,\mathrm{d}x_i,
\end{aligned} \tag{1.72}
$$

which is the covariant transformation (1.22). It is then possible to write the infinitesimal length (1.64) as

$$
\mathrm{d}s^2 = g_{jk}\,\mathrm{d}x^j\,\mathrm{d}x^k = \mathrm{d}x_j\,\mathrm{d}x^j, \tag{1.73}
$$

which is *manifestly invariant under a coordinate transformation*, since it is a contraction of the form (1.27).

For a finite position vector $\boldsymbol{r}$ in a Euclidean manifold, the quantity $\mathbf{e}_k \cdot \boldsymbol{r}$ is the covariant component r_k of the vector $\boldsymbol{r}$,

$$
r_k = \mathbf{e}_j \cdot \boldsymbol{r} = \mathbf{e}_k \cdot r^j \mathbf{e}_j = \mathbf{e}_k \cdot \mathbf{e}_j\, r^j = g_{kj} r^j, \tag{1.74}
$$

and this gives the length of the finite position vector as

$$
r^2 = \boldsymbol{r} \cdot \boldsymbol{r} = r^k r^j \mathbf{e}_k \cdot \mathbf{e}_j = r^k g_{kj} r^j = r^k r_k. \tag{1.75}
$$

In the case of Cartesian coordinates, the direction vectors are both fixed in orientation and orthonormal, resulting in the scalar product (1.58). Since the scalar products of (1.58) define the metric tensor components, it follows that the covariant Cartesian components of a position vector given by $r_j = g_{jk} r^k$ result in r^j and r_j coinciding. In a Euclidean manifold the contravariant and covariant components of a vector in Cartesian coordinates are therefore identical.

In order to clarify the difference between the contravariant and covariant components of a vector, it is useful to consider position vectors in the Euclidean plane $\mathbb{R}^2$ described in *skew coordinates*. Figure 1.4 displays an example of skew coordinates, where the y-axes are *not* orthogonal, instead forming the angle $\phi \neq \pi/2$. The skew coordinates of a point, designated (y^1, y^2), are related through *parallel*

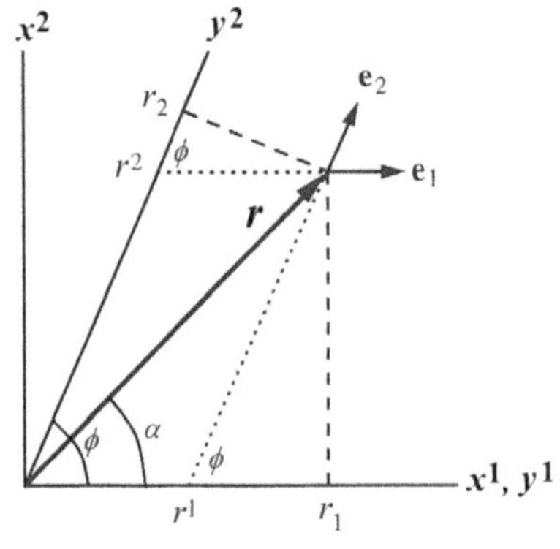

Figure 1.4. Contravariant and covariant position vector components in a skew coordinate system.

projection to the Cartesian coordinates of the same point, designated (x^1, x^2), by the coordinate transformation

$$x^1 = y^1 + \cos\phi\, y^2, \quad x^2 = \sin\phi\, y^2 \iff y^1 = x^1 - \cot\phi\, x^2, \quad y^2 = \csc\phi\, x^2. \tag{1.76}$$

It is useful to express the direction vectors in the skew coordinate system, $\mathbf{e}_1$ and $\mathbf{e}_2$, in terms of the Cartesian direction vectors, $\mathbf{e}_x$ and $\mathbf{e}_y$. This is because the latter are both fixed in orientation and orthonormal. The skew coordinate direction vectors are given by (1.36), so that

$$\begin{aligned} \mathbf{e}_1 &= \frac{\partial x^1}{\partial y^1}\mathbf{e}_x + \frac{\partial x^2}{\partial y^1}\mathbf{e}_y = \mathbf{e}_x, \\ \mathbf{e}_2 &= \frac{\partial x^1}{\partial y^2}\mathbf{e}_x + \frac{\partial x^2}{\partial y^2}\mathbf{e}_y = \cos\phi\, \mathbf{e}_x + \sin\phi\, \mathbf{e}_y. \end{aligned} \tag{1.77}$$

Using the orthonormality of the Cartesian unit vectors, the direction vectors of (1.77) give the metric tensor for the choice of skew coordinates,

$$\begin{aligned} g_{11} &= \mathbf{e}_1 \cdot \mathbf{e}_1 = 1, \quad g_{12} = g_{21} = \mathbf{e}_1 \cdot \mathbf{e}_2 = \cos\phi, \\ g_{22} &= \mathbf{e}_2 \cdot \mathbf{e}_2 = \cos^2\phi + \sin^2\phi = 1. \end{aligned} \tag{1.78}$$

It should be noted that, unlike the Cartesian direction vectors, the skew direction vectors $\mathbf{e}_1$ and $\mathbf{e}_2$ are not orthogonal, since the gridlines to which they are tangent are not orthogonal.

Exercise 1.7. Show that the results of (1.78) can also be obtained from the Cartesian metric tensor by applying the tensor transformation rule (1.67).

Using the metric tensor shows that the covariant components of vector $\boldsymbol{r}$ in figure 1.4 are given by

$$r_1 = g_{11}r^1 + g_{12}r^2 = r^1 + \cos\phi\, r^2, \quad r_2 = g_{21}r^1 + g_{22}r^2 = \cos\phi\, r^1 + r^2. \tag{1.79}$$

Comparing results (1.79) to the diagram shows that the covariant components are indeed the *perpendicular projections* of the vector onto the two skew coordinate axes and coincide with the contravariant components only if $\phi = \pi/2$. Finally, the extension of (1.73) to the finite position vector case gives the magnitude of the vector $\boldsymbol{r}$,

$$\begin{aligned} r^2 = r_j r^j &= r_1 r^1 + r_2 r^2 = (r^1 + \cos\phi\, r^2)r^1 + (\cos\phi\, r^1 + r^2)r^2 \\ &= (r^1)^2 + (r^2)^2 + 2r^1 r^2 \cos\phi, \end{aligned} \tag{1.80}$$

which is formula (1.55) for the addition of two vectors.

Similarly, the metric tensor can be used to define the contravariant version of the direction vectors, denoted $\mathbf{e}^j$, by relating them to the covariant direction vectors,

$$\mathbf{e}_j = g_{jk}\,\mathbf{e}^k. \tag{1.81}$$

Relation (1.81) can be solved for $\mathbf{e}^k$ using the quantities $h^{\ell j}$, which are chosen to have the property

$$h^{\ell j} g_{jk} = \delta^\ell_k. \tag{1.82}$$

The quantities $h^{\ell j}$ are readily found using matrix methods, where they are identified as the elements of the inverse of the metric tensor treated as an $n \times n$ matrix. This method is reviewed in the section on linear algebra in chapter 7. Using (1.82) gives $\mathbf{e}^j$,

$$h^{jk}\,\mathbf{e}_k = h^{jk} g_{k\ell}\,\mathbf{e}^\ell = \delta^j_\ell\,\mathbf{e}^\ell = \mathbf{e}^j. \tag{1.83}$$

The set of direction vectors $\{\mathbf{e}^j\}$ forms what is called the *reciprocal basis*, since these vectors are orthogonal to the covariant direction vectors $\{\mathbf{e}_j\}$. This follows from result (1.83), the metric tensor definition (1.65), and property (1.82) of h^{jk}, all of which combine to give

$$\mathbf{e}^k \cdot \mathbf{e}_j = (h^{k\ell}\mathbf{e}_\ell) \cdot \mathbf{e}_j = h^{k\ell}(\mathbf{e}_\ell \cdot \mathbf{e}_j) = h^{k\ell} g_{\ell j} = \delta^k{}_j. \tag{1.84}$$

Because $\delta^k{}_j$ is invariant under a coordinate transformation, it follows that relation (1.84) must hold after a coordinate transformation. Using the transformation property (1.36) of $\{\mathbf{e}_j\}$ shows that $\{\mathbf{e}^k\}$ obey

$$\begin{aligned} \left(\frac{\partial y^\ell}{\partial x^i}\,\mathbf{e}^i\right) \cdot \mathbf{e}'_k &= \left(\frac{\partial y^\ell}{\partial x^i}\mathbf{e}^i\right) \cdot \left(\frac{\partial x^j}{\partial y^k}\mathbf{e}_j\right) = \left(\frac{\partial y^\ell}{\partial x^i}\,\frac{\partial x^j}{\partial y^k}\right)\mathbf{e}^i. \\ \mathbf{e}_j &= \frac{\partial y^\ell}{\partial x^i}\frac{\partial x^j}{\partial y^k}\delta^i{}_j = \frac{\partial y^\ell}{\partial y^k} = \delta^\ell{}_k. \end{aligned} \tag{1.85}$$

In order for (1.84) to hold, it must be the case that

$$\mathbf{e}'^\ell = \frac{\partial y^\ell}{\partial x^i}\,\mathbf{e}^i, \tag{1.86}$$

which identifies $\{\mathbf{e}^k\}$ as transforming contravariantly.

It is possible to convert both indices of the metric tensor from covariant to contravariant; combining this with (1.82) reveals

$$g^{jk} = h^{j\ell} h^{ki} g_{\ell i} = h^{j\ell} h^{ki} g_{i\ell} = h^{j\ell}\delta^k{}_\ell = h^{jk}. \tag{1.87}$$

The contravariant version of the metric tensor therefore coincides with h^{jk}, so that $g^{ij}g_{jk} = \delta^i{}_k$. It also follows from the symmetry of g_{jk} that $g^{jk} = g^{kj}$. Combining the definition (1.65) of the covariant metric tensor with (1.81) gives

$$g^{jk} = g^{j\ell}g^{ki}g_{\ell i} = g^{j\ell}g^{ki}(\mathbf{e}_\ell \cdot \mathbf{e}_i) = (g^{j\ell}\mathbf{e}_\ell) \cdot (g^{ki}\mathbf{e}_i) = \mathbf{e}^j \cdot \mathbf{e}^k, \tag{1.88}$$

showing that g^{jk} is given by the reciprocal basis scalar product.

As an example, the reciprocal basis will be found for the skew coordinate system depicted in figure 1.4. Using matrix methods (to be reviewed in chapter 7), it can be shown that the skew coordinate metric tensor of (1.78) has the associated $g^{\ell k}$ given by $g^{11} = g^{22} = \csc^2\phi$ and $g^{12} = g^{21} = -\csc^2\phi\cos\phi$. It is straightforward to show that these $g^{\ell k}$ satisfy (1.82) for the metric tensor of (1.78). Using these forms for $g^{\ell k}$ gives the reciprocal basis in terms of the original basis,

$$\begin{aligned} \mathbf{e}^1 &= g^{11}\mathbf{e}_1 + g^{12}\mathbf{e}_2 = \csc^2\phi\,\mathbf{e}_1 - \csc^2\phi\cos\phi\,\mathbf{e}_2, \\ \mathbf{e}^2 &= g^{21}\mathbf{e}_1 + g^{22}\mathbf{e}_2 = -\csc^2\phi\cos\phi\,\mathbf{e}_1 + \csc^2\phi\,\mathbf{e}_2. \end{aligned} \tag{1.89}$$

It is straightforward to show that the reciprocal vectors of (1.89) satisfy (1.84). For example,

$$\mathbf{e}^1\cdot\mathbf{e}_1 = \csc^2\phi\,\mathbf{e}_1\cdot\mathbf{e}_1 - \csc^2\phi\cos\phi\,\mathbf{e}_2\cdot\mathbf{e}_1 = \csc^2\phi\,(1-\cos^2\phi) = 1, \tag{1.90}$$

$$\mathbf{e}^1\cdot\mathbf{e}_2 = \csc^2\phi\,\mathbf{e}_1\cdot\mathbf{e}_2 - \csc^2\phi\cos\phi\,\mathbf{e}_2\cdot\mathbf{e}_2 = \csc^2\phi\,(\cos\phi - \cos\phi) = 0. \tag{1.91}$$

However, the reciprocal vectors are not normalized, since

$$\begin{aligned} \mathbf{e}^1\cdot\mathbf{e}^1 = (e^1)^2 &= \csc^4\phi\,(\mathbf{e}_1\cdot\mathbf{e}_1 + \cos^2\phi\,\mathbf{e}_2\cdot\mathbf{e}_2 - 2\cos\phi\,\mathbf{e}_1\cdot\mathbf{e}_2) = \csc^4\phi \\ &(1-\cos^2\phi) = \csc^2\phi. \end{aligned} \tag{1.92}$$

The magnitude of $\mathbf{e}^1$ is therefore $e^1 = |\csc\phi| = 1/|\sin\phi|$, which is consistent with $g^{11} = \mathbf{e}^1\cdot\mathbf{e}^1 = (e^1)^2 = \csc^2\phi$.

The metric tensor components g_{jk} and those of its inverse g^{jk} can be used to express contractions in multiple forms. For example, a position vector can be written as

$$\boldsymbol{r} = \mathbf{e}_j\,x^j = g_{j\ell}\,\mathbf{e}^\ell g^{kj}x_k = (g^{kj}g_{j\ell})\,\mathbf{e}^\ell x_k = \delta^k{}_\ell\,\mathbf{e}^\ell x_k = \mathbf{e}^k x_k. \tag{1.93}$$

Even if x_k and x^k are different, the two contractions of (1.93) are identical. This gives an important result regarding the scalar product $\boldsymbol{r}\cdot\boldsymbol{r}$, which can be written in three equivalent ways,

$$\begin{aligned} r^2 &= \boldsymbol{r}\cdot\boldsymbol{r} = (\mathbf{e}_j\,x^j)\cdot(\mathbf{e}_k\,x^k) = (\mathbf{e}_j\cdot\mathbf{e}_k)x^jx^k = g_{jk}x^jx^k, \\ r^2 &= \boldsymbol{r}\cdot\boldsymbol{r} = (\mathbf{e}^j\,x_j)\cdot(\mathbf{e}^k\,x_k) = (\mathbf{e}^j\cdot\mathbf{e}^k)x_jx_k = g^{jk}x_jx_k, \\ r^2 &= \boldsymbol{r}\cdot\boldsymbol{r} = (\mathbf{e}_j\,x^j)\cdot(\mathbf{e}^k\,x_k) = (\mathbf{e}_j\cdot\mathbf{e}^k)x^jx_k = \delta^k{}_jx^jx_k = x^jx_j. \end{aligned} \tag{1.94}$$

All three give the same result, although the third version shows that the length of a vector is the contraction of the covariant and contravariant components. This shows that the scalar product of two different vectors can be written as

$$\boldsymbol{r}\cdot\boldsymbol{r}' = r^jr'_j. \tag{1.95}$$

Exercise 1.8. Treating the two-dimensional plane of figure 1.4 as Euclidean, show that the contravariant and covariant components of the vector $\boldsymbol{r}$ are given by

$$\begin{aligned} r^1 &= r\frac{\sin\alpha}{\sin\theta}, & r^2 &= r\frac{\sin(\theta-\alpha)}{\sin\theta}, \\ r_1 &= r\cos\alpha, & r_2 &= r\cos(\theta-\alpha), \end{aligned} \tag{1.96}$$

where r is the Euclidean magnitude of the vector $\boldsymbol{r}$. These immediately verify (1.95), since $r_1 r^1 + r_2 r^2 = r^2$.

The relationship of x^k and x_k carries over into partial derivatives. Using the chain rule gives

$$\frac{\partial}{\partial x_j} = \frac{\partial x^k}{\partial x_j}\frac{\partial}{\partial x^k} = \frac{\partial(g^{k\ell}x_\ell)}{\partial x_j}\frac{\partial}{\partial x^k} = g^{kj}\frac{\partial}{\partial x^k} = g^{kj}\partial_k = \partial^j. \tag{1.97}$$

This shows that $\partial/\partial x_j$ transforms contravariantly.

Another important result follows from considering the partial derivative of a direction vector. An example of this is given by

$$\begin{aligned} \partial_j \mathbf{e}^j &= g_{jk}\partial^k(g^{\ell j}\mathbf{e}_\ell) = (g^{\ell j}g_{jk})\partial^k\mathbf{e}_\ell + g_{jk}(\partial^k g^{\ell j})\mathbf{e}_\ell = \delta^\ell{}_k\partial^k\mathbf{e}_\ell + (\partial_j g^{\ell j})\mathbf{e}_\ell \\ &= \partial^k\mathbf{e}_k + (\partial_j g^{\ell j})\mathbf{e}_\ell. \end{aligned} \tag{1.98}$$

The two forms of the contraction, $\partial_j\mathbf{e}^j$ and $\partial^k\mathbf{e}_k$, are equal only if $\partial_j g^{\ell j} = 0$ for all ℓ. The Levi-Civita symbol defined in (1.5) can undergo a self-contraction using the metric tensor that results in zero. For example, in three dimensions

$$\varepsilon_j^{\ jk} = g_{j\ell}\varepsilon^{jk\ell} = g_{\ell j}\varepsilon^{\ell kj} = -g_{j\ell}\varepsilon^{jk\ell} = -\varepsilon_j^{\ jk}, \tag{1.99}$$

where the relabeling of dummy indices was combined with property (1.66) for the metric, $g_{j\ell} = g_{\ell j}$, and an odd number of permutations for the Levi-Civita symbol, $\varepsilon^{\ell kj} = -\varepsilon^{jk\ell}$. Result (1.99) shows that $\varepsilon_j^{\ jk} = 0$.

Exercise 1.9 Show that $\varepsilon^{\mu\nu\cdots}{}_\mu = 0$, where the number of indices is arbitrary.

1.10 General tensors and tensor densities

The metric tensor transformation property (1.67) serves as the prototype for the transformation property of a *general tensor*. A tensor may depend on position, as the metric tensor does in a non-Euclidean manifold. In a coordinate system a general tensor $\boldsymbol{T}$ has the components $T^{ij\cdots}_{k\ell\ldots}(x)$. The superscripts indicate contravariant indices and the subscripts indicate covariant indices. The general tensor component transforms with the extensions of (1.21) and (1.22),

$$T'^{ij\ldots}_{k\ell\ldots}(y) = \frac{\partial y^i(x)}{\partial x^{i'}}\frac{\partial y^j(x)}{\partial x^{j'}}\cdots\frac{\partial x^{k'}(y)}{\partial y^k}\frac{\partial x^{\ell'}(y)}{\partial y^\ell}\cdots T^{i'j'\ldots}_{k'\ell'\ldots}(x)\bigg|_{x(y)}. \tag{1.100}$$

The *rank* of a tensor is given by the total number of covariant and contravariant indices. If a tensor has r contravariant indices and p covariant indices, it referred to as a type (r, p) tensor. In this context, the direction vector transformation property (1.36) identifies them as *first-rank covariant tensors*, while scalars, such as the differential operator d of (1.25), are *zero-rank tensors*.

There is a class of objects that transforms somewhat differently than (1.100); members of this class are referred to as *tensor densities*. An example is the Levi-Civita symbol of (1.5). This can be seen by contracting both sides of relationship (1.14) with the sequence $\prod_{j=1}^{n} \partial x^{k_j}/\partial y^{\alpha_j}$ and using the chain rule (1.4). The result is

$$\begin{aligned}\frac{1}{J(y)}\frac{\partial x^{k_1}}{\partial y^{\alpha_1}}\cdots\frac{\partial x^{k_n}}{\partial y^{\alpha_n}}\varepsilon_{k_1\ldots k_n} &= \frac{\partial x^{k_1}}{\partial y^{\alpha_1}}\frac{\partial y^{j_1}}{\partial x^{k_1}}\cdots\frac{\partial x^{k_n}}{\partial y^{\alpha_n}}\frac{\partial y^{j_n}}{\partial x^{k_n}}\varepsilon_{j_1\ldots j_n} = \delta^{j_1}{}_{\alpha_1}\cdots \\ &\quad \delta^{j_n}{}_{\alpha_n}\varepsilon_{j_1\ldots j_n} = \varepsilon_{\alpha_1\ldots\alpha_n}.\end{aligned} \tag{1.101}$$

Like the Kronecker delta, the Levi-Civita symbol is a mathematical construct and therefore coordinate independent. Result (1.101) is therefore an *identity* that can be used in defining coordinate invariant expressions involving the Levi-Civita symbol. The left-hand side of identity (1.101) shows that the Levi-Civita symbol can be treated as a covariant tensor of rank n under a transformation, but requires an additional factor of the inverse of the Jacobian of the transformation. This is an important aspect of defining coordinate invariant expressions involving the Levi-Civita symbol, since the factor of the Jacobian that appears can be used to cancel Jacobian factors generated by other quantities. A quantity whose transformation involves the Jacobian of the transformation is referred to as a *tensor density*, and the power of the Jacobian determines its *weight*. Result (1.17) shows that the Jacobian $J(x)$ of the inverse transformation from y-coordinates to x-coordinates is the inverse of the Jacobian $J(y)$ of the transformation from x-coordinates to y-coordinates. This follows from performing the transformation and its inverse successively:

$$\mathrm{d}^n x = J(y)\mathrm{d}^n y = J(y)J(x)\mathrm{d}^n x \implies J(x)J(y) = 1. \tag{1.102}$$

Result (1.102) shows that $J(x) = 1/J(y)$. The covariant components of the Levi-Civita symbol correspond to a tensor density of weight +1, which refers to the single positive power of $J(x)$ appearing on the left-hand side of (1.101).

If a tensor is to represent a physically measurable quantity, it must be expressed in a coordinate-independent way, as in the case of $\mathrm{d}\boldsymbol{r} = \mathrm{d}x^j\mathbf{e}_j$. For a general tensor, this is done by using the *tensor product* for the direction vectors, written $\mathbf{e}_i \otimes \mathbf{e}_j \otimes \cdots \otimes \mathbf{e}_k$. The tensor product is understood to be the product of the directional degrees of freedom. For example, a vector in $\mathbb{R}^n$ can be written $x^j\mathbf{e}_j$, while a second vector in $\mathbb{R}^n$ can be written $y^k \mathbf{e}_k$. Their tensor product, sometimes referred to as a *dyad*, is written

$$\boldsymbol{T} = x^j \mathbf{e}_j \otimes y^k \mathbf{e}_k = x^j y^k \mathbf{e}_j \otimes \mathbf{e}_k \equiv T^{jk} \mathbf{e}_j \otimes \mathbf{e}_k. \tag{1.103}$$

There are n^2 components, $T^{jk} = x^j y^k$, and these components transform as a second-rank contravariant tensor. When coupled with the transformation property (1.36) of $\mathbf{e}_j$, the tensor $\boldsymbol{T}$ defined by (1.103) is coordinate independent. The tensor product of the direction vectors $\mathbf{e}_j \otimes \mathbf{e}_k$ serves as the basis for *linear functions* that map the vectors $\boldsymbol{w}$ into another vector $\boldsymbol{v}$ through the scalar product,

$$\begin{aligned} \boldsymbol{T}(\boldsymbol{w}) &= \boldsymbol{T} \cdot \boldsymbol{w} = \left(T^{jk} \mathbf{e}_j \otimes \mathbf{e}_k\right) \cdot (w^\ell \mathbf{e}_\ell) = \left(T^{jk} w^\ell \mathbf{e}_j\right)(\mathbf{e}_k \cdot \mathbf{e}_\ell) = (g_{k\ell} w^\ell T^{jk})\mathbf{e}_j \\ &= T^{jk} w_k \, \mathbf{e}_j = \boldsymbol{v}, \end{aligned} \tag{1.104}$$

where $v^j = T^{jk} w_k$. Definition (1.104) gives $\boldsymbol{T}(a\boldsymbol{u} + b\boldsymbol{v}) = a\, \boldsymbol{T} \cdot \boldsymbol{u} + b\, \boldsymbol{T} \cdot \boldsymbol{v}$, where $\boldsymbol{u}$ and $\boldsymbol{v}$ are any two vectors and a and b are any two real numbers. This property classifies tensor $\boldsymbol{T}$ as a *linear function of vectors.* Similarly, covariant tensor components can be combined with reciprocal basis vectors in a coordinate-independent manner. For example, the metric tensor can be written

$$\boldsymbol{g} = g_{jk} \, \mathbf{e}^j \otimes \mathbf{e}^k. \tag{1.105}$$

The metric tensor can then be viewed as a scalar function of two vectors. Using (1.84) gives

$$\begin{aligned} \boldsymbol{g}(\mathrm{d}\boldsymbol{r}, \mathrm{d}\boldsymbol{r}) &= (\mathrm{d}x^i \mathbf{e}_i) \cdot (g_{jk} \, \mathbf{e}^j \otimes \mathbf{e}^k) \cdot (\mathrm{d}x^\ell \mathbf{e}_\ell) = g_{jk} \, \mathrm{d}x^i \, \mathrm{d}x^\ell \, \delta^j{}_i \, \delta^k_\ell \\ &= g_{jk} \, \mathrm{d}x^j \, \mathrm{d}x^k = \mathrm{d}s^2. \end{aligned} \tag{1.106}$$

In effect, the metric tensor is a linear function that maps two vectors into a distance.

1.11 Trajectories and tangent spaces

So far, attention has primarily been focused on position vectors, either the infinitesimal version $\mathrm{d}\boldsymbol{r}$ of (1.32) or the finite version $\boldsymbol{r}$ of (1.35). Regardless of whether the manifold is Euclidean or not, the metric tensor relation (1.64) associates a distance *in the manifold* with position vectors, i.e. either infinitesimal position changes or, by integration, finite distance vectors. However, vectors may be present at every point in the manifold that are independent of the choice of an origin. Examples from Newtonian physics include the force acting on a particle or its velocity when it is located at a specific point. Since force and velocity have both a magnitude and direction, they must be represented by a vector *at that point.*

Defining vectors at an arbitrary point starts by defining a *trajectory* $\boldsymbol{r}(t)$ that passes through the point. A trajectory is a continuous sequence of points described by a position vector $\boldsymbol{r}(t)$ that is a differentiable function of the parameter t. This is identical to the trajectory of a point particle in Newtonian mechanics, where the parameter t represents time. In what follows the parameter t will be referred to as the *time*, for the moment ignoring any issues around its units. The trajectory is assumed to pass through the point p of interest in the manifold when the parameter t has the value t_p, so that $\boldsymbol{r}(t_p) = \boldsymbol{r}_p$ is the same position vector of the point p that was used to define the direction vectors (1.30). The position vector is represented in the x-coordinate system by $\boldsymbol{r}(x)$ and the coordinates x^k of the vector tip will now be treated as time-dependent quantities $x^k(t)$. In every other way the trajectory is arbitrary. The rate at which $\boldsymbol{r}(t) = \boldsymbol{r}(x(t))$ changes at p yields the *velocity vector* $\boldsymbol{v}_p$ at the point p, defined by

$$\boldsymbol{v}_p = \left.\frac{\mathrm{d}\boldsymbol{r}(t)}{\mathrm{d}t}\right|_{t_p} = \left.\frac{\partial \boldsymbol{r}(x)}{\partial x^k}\frac{\mathrm{d}x^k(t)}{\mathrm{d}t}\right|_{t_p} = \mathbf{e}_k(x(t_p))\,\dot{x}^k(t_p) \equiv v^k(t_p)\,\mathbf{e}_k(x(t_p)), \tag{1.107}$$

where the notations $v^k(t) = \dot{x}^k(t) \equiv \mathrm{d}x^k(t)/\mathrm{d}t$ and $a^k(t) = \dot{v}^k(t) = \ddot{x}^k(t) \equiv \mathrm{d}^2x^k(t)/\mathrm{d}t^2$ will be used from this point on. The form of v^k, the kth component of the velocity vector at the time t, will depend upon the coordinate system chosen as well as the trajectory. The definition (1.107) shows that the velocity vector at the point p is described in terms of the same direction vectors (1.30) at the point p that are used for the position vectors. Definition (1.107) is also consistent with the Cartan form (1.45) for vectors, since (1.107) can be written as

$$v^k(\boldsymbol{x})\frac{\partial}{\partial x^k} = \left(v^j(\boldsymbol{x})\frac{\partial r^k}{\partial x^j}\right)\frac{\partial}{\partial x^k} = (\boldsymbol{v}(\boldsymbol{x})r^k(\boldsymbol{x}))\frac{\partial}{\partial x^k}, \tag{1.108}$$

although the notation of (1.107) is more suited to Newtonian mechanics.

While the direction vectors $\mathbf{e}_j$ were initially developed for use with position vectors, their utility lies in the fact that they exist *at every point in the manifold*, where they are *tangent* to the jth gridline. In turn, this means that the metric tensor of the manifold *at that point* will determine the magnitude of the velocity vector,

$$v^2 = \boldsymbol{v}\cdot\boldsymbol{v} = v^j(x)v^k(x)\,\mathbf{e}_j(x)\cdot\mathbf{e}_k(x) = g_{jk}(x)v^j(x)v^k(x) = v^j(x)v_j(x). \tag{1.109}$$

Statement (1.109) is true *regardless of whether the manifold is Euclidean or not*. Like the relationship between direction vectors and gridlines, the velocity vector $\boldsymbol{v}(t_p)$ is *tangent* to the trajectory $\boldsymbol{r}(t_p)$ at the point p. This is true since the components of $\boldsymbol{v}(t_p)$ have the same slope as the trajectory in all directions. This follows from

$$\frac{v^j(t_p)}{v^k(t_p)} = \left(\left.\frac{\mathrm{d}x^j(t)}{\mathrm{d}t}\right|_{t_p}\right)\left(\left.\frac{\mathrm{d}x^k(t)}{\mathrm{d}t}\right|_{t_p}\right)^{-1} = \frac{\mathrm{d}x^j(t_p)}{\mathrm{d}x^k(t_p)}, \tag{1.110}$$

which is identical to the slope of the trajectory at point p for all possible values of j and k.

Since there is a nondenumerable infinity of trajectories that pass through point p, the set of all possible tangent vectors at point p given by (1.107) forms a *space of vectors* that are tangent to the manifold $\mathcal{M}$ at the point p. The term *vector space* means that any two vectors belonging to the space can be combined through addition and scalar multiplication, so that a linear combination of any two tangent vectors at a point results in a vector that is also tangent to the point. As a result, the space of vectors defined in this manner is referred to as the *tangent space* at the point p and is denoted by $T_p(\mathcal{M})$. It is important to understand that the vectors in the tangent space do not lie in the manifold and their magnitude is *not* correlated with the distance from the origin to the point. However, the definition of (1.107) shows that the vectors in the tangent space of a point can be expressed in terms of the same basis direction vectors used to describe the position vector, either finite or infinitesimal, at the point. The tangent space therefore has the same dimension as the manifold, which means that any vector in the tangent space at a point can be expanded as a linear combination of the normalized direction vectors at the point. This is because the direction vectors are also tangent to the manifold $\mathcal{M}$ at the point p and therefore form a basis for the tangent vectors. For an n-dimensional manifold the tangent space $T_p(\mathcal{M})$ at each point coincides with $\mathbb{R}^n$, and the tangent point serves as the origin of the tangent space.

The one-dimensional circle and two-dimensional sphere, denoted S^1 and S^2 respectively, have easily visualized tangent spaces. In the case of the circle, the tangent space is the line $\mathbb{R}^1$ tangent to the circle at the point p, while in the case of the sphere, it is a plane $\mathbb{R}^2$ tangent to the sphere at point p. These are depicted in figure 1.5 by *embedding* them in a higher-dimensional space which has the position vector $\boldsymbol{r}$ defining the direction vectors tangent at that point. In each case the tangent vectors at point p can be visualized as an arrow in the tangent space. It should be stressed that the tangent vectors in $T_p(\mathcal{M})$ are defined only at point p and do not extend to other points in the underlying manifold. In the case of $T_p(S^1)$, the polar coordinate direction vector $\mathbf{e}_\theta$ from the polar coordinates of $\mathbb{R}^2$ serves as the single direction vector required to describe a vector in the tangent space. In the case of $T_p(S^2)$, the spherical coordinate direction vectors $\mathbf{e}_\theta$ and $\mathbf{e}_\phi$ of $\mathbb{R}^3$ serve as orthogonal direction vectors for the tangent plane. A vector in $T_p(S^1)$ takes the form $v = v^\theta \mathbf{e}_\theta$, while a vector in $T_p(S^2)$ takes the form $v = v^\theta \mathbf{e}_\theta + v^\phi \mathbf{e}_\phi$. It is possible to perform a coordinate transformation in the manifold and use the new direction vectors as the basis in the tangent space. This transformation also requires the tangent vector components to undergo the contravariant transformation, just as the components of position vectors are transformed. However, the tangent space and the manifold only

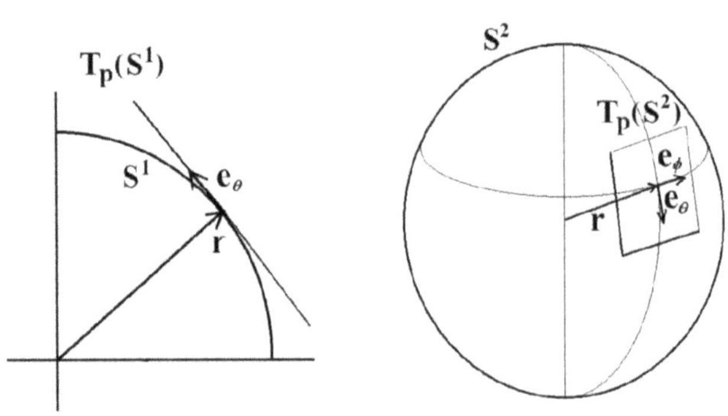

Figure 1.5. The tangent spaces for the circle and the sphere, $T_p(S^1)$ and $T_p(S^2)$.

share the tangent point p, which serves as the origin of the tangent space and the point of definition of the direction vectors.

1.12 The vector product

A second form of vector multiplication, known as the *cross* or *vector product*, defines a vector that is orthogonal to a collection of other vectors. The version of the vector product presented in this section will be specific to three spatial dimensions. A related definition of the vector product for higher dimensions is given in a later section. The vector product is quite often used in three-dimensional quantities, such as angular momentum and the Lorentz force. However, it is less useful in arbitrary dimensions and has largely been superseded by the *wedge product* to be discussed later in this chapter. The definition of the vector product in three dimensions uses the *orthonormal* Cartesian unit vectors $\mathbf{e}_j$,

$$\mathbf{e}_j \times \mathbf{e}_k = \varepsilon_{jk\ell}\, \mathbf{e}^\ell, \tag{1.111}$$

where ε_{jki} is the covariant Levi-Civita symbol defined by (1.5) which obeys the identity (1.101). Because $\varepsilon_{ijk} = -\varepsilon_{jik}$, definition (1.112) has the property that $\mathbf{e}_j \times \mathbf{e}_k = -\mathbf{e}_k \times \mathbf{e}_j$, so that $\mathbf{e}_j \times \mathbf{e}_j = 0$.

In three dimensions the vector product gives rise to the *vector product* of two Euclidean position vectors in Cartesian coordinates,

$$\boldsymbol{x} \times \boldsymbol{y} = x^i y^j (\mathbf{e}_i \times \mathbf{e}_j) = \varepsilon_{ijk} x^i y^j \mathbf{e}^k. \tag{1.112}$$

The vector given by (1.112) is orthogonal to both $\boldsymbol{x}$ and $\boldsymbol{y}$, since $\boldsymbol{x} \cdot (\boldsymbol{x} \times \boldsymbol{y}) = \varepsilon_{ijk} x^i x^j y^k = 0$. If both position vectors $\boldsymbol{x}$ and $\boldsymbol{y}$ lie in the 1-2 plane, so that $x^3 = y^3 = 0$, their vector product is given by

$$\boldsymbol{x} \times \boldsymbol{y} = x^i y^j \varepsilon_{ijk} \mathbf{e}^k = x^i y^j \varepsilon_{ij3} \mathbf{e}^3 = (x^1 y^2 - x^2 y^1)\mathbf{e}^3. \tag{1.113}$$

Expressing the Cartesian components of the two vectors in polar coordinates, $x^1 = r\cos\theta$, $x^2 = r\sin\theta$, $y^1 = r'\cos\theta'$, and $y^2 = r'\sin\theta'$, the vector product (1.113) becomes

$$\boldsymbol{x} \times \boldsymbol{y} = rr'(\cos\theta\sin\theta' - \sin\theta\cos\theta')\mathbf{e}^3 = rr'\sin(\theta' - \theta)\,\mathbf{e}^3. \tag{1.114}$$

If the two vectors are parallel then $\theta' = \theta$ and the vector product vanishes, which is consistent with $\boldsymbol{x} \times \boldsymbol{x} = 0$. Euclidean geometry shows that the magnitude of the vector product (1.114) is the area of the parallelogram formed by the two vectors, while its direction is perpendicular to the plane the two vectors form. For the case of $\mathbf{e}^3 = \mathbf{e}^z$, the vector points in the $+z$ direction if $\theta' - \theta > 0$. This is often referred to as the *right-hand rule*, since it is identical to a right-handed coordinate system where the right thumb aligns along the z-axis when the first two fingers are aligned with the x- and y-axes, respectively.

Defining the three *unsummed* infinitesimal vectors $\mathrm{d}\boldsymbol{r}_j = \mathrm{d}x^{(j)}\mathbf{e}_j$ and using (1.111) gives

$$\mathrm{d}\boldsymbol{r}_i \times \mathrm{d}\boldsymbol{r}_j = \mathrm{d}x^{(i)}\,\mathrm{d}x^{(j)}\,\mathbf{e}_i \times \mathbf{e}_j = \left(\mathrm{d}x^{(i)}\,\mathrm{d}x^{(j)}\,\varepsilon_{ijk}\right)\mathbf{e}^k \equiv \varepsilon_{ijk}\,\mathrm{d}\boldsymbol{S}^k, \tag{1.115}$$

where $\mathrm{d}\boldsymbol{S}^k$ is the kth contravariant component of the *oriented surface element*. This is understood by considering the case $i = 1$ and $j = 2$ for (1.115), which shows that the infinitesimal surface element $\mathrm{d}S^{(3)} = \mathrm{d}S^z = \mathrm{d}x\,\mathrm{d}y$ is a rectangular surface element in the x–y plane. This infinitesimal element is associated with the direction vector $\mathbf{e}^z$, which is the direction *normal* to the surface element in three dimensions. The oriented surface element is written $\mathrm{d}\boldsymbol{S} = \mathrm{d}S\,\hat{\boldsymbol{n}}$, where $\hat{\boldsymbol{n}}$ is the unit vector perpendicular or normal to the plane defining the infinitesimal surface element $\mathrm{d}S$. For example, using the spherical coordinate direction vectors (1.43) and $\mathbf{e}_x \times \mathbf{e}_y = \mathbf{e}_z$, $\mathbf{e}_z \times \mathbf{e}_x = \mathbf{e}_y$, and $\mathbf{e}_z \times \mathbf{e}_y = -\mathbf{e}_x$ shows that

$$\mathbf{e}_\theta \times \mathbf{e}_\phi = r^2 \sin\theta(\sin\theta\cos\theta\,\mathbf{e}_x + \sin\theta\sin\phi\,\mathbf{e}_y + \cos\theta\,\mathbf{e}_z) = r^2\sin\theta\,\mathbf{e}_r. \tag{1.116}$$

The surface element of a sphere is therefore given by

$$\mathrm{d}\boldsymbol{S} = \mathrm{d}\theta\,\mathrm{d}\phi\,(\mathbf{e}_\theta \times \mathbf{e}_\phi) = r^2\sin\theta\,\mathrm{d}\theta\,\mathrm{d}\phi\,\mathbf{e}^r = \mathrm{d}S_{(r)}\,\mathbf{e}^r, \tag{1.117}$$

so that the oriented surface elements of a sphere are radial and perpendicular to the surface of the sphere.

Exercise 1.10. Verify (1.117).

Similarly, definition (1.111) also gives the Cartesian result

$$\mathbf{e}_i \cdot (\mathbf{e}_j \times \mathbf{e}_k) = \varepsilon_{jk\ell}\,\mathbf{e}_i \cdot \mathbf{e}^\ell = \varepsilon_{jk\ell}\,\delta^\ell{}_i = \varepsilon_{jki} = \varepsilon_{ijk}, \tag{1.118}$$

which vanishes if any of the indices on the left-hand side are the same. Result (1.118) is related to the three-dimensional volume element $\mathrm{d}^3x = \mathrm{d}x\,\mathrm{d}y\,\mathrm{d}z$ in the following way. Using (1.118) gives

$$\mathrm{d}\boldsymbol{r}_i \cdot (\mathrm{d}\boldsymbol{r}_j \times \mathrm{d}\boldsymbol{r}_k) = \mathrm{d}x^{(i)}\,\mathrm{d}x^{(j)}\,\mathrm{d}x^{(k)}\,\mathbf{e}_i \cdot (\mathbf{e}_j \times \mathbf{e}_k) = \varepsilon_{ijk}\,\mathrm{d}^3x, \tag{1.119}$$

where $\mathrm{d}^3x = \mathrm{d}x^{(1)}\,\mathrm{d}x^{(2)}\,\mathrm{d}x^{(3)}$ is the infinitesimal volume element in Cartesian coordinates. Relation (1.119) is understood to be an *oriented volume element*, an infinitesimal parallelepiped with an oriented base rectangle of $\mathrm{d}\boldsymbol{r}_j \times \mathrm{d}\boldsymbol{r}_k$ and a height of $\mathrm{d}\boldsymbol{r}_i$. The relation (1.118) can be subjected to a coordinate transformation by contracting it with factors of $\partial x^i/\partial y^\ell$, giving

$$\frac{\partial x^i}{\partial y^\ell}\frac{\partial x^j}{\partial y^m}\frac{\partial x^k}{\partial y^n}\,\mathbf{e}_i \cdot (\mathbf{e}_j \times \mathbf{e}_k) = \frac{\partial x^i}{\partial y^\ell}\frac{\partial x^j}{\partial y^m}\frac{\partial x^k}{\partial y^n}\,\varepsilon_{ijk}. \tag{1.120}$$

Using (1.36) shows that the left-hand side of (1.120) gives the product of the transformed direction vectors, denoted by $\mathbf{e}'_\ell$, while identity (1.101) can be employed on the right-hand side. The result is

$$\mathbf{e}'_\ell \cdot (\mathbf{e}'_m \times \mathbf{e}'_n) = J(y)\, \varepsilon_{\ell mn}, \tag{1.121}$$

which shows that the product of the magnitudes of the transformed direction vectors is given by the Jacobian of the coordinate transformation. Combining $\mathrm{d}\boldsymbol{r}'_j = \mathrm{d}y^{(j)}\mathbf{e}'_{(j)}$ with result (1.121) gives

$$\mathrm{d}\boldsymbol{r}'_\ell \cdot (\mathrm{d}\boldsymbol{r}'_m \times \mathrm{d}\boldsymbol{r}'_n) = \mathrm{d}y^{(\ell)}\,\mathrm{d}y^{(m)}\,\mathrm{d}y^{(n)}\,\mathbf{e}'_\ell \cdot (\mathbf{e}'_m \times \mathbf{e}'_n) = J(y)\,\mathrm{d}^3y\,\varepsilon_{\ell mn}, \tag{1.122}$$

which is the transformed version of (1.119). The appearance of the Jacobian in (1.122) demonstrates the rule (1.17) for the transformation of three-dimensional infinitesimal volume elements. Result (1.121) also shows that the transformed direction vectors are normalized by the Jacobian,

$$\hat{\mathbf{e}}'_\ell \cdot (\hat{\mathbf{e}}'_m \times \hat{\mathbf{e}}'_n) = \frac{\mathbf{e}'_\ell \cdot (\mathbf{e}'_m \times \mathbf{e}'_n)}{J(y)} = \varepsilon_{\ell mn}, \tag{1.123}$$

where the associated unit vectors $\hat{\mathbf{e}}'_j$ are given by

$$\hat{\mathbf{e}}'_j = \mathbf{e}'_j\left(\mathbf{e}'_{(j)} \cdot \mathbf{e}'_{(j)}\right)^{-\frac{1}{2}} = \mathbf{e}'_j\left(g'_{(jj)}\right)^{-\frac{1}{2}}. \tag{1.124}$$

This shows that the Jacobian of the transformation for three dimensions is given by

$$J(y) = \prod_{j=1}^{3} \sqrt{\mathbf{e}'_{(j)} \cdot \mathbf{e}'_{(j)}}. \tag{1.125}$$

> **Exercise 1.11.** Use the spherical coordinate direction vectors (1.43) and expression (1.125) to show that the Jacobian for transforming from three-dimensional Cartesian coordinates to spherical coordinates (1.42) is $J(r, \theta, \phi) = r^2 \sin\theta$. Verify that $\mathbf{e}_r \cdot (\mathbf{e}_\theta \times \mathbf{e}_\phi) = r^2 \sin\theta = J(r, \theta, \phi)$.

The results of this section, such as (1.115) and (1.119), have been limited to three-dimensional spaces. In a later section this limitation is lifted by defining the *wedge product*. However, this requires further developments.

1.13 The gradient

The relationship between direction vectors and partial derivatives is encapsulated in the *gradient*, a key mathematical tool in both differential geometry and field theory. The gradient, denoted ∇, is defined by combining the contravariant reciprocal vectors $\mathbf{e}^k$ of (1.84) with the covariant partial derivatives $\partial/\partial x^k$,

$$\nabla = \mathbf{e}^k(x)\frac{\partial}{\partial x^k} \equiv \mathbf{e}^k(x)\partial_k. \tag{1.126}$$

It is important to note that the gradient uses the reciprocal basis at the same point where the partial derivatives will be evaluated. In the case of a Euclidean manifold with Cartesian coordinates, these reciprocal vectors are the same everywhere. However, in other coordinate systems and in non-Euclidean manifolds, the direction vectors and their reciprocals change with location. As a result, the order of the two components in (1.126) is critical, since it *cannot* be assumed that $\partial_k \mathbf{e}^j = 0$. Because it takes the familiar form (1.27) of a contraction, the gradient is a coordinate invariant *vector operator*. The action of the gradient on a *scalar function* $f(x)$, whose argument x is the coordinates of the point of interest, yields a vector at $\boldsymbol{r}(x)$ given by

$$\nabla f(x) = \mathbf{e}^j(x)\,\frac{\partial f(x)}{\partial x^j} = \mathbf{e}^j(x)\,\partial_j f(x). \tag{1.127}$$

The gradient appears in the application of the differential operator (1.23) to a scalar function $f(x)$ of several variables, which can be written

$$\mathrm{d}f(x) = \mathrm{d}x^k \frac{\partial f(x)}{\partial x^k} = \mathrm{d}x^k\,\delta^j{}_k\,\frac{\partial}{\partial x^j} f(x) = \mathrm{d}x^k \mathbf{e}_k \cdot \mathbf{e}^j \frac{\partial}{\partial x^j} f(x) = \mathrm{d}\boldsymbol{r} \cdot \nabla f(x). \tag{1.128}$$

Using (1.84) shows that

$$\mathbf{e}_k(x) \cdot \nabla f(x) = \mathbf{e}_k(x) \cdot \mathbf{e}^j(x)\,\partial_j f(x) = \delta^j{}_k \partial_j f(x) = \partial_k f(x), \tag{1.129}$$

which is the kth covariant component of $\nabla f(x)$.

In that regard, the differential of a single-valued multivariable function $f(\boldsymbol{r})$ can be combined with the concept of the tangent space to find its extremum points. For example, the extrema of a function $f(x, y)$ of two variables can be found by extending consideration to a three-dimensional Euclidean space and three-dimensional trajectory

$$\boldsymbol{r}(t) = x(t)\hat{\mathbf{e}}_x + y(t)\hat{\mathbf{e}}_y + f(x(t), y(t))\hat{\mathbf{e}}_z, \tag{1.130}$$

where $x(t)$ and $y(t)$ are arbitrary trajectories through the x–y plane. This trajectory traces a path through the surface defined by $f(x, y)$. The function $f(x, y)$ has extrema at those points where the tangent plane of the trajectory is orthogonal to $\hat{\mathbf{e}}_z$, making it parallel to the x–y plane. This gives

$$\hat{\mathbf{e}}_z \cdot \frac{\mathrm{d}\boldsymbol{r}(t)}{\mathrm{d}t} = \dot{x}\frac{\partial f(x, y)}{\partial x} + \dot{y}\frac{\partial f(x, y)}{\partial y} = \frac{\mathrm{d}\boldsymbol{r}}{\mathrm{d}t} \cdot \nabla f(x, y) = 0. \tag{1.131}$$

Since $\boldsymbol{v} = \mathrm{d}\boldsymbol{r}(t)/\mathrm{d}t$ is arbitrary, the extrema of a multivariable function are determined as those points where

$$\nabla f = 0 \implies \mathrm{d}f(\boldsymbol{r}) = \mathrm{d}\boldsymbol{r} \cdot \nabla f(\boldsymbol{r}) = 0, \tag{1.132}$$

which is the usual criterion for an extremum point.

In coordinate systems other than Cartesian, a great deal of confusion regarding units can be avoided by rewriting (1.127) in terms of *unit vectors*. This is done using the metric tensor to swap the indices, as in (1.93), and then normalizing the direction vectors by the replacement $\mathbf{e}_j = e_{(j)}\hat{\mathbf{e}}_j$, where $(e_j)^2 = \mathbf{e}_{(j)} \cdot \mathbf{e}_{(j)}$. This gives

$$\begin{aligned}\nabla f(x) &= \mathbf{e}^k(x)\,\partial_k f(x) = \mathbf{e}_k(x)\,\partial^k f(x) = \hat{\mathbf{e}}_k(x)\big(e_{(k)}(x)\,\partial^k f(x)\big) \\ &= \hat{\mathbf{e}}_k(x) V^k(x) \;\equiv\; \boldsymbol{V}(x).\end{aligned} \tag{1.133}$$

The action of the gradient on a scalar function creates a *vector function* $\boldsymbol{V}(x) = \nabla f(x)$ at the point whose coordinates are x. It is important to note that the vector function $\boldsymbol{V}(x)$ exists at all points in the manifold where the gradient of the function is well defined, and so the gradient creates a *vector field* throughout the manifold once the function $f(x)$ is specified. The components of the vector field, $V^k(x)$, depend on the coordinate system chosen to define the direction vectors. The vector field $\boldsymbol{V}(x)$ is the sum of component vectors directed along the direction vectors, and so the vector field defined by (1.133) is also *tangent* to the manifold at that point, just as the direction vectors are. This is discussed further in the next section.

The form of the gradient depends upon the choice of coordinate system. For consistency, the gradient (1.126) will be expressed in terms of dimensionless unit vectors using the method of (1.133). This gives

$$\nabla = \mathbf{e}^k(x)\,\partial_k = \mathbf{e}_j(x)\,g^{jk}(x)\,\partial_k = \hat{\mathbf{e}}_j(x)\big(e_{(j)}(x)\,g^{jk}(x)\partial_k\big) \;\equiv\; \hat{\mathbf{e}}_j(x)\,\hat{\partial}^j. \tag{1.134}$$

Definition (1.134) is one of the occasions for which there is a sum over j even though (j) occurs in the expression. In the case of Cartesian coordinates in a Euclidean space, the orthonormality of the direction vectors shows that the reciprocal vectors $\mathbf{e}^k$ coincide with the unit vectors $\hat{\mathbf{e}}_k$. This follows from result (1.58), which shows that $g_{jk} = 1$ if $j = k$ and is zero otherwise. This gives both $e_j = (g_{(jj)})^{1/2} = 1$ and $g^{jk} = 1$ if $j = k$, with g^{jk} equal to zero if $j \neq k$. In three dimensions this gives the Cartesian gradient

$$\nabla = \hat{\mathbf{e}}_j\big(e_{(j)}\,g^{jk}\partial_k\big) = \hat{\mathbf{e}}_x\frac{\partial}{\partial x} + \hat{\mathbf{e}}_y\frac{\partial}{\partial y} + \hat{\mathbf{e}}_z\frac{\partial}{\partial z}. \tag{1.135}$$

This Cartesian form extends to a Euclidean manifold with an arbitrary dimension.

For spherical coordinates the direction vectors were found in (1.43). The metric tensor components are

$$\begin{aligned} g_{11} = \mathbf{e}_r\cdot\mathbf{e}_r = e_r^2 = 1, \quad & g_{22} = \mathbf{e}_\theta\cdot\mathbf{e}_\theta = e_\theta^2 = r^2, \\ g_{33} = \mathbf{e}_\phi\cdot\mathbf{e}_\phi &= e_\phi^2 = r^2\sin^2\theta, \end{aligned} \tag{1.136}$$

and all other components are zero. Because the other components are zero, the contravariant or inverse metric tensor components can be found simply by inverting these results. This gives

$$g^{11} = 1, \quad g^{22} = \frac{1}{r^2}, \quad g^{33} = \frac{1}{r^2\sin^2\theta}, \tag{1.137}$$

and all other components are zero. Using (1.136) and (1.137) in (1.134) gives the gradient in spherical coordinates in terms of unit vectors,

$$\begin{aligned}\nabla &= \mathbf{e}^k\partial_k = \mathbf{e}^r\frac{\partial}{\partial r} + \mathbf{e}^\theta\frac{\partial}{\partial\theta} + \mathbf{e}^\phi\frac{\partial}{\partial\phi} = \hat{\mathbf{e}}_j\left(e_{(j)}\, g^{jk}\partial_k\right) \\ &= \hat{\mathbf{e}}_r\frac{\partial}{\partial r} + \hat{\mathbf{e}}_\theta\frac{1}{r}\frac{\partial}{\partial\theta} + \hat{\mathbf{e}}_\phi\frac{1}{r\sin\theta}\frac{\partial}{\partial\phi}.\end{aligned} \tag{1.138}$$

The spherical coordinate unit vectors are given explicitly in terms of the Cartesian unit vectors by

$$\begin{aligned}\hat{\mathbf{e}}_r &= \sin\theta\cos\phi\,\hat{\mathbf{e}}_x + \sin\theta\sin\phi\,\hat{\mathbf{e}}_y + \cos\theta\,\hat{\mathbf{e}}_z = \mathbf{e}_r = \mathbf{e}^r, \\ \hat{\mathbf{e}}_\theta &= \cos\theta\cos\phi\,\hat{\mathbf{e}}_x + \cos\theta\sin\phi\,\hat{\mathbf{e}}_y - \sin\theta\,\hat{\mathbf{e}}_z = \frac{\mathbf{e}_\theta}{r} = r\mathbf{e}^\theta, \\ \hat{\mathbf{e}}_\phi &= -\sin\phi\,\hat{\mathbf{e}}_x + \cos\phi\,\hat{\mathbf{e}}_y = \frac{\mathbf{e}_\phi}{r\sin\theta} = r\sin\theta\,\mathbf{e}^\phi.\end{aligned} \tag{1.139}$$

At each angular position (θ, ϕ) the unit vectors of (1.139) form an orthonormal set. It is useful to note that the covariant components of the gradient in spherical coordinates are given by

$$\partial_1 = \frac{\partial}{\partial r}, \quad \partial_2 = \frac{\partial}{\partial\theta}, \quad \partial_3 = \frac{\partial}{\partial\phi}, \tag{1.140}$$

while the contravariant components are found using the inverse metric tensor (1.137),

$$\begin{aligned}\partial^r = \partial^1 = g^{1j}\partial_j = \frac{\partial}{\partial r}, \quad \partial^\theta = \partial^2 = g^{2j}\partial_j = \frac{1}{r^2}\frac{\partial}{\partial\theta}, \\ \partial^\phi = \partial^3 = g^{3j}\partial_j = \frac{1}{r^2\sin^2\theta}\frac{\partial}{\partial\phi}.\end{aligned} \tag{1.141}$$

Exercise 1.12. Show that the contravariant components of the gradient in spherical coordinates given by (1.141) obey the contravariant transformation law (1.21) by considering the explicit transformation from spherical coordinates to Cartesian coordinates.

Exercise 1.13. Cylindrical coordinates (ρ, ϕ, z) are related to Cartesian coordinates (x, y, z) in three dimensions by

$$\begin{aligned}&x = \rho\cos\phi, \quad y = \rho\sin\phi, \quad z = z \iff \rho = \sqrt{x^2 + y^2}, \quad \phi = \tan^{-1}(y/x), \\ &z = z.\end{aligned} \tag{1.142}$$

Show that the cylindrical coordinate direction vectors are given in terms of Cartesian unit vectors by

$$\mathbf{e}_\rho = \cos\phi\,\mathbf{e}_x + \sin\phi\,\mathbf{e}_y, \quad \mathbf{e}_\phi = -\rho\sin\phi\,\mathbf{e}_x + \rho\cos\phi\,\mathbf{e}_y, \quad \mathbf{e}_z = \mathbf{e}_z. \tag{1.143}$$

Show that a finite Euclidean position vector in cylindrical coordinates is given by

$$\boldsymbol{r} = \rho\,\mathbf{e}_\rho + z\,\mathbf{e}_z, \tag{1.144}$$

and that the gradient in cylindrical coordinates is given in terms of normalized unit vectors by

$$\nabla = \mathbf{e}^k \partial_k = \mathbf{e}^\rho \frac{\partial}{\partial\rho} + \mathbf{e}^\phi \frac{\partial}{\partial\phi} + \mathbf{e}^z \frac{\partial}{\partial z} = \hat{\mathbf{e}}_\rho \frac{\partial}{\partial\rho} + \hat{\mathbf{e}}_\phi \frac{1}{\rho}\frac{\partial}{\partial\phi} + \hat{\mathbf{e}}_z \frac{\partial}{\partial z}. \tag{1.145}$$

1.14 The divergence, the Laplacian, and the curl

A general vector field $\boldsymbol{V}(x)$, not necessarily given by $\nabla f(x)$, can be combined with the gradient in several ways. Expressing the vector field in terms of *unit vectors*, $\boldsymbol{V}(x) = \hat{\mathbf{e}}_k V^k(x)$, the *divergence* of $\boldsymbol{V}(x)$ is a scalar quantity defined by

$$\begin{aligned}\nabla \cdot \boldsymbol{V}(x) &= \mathbf{e}^j\partial_j \cdot (\hat{\mathbf{e}}_k V^k(x)) = (\mathbf{e}^j \cdot \hat{\mathbf{e}}_k)\partial_j V^k(x) + \mathbf{e}^j \cdot \left(\partial_j \hat{\mathbf{e}}_k\right) V^k(x) \\ &= \frac{1}{e_{(k)}} \partial_k V^k(x) + \mathbf{e}^j \cdot \left(\partial_j \hat{\mathbf{e}}_k\right) V^k(x),\end{aligned} \tag{1.146}$$

where the position dependence of the direction vectors is implicit. In that regard, Cartesian coordinate direction vectors are constant and normalized, so that $\partial_j \hat{\mathbf{e}}_k = 0$ and $e_{(k)} = 1$. The Cartesian form of the divergence in three dimensions is therefore given by

$$\begin{aligned}\nabla \cdot \boldsymbol{V}(x) = \partial_k V^k(x) &= \frac{\partial V^1(x)}{\partial x^1} + \frac{\partial V^2(x)}{\partial x^2} + \frac{\partial V^3(x)}{\partial x^3} \\ &= \frac{\partial V^x(x)}{\partial x} + \frac{\partial V^y(x)}{\partial y} + \frac{\partial V^z(x)}{\partial z}.\end{aligned} \tag{1.147}$$

The spherical coordinate version of the divergence requires the use of (1.139) to show that

$$\begin{aligned}&\partial_\theta \hat{\mathbf{e}}_r = \hat{\mathbf{e}}_\theta, \quad \partial_\theta \hat{\mathbf{e}}_\theta = -\hat{\mathbf{e}}_r, \quad \partial_\phi \hat{\mathbf{e}}_r = \sin\theta\,\hat{\mathbf{e}}_\phi, \quad \partial_\phi \hat{\mathbf{e}}_\theta = \cos\theta\,\hat{\mathbf{e}}_\phi, \\ &\partial_\phi \hat{\mathbf{e}}_\phi = -\sin\theta\,\hat{\mathbf{e}}_r - \cos\theta\,\hat{\mathbf{e}}_\theta,\end{aligned} \tag{1.148}$$

and all other derivatives vanish. Combining (1.148) with $e_\theta = r$ and $e_\phi = r\sin\theta$ gives

$$\begin{aligned}\nabla \cdot \boldsymbol{V}(x) = &\frac{\partial V^r(x)}{\partial r} + \frac{1}{r}\frac{\partial V^\theta(x)}{\partial\theta} + \frac{1}{r\sin\theta}\frac{\partial V^\phi(x)}{\partial\phi} + \frac{2}{r}V^r(x) \\ &+ \frac{\cot\theta}{r} V^\theta(x).\end{aligned} \tag{1.149}$$

Exercise 1.14. Show that the divergence in cylindrical coordinates is given by

$$\nabla \cdot V(x) = \frac{\partial V^{\rho}(x)}{\partial \rho} + \frac{1}{\rho}\frac{\partial V^{\phi}(x)}{\partial \phi} + \frac{\partial V^{z}(x)}{\partial z} + \frac{1}{\rho}V^{\rho}(x). \tag{1.150}$$

Using the definition of the gradient (1.126), the *Laplacian operator* is defined as

$$\nabla^2 = \nabla \cdot \nabla = (\mathbf{e}^k \partial_k) \cdot (\mathbf{e}^j \partial_j). \tag{1.151}$$

The Laplacian is a coordinate invariant *scalar operator* that acts on functions. Its action on a function $f(x)$ is given by

$$\begin{aligned} \nabla^2 f(x) &= \nabla \cdot \nabla f(x) = \mathbf{e}^k \cdot \partial_k\big(\mathbf{e}^j \partial_j f(x)\big) \\ &= (\mathbf{e}^k \cdot \mathbf{e}^j)\partial_k \partial_j f(x) + (\mathbf{e}^k \cdot \partial_k \mathbf{e}^j)\partial_j f(x). \end{aligned} \tag{1.152}$$

In the case of Cartesian coordinates, $\partial_k \mathbf{e}^j = 0$ and the second term in (1.152) vanishes. The orthonormality of the Cartesian reciprocal vectors shows that the three-dimensional Laplacian in Cartesian coordinates is

$$\nabla^2 = \frac{\partial^2}{\partial x^2} + \frac{\partial^2}{\partial y^2} + \frac{\partial^2}{\partial z^2}. \tag{1.153}$$

The Cartesian result (1.153) has the same general form in lower or higher dimensions. However, in the case of spherical coordinates, the formulas of (1.139) show that the first term in (1.152) generates nonzero contributions from the following cross terms,

$$\begin{aligned} \mathbf{e}^{\theta} \cdot \partial_{\theta}\mathbf{e}^{r} = \mathbf{e}^{\theta} \cdot \frac{\mathbf{e}_{\theta}}{r} = \frac{1}{r}, \quad \mathbf{e}^{\phi} \cdot \partial_{\phi}\mathbf{e}^{r} = \mathbf{e}^{\phi} \cdot \frac{\mathbf{e}_{\phi}}{r} = \frac{1}{r}, \\ \mathbf{e}^{\phi} \cdot \partial_{\phi}\mathbf{e}^{\theta} = \mathbf{e}^{\phi} \cdot \frac{\cos\theta}{r^2 \sin\theta}\mathbf{e}_{\phi} = \frac{\cot\theta}{r^2}. \end{aligned} \tag{1.154}$$

Using (1.154) in the second term in (1.152) gives the spherical coordinate Laplacian,

$$\begin{aligned} \nabla^2 &= (\mathbf{e}^k \cdot \mathbf{e}^j)\partial_k \partial_j + \left(\mathbf{e}^{\theta} \cdot \partial_{\theta}\mathbf{e}^{r} + \mathbf{e}^{\phi} \cdot \partial_{\phi}\mathbf{e}^{r}\right)\frac{\partial}{\partial r} + (\mathbf{e}^{\phi} \cdot \partial_{\phi}\mathbf{e}^{\theta})\frac{\partial}{\partial \theta} \\ &= \frac{\partial^2}{\partial r^2} + \frac{1}{r^2}\frac{\partial^2}{\partial \theta^2} + \frac{1}{r^2 \sin^2\theta}\frac{\partial^2}{\partial \phi^2} + \frac{2}{r}\frac{\partial}{\partial r} + \frac{\cot\theta}{r^2}\frac{\partial}{\partial \theta}. \end{aligned} \tag{1.155}$$

Exercise 1.15. Use (1.143) in (1.151) to show that the Laplacian in three-dimensional cylindrical coordinates is given by

$$\nabla^2 = \frac{\partial^2}{\partial \rho^2} + \frac{1}{\rho}\frac{\partial}{\partial \rho} + \frac{1}{\rho^2}\frac{\partial^2}{\partial \phi^2} + \frac{\partial^2}{\partial z^2}. \tag{1.156}$$

In three dimensions the *curl* of $V(x)$, denoted by $\nabla \times V(x)$, is defined using the vector product (1.111),

$$\nabla \times V(x) = \mathbf{e}^k \partial_k \times \hat{\mathbf{e}}_j V^j(x) = (\mathbf{e}^k \times \hat{\mathbf{e}}_j)\partial_k V^j(x) + (\mathbf{e}^k \times \partial_k \hat{\mathbf{e}}_j) V^j(x). \tag{1.157}$$

The expressions in (1.157) can be evaluated using the direction vectors in various coordinate systems. In a three-dimensional Euclidean manifold with Cartesian coordinates, the direction vectors obey $\partial_k \hat{\mathbf{e}}_j = 0$ and $\mathbf{e}_j = \hat{\mathbf{e}}_j$, and so in three-dimensional Cartesian coordinates, expression (1.157) is given by

$$\begin{aligned} \nabla \times V(x) &= \hat{\mathbf{e}}^k \times \hat{\mathbf{e}}_j \partial_k V^j(x) = g^{ki}(\hat{\mathbf{e}}_i \times \hat{\mathbf{e}}_j)\partial_k V^j = \varepsilon_{ij\ell} \hat{\mathbf{e}}^\ell g^{ki} \partial_k V^j(x) \\ &= \varepsilon_{ij\ell} \hat{\mathbf{e}}^\ell \partial^i V^j(x) \\ &= \hat{\mathbf{e}}_x \left(\frac{\partial A^z}{\partial y} - \frac{\partial A^y}{\partial z} \right) + \hat{\mathbf{e}}_y \left(\frac{\partial A^x}{\partial z} - \frac{\partial A^z}{\partial x} \right) + \hat{\mathbf{e}}_z \left(\frac{\partial A^y}{\partial x} - \frac{\partial A^x}{\partial y} \right). \end{aligned} \tag{1.158}$$

Exercise 1.16. Use the spherical coordinate direction vectors (1.43) to show that the three-dimensional curl (1.157) in spherical coordinates (1.42) is given by

$$\begin{aligned} \nabla \times V(x) = \hat{\mathbf{e}}_r \frac{1}{r \sin\theta} \left(\frac{\partial}{\partial\theta} (\sin\theta \, V^\phi) - \frac{\partial V^\theta}{\partial\phi} \right) \\ + \hat{\mathbf{e}}_\theta \left(\frac{1}{r \sin\theta} \frac{\partial V^r}{\partial\phi} - \frac{1}{r} \frac{\partial}{\partial r} (r V^\phi) \right) \\ + \hat{\mathbf{e}}_\phi \frac{1}{r} \left(\frac{\partial}{\partial r} (r V^\theta) - \frac{\partial V^r}{\partial\theta} \right). \end{aligned} \tag{1.159}$$

Exercise 1.17. Use the cylindrical coordinate direction vectors (1.143) to show that the three-dimensional curl (1.157) in the cylindrical coordinates of (1.142) is given by

$$\begin{aligned} \nabla \times V(x) = \hat{\mathbf{e}}_\rho \left(\frac{1}{\rho} \frac{\partial V^z}{\partial\phi} - \frac{\partial V^\phi}{\partial z} \right) + \hat{\mathbf{e}}_\phi \left(\frac{\partial V^r}{\partial\phi} - \frac{\partial V^z}{\partial\rho} \right) \\ + \hat{\mathbf{e}}_z \left(\frac{\partial V^\phi}{\partial\rho} - \frac{1}{\rho} \frac{\partial V^\rho}{\partial\phi} + \frac{1}{\rho} V^\phi \right). \end{aligned} \tag{1.160}$$

The curl of the gradient of a scalar function f is given by

$$\nabla \times \nabla f = \mathbf{e}^k \partial_k \times \mathbf{e}^j \partial_j f = (\mathbf{e}^k \times \partial_k \mathbf{e}^j)\partial^j f + \varepsilon_{kji} \mathbf{e}^i \partial^k \partial^j f. \tag{1.161}$$

In the case of Cartesian coordinates, $\partial_k \mathbf{e}^j = 0$. If the function f is differentiable, then $\partial^j \partial^k f = \partial^k \partial^j f$, and it follows that

$$\nabla \times \nabla f = 0. \tag{1.162}$$

This requirement is met for differentiable functions. However, it is worth noting for future reference that $\nabla \times V = 0$ does *not* mean that $V = \nabla f$. Similarly, the divergence of the curl of a vector function is given by

$$\nabla \cdot (\nabla \times V) = \mathbf{e}^i \partial_i \cdot \varepsilon_{jk\ell} \mathbf{e}^j \partial^k V^\ell = \varepsilon_{jk\ell}(\mathbf{e}^i \cdot \partial_i \mathbf{e}^j)\partial^k V^\ell + \varepsilon_{jk\ell}\partial^j \partial^k V^\ell. \tag{1.163}$$

Once again, for the case of Cartesian coordinates and $\partial^j \partial^k V^\ell = \partial^k \partial^j V^\ell$, it follows that

$$\nabla \cdot (\nabla \times V) = 0. \tag{1.164}$$

The divergence and the curl are used in two theorems of great importance in field theory. For the moment, these theorems will be presented in their three-dimensional versions. The first is the *divergence theorem*, which states that the integral of $\nabla \cdot V(x)$ over a volume V bounded by the closed surface S (so that $S = S(V)$) is given by

$$\int_V \mathrm{d}^3 x\, \nabla \cdot V(x) = \oint_{S(V)} \mathrm{d}\boldsymbol{S} \cdot V(x), \tag{1.165}$$

where d$\boldsymbol{S}$ stands for the oriented surface elements that cover the closed surface S. These were discussed in (1.115). The second is *Stokes' theorem*, which relates the *line integral* of $V(x)$ around a closed path P to the curl of $V(x)$ evaluated over *any* surface S bounded by the path,

$$\int_S \mathrm{d}\boldsymbol{S} \cdot \nabla \times V(x) = \oint_{P(S)} \mathrm{d}\boldsymbol{r} \cdot V(\boldsymbol{r}(x)). \tag{1.166}$$

In both cases, (1.165) and (1.166), the integral on the left is found entirely from the vector field *at the boundary*, i.e. either the surface of the volume or the perimeter of the surface.

1.15 Differential forms and the wedge product

In order to lift the limitation of Stokes' theorem to three dimensions, it is necessary to define the relationship between the tangent space of vectors $T_p(\mathcal{M})$ and the set of *linear functions* on the tangent space, denoted by $T_p^*(\mathcal{M})$ and referred to as the *dual space*. The members of $T_p^*(\mathcal{M})$ are scalar functions f of the tangent vectors $\boldsymbol{v}$ that map the vectors into a real number *in a linear manner*. If $\boldsymbol{v}$ and $\boldsymbol{w}$ are two tangent vectors and a and b are two real numbers, linearity requires

$$f(a\boldsymbol{v} + b\boldsymbol{w}) = af(\boldsymbol{v}) + bf(\boldsymbol{w}). \tag{1.167}$$

It is important to find a set of basis functions that allows the dual space members to be expressed as a linear combination, in much the same way that the tangent vectors are a linear combination of the unit vectors. This is resolved by considering a function $f(\boldsymbol{x})$ of the coordinates $\boldsymbol{x}$ for the point p where the tangent space is defined. If v is an infinitesimal tangent vector in the tangent space at some point p with the coordinates $\boldsymbol{x}$, then its length (1.64) is infinitesimal. As a result, the function $f(\boldsymbol{x} + v)$ will differ from $f(\boldsymbol{x})$ by an amount $\mathrm{d}f(v)$. A Taylor series expansion gives

$$\mathrm{d}f(v) = f(\boldsymbol{x} + v) - f(\boldsymbol{x}) = v \cdot \nabla f(\boldsymbol{x}) = Vf(\boldsymbol{x}), \tag{1.168}$$

where V is the vector written in terms of the basis vectors $\partial/\partial x^j$, $V = v \cdot \nabla$. Although it is derived for infinitesimal v, (1.168) is linear in v. Therefore, (1.168) is defined to hold for all vectors in the tangent space at $\boldsymbol{x}$ and all functions f in the dual space. In effect, the definition of (1.168) creates a generalized *inner product* between the members of $T_p(\mathcal{M})$ and the members of $T_p^*(\mathcal{M})$. This inner product is a linear relation, just as the scalar product between two position vectors is a linear relation. In some presentations of differential geometry, (1.168) is written

$$(\mathrm{d}f, v) = \mathrm{d}f[v] = v \cdot \nabla f(\boldsymbol{x}), \tag{1.169}$$

which defines an inner product between the differential $\mathrm{d}f$ in the dual space and the vector v in the tangent space. Viewing it in this manner shows that the differential of a linear function has a projection onto the vector v through definition (1.169).

Picking the function f in (1.168) to be the jth coordinate x^j of point p gives

$$\mathrm{d}x^j[v] = v \cdot \nabla x^j = v^k\, \partial_k(x^j) = v^k\, \delta^j{}_k = v^j. \tag{1.170}$$

Expressing v in components and using the linearity (1.167), result (1.170) gives

$$\mathrm{d}x^j[v] = \mathrm{d}x^j[v^k\mathbf{e}_k] = v^k\, \mathrm{d}x^j[\mathbf{e}_k] = v^j, \tag{1.171}$$

which can only hold if

$$\mathrm{d}x^j[\mathbf{e}_k] = \delta^j{}_k. \tag{1.172}$$

Combining (1.172) and (1.170) with the definition of the differential of a function gives

$$\mathrm{d}f[v] = \partial_j f(\boldsymbol{x})\, \mathrm{d}x^j[v] = v^j(\boldsymbol{x})\, \partial_j f(\boldsymbol{x}) = v(\boldsymbol{x}) \cdot \nabla f(\boldsymbol{x}), \tag{1.173}$$

which reproduces (1.168).

An arbitrary member of the dual space can therefore be written

$$\boldsymbol{\omega} = \omega_j \mathrm{d}x^j, \tag{1.174}$$

and is referred to as a *one-form*. Like the direction vectors $\mathbf{e}_j$ in the case of the tangent space, the dual space is spanned by the basis one-forms $\mathrm{d}x^j$ as long as $\mathrm{d}x^j$ are viewed as functions of vectors defined by (1.172). Since the one-form components ω_j transform covariantly, the one-form $\boldsymbol{\omega}$ is coordinate independent. Its action on a tangent vector v is also coordinate independent,

$$\boldsymbol{\omega}[\boldsymbol{v}] = \omega_j v^k \mathrm{d}x^j[\mathbf{e}_k] = \omega_j v^k \delta^j{}_k = \omega_j v^j. \tag{1.175}$$

Result (1.175) is manifestly linear, satisfying (1.167). The two components appearing in the sum are implicit functions of the position vector $\boldsymbol{r}$ for the point under consideration.

The differential of a function $\mathrm{d}\varphi(\boldsymbol{x})$ gives a one-form with the covariant components $\omega_j = \partial_j\varphi$. This follows from elementary calculus,

$$\mathrm{d}\varphi(\boldsymbol{x}) = \frac{\partial\varphi(\boldsymbol{x})}{\partial x^j}\,\mathrm{d}x^j = \partial_j\varphi(\boldsymbol{x})\,\mathrm{d}x^j. \tag{1.176}$$

As a result, a scalar function $\varphi(\boldsymbol{x})$ of the coordinates is referred to as a *zero-form* and the one-form $\boldsymbol{\omega}(\boldsymbol{x}) = \mathrm{d}\varphi(\boldsymbol{x})$ is referred to as an *exact one-form*. A one-form is identical to the line element from basic calculus, since it takes the form $\boldsymbol{\Omega}\cdot\mathrm{d}\boldsymbol{x}$, where $\boldsymbol{\Omega}$ is a vector function given by $\boldsymbol{\Omega} = \omega_j(\boldsymbol{x})\mathbf{e}^j$ expressed in terms of the covariant components of the vector. The action of the differential d on a zero-form will be generalized to define the *exterior derivative*, which yields an $(n+1)$-form of a given n-form.

The one-form created in (1.176) is a combination of the contravariant components of $\mathrm{d}\boldsymbol{x}$ and the covariant components of $\nabla\varphi$. Using the transformation properties of the two types of vectors shows that the one-form $\mathrm{d}\varphi$ transforms according to

$$\begin{aligned}\mathrm{d}\varphi(\boldsymbol{y}) &= \frac{\partial\varphi(\boldsymbol{y})}{\partial y^i}\,\mathrm{d}y^i = \frac{\partial x^j}{\partial y^i}\frac{\partial y^i}{\partial x^k}\frac{\partial\varphi(\boldsymbol{x})}{\partial x_j}\,\mathrm{d}x^k = \delta^j{}_k\frac{\partial\varphi(\boldsymbol{x})}{\partial x_j}\,\mathrm{d}x^k = \frac{\partial\varphi(\boldsymbol{x})}{\partial x_j}\,\mathrm{d}x^j \\ &= \mathrm{d}\varphi(\boldsymbol{x}).\end{aligned} \tag{1.177}$$

This shows that the one-form is a scalar quantity, since it is unchanged or *invariant* under coordinate transformations regardless of the number of dimensions. It is therefore a coordinate-free quantity.

Just as the one-form is an oriented line element, forms can now be extended to involve infinitesimal oriented area and volume elements in arbitrary dimensions. This was motivated earlier in (1.115) by constructing oriented area elements in three dimensions using the vector product. In differential geometry, infinitesimal oriented area elements are defined by the *wedge product*, sometimes known as the *exterior product*, denoted by $\mathrm{d}x^i \wedge \mathrm{d}x^j$. This product involves the basis elements of the dual space at each point. The wedge product is defined to be antisymmetric, so that, like the Levi-Civita symbol in the vector product,

$$\mathrm{d}x^i \wedge \mathrm{d}x^j = -\mathrm{d}x^j \wedge \mathrm{d}x^i \implies \mathrm{d}x^j \wedge \mathrm{d}x^j = 0. \tag{1.178}$$

In three dimensions the wedge product is defined to be identical to the vector product result (1.115),

$$\mathrm{d}x^i \wedge \mathrm{d}x^j = \varepsilon^{ijk}\,\mathrm{d}S_k, \tag{1.179}$$

where $\mathrm{d}S_k$ is the kth component of the oriented area element perpendicular to both directions i and j. Given two one-forms in three dimensions, $\boldsymbol{\omega}$ and $\boldsymbol{\zeta}$, a *two-form* $\boldsymbol{\xi}$ can be obtained from them using the wedge product, as follows:

$$\boldsymbol{\xi} = \boldsymbol{\omega} \wedge \boldsymbol{\zeta} = \omega_i \zeta_j \, \mathrm{d}x^i \wedge \mathrm{d}x^j = \varepsilon^{ijk} \omega_i \zeta_j \, \mathrm{d}S_k = (\boldsymbol{\omega} \times \boldsymbol{\zeta}) \cdot \mathrm{d}\boldsymbol{S}. \tag{1.180}$$

The vector product of two three-dimensional vectors has appeared, along with the infinitesimal oriented area element $\mathrm{d}\boldsymbol{S}$. This is the two-dimensional extension of the one-form, so it is called a *two-form*. This can be used with the previous results of (1.19) to find the surface element in polar coordinates,

$$\begin{aligned} \mathrm{d}x \wedge \mathrm{d}y &= (\cos\theta \, \mathrm{d}r - r \sin\theta \, \mathrm{d}\theta) \wedge (\sin\theta \, \mathrm{d}r + r\cos\theta \, \mathrm{d}\theta) \\ &= r\cos^2\theta \, \mathrm{d}r \wedge \mathrm{d}\theta - r\sin^2\theta \mathrm{d}\theta \wedge \mathrm{d}r = r(\cos^2\theta + \sin^2\theta)\mathrm{d}r \\ &\quad \wedge \mathrm{d}\theta = r \, \mathrm{d}r \, \mathrm{d}\theta \, \boldsymbol{n}, \end{aligned} \tag{1.181}$$

where $\boldsymbol{n}$ is a unit vector perpendicular to the plane of the surface element in the right-handed sense. This has correctly reproduced the Jacobian (1.13) for the coordinate transformation from Cartesian coordinates to polar coordinates.

While the wedge product of two one-forms gives a two-form, a general two-form $\boldsymbol{\eta}$ in three dimensions is written as

$$\boldsymbol{\eta} = \eta_{ij}(\boldsymbol{x}) \, \mathrm{d}x^i \wedge \mathrm{d}x^j = \eta_{ij}\varepsilon^{ijk}\mathrm{d}S_k. \tag{1.182}$$

The components η_{ij} can be viewed as the covariant components of a second-rank tensor. It is useful to note that only the antisymmetric part of η_{ij} contributes to (1.182). To see this, note that η_{ij} can be written as the sum of an antisymmetric part and a symmetric part,

$$\eta_{ij} = \frac{1}{2}(\eta_{ij} - \eta_{ji}) + \frac{1}{2}(\eta_{ij} + \eta_{ji}) \equiv \eta_{ij}^A + \eta_{ij}^S, \tag{1.183}$$

where $\eta_{ij}^A = -\eta_{ji}^A$ and $\eta_{ij}^S = \eta_{ji}^S$. The symmetric part of η_{ij} drops out of the two-form, since the dummy indices can be relabeled, which gives

$$\eta_{ij}^S \, \mathrm{d}x^i \wedge \mathrm{d}x^j = \eta_{ji}^S \, \mathrm{d}x^j \wedge \mathrm{d}x^i = -\eta_{ij}^S \, \mathrm{d}x^i \wedge \mathrm{d}x^j \implies \eta_{ij}^S \, \mathrm{d}x^i \wedge \mathrm{d}x^j = 0. \tag{1.184}$$

As a result, for a two-form in three dimensions, η_{ij} have only three independent components, since the symmetric part of η_{ij} has six independent components. As long as the η_{ij} transform like the covariant components of a second-rank tensor, the general two-form of (1.182) is invariant under coordinate transformations, regardless of the number of dimensions. This is demonstrated identically to the one-form of (1.177).

So far, the Levi-Civita symbol has been employed in the definition of the vector product and the Jacobian, and it will now play a central role in forms. It was shown in (1.101) that the covariant components of the Levi-Civita symbol correspond to a tensor density of weight +1. Similarly, it is possible to show that the contravariant components correspond to a tensor density of weight −1, since they obey the identity given by

$$\varepsilon^{abc\ldots} = \frac{1}{J(y)} \frac{\partial y^a}{\partial x^{a'}} \frac{\partial y^b}{\partial x^{b'}} \frac{\partial y^c}{\partial x^{c'}} \cdots \varepsilon^{a'b'c'\ldots}. \tag{1.185}$$

Using (1.101) and (1.185) shows that both the transformed and original versions of the Levi-Civita symbol obey the earlier result (1.11),

$$\varepsilon^{abc\ldots}\varepsilon_{abc\ldots} = \varepsilon^{a'b'c'\ldots}\varepsilon_{a'b'c'\ldots} = n!. \tag{1.186}$$

The Jacobian factors cancel and relation (1.186) is invariant under coordinate changes, as it must be.

Exercise 1.18. Verify result (1.186).

The identity (1.185) determines the transformation properties of the curl of a vector **A**, written in three dimensions as $\mathbf{B} = \nabla \times \mathbf{A}$ with components given by $B^k = \varepsilon^{ijk}\partial_i A_j$. Employing the identity (1.185) and the transformation properties of ∇ and **A** shows that $\mathbf{B}'$ has the components

$$\begin{aligned} B'^k(y) &= \frac{1}{J} \frac{\partial y^i}{\partial x^n} \frac{\partial y^j}{\partial x^p} \frac{\partial y^k}{\partial x^r} \varepsilon^{npr} \frac{\partial x^\ell}{\partial y^i} \frac{\partial x^m}{\partial y^j} \frac{\partial A_m}{\partial x^\ell} \\ &= \frac{1}{J} \frac{\partial y^k}{\partial x^r} \varepsilon^{\ell mr} \frac{\partial A_m}{\partial x^\ell} = \frac{1}{J} \frac{\partial y^k}{\partial x^r} B^r(x), \end{aligned} \tag{1.187}$$

so that the curl transforms as a *contravariant pseudovector of weight* −1. In deriving (1.187), **A** was treated as a *true vector*, i.e. one that does not pick up a factor of the Jacobian under transformation. The presence of the Jacobian has a profound effect on a pseudovector when the change of coordinates causes a *reflection* or *change of handedness*, which are coordinate changes of the form $y^i = -x^i$. A true vector changes sign under a reflection or change of handedness, since such a transformation gives $\partial y^i/\partial x^j = -\delta^i{}_j$, yielding $\mathbf{A}'(y) = -\mathbf{A}(x)$. This means that a true vector maintains its orientation when the coordinate system changes handedness. Result (1.187) shows that the curl of a true vector does not change sign under a reflection, since the Jacobian in three dimensions gives $J = -1$, and its sign compensates for the sign change created by $\partial y^i/\partial x^j = -\delta^i{}_j$. This means that a pseudovector, the curl of a true vector, flips its direction under reflection, since the signs of its components in the new coordinate system, which have changed handedness, are the same as in the old coordinate system.

The development of forms continues by defining a *three-form* in three dimensions as

$$V = V_{ijk}\, \mathrm{d}x^i \wedge \mathrm{d}x^j \wedge \mathrm{d}x^k = V_{ijk}\, \varepsilon^{ijk} \mathrm{d}^3x. \tag{1.188}$$

The components V_{ijk} are antisymmetric in a manner identical to the Levi-Civita symbol ε^{ijk}, changing sign under an odd permutation of the indices. In three dimensions a general three-form becomes $V = \varepsilon^{ijk}\, V_{ijk}\, \mathrm{d}^3x$. In three dimensions it is not possible to define a four-form or higher, since such forms will have at least one pair of identical indices and antisymmetry in the indices requires that such a pair should vanish. Using the same two-form arguments, it follows that V_{ijk} transforms as a third-rank covariant tensor, while $\mathrm{d}x^i \wedge \mathrm{d}x^j \wedge \mathrm{d}x^k$ transforms as a third-rank contravariant tensor. It is now possible to show that these transformation properties are consistent with the volume element d^3x that picks up a factor of the Jacobian when the coordinate transformation occurs. The first step is to use (1.186) to write

$$\mathrm{d}^3x = \frac{1}{3!}\varepsilon_{ijk}\, \mathrm{d}x^i \wedge \mathrm{d}x^j \wedge \mathrm{d}x^k. \tag{1.189}$$

Using the contravariant transformation property of the differential and $\mathrm{d}y^\ell \wedge \mathrm{d}y^m \wedge \mathrm{d}y^n = \varepsilon^{\ell mn}\mathrm{d}^3y$ gives

$$\mathrm{d}^3x = \frac{1}{3!}\varepsilon_{ijk}\frac{\partial x^i}{\partial y^\ell}\frac{\partial x^j}{\partial y^m}\frac{\partial x^k}{\partial y^n}\, \mathrm{d}y^\ell \wedge \mathrm{d}y^m \wedge \mathrm{d}y^n = \frac{1}{3!}\varepsilon_{ijk}\varepsilon^{\ell mn}\frac{\partial x^i}{\partial y^\ell}\frac{\partial x^j}{\partial y^m}\frac{\partial x^k}{\partial y^n}\, \mathrm{d}^3y = J\, \mathrm{d}^3y, \tag{1.190}$$

where the Jacobian (1.12) for a three-dimensional volume transformation has appeared, justifying its definition using the *exterior algebra* of exterior forms. Since the volume element d^3x has no direction associated with it and is always positive regardless of the coordinate system, it follows that

$$|J| = \left| \frac{1}{3!}\varepsilon_{ijk}\varepsilon^{\ell mn}\frac{\partial x^i}{\partial y^\ell}\frac{\partial x^j}{\partial y^m}\frac{\partial x^k}{\partial y^n} \right|. \tag{1.191}$$

The volume element is not a true scalar, instead transforming as a pseudoscalar of weight +1.

The *exterior derivative* mentioned earlier in this section can now be defined. The exterior derivative of a zero-form φ was defined to be the exact differential or one-form $\mathrm{d}\varphi = \partial_i\varphi\, \mathrm{d}x^i$. Given the one-form $\boldsymbol{\omega} = \omega_j \mathrm{d}x^j$, the action of the exterior derivative d on $\boldsymbol{\omega}$ is defined as

$$\mathrm{d}\boldsymbol{\omega} = \partial_j\omega_k\, \mathrm{d}x^j \wedge \mathrm{d}x^k, \tag{1.192}$$

so that a two-form is created from a one-form. Using $\mathrm{d}x^j \wedge \mathrm{d}x^k$ for three dimensions gives

$$\mathrm{d}\boldsymbol{\omega} = \varepsilon^{jk\ell}\partial_j\omega_k\, \mathrm{d}S_\ell. \tag{1.193}$$

The quantity $B^k \equiv \varepsilon^{j\ell k}\partial_j A_\ell$ was defined earlier as the kth component of the curl of the vector field $\boldsymbol{A}$, so that

$$\mathrm{d}A = B^k\,\mathrm{d}S_k = (\partial \times A)\cdot \mathrm{d}\mathbf{S}. \tag{1.194}$$

The action of the exterior derivative has changed the one-form, which is a line element, into a two-form, which is a surface or flux element of the curl of the vector field appearing in the one-form. Due to the antisymmetry of the wedge product, it follows that $\mathrm{d}^2\boldsymbol{A} = 0$. This occurs because $\partial_i\partial_j A_k\,\mathrm{d}x^i \wedge \mathrm{d}x^j \wedge \mathrm{d}x^k$ vanishes if $\partial_i\partial_j A_k = \partial_j\partial_i A_k$. Similarly, it follows that $\mathrm{d}^2\varphi = 0$. Since the action of d^2 is zero, d is referred to as a *nilpotent* operator. The exterior derivative can be applied to the wedge product of forms as well, but obeys a generalized Leibniz property. This follows by examining the exterior derivative of the wedge product of two one-forms,

$$\begin{aligned}\mathrm{d}(\boldsymbol{\omega}\wedge\boldsymbol{\chi}) &= \mathrm{d}(\omega_j\chi_k\,\mathrm{d}x^j\wedge\mathrm{d}x^k) = (\partial_i\omega_j\chi_k + \omega_j\partial_i\chi_k)\,\mathrm{d}x^i\wedge\mathrm{d}x^j\wedge\mathrm{d}x^k\\ &= \mathrm{d}\boldsymbol{\omega}\wedge\boldsymbol{\chi} - \boldsymbol{\omega}\wedge\mathrm{d}\boldsymbol{\chi},\end{aligned} \tag{1.195}$$

where $\mathrm{d}x^j \wedge \mathrm{d}x^i = -\mathrm{d}x^i \wedge \mathrm{d}x^j$ creates the minus sign. The wedge product of two *arbitrary* forms $\boldsymbol{\omega}$ and $\boldsymbol{\chi}$ has an exterior derivative given by

$$\mathrm{d}(\boldsymbol{\omega}\wedge\boldsymbol{\chi}) = \mathrm{d}\boldsymbol{\omega}\wedge\boldsymbol{\chi} + (-1)^n\boldsymbol{\omega}\wedge\mathrm{d}\boldsymbol{\chi}, \tag{1.196}$$

where n corresponds to the *degree* of the form $\boldsymbol{\omega}$, which is $n = 0$ for a zero-form, $n = 1$ for a one-form, and so on.

It is straightforward to generalize these results to higher dimensions. In four dimensions a one-form is written

$$\boldsymbol{A} = A_\mu\,\mathrm{d}x^\mu, \tag{1.197}$$

where μ runs from 0 to 3. Continuing in four dimensions the two-form surface element is given by

$$\mathrm{d}x^\mu \wedge \mathrm{d}x^\nu = \varepsilon^{\mu\nu\rho\sigma}\mathrm{d}S_{\rho\sigma}, \tag{1.198}$$

where $\mathrm{d}S_{\rho\sigma} = -\mathrm{d}S_{\sigma\rho}$ corresponds to all the possible two-dimensional surface elements. For example, choosing $\mu = 0$ and $\nu = 1$ gives the oriented surface elements in the y–z plane. Because $\mathrm{d}S_{\sigma\rho}$ is antisymmetric, it has $(4 \times 4 - 4)/2 = 6$ distinct components. These represent the six distinct two-dimensional surfaces possible in a four-dimensional space. This is the four-dimensional counterpart of the three two-dimensional oriented surface elements in three dimensions. Remaining in four dimensions, the three-forms are written

$$\boldsymbol{G} = G_{\mu\nu\rho}\,\mathrm{d}x^\mu \wedge \mathrm{d}x^\nu \wedge \mathrm{d}x^\rho = G_{\mu\nu\rho}\,\varepsilon^{\mu\nu\rho\sigma}\mathrm{d}^3\Sigma_\sigma. \tag{1.199}$$

Here, $\mathrm{d}^3\Sigma_\sigma$ are the four oriented three-volume elements, where $\mathrm{d}^3\Sigma_0 = \mathrm{d}^3x$ is the spatial volume element and $\mathrm{d}^3\Sigma_1 = \mathrm{d}x^0\,\mathrm{d}x^2\,\mathrm{d}x^3$. While there is no four-form in three dimensions, the four-form in four dimensions is given by

$$\mathrm{d}x^\mu \wedge \mathrm{d}x^\nu \wedge \mathrm{d}x^\rho \wedge \mathrm{d}x^\sigma = \varepsilon^{\mu\nu\rho\sigma}\,\mathrm{d}^4x. \tag{1.200}$$

The behavior of d^4x under a change of coordinates to $y^\mu(x)$ is derived identically to (1.190), so that

$$d^4x = |J|\, d^4y, \tag{1.201}$$

where J is the four-dimensional version of expression (1.191). The exterior derivative can be applied to an r-form, $\boldsymbol{\omega}$, to obtain an $(r+1)$-form, $d\boldsymbol{\omega}$, given by

$$d\boldsymbol{\omega} = \partial_\mu \omega_{\nu\cdots\rho}\, dx^\mu \wedge dx^\nu \wedge \cdots \wedge dx^\rho. \tag{1.202}$$

Higher dimensions follow the same pattern. Clearly, if r matches the dimension n of the space, then $d\boldsymbol{\omega} = 0$. Regardless of the dimension of the space, $d^2 = 0$ because of the antisymmetry of the wedge product and the usual assumption $\partial_i \partial_l \omega_{j\cdots k} = \partial_l \partial_i \omega_{j\cdots k}$, so that the exterior derivative remains nilpotent.

1.16 Differential forms and Stokes' theorem

The exterior derivative allows Stokes' theorem to be restated. In three dimensions the standard version of Stokes' theorem states that the vector field $\mathbf{A}$ satisfies

$$\int_S (\nabla \times \mathbf{A}) \cdot d\boldsymbol{S} = \oint_{P(S)} \mathbf{A} \cdot d\boldsymbol{x}, \tag{1.203}$$

where $\boldsymbol{S}$ is a two-dimensional surface and $P(S)$ is the closed path that is the boundary of that surface. If $\boldsymbol{S}$ is a *closed surface* then it has no boundary, and in that case, the right-hand side of (1.203) vanishes. Examining (1.203) shows that the one-form $\mathbf{A}$ is the argument of the right-hand side, while $d\mathbf{A}$ is the argument of the left-hand side. The generalization of Stokes' theorem for an r-form $\boldsymbol{\omega}$ is then given by

$$\int_{\mathcal{M}} d\boldsymbol{\omega} = \int_{\partial\mathcal{M}} \boldsymbol{\omega}, \tag{1.204}$$

where $\mathcal{M}$ is the $(r+1)$-dimensional space over which the $(r+1)$-form $d\boldsymbol{\omega}$ is being integrated, while $\partial\mathcal{M}$ is the r-dimensional boundary of $\mathcal{M}$ over which the r-form $\boldsymbol{\omega}$ is being integrated. The proof of (1.204) in arbitrary dimensions for arbitrary surfaces is beyond the scope of this book, but it is done in a manner similar to the proof of Stokes' theorem in three dimensions. The space $\mathcal{M}$ is covered with a generalized set of triangles, referred to as a *simplicial complex*, that is appropriate to the dimension of $\mathcal{M}$. It is then shown that only the boundary of that covering contributes to the left side, reducing it to the right-hand side. For a space with an unusual topology, such as holes and handles and boundaries, (1.204) leads to a way of characterizing the topology in terms of the classification of forms defined on the space. The interested reader should study *homology and cohomology theory*.

However, in order to clarify the new version of Stokes' theorem, it will be demonstrated that the usual three-dimensional version of the divergence theorem is also contained in (1.204) using the two-form $\boldsymbol{F} = F_{ij}\, dx^i \wedge dx^j$ in the right-hand side of (1.204). Previous results showed that the antisymmetry of F_{ij} allows the two-form to be written in terms of the vector $\boldsymbol{V}$,

$$F_{ij} = \frac{1}{2}\varepsilon_{ij\ell} V^\ell. \tag{1.205}$$

This gives

$$\boldsymbol{F} = F_{ij}\,\mathrm{d}x^i \wedge \mathrm{d}x^j = \frac{1}{2}\varepsilon_{ij\ell}\varepsilon^{ijk}\,V^\ell \mathrm{d}S_k = \delta^k{}_\ell V^\ell\,\mathrm{d}S_k = V^k\,\mathrm{d}S_k = \mathbf{V}\cdot\mathrm{d}\boldsymbol{S}. \tag{1.206}$$

Using the same form for F_{ij} gives

$$\begin{aligned}\mathrm{d}\boldsymbol{F} &= \partial_k F_{ij}\,\mathrm{d}x^k \wedge \mathrm{d}x^i \wedge \mathrm{d}x^j = \partial_k F_{ij}\,\varepsilon^{kij}\mathrm{d}^3x \\ &= \frac{1}{2}\varepsilon^{kij}\varepsilon_{ij\ell}\partial_k V^\ell\,\mathrm{d}^3x = \delta^k{}_\ell\partial_k V^\ell\,\mathrm{d}^3x = \partial_k V^k\,\mathrm{d}^3x = \nabla\cdot\mathbf{V}\,\mathrm{d}^3x.\end{aligned} \tag{1.207}$$

Inserting (1.206) and (1.207) into (1.204) yields the divergence theorem in three dimensions (1.165),

$$\int_{\mathcal{M}} \mathrm{d}^3x\,\nabla\cdot\mathbf{V} = \int_{\partial\mathcal{M}} \mathrm{d}\boldsymbol{S}\cdot\mathbf{V}, \tag{1.208}$$

where $\mathcal{M}$ is a three-dimensional volume and $\partial\mathcal{M}$ is the two-dimensional surface forming the boundary of the three-dimensional volume.

Exercise 1.19. In a two-dimensional Cartesian plane the exterior derivative of a one-form $\boldsymbol{\omega} = \omega_a \mathrm{d}x^a$ is given by $\mathrm{d}\boldsymbol{\omega} = \partial_a\omega_b(x, y)\varepsilon^{ab}\mathrm{d}^2S$, where $\mathrm{d}^2S = \mathrm{d}x\,\mathrm{d}y$. For the case in which $\mathrm{d}\boldsymbol{\omega}$ is integrated over a rectangular patch V of width $2w$ and height $2h$ centered on the origin, show that the integral becomes

$$\begin{aligned}\int_V \mathrm{d}\boldsymbol{\omega} &= \int_{-w}^{w} \mathrm{d}x\,\omega_x(x, -h) + \int_{-h}^{h} \mathrm{d}y\,\omega_y(w, y) + \int_{w}^{-w} \mathrm{d}x\,\omega_x(x, h) \\ &\quad + \int_{h}^{-h} \mathrm{d}y\,\omega_y(-w, y) = \int_{\partial V} \boldsymbol{\omega}.\end{aligned}$$

The second application of Stokes' theorem in two dimensions is trivial, consisting of the line integral of the exterior derivative of a zero-form $\mathrm{d}\varphi = \partial_a\varphi\,\mathrm{d}x^a$. Since $\mathrm{d}\varphi$ is an exact differential the result is

$$\int_P \mathrm{d}\varphi = \varphi(x_f, y_f) - \varphi(x_o, y_o).$$

This is the sum of the zero-form at the oriented endpoints of path P.

Because d is nilpotent, $\boldsymbol{\omega}$ can be replaced in (1.204) with $\boldsymbol{\omega} + \mathrm{d}\boldsymbol{\Lambda}$, where $\boldsymbol{\Lambda}$ is an $(r-1)$-form, and the left-hand side of (1.204) is unchanged, since

$$\mathrm{d}\boldsymbol{\omega} \to \mathrm{d}\boldsymbol{\omega} + \mathrm{d}^2\boldsymbol{\Lambda} = \mathrm{d}\boldsymbol{\omega}. \tag{1.209}$$

This is consistent with the right-hand side being unchanged, since applying Stokes' theorem to the right-hand side shows that its change is given by

$$\int_{\partial\mathcal{M}} \mathrm{d}\boldsymbol{\Lambda} = \int_{\partial^2\mathcal{M}} \boldsymbol{\Lambda} = 0, \tag{1.210}$$

which vanishes, because the boundary of a boundary does not exist.

Finally, Stokes' theorem (1.204) remains true regardless of the dimension. For an n-dimensional space there are $n - 1$ nontrivial applications of the theorem, just as there is one nontrivial application in two dimensions, given by the previous exercise, and two nontrivial applications in three dimensions, the divergence theorem (1.165) and Stokes' theorem (1.166).

1.17 The Lie derivative

It is often important to find the change in a tensor function between two points separated by an infinitesimal position-dependent vector displacement $\epsilon X(\boldsymbol{x})$, where the limit $\epsilon \to 0$ is understood. The starting point has the coordinates x^μ and the goal is to find the change in the tensor function at the point $y^\mu = x^\mu + \epsilon X^\mu(\boldsymbol{x})$. Since ϵ is infinitesimal and the tensor function is assumed to be nonsingular, the change will be proportional to ϵ. It is straightforward to find the change in a scalar function $\varphi(\boldsymbol{x})$. Since the scalar field does not change form at the new point, a Taylor series expansion gives

$$\varphi(\boldsymbol{y}) = \varphi(\boldsymbol{x} + \epsilon X) = \varphi(\boldsymbol{x}) + \epsilon X^\nu \partial_\nu \varphi(\boldsymbol{x}), \tag{1.211}$$

which holds to $O(\epsilon)$. The *Lie derivative* $\mathcal{L}_X$ of a scalar function is therefore defined as

$$\mathcal{L}_X \varphi(\boldsymbol{x}) = \lim_{\epsilon \to 0} \frac{\varphi(\boldsymbol{y}) - \varphi(\boldsymbol{x})}{\epsilon} = X^\nu(\boldsymbol{x}) \partial_\nu \varphi(\boldsymbol{x}) = X(\boldsymbol{x}) \varphi(\boldsymbol{x}), \tag{1.212}$$

where the Cartan notation (1.45) for the vector X is being used. This is simply the directional derivative of $\varphi(\boldsymbol{x})$, and gives the rate of change in $\varphi(\boldsymbol{x})$ along the direction given by X.

In the case of a vector function $V(\boldsymbol{x})$, it may also undergo a change in its components at the new point, and the components will depend on the coordinate system to which they are referred. Finding the actual change of a vector's components between the two points requires finding its components using the *same coordinate system*. This is referred to as pulling the components of the vector $V^\mu(\boldsymbol{y})$ at $\boldsymbol{y}$ back to the original coordinate system $\boldsymbol{x}$ to find the value $V'^\mu(\boldsymbol{x})$ at $\boldsymbol{x}$, and is called the *pullback* of vector V. Once this is done, $V'^\mu(\boldsymbol{x}) - V^\mu(\boldsymbol{x})$ gives the true change of the vector function between the two points. The pullback is found by combining the contravariant vector transformation property (1.21) with $x^\mu = y^\mu - \epsilon X^\mu$, which gives

$$V'^\mu(\boldsymbol{x}) = \frac{\partial x^\mu}{\partial y^\nu} V^\nu(\boldsymbol{y}) = (\delta_\nu^\mu - \epsilon\, \partial_\nu X^\mu(\boldsymbol{x})\,) V^\nu(\boldsymbol{x} + \epsilon X). \tag{1.213}$$

A term of $O(\epsilon^2)$ has been suppressed in (1.213) to write

$$\epsilon \frac{\partial X^\mu(\boldsymbol{x})}{\partial y^\nu} \approx \epsilon \frac{\partial X^\mu(\boldsymbol{x})}{\partial x^\nu}. \tag{1.214}$$

Using a Taylor series expansion in (1.213) gives the $O(\epsilon)$ change in the vector

$$V'^{\mu}(\boldsymbol{x}) - V^{\mu}(\boldsymbol{x}) = \epsilon X^{\nu}(\boldsymbol{x})\partial_{\nu}V^{\mu}(\boldsymbol{x}) - \epsilon V^{\nu}(\boldsymbol{x})\partial_{\nu}X^{\mu}(\boldsymbol{x}). \tag{1.215}$$

From (1.215) the Lie derivative of a vector V along X is given by

$$(\mathcal{L}_X V)^{\mu} = \lim_{\epsilon\to 0}\frac{V'^{\mu}(\boldsymbol{x}) - V^{\mu}(\boldsymbol{x})}{\epsilon} = X^{\nu}(\boldsymbol{x})\partial_{\nu}V^{\mu}(\boldsymbol{x}) - V^{\nu}(\boldsymbol{x})\partial_{\nu}X^{\mu}(\boldsymbol{x}). \tag{1.216}$$

This gives the rate of change of vector V at point $\boldsymbol{x}$ along an infinitesimal displacement in the direction of vector $X(\boldsymbol{x})$. Using the Cartan notation (1.45) for vectors, the Lie derivative of a vector can be written as a *commutator*,

$$\mathcal{L}_X V = XV - VX \equiv [X, V]. \tag{1.217}$$

This commutator is often referred to as a *Lie bracket*. This follows, since the Cartan notation uses

$$X = X^{\mu}\partial_{\mu}, \quad Y = Y^{\mu}\partial_{\mu}, \tag{1.218}$$

so that the action of the Lie bracket $[X, Y]$ on a scalar function f defines a vector Z given by

$$\begin{aligned}[X, Y]f &= X^{\mu}(\partial_{\mu}Y^{\nu})\partial_{\nu}f + X^{\mu}Y^{\nu}\partial_{\mu}\partial_{\nu}f - Y^{\nu}(\partial_{\nu}X^{\mu})\partial_{\mu}f - Y^{\nu}X^{\mu}\partial_{\nu}\partial_{\mu}f \\ &= (X^{\nu}\partial_{\nu}Y^{\mu} - Y^{\nu}\partial_{\nu}X^{\mu})\partial_{\mu}f = Zf,\end{aligned} \tag{1.219}$$

where the vector Z has the components

$$Z^{\mu} = (X^{\nu}\partial_{\nu}Y^{\mu} - Y^{\nu}\partial_{\nu}X^{\mu}). \tag{1.220}$$

The Lie bracket has the property of antisymmetry,

$$[X, Y] = -[Y, X], \tag{1.221}$$

and satisfies the *Jacobi identity*,

$$[[X, Y], Z] + [[Z, X], Y] + [[Y, Z], X] = 0. \tag{1.222}$$

The Jacobi identity follows from direct substitution.

Result (1.217) for $\mathcal{L}_X V$ can also be derived from the action of the Lie derivative on a scalar function φ, given by (1.212). Since $V\varphi = V^{i}\partial_{i}\varphi$ is a scalar, associativity gives

$$\mathcal{L}_X(V\varphi) = X(V\varphi) = (XV)\varphi. \tag{1.223}$$

The Lie derivative obeys the Leibniz property,

$$\mathcal{L}_X(V\varphi) = (\mathcal{L}_X V)\varphi + V(\mathcal{L}_X\varphi) = (\mathcal{L}_X V)\varphi + (VX)\varphi. \tag{1.224}$$

Equating (1.223) and (1.224) immediately gives

$$(\mathcal{L}_X V)\varphi = ([X, V])\varphi. \tag{1.225}$$

The Lie derivative can be generalized to include its action on tensors and forms. For example, if $g_{\mu\nu}$ are the components of the second-rank covariant tensor $\boldsymbol{g}$, its Lie derivative has the components

$$(\mathcal{L}_X\boldsymbol{g})_{\mu\nu}(\boldsymbol{x}) = X^{\rho}(\boldsymbol{x})\partial_{\rho}g_{\mu\nu}(\boldsymbol{x}) + g_{\rho\nu}(\boldsymbol{x})\partial_{\mu}X^{\rho}(\boldsymbol{x}) + g_{\mu\rho}(\boldsymbol{x})\partial_{\nu}X^{\rho}(\boldsymbol{x}). \tag{1.226}$$

Exercise 1.20. Derive (1.226) using the fact that $g_{\mu\nu} Y^{\mu} Z^{\nu}$ is a scalar function, so that

$$\mathcal{L}_X(g_{\mu\nu} Y^{\mu} Z^{\nu}) = X^{\rho} \partial_{\rho}(g_{\mu\nu} Y^{\mu} Z^{\nu}), \tag{1.227}$$

then applying the Leibniz property of the Lie derivative $\mathcal{L}_X$ to $g_{\mu\nu} Y^{\mu} Z^{\nu}$, and finally equating the two forms using result (1.225).

References and recommended further reading

Mathematical monographs that develop the topics of this chapter include

- Y Matsushima 1972 *Differentiable Manifolds* (New York: Dekker)
- P Bamberg, S Sternberg 1988 *A Course in Mathematics for Students of Physics Vol I* (Cambridge: Cambridge University Press)
- Darling R 1994 *Differential Forms and Connections* (Cambridge: Cambridge University Press)
- P Dennery, A Krzywicki 1997 *Mathematics For Physicists* (New York: Dover)
- T Frankel 1997 *The Geometry of Physics: An Introduction* (Cambridge: Cambridge University Press)
- B Kusse, E Westwig 1998 *Mathematical Physics* (New York: Wiley)
- B O'Neill 2006 *Elementary Differential Geometry 2nd edn* (Amsterdam: Elsevier)
- Arfken G, Weber H, Harris F 2013 *Mathematical Methods for Physicists 7th edn* (Amsterdam: Elsevier)
- Hassani S 2013 *Mathematical Physics 2nd edn* (Berlin: Springer–Verlag)
- Hubbard J, Hubbard B 2015 *Vector Calculus, Linear Algebra, and Differential Forms 5th edn* (Ithaca: Matrix Editions)

Chapter 2

Newtonian mechanics and functional methods

In order for the mathematical rules that govern fields and material particles to constitute a complete theory, the physical quantities defined in Newtonian physics, such as momentum and energy, must be generalized to include the contributions from fields. For that reason it is useful to review Newtonian physics in the context of basic differential geometry, and this starts with Newtonian particle trajectories and their relationship to vectors and vector fields. Functional methods are then introduced to formulate the variational approach to Newtonian mechanics. It is the variational approach that can be generalized to field theory.

2.1 Newton's second law

Although tangent vectors have been defined as the velocity associated with a trajectory, the tangent space can be generalized to all vector quantities specific to a point in a manifold. In the context of Newtonian physics this means that both the velocity of a particle and the force acting on the particle are members of the tangent space *at the position of the particle.* In any physical system there is a net force present on the particle at every point in its manifold of motion, even if it is zero. Therefore a *velocity vector field* $\boldsymbol{v}(\boldsymbol{r}(t))$ and a *force vector field* $\boldsymbol{F}(\boldsymbol{r}(t))$ both exist in the tangent space at every position $\boldsymbol{r}(t)$ along the trajectory. The collection of tangent spaces associated with a manifold is referred to as the *tangent bundle.*

For the case of an inertial point mass m, *Newton's second law* relates the total force $\boldsymbol{F}(\boldsymbol{r}(t))$ acting on the particle as it moves along the trajectory $\boldsymbol{r}(t)$ to the particle's instantaneous acceleration $\boldsymbol{a}(\boldsymbol{r}(t)) \equiv \boldsymbol{a}(t) = \mathrm{d}\boldsymbol{v}(t)/\mathrm{d}t = \mathrm{d}^2\boldsymbol{r}(t)/\mathrm{d}t^2$. In this case Newton's second law is stated mathematically as

$$\boldsymbol{F}(\boldsymbol{r}(t)) = m\,\boldsymbol{a}(t) = m\,\frac{\mathrm{d}\boldsymbol{v}(t)}{\mathrm{d}t} = m\frac{\mathrm{d}^2\boldsymbol{r}(t)}{\mathrm{d}t^2}. \tag{2.1}$$

doi:10.1088/978-0-7503-3455-6ch2

© IOP Publishing Ltd 2022

It is important to note that the total force $\boldsymbol{F}$ is assumed to have no *explicit* dependence on time in the coordinate system used in (2.1), only an *implicit* dependence on time through the trajectory $\boldsymbol{r}(t)$. In that sense the coordinate system in use is privileged. This is discussed later when Galilean relativity is defined. Once the net force $\boldsymbol{F}(\boldsymbol{r})$ is stipulated as a function of position, Newton's law yields a second-order differential equation that is solved by the *classical trajectory* $\boldsymbol{r}_s(t)$ of the Newtonian particle. Obtaining a unique solution to (2.1) therefore requires two boundary conditions for $\boldsymbol{r}(t)$. It is important to note that both $\boldsymbol{F}$ and $\boldsymbol{r}$ are vectors, and so they take the coordinate-independent forms $\boldsymbol{F} = F^j\hat{\mathbf{e}}_j$ and $\boldsymbol{r} = r^k\hat{\mathbf{e}}_k$, where the same *dimensionless* unit vectors are used for both, since they share the same tangent space. Inserting these into (2.1) gives

$$F^j(\boldsymbol{r}, t)\,\hat{\mathbf{e}}_j = m\frac{\mathrm{d}^2}{\mathrm{d}t^2}(r^j\hat{\mathbf{e}}_j) = m\left(\ddot{r}^j\hat{\mathbf{e}}_j + 2\dot{r}^j\dot{\hat{\mathbf{e}}}_j + r^j\ddot{\hat{\mathbf{e}}}_j\right). \tag{2.2}$$

If Cartesian coordinates are chosen, so that $\dot{\hat{\mathbf{e}}}_i = \ddot{\hat{\mathbf{e}}}_i = 0$, Newton's second law (2.2) becomes

$$F^j(\boldsymbol{r}, t) = m\,\ddot{x}^j. \tag{2.3}$$

Other coordinate systems require the use of their unit vector forms to determine the time derivatives.

A simple example helps to clarify Newton's law. A two-dimensional trajectory in polar coordinates (1.1) is parameterized by the continuous functions $r = r(t)$ and $\theta = \theta(t)$. Result (1.40) shows that the trajectory in spherical coordinates is given by $\boldsymbol{r}(t) = r(t)\,\hat{\mathbf{e}}_r(t)$. The velocity vector is therefore given by

$$\boldsymbol{v}(t) = \frac{\mathrm{d}\boldsymbol{r}(t)}{\mathrm{d}t} = \frac{\mathrm{d}}{\mathrm{d}t}(r\,\hat{\mathbf{e}}_r) = \dot{r}\,\hat{\mathbf{e}}_r + r\,\dot{\hat{\mathbf{e}}}_r. \tag{2.4}$$

Because θ is a function of time and the Cartesian direction vectors obey $\dot{\hat{\mathbf{e}}}_x = \dot{\hat{\mathbf{e}}}_y = 0$, the radial direction vector (1.38) has the time derivative

$$\begin{aligned} r\,\dot{\hat{\mathbf{e}}}_r &= -r\,\dot{\theta}\sin\theta\,\hat{\mathbf{e}}_x + r\,\dot{\theta}\cos\theta\,\hat{\mathbf{e}}_y \\ &= \dot{\theta}(-r\sin\theta\,\hat{\mathbf{e}}_x + r\cos\theta\,\hat{\mathbf{e}}_y) = r\,\dot{\theta}\left(\frac{\mathbf{e}_\theta}{r}\right) = r\,\dot{\theta}\,\hat{\mathbf{e}}_\theta \equiv v^\theta\hat{\mathbf{e}}_\theta. \end{aligned} \tag{2.5}$$

This gives the velocity tangent vector as

$$\boldsymbol{v}(t) = \dot{r}\,\hat{\mathbf{e}}_r + r\,\dot{\theta}\,\hat{\mathbf{e}}_\theta = v^r\hat{\mathbf{e}}_r + v^\theta\hat{\mathbf{e}}_\theta, \tag{2.6}$$

where $v^r = \dot{r}$ and $v^\theta = r\,\dot{\theta}$. Like (2.5), it follows from (1.39) that

$$\dot{\hat{\mathbf{e}}}_\theta = -\dot{\theta}\,(\cos\theta\,\hat{\mathbf{e}}_x + \sin\theta\,\hat{\mathbf{e}}_y) = -\dot{\theta}\,\hat{\mathbf{e}}_r. \tag{2.7}$$

Exercise 2.1. Result (1.44) showed that the trajectory in spherical coordinates has the form $\boldsymbol{r}(t) = r\,\hat{\mathbf{e}}_r$. Using the relations (1.43) show that it satisfies

$$\frac{\mathrm{d}\boldsymbol{r}(t)}{\mathrm{d}t} = \dot{r}\,\hat{\mathbf{e}}_r + r\,\dot{\hat{\mathbf{e}}}_r = \dot{x}\,\hat{\mathbf{e}}_x + \dot{y}\,\hat{\mathbf{e}}_y + \dot{z}\,\hat{\mathbf{e}}_z. \tag{2.8}$$

Using (2.5) and (2.7), the two-dimensional acceleration vector $\boldsymbol{a}(t)$ is given by

$$\begin{aligned} \boldsymbol{a}(t) = \frac{\mathrm{d}\boldsymbol{v}(t)}{\mathrm{d}t} &= \dot{v}^r \hat{\mathbf{e}}_r + v^r \dot{\hat{\mathbf{e}}}_r + \dot{v}^\theta \hat{\mathbf{e}}_\theta + v^\theta \dot{\hat{\mathbf{e}}}_\theta \\ &= \ddot{r}\, \hat{\mathbf{e}}_r + \dot{r}\, \dot{\hat{\mathbf{e}}}_r + \dot{r}\dot{\theta}\, \hat{\mathbf{e}}_\theta + r\ddot{\theta}\, \hat{\mathbf{e}}_\theta + r\dot{\theta}\, \dot{\hat{\mathbf{e}}}_\theta \\ &= (\ddot{r} - r\dot{\theta}^2)\, \hat{\mathbf{e}}_r + (2\dot{r}\dot{\theta} + r\ddot{\theta})\, \hat{\mathbf{e}}_\theta = a^r\, \hat{\mathbf{e}}_r + a^\theta \hat{\mathbf{e}}_\theta. \end{aligned} \tag{2.9}$$

For the case that the particle is observed to move in a circle of constant radius r, so that $\dot{r} = \ddot{r} = 0$, at a constant angular speed $\omega = \dot{\theta}$, so that $\dot{\omega} = \ddot{\theta} = 0$, the results (2.9) show that there must be a constant *centripetal force* given by the familiar centripetal force formula,

$$F^r = ma^r = -mr\,\dot{\theta}^2 = -m\omega^2 r = -\frac{m(v^\theta)^2}{r}, \tag{2.10}$$

where (2.6) gives the tangential speed $v^\theta = \omega r$. It is useful in understanding fluid dynamics to consider the motion of the constituents of a rotating rigid body. In three dimensions this is accomplished by introducing a vector-valued angular velocity $\boldsymbol{\omega}$, which points along the axis of rotation and has a magnitude given by $\omega^2 = \omega_j \omega^j$. The velocity $\boldsymbol{v}$ of the rotating element of the rigid body at $\boldsymbol{r}$ is given by

$$\boldsymbol{v}(\boldsymbol{r}) = \boldsymbol{\omega} \times \boldsymbol{r}. \tag{2.11}$$

Exercise 2.2. Verify (2.11) by setting $\boldsymbol{\omega} = \omega\, \hat{\mathbf{e}}_z$ and $\boldsymbol{r} = x\, \hat{\mathbf{e}}_x + y\, \hat{\mathbf{e}}_y + z\, \hat{\mathbf{e}}_z$. Show that these choices reproduce the polar coordinate result $\boldsymbol{v} = \omega x\, \hat{\mathbf{e}}_y - \omega y\, \hat{\mathbf{e}}_x = \omega\sqrt{x^2 + y^2}\, \hat{\mathbf{e}}_\theta = \omega r\, \hat{\mathbf{e}}_\theta$.

2.2 Newtonian trajectories and tangent vectors

A scalar function of position, $f(\boldsymbol{r})$, can be combined with a trajectory. The time derivative of the function along the trajectory is given by

$$\frac{\mathrm{d}}{\mathrm{d}t} f(\boldsymbol{r}(t)) = \frac{\mathrm{d}x^k}{\mathrm{d}t} \frac{\partial f(\boldsymbol{r})}{\partial x^k} = v^k \mathbf{e}_k \cdot \mathbf{e}^j \partial_j f(\boldsymbol{r}) = \boldsymbol{v} \cdot \nabla f(\boldsymbol{r}) \equiv \mathbf{V} f(\boldsymbol{r}). \tag{2.12}$$

The tangent space vector $\boldsymbol{v}(t) = \mathrm{d}\boldsymbol{r}(t)/\mathrm{d}t$ has appeared in the derivative of the function $f(\boldsymbol{r})$. Because the trajectory $\boldsymbol{r}(t)$ is arbitrary, the entire tangent space of vectors is associated with the differential operator $\boldsymbol{v} \cdot \nabla = \mathbf{V}$, often referred to as the *velocity gradient*. This is another instance of the equivalence $\{\mathbf{e}_j\} \leftrightarrow \{\partial_j\}$. Differential geometry uses this association by defining the tangent space vector $\boldsymbol{v}(x)$ in terms of the action of the differential operator

$$\mathbf{V} = v^k(x) \frac{\partial}{\partial x^k} \equiv v^k \partial_k. \tag{2.13}$$

This definition of a tangent vector in terms of its action on an arbitrary differentiable function carries all the familiar properties of the vector established earlier in terms of basis vectors and components. This equivalence was developed in chapter 1 when the space of linear functions dual to vectors was introduced in (1.172).

This relationship is used in Newtonian mechanics to introduce the scalar function $U(\boldsymbol{r}, \dot{\boldsymbol{r}})$, known as the *potential energy*, which has the property that the net force $\mathbf{F}$ acting on a particle when it is at point $\boldsymbol{r}$ with a velocity of $\dot{\boldsymbol{r}}$ is given by

$$\mathbf{F}_j(\boldsymbol{r}, \dot{\boldsymbol{r}}) = -\partial_j U(\boldsymbol{r}, \dot{\boldsymbol{r}}) + \frac{\mathrm{d}}{\mathrm{d}\dot{r}^j} U(\boldsymbol{r}, \dot{\boldsymbol{r}}). \tag{2.14}$$

In what follows it will be *assumed* that the net force acting on the particle at point $\boldsymbol{r}$ is *static* according to the observer, so that $\dot{U} = 0$, and U is independent of the velocity $\dot{\boldsymbol{r}}$. This is discussed again when Galilean relativity is introduced and the potential energy becomes explicitly time dependent. The total *mechanical energy* of the particle along the trajectory $\boldsymbol{r}(t)$ is defined as

$$E(t) = \frac{1}{2} m v(t) \cdot v(t) + U(\boldsymbol{r}(t)) \equiv K + U, \tag{2.15}$$

where $v(t)$ is the velocity vector of the trajectory at time t and $K = \frac{1}{2} m v \cdot v \equiv \frac{1}{2} m v^2$ is the *kinetic energy*. The differential of E gives

$$\mathrm{d}E = \mathrm{d}K + \mathrm{d}U = m\frac{\mathrm{d}v(t)}{\mathrm{d}t} \cdot v(t)\mathrm{d}t + \nabla U \cdot \mathrm{d}\boldsymbol{r}(t) = (m\boldsymbol{a}(t) - \boldsymbol{F}(\boldsymbol{r}(t))) \cdot \mathrm{d}\boldsymbol{r}(t), \tag{2.16}$$

where the identification $v(t)\,\mathrm{d}t = \mathrm{d}\boldsymbol{r}(t)$ follows from the definition (1.107). If the trajectory coincides with the classical trajectory that satisfies (2.1), the right-hand side of (2.16) vanishes and the total mechanical energy is conserved, since $\mathrm{d}E = 0$ along the trajectory. For such a case, it follows that $\mathrm{d}K = -\mathrm{d}U$ and the change in the kinetic energy of the particle as it moves from $\boldsymbol{r}_i$ to $\boldsymbol{r}_f$ is given by the *work–energy theorem*,

$$K_f - K_i = \int_{K_i}^{K_f} \mathrm{d}K = -\int_{U_i}^{U_f} \mathrm{d}U = -\int_{\mathbf{r}_i}^{\mathbf{r}_f} \nabla U \cdot \mathrm{d}\boldsymbol{r} = \int_{\mathbf{r}_i}^{\mathbf{r}_f} \boldsymbol{F} \cdot \mathrm{d}\boldsymbol{r} \equiv W, \tag{2.17}$$

where W is the *work* done by the force.

2.3 Newton's first law and Galilean relativity

In Newtonian physics the time t is treated as a parameter that characterizes the motion along a trajectory. It is treated as an *absolute* quantity provided experimentally by a mechanical device generically referred to as a clock. All Newtonian observers of physical phenomena are equipped with identically functioning clocks which remain synchronized throughout their observations of phenomena. Similarly, Newtonian physics measures distances with a second device generically referred to as a meter stick that remains identical throughout all motion, and all observers are equipped with identical meter sticks. Therefore, in Newtonian physics, all observers agree upon the time intervals between events as well as the distances separating material objects at any given moment. For the purpose of comparing or *relating* observations made by two different observers, the origins of two different coordinate systems can be thought of as the positions of two distinct observers watching the motion of some point mass m. The respective coordinate systems serve as *reference frames*, or simply *frames*, for the two observers who record the positions of *events* with respect to the origin as well as the time of occurrence. An example of an event is the presence of an object at a particular location at a particular time. The rules for relating observations made by the two different observers are referred to as *relativity* and these rules play a critical role in determining physical laws.

Newton's first law, the law of inertia, states that an object will remain in *uniform unaccelerated motion* if zero total force acts upon it. For simplicity the manifold for motion will be assumed to be Euclidean, so that finite position vectors are viable. Newton's first law must be consistent with Newton's second law (2.1). If the total force $\boldsymbol{F}$ on an object is zero, then there must be a coordinate system or an observer, referred to as $\mathcal{O}$, such that the position vector of the object with respect to the observer at the origin, denoted $\boldsymbol{r}(t)$, satisfies

$$\frac{\mathrm{d}^2\boldsymbol{r}(t)}{\mathrm{d}t^2} = \frac{\mathrm{d}\boldsymbol{v}(t)}{\mathrm{d}t} = \boldsymbol{a}(t) = 0. \tag{2.18}$$

A second observer, denoted by $\mathcal{O}'$, will agree with $\mathcal{O}$ if the position of the object with respect to the second origin, denoted by $\boldsymbol{r}'(t)$, is such that $\mathrm{d}^2\boldsymbol{r}'(t)/\mathrm{d}t^2 = 0$. Both observers will then agree that Newton's second law is consistent with zero total force. The two frames associated with the two observers, $\mathcal{O}$ and $\mathcal{O}'$, are referred to as *inertial frames*, since *both* are consistent with the law of inertia. Observers $\mathcal{O}$ and $\mathcal{O}'$ are often referred to as *inertial observers*. Since both observers are describing the same object using synchronized clocks and identical meter sticks, consistency with zero force requires that the two inertial frames yield

$$\frac{\mathrm{d}^2\boldsymbol{r}'(t')}{\mathrm{d}t'^2} = \frac{\mathrm{d}^2\boldsymbol{r}(t)}{\mathrm{d}t^2} = 0. \tag{2.19}$$

This requirement is clearly satisfied if

$$\boldsymbol{r}'(t') = \boldsymbol{r}(t) - \boldsymbol{r}_o - \boldsymbol{v}_o t, \quad \boldsymbol{r}(t) = \boldsymbol{r}'(t') + \boldsymbol{r}_o + \boldsymbol{v}_o t', \quad t' = t, \tag{2.20}$$

where $\boldsymbol{r}_o$ and $\boldsymbol{v}_o$ are *constant vectors*. In what follows $\boldsymbol{r}_o$ will be set to zero for simplicity, and this means that the origins of the two coordinate systems coincide at $t = t' = 0$. In anticipation of special relativity, the change of coordinates given by (2.20) is referred to as a *Galilean boost*. This term is used since the transformation boosts the measurements into the frame of the inertial observer $\mathcal{O}'$ moving at a velocity of $\boldsymbol{v}_o$ relative to the inertial observer $\mathcal{O}$. It is important to note that there is no limit on the relative velocity $\boldsymbol{v}_o$.

Although the time parameter t does not change in (2.27), since time is treated as absolute in Newtonian physics, it is useful to distinguish both the second observer's time and space measurements with primes. The Galilean boost of (2.20) assumes that the two clocks are synchronized, so that the measurements of two particles' positions that are made simultaneously by one inertial observer will therefore appear simultaneous to all inertial observers regardless of the presence of force. This assumption is central to Newtonian mechanics, but experimental results in the 19th century proved it to be incorrect. In that regard it is important to note that the Galilean boost of (2.20) results in no difference in time intervals measured by the two observers, since $\mathrm{d}t' = \mathrm{d}t$, so that the two observers agree on the action of the time derivative,

$$\frac{\mathrm{d}}{\mathrm{d}t}F(t) = \frac{\mathrm{d}}{\mathrm{d}t'}F(t'). \tag{2.21}$$

It is important to understand the effect of a Galilean boost on both the spatial gradient and the partial time derivative. The new form of the spatial gradient is found using the usual covariant transformation of the partial derivative combined with the form (2.20) of the Galilean boost,

$$\frac{\partial}{\partial r^j} = \frac{\partial r'^k}{\partial r^j}\frac{\partial}{\partial r'^k} + \frac{\partial t'}{\partial r^j}\frac{\partial}{\partial t'} = \delta^k{}_j\frac{\partial}{\partial r'^k} = \frac{\partial}{\partial r'^j} \implies \nabla = \nabla'. \tag{2.22}$$

However, the covariant transformation of the partial time derivative yields a more complicated result

$$\frac{\partial}{\partial t} = \frac{\partial t'}{\partial t}\frac{\partial}{\partial t'} + \frac{\partial r'^j}{\partial t}\frac{\partial}{\partial r'^j} = \frac{\partial}{\partial t'} - \boldsymbol{v}_o \cdot \nabla'. \tag{2.23}$$

As a result, if the two inertial observers are comparing the action of the partial time derivative on an observation, they must correct for their relative motion by using (2.23). This will come into play later when the action of a Galilean boost on the conservation of energy is discussed.

The Galilean boost (2.20) can be generalized to include the mixed tensor quantity $\boldsymbol{R}$, defined as

$$\boldsymbol{R} = R^j_k\, \mathbf{e}_j \otimes \mathbf{e}^k, \tag{2.24}$$

as long as the components of $\boldsymbol{R}$ satisfy *orthogonality*,

$$R_j{}^\ell R^j_k = \delta^\ell_k, \tag{2.25}$$

and are time independent, so that $\mathrm{d}\boldsymbol{R}/\mathrm{d}t = 0$. The tensor indices are raised and lowered by the usual procedure,

$$R_j{}^\ell = g_{jk}g^{i\ell}R^k_i. \tag{2.26}$$

Using $\boldsymbol{R}$ the Galilean boost (2.20) is generalized to

$$\boldsymbol{r}'(t') = \boldsymbol{R}\cdot\boldsymbol{r}(t) - \boldsymbol{v}_o t, \quad t' = t, \tag{2.27}$$

which is referred to as a *Galilean transformation*. The requirement of orthogonality preserves position vector lengths for the case where $\boldsymbol{v}_o = \boldsymbol{r}_o = 0$. This follows from (1.73), which gives

$$\boldsymbol{r}' \cdot \boldsymbol{r}' = r'^j r'_j = R^j{}_k r^k R_j{}^\ell r_\ell = \delta_k^\ell r^k r_\ell = r^k r_k = \boldsymbol{r} \cdot \boldsymbol{r}. \tag{2.28}$$

Result (2.28) shows that the finite position vector's length, given by (1.75), is the same according to both inertial observers as long as they are stationary with respect to each other and have the same origin. This is discussed in greater detail in chapter 7, where the rotation group is analyzed. The form of (2.27) allows the possibility of rotating the coordinate system with an orthogonal transformation denoted by $\boldsymbol{R}$. The tensor $\boldsymbol{R}$ will be of more importance in later chapters in determining the angular momentum content of the fields. For simplicity, in many applications of Galilean relativity, the transformation (2.27) will be restricted to the form of a boost given by (2.20).

While the Galilean transformations yield a family of inertial frames where there will also be zero total force, in 1632 Galileo argued that consistency with experiments required that the *universal* laws of motion take the *same form* in all inertial frames. In that regard, Newton's second law of motion is formulated to be consistent with what is now called the *Galilean principle*. In modern terms the ability to change coordinate systems in specific ways and not affect the form of the equations of motion is referred to as a *symmetry*. In that sense, the Galilean transformation (2.27) is a symmetry of Newtonian physics *based on the experimental evidence available in Newton's time*. However, this does not mean that the force acting on a material object is independent of the frame chosen to represent it. Restricting attention to the Galilean boost (2.20) assumes that the origins and axes of the two coordinate systems coincide at $t = 0$. The origin of the $\mathcal{O}'$ system, given by $\boldsymbol{r}' = 0$, is located at $\boldsymbol{r}(t) = \boldsymbol{v}_o t$ at any later time *according to* $\mathcal{O}$. As a result, $\boldsymbol{v}_o$ is identified as the velocity of observer $\mathcal{O}'$ *relative* to observer $\mathcal{O}$. For simplicity, it will be *assumed* that the total force $\boldsymbol{F}$ is *static* at every point *according to* $\mathcal{O}$. This means that the force at position $\boldsymbol{r}$ has no explicit time dependence. Using Newton's second law, the two inertial observers will claim that a point mass follows the trajectories $\boldsymbol{r}(t)$ and $\boldsymbol{r}'(t')$ respectively, and these satisfy the second-order differential equations given by

$$\begin{aligned} \mathcal{O} &\Longrightarrow m\frac{\mathrm{d}^2\boldsymbol{r}(t)}{\mathrm{d}t^2} = \boldsymbol{F}(\boldsymbol{r}(t)), \\ \mathcal{O}' &\Longrightarrow m\frac{\mathrm{d}^2\boldsymbol{r}'(t')}{\mathrm{d}t'^2} = m\frac{\mathrm{d}^2}{\mathrm{d}t^2}(\boldsymbol{r}(t) - \boldsymbol{v}_o t) = m\frac{\mathrm{d}^2\boldsymbol{r}(t)}{\mathrm{d}t^2} = \boldsymbol{F}(\boldsymbol{r}(t)) \\ &\qquad = \boldsymbol{F}(\boldsymbol{r}'(t') + \boldsymbol{v}_o t'), \end{aligned} \tag{2.29}$$

where $\boldsymbol{F}$ is the *same function* for both observers. In Newtonian physics it is not necessary to distinguish t and t', but it draws attention to the idea that events can be measured by separate observers in relative motion. It is important to note that the assumption that $\mathcal{O}$ observes a static force field means that another *inertial observer* will observe *the force field in motion* and it will therefore have an explicit time dependence. A Galilean boost therefore does *not* leave Newton's second law unchanged if a force is acting on the object. Instead, Newton's second law in the $\mathcal{O}'$ frame is changed to (2.29), so that the second observer records a force acting on the object that is consistent with the observer's motion relative to the *agent of the force*. The agent of the force is typically another material body creating the force that is stationary relative to the inertial mass under observation.

However, Newton's second law shows that a *noninertial observer* will disagree with the form of the force measured by an inertial observer. This is demonstrated by considering two observers in *constant relative acceleration*. For the case of two observers $\mathcal{O}$ and $\mathcal{O}'$ in constant relative acceleration $\boldsymbol{a}_o$, such that their origins and axes coincide at $t = 0$, the trajectory $\boldsymbol{r}'(t)$ of an object according to $\mathcal{O}'$ compared to the trajectory $\boldsymbol{r}(t)$ observed by $\mathcal{O}$ is given by

$$\boldsymbol{r}'(t') = \boldsymbol{r}(t) - v_o t - \frac{1}{2}\boldsymbol{a}_o t^2, \quad t = t'. \tag{2.30}$$

This transformation results in observer $\mathcal{O}$ claiming that the origin of the $\mathcal{O}'$ reference frame is located at $v_o t + \frac{1}{2}\boldsymbol{a}_o t^2$ at time t. The transformation (2.30) is not a Galilean transformation, so that the two observers will disagree on the force acting on the point mass. This is demonstrated by applying (2.30) to Newton's second law, which gives

$$\boldsymbol{F}'(\boldsymbol{r}'(t')) = m\frac{\mathrm{d}^2\boldsymbol{r}'(t)}{\mathrm{d}t'^2} = m\frac{\mathrm{d}^2\boldsymbol{r}(t)}{\mathrm{d}t^2} - m\boldsymbol{a}_o = \boldsymbol{F}(\boldsymbol{r}(t)) - m\boldsymbol{a}_o. \tag{2.31}$$

To observer $\mathcal{O}'$ there appears to be an additional force, given by $-m\boldsymbol{a}_o$, acting on *all* inertial masses m, regardless of their mass.

Result (2.31) provides a simple but profound insight into gravitation. In the Newtonian version of gravitation the force of attraction on a point mass m at a distance r from a point mass M located at the origin is given by

$$\boldsymbol{F}_G = -\frac{GmM}{r^2}\hat{\mathbf{e}}_r = -m\left(\frac{GM}{r^2}\hat{\mathbf{e}}_r\right) \equiv m\boldsymbol{g}. \tag{2.32}$$

The Eötvös experiments in the late 19th century and numerous subsequent experiments are consistent with the *gravitational masses m* and *M* appearing in (2.32) being identical to the inertial masses appearing in Newton's second law. Of course, this result is within the limits of accuracy for the measuring devices, which is currently one part in 10^{15}. According to an observer $\mathcal{O}$ that is stationary with respect to mass M, *all* point masses m at a distance r will fall toward mass M with the *same initial acceleration* $\boldsymbol{g} = -GM\hat{\mathbf{e}}_r/r^2(0)$ since

$$\mathcal{O} \Longrightarrow m\left.\frac{\mathrm{d}^2\boldsymbol{r}}{\mathrm{d}t^2}\right|_{t=0} = m\boldsymbol{a}(0) = m\boldsymbol{g} \Longrightarrow \boldsymbol{a}(0) = \boldsymbol{g}; \tag{2.33}$$

again, this is in agreement with the perhaps apocryphal experiments of Galileo. This is referred to as the *weak principle of equivalence*. For the case that $r(0)$ is large compared to the distance mass m is allowed to fall, the acceleration will remain virtually constant. The radius vector for m is then given by $\boldsymbol{r}(t) = \boldsymbol{r}_0 + \frac{1}{2}\boldsymbol{g}t^2$. Treating a second observer $\mathcal{O}'$ as a point mass, the second observer $\mathcal{O}'$ will also fall with the same acceleration $\boldsymbol{g}$ according to $\mathcal{O}$. Using (2.31) this *freely falling* observer will therefore claim that all other freely falling objects are experiencing the net force

$$F'(\boldsymbol{r}'(t)) = \boldsymbol{F}(\boldsymbol{r}(t)) - m\boldsymbol{g} = m\boldsymbol{g} - m\boldsymbol{g} = 0. \tag{2.34}$$

In other words, in a freely falling frame there appears to be no force of gravitation, a state often referred to as *weightlessness*. This occurs in the case of an object orbiting another under the influence of the Newtonian gravitational force (2.32). For such a case, (2.9) shows that if the orbiting object is freely falling toward the other with a constant centripetal acceleration of $a_c = v_\theta^2/r = GM/r^2$, then there is zero net radial force on the orbiting object. This observation is also referred to as the *weak principle of equivalence*, in this case stating that the force of gravitation cannot be distinguished from an accelerated frame of reference through *local experiments*. The latter restriction is important, since the acceleration due to gravity will typically have a spatial dependence which can be detected if the measurements are not local, whereas a *uniformly accelerated* reference frame does not have a spatial dependence. The existence of a local frame of reference where gravitation disappears is critical to developing Einstein's general theory of relativity.

Galilean relativity can be extended to velocity addition, and for that case it is notationally more convenient to designate observers with a subscript. The velocity of a particle measured relative to the $\mathcal{O}_A$ origin is given by $\boldsymbol{v}_A = \mathrm{d}\boldsymbol{r}_A(t)/\mathrm{d}t$, while the velocity of the same particle measured relative to the observer $\mathcal{O}_B$ origin is given by $\boldsymbol{v}_B = \mathrm{d}\boldsymbol{r}_B(t)/\mathrm{d}t$. For the case that the two observers are related by the Galilean boost of (2.20), these two velocities are related according to

$$\begin{aligned} \boldsymbol{v}_{AB} &\equiv \frac{\mathrm{d}\boldsymbol{r}_A(t)}{\mathrm{d}t} - \frac{\mathrm{d}\boldsymbol{r}_B(t)}{\mathrm{d}t} = \boldsymbol{v}_A - \boldsymbol{v}_B = \boldsymbol{v}_o \implies \boldsymbol{v}_{BA} \\ &= \frac{\mathrm{d}\boldsymbol{r}_B(t)}{\mathrm{d}t} - \frac{\mathrm{d}\boldsymbol{r}_A(t)}{\mathrm{d}t} = \boldsymbol{v}_B - \boldsymbol{v}_A = -\boldsymbol{v}_{AB}. \end{aligned} \tag{2.35}$$

The relative velocity $\boldsymbol{v}_{AB}$ is understood to be the velocity of observer A relative to observer B, and result (2.35) shows that $\boldsymbol{v}_{BA}$ is equal and opposite to $\boldsymbol{v}_{AB}$. It also shows that the velocity of the particle measured by $\mathcal{O}_A$ is the velocity of the particle measured by $\mathcal{O}_B$ plus the velocity of observer $\mathcal{O}_B$ relative to observer $\mathcal{O}_A$.

Velocities in Newtonian physics are often measured relative to a third inertial frame, designated $\mathcal{O}_C$. For such a case the relative velocity of two observers can be found from the velocities they observe for particle motion by using the particle motion observed in the third inertial frame,

$$\begin{aligned} \boldsymbol{v}_{AB} &= \frac{\mathrm{d}\boldsymbol{r}_A}{\mathrm{d}t} - \frac{\mathrm{d}\boldsymbol{r}_B}{\mathrm{d}t} = \left(\frac{\mathrm{d}\boldsymbol{r}_A}{\mathrm{d}t} - \frac{\mathrm{d}\boldsymbol{r}_C}{\mathrm{d}t}\right) + \left(\frac{\mathrm{d}\boldsymbol{r}_C}{\mathrm{d}t} - \frac{\mathrm{d}\boldsymbol{r}_B}{\mathrm{d}t}\right) = \boldsymbol{v}_{AC} + \boldsymbol{v}_{CB} \\ &= \boldsymbol{v}_{AC} - \boldsymbol{v}_{BC}. \end{aligned} \tag{2.36}$$

Result (2.36) is the Galilean rule for velocity addition, which demonstrates that two inertial observers can refer their velocities to a third *reference frame* as long as that third frame is also an inertial frame. The Earth is commonly used as such a reference frame in elementary physics, but it is not a true inertial frame, since the Earth is both orbiting the Sun and rotating about an axis relative to the *fixed stars*. Both of these rotational motions involve a centripetal acceleration, and therefore a reference frame embedded in the Earth is not a true Galilean inertial frame. In Newtonian mechanics the acceleration associated with a reference frame that is rotating relative to a true inertial frame creates apparent forces, or *pseudoforces*, such as the *Coriolis force*, similarly to the appearance of the pseudoforce $-m\boldsymbol{a}_o$ in (2.31). This is resolved in Newtonian mechanics by the third law of motion, which defines a *true force* as a member of an action–reaction pair between two material objects. These are distinguished from pseudoforces that do not possess an *agent*, which is another material body, that creates them. Such pseudoforces are discussed in classical mechanics texts.

2.4 Functionals and the calculus of variations

The concept of the *functional* plays a central role in the modern formulation of Newtonian mechanics and classical field theory. A functional is a means to map a collection of functions and their derivatives into a real or complex number. The calculus of variations was developed to understand the behavior of functionals and to determine those functions for which the associated functional attains an extremum value.

Given a function $f(x)$ of the independent variable x, the functional $F[f, f']$ denotes a method to obtain a real or complex number from this function and typically its first derivative, denoted by $f'(x) = \mathrm{d}f(x)/\mathrm{d}x$. Although the function depends on x, the functional does not. The simplest functional consists of an integral involving $f(x)$ and $f'(x)$, and is written

$$F[f, f'] = \int_{x_i}^{x_f} \mathrm{d}x\, \mathcal{L}(x, f(x), f'(x)), \tag{2.37}$$

where $\mathcal{L}$ represents a collection of powers of $f(x)$ and $f'(x)$ with a possible separate dependence on x, while x_i and x_f are the limits of integration. A variation of the function f is given by $\delta f(x)$, and satisfies $\delta f(x_i) = \delta f(x_f) = 0$. It is an infinitesimal function, so that $(\delta f(x))^2 = 0$ if $x \in (x_i, x_f)$. If $f_s(x)$ is the function that extremizes (2.37), then the *variation of the functional* F around the extremum value created by the *variation of the function* $f = f_s + \delta f$ is given by

$$\begin{aligned}\delta F[f_s, f_s'] &= F[f_s + \delta f, f_s' + \delta f'] - F[f_s, f_s'] \\ &= \int_{x_i}^{x_f} \mathrm{d}x \left(\mathcal{L}(x, f_s + \delta f, f_s' + \delta f') - \mathcal{L}(x, f_s, f_s')\right),\end{aligned} \tag{2.38}$$

where $\delta f'(x) = \mathrm{d}(\delta f(x))/\mathrm{d}x$.

Definition (2.38) is the functional analogy of the infinitesimal differential of a function of two variables, $\mathrm{d}f(x, y)$, as long as f and f' are treated as separate degrees of freedom. Result (1.132) showed that the extrema of a function occurred at $\mathrm{d}f = 0$, and so the extrema of the functional $F[f, f']$ will occur for those functions f_s such that $\delta F[f_s, f_s'] = 0$. Expanding the integrand in (2.38) in a Taylor series in δf and using the infinitesimal nature of δf gives

$$\begin{aligned}\delta F[f_s, f_s'] &= \int_{x_i}^{x_f} \mathrm{d}x \left(\frac{\partial \mathcal{L}(x, f, f')}{\partial f(x)} \delta f(x) + \frac{\partial \mathcal{L}(x, f, f')}{\partial f'(x)} \delta f'(x) \right) \Bigg|_{f=f_s} \\ &= \int_{x_i}^{x_f} \mathrm{d}x \left(\frac{\partial \mathcal{L}(x, f, f')}{\partial f(x)} \delta f(x) + \frac{\partial \mathcal{L}(x, f, f')}{\partial f'(x)} \frac{\mathrm{d}\delta f(x)}{\mathrm{d}x} \right) \Bigg|_{f=f_s} \\ &= \frac{\partial \mathcal{L}(x, f_s, f_s')}{\partial f_s'(x)} \delta f(x) \Bigg|_{x_i}^{x_f} + \int_{x_i}^{x_f} \mathrm{d}x \\ &\quad \left(\frac{\partial \mathcal{L}(x, f, f')}{\partial f(x)} - \frac{\mathrm{d}}{\mathrm{d}x} \frac{\partial \mathcal{L}(x, f, f')}{\partial f'(x)} \right) \Bigg|_{f=f_s} \delta f(x) \\ &= \int_{x_i}^{x_f} \mathrm{d}x \left(\frac{\partial \mathcal{L}(x, f, f')}{\partial f(x)} - \frac{\mathrm{d}}{\mathrm{d}x} \frac{\partial \mathcal{L}(x, f, f')}{\partial f'(x)} \right) \Bigg|_{f=f_s} \delta f(x) = 0,\end{aligned} \tag{2.39}$$

where the boundary conditions on δf, $\delta f(x_f) = \delta f(x_i) = 0$, allow the surface terms from the integration by parts to be dropped. Because $\delta f(x)$ is an arbitrary variation, an extremal function f_s must therefore satisfy the differential equation generated by

$$\left(\frac{\partial \mathcal{L}(x, f, f')}{\partial f(x)} - \frac{\mathrm{d}}{\mathrm{d}x} \frac{\partial \mathcal{L}(x, f, f')}{\partial f'(x)} \right) \Bigg|_{f=f_s} = 0, \tag{2.40}$$

which is referred to as the *Euler–Lagrange equation.*

A simple application is to identify *geodesics*, paths of minimum length between two points, in the Euclidean plane. Using the Euclidean metric (1.68) in Cartesian coordinates, the length L of an arbitrary path P between two arbitrary points, (x_1, y_1) and (x_2, y_2), is given by

$$L = \int_P \mathrm{d}s = \int_P \sqrt{\mathrm{d}x^2 + \mathrm{d}y^2} = \int_{x_1}^{x_2} \mathrm{d}x \sqrt{1 + \left(\frac{\mathrm{d}y(x)}{\mathrm{d}x} \right)^2}, \tag{2.41}$$

where the boundary conditions on the function $y(x)$ are $y(x_1) = y_1$ and $y(x_2) = y_2$ with $x_2 - x_1 \neq 0$. This identifies the argument of the functional as

$$\mathcal{L}(x, y(x), y'(x)) = \sqrt{1 + (y'(x))^2}. \tag{2.42}$$

The Euler–Lagrange equation for the extremum path gives

$$
\begin{aligned}
-\frac{\mathrm{d}}{\mathrm{d}x}\frac{\partial\mathcal{L}(x, y(x), y'(x))}{\partial y'(x)} &= -\frac{\mathrm{d}}{\mathrm{d}x}\left(\frac{y'(x)}{\sqrt{1 + y'^2(x)}}\right) = 0 \\
&\Longrightarrow \frac{y'(x)}{\sqrt{1 + y'^2(x)}} = c,
\end{aligned}
\tag{2.43}
$$

where c is a constant that is independent of x. Equation (2.43) is easily solved,

$$
y'(x) = \frac{c}{\sqrt{1 - c^2}} \equiv m \Longrightarrow y(x) = mx + b, \tag{2.44}
$$

where b is another constant. Result (2.44) is the well-known equation for a straight line in Cartesian coordinates, with the two arbitrary constants m and b chosen to match the two boundary conditions. A similar analysis is used in chapter 10 to find the equation for geodesics in a non-Euclidean manifold.

2.4.1 The Dirac delta

The Dirac delta is commonly used in field theory to represent a point mass or charge. Defining it begins with the Heaviside step function $\theta(x)$, which is defined as

$$
\theta(x) = \begin{cases} 1 & \text{if } x > 0 \\ 0 & \text{if } x < 0 \end{cases}. \tag{2.45}
$$

If $a < x_o < b$, the step function has an integral given by

$$
\int_a^b \mathrm{d}x\, \theta(x - x_o) = \int_{x_o}^b \mathrm{d}x = b - x_o. \tag{2.46}
$$

Defining $\theta(0) = \frac{1}{2}\lim\limits_{\epsilon\to 0}(\theta(\epsilon) + \theta(-\epsilon))$ results in $\theta(0) = 1/2$. Using the properties of the integral, it follows that such a function has a derivative that formally obeys

$$
\lim_{\epsilon\to 0^+}\int_{a-\epsilon}^{a+\epsilon} \mathrm{d}x\, \frac{\mathrm{d}}{\mathrm{d}x}\theta(x - a) = \lim_{\epsilon\to 0^+}(\theta(\epsilon) - \theta(-\epsilon)) = \lim_{\epsilon\to 0^+}(1 - 0) = 1. \tag{2.47}
$$

A continuous function would not exhibit such behavior. The derivative of the step function $\theta(x - a)$ is therefore a singular mathematical object, known as the *Dirac delta* and denoted by $\delta(x - a)$, so that

$$
\delta(x - a) = \frac{\mathrm{d}}{\mathrm{d}x}\theta(x - a). \tag{2.48}
$$

For $a < b$ the definition (2.48) combined with result (2.47) gives

$$
\int_a^b \mathrm{d}x\, \delta(x - x_o) = \begin{cases} 1 & \text{if } a < x_o \\ 0 & \text{if } x_o < a \text{ or } x_o > b \end{cases}. \tag{2.49}
$$

The reader should note that different definitions of the Dirac delta are possible, and the area of mathematics known as *distribution theory* deals with a formal analysis of such singular functions. For $a < x_o < b$, the inclusion of a continuous *test function* $f(x)$ in the integrand is evaluated using an integration by parts to obtain

$$\begin{aligned}\int_a^b \mathrm{d}x\, f(x)\, \delta(x - x_o) &= \int_a^b \mathrm{d}x\, f(x)\frac{\mathrm{d}}{\mathrm{d}x}\theta(x - x_o)\\ &= \int_a^b \mathrm{d}x \left\{\frac{\mathrm{d}}{\mathrm{d}x}[f(x)\theta(x - x_o)] - \theta(x - x_o)\frac{\mathrm{d}f(x)}{\mathrm{d}x}\right\}\\ &= f(b)\,\theta(b - x_o) - f(a)\theta(a - x_o)\\ &\qquad - \int_{x_o}^b \mathrm{d}x\, \theta(x - x_o)\frac{\mathrm{d}f(x)}{\mathrm{d}x}\\ &= f(b) - \int_{x_o}^b \mathrm{d}x\, \frac{\mathrm{d}f(x)}{\mathrm{d}x} = f(b) - f(b) + f(x_o)\\ &= f(x_o).\end{aligned} \tag{2.50}$$

If $f(x) = x - x_o$ then $(x - x_o)\delta(x - x_o) = 0$ when integrated, and so $(x - x_o)\,\delta(x - x_o)$ can be treated as zero as long as there are no additional factors present in the integrand that are singular at x_o.

The Dirac delta has a number of important additional properties. Assuming $b > a$, the change of variables $x \to -y$ shows that

$$\begin{aligned}\int_{-b}^b \mathrm{d}x\, f(x)\, \delta(-x + a) \quad &\longrightarrow \quad -\int_b^{-b} \mathrm{d}y\, f(-y)\, \delta(y + a)\\ &= \int_{-b}^b \mathrm{d}y\, f(-y)\, \delta(y + a) = f(a)\\ &\Longrightarrow \quad \delta(\pm(x - a)) = \delta(x - a).\end{aligned} \tag{2.51}$$

The Dirac delta can be integrated by parts, meaning that for $-b < y < b$, the derivative of a function obeys

$$\begin{aligned}\frac{\mathrm{d}f(y)}{\mathrm{d}y} = \frac{\mathrm{d}}{\mathrm{d}y}\int_{-b}^b \mathrm{d}x\, f(x)\delta(x - y) &= \int_{-b}^b \mathrm{d}x\, f(x)\frac{\mathrm{d}}{\mathrm{d}y}\delta(x - y)\\ &= -\int_{-b}^b \mathrm{d}x\, f(x)\frac{\mathrm{d}}{\mathrm{d}x}\delta(x - y)\\ = -\int_{-b}^b \mathrm{d}x\, \frac{\mathrm{d}}{\mathrm{d}x}(f(x)\delta(x - y)) &+ \int_{-b}^b \mathrm{d}x\, \frac{\mathrm{d}f(x)}{\mathrm{d}x}\,\delta(x - y) = \frac{\mathrm{d}f(y)}{\mathrm{d}y},\end{aligned} \tag{2.52}$$

where the surface terms vanish, since $-b < y < b$. This shows the validity of the identification

$$\frac{\mathrm{d}}{\mathrm{d}y}\delta(x - y) = -\frac{\mathrm{d}}{\mathrm{d}x}\delta(x - y). \tag{2.53}$$

Another important property is $\delta(ax) = \delta(x)/|a|$, where $|a|$ is the absolute value of the real constant a. The demonstration of this begins by noting that if $a < 0$ then

$a = -|a|$, while if $a > 0$ then $a = |a|$. Combining (2.51) with the two possibilities, $a = \pm|a|$, and the change of variable $x = z/|a|$ demonstrates the property

$$\begin{aligned}&\int_{-b}^{b} \mathrm{d}x\, f(x)\delta(\pm|a|\, x)\\ &= \int_{-b}^{b} \mathrm{d}x\, f(x)\delta(|a|\, x) = \frac{1}{|a|}\int_{-|a|\, b}^{|a|\, b} \mathrm{d}z\, f(\pm z/|a|)\delta(z) = \frac{1}{|a|}f(0).\end{aligned} \tag{2.54}$$

It is common to treat the Dirac delta as a function that is essentially zero everywhere except where its argument vanishes, and there it is effectively infinite, since it must have a value of $\approx(\mathrm{d}x)^{-1}$. There are pitfalls in doing this, since this is not the definition of a standard function. However, visualizing the function in this manner means that $\delta(f(x))$ is zero everywhere except where the function $f(x)$ vanishes. Therefore, if $f(x)$ has a simple zero at x_o, a Taylor series expansion around x_o in the vicinity of x_o gives

$$\begin{aligned}\delta(f(x)) &= \delta(f(x_o + x - x_o)) = \delta\left(f(x_o) + (x - x_o)\left.\frac{\mathrm{d}f(x)}{\mathrm{d}x}\right|_{x_o}\right)\\ &= \delta\left((x - x_o)\frac{\mathrm{d}f(x_o)}{\mathrm{d}x_o}\right).\end{aligned} \tag{2.55}$$

The previous result (2.54) gives

$$\delta(f(x)) = \sum_{x_o} \frac{\delta(x - x_o)}{|\mathrm{d}f(x_o)/\mathrm{d}x_o|}, \tag{2.56}$$

where the sum is over all the simple zeroes x_o of the function $f(x)$.

The Dirac delta can be generalized to higher dimensions. In Cartesian coordinates the three-dimensional version is given by

$$\delta^3(\boldsymbol{r} - \boldsymbol{r}_o) = \delta(x - x_o)\,\delta(y - y_o)\,\delta(z - z_o), \tag{2.57}$$

so that the Cartesian volume integral is given by

$$\begin{aligned}&\int_V \mathrm{d}^3x\, f(\boldsymbol{r})\,\delta^3(\boldsymbol{r} - \boldsymbol{r}_o)\\ &= \int_{V_x} \mathrm{d}x \int_{V_y} \mathrm{d}y \int_{V_z} \mathrm{d}z\, f(\boldsymbol{r})\,\delta(x - x_o)\,\delta(y - y_o)\,\delta(z - z_o) = f(\boldsymbol{r}_o),\end{aligned} \tag{2.58}$$

if the volume $V = V_x V_y V_z$ contains the point defined by the position vector $\boldsymbol{r}_o$ and is zero otherwise.

The multidimensional Dirac delta can be expressed in other coordinate systems by recalling that $\mathrm{d}^n x$ acquires a factor of the Jacobian when coordinate systems are changed. This means that the Dirac delta in the new coordinate system must have an inverse factor of the Jacobian. For example, changing from Cartesian coordinates to polar coordinates in two dimensions, $x = r\cos\theta$ and $y = r\sin\theta$, generates the

Jacobian (1.13), given by $J = r$. As a result, the two-dimensional Dirac delta in polar coordinates is given by

$$\begin{aligned} &\delta^2(\mathbf{r} - \mathbf{r}_o) = \delta(x - x_o)\,\delta(y - y_o) = \frac{1}{r}\delta(r - r_o)\,\delta(\theta - \arctan(y_o/x_o)) \\ &\Longrightarrow \int_S r\,\mathrm{d}r\,\mathrm{d}\theta\, f(\mathbf{r})\,\delta^2(\mathbf{r} - \mathbf{r}_o) = \int_S \mathrm{d}r\,\mathrm{d}\theta\, f(\mathbf{r})\,\delta(r - r_o)\,\delta(\theta - \arctan(y_o/x_o)) \\ &= f(\mathbf{r}_o), \end{aligned} \tag{2.59}$$

where $r_o = \sqrt{x_o^2 + y_o^2}$.

A general change from Cartesian coordinates $x^k = x^k(y)$ to the y-coordinate system is required to be invertible, so that $y^j = y^j(x)$ are the coordinates in the new system. The Dirac delta in the n-dimensional Cartesian x-coordinate system is written in terms of the Cartesian coordinates of the point $\boldsymbol{x}_o$, so that

$$\delta^n(\boldsymbol{x} - \boldsymbol{x}_o) = \prod_{k=1}^{n} \delta(x^k - x_o^k). \tag{2.60}$$

In the y-coordinate system, where the coordinates of the point $\boldsymbol{x}_o$ are given by $y_o^k(x_o)$, the Dirac delta is given by

$$\delta^n(\boldsymbol{x} - \boldsymbol{x}_o) = \frac{1}{J(y)} \prod_{k=1}^{n} \delta(y^k - y_o^k) = \frac{1}{J(y)} \delta^n(\boldsymbol{y} - \boldsymbol{y}_o), \tag{2.61}$$

where $J(y)$ is the Jacobian (1.12) that appears in the transformed volume element (1.17), $\mathrm{d}^n x = J(y)\,\mathrm{d}^n y$.

2.4.2 The functional derivative and differential

In order to generalize the functional differential $\delta F[f]$ of (2.38) to an arbitrary functional, the functional derivative $\delta F[f]/\delta f(x)$ at point x is defined so that

$$\delta F[f] = \lim_{\delta f \to 0} (F[f + \delta f] - F[f]) = \lim_{\delta f \to 0} \int \mathrm{d}x\, \frac{\delta F[f]}{\delta f(x)}\, \delta f(x). \tag{2.62}$$

Definition (2.62) treats the function $f(x)$ at each point x as an independent variable, so that $\partial/\partial x$ and $\mathrm{d}x$ are extended to functional variables $\delta/\delta f(x)$ and $\delta f(x)$. It is important to note that the integral on the right-hand side of (2.62) is taken over the entire manifold where $f(x)$ is defined, so that definition (2.62) can be viewed as the case where the sum over the independent variables $\mathrm{d}x^j$ in the multivariable differential d defined by (1.25) becomes infinite. By setting $\delta f(x) = \epsilon\, \delta(x - x')$, where $\epsilon \to 0$ is understood, definition (2.62) gives

$$\begin{aligned}\delta F[f] &= \lim_{\epsilon\to 0}(F[f(x) + \epsilon\, \delta(x - x')] - F[f(x)]) \\ &= \lim_{\epsilon\to 0}\int \mathrm{d}x\, \epsilon\, \frac{\delta F[f]}{\delta f(x)}\, \delta(x - x') = \lim_{\epsilon\to 0}\epsilon\, \frac{\delta F[f]}{\delta f(x')}.\end{aligned} \tag{2.63}$$

Result (2.63) therefore gives

$$\frac{\delta F[f]}{\delta f(x')} = \lim_{\epsilon\to 0}\frac{(F[f(x) + \epsilon\, \delta(x - x')] - F[f(x)])}{\epsilon}, \tag{2.64}$$

a result similar to the definition of the standard derivative and referred to as the *Volterra variational derivative*. An alternative and equivalent definition of the functional derivative is given by

$$\frac{\mathrm{d}}{\mathrm{d}s}F[f + s\delta f]_{s=0} = \int \mathrm{d}x\, \frac{\delta F[f]}{\delta f(x)}\, \delta f(x). \tag{2.65}$$

As examples, (2.64) gives two results similar to those of basic calculus,

$$\frac{\delta}{\delta f(x)}\int_{-\infty}^{\infty} \mathrm{d}y\, f^{n}(y) = nf^{n-1}(x), \tag{2.66}$$

$$\frac{\delta}{\delta f(x)}\exp\left(\int_{-\infty}^{\infty} \mathrm{d}y\, K(y)f(y)\right) = K(x)\exp\left(\int_{-\infty}^{\infty} \mathrm{d}y\, K(y)f(y)\right). \tag{2.67}$$

These results and definitions can be generalized to higher-order functional derivatives and to higher-dimensional manifolds. However, some subtleties can occur. For example, if the function f appearing in the functional is a function of two variables, so that $f = f(x, y)$, the functional may be a mapping into a function of x, so that the functional $F[f]$ is written $F[f](x)$. For such a case, the functional derivative of $F[f](x)$ will yield a delta function factor. This follows from extending definition (2.64),

$$\frac{\delta F[f](x)}{\delta f(z, w)} = \lim_{\epsilon\to 0}\frac{1}{\epsilon}(F[f(x, y) + \epsilon\delta(x - z)\delta(y - w)](x) - F[f(x, y)](x)). \tag{2.68}$$

For example, if the functional is given by

$$F[f(x, y)](x) = \int \mathrm{d}y\, \frac{1}{2}f^{2}(x, y), \tag{2.69}$$

then (2.68) yields

$$\frac{\delta F[f](x)}{\delta f(z, w)} = \delta(x - z)f(x, w). \tag{2.70}$$

It should be noted that this delta function factor does not occur for the functional

$$F[f](x) = \int \mathrm{d}y\, f(y)\sin(k(y - x)), \tag{2.71}$$

which uses (2.62) to find

$$\frac{\delta F[f](x)}{\delta f(z)} = \sin(k(z - x)). \tag{2.72}$$

Because definition (2.64) obeys the Leibniz or product rule, it yields results similar in form to the derivatives of functions, but with the added feature that the argument of the function appears in the final result. In the case of a functional of a functional, $F[G[f](z)]$, it is assumed that the functional $G[f](z)$ may also depend on the independent variable z. Such a possibility can occur as the limit of an integral or as the argument of an independent function in the functional. For such a case, definition (2.64) gives rise to the *functional chain rule*,

$$\frac{\delta F[G[f]]}{\delta f(x)} = \int_{-\infty}^{\infty} \mathrm{d}y\, \frac{\delta F[G[f]]}{\delta G[f](y)} \frac{\delta G[f](y)}{\delta f(x)}. \tag{2.73}$$

Exercise 2.3. Consider the functional

$$F[f] = \int_{-\infty}^{\infty} \mathrm{d}z \left[\frac{1}{2}\int_{-\infty}^{\infty} \mathrm{d}y\, f^2(y) \exp\left\{-\frac{1}{2}\alpha(y - z)^2\right\}\right]^2, \tag{2.74}$$

where α is real and positive. Use (2.73) to show that

$$\frac{\delta F[f]}{\delta f(x)} = \sqrt{\frac{\pi}{\alpha}}\, f(x) \int_{-\infty}^{\infty} \mathrm{d}z\, f^2(z) \exp\left\{-\frac{1}{4}\alpha(x - z)^2\right\}. \tag{2.75}$$

The chain rule (2.73) also holds for the case that G is simply a function, $G = g(z)$, but defining the quantity $\delta g(z)/\delta f(x)$ will require the introduction of Fourier transforms. It should be pointed out that such a quantity must satisfy the identity

$$\frac{\delta f(x)}{\delta f(y)} = \lim_{\epsilon \to 0} \frac{f(x) + \epsilon\delta(x - y) - f(x)}{\epsilon} = \delta(x - y), \tag{2.76}$$

which follows from (2.64). It is also often useful to represent a functional in terms of a functional Taylor series, which takes a form quite similar to the Taylor series for functions. For the case that $F[f]$ is well-behaved in the neighborhood of $f(x) = 0$, it can be expanded around $f(x) = 0$, and the functional Taylor series takes the form

$$F[f] = \sum_{n=0}^{\infty} \frac{1}{n!} \int_{-\infty}^{\infty} \mathrm{d}x_1 \cdots \int_{-\infty}^{\infty} \mathrm{d}x_n\, f(x_1) \cdots f(x_n) \frac{\delta^n F[f]}{\delta f(x_1) \cdots \delta f(x_n)}\Bigg|_{f=0}. \tag{2.77}$$

The validity of (2.77) is readily demonstrated at each order by using (2.64). It is also possible to use (2.77) to expand a functional around a non-zero function.

2.4.3 Complex vectors, Fourier transforms, and the Dirac delta

Complex analysis plays an extremely important role in field theory and physics in general. A complex number $z = x + iy$ consists of two real numbers, x and y, and the imaginary number i, which satisfies $i^2 = -1$. The two values x and y are known respectively as the *real part* of z and the *imaginary part* of z. Two complex numbers are equal only if their real and imaginary parts are identical. The *complex conjugate* of z is denoted by z^* and is given by

$$z^* = (x + iy)^* = x - iy, \tag{2.78}$$

so that $i^* = -i$. It follows that

$$zz^* = (x + iy)(x - iy) = x^2 + y^2, \tag{2.79}$$

so that $zz^* = |z|^2$, where $|z| = +\sqrt{x^2 + y^2}$ is known as the *modulus* of z. A complex number can be expressed in polar form using the Euler representation

$$z = re^{i\theta}, \tag{2.80}$$

where both r and θ are both real variables. The Euler representation gives

$$z^* = re^{-i\theta} \implies zz^* = r^2. \tag{2.81}$$

Comparing (2.79) to (2.81) immediately gives

$$\left|z\right| = r = \sqrt{x^2 + y^2}. \tag{2.82}$$

Using the Taylor series expansion for the exponential $e^{i\theta}$ gives the generalized Euler identity,

$$e^{i\theta} = \sum_{n=0}^{\infty} \frac{1}{n!}(i\theta)^n = \sum_{j=0}^{\infty}(-1)^j \frac{\theta^{2j}}{(2j)!} + i\sum_{j=1}^{\infty}(-1)^{j-1}\frac{\theta^{2j-1}}{(2j-1)!} = \cos\theta + i\sin\theta, \tag{2.83}$$

from which it follows that $x = r\cos\theta$ and $y = r\sin\theta$, so that $\tan\theta = y/x$. The angle θ is referred to as the *phase angle* of the complex number. A complex number can be visualized as a point in the *complex plane*, given by using x and y as the Cartesian coordinates of the point. The Euler representation is then equivalent to the polar coordinates of z. A complex function $f(z) = f(x + iy)$ will return a complex number, which is another point in the complex plane. As a result, a complex function can be written $f(z) = u(x, y) + iv(x, y)$.

The concept of the vector can be extended to the case that its components are complex valued. While position vectors $\boldsymbol{r}$ must have real components, the concept of the complex vector $\boldsymbol{z}$ is useful in dealing with complex-valued functions. The first step is to generalize the real vector Cartesian scalar product for Euclidean manifolds (1.59) to the case of vectors with n complex components. For the case that the components of the two complex vectors $\boldsymbol{z}$ and $\boldsymbol{w}$ are z^k and w^k, respectively, the scalar product (1.59) of the two vectors is generalized to the *inner product*, denoted by $(\boldsymbol{w}, \boldsymbol{z})$, and given by

$$(w, z) = \sum_{k=1}^{n} w^{k*} z^k \implies (w, z)^* = \sum_{k=1}^{n} z^{k*} w^k = (z, w). \tag{2.84}$$

For the case that all components are real, (2.84) reproduces (1.59). However, in general, (2.84) will yield a complex number. The exception is the inner product of a vector with itself, which is given by

$$(z, z) = \sum_{k=1}^{n} z^{k*} z^k = \sum_{k=1}^{n} |z^k|^2 \equiv |z|^2 \geqslant 0. \tag{2.85}$$

Because (2.85) is positive-definite, $|z| = \sqrt{(z, z)}$ is a real positive number referred to as the *norm* or *magnitude* of the complex vector z. The definitions (2.84) and (2.85) satisfy the *Cauchy–Schwarz inequality*,

$$|w|^2 |z|^2 \geqslant |(w, z)|^2. \tag{2.86}$$

This relation is proved by first defining the vector $v = w - \lambda z$, where λ is an *arbitrary* complex number. Multiplying z by λ is understood to mean that the components of λz are λz^k. Combining the fact that the magnitude of this vector is positive-definite, $(v, v) = |v|^2 = |w - \lambda z|^2 \geqslant 0$, with the definition of the inner product (2.84) gives

$$(v, v) = |w|^2 + \lambda^* \lambda \, |z|^2 - \lambda \, (z, w) - \lambda^* \, (w, z) \geqslant 0. \tag{2.87}$$

Since the inequality (2.87) holds for any value of λ, choosing $\lambda = (w, z)/|w|^2$ and $\lambda^* = (w, z)^*/|w|^2 = (z, w)/|w|^2$ causes (2.87) to yield $|z|^2 |w|^2 \geqslant (z, w)(w, z) = |(w, z)|^2$, thereby proving (2.86). Restricting (2.86) to real two-dimensional position vectors separated by the angle θ and using the equivalence between (1.59) and (1.61) shows that (2.86) reduces to the obvious inequality $\cos^2 \theta \leqslant 1$.

By analogy the inner product of two complex-valued functions of the real variable x, denoted by $f(x)$ and $g(x)$, can be defined to be identical in nature to the complex vector inner product (2.84). If the interval where the functions of x are defined or are of interest is $(-L/2, L/2)$, the infinitesimal subinterval $\Delta x = L/N$ is defined, where N is an arbitrarily large integer such that the limit $N \to \infty$ will be taken. For an integer j, such that $0 \leqslant j \leqslant N$, the value of the function at $x = -L/2 + j \, \Delta x$ is defined as $f_j = f(-L/2 + j\Delta x)$ and treated as the jth component of the complex vector f. In complete analogy to complex vectors, the inner product of the two functions, denoted by (f, g), will be proportional to $\sum_{j=0}^{N} f_j^* g_j$. The definition is completed by forming a Riemann sum and taking the limit $N \to \infty$, so that $\Delta x \to \mathrm{d}x$. This defines the inner product of two complex functions as a definite integral,

$$(f, g) = \lim_{N \to \infty} \sum_{j=0}^{N} \frac{\Delta x}{L} f_j^* \, g_j = \frac{1}{L} \int_{-L/2}^{L/2} \mathrm{d}x \, f^*(x) \, g(x). \tag{2.88}$$

Attention will be restricted to functions that are *square integrable* and *piecewise continuous* over the interval $(-L/2, L/2)$. The requirement of square integrability is equivalent to restricting the norm of the function $|f|$ to be finite. For such a case, the inner product (2.88) is well-defined and the interval of interest can be broken up into subintervals where the function under consideration is continuous. This allows the introduction of the following periodic *basis functions*,

$$f_n(x) = \exp\left(\frac{i2\pi nx}{L}\right), \tag{2.89}$$

where n is an arbitrary integer. Using (2.88) shows that this set of functions is *orthonormal*, since their inner product is given by

$$(f_n, f_m) = \frac{1}{L}\int_{-L/2}^{L/2} \mathrm{d}x \exp\left(\frac{i2\pi(m-n)x}{L}\right) = \begin{cases} 1 & \text{if } m = n \\ 0 & \text{if } m \neq n \end{cases}. \tag{2.90}$$

Result (2.90) follows, since $e^{\pm i\pi(m-n)} = (-1)^{(m-n)}$. This is the functional counterpart of the Cartesian unit vector scalar product (1.58).

The orthonormal functions $f_n(x)$ of (2.89) are an integer power of the modulus one complex number $z(x) = \exp(i2\pi x/L)$. The orthonormal functions are also periodic, since $z(x + L) = z(x)$. The Weierstrass approximation theorem can be extended to show that any piecewise continuous complex function that is periodic over the interval $(-L/2, L/2)$ can be represented as a polynomial in $z(x)$ and $z^*(x)$. In this sense, the functions appearing in (2.89) form a *complete set* of orthonormal functions, and can serve as a *basis* to express other functions. The orthonormal functions f_n therefore play the role that unit vectors did for position vectors. An arbitrary square integrable piecewise continuous complex function $f(x)$ can be expressed as the polynomial

$$f(x) = a_0 + \sum_{k=1}^{\infty} a_k z^k + \sum_{k=1}^{\infty} b_k z^{*k} = \sum_{k=-\infty}^{\infty} a_k e^{i2\pi kx/L}, \tag{2.91}$$

where $a_{-k} = b_k$. The sum (2.91) is referred to as a *Fourier series*. The coefficients a_k are a set of constants that depend on the function $f(x)$, and can be found by using (2.89), which gives

$$\begin{aligned}(f_j, f) &= \frac{1}{L}\int_{-L/2}^{L/2} \mathrm{d}y\, f(y) e^{-i2\pi jy/L} = \sum_{k=-\infty}^{\infty} a_k \frac{1}{L}\int_{-L/2}^{L/2} \mathrm{d}y\; e^{i2\pi(k-j)y/L} \\ &= \sum_{k=-\infty}^{\infty} a_k \delta_{jk} = a_j.\end{aligned} \tag{2.92}$$

Result (2.92) is the functional counterpart to (1.74), where $r^j = \mathbf{e}^j \cdot \boldsymbol{r}$ gave the jth component of the vector $\boldsymbol{r}$. This allows a_j to be viewed as the jth *Fourier component* of the function $f(x)$. Because the basis functions are periodic, the Fourier series (2.91) gives a function $f(x)$ that is periodic over the x axis, so that if n is any integer then $f(x) = f(x + nL)$. Combining (2.91) and result (2.92) and interchanging the order of summation and integration yields

$$f(x) = \sum_{k=-\infty}^{\infty} a_k e^{i2\pi kx/L} = \int_{-L/2}^{L/2} \mathrm{d}y\, f(y)\left(\frac{1}{L}\sum_{k=-\infty}^{\infty} e^{i2\pi k(x-y)/L}\right). \tag{2.93}$$

Comparing (2.93) to (2.50) shows that

$$\frac{1}{L}\sum_{k=-\infty}^{\infty}\exp\left(i\frac{2\pi k}{L}(x-y)\right)=\delta(x-y), \tag{2.94}$$

which gives an alternate representation of the Dirac delta for the interval $(-L/2, L/2)$.

If the function $f(x)$ is not periodic, it can still be represented using a Fourier series by allowing the range of periodicity L to become infinite. For large L, the variable $p = 2\pi k/L$ is defined, so that $\Delta p = 2\pi\ \Delta k/L \Longrightarrow \Delta k = \Delta p\ L/2\pi$. Since $\Delta k = 1$, this factor can be inserted into the sum (2.94) and the limit $L \to \infty$ taken, resulting in the integral representation of the non-periodic Dirac delta,

$$\delta(x-y)=\lim_{L\to\infty}\frac{1}{2\pi}\sum_{p=-\infty}^{\infty}\Delta p\ e^{ip(x-y)}=\int_{-\infty}^{\infty}\frac{\mathrm{d}p}{2\pi}e^{ip(x-y)}, \tag{2.95}$$

where p has units of $1/L$. Result (2.95) gives an integral representation of a non-periodic function,

$$\begin{aligned} f(x) &= \int_{-\infty}^{\infty}\mathrm{d}y\, f(y)\,\delta(x-y)=\int_{-\infty}^{\infty}\frac{\mathrm{d}p}{2\pi}\int_{-\infty}^{\infty}\mathrm{d}y\, f(y)\, e^{ip(x-y)} \\ &= \int_{-\infty}^{\infty}\frac{\mathrm{d}p}{2\pi}\tilde{f}(p)\, e^{ipx}, \end{aligned} \tag{2.96}$$

where $\tilde{f}(p)$, referred to as the *Fourier transform* of $f(x)$, is given by

$$\tilde{f}(p)=\int_{-\infty}^{\infty}\mathrm{d}y\, f(y)\, e^{-ipy}. \tag{2.97}$$

Using the Fourier representation (2.96) in (2.97) shows that

$$\tilde{f}(p)=\int_{-\infty}^{\infty}\mathrm{d}y\int_{-\infty}^{\infty}\frac{\mathrm{d}k}{2\pi}\tilde{f}(k)\, e^{i(k-p)y}=\int_{-\infty}^{\infty}\mathrm{d}k\,\tilde{f}(k)\int_{-\infty}^{\infty}\frac{\mathrm{d}y}{2\pi}\, e^{i(k-p)y}, \tag{2.98}$$

which immediately gives the inverse Dirac delta expression,

$$\delta(k-p)=\int_{-\infty}^{\infty}\frac{\mathrm{d}y}{2\pi}\, e^{i(k-p)y}. \tag{2.99}$$

Results (2.95) and (2.97) are easily extended to multidimensional manifolds. For example, in three spatial dimensions, the Dirac delta is given by

$$\begin{aligned} \delta^3(\mathbf{r}-\mathbf{r}') &= \int_{-\infty}^{\infty}\frac{\mathrm{d}p_x}{2\pi}\, e^{ip_x(x-x')}\int_{-\infty}^{\infty}\frac{\mathrm{d}p_y}{2\pi}\, e^{ip_y(y-y')}\int_{-\infty}^{\infty}\frac{\mathrm{d}p_z}{2\pi}\, e^{ip_z(z-z')} \\ &= \int_{-\infty}^{\infty}\frac{\mathrm{d}^3p}{(2\pi)^3}\, e^{i\mathbf{p}\cdot(\mathbf{r}-\mathbf{r}')}, \end{aligned} \tag{2.100}$$

so that $p \to \mathbf{p}$, a three-dimensional vector with units of inverse length known as the *wave vector*. Similarly, in three dimensions, expression (2.99) becomes

$$\delta^3(\boldsymbol{k} - \boldsymbol{p}) = \int_{-\infty}^{\infty} \frac{\mathrm{d}^3 r}{(2\pi)^3} e^{i(\boldsymbol{k}-\boldsymbol{p})\cdot \boldsymbol{r}}. \tag{2.101}$$

It is now possible to define the functional derivative of a function $f(x)$ with respect to another function $g(x)$ in terms of their Fourier components $\tilde{f}(k)$ and $\tilde{g}(k)$,

$$\frac{\delta f(x)}{\delta g(y)} = \int \frac{\mathrm{d}p}{2\pi} \frac{\partial \tilde{f}(p)}{\partial \tilde{g}(p)} e^{ip(x-y)} = \int \frac{\mathrm{d}p}{2\pi} \frac{\partial \tilde{f}(p)}{\partial p} \frac{\partial p}{\delta \tilde{g}(p)} e^{ip(x-y)}. \tag{2.102}$$

For the case that $\tilde{f}(p) = \tilde{g}(p)$, so that $f(x) = g(x)$ and $\partial \tilde{f}(p)/\partial \tilde{g}(p) = 1$, the definition (2.102) correctly reduces to the result given by (2.76). Definition (2.102) assumes that the expression for $\tilde{g}(p)$ is uniquely invertible, so that p can be expressed in terms of $\tilde{g}(p)$, allowing $\partial p/\partial \tilde{g}(p)$ to be found. If this is not the case, the functional derivative of (2.102) may be multivalued or undefined.

2.5 The action approach to Newtonian mechanics

Modern Newtonian mechanics and classical field theory are formulated in terms of a *variational* or *action principle*. This approach has the advantage of revealing the relationship between conservation laws, such as energy and angular momentum, and the *symmetries* of the system's action, such as translational or rotational invariance. This relationship will be defined and explored in detail throughout this book. The equations of motion, which generalize Newton's second law (2.1), are then generated by applying the techniques of the calculus of variations to *extremize* the action. For simplicity, it is assumed that a mass m point particle is moving under the influence of a force associated with a scalar potential energy $U(\boldsymbol{r}, \dot{\boldsymbol{r}})$. The functions to be determined are the real-valued time-dependent position vectors of the particle, denoted by $\boldsymbol{r}(t)$, referred to as trajectories. The parameter t corresponds to time, and the components of the position vector are denoted by $r^i(t)$, where i indexes the coordinates for the particular choice of coordinate system. The action approach to classical mechanics constructs the *action functional* S from the position vector $\boldsymbol{r}(t)$ and its first derivative, denoted by $\dot{\boldsymbol{r}}(t) = \mathrm{d}\boldsymbol{r}(t)/\mathrm{d}t$, so that $S = S[\boldsymbol{r}(t), \dot{\boldsymbol{r}}(t)]$. The general action functional for classical mechanics is written

$$S[\boldsymbol{r}(t), \dot{\boldsymbol{r}}(t)] = \int_{t_i}^{t_f} \mathrm{d}t\, \mathcal{L}(\boldsymbol{r}(t), \dot{\boldsymbol{r}}(t), t), \tag{2.103}$$

where t_i and t_f are the initial and final times, respectively, for the motion of the particle under consideration and $\mathcal{L}$ is a *real-valued* quantity commonly known as the *Lagrangian density*, or more commonly just the *Lagrangian*. For the sake of notational brevity the position vector functions will often be written $\boldsymbol{r}$, with the time dependence suppressed, so that $\boldsymbol{r}$ is shorthand for $\boldsymbol{r}(t)$. In the most general case, $\mathcal{L}$ may also have an explicit time dependence. However, in typical elementary applications, the Lagrangian takes the general form

$$\mathcal{L}(\boldsymbol{r}, \dot{\boldsymbol{r}}) = K(\boldsymbol{r}, \dot{\boldsymbol{r}}) - U(\boldsymbol{r}, \dot{\boldsymbol{r}}), \tag{2.104}$$

where $K = \frac{1}{2}m\dot{\boldsymbol{r}} \cdot \dot{\boldsymbol{r}}$ is the kinetic energy familiar from (2.15) and $U(\boldsymbol{r}, \dot{\boldsymbol{r}})$ is the potential energy of the particle specific to the physical system under consideration. For this form of the Lagrangian there is no explicit time dependence and $\partial\mathcal{L}/\partial t = 0$. It is important to note that *the physical units of the action are energy·time*, which are the same units as those of Planck's constant, $\hbar$. This is true for all systems that can be understood using an action approach.

The derivation of the Euler–Lagrange equation (2.40) is readily adapted to provide a differential equation which the *trajectories* $\boldsymbol{r}(t)$ must solve in order to make the action functional (2.103) an extremum. After using the definition of the functional derivative and performing an integration by parts, it follows that

$$\begin{aligned}\delta S[\boldsymbol{r}, \dot{\boldsymbol{r}}] &= \int_{t_i}^{t_f} \mathrm{d}t \left(\frac{\delta S[\boldsymbol{r}, \dot{\boldsymbol{r}}]}{\delta \boldsymbol{r}(t)} \cdot \delta \boldsymbol{r}(t) + \frac{\delta S[\boldsymbol{r}, \dot{\boldsymbol{r}}]}{\delta \dot{\boldsymbol{r}}(t)} \cdot \delta \dot{\boldsymbol{r}}(t) \right) = 0 \\ \Longrightarrow &\int_{t_i}^{t_f} \mathrm{d}t \left(\frac{\partial \mathcal{L}(\boldsymbol{r}(t), \dot{\boldsymbol{r}}(t))}{\partial \boldsymbol{r}(t)} - \frac{\mathrm{d}}{\mathrm{d}t} \frac{\partial \mathcal{L}(\boldsymbol{r}(t), \dot{\boldsymbol{r}}(t))}{\partial \dot{\boldsymbol{r}}(t)} \right) \cdot \delta \boldsymbol{r}(t) = 0.\end{aligned} \tag{2.105}$$

Since $\delta\boldsymbol{r}(t)$ is arbitrary, the expression in parentheses must vanish along an extremum trajectory $\boldsymbol{r}_s(t)$,

$$\left(\frac{\partial \mathcal{L}(\boldsymbol{r}(t), \dot{\boldsymbol{r}}(t))}{\partial r^k(t)} - \frac{\mathrm{d}}{\mathrm{d}t} \frac{\partial \mathcal{L}(\boldsymbol{r}(t), \dot{\boldsymbol{r}}(t))}{\partial \dot{r}^k(t)} \right) \Bigg|_{\boldsymbol{r}_s(t)} = 0, \tag{2.106}$$

and (2.106) generalizes Newton's equation of motion (2.1).

In addition to the Euler–Lagrange equation (2.106), the functional approach allows the definition of the covariant momentum component p_k, *canonically conjugate* to r^k, as

$$p_k = \frac{\delta S[\boldsymbol{r}, \dot{\boldsymbol{r}}]}{\delta \dot{r}^k} = \frac{\partial \mathcal{L}(\boldsymbol{r}, \dot{\boldsymbol{r}}, t)}{\partial \dot{r}^k}. \tag{2.107}$$

It is assumed that this expression can be inverted to find $\dot{r}^k$ in terms of $\boldsymbol{p}$ and $\boldsymbol{r}$, so that $\dot{r}^k = \dot{r}^k(\boldsymbol{p}, \boldsymbol{r})$. Using (2.107) for the canonically conjugate momentum and working in Cartesian coordinates, the generalization of mechanical energy, referred to as the *Hamiltonian* H, is given by a Legendre transformation,

$$H(\boldsymbol{p}, \boldsymbol{r}) = p_k \dot{r}^k - \mathcal{L}(\boldsymbol{r}, \dot{\boldsymbol{r}}(\boldsymbol{p}, \boldsymbol{r}), t) = \boldsymbol{p} \cdot \dot{\boldsymbol{r}}(\boldsymbol{p}, \boldsymbol{r}) - \mathcal{L}(\boldsymbol{r}, \dot{\boldsymbol{r}}(\boldsymbol{p}, \boldsymbol{r}), t). \tag{2.108}$$

Using the definition (2.108) the Lagrangian can be rewritten in its *phase-space version*,

$$\mathcal{L}(\boldsymbol{p}, \boldsymbol{r}) = \boldsymbol{p} \cdot \dot{\boldsymbol{r}}(\boldsymbol{p}, \boldsymbol{r}) - H(\boldsymbol{p}, \boldsymbol{r}), \tag{2.109}$$

and used in the definition (2.103) to obtain a second version of the action functional,

$$S[\boldsymbol{p}, \boldsymbol{r}] = \int_{t_i}^{t_f} \mathrm{d}t\, \mathcal{L}(\boldsymbol{p}, \boldsymbol{r}) = \int_{t_i}^{t_f} \mathrm{d}t\, (\boldsymbol{p} \cdot \dot{\boldsymbol{r}}(\boldsymbol{p}, \boldsymbol{r}) - H(\boldsymbol{p}, \boldsymbol{r})). \tag{2.110}$$

Demanding that $\delta S[\boldsymbol{p}, \boldsymbol{r}] = 0$ and integrating by parts gives

$$\begin{aligned}
\delta S[\boldsymbol{p}, \boldsymbol{r}] &= \int_{t_i}^{t_f} \mathrm{d}t \left(\frac{\delta S[\boldsymbol{p}, \boldsymbol{r}]}{\delta \boldsymbol{p}(t)} \cdot \delta \boldsymbol{p}(t) + \frac{\delta S[\boldsymbol{p}, \boldsymbol{r}]}{\delta \boldsymbol{r}(t)} \cdot \delta \boldsymbol{r}(t) \right) \\
&= \int_{t_i}^{t_f} \mathrm{d}t \left\{ \left(\dot{\boldsymbol{r}}(t) - \frac{\partial H(\boldsymbol{p}, \boldsymbol{r})}{\partial \boldsymbol{p}(t)} \right) \cdot \delta \boldsymbol{p}(t) + \boldsymbol{p}(t) \cdot \delta \dot{\boldsymbol{r}}(t) \right. \\
&\qquad \left. - \frac{\partial H(\boldsymbol{p}, \boldsymbol{r})}{\partial \boldsymbol{r}(t)} \cdot \delta \boldsymbol{r}(t) \right\} \\
&= \int_{t_i}^{t_f} \mathrm{d}t \left\{ \left(\dot{\boldsymbol{r}}(t) - \frac{\partial H(\boldsymbol{p}, \boldsymbol{r})}{\partial \boldsymbol{p}(t)} \right) \cdot \delta \boldsymbol{p}(t) - \left(\dot{\boldsymbol{p}}(t) + \frac{\partial H(\boldsymbol{p}, \boldsymbol{r})}{\partial \boldsymbol{r}(t)} \right) \cdot \delta \boldsymbol{r}(t) \right\} \\
&= 0.
\end{aligned} \tag{2.111}$$

Because $\delta \boldsymbol{r}(t)$ and $\delta \boldsymbol{p}(t)$ are independent variations, this yields *Hamilton's equations of motion* for the extremal solutions $\boldsymbol{r}_s$ and $\boldsymbol{p}_s$,

$$\dot{\boldsymbol{r}}_s(t) = \left. \frac{\partial H(\boldsymbol{p}, \boldsymbol{r})}{\partial \boldsymbol{p}(t)} \right|_{\boldsymbol{r}_s, \boldsymbol{p}_s}, \qquad \dot{\boldsymbol{p}}_s(t) = - \left. \frac{\partial H(\boldsymbol{p}, \boldsymbol{r})}{\partial \boldsymbol{r}(t)} \right|_{\boldsymbol{r}_s, \boldsymbol{p}_s}. \tag{2.112}$$

In a general n-dimensional system, the equations of (2.112) comprise a set of $2n$ coupled first-order differential equations that can be combined to recreate the n second-order Euler–Lagrange equations of (2.106). In Cartesian coordinates, (2.112) can be written in component form as

$$\dot{r}_s^k(t) = \left. \frac{\partial H(\boldsymbol{p}, \boldsymbol{r})}{\partial p_k(t)} \right|_{\boldsymbol{r}_s, \boldsymbol{p}_s}, \qquad \dot{p}_{ks}(t) = - \left. \frac{\partial H(\boldsymbol{p}, \boldsymbol{r})}{\partial r^k(t)} \right|_{\boldsymbol{r}_s, \boldsymbol{p}_s}. \tag{2.113}$$

The time derivative of a general Hamiltonian is given by

$$\frac{\mathrm{d}H(\boldsymbol{p}, \boldsymbol{r}, t)}{\mathrm{d}t} = \frac{\partial H(\boldsymbol{p}, \boldsymbol{r}, t)}{\partial p_k} \dot{p}_k + \frac{\partial H(\boldsymbol{p}, \boldsymbol{r}, t)}{\partial r^k} \dot{r}^k + \frac{\partial H(\boldsymbol{p}, \boldsymbol{r}, t)}{\partial t}. \tag{2.114}$$

If (2.114) is evaluated using the solutions to (2.112), it follows that

$$\frac{\mathrm{d}H(\boldsymbol{p}_s, \boldsymbol{r}_s, t)}{\mathrm{d}t} = \frac{\partial H(\boldsymbol{p}_s, \boldsymbol{r}_s, t)}{\partial t}, \tag{2.115}$$

and energy is conserved unless the Hamiltonian is explicitly time dependent.

However, in the event of a Galilean boost, a static potential energy becomes explicitly time dependent,

$$U(\boldsymbol{r}(t)) = U(\boldsymbol{r}'(t) + v_o t). \tag{2.116}$$

The resolution of this question is obtained by considering the effect of a Galilean boost on the action of a free particle. For such a case, the action transforms to

$$
\begin{aligned}
S[\boldsymbol{r}, \dot{\boldsymbol{r}}] &= \int_0^T \mathrm{d}t \, \frac{1}{2} m \dot{\boldsymbol{r}} \cdot \dot{\boldsymbol{r}} = \int_0^T \mathrm{d}t \, \frac{1}{2} m (\dot{\boldsymbol{r}}' + \boldsymbol{v}_o) \cdot (\dot{\boldsymbol{r}}' + \boldsymbol{v}_o) \\
&= \frac{1}{2} m v_o^2 T + m \boldsymbol{v}_o \cdot (\boldsymbol{r}'(T) - \boldsymbol{r}'(0)) + \int_0^T \mathrm{d}t \, \frac{1}{2} m \dot{\boldsymbol{r}}' \cdot \dot{\boldsymbol{r}}',
\end{aligned}
\tag{2.117}
$$

which occurs since $\boldsymbol{v}_o$ is constant in time and therefore $m\boldsymbol{v}_o \cdot \dot{\boldsymbol{r}}' = \mathrm{d}(m\boldsymbol{v}_o \cdot \boldsymbol{r}')/\mathrm{d}t$. The first two terms in (2.117) are therefore constants and contribute nothing to the variation of the transformed action. As a result, the kinetic energy effectively transforms under a Galilean boost according to

$$
\frac{1}{2} m \dot{\boldsymbol{r}} \cdot \dot{\boldsymbol{r}} \to \frac{1}{2} m \dot{\boldsymbol{r}}' \cdot \dot{\boldsymbol{r}}'. \tag{2.118}
$$

In the terminology of the action approach, the Galilean boost is said to be *a symmetry* of the free action.

As a result of (2.118), the Lagrangian density (2.104) transforms to

$$
\mathcal{L}(\boldsymbol{r}, \dot{\boldsymbol{r}}) \to \mathcal{L}'(\boldsymbol{r}', \dot{\boldsymbol{r}}') = \frac{1}{2} m \dot{\boldsymbol{r}}' \cdot \dot{\boldsymbol{r}}' - U(\boldsymbol{r}' - \boldsymbol{v}_o t). \tag{2.119}
$$

The Euler–Lagrange equation for (2.119) gives

$$
\frac{\partial \mathcal{L}'}{\partial r'^k} - \frac{\mathrm{d}}{\mathrm{d}t} \frac{\partial \mathcal{L}'}{\partial \dot{r}'^k} = -\frac{\partial}{\partial r_k'} U(\boldsymbol{r}' + \boldsymbol{v}_o t) - m \frac{\mathrm{d}^2 r'^k}{\mathrm{d}t^2} = F^k(\boldsymbol{r}' + \boldsymbol{v}_o t) - m \frac{\mathrm{d}^2 r'^k}{\mathrm{d}t^2} = 0, \tag{2.120}
$$

which is precisely the boosted version of Newton's second law found in (2.29). The Hamiltonian obtained from the boosted Lagrangian is given by

$$
H'(\boldsymbol{p}', \boldsymbol{r}') = \frac{1}{2m} \boldsymbol{p}' \cdot \boldsymbol{p}' + U(\boldsymbol{r}' + \boldsymbol{v}_o t). \tag{2.121}
$$

Exercise 2.4. Verify (2.121) and show that Hamilton's equations give

$$
\begin{aligned}
\dot{\boldsymbol{r}}' &= \frac{\partial H'(\boldsymbol{p}', \boldsymbol{r}')}{\partial \boldsymbol{p}'} = \frac{\boldsymbol{p}'}{m}, \\
\dot{\boldsymbol{p}}' &= -\nabla' H'(\boldsymbol{p}', \boldsymbol{r}') = -\nabla' U(\boldsymbol{r}' + \boldsymbol{v}_o t) = \boldsymbol{F}(\boldsymbol{r}' + \boldsymbol{v}_o t).
\end{aligned}
\tag{2.122}
$$

The Hamiltonian (2.121) is the energy of mass m as observed by $\mathcal{O}'$, whereas (2.108) is the energy of mass m as observed by $\mathcal{O}$. In this case the two observers are distinguished by the fact that $\mathcal{O}$ sees a static potential, while $\mathcal{O}'$ observes a potential that is moving with a velocity of $\boldsymbol{v}_o$. Whether or not energy is conserved is agreed upon by both observers, since the partial derivative with respect to time transforms according to (2.23), which gives

$$\frac{\partial H(\boldsymbol{p}, \boldsymbol{r}, t)}{\partial t} = \left(\frac{\partial}{\partial t'} - \boldsymbol{v}_o \cdot \nabla'\right) H'(\boldsymbol{p}', \boldsymbol{r}' - \boldsymbol{v}_o t', t'). \tag{2.123}$$

If the left-hand side of (2.123) vanishes, then so does the right-hand side. As a result, if total mechanical energy is conserved according to one inertial observer, it is conserved according to all other inertial observers.

Another useful definition that can be generalized to field theory is the *Poisson bracket*. If $F(\boldsymbol{p}, \boldsymbol{r})$ and $G(\boldsymbol{p}, \boldsymbol{r})$ are two functions of the n canonically conjugate pairs, then their Poisson bracket is defined as

$$\{F, G\} = \sum_{k=1}^{n} \left(\frac{\partial F(\boldsymbol{p}, \boldsymbol{r})}{\partial r^k} \frac{\partial G(\boldsymbol{p}, \boldsymbol{r})}{\partial p_k} - \frac{\partial G(\boldsymbol{p}, \boldsymbol{r})}{\partial r^k} \frac{\partial F(\boldsymbol{p}, \boldsymbol{r})}{\partial p_k} \right) \Bigg|_{r_s, p_s}. \tag{2.124}$$

The Poisson bracket is antisymmetric, since $\{F, G\} = -\{G, F\}$. Using Hamilton's equations in component form (2.113), it follows that

$$\begin{aligned} \{H, F\} &= \sum_{k=1}^{n} \left(\frac{\partial F(\boldsymbol{p}, \boldsymbol{r})}{\partial r^k} \frac{\partial H(\boldsymbol{p}, \boldsymbol{r})}{\partial p_k} - \frac{\partial H(\boldsymbol{p}, \boldsymbol{r})}{\partial r^k} \frac{\partial F(\boldsymbol{p}, \boldsymbol{r})}{\partial p_k} \right) \Bigg|_{r_s, p_s} \\ &= \sum_{k=1}^{n} \left(\frac{\partial F(\boldsymbol{p}, \boldsymbol{r})}{\partial r^k} \dot{r}^k + \frac{\partial F(\boldsymbol{p}, \boldsymbol{r})}{\partial p_k} \dot{p}_k \right) \Bigg|_{r_s, p_s} = \frac{\mathrm{d}}{\mathrm{d}t} F(\boldsymbol{p}, \boldsymbol{r}). \end{aligned} \tag{2.125}$$

References and recommended further reading

Monographs on Newtonian mechanics include

- H Goldstein, C Poole, and J Safko 2000 *Classical Mechanics 3rd edn* (San Francisco: Addison–Wesley)
- S Thornton and J Marion 2004 *Classical Dynamics of Particles and Systems* (Cengage)

Monographs presenting mathematical aspects of the calculus of variations include

- C Lanczos 1970 *The Variational Principles of Mechanics 4th edn* (New York: Dover)
- D Lovelock and H Rund 1989 *Tensors, Differential Forms, and Variational Principles* (New York: Dover)

Monographs on classical field theory that discuss variational techniques include

- D Soper 2008 *Classical Field Theory* (New York: Dover)
- H Nastase 2019 *Classical Field Theory* (Cambridge: Cambridge University Press)

Monographs focusing on complex variables and analysis include

- T Gamelin 2001 *Complex Analysis* (New York: Springer–Verlag)

- J Brown and R Churchill 2009 *Complex Variables and Applications 9th edn* (New York: McGraw–Hill)

Monographs presenting the mathematical aspects of functional methods include

- R Courant and D Hilbert 1989 *Methods of Mathematical Physics Vol. II* (New York: Wiley)
- F Byron and R Fuller 1992 *Mathematics of Classical and Quantum Physics* (New York: Dover)

IOP Publishing

Classical Field Theory and the Stress-Energy Tensor
(Second Edition)

Mark S Swanson

Chapter 3

Basic field theory

In this chapter the fundamental concepts and tools for formulating and analyzing field theory are developed by applying Newtonian mechanics to a uniform horizontally stretched string. In particular, analyzing the stretched string yields the generalization of the Newtonian concepts of energy and momentum to continuous systems, encapsulated in the stress–energy tensor. It also quantifies the emergence of waves, representing the motion of momentum and energy through the medium of the string. In this simple case, the field quantifies the disturbance of the stretched string away from its equilibrium or zero kinetic energy configuration. The advantage of this simple example is that it merges basic Newtonian mechanics and the concept of the field.

3.1 The mechanical properties of a stretched string

The mechanical properties of a uniform stretched string are critical to formulating its behavior when it is disturbed from its equilibrium configuration. Consideration begins with a uniform string of mass M consisting of two segments, each with a length L. The string is horizontally stretched between two end points by an arbitrary but uniform tension F . Since the tension creates a force that has an opposite direction at each end of the string, it is understood that F is the *magnitude* of the force at each end. The agent creating this tension is external, such as weights attached over a pulley at each end of the string or the tuning key of a guitar. It is assumed that the string can stretch slightly but that the tension remains *constant* during this process. This is an approximation since the tension in the string will change as it is stretched or a load is placed on it. This is discussed again in the last two sections of this chapter, where the weight of the string itself results in a varying tension in the string and affects the static shape of the string. As depicted in figure 3.1, the potential energy of the string is revealed by suspending a mass m from

doi:10.1088/978-0-7503-3455-6ch3

© IOP Publishing Ltd 2022

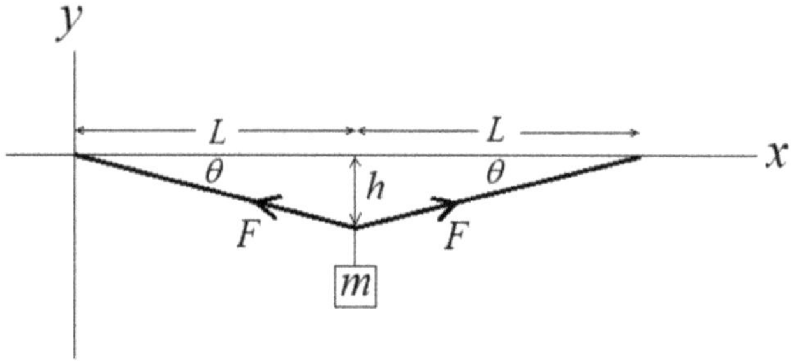

Figure 3.1. The mechanical response of the stretched string.

its midpoint and allowing it to descend to a distance h below the horizontal equilibrium shape of the string.

As the mass moves through the vertical element $\mathrm{d}y$ at position y, the net vertical force acting on the string is $2F \sin\theta - mg$. Using $\sin\theta = y/\sqrt{L^2 + y^2}$, the work done by this force on the string is given by

$$W = \int_0^{-h} \mathrm{d}y\, (2F\sin\theta - mg) = \int_0^{-h} \mathrm{d}y \left(\frac{2Fy}{\sqrt{L^2 + y^2}} - mg \right) = mgh \\ + 2F\left(\sqrt{L^2 + h^2} - L\right). \tag{3.1}$$

This gives the potential energy of the configuration as a function of $y = h$,

$$U(y) = mgy + 2F\left(\sqrt{L^2 + y^2} - L\right). \tag{3.2}$$

The first term is the familiar gravitational potential of the attached mass. The second term can be viewed as the potential energy that arises from the stretching of the two segments. This follows by noting that the factor $\Delta L = \sqrt{L^2 + y^2} - L$ is the change of length in each of the two segments. This assumes that the tension F remains constant as the segment changes length. A more realistic treatment in a later section allows the tension to vary in the string. Since the two segments are identical, the potential energy of each straight segment of the stretched string is $U = F\,\Delta L$, where ΔL is the change in length of the string segment from its equilibrium length of L. It follows from (3.2) that the equilibrium position of the system depicted in figure 3.1 is given by

$$\frac{\partial U}{\partial y} = 0 \implies \frac{2Fy}{\sqrt{L^2 + y^2}} + mg = 0 \implies \frac{y}{\sqrt{L^2 + y^2}} = \sin\theta = -\frac{mg}{2F}, \tag{3.3}$$

which is the usual elementary result for the equilibrium angle of the configuration. This result assumes that $F \geqslant mg$, since otherwise an unphysical result occurs and the string either breaks or increases in tension. The potential energy of an *individual* string segment of original length L can be expressed as a function of the difference in height y between its two ends,

$$U(y) = F\ \Delta L = F\left(\sqrt{L^2 + y^2} - L\right). \tag{3.4}$$

In the absence of the external mass m the equilibrium configuration of the string segment is $y = 0$.

3.2 The stretched string as a field theory

In this section some important aspects of field theory will be derived by considering the response of a stretched string with no weights attached when disturbed away from equilibrium. The string has a length L with one end fixed at $x = 0$ and the other at $x = L$. It has a total mass M with a uniform mass density $\mu = M/L$ and negligible thickness. It is stretched taut with a constant tension F throughout. For analysis, the continuous string is broken into a sequence of N small straight segments of length $\Delta x = L/N$, where N will become arbitrarily large. Each of the N segments has a mass $m = \mu\ \Delta x$, which is treated as a point mass. In the limit of vanishing size for the segments, the segment and the point mass coincide. Each segment is free to move only perpendicularly to the x-axis in the y-direction and is subject to Newtonian dynamics. This approach goes by the general name of *continuum mechanics* and is often used in the study of extended objects or materials.

The string segments are labeled sequentially with positive integers j that range from 0 to $N - 1$, so that the horizontal position of the jth segment is $x_j = j\ \Delta x$. Since the string segments are free to move in only one vertical direction, each segment has a vertical coordinate, $\varphi_j(t)$, that measures the distance the jth string segment center has deviated vertically from the string's undisturbed or equilibrium shape. It follows that the jth segment has the kinetic energy

$$K_j = \frac{1}{2} m \dot{\varphi}_j^{\,2} = \Delta x\ \frac{1}{2} \mu \dot{\varphi}_j^{\,2}. \tag{3.5}$$

The height difference between the ends of the jth segment at time t is given by $\Delta\varphi_j = \varphi_{j+1}(t) - \varphi_j(t)$. Using (3.4) and setting $y = y_j$ gives the potential energy U_j of the jth string segment,

$$U_j = F\ \Delta L_j = F\left(\sqrt{(\Delta x)^2 + (\Delta\varphi_j)^2} - \Delta x\right) = F\Delta x\left(\sqrt{1 + \left(\frac{\Delta\varphi_j}{\Delta x}\right)^2} - 1\right). \tag{3.6}$$

For that case that the angular deviation from equilibrium, $\Delta\varphi_j/\Delta x = \tan\theta_j$, is small for each segment, the Taylor series expansion $\sqrt{1 + x} \approx 1 + \frac{1}{2}x$ for small x can be used. The potential energy (3.6) becomes

$$U_j \approx \Delta x\ \frac{1}{2} F \left(\frac{\Delta\varphi_j}{\Delta x}\right)^2. \tag{3.7}$$

Approximation (3.7) will be discarded in the last section of this chapter and the full expression (3.6) will be employed for the potential energy.

The action for the total string is obtained by treating the collection of segments as N point masses, each contributing a kinetic and a potential energy. Using (3.7) and summing the individual actions for each segment yields the action for small disturbances,

$$\mathcal{S}[\varphi, \dot{\varphi}] = \sum_{j=0}^{N-1} \int_0^T \mathrm{d}t \, (K_j - U_j) = \sum_{j=0}^{N-1} \Delta x \int_0^T \mathrm{d}t \left(\frac{1}{2}\mu \, \dot{\varphi}_j^{\,2} - \frac{1}{2}F \left(\frac{\Delta \varphi_j}{\Delta x} \right)^2 \right). \quad (3.8)$$

The next step is to replace $\varphi_j(t)$ by introducing a *scalar field variable* $\varphi(x_j, t)$, which gives the height of the string's disturbance at position x_j at time t. Assuming that this variable is differentiable allows the Taylor series expansion

$$\Delta\varphi_j(t) = \varphi(x_{j+1}, t) - \varphi(x_j, t) = \varphi(x_j + \Delta x, t) - \varphi(x_j, t) \approx \Delta x \, \frac{\partial \varphi(x_j, t)}{\partial x_j}. \quad (3.9)$$

For small disturbances, action (3.8) becomes

$$\mathcal{S} = \sum_{j=0}^{N-1} \Delta x \int_0^T \mathrm{d}t \left(\frac{1}{2}\mu \left(\frac{\partial \varphi(x_j, t)}{\partial t} \right)^2 - \frac{1}{2}F \left(\frac{\partial \varphi(x_j, t)}{\partial x_j} \right)^2 \right). \quad (3.10)$$

The final step is to take the $N \to \infty$ *continuum limit*, so that $x_j = j\,\Delta x \to x$ and $\sum_j \Delta x \to \int \mathrm{d}x$. The discrete action of (3.10) becomes an integral over both space x and time t and, for this specific case, is written

$$S[\varphi, \partial\varphi] = \int_0^L \mathrm{d}x \int_0^T \mathrm{d}t \left(\frac{1}{2}\mu \left(\frac{\partial \varphi}{\partial t} \right)^2 - \frac{1}{2}F \left(\frac{\partial \varphi}{\partial x} \right)^2 \right) \equiv \int_{LT} \mathrm{d}x \, \mathrm{d}t \, \mathcal{L}(\varphi, \partial\varphi). \quad (3.11)$$

For the sake of streamlining notation, φ is understood to stand for $\varphi(x, t)$ and $\partial\varphi$ is understood to be shorthand for the set of partial derivatives of $\varphi(x, t)$. The Lagrangian density $\mathcal{L}(\varphi, \partial\varphi)$ that appears in the action is a function of both position and time, x and t, and the integration in the action is over the *volume of space and time LT*. Although the specific stretched string Lagrangian density in (3.11),

$$\mathcal{L}(\varphi, \partial\varphi) = \frac{1}{2}\mu \left(\frac{\partial \varphi}{\partial t} \right)^2 - \frac{1}{2}F \left(\frac{\partial \varphi}{\partial x} \right)^2, \quad (3.12)$$

has only derivatives of φ, a general scalar Lagrangian may depend on both the field φ and its derivatives $\partial\varphi$, and so it is written $\mathcal{L}(\varphi, \partial\varphi)$. Actions with higher-order derivatives have proven to be problematic and will not be considered in this text. The important distinction between point particle mechanics and fields is that x no longer characterizes a trajectory, and therefore has no implicit time dependence. Rather,

x is simply a location where the field disturbance φ can occur. The time dependence is now present in the field variable $\varphi(x, t)$, which, in this case, is a coordinate measuring the vertical disturbance of the string from equilibrium. Since $\mathcal{L}$ has units of force F in this specific case, action (3.11) has units of $F\ \mathrm{d}x\ \mathrm{d}t$, which gives the correct units of energy × time for the action. This is identical to the units of $\hbar$ and is the same units as those of single-particle action (2.103).

In this particular case the action is characterized by a Lagrangian density $\mathcal{L}(\varphi, \partial\varphi)$ that consists of a *kinetic energy density* or *kinetic energy per unit length*,

$$K(x, t) = \frac{1}{2}\mu\left(\frac{\partial\varphi}{\partial t}\right)^2, \tag{3.13}$$

and a *potential energy density*,

$$U(x, t) = \frac{1}{2}F\left(\frac{\partial\varphi}{\partial x}\right)^2. \tag{3.14}$$

These two densities generalize the Newtonian concepts of kinetic and potential energy into the continuum of the string and will be employed to generalize the Newtonian concepts of momentum and force.

3.3 The Euler–Lagrange equation for the stretched string

The next step is to adapt the variational techniques of (2.105) to the field theory action (3.11). This will determine the equation of motion for the field φ that will extremize the action. As in the case of the point particle (2.105), the equation of motion for the function $\varphi_s(x, t)$ that extremizes the action (3.11) is found by demanding that $\delta S[\varphi_s, \partial\varphi_s] = 0$. The variation of the action is found using an infinitesimal variation of the field $\varphi_s \to \varphi_s + \delta\varphi$ that satisfies the boundary conditions

$$\delta\varphi(x, 0) = \delta\varphi(x, T) = 0. \tag{3.15}$$

These boundary conditions on the variation are consistent with the initial and final shapes of the string being *fixed*, so that $\varphi_s(x, 0) = I(x)$ and $\varphi_s(x, T) = F(x)$. As a result of these boundary conditions on the shape, there can be no variation of the field φ at the initial and final times, and this yields the conditions of (3.15). Specifying the initial and final shape of the string, $\varphi_s(x, 0)$ and $\varphi_s(x, T)$, is referred as *Dirichlet boundary conditions*. In addition, fixing the end points of the strings places a restriction on the variations given by

$$\delta\varphi(0, t) = \delta\varphi(L, t) = 0, \tag{3.16}$$

which must hold, since the end points of the string are fixed and cannot be varied at any time. In the case of Dirichlet boundary conditions, this corresponds to boundary conditions on the end points given by

$$\frac{\partial\varphi(0, t)}{\partial t} = \frac{\partial\varphi(L, t)}{\partial t} = 0. \tag{3.17}$$

For fixed end points the initial and final shapes of the string must therefore satisfy

$$I(0) = I(L) = F(0) = F(L) = 0. \tag{3.18}$$

This combination of conditions is referred to as the *Dirichlet boundary conditions* for the fixed string.

The variation of the action is given by

$$\delta S[\varphi_s, \partial\varphi_s] = \int_{LT} \mathrm{d}x\, \mathrm{d}t \left(\delta\varphi(x, t)\frac{\delta S[\varphi, \partial\varphi]}{\delta\varphi(x, t)}\bigg|_{\varphi_s, \partial\varphi_s} + \delta\partial_\mu\varphi(x, t)\frac{\delta S[\varphi, \partial\varphi]}{\delta\partial_\mu\varphi(x, t)}\bigg|_{\varphi_s, \partial\varphi_s} \right) = 0. \tag{3.19}$$

Relation (3.19) uses the summation convention for Greek subscripts μ that ranges from zero to one, where $\partial_0 = \partial/\partial t$ and $\partial_1 = \partial/\partial x$. The remainder of this book will observe this convention, with the range of spatial subscripts determined by the spatial dimension of the system under consideration. This begins the elevation of time t to a coordinate, although considerable further development is required to justify such a step. Using $\delta\partial_\mu\varphi(x, t) = \partial_\mu\delta\varphi(x, t)$, integrating by parts, and using the boundary conditions (3.15) and (3.16) on $\delta\varphi$ gives

$$\begin{aligned} \delta S[\varphi_s, \partial\varphi_s] &= \int_{LT} \mathrm{d}x\, \mathrm{d}t \left\{ \partial_\mu\left(\delta\varphi \frac{\partial\mathcal{L}(\varphi, \partial\varphi)}{\partial(\partial_\mu\varphi)} \right)\bigg|_{\varphi_s, \partial\varphi_s} \right. \\ &\quad \left. + \delta\varphi\left(\frac{\partial\mathcal{L}(\varphi, \partial\varphi)}{\partial\varphi} - \partial_\mu\frac{\partial\mathcal{L}(\varphi, \partial\varphi)}{\partial(\partial_\mu\varphi)} \right)\bigg|_{\varphi_s, \partial\varphi_s} \right\} \\ &= \int_{LT} \mathrm{d}x\, \mathrm{d}t\, \delta\varphi\left(\frac{\partial\mathcal{L}(\varphi, \partial\varphi)}{\partial\varphi} - \partial_\mu\frac{\partial\mathcal{L}(\varphi, \partial\varphi)}{\partial(\partial_\mu\varphi)} \right)\bigg|_{\varphi_s, \partial\varphi_s} = 0. \end{aligned} \tag{3.20}$$

For the case that $\delta\varphi(x, t)$ is arbitrary the extremum function φ_s must satisfy

$$\left(\frac{\partial\mathcal{L}(\varphi, \partial\varphi)}{\partial\varphi} - \partial_\mu\frac{\partial\mathcal{L}(\varphi, \partial\varphi)}{\partial(\partial_\mu\varphi)} \right)\bigg|_{\varphi_s, \partial\varphi_s} = 0, \tag{3.21}$$

which is the Euler–Lagrange equation for a *scalar field* φ. In the case of the stretched string the field variable φ is a scalar, because the string is assumed to rise and fall in a single direction. For the case of the stretched string Lagrangian (3.11), the Euler–Lagrange equation (3.21) gives the *equation of motion*

$$F\frac{\partial^2\varphi_s(x, t)}{\partial x^2} - \mu\frac{\partial^2\varphi_s(x, t)}{\partial t^2} = F\left(\frac{\partial^2\varphi_s(x, t)}{\partial x^2} - \frac{1}{\tilde{v}^2}\frac{\partial^2\varphi_s(x, t)}{\partial t^2} \right) = 0, \tag{3.22}$$

where $\tilde{v} = \sqrt{F/\mu}$ has units of length per time, i.e. velocity.

3.4 Solving the Euler–Lagrange equation

Obtaining a *unique* solution to (3.22) requires an appropriate set of *boundary conditions*. In general, the necessary boundary conditions follow from the fact that the equation of motion (3.22) is a second-order hyperbolic partial differential equation. In order to understand what constitutes sufficient *boundary conditions* to specify a unique solution, the string can be viewed as a sequence of point masses or segments, each of which is governed by Newton's equation of motion. Because Newton's second law is a second-order differential equation, the jth segment requires two boundary conditions. In the case of Dirichlet boundary conditions the height φ_j of the jth segment is specified at $t = 0$ and $t = T$. In the continuum limit for the stretched string, Dirichlet boundary conditions therefore correspond to the *spatial shape* of the string at those two times. As discussed earlier Dirichlet boundary conditions correspond to specifying $\varphi(x, 0) = I(x)$ and $\varphi(x, T) = F(x)$.

Just as other boundary conditions are possible for the motion of a point mass, the same is true for fields. Cauchy boundary conditions specify the initial shape of the string, $\varphi(x, 0) = I(x)$, and the initial transverse velocity of the string, $\dot{\varphi}(x, 0) = V(x)$. Neumann boundary conditions specify the string's initial velocity, $\dot{\varphi}(x, 0) = V_i(x)$, and its final velocity, $\dot{\varphi}(x, T) = V_f(x)$. It is important to note that Neumann and Cauchy boundary conditions *do not* require that the end points of the string remain fixed. Such a situation corresponds to a physical situation where the string is free to slide vertically on a post at the two end positions. This is discussed again in chapter 11 in the context of string theory. As a result, Dirichlet conditions are used for the fixed endpoint situation. Since the string disturbance $\varphi(x, t)$ cannot extend beyond the string, it must also be the case that $\varphi(0, t) = \varphi(L, t) = 0$ and $\dot{\varphi}(0, t) = \dot{\varphi}(L, t) = 0$ for all t.

The differential operator appearing in (3.22), denoted by

$$\Box_{\tilde{v}} \equiv \frac{\partial^2}{\partial x^2} - \frac{1}{\tilde{v}^2}\frac{\partial^2}{\partial t^2}, \tag{3.23}$$

is known as the *d'Alembertian*. Since F and $\tilde{v}$ are constants, it follows that (3.22) can be rewritten as

$$\Box_{\tilde{v}}\varphi_s(x, t) = \left(\frac{\partial}{\partial x} + \frac{1}{\tilde{v}}\frac{\partial}{\partial t}\right)\left(\frac{\partial}{\partial x} - \frac{1}{\tilde{v}}\frac{\partial}{\partial t}\right)\varphi_s(x, t) = 0. \tag{3.24}$$

Equation (3.24) can easily be solved by introducing two new independent variables, $y_- = x + \tilde{v}t$ and $y_+ = x - \tilde{v}t$. The subscripts are chosen to represent the direction that a fixed value for the new variable appears to move in terms of the old variable. For example, $x = y_- - \tilde{v}t$ shows that a fixed value of y_- is moving leftward, or in the negative x direction, at a speed $\tilde{v}$ as time passes. The transformation properties of the partial derivatives give

$$\begin{aligned} \frac{\partial}{\partial x} &= \frac{\partial y_+}{\partial x}\frac{\partial}{\partial y_+} + \frac{\partial y_-}{\partial x}\frac{\partial}{\partial y_-} = \frac{\partial}{\partial y_+} + \frac{\partial}{\partial y_-}, \\ \frac{1}{\tilde{v}}\frac{\partial}{\partial t} &= \frac{1}{\tilde{v}}\frac{\partial y_+}{\partial t}\frac{\partial}{\partial y_+} + \frac{1}{\tilde{v}}\frac{\partial y_-}{\partial t}\frac{\partial}{\partial y_-} = \frac{\partial}{\partial y_-} - \frac{\partial}{\partial y_+}, \end{aligned} \tag{3.25}$$

with the result that

$$\begin{aligned}\frac{1}{2}\left(\frac{\partial}{\partial x} - \frac{1}{\tilde{v}}\frac{\partial}{\partial t}\right) &= \frac{\partial}{\partial y_+},\\ \frac{1}{2}\left(\frac{\partial}{\partial x} + \frac{1}{\tilde{v}}\frac{\partial}{\partial t}\right) &= \frac{\partial}{\partial y_-}.\end{aligned} \tag{3.26}$$

Using these new variables, (3.24) takes the form

$$\frac{1}{4}\Box_{\tilde{v}}\varphi_s(x, t) = \frac{\partial^2}{\partial y_+ \partial y_-}\varphi_s(y_+, y_-) = 0. \tag{3.27}$$

Result (3.27) shows that an *arbitrary* function of only one of these variables, either y_+ or y_-, solves (3.27). This follows from the fact that the two partial derivatives commute,

$$\frac{\partial^2}{\partial y_+ \partial y_-} = \frac{\partial}{\partial y_+}\frac{\partial}{\partial y_-} = \frac{\partial}{\partial y_-}\frac{\partial}{\partial y_+} = \frac{\partial^2}{\partial y_- \partial y_+}, \tag{3.28}$$

and the fact that y_+ and y_- are independent variables, so that

$$\frac{\partial \varphi(y_+)}{\partial y_-} = \frac{\partial \varphi(y_-)}{\partial y_+} = 0. \tag{3.29}$$

Solutions to (3.27) therefore take the form $\varphi_s(x, t) = \varphi(y_\pm) = \varphi(x \pm \tilde{v}t)$, where φ is an arbitrary function that can be chosen to match the boundary conditions. For the choice $\varphi(x + \tilde{v}t)$ the shape given by φ is moving to the left or negative x direction as time passes, since a specific function argument $x + \tilde{v}t = x_o$ occurs at the left-moving position $x = x_o - \tilde{v}t$. Similarly, the choice $\varphi(x - \tilde{v}t)$ gives a shape that is moving to the right or in the positive x direction. In both cases the shape is moving at the constant speed $\tilde{v}$, known as the *velocity of propagation.*

For the case of the stretched string the velocity of propagation is determined by the mechanical properties of the string, since $\tilde{v} = \sqrt{F/\mu}$. The solutions to (3.22) therefore represent a *disturbance* from equilibrium in the *medium* of the string that *propagates* along the string. A propagating disturbance in a medium is the original definition of a *wave* and accurately describes many *material waves*, such as water waves and sound waves. In the case of the string the motion of the medium is perpendicular or *transverse* to the direction of the wave as it passes a point, and so it is referred to as a *transverse wave*. In some cases, such as sound, the motion of the medium is parallel to the direction of the wave as it passes a point and such waves are referred to as *longitudinal*. In both cases the wave disturbance is governed by (3.22), and so (3.22) is commonly referred to as the *wave equation*. While (3.22) was obtained for the case of a stretched string, the wave equation and its modified forms occur in many physical systems governed by field theories. It is therefore important to investigate its basic properties.

The d'Alembertian operator appearing in the wave equation (3.22) is a *linear operator*. This means that two solutions to (3.22), denoted by $\varphi_1(x, t)$ and $\varphi_2(x, t)$, can be used to create a third solution to (3.22) through the linear combination $\varphi(x, t) = a\,\varphi_1(x, t) + b\,\varphi_2(x, t)$, where a and b are two constants. This is often referred to as the *principle of superposition* and gives rise to many important effects, such as *interference* and *diffraction*, when waves overlap. Using the principle of superposition allows a very general solution of (3.22) to be written as a linear combination of left- and right-moving solutions, so that

$$\varphi_s(x, t) = \varphi_-(x + \tilde{v}t) + \varphi_+(x - \tilde{v}t). \tag{3.30}$$

Each of the left- and right-moving solutions in (3.30) can themselves be general linear superpositions of left- and right-moving solutions. This will be discussed when Fourier methods are presented.

An important *real* solution of (3.22) is the *simple harmonic wave*. These are solutions to (3.22) with the forms

$$\varphi_s(x, t) = \begin{cases} A\sin(kx \pm \omega t + \phi) \\ A\cos(kx \pm \omega t + \phi) \end{cases}. \tag{3.31}$$

The simple harmonic wave is a solution to the wave equation (3.22) as long as $k^2 - (\omega^2/\tilde{v}^2) = 0$, which is solved if $\omega/k = \pm\tilde{v}$. The simple harmonic wave therefore has the expected form of left- and right-moving waves, since this relation gives $kx \pm \omega t = k(x \pm (\omega/k)t) = k(x \pm \tilde{v}t)$. The constant A is the *amplitude* or magnitude of the wave, k is the *wave number*, ω is the *angular frequency*, and ϕ is the *phase* of the wave. If ϕ is replaced with $\phi + \pi/2$, the sine function in (3.31) becomes the cosine function in (3.31). The choice of sine or cosine function is typically dictated by boundary conditions, although they are often interchangeable with the correct choice of phase. The simple harmonic wave is an example of a *periodic wave*, which has a recurring shape. Since the sine and cosine functions have a period of 2π, the simple harmonic wave recurs over the distance $2\pi/k = \lambda$, which is known as the *wavelength*. Similarly, the amount of time T required for one wavelength to pass a point, known as the *period*, is given by $T = 2\pi/\omega$. The number of wavelengths passing a point per unit time is known as the *frequency* f, and is given by $f = 1/T = \omega/2\pi$, so that the simple harmonic wave satisfies

$$f\lambda = \omega/k = \tilde{v}. \tag{3.32}$$

If complex solutions represent a physical phenomenon, the wave equation also has the complex-valued harmonic solutions given by

$$\varphi_s(x, t) = A\exp\{i(kx \pm \omega t + \phi)\}. \tag{3.33}$$

The same periodicity holds for (3.33) as for (3.31).

If equal-amplitude and -wavelength left- and right-moving simple harmonic waves are present in the string, they will combine according to the principle of superposition. Using either the sine or cosine functions of (3.31) with a phase angle of $\phi = 0$ gives the following solutions:

$$\begin{aligned}\varphi_s(x, t) &= \varphi_-(x + vt) + \varphi_+(x - vt) = A \sin(kx + \omega t) + A \sin(kx - \omega t) \\ &= 2A \sin(kx) \cos(\omega t),\end{aligned} \tag{3.34}$$

$$\begin{aligned}\varphi_s(x, t) &= \varphi_-(x + vt) + \varphi_+(x - vt) = A \cos(kx - \omega t) - A \cos(kx + \omega t) \\ &= 2A \sin(kx) \sin(\omega t),\end{aligned} \tag{3.35}$$

where the trigonometric identities $\sin(\alpha \pm \beta) = \sin\alpha \cos\beta \pm \cos\alpha \sin\beta$ and $\cos(\alpha \pm \beta) = \cos\alpha \cos\beta \mp \sin\alpha \sin\beta$ were used. For the case of a length L string fixed at both $x = 0$ and $x = L$, the boundary conditions include $\varphi_s(0, t) = \varphi_s(L, t) = 0$. Solutions (3.34) and (3.35) automatically satisfy the boundary condition $\varphi_s(0, t) = 0$. However, satisfying the boundary condition $\varphi_s(L, t) = 0$ requires that $k = j\pi/L \equiv k_j$, where j is a positive integer, since only then does $\sin(k_j L) = \sin(j\pi) = 0$. A standing wave in a finite string can therefore have only one of the discrete wavelengths and frequencies determined by the integer j,

$$\lambda_j = \frac{2\pi}{k_j} = \frac{2L}{j} \implies f_j = \frac{\tilde{v}}{\lambda_j} = \frac{j}{2L}\sqrt{\frac{F}{\mu}}. \tag{3.36}$$

Such a set of fixed wavelengths and frequencies is often referred to as the *normal modes* of the system.

At the locations that satisfy $k_j x_n = n\pi$, where n is an arbitrary positive or negative integer, solutions (3.34) and (3.35) are zero at all times. These locations are stationary and referred to as *nodes*, giving rise to the name *standing wave* for this phenomenon. This occurs because the two waves *interfere destructively* at the nodes of the standing wave. Adjacent nodes are separated by the distance $x_{n+1} - x_n = \pi/k_j = \frac{1}{2}\lambda_j = L/j$. At the locations exactly midway between two nodes the string performs simple harmonic oscillation with an amplitude of $2A$ and a frequency $f_j = j\tilde{v}/2L$. These locations are referred to as *antinodes*. The antinodes correspond to places where the two waves *interfere constructively*. It is straightforward to see that the velocity at each end is also zero, $\dot{\varphi}_s(0, t) = \dot{\varphi}_s(L, t) = 0$, at all times for either form (3.34) or (3.35). The standing wave is depicted in figure 3.2 for $j = 3$. In a stringed musical instrument, such as a piano or violin, the frequency $f_1 = (\sqrt{F/\mu})/2L$ determines the basic musical note of the string, while the multiples f_j are referred to as octaves or overtones. The full response of the string in a musical instrument is a combination of the basic frequency and a mixture of overtones determined by the acoustic properties of the instrument.

By adapting (2.91), the full solution to (3.22) consistent with all boundary conditions can be expressed as a Fourier series. The Cauchy boundary conditions for the string can be expressed as the shape of the string $I(x)$ and the vertical velocity of the string $V(x)$ at the initial time $t = 0$,

$$\varphi_s(x, 0) = I(x), \quad \dot{\varphi}_s(x, 0) = V(x), \tag{3.37}$$

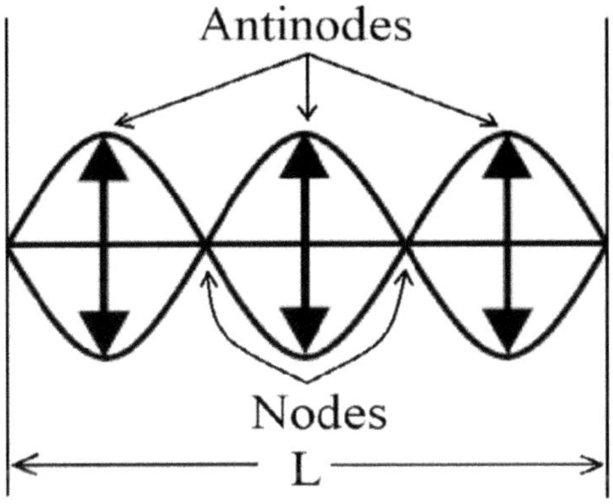

Figure 3.2. The standing wave for $j = 3$.

where both $I(x)$ and $V(x)$ satisfy the required boundary conditions $I(0) = I(L) = V(0) = V(L) = 0$, but are otherwise *arbitrary* piecewise continuous functions. If these conditions are met, the boundary functions can be expanded in a Fourier series using the spatial part of the standing-wave solutions, $\varphi_j = \sin(k_j x)$, present in both (3.34) and (3.35),

$$I(x) = \frac{2}{L}\sum_{j=1}^{\infty} \tilde{I}_j \sin(k_j x), \quad V(x) = \frac{2}{L}\sum_{j=1}^{\infty} \tilde{V}_j \sin(k_j x). \tag{3.38}$$

The series in (3.38) automatically satisfy the boundary conditions at $x = 0$ and $x = L$. The coefficients in the Fourier series are found by using the *orthonormality* of the $\sin(k_j x)$ functions over the interval $(0, L)$,

$$\frac{2}{L}\int_0^L \mathrm{d}x \sin(k_j x) \sin(k_\ell x) = \delta_{j\ell}. \tag{3.39}$$

This orthonormality gives the Fourier coefficients in terms of spatial integrals,

$$\tilde{I}_j = \int_0^L \mathrm{d}x\, I(x)\sin(k_j x), \quad \tilde{V}_j = \int_0^L \mathrm{d}x\, V(x)\sin(k_j x). \tag{3.40}$$

In effect, the $\sin(k_j x)$ functions act as unit vectors in the space of possible solutions, while an arbitrary solution can be expanded in terms of the Fourier coefficients, which act as the vector components.

The solution φ_s for such general boundary conditions is written using a Fourier series that combines the solutions of both (3.34) and (3.35),

$$\varphi_s(x, t) = \sum_{j=1}^{\infty}\left(\tilde{\varphi}_j \cos(\omega_j t) + \tilde{\varphi}'_j \sin(\omega_j t)\right)\sin(k_j x), \tag{3.41}$$

where $\omega_j = \tilde{v}k_j = j\pi\tilde{v}/L$. Since $\omega_j^2 = \tilde{v}^2 k_j^2$, it follows that $\Box_{\tilde{v}}\varphi_s = 0$. As in the case of the standing waves (3.34) and (3.35), the discrete wave numbers $k_j = j\pi/L$ used in

(3.41) allow φ_s to satisfy the boundary conditions $\varphi_s(0, t) = \varphi_s(L, t) = 0$ and $\dot{\varphi}_s(0, t) = \dot{\varphi}_s(L, t) = 0$. The Cauchy boundary conditions (3.38) are satisfied by matching the Fourier coefficients in (3.41) to the Fourier coefficients in (3.40). Setting $t = 0$ in (3.41) as well as in its time derivative gives

$$\begin{aligned} \varphi_s(x, 0) = I(x) &\implies \tilde{\varphi}_j = \frac{2}{L}\tilde{I}_j, \\ \dot{\varphi}_s(x, 0) = V(x) &\implies \tilde{\varphi}_j' = \frac{2}{\omega_j L}\tilde{V}_j. \end{aligned} \tag{3.42}$$

Exercise 3.1. Solve (3.22) for the Dirichlet boundary conditions $\varphi_s(x, 0) = I(x)$ and $\varphi_s(x, T) = F(x)$. Assume that the Fourier components of $I(x)$ and $F(x)$ satisfy the periodicity requirement $\cos(\omega_j T)\tilde{I}_j = \tilde{F}_j$ if $\sin(\omega_j T) = 0$.

For the case of an infinitely long string, the wavelengths and frequencies of the waves become continuous rather than the discrete wavelengths and frequencies of the standing waves in a finite-length string. The solution to the wave equation that is consistent with the Cauchy boundary conditions (3.38) is found using the method of Fourier transforms (2.96). The Cauchy boundary conditions are expressed as Fourier integrals,

$$I(x) = \int_{-\infty}^{\infty} \frac{dk}{2\pi}\, \tilde{I}(k) e^{ikx}, \quad V(x) = \int_{-\infty}^{\infty} \frac{dk}{2\pi}\, \tilde{V}(k) e^{ikx}, \tag{3.43}$$

where it is assumed that $\tilde{I}^*(-k) = \tilde{I}(k)$ and $\tilde{V}^*(-k) = \tilde{V}(k)$, so that both $I(x)$ and $V(x)$ are real. The solution is written as the general Fourier integral

$$\varphi_s(x, t) = \varphi_+(x - \tilde{v}t) + \varphi_-(x + \tilde{v}t) = \int \frac{dk}{2\pi} \left(\tilde{\varphi}_+(k)\, e^{ik(x-\tilde{v}t)} + \tilde{\varphi}_-(k)\, e^{ik(x+\tilde{v}t)}\right), \tag{3.44}$$

where $\tilde{\varphi}_\pm^*(-k) = \tilde{\varphi}_\pm(k)$ ensures that (3.44) is real valued. Satisfying the boundary conditions reduces to solving the system of equations

$$\begin{aligned} \varphi(x, 0) = I(x) &\implies \tilde{\varphi}_-(k) + \tilde{\varphi}_+(k) = \tilde{I}(k), \\ \dot{\varphi}(x, 0) = V(x) &\implies ik\tilde{v}\left(\tilde{\varphi}_-(k) - \tilde{\varphi}_+(k)\right) = \tilde{V}(k), \end{aligned} \tag{3.45}$$

which gives

$$\tilde{\varphi}_+(k) = \frac{1}{2}\left(\tilde{I}(k) + i\frac{\tilde{V}(k)}{k\tilde{v}}\right), \quad \tilde{\varphi}_-(k) = \frac{1}{2}\left(\tilde{I}(k) - i\frac{\tilde{V}(k)}{k\tilde{v}}\right). \tag{3.46}$$

Inverting these Fourier transforms gives the left- and right-moving solutions $\varphi_\pm(x \mp \tilde{v}t)$ that combine in (3.44) to satisfy the wave equation (3.22) and the boundary conditions (3.38). The reality conditions $\tilde{I}^*(-k) = \tilde{I}(k)$ and

$\tilde{V}^*(-k) = \tilde{V}(k)$ ensure that the reality conditions $\tilde{\varphi}^*_\pm(-k) = \tilde{\varphi}_\pm(k)$ also hold. Fourier solutions such as (3.41) and (3.44) are often referred to as *wave packets*, since they are a superposition of different wavelengths, which is effectively a packet of component waves. The reader is reminded that the principle of superposition holds because the equation of motion $\Box_{\tilde{v}}\varphi = 0$ is a linear partial differential equation.

3.5 Galilean relativity and wave solutions

Action (3.11) was derived using a coordinate system embedded in the stretched string, the *rest frame* of the string, denoted by $\mathcal{O}$. It is instructive to understand the behavior of the wave equation under the Galilean transformation of (2.27). The one-dimensional Galilean boost (2.20) and its inverse are given by

$$x' = x - v_o t, \quad x = x' + v_o t', \quad t' = t. \tag{3.47}$$

In the $\mathcal{O}'$ frame the medium of the string appears to be moving with a constant velocity $-v_o$. This follows from the fact that the origin in the rest frame of the string, $x = 0$, is at $x' = -v_o t'$ in the $\mathcal{O}'$ frame. If v_o is positive the medium of the string is moving leftward according to $\mathcal{O}'$, while if v_o is negative it appears to be moving rightward according to $\mathcal{O}'$. Combining (3.47) with the chain rule for partial derivatives gives

$$\begin{aligned}\frac{\partial}{\partial x} &= \frac{\partial x'}{\partial x}\frac{\partial}{\partial x'} + \frac{\partial t'}{\partial x}\frac{\partial}{\partial t'} = \frac{\partial}{\partial x'},\\ \frac{\partial}{\partial t} &= \frac{\partial x'}{\partial t}\frac{\partial}{\partial x'} + \frac{\partial t'}{\partial t}\frac{\partial}{\partial t'} = -v_o\frac{\partial}{\partial x'} + \frac{\partial}{\partial t'}.\end{aligned} \tag{3.48}$$

Using (3.48) in (3.24) along with the scalar property $\varphi_s(x', t') = \varphi_s(x, t)$ shows that the transformed wave equation is

$$\begin{aligned}\Box_{\tilde{v}}\varphi_s(x, t) &= \left(\frac{\partial}{\partial x} + \frac{1}{\tilde{v}}\frac{\partial}{\partial t}\right)\left(\frac{\partial}{\partial x} - \frac{1}{\tilde{v}}\frac{\partial}{\partial t}\right)\varphi_s(x, t)\\ &= \left[\left(1 - \frac{v_o}{\tilde{v}}\right)\frac{\partial}{\partial x'} + \frac{1}{v}\frac{\partial}{\partial t'}\right]\left[\left(1 + \frac{v_o}{\tilde{v}}\right)\frac{\partial}{\partial x'} - \frac{1}{\tilde{v}}\frac{\partial}{\partial t'}\right]\varphi_s(x', t')\\ &= \left(1 - \frac{v_o^2}{\tilde{v}^2}\right)\left(\frac{\partial}{\partial x'} + \frac{1}{(\tilde{v}-v_o)}\frac{\partial}{\partial t'}\right)\left(\frac{\partial}{\partial x'} - \frac{1}{(\tilde{v}+v_o)}\frac{\partial}{\partial t'}\right)\varphi_s(x', t') = 0.\end{aligned} \tag{3.49}$$

Subjecting (3.49) to the same analysis as that used for (3.24) shows that its solutions are right-moving waves with a velocity of propagation $\tilde{v}-v_o$ and left-moving waves with a velocity of propagation $\tilde{v}+v_o$. The $\mathcal{O}'$ observer will claim that the waves in the moving medium of the string have different velocities in different directions. This is consistent with Galilean relativity since the inertial observer $\mathcal{O}'$ sees the medium of the string moving at the velocity $-v_o$, so that a left-moving wave will have the relative velocity $-\tilde{v}-v_o = -(\tilde{v}+v_o)$, while a right-moving wave will have the relative velocity $\tilde{v}-v_o$. These are the same results as those obtained by using the inverse of transformation (3.47) and the scalar property of φ_s to transform a right-moving wave solution, $\varphi_s(x - \tilde{v}t) = \varphi_s(x' - (\tilde{v}-v_o)t')$, and a left-moving wave solution, $\varphi_s(x + \tilde{v}t) = \varphi_s(x' + (\tilde{v}+v_o)t')$.

This reveals an important aspect of the d'Alembertian and its wave solutions. In the case of waves traveling through the medium of a stretched string, the equation of motion (3.22) was derived using an action developed in the *rest frame of the medium.* The solutions in another inertial frame were obtained in (3.49) by applying the Galilean transformation to this equation of motion and solving it. This shows that the d'Alembertian is *not* invariant under a Galilean transformation. This result also follows from performing a Galilean transformation on the action (3.11). The Jacobian associated with (3.47) is unity, so that $\mathrm{d}x\,\mathrm{d}t = \mathrm{d}x'\,\mathrm{d}t'$. The action becomes

$$S[\varphi, \partial'\varphi] = \int_V \mathrm{d}x'\,\mathrm{d}t'\,\mathcal{L}'(\partial'\varphi, \varphi), \tag{3.50}$$

where, using $F = \mu\tilde{v}^2$, shows that the transformed Lagrangian is given by

$$\begin{aligned}\mathcal{L}'(\varphi, \partial'\varphi) &= \frac{1}{2}\mu\left(\frac{\partial\varphi(x', t')}{\partial t'}\right)^2 \\ &- \mu v_o\frac{\partial\varphi(x', t')}{\partial x'}\frac{\partial\varphi(x', t')}{\partial t'} - \frac{1}{2}\mu(\tilde{v}^2 - v_o^2)\left(\frac{\partial\varphi(x', t')}{\partial x'}\right)^2.\end{aligned} \tag{3.51}$$

A key observation is that the Lagrangian has changed *form* under a Galilean transformation, so that Lagrangian (3.11) is *not* a scalar under the Galilean transformation. The Euler–Lagrange equation associated with (3.51) can be written

$$\begin{aligned}\left.\left(\frac{\partial\mathcal{L}'}{\partial\varphi} - \partial'_\mu\frac{\partial\mathcal{L}'}{\partial(\partial'_\mu\varphi)}\right)\right|_{\varphi_s,\partial\varphi_s} &= -\frac{\partial}{\partial t'}\left(\mu\frac{\partial\varphi_s}{\partial t'} - \mu v_o\frac{\partial\varphi_s}{\partial x'}\right) \\ &\quad + \frac{\partial}{\partial x'}\left(\mu(\tilde{v}^2 - v_o^2)\frac{\partial\varphi_s}{\partial x'} - \mu v_o\frac{\partial\varphi_s}{\partial t'}\right) \\ = \mu\tilde{v}^2\left(1 - \frac{v_o^2}{\tilde{v}^2}\right)\left(\frac{\partial}{\partial x'} + \frac{1}{(\tilde{v}-v_o)}\frac{\partial}{\partial t'}\right)&\left(\frac{\partial}{\partial x'} - \frac{1}{(\tilde{v}+v_o)}\frac{\partial}{\partial t'}\right)\varphi_s = 0,\end{aligned} \tag{3.52}$$

which, up to the factor $\mu\tilde{v}^2 = F$, is identical to (3.49). The Euler–Lagrange equation obtained from the transformed action is therefore identical to the one obtained by transforming the equation of motion (3.22).

Exercise 3.2. Show that the solutions to (3.52) take the form $\varphi_s(x' + (v_o \pm \tilde{v})t)$ as long as $v_o \neq \pm\tilde{v}$, but that the solutions are undefined at $v_o = \pm\tilde{v}$.

Prior to the experimental study of electromagnetic waves, all known wave phenomena exhibited the property that an observer in motion *relative to the wave medium* observed velocities of propagation that obeyed Galilean relativity. This reinforced both the concept of waves as a disturbance that propagates through a medium and the belief that Galilean relativity governed the observed velocities of propagation. It will be seen in chapter 6 that Maxwell's equations give rise to an *electromagnetic wave* governed by the d'Alembertian equation with a velocity of propagation that matches the speed of light. However, it was experimentally found that observers in relative motion measured the *same velocity of light in all directions*, in direct contradiction to Galilean relativity. This was ultimately resolved by simultaneously replacing Galilean relativity with special relativity and abandoning the concept of a medium for light waves.

3.6 Momentum and energy in field theories

Because each segment of the stretched string has kinetic and potential energies when disturbed away from equilibrium, the waves associated with the string represent the motion of energy and momentum through the medium of the string. Newtonian mechanics reveals how these quantities manifest themselves in a field theory by viewing the string as a collection of point particles.

Combining the definition (2.107) of the canonical momentum with the discrete action (3.8) shows that the momentum canonically conjugate to coordinate φ_j of the jth segment is

$$p_j = \frac{\delta S[\varphi, \partial\varphi]}{\delta\dot{\varphi}_j(t)} = \frac{\partial \mathcal{L}(\varphi, \partial\varphi)}{\partial\dot{\varphi}_j} = \Delta x\ \mu\dot{\varphi}_j. \tag{3.53}$$

This is consistent with the mass of the jth segment being given by $m = \mu\ \Delta x$ and its velocity in the transverse direction being given by $\dot{\varphi}_j$. The Hamiltonian for the stretched string is found by applying the particle definition (2.108) for each segment and definitions (3.5) and (3.7) and summing over all N segments. The result is

$$H(t) = \sum_{j=0}^{N-1}\left(p_j\dot{\varphi}_j - K_j + U_j\right) = \sum_{j=0}^{N-1}\left(p_j\dot{\varphi}_j - \frac{1}{2}\mu\ \Delta x\ \dot{\varphi}_j^{\,2} + \frac{1}{2}\frac{F}{\Delta x}(\varphi_{j+1} - \varphi_j)^2\right). \tag{3.54}$$

The *canonical momentum density* of the jth string segment is then defined as

$$\pi_j \equiv \frac{p_j}{\Delta x} = \mu\dot{\varphi}_j, \tag{3.55}$$

which is consistent with (3.53). In the continuum limit expression (3.55) becomes

$$\pi(x, t) = \mu\dot{\varphi}(x, t), \tag{3.56}$$

and gives the momentum per unit length in the string in the direction of the field velocity $\dot{\varphi}$ at point x at time t. Combining (3.55) with relation (3.53) allows (3.54) to be written

$$H(t) = \sum_{j=0}^{N-1} \Delta x \left(\frac{\pi_j^2(t)}{2\mu} + \frac{1}{2} F \left(\frac{\varphi_{j+1}(t) - \varphi_j(t)}{\Delta x} \right)^2 \right). \tag{3.57}$$

Taking the continuum limit $N \to \infty$ gives $\pi_j(t) \to \pi(x, t) = \mu\dot{\varphi}(x, t)$ and, using both (3.13) and (3.14), expression (3.57) then becomes

$$\begin{aligned} H(t) &= \int_0^L \mathrm{d}x \left\{ \frac{\pi^2(x, t)}{2\mu} + \frac{1}{2} F \left(\frac{\partial \varphi(x, t)}{\partial x} \right)^2 \right\} \\ &= \int_0^L \mathrm{d}x \, (K(x, t) + U(x, t)) \equiv \int_0^L \mathrm{d}x \, \mathcal{H}(\pi, \varphi), \end{aligned} \tag{3.58}$$

where $\mathcal{H}(\pi, \varphi)$ is referred to as the *Hamiltonian density*, since it has units of energy per unit length. The Hamiltonian density $\mathcal{H}$ obtained from (3.58) is given by

$$\mathcal{H}(\pi, \varphi) = \frac{\pi^2(x, t)}{2\mu} + \frac{1}{2} F \left(\frac{\partial \varphi(x, t)}{\partial x} \right)^2. \tag{3.59}$$

For the case of a stretched string, $\mathcal{H}(\pi, \varphi)$ is the total mechanical energy density of the string, which is a combination of the kinetic energy density of its transverse motion and the potential energy density of its deviation from equilibrium.

While (3.59) was found by taking the continuum limit of the Newtonian particle expression, it is critical to generalize the definitions of the momentum density and the Hamiltonian density in such a way that they can be directly obtained from the continuum action (3.11). The canonical momentum density can be found from (3.11) by adapting (2.107) and noting that the continuum action (3.11) gives a result identical to (3.56) by using the functional derivative,

$$\pi(x, t) = \frac{\delta S[\varphi, \partial\varphi]}{\delta \dot{\varphi}(x, t)} = \frac{\partial \mathcal{L}(\varphi, \partial\varphi)}{\partial \dot{\varphi}(x, t)} = \mu \dot{\varphi}(x, t). \tag{3.60}$$

In the case of the stretched string, where $\varphi(x, t)$ is the vertical displacement at point x at time t, it is clear that (3.60) gives the vertical momentum density. As a result, the first step in (3.60) will be used to define the canonical momentum density *for the case of an arbitrary field theory*. This gives the general expression

$$\pi(x, t) = \frac{\delta S[\varphi, \partial\varphi]}{\delta \dot{\varphi}(x, t)} = \pi(\partial\varphi, \varphi). \tag{3.61}$$

For the case of a general field theory action it is assumed that (3.61) can be solved to give $\dot{\varphi}$ in terms of π, $\nabla\varphi$, and φ, with the solution denoted $\dot{\varphi} = \dot{\varphi}(\pi, \varphi, \nabla\varphi)$. This will require further discussion in later chapters, particularly for the case of electromagnetism. The continuum field theory version of the Legendre transformation (2.108) is then given by

$$\mathcal{H}(\pi, \varphi) = (\pi\dot{\varphi} - \mathcal{L}(\varphi, \partial\varphi)) \big|_{\dot{\varphi}(\pi, \varphi, \nabla\varphi)}, \tag{3.62}$$

where it is understood that the result of solving (3.61) for $\dot{\varphi}$ in terms of π, $\nabla\varphi$, and φ is substituted for all occurrences of $\dot{\varphi}$. Using result (3.56) gives $\dot{\varphi} = \pi/\mu$, and combining it with the Lagrangian density (3.11) shows that definition (3.62) does indeed reproduce the Hamiltonian density (3.59) for the stretched string.

The Legendre transformation (3.62) can be reversed to define the Hamiltonian version of the Lagrangian density in terms of π and φ,

$$\mathcal{L}(\pi, \varphi, \nabla\varphi) = \pi\dot{\varphi}(\pi, \varphi, \nabla\varphi) - \mathcal{H}(\pi, \varphi). \tag{3.63}$$

Using (3.63) to define the action allows the field theory version of Hamilton's equations (2.112) to be derived by varying the action around the extremum values φ_s and π_s. In what follows quantities associated with a solution to the equation of motion will be indicated by using the subscript s, so that $\mathcal{H}_s(x, t) \equiv \mathcal{H}(\pi_s, \varphi_s, \nabla\varphi_s)$. The variation of the stretched string action is then given by

$$\delta S[\pi_s, \varphi_s] = \int_V \mathrm{d}^2x\left(\dot{\varphi}_s\, \delta\pi + \pi_s\, \delta\dot{\varphi} - \frac{\partial\mathcal{H}_s}{\partial\pi_s}\,\delta\pi - \frac{\partial\mathcal{H}_s}{\partial\varphi_s}\,\delta\varphi - \frac{\partial\mathcal{H}_s}{\partial(\partial\varphi_s/\partial x)}\,\delta\left(\frac{\partial\varphi}{\partial x}\right)\right) = 0, \tag{3.64}$$

where $\partial\varphi_s/\partial x$ is present in the Hamiltonian density $\mathcal{H}_s$, while $\dot{\varphi}_s$ is absent due to the substitution $\dot{\varphi}_s = \pi_s/\mu$ from (3.61). The particular form (3.64) reflects the assumption that the field theory under consideration has only first derivatives of φ_s in the Hamiltonian density $\mathcal{H}_s$. Using $\delta(\partial\varphi/\partial x) = \partial(\delta\varphi)/\partial x$, $\delta\dot{\varphi} = \partial(\delta\varphi)/\partial t$, integrating by parts, and applying the boundary conditions $\delta\varphi(0, t) = \delta\varphi(L, t) = \delta\varphi(x, 0) = \delta\varphi(x, T) = 0$ allows (3.64) to be written

$$\delta S[\pi_s, \varphi_s] = \int_V \mathrm{d}^2x\left[\left(\dot{\varphi}_s - \frac{\partial\mathcal{H}_s}{\partial\pi_s}\right)\delta\pi - \left(\dot{\pi}_s + \frac{\partial\mathcal{H}_s}{\partial\varphi_s} - \frac{\partial}{\partial x}\frac{\partial\mathcal{H}_s}{\partial(\partial\varphi_s/\partial x)}\right)\delta\varphi\right] = 0. \tag{3.65}$$

Since $\delta\varphi$ and $\delta\pi$ are independent variations, $\delta S[\pi_s, \varphi_s] = 0$ yields the field theory versions of Hamilton's equations (2.112),

$$\dot{\varphi}_s(x, t) = \frac{\partial\mathcal{H}_s(x, t)}{\partial\pi_s(x, t)}, \tag{3.66}$$

$$\dot{\pi}_s(x, t) = \frac{\partial}{\partial x}\frac{\partial\mathcal{H}_s(x, t)}{\partial(\partial\varphi_s(x, t)/\partial x)} - \frac{\partial\mathcal{H}_s(x, t)}{\partial\varphi_s(x, t)}. \tag{3.67}$$

Applying (3.66) to the string Hamiltonian density (3.59) reproduces result (3.60), $\dot{\varphi}_s = \pi_s/\mu$. Applying (3.67) to (3.59) and using (3.60) gives

$$\begin{aligned}\dot{\pi}_s(x, t) &= \mu\frac{\partial^2\varphi_s(x, t)}{\partial t^2} = \frac{\partial}{\partial x}\left(F\,\frac{\partial\varphi_s(x, t)}{\partial x}\right)\\ \Longrightarrow\ F\left(\frac{\partial^2}{\partial x^2} - \frac{\mu}{F}\frac{\partial^2}{\partial t^2}\right)\varphi_s(x, t) &= F\left(\frac{\partial^2}{\partial x^2} - \frac{1}{\tilde{v}^2}\frac{\partial^2}{\partial t^2}\right)\varphi_s(x, t) = 0,\end{aligned} \tag{3.68}$$

which is identical to the Euler–Lagrange equation (3.22).

One of the important outgrowths of defining the Hamiltonian density (3.62) is the quantification of the flow of energy and momentum associated with the wave solutions $\varphi_s(x, t)$ to (3.22). For the case of the stretched string, this can be quantified by examining the arrival of a wave disturbance $\varphi_s(x, t)$ at a section of the string that is initially at equilibrium and therefore has $\mathcal{H}_s = 0$. During the infinitesimal time interval Δt a right-moving wave will travel the distance $\Delta x = \tilde{v}\,\Delta t$. The amount of energy delivered to the interval between x and $x + \Delta x$ at time t is $\Delta E_s(x, t) = \mathcal{H}_s(x, t)\,\Delta x = \tilde{v}\,\mathcal{H}_s(x, t)\,\Delta t$, where $\mathcal{H}_s(x, t)$ is the energy density of the wave arriving at x at the time t. The *instantaneous power* $P_s(x, t)$ of the arriving wave passing point x at time t is therefore given by the rate of energy delivery,

$$P_s(x, t) = \frac{\partial E_s(x, t)}{\partial t} = \lim_{\Delta t \to 0} \frac{\Delta E_s(x, t)}{\Delta t} = \tilde{v}\,\mathcal{H}_s(x, t). \tag{3.69}$$

Because all wave phenomena are associated with the flow of energy, the power relationship (3.69) will hold as long as the Hamiltonian density is evaluated using a traveling wave solution to the field theory.

Several important relations are revealed by evaluating the derivatives of the Hamiltonian density (3.59) using an arbitrary solution of the equation of motion (3.22) and relation (3.56). The time derivative gives

$$\begin{aligned} \frac{\partial \mathcal{H}_s}{\partial t} &= \frac{\partial \pi_s}{\partial t}\frac{\pi_s}{\mu} + F\frac{\partial \varphi_s}{\partial x}\frac{\partial^2 \varphi_s}{\partial x\,\partial t} = \mu\frac{\partial^2 \varphi_s}{\partial t^2}\frac{\partial \varphi_s}{\partial t} + F\frac{\partial \varphi_s}{\partial x}\frac{\partial^2 \varphi_s}{\partial x\,\partial t} \\ &= \left(\mu\frac{\partial^2 \varphi_s}{\partial t^2} - F\frac{\partial^2 \varphi_s}{\partial x^2}\right)\frac{\partial \varphi_s}{\partial t} + \frac{\partial}{\partial x}\left(F\frac{\partial \varphi_s}{\partial x}\frac{\partial \varphi_s}{\partial t}\right). \end{aligned} \tag{3.70}$$

Because φ_s satisfies the wave equation (3.22), the first term on the second line of (3.70) vanishes, so that (3.70) reduces to

$$\frac{\partial \mathcal{H}_s(x, t)}{\partial t} + \tilde{v}^2\frac{\partial \mathcal{P}_s(x, t)}{\partial x} = 0, \tag{3.71}$$

where the relation $F = \mu\tilde{v}^2$ was used and the quantity $\mathcal{P}_s(x, t)$ is defined as

$$\mathcal{P}_s(x, t) \equiv -\mu\,\frac{\partial \varphi_s(x, t)}{\partial t}\frac{\partial \varphi_s(x, t)}{\partial x} = -\pi_s(x, t)\,\frac{\partial \varphi_s(x, t)}{\partial x}. \tag{3.72}$$

Result (3.71) is a *conservation law* for the *total energy* $H_s(t)$ present in a string wave solution, which is defined for the string wave as

$$H_s(t) = \int_0^L \mathrm{d}x\,\mathcal{H}_s(x, t). \tag{3.73}$$

Using (3.71) gives

$$\begin{aligned} \frac{\mathrm{d}H_s(t)}{\mathrm{d}t} &= \int_0^L \mathrm{d}x\,\frac{\partial \mathcal{H}_s(x, t)}{\partial t} = -\tilde{v}^2\int_0^L \mathrm{d}x\,\frac{\partial \mathcal{P}_s(x, t)}{\partial x} \\ &= -\tilde{v}^2(\mathcal{P}_s(L, t) - \mathcal{P}_s(0, t)) = 0. \end{aligned} \tag{3.74}$$

The final step follows from the boundary conditions $\dot{\varphi}_s(L, t) = \dot{\varphi}_s(0, t) = 0$, so that $\mathcal{P}_s(L, t) = \mathcal{P}_s(0, t) = 0$.

Similarly, applying the spatial derivative to the Hamiltonian density and using $\pi_s = \mu\dot{\varphi}$ gives

$$\begin{aligned} \frac{\partial \mathcal{H}_s}{\partial x} &= \frac{\partial \pi_s}{\partial x}\frac{\pi_s}{\mu} + F\frac{\partial \varphi_s}{\partial x}\frac{\partial^2 \varphi_s}{\partial x^2} = \mu\frac{\partial^2 \varphi_s}{\partial t\, \partial x}\frac{\partial \varphi_s}{\partial t} + F\frac{\partial \varphi_s}{\partial x}\frac{\partial^2 \varphi_s}{\partial x^2} \\ &= \left(F\frac{\partial^2 \varphi_s}{\partial x^2} - \mu\frac{\partial^2 \varphi_s}{\partial t^2}\right)\frac{\partial \varphi_s}{\partial x} + \frac{\partial}{\partial t}\left(\mu\frac{\partial \varphi_s}{\partial t}\frac{\partial \varphi_s}{\partial x}\right). \end{aligned} \tag{3.75}$$

Again, the first term on the second line of (3.75) vanishes, and definition (3.72) allows (3.75) to be written

$$\frac{\partial \mathcal{P}_s(x, t)}{\partial t} = -\frac{\partial \mathcal{H}_s(x, t)}{\partial x}. \tag{3.76}$$

The next step is demonstrate that (3.76) is the generalization of Newton's second law for point particles, $\dot{\boldsymbol{p}} = -\nabla U(\boldsymbol{r})$, to waves.

Insight into the nature of $\mathcal{P}_s$ as defined by (3.72) can be obtained from the specific case of the stretched string. It starts by noting that $\partial\varphi_s/\partial x$ is dimensionless for the string. As a result, the physical units of $\mathcal{P}_s$ are the same as those of the canonical momentum density, $\pi_s = \mu\, \partial\varphi_s/\partial t$. Further insight is gained by recalling that the solutions φ_s of the equation of motion fall into two general categories: right movers $\varphi_+(x - \tilde{v}t)$ and left movers $\varphi_-(x + \tilde{v}t)$. In the rest frame of the string, (3.25) and (3.29) show that

$$\begin{aligned} \pi_+(x - \tilde{v}t) &= \mu\frac{\partial \varphi_+(x - \tilde{v}t)}{\partial t} = -\mu\tilde{v}\,\frac{\partial \varphi_+(x - \tilde{v}t)}{\partial x}, \\ \pi_-(x + \tilde{v}t) &= \mu\frac{\partial \varphi_-(x + \tilde{v}t)}{\partial t} = +\mu\tilde{v}\,\frac{\partial \varphi_-(x + \tilde{v}t)}{\partial x}. \end{aligned} \tag{3.77}$$

Returning (3.77) to (3.72) gives

$$\mathcal{P}_\pm = -\pi_\pm\frac{\partial \varphi_\pm}{\partial x} = \pm\mu\tilde{v}\left(\frac{\partial \varphi_\pm}{\partial x}\right)^2. \tag{3.78}$$

The rest-frame result (3.78) shows that a right-moving wave corresponds to a positive value for $\mathcal{P}_s$, while a left-moving wave corresponds to a negative value for $\mathcal{P}_s$. Since $\mathcal{P}_s$ has units of momentum, this demonstrates that there is a momentum density traveling with the wave at the same velocity as the wave. Further insight is obtained by noting that $\partial\varphi_s/\partial x$ is the spatial slope of the string and therefore is the same as $\tan\theta$, where θ is the angle of the string with respect to the x-axis. For the case of small-amplitude waves, it follows that $\partial\varphi_s/\partial x = \tan\theta \approx \sin\theta$. Since the canonical momentum is, by assumption, perpendicular to the x-axis, it follows that $-\pi_s\, \partial\varphi_s/\partial x \approx -\pi_s \sin\theta$ is the component of the canonical momentum *tangent to the string* and always points in the same direction as the wave motion.

For waves in the string, $\mathcal{P}_s$ therefore represents the *longitudinal momentum density* and $-\partial\mathcal{H}_s(x, t)/\partial x$ represents the *longitudinal force density* in the string. It is important to note that both the kinetic energy density and the potential energy density contribute to (3.78). The two results of (3.71) and (3.76) represent conservation laws. The next step is to connect (3.78) to the energy density of a wave solution. Both results of (3.77) give the rest-frame result

$$\left(\frac{\partial\varphi_\pm}{\partial t}\right)^2 = \tilde{v}^2\left(\frac{\partial\varphi_\pm}{\partial x}\right)^2. \tag{3.79}$$

Using (3.79) in the string Hamiltonian density (3.59) along with $F = \mu\tilde{v}^2$ gives

$$\mathcal{H}_\pm = \frac{1}{2}\mu\left(\frac{\partial\varphi_\pm}{\partial t}\right)^2 + \frac{1}{2}F\left(\frac{\partial\varphi_\pm}{\partial x}\right)^2 = \frac{1}{2}\mu\tilde{v}^2\left(\frac{\partial\varphi_\pm}{\partial x}\right)^2 + \frac{1}{2}\mu\tilde{v}^2\left(\frac{\partial\varphi_\pm}{\partial x}\right)^2 = \mu\tilde{v}^2\left(\frac{\partial\varphi_\pm}{\partial x}\right)^2, \tag{3.80}$$

which is always a positive quantity. Substituting (3.80) into (3.78) gives

$$\mathcal{P}_\pm = \pm\frac{\mathcal{H}_\pm}{\tilde{v}}. \tag{3.81}$$

Although result (3.81) has been derived in the context of a mechanical wave in a stretched string, it will be shown to hold in other field theories where there are disturbances traveling at a constant velocity.

3.7 The stress–energy tensor

Results (3.71), (3.72), and (3.76) were obtained for the specific case of the stretched string. Results (3.78) and (3.79) used specific properties of solutions to the equation of motion in the *rest frame* of the string. It is therefore natural to ask whether there are underlying structures in general field theories that will give rise to similar results for other physical systems. While a complete answer to this must await further developments, it is an important start to note that there is a *space-time quantity*, denoted by $T_\nu^{\ \mu}$, that encapsulates both results for the scalar field φ of the stretched string. For a scalar field theory governed by the Lagrangian density $\mathcal{L}(\varphi, \partial\varphi)$, this quantity is defined as

$$T_\nu^{\ \mu}(\varphi) = \frac{\partial\mathcal{L}(\varphi, \partial\varphi)}{\partial(\partial_\mu\varphi)}\partial_\nu\varphi - \delta_\nu^{\ \mu}\mathcal{L}(\varphi, \partial\varphi), \tag{3.82}$$

where the reader is reminded that the Greek indices in (3.82) range from zero to one for the case of the one-dimensional stretched string, so that

$$\begin{aligned}\partial_0 &\equiv \frac{\partial}{\partial t},\\ \partial_1 &\equiv \frac{\partial}{\partial x}.\end{aligned} \tag{3.83}$$

Form (3.82) is easily generalized to higher-dimensional spatial manifolds and the presence of more field variables. In doing so it will be assumed that $\mathcal{L}$ has no explicit dependence on x. Instead, there is only an implicit dependence on x through $\varphi(x, t)$ and $\partial_\mu\varphi(x, t)$. If T^μ_ν is evaluated using a solution $\varphi_s(x, t)$ to the Euler–Lagrange equation (3.21), it follows from the chain rule that

$$\begin{aligned}
\partial_\mu T^\mu_\nu(\varphi_s) &= \left(\partial_\mu \frac{\partial\mathcal{L}(\varphi_s, \partial\varphi_s)}{\partial(\partial_\mu\varphi_s)}\right)\partial_\nu\varphi_s + \frac{\partial\mathcal{L}(\varphi_s, \partial\varphi_s)}{\partial(\partial_\mu\varphi_s)}\partial_\mu\partial_\nu\varphi_s - \partial_\nu\mathcal{L}(\varphi_s, \partial\varphi_s) \\
&= \left(\frac{\partial\mathcal{L}(\varphi_s, \partial\varphi_s)}{\partial\varphi_s}\right)\partial_\nu\varphi_s + \frac{\partial\mathcal{L}(\varphi_s, \partial\varphi_s)}{\partial(\partial_\mu\varphi_s)}\partial_\nu\partial_\mu\varphi_s - \frac{\partial\mathcal{L}(\varphi_s, \partial\varphi_s)}{\partial\varphi_s}\partial_\nu\varphi_s \\
&\quad - \frac{\partial\mathcal{L}(\varphi_s, \partial\varphi_s)}{\partial(\partial_\mu\varphi_s)}\partial_\nu\partial_\mu\varphi_s = 0.
\end{aligned} \tag{3.84}$$

Result (3.84) will now be shown to encapsulate both results (3.71) and (3.76). This starts by noting that definition (3.82) contains the Hamiltonian density (3.62) and the momentum density (3.72), which follows by combining (3.82) with the definitions (3.61), (3.62), and (3.72) to find

$$\begin{aligned}
T^0_0(\varphi_s) &= \frac{\partial\mathcal{L}(\varphi_s, \partial\varphi_s)}{\partial\dot{\varphi}_s}\dot{\varphi}_s - \mathcal{L}(\varphi_s, \partial\varphi_s) = \pi_s\dot{\varphi}_s - \mathcal{L}(\varphi_s, \partial\varphi_s) = \mathcal{H}_s(x, t), \\
T^0_1(\varphi_s) &= \frac{\partial\mathcal{L}(\varphi_s, \partial\varphi_s)}{\partial\dot{\varphi}_s}\frac{\partial\varphi_s}{\partial x} = \pi_s\frac{\partial\varphi_s}{\partial x} = -\mathcal{P}_s(x, t).
\end{aligned} \tag{3.85}$$

For the case that the specific string Lagrangian (3.12) is used in (3.82), it follows that

$$\begin{aligned}
T^1_0(\varphi_s) &= \frac{\partial\mathcal{L}(\varphi_s, \partial\varphi_s)}{\partial(\partial\varphi_s/\partial x)}\dot{\varphi}_s = -F\frac{\partial\varphi_s}{\partial x}\dot{\varphi}_s = \frac{F}{\mu}\mathcal{P}_s(x, t) = \tilde{v}^2\mathcal{P}_s(x, t), \\
T^1_1(\varphi_s) &= \frac{\partial\mathcal{L}(\varphi_s, \partial\varphi_s)}{\partial(\partial\varphi_s/\partial x)}\frac{\partial\varphi_s}{\partial x} - \mathcal{L}(\varphi_s, \partial\varphi_s) = -\frac{1}{2}F\left(\frac{\partial\varphi_s}{\partial x}\right)^2 - \frac{1}{2}\mu\dot{\varphi}_s^2 \\
&= -\mathcal{H}_s(x, t).
\end{aligned} \tag{3.86}$$

Using (3.85) and (3.86) in (3.84) gives two equations,

$$\partial_\mu T^\mu_0 = \partial_0 T^0_0 + \partial_1 T^1_0 = \frac{\partial\mathcal{H}_s}{\partial t} + \tilde{v}^2\frac{\partial\mathcal{P}_s}{\partial x} = 0, \tag{3.87}$$

$$\partial_\mu T^\mu_1 = \partial_0 T^0_1 + \partial_1 T^1_1 = -\frac{\partial\mathcal{P}_s}{\partial t} - \frac{\partial\mathcal{H}_s}{\partial x} = 0. \tag{3.88}$$

Results (3.87) and (3.88) are identical to the previous results (3.71) and (3.76).

For the case of the stretched string result (3.87) is understood to represent the conservation of energy plus momentum, since integrating it over any subvolume of the string gives the rate of change of the energy E in the subvolume,

$$\frac{\partial E}{\partial t} = \int_a^b \mathrm{d}x\, \partial_0\mathcal{H}_s = \int_a^b \mathrm{d}x\, \partial_0 T^0_0 = -\int_a^b \mathrm{d}x\, \partial_1 T^1_0 = -T^1_0(b) + T^1_0(a). \tag{3.89}$$

For the case of the string, (3.86) shows that the change in the energy present in the subvolume of the string is given by the endpoint contributions of

$$T_0^1 = \tilde{v}^2 \mathcal{P}_s. \tag{3.90}$$

This relates the energy in the string to the momentum flowing through the string. For that reason, the momentum density of the field will be identified as

$$\mathcal{P}_s^j = \frac{1}{\tilde{v}^2} T_0^j. \tag{3.91}$$

Alternatively, it follows from (3.85) that

$$T_j^0 = -\mathcal{P}_{sj}, \tag{3.92}$$

which also allows the identification of the field momentum.

Definition (3.82) has the advantage that it allows T_ν^μ to be derived from a general Lagrangian in such a way that it automatically obeys $\partial_\mu T_\nu^\mu(\varphi_s) = 0$. Furthermore, it allows the momentum density of the field to be generalized to a case where the spatial manifold is described by n-dimensional position vectors $\boldsymbol{r}$,

$$\mathcal{P}_s^j(\boldsymbol{r}, t) = \frac{1}{\tilde{v}^2} T_0^j(\varphi_s) = \frac{1}{\tilde{v}^2} \frac{\partial \mathcal{L}(\varphi_s, \partial\varphi_s)}{\partial(\partial\varphi_s(\boldsymbol{r}, t)/\partial x^j)} \dot{\varphi}_s(\boldsymbol{r}, t), \tag{3.93}$$

so that an arbitrary number of spatial dimensions can now be accommodated.

Result (3.84) then provides a valuable set of *conservation laws* specific to the theory under consideration. To see this, (3.84) is written $\partial_0 T^0{}_\nu + \partial_j T^j{}_\nu = 0$, where j ranges over the n spatial coordinates. Integrating this result over a *spatial* volume V and using the divergence theorem (1.165) gives

$$\frac{\partial}{\partial t} \int_V \mathrm{d}^n x \; T_\nu^0 = -\int_V \mathrm{d}^n x \; \partial_i T_\nu^i = -\int_{S(V)} \mathrm{d}S_i \; T_\nu^i, \tag{3.94}$$

resulting in a surface integral on the right-hand side of (3.94). This surface integral represents the flow or *flux* of a physical quantity, such as momentum for the stretched string, through the surface bounding the volume V. This shows that the change over time of the quantity in the volume integral of T_ν^0 is identical to the flux of the quantity T_ν^i through the closed surface of that volume. Result (3.94) shows that this surface flux accounts for the change in the quantity T_ν^0 contained in the interior of the volume V. This is the essence of a conservation law. In his seminal work on the behavior of continuous media, Cauchy described the generalized force per unit area or *stress* acting on the medium in terms of a second-rank tensor in three spatial dimensions known as the *Cauchy stress tensor*. In the case of the stretched string, the space-time object T_ν^μ encapsulates the effect of a stress, the force per unit length given by $\partial_1 T_1^1 = -\partial\mathcal{H}_s(x, t)/\partial x$, on the rate of change in the momentum density, $\partial_0 T_1^0 = -\partial\mathcal{P}_s/\partial t$, present in the wave solution. Definition (3.82) therefore serves as the precursor to the *stress–energy tensor*, often also referred to as the *energy–momentum tensor*, associated with relativistic field theories.

While it is tempting to treat the indices of quantity (3.82) as the components of a second-rank mixed tensor, in a non-relativistic field theory there are a number of barriers to doing so. First, T_i^0 and T_0^i have different physical units, since $\dot{\varphi}$ and $\partial_i\varphi$ differ by a factor from the units of velocity. Therefore, T_i^0 and T_0^i will typically differ by a factor of velocity squared that is indeterminate for the general expression (3.82). Identifying this velocity requires the details of the equation of motion derived from the specific Lagrangian density. A second impediment to identifying T_ν^μ as the components of a second-rank mixed tensor is the indeterminate relationship of T_ν^μ to $T_{\mu\nu}$ or $T^{\mu\nu}$, i.e. it is unclear how the contravariant and covariant components of such a tensor are related. Result (1.74) showed that this relationship requires knowledge of the metric tensor $\boldsymbol{g}$, and for this case, such a metric tensor would require treating both space and time as coordinates with unit vectors. While this approach will be developed when special relativity is considered in chapter 6, Newtonian physics treats position as a *coordinate* and time as a *parameter*. Although position and time both appear in the Galilean transformation (2.27), only the spatial manifold has a well-defined metric in Newtonian physics. It is also not possible to use the Euler–Lagrange equation to show that $\partial^\nu T_\nu^\mu = 0$ from definition (3.82). This is a reflection of the lack of a relationship between ∂^0 and ∂_0. For the moment, the expression $\partial_\mu T_\nu^\mu = 0$ will be treated as the correct generalization of the conservation and stress laws to an *arbitrary scalar field theory* $\varphi(\boldsymbol{r}, t)$ involving both position and time.

These impediments can be made concrete by recalling that the Lagrangian density (3.11) for the stretched string is *not* a scalar under a Galilean transformation. Instead, it transforms from (3.11) to (3.51).

Exercise 3.3. Using (3.51) show that an observer $\mathcal{O}'$ moving to the right relative to the string medium with speed v_o will claim that the T_0^0 and T_0^1 components of (3.82) are given by

$$\begin{aligned} T_0^0(\mathcal{L}') &= \frac{1}{2}\mu\left(\frac{\partial\varphi(x', t')}{\partial t'}\right)^2 + \frac{1}{2}\mu(\tilde{v}^2 - v_o^2)\left(\frac{\partial\varphi(x', t')}{\partial x'}\right)^2, \\ T_0^1(\mathcal{L}') &= -\mu v_o\left(\frac{\partial\varphi(x', t')}{\partial t'}\right)^2 - \mu(\tilde{v}^2 - v_o^2)\left(\frac{\partial\varphi(x', t')}{\partial x'}\right)\left(\frac{\partial\varphi(x', t')}{\partial t'}\right). \end{aligned} \tag{3.95}$$

Show that the components of (3.95) satisfy

$$\partial'_\mu T_0^\mu(\varphi(x', t')) = \frac{\partial}{\partial t'}T_0^0(\mathcal{L}') + \frac{\partial}{\partial x'}T_0^1(\mathcal{L}') = 0, \tag{3.96}$$

as long as the function $\varphi(x', t')$ takes the form of a solution to (3.52), $\varphi_s(x' + (v_o \pm \tilde{v})t')$, where φ_s is an arbitrary function.

3.8 Static sources and Green's function techniques

In this section the result of attaching a static *source of disturbance* to the stretched string is analyzed from a field theory point of view. In its simplest version, a static pointlike mass m is fixed to the string at the horizontal position $x = a$, where $0 < a < L$. The weight mg of the point mass causes the string to sag, resulting in a displacement φ of the string away from equilibrium. The point particle therefore acts as a *source* of the field φ, which measures the vertical displacement of the string away from horizontal. This mechanism is similar to those that occur in other field theories, where point masses and charges act as a source of the field in the space around them, allowing the field to transmit force and energy to other masses or charges.

This problem can be solved using elementary physics techniques and the result can then be compared to that obtained using field theory techniques. It is assumed the string sags by a distance h under the weight mg, while its ends remain fixed and its two segments remain straight. As mentioned in the first section of this chapter, the two straight segments depicted in figure 3.3 will require *different tensions* to be at equilibrium. If the tension in the right segment is F, then the tension F' in the left segment must satisfy the relation

$$F' \cos\theta_1 = F \cos\theta_2 \implies F' = F\frac{\cos\theta_2}{\cos\theta_1} = F\left(\frac{L}{a} - 1\right)\sqrt{\frac{a^2 + h^2}{(L-a)^2 + h^2}}. \tag{3.97}$$

Clearly, if $a \approx L/2$, then $F' \approx F$. However, as $a \to 0$ or $a \to L$, a significant difference will occur between the tensions in the two segments. Because field theory action (3.11) was developed under the assumption that the tension was constant throughout the string, such cases will not be considered in order to compare the results of the two methods. The sum of the two vertical components supplied by the string's tensions F and F' must match the weight mg. It is also important to note that the weight of the string itself is being ignored, so this result will be valid only if $\mu L \ll m$. For a small downward displacement h the left segment forms an angle ϕ_1 that satisfies $\sin\phi_1 \approx h/a$, while the right segment is at an angle ϕ_2, satisfying $\sin\phi_2 \approx h/(L-a)$. The two vertical components of the string's tension are therefore

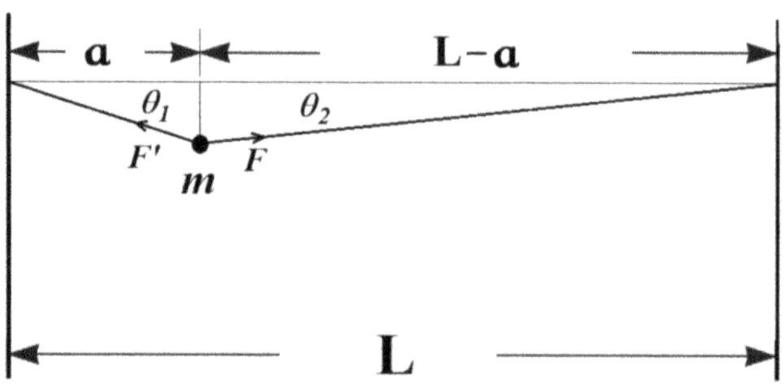

Figure 3.3. The stretched string with a point mass attached.

$F' \sin\phi_1 = F'h/a$ and $F \sin\phi_2 = Fh/(L - a)$. Assuming a is such that $F' \approx F$, the law of statics gives the distance the mass sags downward by as

$$\frac{Fh}{a} + \frac{Fh}{L - a} = mg \implies h = \frac{mg}{F}a\left(1 - \frac{a}{L}\right), \tag{3.98}$$

which is valid only for $a \approx L/2$ and $mg \ll F$.

The same physical situation can be modeled using a field theory treatment by adding a *source term* to the Lagrangian density (3.11). A point mass m at the location $x = a$ can be modeled by the mass density $\rho(x) = m\,\delta(x - a)$, where $\delta(x - a)$ is the Dirac delta. In that regard, the Dirac delta $\delta(x - a)$ can be viewed as a function that is sharply peaked at the point $x = a$ and which has units of inverse length. Recalling that $\varphi(x, t)$ measures the relative vertical height of the string at point x, the potential energy density associated with $\rho(x)$ is nothing more than the standard gravitational potential energy density $U_g(x, t) = \rho(x)\,g\,\varphi(x, t)$. Since it is a potential energy density, the Lagrangian density of (3.11) is modified by subtracting this additional potential energy density, giving

$$\mathcal{L} = \frac{1}{2}\mu\left(\frac{\partial\varphi}{\partial t}\right)^2 - \frac{1}{2}F\left(\frac{\partial\varphi}{\partial x}\right)^2 - \rho g\varphi. \tag{3.99}$$

The Euler–Lagrange equation for the field in the presence of the point source is therefore modified to

$$- \partial_\mu\frac{\partial\mathcal{L}(\varphi_s, \partial\varphi_s)}{\partial(\partial_\mu\varphi_s)} + \frac{\partial\mathcal{L}(\varphi_s, \partial\varphi_s)}{\partial\varphi_s} = 0 \implies \left(F\frac{\partial^2}{\partial x^2} - \mu\frac{\partial^2}{\partial t^2}\right)\varphi_s(x, t) - mg\,\delta(x - a) = 0. \tag{3.100}$$

In order to compare the results of solving (3.100) to result (3.98), a time-independent or *static solution* φ_s to (3.100) is sought, so that (3.100) becomes

$$\frac{\partial^2\varphi_s(x, a)}{\partial x^2} = \frac{mg}{F}\,\delta(x - a). \tag{3.101}$$

The solution to (3.101) must be a continuous function of x, since the string cannot break, and it must satisfy the boundary conditions on the string, $\varphi(0) = \varphi(L) = 0$. The key to finding the solution to (3.101) is the property of the Heaviside step function $\theta(x)$ given by (2.45). The solution to (3.101) consistent with continuity and the boundary conditions is given by

$$\varphi_s(x, a) = -\frac{mg}{F}\left\{\left(1 - \frac{a}{L}\right)x\,\theta(a - x) + \frac{a}{L}(L - x)\,\theta(x - a)\right\}. \tag{3.102}$$

This is demonstrated by using property (2.48) of the Heaviside step function, $\partial\theta(x - a)/\partial x = \delta(x - a)$ and $\partial\theta(a - x)/\partial x = -\delta(x - a)$, to show that the first derivative of (3.102) is given by

$$\frac{\partial\varphi_s(x,a)}{\partial x} = -\frac{mg}{F}\left\{\left(1-\frac{a}{L}\right)\theta(a-x) - \frac{a}{L}\,\theta(x-a) - (x-a)\,\delta(x-a)\right\}. \quad (3.103)$$

Since the Dirac delta obeys $(x - a)\,\delta(x - a) = 0$, the third term in (3.103) vanishes. Using the Heaviside property (2.48) again shows that the second derivative of (3.102) satisfies,

$$\frac{\partial^2\varphi_s(x,a)}{\partial x^2} = \frac{mg}{F}\left\{\left(1-\frac{a}{L}\right)\delta(x-a) + \frac{a}{L}\delta(x-a)\right\} = \frac{mg}{F}\delta(x-a), \quad (3.104)$$

which is the required relation (3.101). Recalling that (3.102) is the height of the string in the presence of the point source at $x = a$, the solution φ_s describes two straight segments of string that start with zero height at $x = 0$ and $x = L$ and meet at $x = a$, where the height is $\varphi_s(x = a, a) = -mga(L - a)/FL$, exactly matching the result (3.98) obtained from the statics method and the restriction $x \approx L/2$. Rather than simply demonstrating that (3.102) satisfies (3.101), it is possible to derive (3.102) using the methods of complex variables. This will be discussed in greater detail in chapter 8.

Result (3.103) shows that the energy of the point mass and string receives a contribution from the tension in the string. Integrating by parts, combined with the boundary conditions on φ_s and (3.101), gives

$$\begin{aligned} E_F &= \int_0^L \mathrm{d}x\,\frac{1}{2}F\left(\frac{\partial\varphi_s}{\partial x}\right)^2 = \frac{1}{2}F\int_0^L \mathrm{d}x\,\frac{\partial}{\partial x}\left(\varphi_s\frac{\partial\varphi_s}{\partial x}\right) - \frac{1}{2}F\int_0^L \mathrm{d}x\,\varphi_s\frac{\partial^2\varphi_s}{\partial x^2} \\ &= -\frac{1}{2}mg\int_0^L \mathrm{d}x\,\varphi_s(x)\delta(x-a) = -\frac{1}{2}mg\varphi_s(a) = \frac{m^2g^2a}{2F}\left(1-\frac{a}{L}\right), \end{aligned} \quad (3.105)$$

where the Heaviside property $\theta(0) = 1/2$ was used. The contribution of the gravitational potential energy is

$$E_g = mg\phi_s(a) = -\frac{m^2g^2a}{F}\left(1-\frac{a}{L}\right), \quad (3.106)$$

where the mass of the string has been ignored. The total mechanical energy of the static point mass and the stretched string is therefore given by

$$E = E_F + E_g = -\frac{m^2g^2a}{2F}\left(1-\frac{a}{L}\right). \quad (3.107)$$

The total energy is negative, since work will be required to restore the string and the point mass to the static horizontal equilibrium configuration.

The usefulness of (3.102) can be extended by noting that it enables a solution for the shape of the string in the event that an *arbitrary* mass density $\rho(x)$ is attached to the string. *Assuming the tension in the string remains uniform*, (3.102) can be used to solve the static Euler–Lagrange equation given by

$$\frac{\partial^2 \varphi_p(x)}{\partial x^2} = \frac{g}{F}\rho(x). \tag{3.108}$$

This is accomplished by noting that (3.102) provides a *Green's function* $G(x, a)$, which for the case of the stretched string is defined as the function that satisfies

$$\frac{\partial^2}{\partial x^2}G(x, a) = \frac{g}{F}\delta(x - a). \tag{3.109}$$

The solution to (3.109) is obtained from (3.102),

$$G(x, a) = \frac{1}{m}\varphi_s(x, a). \tag{3.110}$$

Result (3.110) satisfies both (3.109) *and* the boundary conditions $G(0, a) = G(L, a) = 0$. The Green's function is defined to yield a Dirac delta when the differential operator of the equation of motion acts upon it in a manner that is consistent with the boundary conditions, and in the static case of the string, this differential operator is simply $\partial^2/\partial x^2$.

The solution φ_p to (3.108) that is consistent with the boundary conditions is then given by

$$\varphi_p(x) = \int_0^L \mathrm{d}a\; G(x, a)\rho(a). \tag{3.111}$$

This follows from using (3.109) and the property of the Dirac delta to obtain

$$\frac{\partial^2}{\partial x^2}\varphi_p(x) = \int_0^L \mathrm{d}a\; \frac{\partial^2}{\partial x^2}G(x, a)\rho(a) = \frac{g}{F}\int_0^L \mathrm{d}a\; \delta(x - a)\,\rho(a) = \frac{g}{F}\rho(x), \tag{3.112}$$

so that $\varphi_p(x)$ satisfies (3.108) as well as the boundary conditions $\varphi_p(0) = \varphi_p(L) = 0$ for an arbitrary mass density $\rho(x)$. Green's function techniques are commonly used to solve field theories when interactions are present. For the case of time-dependent sources their construction in a manner consistent with the boundary conditions is intimately tied to the use of Fourier integrals and complex analysis. As mentioned earlier, a more detailed presentation of this method is given in chapter 8.

As a simple example, (3.111) can be applied to the case of a string with a uniform mass per unit length of μ to determine the shape of the string due to its weight. Using (3.102) in (3.110) and evaluating (3.111) gives the solution

$$\begin{aligned} \varphi_s(x) = \mu\int_0^L \mathrm{d}a\; G(x, a) = -\frac{\mu g}{F}\left\{\int_0^x \mathrm{d}a\; a\left(1 - \frac{x}{L}\right)\right. \\ \left. + \int_x^L \mathrm{d}a\; x\left(1 - \frac{a}{L}\right)\right\} = -\frac{\mu g}{2F}(L - x)x. \end{aligned} \tag{3.113}$$

Result (3.113) is a parabola that is zero at $x = 0$ and $x = L$. It extends to its maximum distance below the horizontal of $-\mu g L^2/8F$ at the center of the string,

$x = L/2$. For the case that $\mu g L^2/8F \ll L$, the weight of the string has a negligible effect on its shape.

Exercise 3.4. Consider the case where two equal point masses m are located on the string at $x = a$ and $x = L - a$, where $a < L/2$. Use (3.111) and $\rho(x) = m\,\delta(x - a) + m\,\delta(x - L + a)$ to show that the height of the two point masses is given by

$$\varphi_s(x = a) = \varphi_s(x = L - a) = -\frac{mga}{F}. \tag{3.114}$$

Result (3.114) can be used to model force transmission by a field. In order to do this, results (3.105) and (3.106) are extended to the case that there are two point masses. It is straightforward to show that the total energy of the two-mass and string configuration is given by

$$E_T = E_F + E_g = \frac{1}{2}mg(\varphi_s(x = a) + \varphi_s(x = L - a)) = -\frac{m^2g^2a}{F}. \tag{3.115}$$

Exercise 3.5. Verify result (3.115).

The total energy (3.115) of the two-mass system is negative and becomes more negative as a increases. It follows that the mass at $x = a$ experiences a constant horizontal force given by

$$F(a) = -\frac{\partial E_T}{\partial a} = \frac{m^2g^2}{F}, \tag{3.116}$$

which is attractive toward the mass at $x = L - a$, since it is positively directed. Similarly, the mass at $x = L - a$ experiences a force given by

$$F(L - a) = -\frac{\partial E_T}{\partial(L - a)} = -\frac{\partial a}{\partial(L - a)}\frac{\partial E_T}{\partial a} = \frac{\partial E_T}{\partial a} = -\frac{m^2g^2}{F}, \tag{3.117}$$

which is attractive toward the mass at $x = a$, since it is negatively directed. This shows that the field φ acts to transmit a force of attraction between the two point masses. This is similar to the mechanism for the transmission of electric force between charges and is discussed again in chapter 9.

3.9 The catenary, the Beltrami identity, and constraints

As was the case in the previous section, field theories are often analyzed for static solutions. These solutions provide useful information about the behavior of the medium and the nature of disturbances away from equilibrium that do not propagate. These solutions typically occur in field theories where there are *interaction terms* in the Lagrangian density. In the case of a *stretched string* considered in the previous sections, the string was assumed to maintain both a uniform mass density μ and a uniform tension F. Clearly, this is valid only if the disturbance away from equilibrium is small, which was the assumption made in deriving the potential (3.7). In the previous section the gravitational interaction term $V(\varphi) = \mu g \varphi(x)$ was added to the potential energy density. Its presence led to the stationary shape of a parabola for the string, described by equation (3.113). However, this result was contingent on the assumptions of uniform tension and uniform mass density, which in turn require only small deviations from equilibrium.

The case of large deviations from equilibrium can be considered for the case of a *suspended string*. The tension will not be uniform as the string sags under the effect of its weight. Instead, the tension will vary with position and will be determined by the equations of motion. This is depicted in figure 3.4.

In this section the suspended string will be assumed to maintain the same total length ℓ as it sags. Of course, this requires that $\ell \geqslant L$, where L is the horizontal distance between the two end points. However, this assumption allows the mass density $\mu = M/\ell$, where M is the total mass of the string, to be treated as uniform throughout the string. It also fixes the potential energy associated with stretching a segment of the string, $U - F\,\Delta\ell$, at zero. The forces acting on each segment of the suspended string are only tension and the gravitational force. For such a case it will be shown by several methods that the stationary shape of the string deviates from the simple parabola of (3.113) to the related shape of a *catenary*,

$$\varphi_s(x) = -\frac{F}{\mu g}\left[\cosh\left(\frac{\mu g}{F}\left(x - \frac{L}{2}\right)\right) - \cosh\left(\frac{\mu g L}{2F}\right)\right]. \tag{3.118}$$

In expression (3.118), F is the tension at $x = L/2$, which is the point where the slope of the string, $\mathrm{d}\varphi_s(x)/\mathrm{d}x = \tan\theta(x)$, vanishes. Since only static solutions are being considered and the string is one-dimensional, $\mathrm{d}/\mathrm{d}x$ and $\partial/\partial x$ are interchangeable.

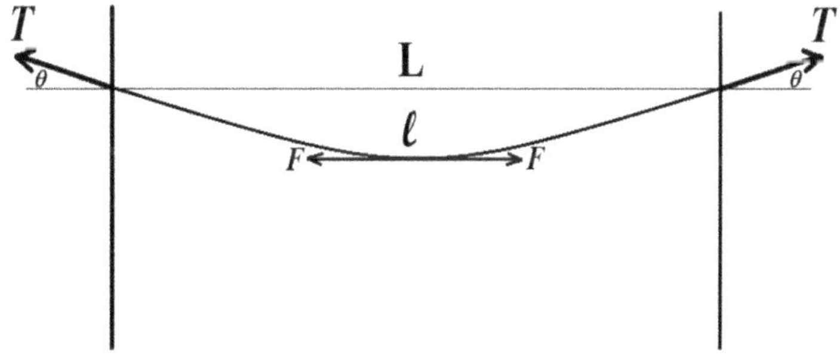

Figure 3.4. The suspended string with variable tension and fixed length and density.

Exercise 3.6. Show that (3.118) becomes the parabola of (3.113) in the limit $\mu g/F \to 0$ by using the approximation for the hyperbolic cosine, $\cosh\alpha \approx 1 + \frac{1}{2}\alpha^2$, for small α.

The catenary (3.118) can be derived by applying the Newtonian criteria for static equilibrium. This requires the tension to depend on position in order to support the weight of the string. The analysis starts by noting that the angle of the slope at point x, denoted by $\theta(x)$, is available from the relationship $\mathrm{d}\varphi_s(x)/\mathrm{d}x = \tan\theta(x)$. This gives $\cos\theta(x)$ from the trigonometric identity,

$$\sqrt{1 + \left(\frac{\mathrm{d}\varphi_s(x)}{\mathrm{d}x}\right)^2} = \sqrt{1 + \tan^2\theta(x)} = \frac{1}{\cos\theta(x)}. \tag{3.119}$$

Similarly, $\sin\theta(x)$ is given by

$$\sin\theta(x) = \tan\theta(x)\cos\theta(x) = \frac{\mathrm{d}\varphi_s(x)}{\mathrm{d}x}\left(1 + \left(\frac{\mathrm{d}\varphi_s(x)}{\mathrm{d}x}\right)^2\right)^{-1/2}. \tag{3.120}$$

Static equilibrium of an arbitrary segment of the string requires that the tension $F(x)$ at $x \neq L/2$ must yield a horizontal component equal and opposite to F. Using (3.119) yields $F(x)$ in terms of F and $\cos\theta(x)$,

$$F(x)\cos\theta(x) = F \implies F(x) = \frac{F}{\cos\theta(x)} = F\sqrt{1 + \left(\frac{\mathrm{d}\varphi_s(x)}{\mathrm{d}x}\right)^2}. \tag{3.121}$$

At $x = L/2$ the string is horizontal so that $\mathrm{d}\varphi_s(x)/\mathrm{d}x = 0$, and the tension possesses no vertical component. As a result, the vertical component of the tension at $x \neq L/2$, given by $F(x)\sin\theta(x)$, must be equal and opposite to the weight of the segment from $L/2$ to x. The weight of this segment is given by $\mu g s(x)$, where $s(x)$ is the length of the segment, given by

$$s(x) = \int_{L/2}^{x} \mathrm{d}s = \int_{L/2}^{x} \frac{\mathrm{d}z}{\cos\theta(z)} = \int_{L/2}^{x} \mathrm{d}z\sqrt{1 + \left(\frac{\mathrm{d}\varphi_s(z)}{\mathrm{d}z}\right)^2}. \tag{3.122}$$

The requirement for vertical equilibrium of the segment is obtained by combining (3.120), (3.121), and (3.122), which gives

$$F(x)\sin\theta(x) = F\frac{\mathrm{d}\varphi_s(x)}{\mathrm{d}x} = \mu g s(x) = \mu g\int_{L/2}^{x} \mathrm{d}z\sqrt{1 + \left(\frac{\mathrm{d}\varphi_s(z)}{\mathrm{d}z}\right)^2}. \tag{3.123}$$

Differentiating (3.123) gives the *second-order catenary equation*,

$$\frac{\mathrm{d}^2\varphi_s(x)}{\mathrm{d}x^2} = \frac{\mu g}{F}\sqrt{1 + \left(\frac{\mathrm{d}\varphi_s(x)}{\mathrm{d}x}\right)^2}. \tag{3.124}$$

It is straightforward to show that (3.118) solves (3.124) in a manner that is consistent with the boundary conditions $\varphi_s(0) = \varphi_s(L) = 0$ and $\mathrm{d}\varphi_s(x)/\mathrm{d}x\,|_{x=L/2} = 0$.

Using (3.118) in relation (3.121) shows that the tension at each end of the string is given by

$$T = F(0) = F(L) = F\sqrt{1 + \sinh^2\left(\frac{\mu g L}{2F}\right)} = F\cosh\left(\frac{\mu g L}{2F}\right). \tag{3.125}$$

The total length ℓ of the string is given by (3.122),

$$\begin{aligned}\ell = 2s(L) &= 2\int_{L/2}^{L} \mathrm{d}z\,\sqrt{1 + \left(\frac{\mathrm{d}\varphi_s(z)}{\mathrm{d}z}\right)^2} = 2\int_{L/2}^{L} \mathrm{d}z\,\cosh\left(\frac{\mu g}{F}\left(z - \frac{L}{2}\right)\right)\\ &= \frac{2F}{\mu g}\sinh\left(\frac{\mu g L}{2F}\right).\end{aligned} \tag{3.126}$$

Result (3.126) determines F from the three independent quantities μ, ℓ, and L. Using the Taylor series for the hyperbolic sine, $\sinh\alpha = \alpha + \alpha^3/6 + \ldots$, shows that the $\mu g L/2F \to 0$ limit of (3.126) corresponds to the length $\ell = L$. Combining (3.123) with (3.126) gives

$$F(0)\sin\theta(0) = F\frac{\mathrm{d}\varphi_s}{\mathrm{d}x} = F\sinh\left(\frac{\mu g L}{2F}\right) = \frac{1}{2}\mu g \ell, \tag{3.127}$$

which shows that the vertical component of the tension at the end points is one-half of the string weight.

It is instructive to demonstrate how result (3.118) is derived using a variational approach. The first step is to demonstrate the Beltrami identity, which is particularly useful for finding static solutions to the Euler–Lagrange equation in one spatial dimension where the Lagrangian involves only $\varphi(x)$ and $\mathrm{d}\varphi(x)/\mathrm{d}x$. For the case of a static solution $\varphi_s(x)$, the Euler–Lagrange equation (3.21) becomes

$$\frac{\partial\mathcal{L}(\varphi_s, \varphi_s')}{\partial\varphi_s(x)} - \frac{\mathrm{d}}{\mathrm{d}x}\frac{\partial\mathcal{L}(\varphi_s, \varphi_s')}{\partial\varphi_\mu'(x)} = 0, \tag{3.128}$$

where the notation $\varphi_s'(x) = \mathrm{d}\varphi_s(x)/\mathrm{d}x$ and $\varphi_s''(x) = \mathrm{d}^2\varphi_s(x)/\mathrm{d}x^2$ is now in use. The Beltrami identity for static solutions follows from (3.128) and the observation that

$$\begin{aligned}\frac{\mathrm{d}}{\mathrm{d}x}\left(\varphi_s'(x)\frac{\partial\mathcal{L}(\varphi_s, \varphi_s')}{\partial\varphi_s'(x)}\right) &= \varphi_s''(x)\frac{\partial\mathcal{L}(\varphi_s, \varphi_s')}{\partial\varphi_s'(x)} + \varphi_s'(x)\frac{\mathrm{d}}{\mathrm{d}x}\frac{\partial\mathcal{L}(\varphi_s, \varphi_s')}{\partial\varphi_s'(x)}\\ &= \varphi_s''(x)\frac{\partial\mathcal{L}(\varphi_s, \varphi_s')}{\partial\varphi_s'(x)} + \varphi_s'(x)\frac{\partial\mathcal{L}(\varphi_s, \varphi_s')}{\partial\varphi_s(x)} = \frac{\mathrm{d}}{\mathrm{d}x}\mathcal{L}(\varphi_s, \varphi_s').\end{aligned} \tag{3.129}$$

Result (3.129) immediately gives the Beltrami identity,

$$\frac{\mathrm{d}}{\mathrm{d}x}\left(\mathcal{L}(\varphi_s, \varphi_s') - \varphi_s'(x)\frac{\partial\mathcal{L}(\varphi_s, \varphi_s')}{\partial\varphi_s'(x)}\right) = 0. \tag{3.130}$$

The Beltrami identity is structurally similar to the conservation of the Hamiltonian in a classical point particle system with x playing the role of time t. Similarly to the Hamiltonian, its value lies in reducing the original second-order static Euler–Lagrange equation to a first-order differential equation, since (3.130) gives

$$\mathcal{L}(\varphi_s, \varphi_s') - \varphi_s'(x)\frac{\partial\mathcal{L}(\varphi_s, \varphi_s')}{\partial\varphi_s'(x)} = k, \tag{3.131}$$

where k is a constant.

The catenary solution (3.118) can be derived using the Beltrami identity and the static action

$$\int_0^L \mathrm{d}x\, \mathcal{L}(\varphi_s, \varphi_s', \lambda) = \int_0^L \mathrm{d}x\left\{\lambda\left(\sqrt{1 + \left(\varphi_s'(x)\right)^2} - \frac{\ell}{L}\right) - \mu g\varphi_s(x)\sqrt{1 + \left(\varphi_s'(x)\right)^2}\right\}. \tag{3.132}$$

The term in the Lagrangian of (3.132) proportional to λ implements the constraint that the length of the string is ℓ. The quantity λ is a *Lagrange multiplier* and, as in point particle mechanics, it is treated as a variable. In this case it has no spatial or time dependence and has units of force. Its Euler–Lagrange equation gives

$$\int_0^L \mathrm{d}x\, \frac{\partial}{\partial\lambda}\mathcal{L}(\varphi_s, \varphi_s', \lambda) = 0 \implies \int_0^L \mathrm{d}x\, \sqrt{1 + \left(\varphi_s'(x)\right)^2} = \ell, \tag{3.133}$$

which simply enforces the length constraint. The action is therefore independent of the value of λ. However, λ is present in the equation of motion for φ_s and can be chosen to satisfy the boundary conditions as long as it is a constant that is independent of x and t. The second term in (3.132) is the gravitational potential energy with a correction to the length of the segment $\mathrm{d}s(x)$ at x due to the slope of the string. This follows from $\mathrm{d}s(x) = \mathrm{d}x/\cos\theta(x)$, where $\cos\theta(x)$ is given by (3.119). The mass of the infinitesimal segment at x is therefore given by $\mathrm{d}m(x) = \mu\, \mathrm{d}s(x) = \mu\, \mathrm{d}x/\cos\theta(x)$. It should be reiterated that there is no energy contribution to (3.132) from the tension in the suspended string, since it retains its fixed length.

Using variational techniques with (3.132) determines the shape of the string that extremizes the gravitational potential energy that is consistent with a constant length and the boundary conditions. Applying the Beltrami identity (3.130) to the Lagrangian appearing in (3.132) gives the first-order differential equation

$$\mathcal{L}(\varphi_s, \varphi_s', \lambda) - \varphi_s'(x)\frac{\partial\mathcal{L}(\varphi_s, \varphi_s', \lambda)}{\partial\varphi_s'(x)} = \frac{\lambda - \mu g\varphi_s(x)}{\sqrt{1 + (\varphi_s'(x))^2}} - \frac{\lambda\ell}{L} = k. \tag{3.134}$$

Defining the new variable $\psi_s(x)$, where

$$\varphi_s(x) = \psi_s(x) + \frac{\lambda}{\mu g}, \tag{3.135}$$

transforms equation (3.134) to

$$-\frac{\mu g \psi_s(x)}{\sqrt{1+\left(\psi_s'(x)\right)^2}} = k + \frac{\lambda \ell}{L} \equiv F. \tag{3.136}$$

The constant F has units of force and will be determined from the length constraint (3.133). Either by solving (3.136) for ψ_s' and then integrating or by simply using the properties of the hyperbolic cosine and sine, it follows that the solution to (3.136) is given by

$$\psi_s(x) = -\frac{F}{\mu g}\cosh\left(\frac{\mu g}{F}(x - x_o)\right), \tag{3.137}$$

where x_o is a constant. Using (3.135) gives the original variable $\varphi_s(x)$ as

$$\varphi_s(x) = -\frac{F}{\mu g}\cosh\left(\frac{\mu g}{F}(x - x_o)\right) + \frac{\lambda}{\mu g}. \tag{3.138}$$

Satisfying the two boundary conditions $\varphi_s(0) = \varphi_s(L) = 0$ requires that $x_o = L/2$ and

$$\lambda = F\cosh\left(\frac{\mu g L}{2F}\right). \tag{3.139}$$

Using (3.138) in the length constraint (3.133) gives

$$\begin{aligned}\ell &= \int_0^L \mathrm{d}x\sqrt{1+\sinh^2\left(\frac{\mu g}{F}\left(x - \frac{L}{2}\right)\right)} = \int_0^L \mathrm{d}x\cosh\left(\frac{\mu g}{F}\left(x - \frac{L}{2}\right)\right)\\ &= \frac{2F}{\mu g}\sinh\left(\frac{\mu g L}{2F}\right),\end{aligned} \tag{3.140}$$

which determines F in terms of ℓ, μ, and L identically to the statics result (3.126). Variational techniques have therefore reproduced the catenary solution (3.118), although they do not immediately identify F as the tension in the string at its midpoint.

Exercise 3.7. Noting that $M = \mu\ell$ is the mass of the string, use relation (3.140) to show that F is determined by the equivalent relation

$$\frac{Mg}{2F} = \sinh\left(\frac{MgL}{2F\ell}\right). \tag{3.141}$$

For the case that $L = \ell$, show that (3.141) requires $Mg/2F \to 0$.

3.10 Functional derivatives and Poisson brackets

The total energy (3.58) of the string configuration at time t is given by the *spatial integral* of the energy density $\mathcal{H}(x, t)$. A similar result for the total momentum of the configuration is given by the spatial integral of the momentum density $\mathcal{P}(x, t)$ given by (3.72). This means that the total energy and momentum of the string configuration are functionals of the field variable $\varphi(x, t)$ and its canonical momentum $\pi(x, t)$. In developing field theories, it is useful to define a form for the functional derivative that is consistent with the spatial functional. This enables the Poisson bracket formalism (2.124) to be used and results such as (2.125) to be expressed in terms of functional derivatives and extended to continuum field theory.

The starting point is to note that the string Hamiltonian was initially developed as the sum of individual segments treated by variables using the definitions for the kinetic energy K_j in (3.5) and the potential energy U_j in (3.7) and summed in (3.57). In that regard, the field variables $\varphi_j(t)$ and $\pi_k(t)$ are independent, satisfying

$$\begin{aligned}\frac{\partial\varphi_k(t)}{\partial\varphi_j(t)} &= \delta^j{}_k,\\ \frac{\partial p_k(t)}{\partial p_j(t)} &= \delta^j{}_k,\\ \frac{\partial\varphi_k(t)}{\partial p_j(t)} &= \frac{\partial p_k(t)}{\partial\varphi_j(t)} = 0.\end{aligned} \tag{3.142}$$

Prior to taking the continuum limit $\Delta x \to 0$, the discrete Hamiltonian of (3.57) has the form

$$H[\varphi, \pi](t) = \sum_{j=0}^{N-1} \Delta x \left(\frac{1}{2}\frac{\pi_j^2(t)}{\mu} + \frac{1}{2}F\left(\frac{\varphi_{j+1}(t) - \varphi_j(t)}{\Delta x}\right)^2 \right). \tag{3.143}$$

For the case that $x' = x_k$, it is now possible to define the action of the functional derivative in terms of the partial derivative by defining the functional derivative as

$$\begin{aligned}\frac{\delta H[\varphi, \pi](t)}{\delta\pi(x', t)} &= \left(\frac{1}{\Delta x}\frac{\partial}{\partial\pi_k(t)}\right)\sum_{j=0}^{N-1} \Delta x \left(\frac{1}{2}\frac{\pi_j^2(t)}{\mu} + \frac{1}{2}F\left(\frac{\varphi_{j+1}(t) - \varphi_j(t)}{\Delta x}\right)^2 \right)\\ &= \frac{\pi_k(t)}{\mu} = \frac{\pi(x', t)}{\mu}.\end{aligned} \tag{3.144}$$

The functional derivative for the case of time-dependent spatial functionals is therefore given by

$$\frac{\delta}{\delta\pi(x_k, t)} = \frac{1}{\Delta x}\frac{\partial}{\partial\pi(x_k, t)}. \tag{3.145}$$

Using this definition with (3.142) gives

$$\frac{\delta p_j(t)}{\delta p_k(t)} = \frac{1}{\Delta x}\frac{\partial p_j(t)}{\partial p_k(t)} = \frac{1}{\Delta x}\delta_j^k. \tag{3.146}$$

In the limit $\Delta x \to 0$, the right-hand side of (3.146) changes to

$$\lim_{\Delta x \to 0} \frac{1}{\Delta x} \delta^k{}_j = \delta(x_k - x_j), \tag{3.147}$$

becoming the spatial Dirac delta function. This can be shown to be the case by examining the sum

$$\sum_{j=0}^{N-1} \Delta x \left(\frac{1}{\Delta x} \delta^k{}_j \right) = 1, \tag{3.148}$$

which is identical to the behavior of the delta function,

$$\int_0^L \mathrm{d}x \, \delta(x - x_k) = 1. \tag{3.149}$$

In effect, (3.146) is a *discrete* version of the delta function which is singular at $j = k$ as $\Delta x \to 0$ and zero elsewhere. As a result, in the continuum limit where $\Delta x \to 0$, it follows that

$$\begin{aligned} \frac{\delta \varphi(x, t)}{\delta \varphi(y, t)} &= \frac{\delta \pi(x, t)}{\delta \pi(y, t)} = \delta(x - y), \\ \frac{\delta \varphi(x, t)}{\delta \pi(y, t)} &= \frac{\delta \pi(x, t)}{\delta \pi(y, t)} = 0. \end{aligned} \tag{3.150}$$

This allows the definition of the continuum functional derivative for spatial functionals similarly to (2.64):

$$\frac{\delta F[\varphi, \pi](t)}{\delta \varphi(x, t)} = \lim_{\epsilon \to 0} \frac{1}{\epsilon} (F[\varphi(y, t) + \epsilon \delta(y - x), \pi(y, t)] - F[\varphi(y, t), \pi(y, t)]). \tag{3.151}$$

This definition of the functional derivative is used in the continuum limit of expressions such as (3.143).

Exercise 3.8. For the case that the functional is given by

$$F[\varphi](t) = \int \mathrm{d}x \, \frac{1}{2} \left(\frac{\partial \varphi(x, t)}{\partial x} \right)^2, \tag{3.152}$$

show that definition (3.151) gives

$$\frac{\delta F[\varphi](t)}{\delta \varphi(y, t)} = - \frac{\partial^2 \varphi(y, t)}{\partial y^2}. \tag{3.153}$$

The use of definition (3.151) allows the Poisson bracket (2.124) to be adapted to field theory. If $A[\varphi, \pi](t)$ and $B[\varphi, \pi](t)$ are two spatial functionals of the field variables, then their Poisson bracket is defined as

$$\begin{aligned}&\{A[\varphi, \pi](t), B[\varphi, \pi](t)\}\\&= \int \mathrm{d}x \left(\frac{\delta A[\varphi, \pi](t)}{\delta \varphi(x, t)} \frac{\delta B[\varphi, \pi](t)}{\delta \pi(x, t)} - \frac{\delta A[\varphi, \pi](t)}{\delta \pi(x, t)} \frac{\delta B[\varphi, \pi](t)}{\delta \varphi(x, t)} \right) \Bigg|_{\varphi=\varphi_s},\end{aligned} \tag{3.154}$$

where it is assumed that there is one spatial degree of freedom for the fields and the Poisson bracket is evaluated using a solution φ_s to the Euler–Lagrange equation of motion. For example, using the string variable $A[\varphi, \pi](t) = \varphi(y, t)$ and the string Hamiltonian $B[\varphi, \pi](t) = H(t)$, where the Hamiltonian is the continuum form of (3.143), given by

$$H[\varphi, \pi](t) = \int \mathrm{d}y \left(\frac{\pi^2(y, t)}{\mu} + F\left(\frac{\partial \varphi(y, t)}{\partial y} \right)^2 \right), \tag{3.155}$$

gives the Poisson bracket

$$\{\varphi(z, t), H[\varphi, \pi](t)\} = \int \mathrm{d}x\ \delta(x - z) \frac{\pi(x, t)}{\mu} = \frac{\pi(z, t)}{\mu} = \dot{\varphi}(z, t), \tag{3.156}$$

where relation (3.55) was used. This shows that the string Hamiltonian (3.155) is associated with the time evolution of the field φ, just as the Hamiltonian in mechanics gives the time derivative of position when used in the Poisson bracket. Similarly, using (3.153), it follows that

$$\begin{aligned}\{\pi_s(y, t), H[\varphi, \pi](t)\} &= \int \mathrm{d}x\ \delta(x - y)\, F\, \frac{\partial^2 \varphi_s(x, t)}{\partial x^2} = \frac{F}{v_p^2} \ddot{\varphi}_s(y, t) = \mu \ddot{\varphi}_s(y, t)\\&= \dot{\pi}_s(y, t).\end{aligned} \tag{3.157}$$

> **Exercise 3.9.** Show that
>
> $$\{\varphi_s(y, t), \pi_s(z, t)\} = \delta(y - z). \tag{3.158}$$

The Poisson bracket for classical theory carries over into quantum field theory, where it becomes the commutator of the fields and their canonical momenta. With slight modifications, the Poisson bracket formalism can be adapted to all the field theories that will be treated in this book.

References and recommended further reading

Monographs that cover the topic of continuum mechanics include

- Lanczos C 1970 *The Variational Principles of Mechanics* 4th edn (New York: Dover)
- Goldstein H, Poole C and Safko J 2000 *Classical Mechanics* 3rd edn (San Francisco, CA: Addison-Wesley)
- Thornton S and Marion J 2004 *Classical Dynamics of Particles and Systems* (Boston, MA: Cengage)
- Soper D 2008 *Classical Field Theory* (New York: Dover)

Monographs dealing exclusively with continuum mechanics include

- Truesdell C and Noll W 2003 *The Non-Linear Field Theories of Mechanics* 3rd edn (Berlin: Springer)

Monographs that focus on solving the wave equation include

- Freegarde T 2013 *Introduction to the Physics of Waves* (Cambridge: Cambridge University Press)

Mathematical monographs presenting methods for solving partial differential equations include

- Courant R and Hilbert D 1989 *Methods of Mathematical Physics* vol II (New York: Wiley)
- Byron F and Fuller R 1992 *Mathematics of Classical and Quantum Physics* (New York: Dover)

IOP Publishing

Classical Field Theory and the Stress–Energy Tensor (Second Edition)

Mark S Swanson

Chapter 4

Newtonian fluid dynamics

A fluid is, by definition, a substance that flows, which means that it responds to pressure differences and forces with the motion of its constituent mass. Despite being a familiar phenomenon on the everyday level, fluid dynamics has remained a mathematically difficult and incomplete subject since the seminal work of Euler in the 18th century. Using Euler's approach, fluid dynamics consistent with Newtonian physics can be developed by treating the velocity $\boldsymbol{v}$ of the fluid at position $\boldsymbol{r}$ at time t as the *velocity field* $\boldsymbol{v}(\boldsymbol{r}, t)$. This allows a straightforward derivation of the *Euler equation* and the *Navier–Stokes equation*, with their numerous applications to fluid flow. However, such an approach ignores the *atomic or microscopic properties* of the matter making up the fluid, instead using the *macroscopic or bulk properties* of the fluid, such as density, viscosity, and bulk modulus. The alternative is to use a statistical approach in which a distribution function $f(\boldsymbol{v}, \boldsymbol{r}, t)$ gives the number of identical pointlike atoms or molecules with the velocity $\boldsymbol{v}$ at point $\boldsymbol{r}$ at time t. Such a microscopic statistical approach leads to the *Boltzmann–Vlasov equation* and allows the bulk properties of fluids, such as pressure and viscosity, to be defined in terms of the distribution function augmented by the mean free path between atomic collisions. Such an approach also allows the eventual derivation of the Navier–Stokes equation by using the *mean or average velocity* of the atoms as the Eulerian velocity field $\boldsymbol{v}(\boldsymbol{r}, t)$. In this monograph the Newtonian approach to fluid dynamics will be followed, since, like the stretched string considered in chapter 3, it allows an action approach to deriving the equations of motion for fluid flow. In turn, this allows the stress–energy tensor for perfect fluids to be defined from the action, in keeping with the central themes of this monograph. The diligent reader should consult the bibliography to find one of the excellent monographs where the statistical approach is presented.

doi:10.1088/978-0-7503-3455-6ch4 © IOP Publishing Ltd 2022

4.1 Fluid flow from Newtonian physics

At the macroscopic level, the two generic forms for a fluid are liquid and gas. A liquid is characterized by maintaining its volume while flowing, while a gas is characterized by expanding to fill the volume available to it. These two *phases* are governed by the *equation of state* for the substance. In the macroscopic approach to fluid flow, liquids are often characterized by being incompressible, so that their density remains constant, while gases are compressible and have variable densities. The macroscopic approach to fluid flow therefore requires using the *scalar density field*, $\rho(\boldsymbol{r}, t)$, which gives the mass per unit volume at position $\boldsymbol{r}$ at time t. It describes the flow in terms of a *velocity* or *vector field*, $\boldsymbol{v}(\boldsymbol{r}, t)$, which gives the velocity of the fluid at position $\boldsymbol{r}$ at time t. For simplicity, the flow occurs in a three-dimensional Euclidean manifold and Cartesian coordinates are typically employed. Using the summation convention and the Cartesian direction vectors $\mathbf{e}_j$ gives $\boldsymbol{r} = \mathbf{e}_j r^j$. The velocity field is also expressed in terms of the direction vectors at $\boldsymbol{r}$ in the manifold, so that $\boldsymbol{v}(\boldsymbol{r}, t) = \mathbf{e}_j v^j(\boldsymbol{r}, t)$. A scalar pressure field, $P(\boldsymbol{r}, t)$, is present in the fluid, as well as the possibility of other forces, such as gravity and internal friction characterized by viscosity. The viscosity of a fluid converts the mechanical energy of flow into internal heat, and the pressure field $P(\boldsymbol{r}, t)$ requires knowledge of the *equation of state* for the material making up the fluid. The dependence on the thermodynamic properties of the substance that is flowing complicates the analysis considerably.

Macroscopic fluid dynamics is formulated by considering an infinitesimal volume element of fluid, given by $\mathrm{d}^3r = \mathrm{d}x\,\mathrm{d}y\,\mathrm{d}z$ in Cartesian coordinates, located at the position $\boldsymbol{r}$. The element of mass $\mathrm{d}m(\boldsymbol{r}, t)$ contained in the volume element located at $\boldsymbol{r}$ at time t is obtained from the density $\rho(\boldsymbol{r}, t)$ at position $\boldsymbol{r}$ at time t as

$$\mathrm{d}m(\boldsymbol{r}, t) = \rho(\boldsymbol{r}, t)\,\mathrm{d}^3r. \tag{4.1}$$

The acceleration of the mass element at position $\boldsymbol{r}$ at time t is denoted by $\boldsymbol{a}(\boldsymbol{r}, t)$, while the total force acting on the mass element at $\boldsymbol{r}$ at time t is denoted by $\boldsymbol{F}(\boldsymbol{r}, t)$. Newton's second law relates the acceleration to the total force according to

$$\boldsymbol{F}(\boldsymbol{r}, t) = \mathrm{d}m(\boldsymbol{r}, t)\,\boldsymbol{a}(\boldsymbol{r}, t) = \rho(\boldsymbol{r}, t)\,\boldsymbol{a}(\boldsymbol{r}, t)\,\mathrm{d}^3r. \tag{4.2}$$

The goal is to express (4.2) in terms of the velocity field $\boldsymbol{v}(\boldsymbol{r}, t)$, and this requires expressing the acceleration of the mass element in terms of $\boldsymbol{v}(\boldsymbol{r}, t)$. It is tempting to identify the acceleration of the mass element $\boldsymbol{a}$ with the partial derivative of velocity $\boldsymbol{v}$ with respect to time. However, doing so overlooks the fact that the mass element has moved to a *new position* $\boldsymbol{r}' = \boldsymbol{r} + \boldsymbol{v}(\boldsymbol{r}, t)\,\Delta t$ after the time interval Δt has passed. Computing the acceleration of the mass element $\boldsymbol{a}(\boldsymbol{r}, t)$ from the velocity field $\boldsymbol{v}(\boldsymbol{r}, t)$ therefore requires using its new position to find its final velocity, which is given by $\boldsymbol{v}(\boldsymbol{r}', t')$. This gives the instantaneous acceleration of the mass element,

$$\begin{aligned} \boldsymbol{a}(\boldsymbol{r}, t) &= \lim_{\Delta t \to 0} \frac{\boldsymbol{v}(\boldsymbol{r}', t') - \boldsymbol{v}(\boldsymbol{r}, t)}{\Delta t} = \lim_{\Delta t \to 0} \frac{\boldsymbol{v}(\boldsymbol{r} + \boldsymbol{v}(\boldsymbol{r}, t)\, \Delta t, t + \Delta t) - \boldsymbol{v}(\boldsymbol{r}, t)}{\Delta t} \\ &= \left(\frac{\partial}{\partial t} + \boldsymbol{v}(\boldsymbol{r}, t) \cdot \nabla\right) \boldsymbol{v}(\boldsymbol{r}, t) = \left(\frac{\partial}{\partial t} + v^j(\boldsymbol{r}, t)\, \partial_j\right) \boldsymbol{v}(\boldsymbol{r}, t). \end{aligned} \tag{4.3}$$

The term in the parentheses in (4.3) is referred to as the *convective derivative* and is typically denoted

$$\frac{\mathrm{D}}{\mathrm{D}t} \equiv \frac{\partial}{\partial t} + \boldsymbol{v}(\boldsymbol{r}, t) \cdot \nabla. \tag{4.4}$$

If $\boldsymbol{v}(\boldsymbol{r}, t) = 0$ then the convective derivative coincides with the partial derivative with respect to time. Newton's second law (4.2) becomes

$$\boldsymbol{F}(\boldsymbol{r}, t) = \rho(\boldsymbol{r}, t)\, \boldsymbol{a}(\boldsymbol{r}, t)\, \mathrm{d}^3 r = \rho(\boldsymbol{r}, t) \left(\frac{\partial}{\partial t} + \boldsymbol{v}(\boldsymbol{r}, t) \cdot \nabla\right) \boldsymbol{v}(\boldsymbol{r}, t)\, \mathrm{d}^3 r. \tag{4.5}$$

The vector quantity $(\boldsymbol{v}(\boldsymbol{r}, t) \cdot \nabla)\boldsymbol{v}(\boldsymbol{r}, t)$ appearing in (4.5) is most easily manipulated by using the summation convention in Cartesian coordinates,

$$(\boldsymbol{v}(\boldsymbol{r}, t) \cdot \nabla)\boldsymbol{v}(\boldsymbol{r}, t) = v^k(\boldsymbol{r}, t)\, \partial_k\big(\mathbf{e}_j v^j(\boldsymbol{r}, t)\big) = v^k(\boldsymbol{r}, t)\, \partial_k v^j(\boldsymbol{r}, t)\, \mathbf{e}_j, \tag{4.6}$$

where the property of Cartesian direction vectors $\partial_k \mathbf{e}_j = 0$ was used. The jth component of (4.6) is written $v^k \partial_k v^j$, where the arguments of the velocity components v^k are suppressed for notational simplicity. This quantity can be rewritten using two identities. The first is given by

$$\frac{1}{2}\partial_j(\boldsymbol{v} \cdot \boldsymbol{v}) = \frac{1}{2}\partial_j(v^k v_k) = \frac{1}{2} v^k \partial_j v_k + \frac{1}{2} v_k \partial_j v^k = v^k \partial_j v_k. \tag{4.7}$$

Exercise 4.1. Demonstrate the last step of (4.7) by using the property of the Cartesian coordinate metric tensor in a Euclidean space $\partial_j v^k = \partial_j(g^{k\ell} v_\ell) = g^{k\ell}\partial_j v_\ell$.

A second important identity is provided by examining the jth component of $(\nabla \times \boldsymbol{v}) \times \boldsymbol{v}$, which is given by using the expressions (1.112) and (1.158) for the vector product and the curl,

$$\begin{aligned} [(\nabla \times \boldsymbol{v}) \times \boldsymbol{v}]_j &= \varepsilon_{jk\ell}(\nabla \times \boldsymbol{v})^k v^\ell = \varepsilon_{jk\ell}\, \varepsilon^{kmn}(\partial_m v_n) v^\ell \\ &= \left(\delta^m{}_\ell \delta^n{}_j - \delta^m{}_j \delta^n{}_\ell\right) v^\ell \partial_m v_n = v^\ell \partial_\ell v_j - v^\ell \partial_j v_\ell, \end{aligned} \tag{4.8}$$

where the three-dimensional Levi-Civita identity (1.7) was used. Combining (4.7) and (4.8) gives

$$
\begin{aligned}
v^k \partial_k v_j &= \frac{1}{2}\partial_j(v \cdot v) + [(\nabla \times v) \times v]_j \implies \\
(v \cdot \nabla)v &= \frac{1}{2}\nabla(v^2) + (\nabla \times v) \times v.
\end{aligned} \tag{4.9}
$$

The curl of the velocity field,

$$\boldsymbol{\Omega} = \nabla \times v, \tag{4.10}$$

is known as the *vorticity* of the fluid. By virtue of identity (1.164) the vorticity satisfies

$$\nabla \cdot \boldsymbol{\Omega} = 0. \tag{4.11}$$

Understanding the physical meaning of the vorticity (4.10) begins by noting that the curl of the velocity has the physical units of inverse time, as does angular velocity. This relationship can be clarified by considering the case where the fluid undergoes rigid-body rotation around the z-axis with a constant angular velocity $\boldsymbol{\omega} = \omega\, \hat{\mathbf{e}}_z$. Result (2.11) and the subsequent exercise showed that this is best analyzed using cylindrical coordinates. The velocity of the fluid is tangent to circles around the z-axis with the only nonzero component given by $v^\phi \hat{\mathbf{e}}_\phi = \omega r\, \hat{\mathbf{e}}_\phi$, where r rather than ρ denotes the perpendicular distance to the z-axis to avoid confusion with the fluid density. Using (1.160) to find the curl of v in cylindrical coordinates gives

$$\boldsymbol{\Omega} = \nabla \times v = \frac{1}{r}\frac{\partial}{\partial r}(r v^\phi)\hat{\mathbf{e}}_z = \frac{1}{r}\frac{\partial}{\partial r}(\omega r^2)\hat{\mathbf{e}}_z = 2\omega\, \hat{\mathbf{e}}_z. \tag{4.12}$$

Since $\frac{1}{2}\boldsymbol{\Omega}$ is the local angular velocity in the fluid in this case, a nonzero $\boldsymbol{\Omega}$ indicates the presence of local rotation or vorticity in the fluid. If the vorticity vanishes the fluid flow is said to be *irrotational*. This is consistent with Stokes' theorem (1.166), which gives

$$\oint_{P(S)} \mathrm{d}\boldsymbol{r} \cdot v = \int_S \mathrm{d}\boldsymbol{S} \cdot \nabla \times v = \int_S \mathrm{d}\boldsymbol{S} \cdot \boldsymbol{\Omega}, \tag{4.13}$$

where P is the perimeter of the two-dimensional surface S. The line integral of v around the closed path P is sometimes referred to as the *circulation* of the velocity field. This is made explicit in the following exercise.

Exercise 4.2. Using cylindrical coordinates, where r is the radial distance from the z-axis, show that the uniform fluid velocity $v = v_o\, \hat{\mathbf{e}}_\phi$ gives $\boldsymbol{\Omega}(r) = \nabla \times v = v_o \hat{\mathbf{e}}_z / r$, where v_o is a constant velocity.

The next step in developing (4.5) is to identify contributions to the total force $\boldsymbol{F}(\boldsymbol{r}, t)$ acting on the fluid element. The first contribution arises from the action of pressure on the element. Pressure represents the force per unit area present in the

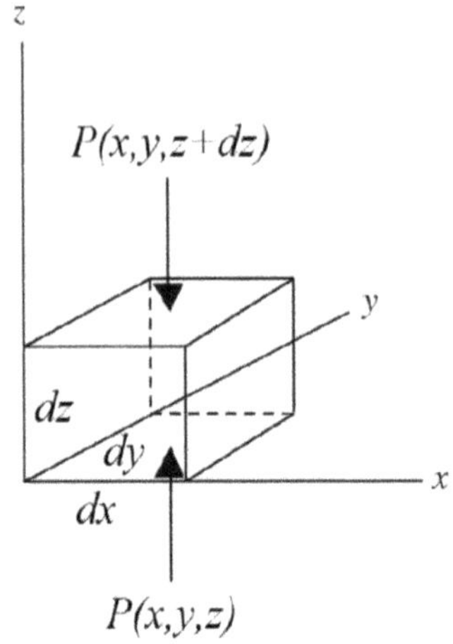

Figure 4.1. The force created by pressure on the mass element in the z direction.

fluid and the force created by pressure on the mass element results from the *pressure difference* across the volume element $\mathrm{d}^3r = \mathrm{d}x\,\mathrm{d}y\,\mathrm{d}z$. To see this, attention is momentarily restricted to the z direction and the volume element is visualized as a cube, as depicted in figure 4.1. There is a force in the $+z$ direction due to pressure acting on the bottom side of the mass element, and this is given by $F_{zB} = P(x, y, z, t)\,\mathrm{d}S_z = P(x, y, z, t)\,\mathrm{d}x\,\mathrm{d}y$. There is also a force due to pressure in the $-z$ direction acting on the top side of the mass element given by $F_{zT} = -P(x, y, z + \mathrm{d}z)\,\mathrm{d}x\,\mathrm{d}y = -P(x, y, z + \mathrm{d}z, t)\,\mathrm{d}S_z$. The net force in the $+z$ direction due to pressure is therefore

$$\begin{aligned} F_{zP}(\boldsymbol{r}, t) &= F_{zB} + F_{zT} = P(x, y, z, t)\,\mathrm{d}y\,\mathrm{d}z - P(x, y, z + \mathrm{d}z, t)\,\mathrm{d}y\,\mathrm{d}z \\ &= -\frac{\partial P(\boldsymbol{r}, t)}{\partial z}\,\mathrm{d}x\,\mathrm{d}y\,\mathrm{d}z = -\frac{\partial P(\boldsymbol{r}, t)}{\partial z}\,\mathrm{d}^3r. \end{aligned} \tag{4.14}$$

Similar expressions follow for the other two directions. As a result, the force due to pressure on the mass element, denoted by $\boldsymbol{F}_P(\boldsymbol{r}, t)$, can be written in terms of the gradient of the pressure,

$$\boldsymbol{F}_P(\boldsymbol{r}, t) = -\nabla P(\boldsymbol{r}, t)\,\mathrm{d}^3r. \tag{4.15}$$

The relationship of the pressure to other macroscopic variables, such as temperature and density, requires the equation of state for the fluid.

The force of gravity on the element is given by $\boldsymbol{F}_G(\boldsymbol{r}, t) = \mathrm{d}m(\boldsymbol{r}, t)\,\boldsymbol{g}(\boldsymbol{r}, t)$, where $\boldsymbol{g}(\boldsymbol{r}, t)$ is the local acceleration of gravity. For the case of Newtonian gravity, the local acceleration of gravity $\boldsymbol{g}(\boldsymbol{r}, t)$ can be derived from a scalar gravitational potential $\phi_g(\boldsymbol{r}, t)$ as $\boldsymbol{g}(\boldsymbol{r}, t) = -\nabla\phi_g(\boldsymbol{r}, t)$. As a result, the force of gravity on the mass element is given by

$$\boldsymbol{F}_G(\boldsymbol{r}, t) = \mathrm{d}m(\boldsymbol{r}, t)\, \boldsymbol{g}(\boldsymbol{r}, t) = -\rho(\boldsymbol{r}, t)\, \nabla\phi_g(\boldsymbol{r}, t)\, \mathrm{d}^3r. \tag{4.16}$$

For the moment the action of all the remaining forces, $\boldsymbol{F}'(\boldsymbol{r}, t)$, will be written as $\boldsymbol{F}'(\boldsymbol{r}, t) = \boldsymbol{f}(\boldsymbol{r}, t)\, \mathrm{d}^3r$, where $\boldsymbol{f}(\boldsymbol{r}, t)$ is also assumed to be a *force density*. The total force on the mass element is, suppressing the position and time dependence,

$$\boldsymbol{F}_T = \left(\boldsymbol{f} - \nabla P - \rho\, \nabla\phi_g\right)\mathrm{d}^3r. \tag{4.17}$$

Combining (4.3) with (4.17) in Newton's law (4.5) and canceling the volume element factor results in

$$\rho\left(\frac{\partial}{\partial t} + \boldsymbol{v} \cdot \nabla\right)\boldsymbol{v} = \boldsymbol{f} - \nabla P - \rho\, \nabla\phi_g. \tag{4.18}$$

Relation (4.18) is known as the *Euler equation* in recognition of Euler's ground-breaking work on fluid flow in the 18th century. It provides three *first-order partial differential equations*, but contains at least four unknowns, the density ρ and the three components of the velocity field $\boldsymbol{v}$. It is also *nonlinear* since it possesses a quadratic term in the velocity. As a result, unlike the stretched string result (3.30), solutions to (4.18) do not obey the principle of superposition, and this increases the difficulty in solving (4.18).

The presence of pressure in (4.18) often requires knowledge of the equation of state for the fluid, and it is important to note that the pressure $P(\boldsymbol{r}, t)$ has been defined *locally* in each element of the fluid, since it is the spatial variation of the pressure that contributes to (4.18). To complicate matters further, the equation of state for the fluid can result in the pressure being dependent on the absolute temperature T, which can also vary with position. For example, the perfect gas law is given by $P(\boldsymbol{r}, t) = \rho(\boldsymbol{r}, t)R\, T(\boldsymbol{r}, t)$, where R is the perfect gas constant. It is also necessary to assume that the equation of state holds locally within an element of the fluid. As a result, the temperature may not be uniform throughout the body of the fluid, which means that the fluid is not in thermal equilibrium and heat conduction is occurring. It follows that solving (4.18) may therefore require the inclusion of thermodynamic processes where mechanical energy and heat are exchanged locally as a result of temperature gradients, resulting in non-equilibrium processes within the fluid and between the fluid and its environment. Modeling these processes require further properties of the fluid, such as its specific heat and thermal conductivity, as well as specifying its environment. This is discussed in somewhat more detail in the coming sections dealing with viscosity and thermodynamics. The upshot is that modeling fluid flow often results in a complicated and in many cases *insolvable set of equations*. In order to make progress in extracting analytical information from (4.18), it is often necessary to make simplifying assumptions, such as the presence of uniform pressure and temperature, the incompressiblity of the fluid, the absence of angular momentum in the flow, or that the flow is reversible. These simplifying assumptions will be discussed in the section on applications.

An alternative form for Euler's equation is obtained by using the identity (4.9), so that (4.18) becomes

$$\rho \frac{\partial \boldsymbol{v}}{\partial t} + \frac{1}{2}\rho \nabla(v^2) + \rho \boldsymbol{\Omega} \times \boldsymbol{v} = -\nabla P - \rho \nabla\phi_g + \boldsymbol{f}. \tag{4.19}$$

Result (4.19) will be referred to as the *Navier–Stokes equation*, although the full Navier–Stokes equation typically includes the contribution of viscosity to the force $\boldsymbol{f}$. Like the Euler equation, solving the Navier–Stokes equation requires more information regarding the properties of the fluid under consideration.

4.2 The equation of continuity

Solving the Euler or Navier–Stokes equations typically requires additional mathematical relationships. One such additional relationship is obtained by considering the rate at which mass is changing in a volume. Mass is conserved in almost all terrestrial fluids, but it is not conserved in the event that nuclear processes are taking place. As a result, mass conservation does not hold in the core of a main sequence star where nuclear fusion converts mass into thermal energy and radiation. The Navier–Stokes equation (4.19) can be supplemented with a second equation involving the density $\rho(\boldsymbol{r}, t)$, the velocity field $\boldsymbol{v}(\boldsymbol{r}, t)$, and the rate $\tau(\boldsymbol{r}, t)$ at which mass is being destroyed or created. If mass is being created at $\boldsymbol{r}$ then $\tau(\boldsymbol{r}, t)$ is positive and the volume element at $\boldsymbol{r}$ serves as a *source* of mass creation. On the other hand, mass destruction corresponds to a negative value for $\tau(\boldsymbol{r}, t)$, and the volume element serves as a *sink* where mass is disappearing.

The mathematical relation involving ρ, $\boldsymbol{v}$, and τ is motivated by looking at the simple case of fluid flowing at a uniform velocity v in a cylindrical pipe with the cross-sectional surface area S. After the time interval Δt a cylinder of fluid with the volume $\Delta V = Sv\,\Delta t$ will have passed a point in the pipe. For this simple case, the *flow rate* of the fluid is therefore defined as

$$\mathcal{R} = \frac{\Delta V}{\Delta t} = vS, \tag{4.20}$$

and has the units of volume per unit time. This concept can be generalized to the case of an arbitrary surface element $\mathrm{d}\boldsymbol{S}$, with components defined by (1.115), at point $\boldsymbol{r}$ where the fluid velocity field is $\boldsymbol{v}(\boldsymbol{r}, t)$. The differential flow rate *through a surface element* is given by

$$\mathrm{d}\mathcal{R}(\boldsymbol{r}, t) = \boldsymbol{v}(\boldsymbol{r}, t) \cdot \mathrm{d}\boldsymbol{S} = v(\boldsymbol{r}, t)\cos\theta\,\mathrm{d}S, \tag{4.21}$$

where θ is the angle between $\boldsymbol{v}(\boldsymbol{r}, t)$ and $\mathrm{d}\boldsymbol{S}$. This reproduces result (4.20) for the case that $\theta = 0$ and $\boldsymbol{v}$ is constant over the cross-sectional area of the total surface, and gives the differential flow rate *through the surface element*.

The rate of change in the mass m inside an infinitesimal volume element is given by the difference in the mass flowing into and out of the surface of the volume as well as the rate at which mass is created or destroyed in the volume. This is depicted using Cartesian coordinates in figure 4.2 for the case that the velocity field has a z-component. The rate at which mass enters the infinitesimal volume through the bottom surface element $\mathrm{d}S_z = \mathrm{d}x\,\mathrm{d}y$ is given by

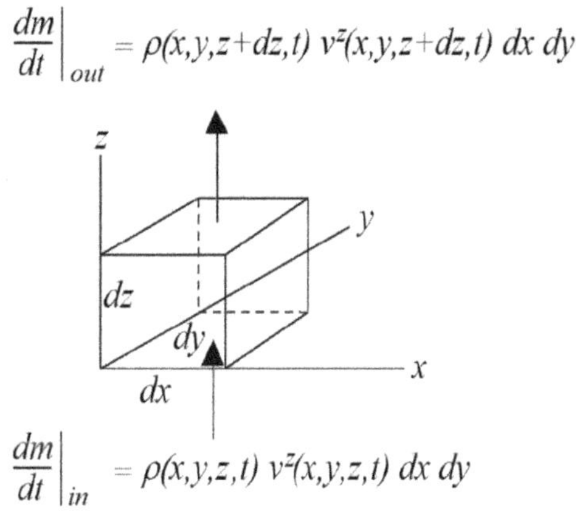

Figure 4.2. The mass flow in the z direction through an infinitesimal volume element.

$$\left.\frac{\mathrm{d}m}{\mathrm{d}t}\right|_{\mathrm{in}} = \rho(x,\, y,\, z,\, t)v^z(x,\, y,\, z,\, t)\,\mathrm{d}x\,\mathrm{d}y, \tag{4.22}$$

while the mass exiting the infinitesimal volume at the top surface element is given by

$$\left.\frac{\mathrm{d}m}{\mathrm{d}t}\right|_{\mathrm{out}} = \rho(x,\, y,\, z + \mathrm{d}z,\, t)v^z(x,\, y,\, z + \mathrm{d}z,\, t)\,\mathrm{d}x\,\mathrm{d}y. \tag{4.23}$$

The net rate of mass change in the infinitesimal volume due to the z-component of the velocity field and $\tau(\boldsymbol{r},\, t)$ is given by

$$\begin{aligned}\frac{\mathrm{d}m}{\mathrm{d}t} &= \left.\frac{\mathrm{d}m}{\mathrm{d}t}\right|_{\mathrm{in}} - \left.\frac{\mathrm{d}m}{\mathrm{d}t}\right|_{\mathrm{out}} + \tau(\boldsymbol{r},\, t)\,\mathrm{d}^3r \\ &= (\rho(x,\, y,\, z,\, t)v^z(x,\, y,\, z,\, t) - \rho(x,\, y,\, z + \mathrm{d}z,\, t) \\ &\qquad v^z(x,\, y,\, z + \mathrm{d}z,\, t))\mathrm{d}x\,\mathrm{d}y + \tau(\boldsymbol{r},\, t)\mathrm{d}^3r \\ &= -\left(\frac{\partial\rho(\boldsymbol{r},\, t)}{\partial z}v^z(\boldsymbol{r},\, t) + \rho(\boldsymbol{r},\, t)\frac{\partial v^z(\boldsymbol{r},\, t)}{\partial z}\right)\mathrm{d}x\,\mathrm{d}y\,\mathrm{d}z + \tau(\boldsymbol{r},\, t)\,\mathrm{d}^3r \\ &= \left(-\frac{\partial}{\partial z}(\rho(\boldsymbol{r},\, t)v^z(\boldsymbol{r},\, t)) + \tau(\boldsymbol{r},\, t)\right)\mathrm{d}^3r,\end{aligned} \tag{4.24}$$

where a Taylor series expansion to $O(\mathrm{d}z)$ was used. For the case that the velocity field has all three components, expression (4.24) becomes

$$\begin{aligned}\frac{\mathrm{d}m}{\mathrm{d}t} &= \left(-\frac{\partial}{\partial x^j}(\rho(\boldsymbol{r},\, t)v^j(\boldsymbol{r},\, t)) + \tau(\boldsymbol{r},\, t)\right)\mathrm{d}^3r \\ &= (-\nabla\cdot(\rho(\boldsymbol{r},\, t)\boldsymbol{v}(\boldsymbol{r},\, t)) + \tau(\boldsymbol{r},\, t))\mathrm{d}^3r.\end{aligned} \tag{4.25}$$

The change in a mass M in a *finite volume* V is obtained by summing the infinitesimal changes of (4.25),

$$\begin{aligned}\frac{\mathrm{d}M}{\mathrm{d}t} &= \int_V \mathrm{d}^3r\,(-\nabla\cdot(\rho(\boldsymbol{r},t)\boldsymbol{v}(\boldsymbol{r},t)) + \tau(\boldsymbol{r},t))\\ &= -\oint_{S(V)} \rho(\boldsymbol{r},t)\boldsymbol{v}(\boldsymbol{r},t)\cdot\mathrm{d}\boldsymbol{S} + \int_V \mathrm{d}^3r\,\tau(\boldsymbol{r},t),\end{aligned} \tag{4.26}$$

where the second step follows from the divergence theorem (1.165). The product $\boldsymbol{J}(\boldsymbol{r},t) = \rho(\boldsymbol{r},t)\boldsymbol{v}(\boldsymbol{r},t)$ can be viewed as a *mass current density* and has units of mass per unit time per unit area. It characterizes the flow of mass through the surface enclosing the volume of fluid.

A second expression for the rate of mass change in the volume V is given by

$$\frac{\mathrm{d}M}{\mathrm{d}t} = \frac{\mathrm{d}}{\mathrm{d}t}\int_V \mathrm{d}^3r\,\rho(\boldsymbol{r},t) = \int_V \mathrm{d}^3r\,\frac{\partial\rho(\boldsymbol{r},t)}{\partial t}. \tag{4.27}$$

The two integrals (4.26) and (4.27) must be equal, and this gives

$$\int_V \mathrm{d}^3r\left(\frac{\partial\rho(\boldsymbol{r},t)}{\partial t} + \nabla\cdot(\rho(\boldsymbol{r},t)\,\boldsymbol{v}(\boldsymbol{r},t)) - \tau(\boldsymbol{r},t)\right) = 0. \tag{4.28}$$

Since the volume of integration V in (4.28) is arbitrary, the integrand must vanish, giving the *equation of continuity*,

$$\frac{\partial\rho(\boldsymbol{r},t)}{\partial t} + \nabla\cdot(\rho(\boldsymbol{r},t)\,\boldsymbol{v}(\boldsymbol{r},t)) = \tau(\boldsymbol{r},t). \tag{4.29}$$

The case $\tau(\boldsymbol{r},t) = 0$ corresponds to the conservation of mass. For the case that $\tau(\boldsymbol{r},t)$ vanishes relation (3.71) is the prototype of a three-dimensional conservation law, similar to the energy–momentum conservation law (3.71) found for the stretched string. A similar expression will be found for the conservation of electric charge in chapter 9.

The compressibility of a macroscopic substance is characterized by its *bulk modulus, B*. If the change in external pressure $\mathrm{d}P$ at the surface of a volume V of the material results in a change of volume $\mathrm{d}V$, then the bulk modulus B is the constant of proportionality between $\mathrm{d}P$ and the fractional change in volume $\mathrm{d}V/V$,

$$\mathrm{d}P = -B\,\frac{\mathrm{d}V}{V}. \tag{4.30}$$

The minus sign is present to yield a decrease in the volume in response to an increase in the *external pressure*. The limit $B \to \infty$ corresponds to $\mathrm{d}V/V \to 0$, signifying a material that retains its volume and is said to be *incompressible*. The bulk modulus has the same units as pressure, which in the SI system of units is the pascal (Pa). Liquid H_2O has a bulk modulus of $B = 2.2 \times 10^9$ Pa, while air at the surface of the Earth has $B \approx 1.4 \times 10^5$ Pa. The bulk modulus formula (4.30) gives the increase in pressure ΔP *acting on the fluid* required to generate a 1% fractional volume change $\Delta V/V = -0.01$ when the two substances are compressed,

$$\Delta P_{\text{water}} = -\, B_{\text{water}}\frac{\Delta V}{V} = 2.2 \times 10^7 \text{ Pa},$$
$$\Delta P_{\text{air}} = -\, B_{\text{air}}\frac{\Delta V}{V} = 1400 \text{ Pa}. \tag{4.31}$$

This shows that a 1% compression of liquid H_2O requires a change in pressure that is $\approx 15\,000$ times greater than that required by air under terrestrial conditions. Result (4.31) shows that it is therefore possible to treat some fluids as *incompressible* for a large range of conditions. This applies to many liquids under normal terrestrial conditions, allowing liquids to be characterized as maintaining their volume as they flow. Of course, it also shows that a fluid in gaseous form readily undergoes compression and expansion. The definition of bulk modulus can be expressed in terms of a change in the density ρ of the fluid. If M is the total mass of the fluid, then $V = M/\rho$ and $\mathrm{d}V = -M\mathrm{d}\rho/\rho^2$, so that bulk modulus formula (4.30) becomes

$$B = \rho\frac{\mathrm{d}P}{\mathrm{d}\rho}. \tag{4.32}$$

For a perfect gas $P = \rho RT$, and this gives $B = \rho RT = P$ for the case that the temperature is held constant.

If the fluid can be treated as incompressible then $\mathrm{d}V = 0$ and the density $\rho = M/V$ can be treated as a constant during flow. For such a case both $\partial\rho/\partial t$ and $\nabla\rho$ will vanish, and the equation of continuity (3.71) reduces to

$$\rho\nabla \cdot \boldsymbol{v}(\boldsymbol{r}, t) = 0 \implies \nabla \cdot \boldsymbol{v}(\boldsymbol{r}, t) = 0. \tag{4.33}$$

Using the divergence theorem (1.165) for an incompressible fluid gives

$$\int_V \mathrm{d}^3r\, \nabla \cdot \boldsymbol{v}(\boldsymbol{r}, t) = \oint_{S(V)} \boldsymbol{v}(\boldsymbol{r}, t) \cdot \mathrm{d}\boldsymbol{S} = \oint_{S(V)} \mathrm{d}\mathcal{R}(\boldsymbol{r}, t) = \mathcal{R}_S(t) = 0, \tag{4.34}$$

where V is an arbitrary volume of the fluid. The *net* flow rate $\mathcal{R}_S(t)$ of mass through the surface S enclosing a volume V is zero if the fluid is incompressible. This simply means that the mass of fluid entering the volume is the same as the mass of fluid exiting the volume in the same time interval.

4.3 Viscosity

A dissipative force of great importance in the flow of fluids is quantified by the *viscosity* of the fluid. When adjacent layers of a fluid move at different speeds, referred to as *shearing*, a dissipative force akin to friction appears between the layers that opposes their motion. This property of the fluid is referred to as *shearing viscosity*. At the microscopic level this viscous force is created by the transport of momentum via atomic diffusion between adjacent layers of the flowing fluid and can be modeled in the statistical approach to fluids in terms of atomic collisions converting mechanical energy into the random energy of heat. In the macroscopic description of fluids it is necessary to introduce a shearing viscosity coefficient η,

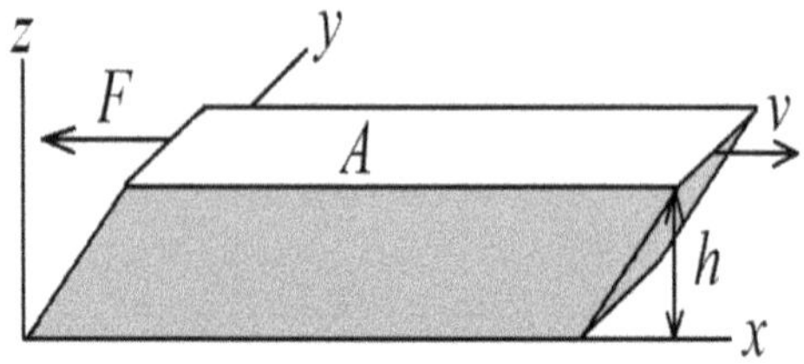

Figure 4.3. The force F of shearing viscosity that opposes motion for a layer of fluid thickness h and area A.

which is measured by separating two perfectly smooth plates of area A by a vertical distance h and placing a thin layer of the fluid between them.

If a constant F is applied *tangentially* to the top plate, it will reach a terminal velocity v that is proportional to F as it slides on the layer of fluid separating it from the stationary bottom plate. Similarly, if the top plate has the velocity v parallel to the bottom plate, there is a force F of resistance to the motion acting *tangentially* on the top plate that is found to be proportional to v. It is the latter situation that is depicted in figure 4.3. For a *Newtonian fluid* the coefficient of shearing viscosity η is defined as

$$\eta = \frac{Fh}{vA}, \tag{4.35}$$

and is roughly constant for fixed temperatures. There are many materials, such as plastics and suspensions, for which (4.35) does not hold, and these are referred to as *non-Newtonian fluids*. These will not be considered here. The units of η are the same as momentum per unit area, which is consistent with viscosity being the result of diffusion of momentum between sliding layers of fluid. Of course, the tangential drag force F due to viscosity is opposite to the direction of the upper plate's motion. For the case that the top plate has the mass m and slides with an initial velocity v_o, Newton's second law shows that, in the absence of other forces, viscosity causes exponentially damped motion,

$$\frac{\mathrm{d}v}{\mathrm{d}t} = a = \frac{F}{m} = -\left(\frac{\eta A}{mh}\right)v \quad \Longrightarrow \quad v = v_o \exp\left(-\frac{\eta A}{mh}t\right), \tag{4.36}$$

and so kinetic energy and momentum are being lost if η is nonzero. This situation also occurs in elementary mechanics problems when kinetic friction is present and acting in opposition to an object's motion. Result (4.36) does not indicate how to account for the loss of kinetic energy, only that nonzero viscosity results in it. Like kinetic friction in mechanics, the presence of viscosity prevents an action approach to analyzing fluid flow, since this would require the inclusion of the exchange of heat with the fluid's environment as well as the flow of heat within the fluid due to temperature gradients. However, the effects of viscosity can be included in the Navier–Stokes equation (4.19) as a contribution to the unspecified force density $\boldsymbol{f}$, just as the force of kinetic friction can be included as part of the total force in Newton's second law.

In order to be included in the force $\boldsymbol{f}$ in the Navier–Stokes equation, the local force density associated with viscosity must be derived. Since viscosity creates a force on the surfaces of layers of fluid sliding at different speeds, it is reasonable to describe this by introducing a viscosity tensor $\mathcal{V}(\boldsymbol{v})$, where $\boldsymbol{v}$ stands for $\boldsymbol{v}(\boldsymbol{r}, t)$, the velocity at position $\boldsymbol{r}$ at time t. The viscous forces on the surfaces $\mathrm{d}S_k$ of an element of fluid d^3r at $\boldsymbol{r}$ at time t are obtained from its components according to

$$\mathrm{d}F_{\mathcal{V}}^j(\boldsymbol{v}) = \mathcal{V}^{jk}(\boldsymbol{v})\, \mathrm{d}S_k. \tag{4.37}$$

Expression (4.37) can then be converted to a force density by applying the divergence theorem (1.165) to the total viscous force $F_{\mathcal{V}}^j$ acting on the surface S of the volume V,

$$F_{\mathcal{V}}^j = \oint_{S(V)} \mathrm{d}F_{\mathcal{V}}^j(\boldsymbol{v}) = \oint_{S(V)} \mathrm{d}S_k\, \mathcal{V}^{jk}(\boldsymbol{v}) = \int_V \mathrm{d}^3r\, \partial_k \mathcal{V}^{jk}(\boldsymbol{v}). \tag{4.38}$$

From (4.38) it follows that the force density due to viscosity is given by

$$f_{\mathcal{V}}^j(\boldsymbol{v}) = \partial_k \mathcal{V}^{jk}(\boldsymbol{v}). \tag{4.39}$$

The next step is to identify the components $\mathcal{V}^{jk}(\boldsymbol{v})$ that are consistent with the nature of viscosity. This begins by considering the elementary description of the viscous force given by (4.35). Replacing h with $\mathrm{d}z$, A with $\mathrm{d}S^z$, v with $\mathrm{d}v^j$, and F with $-\mathrm{d}F^j$, (4.35) becomes a Cartesian coordinate differential relationship first proposed by Newton,

$$F_{\mathcal{V}} = \frac{\eta v A}{h} \quad \Longrightarrow \quad \mathrm{d}F_{\mathcal{V}}^j = \eta \frac{\partial v^j}{\partial z}\, \mathrm{d}S^z. \tag{4.40}$$

Expression (4.40) is indicative of how to formulate the viscosity tensor in terms of a velocity gradient, but it has several flaws. The first is that it gives a nonzero result for the case that the fluid is undergoing rigid-body rotation. Using (2.11), $\boldsymbol{v} = \boldsymbol{\omega} \times \boldsymbol{r}$, for the velocity of the rigid-body element gives

$$v^j = \varepsilon^{j\ell i} \omega_\ell x_i \quad \Longrightarrow \quad \frac{\partial v^j}{\partial x_k} = \varepsilon^{j\ell k}\, \omega_\ell, \tag{4.41}$$

which is nonzero for many cases of $\boldsymbol{\omega}$. Since there is no sliding of fluid layers in a rigid body, no viscous force should be present for this case. As a result, expression (4.40) is incomplete.

The solution is to introduce the *strain rate tensor* $\boldsymbol{\sigma}(\boldsymbol{v})$, which has the symmetric components

$$\sigma^{jk}(\boldsymbol{v}) = \frac{\partial v^j(\boldsymbol{r}, t)}{\partial x_k} + \frac{\partial v^k(\boldsymbol{r}, t)}{\partial x_j} = \partial^k v^j(\boldsymbol{r}, t) + \partial^j v^k(\boldsymbol{r}, t). \tag{4.42}$$

In what follows $\boldsymbol{v}$ will once again be shorthand for $\boldsymbol{v}(\boldsymbol{r}, t)$ and v^j shorthand for $v^j(\boldsymbol{r}, t)$. The presence of the second velocity gradient in (4.42) causes it to vanish for rigid-body rotation. This follows by using (4.41) and $\partial^j = \partial/\partial x_j$ to obtain

$$\partial^k v^j = \frac{\partial v^j}{\partial x_k} = \varepsilon^{j\ell k}\omega_\ell, \quad \partial^j v^k = \frac{\partial v^k}{\partial x_j} = \varepsilon^{k\ell j}\omega_\ell = -\varepsilon^{j\ell k}\omega_\ell, \tag{4.43}$$

which shows that the sum of the two terms in the strain rate tensor (4.42) is zero for rigid-body motion. The strain rate tensor is also *linear* in the velocity gradients, sharing that property with the Newtonian differential relation (4.40). As a result, the strain rate tensor $\boldsymbol{\sigma}(\boldsymbol{v})$ provides the basic building block for viscosity in Newtonian fluids.

Since $\sigma^{jk}(\boldsymbol{v})$ vanishes in the absence of velocity gradients and for the case of rigid-body motion, it is possible to define a scalar quantity that has these same properties by forming the *trace* of $\boldsymbol{\sigma}(\boldsymbol{v})$. Denoted by $\tau(\boldsymbol{v})$, this is given by

$$\tau(\boldsymbol{v}) = \sigma^j{}_j(\boldsymbol{v}) = \partial_j v^j + \partial^j v_j = 2\,\partial_j v^j = 2\,\nabla \cdot \boldsymbol{v}. \tag{4.44}$$

The divergence of the rigid-body velocity components of (4.41) does indeed vanish, since it is given by

$$\partial_j v^j = \varepsilon^{j\ell k}\omega_\ell\, g_{kj} = \varepsilon^{j\ell}{}_j\omega_\ell = 0, \tag{4.45}$$

where (1.99) established that $\varepsilon^{j\ell}{}_j = 0$. Because the divergence in (4.44) is a contraction, it is invariant under coordinate transformations and is therefore a scalar quantity, just as pressure is. Finally, the components of the viscosity tensor associated with $\tau(\boldsymbol{v})$ must also be symmetric, like $\sigma^{jk}(\boldsymbol{v})$, and so it must be proportional to g^{jk}.

Using $\boldsymbol{\sigma}$ and τ there are two possible extensions of (4.40) that have the required behavior of vanishing for both rigid-body motion and the absence of velocity gradients while remaining *linear* in the velocity gradients. Continuing the notation $\boldsymbol{v} \equiv \boldsymbol{v}(\boldsymbol{r}, t)$, these are combined in the expression for the viscous force,

$$\begin{aligned} \mathrm{d}F^j_{\mathcal{V}}(\boldsymbol{v}) &= \mathcal{V}^{jk}(\boldsymbol{v})\mathrm{d}S_k = \left(\eta\,\sigma^{jk}(\boldsymbol{v}) + \frac{1}{2}\mu\, g^{jk}\tau(\boldsymbol{v})\right)\mathrm{d}S_k \\ &= \eta(\partial^k v^j + \partial^j v^k)\mathrm{d}S_k + \mu\nabla\cdot\boldsymbol{v}\, g^{jk}\mathrm{d}S_k, \end{aligned} \tag{4.46}$$

where η and μ represent two possible coefficients of viscosity. Expression (4.46) can be rewritten by separating a contribution to the μ viscosity from the trace of the η viscosity. The force element of (4.46) is rewritten

$$\begin{aligned} \mathrm{d}F^j_{\mathcal{V}}(\boldsymbol{v}) &= \eta\left(\partial^k v^j + \partial^j v^k - \frac{2}{3}g^{jk}\nabla\cdot\boldsymbol{v}\right)\mathrm{d}S_k + \left(\frac{2}{3}\eta + \mu\right)\nabla\cdot\boldsymbol{v}\, g^{jk}\mathrm{d}S_k \\ &\equiv \eta\,\mathcal{S}^{jk}(\boldsymbol{v})\mathrm{d}S_k + \left(\zeta - \frac{1}{3}\eta\right)\mathcal{B}^{jk}(\boldsymbol{v})\mathrm{d}S_k, \end{aligned} \tag{4.47}$$

where $\zeta = \eta + \mu$. This rearrangement yields the *shear viscosity tensor* $\mathcal{S}(\boldsymbol{v})$, which has the components

$$\mathcal{S}^{jk}(\boldsymbol{v}) = \eta\left(\partial^k v^j + \partial^j v^k - \frac{2}{3}g^{jk}\nabla\cdot\boldsymbol{v}\right). \tag{4.48}$$

The shear viscosity tensor has the important property that its trace is zero, since $g^j{}_j = \delta^j{}_j = 3$, so that $\mathcal{S}^j{}_j(v) = 0$. As a result, there is no *new* nonzero scalar that can obtained from it. The second tensor, $\mathcal{B}(v)$, has the components

$$\mathcal{B}^{jk}(v) = \left(\zeta - \frac{1}{3}\eta\right) g^{jk}\nabla \cdot v. \tag{4.49}$$

Result (4.33) showed that $\nabla \cdot v$ is associated with the compressibility of the fluid. For that reason it related to the bulk modulus B defined by (4.30), and so it is often referred to as the *bulk or volume viscosity tensor*. It is important to note that the trace of (4.49) is $\mathcal{B}^j{}_j = (3\zeta - \eta)\nabla \cdot v$, which also provides no *new* nonzero scalar. The shear viscosity and bulk viscosity tensors are therefore the two unique tensors obtained from $\boldsymbol{\sigma}(v)$ that are both *linear* in the velocity gradients and vanish for rigid-body motion.

The force density $f^j_{\mathcal{S}}(v)$ associated with the shear viscosity tensor $\mathcal{S}(v)$ is obtained by applying the result of the divergence theorem (4.39), which identifies $f^j_{\mathcal{S}}(v)$ as

$$\begin{aligned} f^j_{\mathcal{S}}(v) &= \partial_k \mathcal{S}^{jk}(v) = \partial_k\left(\eta\, \partial^k v^j + \eta\, \partial^j v^k - \frac{2}{3} g^{jk}\eta\, \nabla \cdot v\right) \\ &= \eta\left(\nabla^2 v^j + \frac{1}{3}\partial^j \nabla \cdot v\right), \end{aligned} \tag{4.50}$$

where it was assumed that $\partial_k \eta = 0$. Similarly, the divergence theorem (4.39) identifies the force density $f^j_{\mathcal{B}}(v)$ associated with the bulk viscosity tensor $\mathcal{B}(v)$ as

$$f^j_{\mathcal{B}}(v) = \partial_k \mathcal{B}^{jk}(v) = \partial_k\left(\left(\zeta - \frac{1}{3}\eta\right) g^{jk}\, \nabla \cdot v\right) = \left(\zeta - \frac{1}{3}\eta\right) \partial^j \nabla \cdot v, \tag{4.51}$$

where it was assumed that $\partial_k \zeta = 0$. Combining (4.50) and (4.51) with $\zeta = \eta + \mu$ gives the viscosity tensor components,

$$\mathcal{V}^{jk}(v) = \eta(\partial^k v^j + \partial^j v^k) + \mu\, g^{jk}\nabla \cdot v. \tag{4.52}$$

The total force density due to viscosity appearing in the Navier–Stokes equation is then given by

$$f^j_{\mathcal{V}}(v) = f^j_{\mathcal{S}}(v) + f^j_{\mathcal{B}}(v) = \partial_k \mathcal{V}^{jk}(v) = \eta\, \nabla^2 v^j + (\eta + \mu)\, \partial^j \nabla \cdot v. \tag{4.53}$$

The Navier–Stokes equation (4.19) for a viscous compressible Newtonian fluid flowing in the presence of gravitation and an additional force density $\boldsymbol{f}$ is given by

$$\begin{aligned} &\rho \frac{\partial v}{\partial t} + \frac{1}{2}\rho\, \nabla(v^2) + \rho\, \boldsymbol{\Omega} \times v + \nabla P \\ &= -\rho\, \nabla\phi_g + \boldsymbol{f} + \eta\, \nabla^2 v + (\eta + \mu)\, \nabla(\nabla \cdot v) = 0, \end{aligned} \tag{4.54}$$

where $\boldsymbol{\Omega} = \nabla \times v$.

Relation (4.54) is typically combined with the equation of continuity (3.71) as well as the equation of state for the specific fluid being analyzed. In the presence of viscosity the Navier–Stokes equation has become a second-order partial differential equation, even if $\boldsymbol{f} = 0$, requiring twice as many boundary conditions as the original version (4.19). It is also important to note that three terms in (4.54) are the gradients of *scalar quantities*, and these are v^2, P, and $\nabla \cdot \boldsymbol{v}$. In the case of an incompressible fluid, the term $\rho \nabla \phi_g = \nabla(\rho \phi_g)$ is also the gradient of a scalar quantity. In addition, the second and third terms are quadratic in $\boldsymbol{v}$, so that (4.54) is a *second-order nonlinear differential equation.* The presence of viscosity also creates dissipative forces that prevent the conservation of mechanical energy from being invoked and may require thermodynamical considerations to complete the analysis.

4.4 The Navier–Stokes equation and the stress–energy tensor

Assuming that the equation of continuity (3.71) holds with $\sigma = 0$, it can be combined with the Navier–Stokes equation (4.54) to reveal a tensor version of the Navier–Stokes equation. This starts by noting that the equation of continuity (3.71) gives

$$\begin{aligned} \rho \frac{\partial v_j}{\partial t} &= \frac{\partial}{\partial t}(\rho v_j) - v_j \frac{\partial \rho}{\partial t} = \frac{\partial}{\partial t}(\rho v_j) + v_j \partial_k(\rho v^k) \\ &= \frac{\partial}{\partial t}(\rho v_j) + \partial_k(\rho v^k v_j) - \rho v^k \partial_k v_j. \end{aligned} \tag{4.55}$$

Adding the Euclidean identity (4.7), $\frac{1}{2}\rho\, \partial_j(v^2) = \rho\, v^k \partial_j v_k$, to both sides of (4.55) gives

$$\rho \frac{\partial v_j}{\partial t} + \frac{1}{2}\rho\, \partial_j(v^2) = \frac{\partial}{\partial t}(\rho v_j) + \partial_k\left(\rho v_j v^k\right) - \rho\left(v^k \partial_k v_j - v^k \partial_j v_k\right). \tag{4.56}$$

The last term on the right-hand side of (4.56) matches the jth component of the vorticity identity (4.8), $\rho\, (\boldsymbol{\Omega} \times \boldsymbol{v})_j = \rho(v^k \partial_k v_j - v^k \partial_j v_k)$. Using this identification in (4.56) yields the identity

$$\rho\, \frac{\partial v_j}{\partial t} + \frac{1}{2}\rho\, \partial_j v^2 + \rho\, (\boldsymbol{\Omega} \times \boldsymbol{v})_j = \frac{\partial}{\partial t}(\rho v_j) + \partial_k\left(\rho v^k v_j\right). \tag{4.57}$$

Adding the gradient of the pressure $\partial_j P = \partial_k(\delta_j^k P)$ to both sides of (4.57) gives

$$\rho\, \frac{\partial v_j}{\partial t} + \frac{1}{2}\rho\, \partial_j v^2 + \rho\, (\boldsymbol{\Omega} \times \boldsymbol{v})_j + \partial_j P = \frac{\partial}{\partial t}(\rho v_j) + \partial_k\left(\rho v^k v_j + \delta_j^k P\right). \tag{4.58}$$

The left-hand side of (4.58) matches the left-hand side of (4.54). As a result, for the case that ϕ_g, η, μ, and $\boldsymbol{f}$ all vanish, the Navier–Stokes equation (4.54) can be written

$$\pm\left(\frac{\partial}{\partial t}(\rho v_j) + \partial_k\left(\rho v^k v_j + \delta_j^k P\right)\right) = 0, \tag{4.59}$$

where either choice of sign is consistent with the Navier–Stokes equation. Equation (4.59) has the same general form as the string stress–energy tensor conservation law (3.94). Because dissipative forces such as viscosity and processes such as radiative cooling have been ignored, (4.59) holds only for a *perfect* or *ideal fluid* where there is no heat flow. The perfect fluid stress–energy tensor components $T^{\mu}{}_{j}$ and $T^{\mu j}$ are therefore identified as

$$\begin{aligned} T^0_j &= -\rho v_j = -\mathcal{P}_j, \\ T^k_j &= -\rho v^k v_j - \delta^k_j P, \end{aligned} \tag{4.60}$$

where $\mathcal{P}_j$ is the momentum density of the flowing fluid. The negative sign in (4.59) was chosen for consistency with (3.92). In anticipation of special relativity the jth spatial components are raised by simply eliminating the minus sign,

$$\begin{aligned} T^{0j} &= \rho v^j = \mathcal{P}^j, \\ T^{kj} &= \rho v^k v^j + g^{kj} P, \end{aligned} \tag{4.61}$$

where g^{kj} is the metric tensor of the manifold. The Navier–Stokes equation (4.19) for a perfect fluid can therefore be written

$$\partial_\mu T^{\mu j} = \partial_0 T^{0j} + \partial_k T^{kj} = 0, \tag{4.62}$$

or equivalently for the covariant components as

$$\partial_\mu T^\mu_j = \partial_0 T^0_j + \partial_k T^k_j = 0. \tag{4.63}$$

However, further efforts will be required to identify T^{00} and T^{j0}. As in the case of the string stress–energy tensor, it is not possible to define the metric tensor components g^{00} and g^{j0} in Newtonian physics, and so raising and lowering the indices involving time is not possible. If it were, it would be possible to deduce $T^0{}_0$ from the mixed components (4.62) through the relation

$$T^{0j} = g^{j\mu} T^0_\mu = g^{j0} T^0_0 + g^{jk} T^0{}_k. \tag{4.64}$$

As it is, consistency between (4.64) and (4.60) requires that $g^{j0} = 0$, so that $T^{0j} = \rho g^{jk} v_k$ when $T^0_k = \rho v_k$.

Insight into the spatial components T^{kj} for a fluid is obtained by considering the force or *stress* on the fluid contained in a volume V. The total momentum $\boldsymbol{p}$ in the volume V is obtained by integrating $\rho \boldsymbol{v}$. The stress-tensor version (4.62) of the Navier–Stokes equation therefore gives the components of the net force $\boldsymbol{F}$ on the volume as

$$\begin{aligned} F^j &= \frac{\mathrm{d}p^j}{\mathrm{d}t} = \int_V \mathrm{d}^3 r \, \frac{\partial}{\partial t} T^{0j} = -\int_V \mathrm{d}^3 r \, \partial_k T^{kj} \\ &= -\oint_{S(V)} \mathrm{d}S_k \, T^{kj} = -\oint_{S(V)} \mathrm{d}S_k \, (g^{kj} P + \rho v^k v^j), \end{aligned} \tag{4.65}$$

where the divergence theorem (1.165) was used. For a perfect fluid the stress-tensor components T^{kj} show that the net force on the fluid in the volume comes from two sources at the surface of the volume. The first is the inward force element

$\mathrm{d}F_P^j = -Pg^{kj}\mathrm{d}S_k = -P\,\mathrm{d}S^j$ resulting from the pressure on the surface element $\mathrm{d}S^j$. The second source of force is the change in momentum per unit time in a volume due to the flow of fluid through the surface of the volume. The momentum in the *j*th direction that is carried by the fluid per unit time across the surface element out of the volume contributes to $\mathrm{d}p^j/\mathrm{d}t$ and therefore F^j. The momentum carried through a surface element per unit time is proportional to $\mathrm{d}\mathcal{R}$ and creates the force element $\mathrm{d}F_v^j = -\rho v^j\,\mathrm{d}\mathcal{R} = -\rho v^j\,\boldsymbol{v}\cdot\mathrm{d}\boldsymbol{S} = -\rho v^j v^k\,\mathrm{d}S_k$. If $\mathrm{d}\mathcal{R}$ is positive, then the momentum $\rho v^j\,\mathrm{d}\mathcal{R}$ is being lost by the volume element and the force on the volume element is in the opposite direction to that of ρv^j. This is precisely the contribution of the second term in the surface integral of (4.65).

Exercise 4.3. The perfect fluid stress–energy tensor can be amended to accommodate the presence of gravitation for the case of an incompressible fluid. Show that the spatial components are then given by

$$T^{jk}(\boldsymbol{r},\,t) = \rho v^j(\boldsymbol{r},\,t)v^k(\boldsymbol{r},\,t) + g^{jk}(P(\boldsymbol{r},\,t) - \rho\phi_g(\boldsymbol{r},\,t)). \tag{4.66}$$

Result (4.33) shows that the force density proportional to $\partial^j\nabla\cdot\boldsymbol{v}$ appearing in both the viscosity terms (4.50) and (4.51) vanishes for an incompressible fluid. If it is nonzero, it is the gradient of a scalar quantity and so it enters the Navier–Stokes equation (4.19) in a manner identical to the pressure gradient ∇P. This is consistent with the units of $\mu\nabla\cdot\boldsymbol{v}$, which are the same as those of pressure. The bulk viscosity contribution becomes important in understanding shock waves where discontinuous pressure and compression takes place in the fluid. It is often the case that the bulk viscosity coefficient μ is much larger than the shear viscosity coefficient η. Since the total viscous force density (4.53) can be written $f_\mathcal{V}^j = \partial_k\mathcal{V}^{jk}$, it can be incorporated into the stress-tensor version (4.62) of the Navier–Stokes equation for a perfect fluid. For the case that $\phi_g = 0$ and $\boldsymbol{f} = 0$, the spatial components of the stress tensor for an *imperfect fluid* are then given by

$$T^{jk}(\boldsymbol{r},\,t) = \rho v^j(\boldsymbol{r},\,t)v^k(\boldsymbol{r},\,t) + g^{jk}P(\boldsymbol{r},\,t) - \mathcal{V}^{jk}(\boldsymbol{v}(\boldsymbol{r},\,t)), \tag{4.67}$$

where the implicit dependence $\boldsymbol{v} = \boldsymbol{v}(\boldsymbol{r},\,t)$ is displayed. The other component is still given by $T^{0k} = \rho v^k$, so that the Navier–Stokes equation for the case of an imperfect fluid retains the form $\partial_\mu T^{\mu k} = 0$.

4.5 Basic solutions to the Navier–Stokes equation

Although a general analytic solution to the Navier–Stokes equation (4.54) for an imperfect fluid is currently unavailable, it is possible to make simplifying assumptions that allow it to be solved for special situations. It is often the case that the fluid can be treated as incompressible, particularly liquids with a large bulk modulus. It is also possible to limit consideration to a steady or time-independent flow.

This section will consider only simple cases and will avoid unresolved issues such as turbulent flow.

4.5.1 The pressure–depth relation for a static fluid

The simplest meaningful version of the Navier–Stokes equation is the case of a static fluid in the presence of a uniform gravitational acceleration g and the absence of additional forces $\boldsymbol{f}$. Since the fluid is static, $v = 0$ everywhere and there is no vorticity or viscous force. The gravitational potential for a uniform gravitational acceleration g is given by $\phi_g(z) = gz$, where z has been chosen as the vertical direction. For such a case the derivatives of ϕ_g vanish except in the z direction. This means that the pressure P is solely a function of z, so that the z-component of the Navier–Stokes equation becomes

$$\frac{\partial P(z)}{\partial z} = -\rho g. \tag{4.68}$$

Initially it will be assumed that ρ is uniform. Relation (4.68) is a first-order differential equation, and so it requires one boundary condition. This is usually chosen to be $P(0) = P_o$. The solution to (4.68) is then given by

$$P(z) = P_o - \rho g z. \tag{4.69}$$

If $z = 0$ corresponds to the surface of a body of water, then SI units give $\rho = 10^3\,\text{kgm}^{-3}$, while $g = 9.8\,\text{ms}^{-2}$ and $P_o = 1.0 \times 10^5\,\text{Pa}$ at the surface of the Earth. It is important to check the validity of the assumption of incompressibility. It was determined in (4.31) that a 1% compression of water requires a change in pressure of $\Delta P = 2.2 \times 10^7$ Pa. Result (4.69) shows that this pressure increase occurs at

$$z = -\frac{\Delta P}{\rho g} \approx 2.2 \times 10^3\,\text{m}. \tag{4.70}$$

Only below a depth of 2.2 km does the compressibility of water begin to affect the solution (4.69) of the Navier–Stokes equation.

However, the density of a gas is much more sensitive to pressure changes. Result (4.31) showed that $\Delta P \approx -1400$ Pa results in a 1% increase in the volume of air, which has a density of $\approx 1.3\,\text{kgm}^{-3}$ near the surface of the Earth. Applying (4.69) to air shows that

$$z = -\frac{\Delta P}{\rho g} \approx 100\,\text{m}, \tag{4.71}$$

so that the density of air will begin to decrease for a height increase of 100 m above the surface of the Earth. For such a case, result (4.69) will not hold for all but small height changes. It is necessary to invoke the equation of state for air to relate pressure to density and temperature by assuming that air can be treated as a perfect gas, so that $P(z) = \rho(z)R\,T(z)$, where $R = 287\ \mathrm{Jkg^{-1}\,°K}$ in SI units. Using $\rho(z) = P(z)/R\,T(z)$, the Navier–Stokes equation (4.68) becomes

$$\frac{\partial P(z)}{\partial z} = -\frac{gP(z)}{R\,T(z)}. \tag{4.72}$$

In order to solve (4.72) it is necessary to deal with a possible temperature gradient in the ideal gas. Doing so requires further information regarding the environment of the gas, since the presence of a temperature gradient is accompanied by the flow of heat from hot to cold. The simplest case is to treat the fluid as *isothermal*, so that T is a constant. For such a case, (4.72) is easily solved. Denoting $P(0) = P_o$, the pressure–height relation is obtained, as follows:

$$P(z) = P_o \exp\left(-\frac{gz}{R\,T}\right), \tag{4.73}$$

showing that the pressure drops exponentially. In SI units for air at standard temperature and pressure, $RT/g \approx 8400\ \mathrm{m}$. This predicts that the atmospheric pressure drops by a factor of $e^{-0.95} \approx 0.39$ at an altitude of 8.0 km, whereas the observed drop factor is 0.35 at 8.0 km.

Exercise 4.4. It is more realistic to include a temperature gradient in (4.72). In the Earth's troposphere the temperature drops linearly up to a height of $\approx$10 km, so that $T = T_o - \lambda z$, where $T_o = 288°\mathrm{K}$ is the average surface temperature and the tropospheric *lapse rate* is $\lambda \approx 0.006\,5\ °\mathrm{Km^{-1}}$. Show that the pressure–height relation becomes

$$P = P_o\left(1 - \frac{\lambda z}{T_o}\right)^{(g/R\lambda)}. \tag{4.74}$$

Show that the $\lambda \to 0$ limit of (4.74) is (4.73). Show that result (4.74) predicts a drop in pressure by a factor of 0.35 at 8.0 km, which is the observed value.

4.5.2 The Bernoulli relation

The Navier–Stokes equation is greatly simplified by making the following assumptions about the fluid and the velocity field. The velocity field is assumed to be *steady*, so that $\partial v/\partial t = 0$. The fluid is assumed to be *incompressible*, so that $\nabla \cdot v = 0$ and ρ is uniform and constant. The velocity field is assumed to be *irrotational*, so that $\boldsymbol{\Omega} = 0$. The fluid is assumed to be *nonviscous*, so that $\nu = 0$. Finally, all other forces except for gravitation are assumed to be absent. For this rather sizable set of assumptions, the Navier–Stokes equation (4.19) can be written

$$\nabla\left(\frac{1}{2}\rho v^2(\mathbf{r}) + P(\mathbf{r}) + \rho\phi_g(\mathbf{r})\right) = 0. \tag{4.75}$$

This equation is satisfied by

$$\frac{1}{2}\rho v^2(\boldsymbol{r}) + P(\boldsymbol{r}) + \rho\phi_g(\boldsymbol{r}) = c, \tag{4.76}$$

where c is a constant such that $\nabla c = 0$. The sum of the three terms in (4.76) is therefore independent of position, subject to the restrictions that $\frac{1}{2}\rho v^2(\boldsymbol{r}) \geqslant 0$ and $P(\boldsymbol{r}) \geqslant 0$. An example of applying the Bernoulli relation is the steady flow of an incompressible fluid of density ρ in a horizontal pipe of varying cross section, as depicted in figure 4.4.

At the entrance the cross-sectional area, pressure, and fluid velocity are A_1, P_1, and v_1, respectively, while at the exit they are A_2, P_2, and v_2. Since the fluid is incompressible the flow rate (4.20) is constant throughout the pipe, and this gives

$$A_1 v_1 = A_2 v_2 \implies v_2 = \frac{A_1 v_1}{A_2}. \tag{4.77}$$

If $A_2 < A_1$ the fluid velocity v_2 at the exit is greater than the entrance velocity v_1. Since the pipe is horizontal the influence of gravity can be ignored. The Bernoulli relation gives

$$P_1 - P_2 = \frac{1}{2}\rho(v_2^2 - v_1^2) = \frac{1}{2}\rho v_1^2\left(\frac{A_1^2}{A_2^2} - 1\right), \tag{4.78}$$

showing that the increased fluid velocity is accompanied by a drop in pressure. This phenomenon also occurs with wings, where the aerofoil shape results in air traveling faster over the top of the wing than under the bottom of the wing. The pressure difference created by the different air velocities results in the creation of an upward force known as *lift* on the wing. If the velocities over the top and bottom surfaces are v_2 and v_1, respectively, while the wing has a horizontal surface area A, the lift is given by

$$F = (P_1 - P_2)A = \frac{1}{2}\rho A(v_2^2 - v_1^2). \tag{4.79}$$

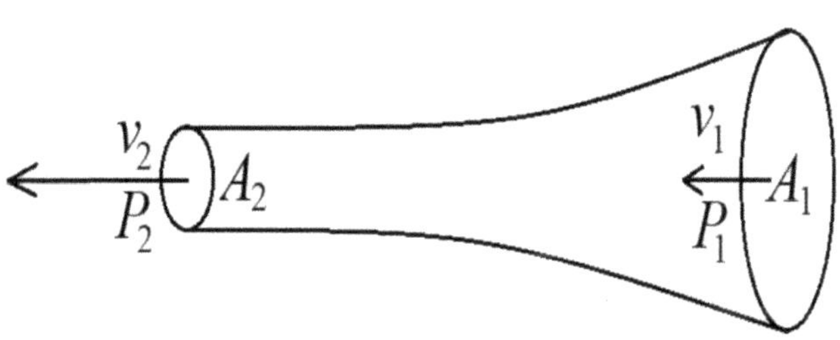

Figure 4.4. Fluid flow in a pipe of varying cross section.

Exercise 4.5. Consider a cylindrical bucket of cross-sectional area A_b holding stationary water of depth h. A small hole of area A is created at the bottom of the bucket. At both the top of the bucket and at the hole the fluid is at atmospheric pressure P_a. Ignoring viscosity and assuming the Bernoulli relation holds throughout, show that the time T required for the bucket to empty is given by

$$T = \frac{A_b}{A}\sqrt{\frac{2h}{g}}. \tag{4.80}$$

4.5.3 The velocity vortex

A very illustrative example of a rotating fluid is a *cylindrically symmetric vortex* oriented along the z-axis. In order to prevent confusion with the density ρ, the cylindrical coordinates will be denoted by (r, ϕ, z). The velocity field of the fluid is assumed to be

$$v(r) = \frac{c}{2\pi r}\,\hat{\mathbf{e}}_\phi, \tag{4.81}$$

where the constant c has the units of area per unit time. The velocity field (4.81) is time-independent and symmetric around the z-axis. However, it is singular at $r = 0$ and therefore unrealistic for small r. Since the velocity is tangent to circles around the z-axis, it models a time-independent circular motion in the fluid around the z-axis. The next step is to determine the vorticity $\boldsymbol{\Omega} = \nabla \times v$. Taking the curl in cylindrical coordinates, the z-component Ω^z for $r \neq 0$ is given by

$$\Omega^z = \frac{1}{r}\frac{\partial}{\partial r}(rv^\phi) = \frac{1}{r}\frac{\partial}{\partial r}\left(\frac{c}{2\pi}\right) = 0. \tag{4.82}$$

However, the singularity in the velocity field at $r = 0$ results in a nonzero vorticity at $r = 0$. This follows from applying Stokes' theorem (1.166) to a circular path P of *arbitrary* radius R around the z-axis, described by the line element $\mathrm{d}\boldsymbol{r} = R\,\mathrm{d}\phi\,\hat{\mathbf{e}}_\phi$. This gives

$$\int_{S(P)} \mathrm{d}\boldsymbol{S}\cdot\boldsymbol{\Omega} = \int_{S(P)} \mathrm{d}\boldsymbol{S}\cdot(\nabla\times v) = \oint_P \mathrm{d}\boldsymbol{r}\cdot v(R) = \int_0^{2\pi} \mathrm{d}\phi\,\frac{c}{2\pi} = c, \tag{4.83}$$

for a surface S which has the circle P as its boundary. Because R is arbitrary, this shows that the singularity at $r = 0$ results in a nonzero vorticity oriented perpendicularly to the circle defined by the path and given by

$$\boldsymbol{\Omega} = \frac{c}{2\pi r}\,\delta(r)\,\hat{\mathbf{e}}_z = c\,\delta^2(r)\,\hat{\mathbf{e}}_z, \tag{4.84}$$

where $\delta^2(r)$ is the two-dimensional delta function. Result (4.84) describes a vorticity that is a singular filament along the z-axis where the tangential velocity of the fluid becomes infinite, in effect, a whirlpool of zero radius. Because (4.84) is independent of z the vorticity satisfies the necessary condition $\nabla \cdot \mathbf{\Omega} = 0$.

Exercise 4.6. Show that (4.84) is consistent with result (4.83).

It is now possible to combine the time-independent velocity field (4.81) with the Euler equation (4.18) for the case that the density ρ is constant and gravitational force is ignored. Suppressing other possible forces $\boldsymbol{f}$ and using $r\nabla = \hat{\mathbf{e}}_\phi(\partial/\partial\phi)$, the Euler equation becomes

$$-\nabla P = \rho(v \cdot \nabla)v = \rho v^\phi \frac{1}{r}\frac{\partial v}{\partial \phi} = \rho\left(\frac{c}{2\pi r}\right)\left(\frac{1}{r}\frac{\partial}{\partial \phi}\right)\left(\frac{c}{2\pi r}\hat{\mathbf{e}}_\phi\right) = \frac{\rho c^2}{4\pi^2 r^3}\frac{\partial \hat{\mathbf{e}}_\phi}{\partial \phi}$$
$$= -\frac{\rho c^2}{4\pi^2 r^3}\hat{\mathbf{e}}_r, \tag{4.85}$$

where the polar coordinate direction vectors (1.38) and (1.39) show that $\partial\hat{\mathbf{e}}_\phi/\partial\phi = -\hat{\mathbf{e}}_r$. Result (4.85) shows that the pressure is solely a function of r, so that the radial component of the pressure gradient in the presence of the vortex (4.81) is given by

$$\frac{\partial P(r)}{\partial r} = \frac{\rho c^2}{4\pi^2 r^3} \implies \mathrm{d}P = \frac{\rho c^2}{4\pi^2 r^3}\mathrm{d}r. \tag{4.86}$$

The first-order differential equation (4.86) is given the boundary condition $P(\infty) = P_0$ and integrated for $R > 0$ to obtain

$$P_0 - P(R) = \int_{P(R)}^{P(\infty)} \mathrm{d}P = \int_R^\infty \mathrm{d}r\, \frac{\rho c^2}{4\pi^2 r^3} = \frac{\rho c^2}{8\pi^2 R^2}, \tag{4.87}$$

so that

$$P(R) = P_0 - \frac{\rho c^2}{8\pi^2 R^2}. \tag{4.88}$$

This shows that the vortex can result in a significant pressure drop in its vicinity, a phenomenon observed in the vicinity of tornadoes, where the air pressure difference creates sufficient outward force on structures to explode them. Since pressure cannot be negative, the simple model provided by (4.81) breaks down for $R < (c/2\pi)\sqrt{\rho/P_0}$. It is worth noting that result (4.88) is consistent with the Bernoulli relation, since the vorticity vanishes for $r > 0$. Result (4.88) can be written as

$$P_0 = P(R) + \frac{\rho c^2}{8\pi^2 R^2} = P(R) + \frac{1}{2}\rho(v^\phi(R))^2, \tag{4.89}$$

which is the Bernoulli relation (4.76) in the absence of gravitational force.

Exercise 4.7. Consider an infinitesimal cylindrical shell of radius $r > 0$, thickness $\mathrm{d}r$, and height H around the z-axis, so that it has the volume $\mathrm{d}V = 2\pi r H \mathrm{d}r$ and the lateral surface area $S = 2\pi r H$. If $\mathrm{d}P$ is the pressure difference across $\mathrm{d}r$, show that the $\mathrm{d}P$ given by (4.86) generates the necessary centripetal force for the cylindrical shell of rotating fluid,

$$\mathrm{d}F_c = S\,\mathrm{d}P = \frac{\mathrm{d}m\, v^2}{r}, \tag{4.90}$$

where $\mathrm{d}m = \rho\,\mathrm{d}V$ and v is given by (4.81).

The angular momentum in an element of fluid in the presence of the vortex can be found for the case that the density ρ is a constant. The element of angular momentum is given by

$$\mathrm{d}\boldsymbol{L} = \boldsymbol{r} \times \mathrm{d}\boldsymbol{p} = \boldsymbol{r} \times (\mathrm{d}m\, \boldsymbol{v}). \tag{4.91}$$

For a constant density ρ the mass element is given by $\mathrm{d}m = \rho\, \mathrm{d}^3r$. Combining relation (4.91) with (4.81) gives the z-component of angular momentum in the volume element,

$$\mathrm{d}L_z = r v^\phi \mathrm{d}m = \rho v^\phi r\, \mathrm{d}^3r = \frac{\rho c}{2\pi}\,\mathrm{d}^3r \;\equiv\; \ell_z\, \mathrm{d}^3r. \tag{4.92}$$

The vortex is therefore characterized by a constant angular momentum per unit volume $\ell_z = \rho c/2\pi$ in the fluid around the z-axis. This allows the constant c to be written $c = 2\pi\ell_z/\rho$ and the vorticity (4.84) becomes

$$\boldsymbol{\Omega} = \frac{\ell_z}{\rho r}\delta(r)\,\hat{\mathbf{e}}_z, \tag{4.93}$$

while the pressure is given by

$$P(R) = P_0 - \frac{\ell_z^2}{2\rho R^2}. \tag{4.94}$$

Exercise 4.8. Repeat the analysis of this section for the case that the velocity field is described by

$$v = \frac{v_0 R}{r}\theta(r - R)\,\hat{\mathbf{e}}_\phi, \tag{4.95}$$

where v_o is a constant velocity and R is the radius of the vortex, which is defined as the radial distance at which $P(R) = 0.9\, P_0$. Show that $\boldsymbol{\Omega} = v_o\, \delta(r - R)\, \hat{\mathbf{e}}_z$ and that $v_o = \sqrt{P_0/5\rho}$. Air has $\rho = 1.2$ kg/m^3 and atmospheric pressure is $P_0 = 1.0 \times 10^5$ Pa, which gives $v_o \approx 130$ m/s or 470 km/hr.

4.5.4 Vorticity transport and stability

It was shown in (4.33) that an incompressible fluid, where ρ is constant and uniform, satisfies both $\partial\rho/\partial t = 0$ and $\nabla\rho = 0$ as well as $\nabla \cdot \boldsymbol{v} = 0$. In the discussion of the Navier–Stokes equation (4.54) that included the force of viscosity, it was pointed out that three terms are the gradient of a scalar quantity, with a fourth for the case of an incompressible fluid. It was also shown in (1.162) that the curl of the gradient of a differential function vanishes. Combining this result with the assumption $\nabla\rho = 0$ for an incompressible fluid and taking the curl of the Navier–Stokes equation (4.54) with an arbitrary additional force density $\boldsymbol{f}$ present, the result is

$$\rho\, \frac{\partial(\nabla \times \boldsymbol{v})}{\partial t} + \rho\, \nabla \times (\boldsymbol{\Omega} \times \boldsymbol{v}) - \eta\nabla^2\nabla \times \boldsymbol{v} - \nabla \times \boldsymbol{f} = 0, \tag{4.96}$$

where it was assumed that η is also a constant. Using $\boldsymbol{\Omega} = \nabla \times \boldsymbol{v}$, equation (4.96) becomes

$$\rho\, \frac{\partial\boldsymbol{\Omega}}{\partial t} + \rho\, \nabla \times (\boldsymbol{\Omega} \times \boldsymbol{v}) - \eta\nabla^2\boldsymbol{\Omega} - \nabla \times \boldsymbol{f} = 0, \tag{4.97}$$

which is known as the *vorticity transport equation*. It is possible to rewrite (4.97) by using the identity

$$\nabla \times (\boldsymbol{\Omega} \times \boldsymbol{v}) = \boldsymbol{\Omega}\, \nabla \cdot \boldsymbol{v} - \boldsymbol{v}\, \nabla \cdot \boldsymbol{\Omega} + (\boldsymbol{v} \cdot \nabla)\, \boldsymbol{\Omega} - (\boldsymbol{\Omega} \cdot \nabla)\, \boldsymbol{v}. \tag{4.98}$$

Exercise 4.9. Use identity (1.7) to prove identity (4.98).

For an incompressible fluid $\nabla \cdot \boldsymbol{v} = 0$, while (4.11) showed that the vorticity obeys $\nabla \cdot \boldsymbol{\Omega} = 0$. Combining these with (4.98) allows equation (4.96) to be written

$$\frac{D\boldsymbol{\Omega}}{Dt} = \left(\frac{\partial}{\partial t} + \boldsymbol{v} \cdot \nabla\right)\boldsymbol{\Omega} = (\boldsymbol{\Omega} \cdot \nabla)\, \boldsymbol{v} + \frac{\eta}{\rho}\, \nabla^2\boldsymbol{\Omega} + \frac{1}{\rho}\nabla \times \boldsymbol{f}, \tag{4.99}$$

where the convective derivative D/Dt was defined in (4.4). The constant $\nu \equiv \eta/\rho$ has units of area per unit time.

If the fluid is not viscous, then $\nu = 0$ and the vorticity transport equation (4.97) becomes

$$\frac{\partial \boldsymbol{\Omega}}{\partial t} = \nabla \times \left(\frac{\boldsymbol{f}}{\rho} - \boldsymbol{\Omega} \times \boldsymbol{v} \right). \tag{4.100}$$

At a given location in the fluid the time evolution of $\boldsymbol{\Omega}$ therefore depends on the sign of the term on the right-hand side of (4.100). If $\nabla \times \boldsymbol{f} = 0$ and $\boldsymbol{\Omega} = 0$ everywhere in an incompressible fluid at some time t, then $\partial \boldsymbol{\Omega}/\partial t = 0$ and $\boldsymbol{\Omega}$ will remain zero. However, if $\boldsymbol{\Omega}$ or $\boldsymbol{f}$ is present in the fluid in such a way that the right-hand side is nonzero at some location, then the vorticity will increase as time passes at those locations where the right-hand side is positive or alternatively decrease as time passes at those locations where the right-hand side is negative. The situation of zero vorticity is therefore *unstable* in the presence of other forces $\boldsymbol{f}$. Even if $\nabla \times \boldsymbol{f} = 0$, the vorticity itself can cause the vorticity to change as time passes.

The presence of viscosity results in the disappearance of vorticity, since viscosity acts as a force opposing the differentially rotating layers of fluid. This occurs in the presence of the vortex motion (4.81) that gives rise to the vorticity (4.84). Using those results gives

$$\begin{aligned} (\boldsymbol{v} \cdot \nabla)\Omega^z &= \frac{c}{2\pi r^2}\frac{\partial}{\partial \phi}\left(\frac{c}{2\pi r}\,\delta(r) \right) = 0, \\ (\boldsymbol{\Omega} \cdot \nabla)v^\phi &= \frac{c}{2\pi r}\,\delta(r)\frac{\partial}{\partial z}\left(\frac{c}{2\pi r} \right) = 0. \end{aligned} \tag{4.101}$$

Using these results in the version of the vorticity transport equation given by (4.99), along with the assumption that $\nabla \times \boldsymbol{f} = 0$, results in the following equation for the time evolution of the vorticity,

$$\frac{\partial \Omega^z}{\partial t} = \nu \nabla^2 \Omega^z. \tag{4.102}$$

Equation (4.102) is known as the *diffusion equation*. Equation (4.102) can be solved under the assumption that the initial vorticity at $t = 0$ is given by (4.84). The solution is given by

$$\Omega^z(r,\, t) = \left(\frac{c}{4\pi \nu t} \right) \exp\left(-\frac{r^2}{4\nu t} \right). \tag{4.103}$$

Exercise 4.10. Using the radial part of the Laplacian in cylindrical coordinates given by (1.156),

$$\nabla_r^2 = \frac{\partial^2}{\partial r^2} + \frac{1}{r}\frac{\partial}{\partial r}, \tag{4.104}$$

verify that (4.103) solves (4.102).

The final step is to show that

$$\lim_{t\to 0^+}\Omega^z(r, t) = \frac{c}{2\pi r}\delta(r), \tag{4.105}$$

where $\delta(r)/2\pi r = \delta^2(r)$ is the two-dimensional delta function appearing in (4.84). This is easily demonstrated by integrating (4.103) over a circular surface of radius R with the surface element $\mathrm{d}S^z = r\,\mathrm{d}r\,\mathrm{d}\phi$ and then taking the limit from positive t. This gives

$$\lim_{t\to 0^+}\int_0^{2\pi}\mathrm{d}\phi\int_0^R \mathrm{d}r\,\frac{cr}{4\pi\nu t}\exp\!\left(-\frac{r^2}{4\nu t}\right) = -c\lim_{t\to 0^+}\left(\exp\!\left(-\frac{R^2}{4\nu t}\right) - 1\right) = c. \tag{4.106}$$

Since R can be chosen to be arbitrarily small, this verifies (4.105). Examining (4.103) shows that the vortex begins as a singular filament along the z-axis and disappears through diffusion as time passes. As νt increases, the exponential tends to one for ever larger values of r, while the overall factor tends to zero, since it is inversely proportional to νt. This describes the initial singularity at $r = 0$ collapsing outward as time passes. The argument used in (4.106) also shows that the limit $\nu \to 0^+$ produces the *time-independent result* (4.105). This demonstrates that viscosity is the agent causing the vorticity to decay, since in the absence of viscosity the vorticity remains concentrated along the z-axis as time passes.

4.5.5 Viscous flow rate

The steady horizontal one-directional flow of an incompressible viscous fluid in a cylindrical pipe of uniform radius R and length L further demonstrates the dissipative effect of viscosity. For simplicity, gravitation will be ignored and the flow is assumed to be cylindrically symmetric along the horizontal z-axis. Using cylindrical coordinates (r, ϕ, z), the cylindrical symmetry of the flow is satisfied if $\partial v/\partial\phi = 0$. The flow is assumed to be steady and *smooth*, so that $v = v(r)\,\hat{\mathbf{e}}_z$ is a differentiable function of r with no dependence on z. This is often referred to as *laminar flow*, since the fluid flow is modeled by cylindrical layers sliding under the dual forces of pressure and viscosity. This assumption is consistent with incompressibility, since it gives

$$\nabla\cdot v = \frac{\partial v(r)}{\partial z} = 0. \tag{4.107}$$

The boundary condition $v_z(R) = 0$ is chosen to reflect the fluid's adhesion to the inner surface of the pipe. Finally, the pressure in the fluid between the two ends of the length L pipe is modeled as dropping linearly in the $+z$ direction,

$$P(z) = P(0) - \frac{\Delta P}{L}z, \tag{4.108}$$

where $\Delta P = P(0) - P(L)$, with ΔP assumed to be positive. Using the identity (4.9) in conjunction with the form for $\boldsymbol{v}$ gives

$$\rho \nabla v^2 + \rho\, \boldsymbol{\Omega} \times \boldsymbol{v} = \rho(\boldsymbol{v} \cdot \nabla)\boldsymbol{v} = \rho\, v(r)\frac{\partial v(r)}{\partial z}\, \hat{\mathbf{e}}_z = 0. \tag{4.109}$$

Exercise 4.11. Show that the vorticity of the flow is given by

$$\boldsymbol{\Omega}(r) = -\frac{\partial v(r)}{\partial r}\, \hat{\mathbf{e}}_\phi. \tag{4.110}$$

It follows from (4.110) and $\hat{\mathbf{e}}_\phi \times \hat{\mathbf{e}}_z = \hat{\mathbf{e}}_r$ that

$$\frac{1}{2}\rho \nabla v^2 = \rho\, \frac{\partial v(r)}{\partial r}\, v(r)\, \hat{\mathbf{e}}_r = \rho\, \frac{\partial v(r)}{\partial r}\hat{\mathbf{e}}_\phi \times v(r)\hat{\mathbf{e}}_z = -\rho\, \boldsymbol{\Omega} \times \boldsymbol{v}, \tag{4.111}$$

explicitly verifying that (4.109) holds.

Combining these results shows that the Navier–Stokes equation (4.54) becomes

$$\frac{\partial}{\partial z}P - \nu \nabla^2 v(r) = 0 \implies -\frac{\Delta P}{L} - \nu\left(\frac{\partial^2}{\partial r^2} + \frac{1}{r}\frac{\partial}{\partial r}\right)v(r) = 0, \tag{4.112}$$

where the cylindrical coordinate version of the Laplacian (1.156) was used. Direct substitution shows that the solution to (4.112) takes the form $v(r) = \alpha + \beta r^2$, where the constant β is given by $\beta = -\Delta P/4\eta L$. The boundary condition $v_z(R) = 0$ fixes the constant α to be given by $\alpha = -\beta R^2 = \Delta P R^2/4\eta L$, so that

$$v(r) = \frac{\Delta P}{4\eta L}(R^2 - r^2). \tag{4.113}$$

The net flow rate $\mathcal{R}$ out of the pipe is found by integrating $\mathrm{d}\mathcal{R} = \boldsymbol{v} \cdot \mathrm{d}\boldsymbol{S} = v(r)\, \mathrm{d}S_z$, where $\mathrm{d}S_z = 2\pi r\, \mathrm{d}r$, over the circular cross section S of the cylinder to find

$$\mathcal{R} = \int_S \mathrm{d}S_z\, v(r) = \frac{\Delta P}{4\eta L}\int_0^R \mathrm{d}r\, 2\pi r(R^2 - r^2) = \frac{\pi\, \Delta P\, R^4}{8\eta L}, \tag{4.114}$$

which is known as *Poiseuille's formula.*

A nonzero pressure difference means that mechanical work is being done on the fluid, since it creates a net force in the direction of the pressure drop and in the same direction that the fluid is flowing. This follows from finding the element of work $\mathrm{d}W_e$ done per unit time $\mathrm{d}t$ on the cylindrical shell, which has a surface area $\mathrm{d}S_z = 2\pi r\, \mathrm{d}r$. The pressure difference between the ends of the pipe creates the force element $\mathrm{d}F(r) = \Delta P\, \mathrm{d}S_z = \Delta P\, 2\pi r\, \mathrm{d}r$ on the shell. As in the Newtonian formula $P = \boldsymbol{F} \cdot \boldsymbol{v}$

for the rate at which a force $\boldsymbol{F}$ does work on an object moving with the instantaneous velocity $\boldsymbol{v}$, the rate at which this force element does work is given by

$$\frac{\mathrm{d}W_e(r)}{\mathrm{d}t} = \mathrm{d}F(r)\, v(r) = \Delta P\, 2\pi r\, \mathrm{d}r\, v(r) = \frac{\pi \Delta P^2}{2\eta L}\mathrm{d}r\, r(R^2 - r^2). \tag{4.115}$$

Expression (4.115) is easily integrated to find the total rate $\mathrm{d}W_t/\mathrm{d}t$ at which the pressure difference is doing mechanical work,

$$\frac{\mathrm{d}W_t}{\mathrm{d}t} = \int_0^R \mathrm{d}r\, \frac{\mathrm{d}W_e(r)}{\mathrm{d}t} = \frac{\pi \Delta P^2}{2\eta L}\int_0^R \mathrm{d}r\, r(R^2 - r^2) = \frac{\pi\, \Delta P^2\, R^4}{8\eta L} = \Delta P\, \mathcal{R}. \tag{4.116}$$

However, since the flow rate remains constant, none of this mechanical work is appearing as kinetic energy in the fluid. Mechanical energy is therefore being lost through the dissipative effect of viscosity, requiring thermodynamic considerations to account for the production of heat by the flow. This requires further information, such as the heat capacity and thermal conductivity of the fluid as well as the temperature of the pipe.

4.5.6 The Reynolds number

The *scaling behavior* of field theories can be a useful tool in analyzing them. In the case of an incompressible viscous fluid the Navier–Stokes equation (4.54) and the vorticity transport equation (4.97) can undergo scaling if the physical situation under consideration has a characteristic length L and a characteristic velocity V. For example, a sphere of diameter L may be immersed in a viscous fluid of fixed density ρ flowing past it with an average relative speed V. Using these as *length, density, and velocity scales*, the Navier–Stokes equation of (4.54) can be written in terms of *dimensionless quantities*, denoted with a prime, by the replacements $\boldsymbol{v} = V\boldsymbol{v}'$, $\boldsymbol{r} = L\boldsymbol{r}'$, $\partial/\partial t = (V/L)\partial/\partial t'$, $\nabla = (1/L)\nabla'$, $\boldsymbol{\Omega} = (V/L)\boldsymbol{\Omega}'$, $P = \rho V^2 P'$, $\phi_g = V^2\phi_g'$, and $\boldsymbol{f} = (\rho V^2/L)\boldsymbol{f}'$. Using $\nabla \cdot \boldsymbol{v} = 0$, the Navier–Stokes equation becomes

$$\left(\frac{\rho V^2}{L}\right)\left(\frac{\partial \boldsymbol{v}'}{\partial t'} + \frac{1}{2}\nabla'(v'^2) + \boldsymbol{\Omega}' \times \boldsymbol{v}' + \nabla' P' + \nabla'\phi_g' - \left(\frac{\eta}{\rho V L}\right)\nabla'^2\boldsymbol{v}' - \boldsymbol{f}'\right) = 0. \tag{4.117}$$

The dimensionless quantity $R \equiv \rho V L/\eta$ is known as the *Reynolds number* of the system. Dropping the overall factor of $\rho V^2/L$, the Navier–Stokes equation for an incompressible viscous fluid has become

$$\frac{\partial \boldsymbol{v}'}{\partial t'} + \frac{1}{2}\nabla'(v'^2) + \boldsymbol{\Omega}' \times \boldsymbol{v}' + \nabla' P' + \nabla'\phi_g' - \frac{1}{R}\nabla'^2\boldsymbol{v}' - \boldsymbol{f}' = 0. \tag{4.118}$$

Exercise 4.12. Show that the vorticity transport equation (4.97) can be scaled to

$$\frac{\partial \mathbf{\Omega}'}{\partial t'} + \nabla' \times (\mathbf{\Omega}' \times v') = \frac{1}{R}\nabla'^2\mathbf{\Omega}' + \nabla' \times \boldsymbol{f}'. \tag{4.119}$$

The specific value of R and the boundary conditions for the system determine the behavior of the velocity field and the vorticity. Scaling the two equations has revealed that the characteristic length and velocity will combine with the viscosity and density of the fluid in a *dimensionless* way in the solution of the two original equations (4.54) and (4.97). Therefore, if two systems share the same Reynolds number and the same boundary conditions, then their velocity field and vorticity will behave identically. This allows the experimental analysis of a small system and extrapolation of the result to a similar large system with the same Reynolds number. This is the principle used in wind tunnels, where scaled-down models of aircraft and wings are tested for their behavior in flight. For such a case, L is given by the flow distance across the actual wing and V is the aircraft's expected average airspeed. If L_m is the wing size of the model, then wind-tunnel testing must occur with an airspeed $V_m = LV/L_m$. For aircraft designed to fly at high speeds the wind-tunnel velocity required for testing the smaller model is often not feasible. For such cases, it is sometimes possible to increase the wind-tunnel air density or replace air with a denser ρ_m but more viscous η_m fluid, such as water, in the testing of the model. The flow speed required for testing is then given by $V_m = \rho V L \eta_m / L_m \rho_m \eta$.

The nature of the flow described by the scaled Navier–Stokes equation is highly dependent on the value of the Reynolds number. If the Reynolds number is relatively small, which occurs in viscous fluids at low flow velocities, then the right-hand side of (4.119) dominates and the vorticity satisfies $\nabla^2\mathbf{\Omega} \approx 0$. For many boundary conditions this equation only has the solution $\mathbf{\Omega} = 0$, showing that viscous fluids flowing at low velocities often do so irrotationally, so that laminar flow persists. As R increases with high flow speeds and larger length scales, the left-hand side of (4.119) becomes important and the interaction of vorticity with the velocity field dominates the dynamics. This results in a much more complicated behavior of the vorticity and velocity field known as *turbulence*. Turbulence is characterized by large variations of the velocity field over small distances as well as the appearance of local angular momentum in the form of localized vortices, referred to as *eddies*, so that the flow appears chaotic. This explains why low-viscosity fluids, such as air and water, exhibit turbulent behavior under a wide variety of circumstances. Numerical analysis shows that for $R < 2000$ laminar flow is stable, while for $R > 3500$ turbulent flow sets in. Turbulence is characterized by a smooth fluid flow splitting into eddies. These eddies generate yet smaller eddies, giving rise to a process known as the *energy cascade* that eventually dissipates energy from the original laminar flow due to viscous forces at small scales. These eddies interact nonlinearly with one another, giving rise to a mathematically intractable situation. On the large scale, energy dissipates gradually, while on the small scale, the motion appears chaotic. Analyzing turbulent flow remains an open and important area of research beyond the scope of this text. The interested reader should consult the references.

4.6 Homentropic flow

Because the action approach to dynamics is predicated on mechanical energy considerations, the presence of dissipative forces renders formulating an action approach difficult if not outright impossible. It was seen in several previous examples that the presence of viscosity removes mechanical energy from the fluid with no indication of how the mechanical energy is lost. This is identical to the effect that the dissipative force of kinetic friction has on the kinetic energy of a moving object. In typical physical systems the loss of mechanical energy is accompanied by the appearance of heat and temperature change. If δQ heat flows into or out of the fluid or its subelements, then thermodynamic variables, such as temperature T, pressure P, and entropy S, are needed to specify how the conversion between heat and mechanical energy occurs. Stating these relations requires such properties as the thermal conductivity, heat capacity, and equation of state for the fluid, as well as the nature of the fluid's environment. For that reason dissipative fluid flow can become extremely complicated. This section will limit consideration to a perfect fluid undergoing constant-entropy flow. It will be shown that a constant-entropy flow is consistent with ignoring the presence of viscosity, since the effects of heat production are suppressed. This allows an action approach to fluid flow, since the conserved mechanical energy can be used in formulating the action. Of course, it comes at the expense of ignoring important aspects of a real fluid, instead describing a fluid Feynman referred to as *dry water*.

It is important to state this in thermodynamic terms This starts by considering a volume $V(\boldsymbol{r})$ of fluid, referred to as an *element of fluid*, located at position $\boldsymbol{r}$ at time t. This volume element may be infinitesimal, but this restriction is not necessary to state its thermodynamic behavior. The element of fluid is characterized thermodynamically by its mass M, its absolute temperature $T(\boldsymbol{r})$, its entropy $S(\boldsymbol{r})$, its pressure $P(\boldsymbol{r})$, and its total energy $U(\boldsymbol{r})$. The total energy $U(\boldsymbol{r})$ includes contributions from both heat and mechanical energy. It is assumed that this element of fluid is at thermal equilibrium internally, which means that its temperature $T(\boldsymbol{r})$ is the same throughout. The mass of fluid in the volume element $V(\boldsymbol{r})$ is assumed to be conserved, but the volume $V(\boldsymbol{r})$, entropy $S(\boldsymbol{r})$, and total energy $U(\boldsymbol{r})$ may change as the result of thermodynamic processes. The volume may absorb or lose the heat $\delta Q(\boldsymbol{r})$ *and* the volume of fluid may do or undergo the mechanical work $\delta W(\boldsymbol{r})$. For such a case, the *first law of thermodynamics* states that the exact change in the total energy $\mathrm{d}U(\boldsymbol{r})$ is given by

$$\mathrm{d}U(\boldsymbol{r}) = \delta Q(\boldsymbol{r}) - \delta W(\boldsymbol{r}). \tag{4.120}$$

This encapsulates the idea that both heat and mechanical work can contribute to the change in the internal energy of the fluid element.

The quantities $\delta Q(\boldsymbol{r})$ and $\delta W(\boldsymbol{r})$ are *not* exact differentials, since their values depend on the details of the thermodynamic processes that are occurring. However, if the fluid element *remains* at the temperature $T(\boldsymbol{r})$ throughout a process, then the heat absorbed is given by

$$\delta Q(\boldsymbol{r}) = T(\boldsymbol{r})\,\mathrm{d}S(\boldsymbol{r}), \tag{4.121}$$

where $\mathrm{d}S(\boldsymbol{r})$ is the exact change in the entropy of the fluid element. Similarly, if the fluid element *remains* at the pressure $P(\boldsymbol{r})$ throughout a process, the work done by the fluid is given by

$$\delta W(\boldsymbol{r}) = P(\boldsymbol{r})\,\mathrm{d}V(\boldsymbol{r}), \tag{4.122}$$

where $\mathrm{d}V(\boldsymbol{r})$ is the exact change in the volume of the fluid element. Combining (4.121) and (4.122) with the first law (4.120) for these two specific cases gives

$$\mathrm{d}U(\boldsymbol{r}) = T(\boldsymbol{r})\,\mathrm{d}S(\boldsymbol{r}) - P(\boldsymbol{r})\,\mathrm{d}V(\boldsymbol{r}). \tag{4.123}$$

This reveals that U can be treated as a function of S and V. For notational convenience the dependence of S and V on $\boldsymbol{r}$ will be suppressed. The chain rule then gives

$$\begin{aligned} \mathrm{d}U(S, V) &= \frac{\partial U(S, V)}{\partial S}\mathrm{d}S + \frac{\partial U(S, V)}{\partial V}\mathrm{d}V = T\,\mathrm{d}S - P\,\mathrm{d}V \\ &\Longrightarrow T = \frac{\partial U(S, V)}{\partial S}, \quad P = -\frac{\partial U(S, V)}{\partial V}. \end{aligned} \tag{4.124}$$

In addition to the relations of (4.124), the element of fluid will obey an *equation of state*, which relates the absolute pressure $P(\boldsymbol{r})$ to the volume $V(\boldsymbol{r})$, the mass M, and the absolute temperature $T(\boldsymbol{r})$.

Thermodynamic processes are also characterized by the net change ΔS_T in the entropy of the system and its environment. In this case, it is given by the sum of the change in the entropy of the fluid element $\Delta S(\boldsymbol{r})$ *and* its surrounding environment ΔS_e. The second law of thermodynamics states that the net change in entropy satisfies

$$\Delta S_T = \Delta S(\boldsymbol{r}) + \Delta S_e \geqslant 0. \tag{4.125}$$

Among many other things, this encapsulates the result that, in the absence of work, heat naturally flows from hot to cold. This follows by considering two large heat reservoirs at fixed temperatures of T_1 and T_2, respectively; heat δQ leaves the first reservoir and arrives at the second without performing mechanical work. The second law immediately gives

$$\Delta S_1 + \Delta S_2 = -\frac{\delta Q}{T_1} + \frac{\delta Q}{T_2} \geqslant 0 \implies T_1 \geqslant T_2. \tag{4.126}$$

This shows that the flow of heat is associated with temperature gradients. If mechanical energy is being converted into heat, this heat will then flow from a hotter element to a cooler element. In the case of the two reservoirs exchanging heat, a growth in entropy occurs if their temperatures are not identical. A thermodynamic process is *irreversible* if $\Delta S_T > 0$, and this occurs if there are temperature gradients in the fluid.

As the fluid element flows some of its mechanical energy may be converted into heat as the result of dissipative processes, such as viscosity. If the heat remains in the fluid element it results in an increase of entropy, according to (4.121). If some of the heat flows into a cooler neighboring fluid element, then (4.126) shows that the total change in the entropy of the two elements combines for a net increase in entropy. In both of these cases, the presence of dissipative processes is accompanied by an increase in entropy. As a result, ignoring dissipative processes requires that $\mathrm{d}S(\boldsymbol{r}) = 0$ throughout the fluid as it flows. Such a fluid flow is said to be *homentropic*, and it has two consequences. The first is that the flow is *reversible*, since there is no change in the net entropy of the fluid or its environment. Since $\mathrm{d}S(\boldsymbol{r}) = 0$, the first law (4.120) becomes

$$\mathrm{d}U(\boldsymbol{r}) = -P(\boldsymbol{r})\,\mathrm{d}V(\boldsymbol{r}). \tag{4.127}$$

The second consequence of homentropic flow is that (4.127) reveals that the energy of the fluid element changes only if *mechanical work* $\delta W(\boldsymbol{r})$ occurs, and this is associated with a change in volume for the element. Since only *mechanical processes* are occurring during homentropic flow, an action for the flow can be formulated. The remainder of this section will therefore consider homentropic flow. This limitation overlooks the many important contributions of dissipative processes to fluid flow, but it allows an action approach based on mechanical energy considerations. It also allows a derivation of the associated stress–energy tensor and a comparison to the fluid stress–energy tensor (4.60) inferred from the Navier–Stokes equation earlier.

Because the mass M of the fluid element is *assumed* to remain constant, the change in volume can be recast as a change in the density $\rho(\boldsymbol{r})$ of the fluid element using $V(\boldsymbol{r}) = M/\rho(\boldsymbol{r})$,

$$V(\boldsymbol{r}) = \frac{M}{\rho(\boldsymbol{r})} \implies \mathrm{d}V(\boldsymbol{r}) = \mathrm{d}\left(\frac{M}{\rho(\boldsymbol{r})}\right) = -\frac{M}{\rho^2(\boldsymbol{r})}\,\mathrm{d}\rho(\boldsymbol{r}). \tag{4.128}$$

The factor of M that has appeared in (4.128) can be removed by expressing the energy $U(\boldsymbol{r})$ in terms of the energy density $u(\boldsymbol{r})$ of the fluid element, defined as $u(\boldsymbol{r}) = U(\boldsymbol{r})/V(\boldsymbol{r})$. This definition and (4.128) give

$$\begin{aligned}\mathrm{d}U(\boldsymbol{r}) &= \mathrm{d}(u(\boldsymbol{r})V(\boldsymbol{r})) = V(\boldsymbol{r})\,\mathrm{d}u(\boldsymbol{r}) + u(\boldsymbol{r})\,\mathrm{d}V(\boldsymbol{r})\\ &= \frac{M}{\rho(\boldsymbol{r})}\,\mathrm{d}u(\boldsymbol{r}) - u(\boldsymbol{r})\,\frac{M}{\rho^2(\boldsymbol{r})}\mathrm{d}\rho(\boldsymbol{r}).\end{aligned} \tag{4.129}$$

Returning (4.128) and (4.129) to (4.127) shows that homentropic flow obeys the relation

$$\rho(\boldsymbol{r})\,\mathrm{d}u(\boldsymbol{r}) = (u(\boldsymbol{r}) + P(\boldsymbol{r}))\mathrm{d}\rho(\boldsymbol{r}). \tag{4.130}$$

The energy density $u(\boldsymbol{r})$ for homentropic flow is therefore a function solely of the density $\rho(\boldsymbol{r})$. The first law for homentropic flow (4.130) can therefore be equivalently expressed as

$$\rho(\boldsymbol{r})\,\frac{\partial u(\boldsymbol{r})}{\partial \rho(\boldsymbol{r})} = u(\boldsymbol{r}) + P(\boldsymbol{r}), \tag{4.131}$$

and this holds in each fluid element.

In thermodynamics the *enthalpy* is defined as

$$H = U + PV, \tag{4.132}$$

so that the *enthalpy density* $h(\boldsymbol{r})$ of the fluid element is given by

$$h(\boldsymbol{r}) = \frac{H(\boldsymbol{r})}{V(\boldsymbol{r})} = \frac{U(\boldsymbol{r})}{V(\boldsymbol{r})} + P(\boldsymbol{r}) = u(\boldsymbol{r}) + P(\boldsymbol{r}) = \rho(\boldsymbol{r})\,\frac{\partial u(\boldsymbol{r})}{\partial \rho(\boldsymbol{r})}, \tag{4.133}$$

where relation (4.131) was used. The energy density of the fluid element is therefore given by

$$u(\boldsymbol{r}) = \rho(\boldsymbol{r})\,\frac{\partial u(\boldsymbol{r})}{\partial \rho(\boldsymbol{r})} - P(\boldsymbol{r}) \implies \rho(\boldsymbol{r})\,\frac{\partial u(\boldsymbol{r})}{\partial \rho(\boldsymbol{r})} = u(\boldsymbol{r}) + P(\boldsymbol{r}) = h(\boldsymbol{r}). \tag{4.134}$$

Because $u(\boldsymbol{r}) = u(\rho(\boldsymbol{r}))$, it follows from the chain rule that the gradient of $u(\boldsymbol{r})$ can be written

$$\nabla u(\boldsymbol{r}) = \frac{\partial u(\rho(\boldsymbol{r}))}{\partial \rho(\boldsymbol{r})}\nabla \rho(\boldsymbol{r}). \tag{4.135}$$

These two results give a useful relation for homentropic flow, again suppressing the $\boldsymbol{r}$ dependence,

$$\rho\nabla\frac{\partial u}{\partial \rho} = \nabla\left(\rho\frac{\partial u}{\partial \rho}\right) - \frac{\partial u}{\partial \rho}\nabla\rho = \nabla h - \nabla u = \nabla(u + P) - \nabla u = \nabla P. \tag{4.136}$$

The important relations (4.135) and (4.136) were derived under the assumption that the homentropic flow respects conservation of mass as expressed by the *equation of continuity* (3.71).

4.7 The action formulation for homentropic flow

Like the stretched string of chapter 3, the Lagrangian density for fluids is defined using *scalar functions* representing mechanical energy. Attention will be limited to homentropic flow, so that the energy density u is a function solely of ρ. The kinetic energy density of the fluid is $\frac{1}{2}\rho v^2 = \frac{1}{2}\rho v_j v^j$, while $u(\rho)$ plays the role of a potential energy density. The action functional, denoted by I to distinguish it from the entropy S, will also include the gravitational potential $\rho\phi_g$, so that it is initially written as

$$I[\rho, v] = \int_V \mathrm{d}^3 r \int_0^T \mathrm{d}t \left(\frac{1}{2}\rho v^2 - u - \rho\phi_g\right) \equiv \int_V \mathrm{d}^3 r \int_0^T \mathrm{d}t\, \mathcal{L}(\rho, v), \tag{4.137}$$

where the integral is over a spatial volume V specified by the boundary conditions for the fluid. In addition, the action should be consistent with the equation of continuity (3.71), given by

$$\frac{\partial \rho}{\partial t} + \nabla \cdot (\rho \boldsymbol{v}) = 0. \tag{4.138}$$

The action for homentropic flow enforces (4.138) through the method of constraints. The method for enforcing constraints through a Lagrange multiplier was introduced in chapter 3 for the hanging string action (3.132). In that case the Lagrange multiplier λ was a constant. In the case of fluid flow, enforcing the equation of continuity as a constraint requires the introduction of a scalar field $\varphi(\boldsymbol{r}, t)$ that acts as a Lagrange multiplier. The Lagrangian density of (4.137) is modified to read

$$\mathcal{L}(\rho, \boldsymbol{v}) \to \mathcal{L}(\rho, \boldsymbol{v}, \varphi) = \frac{1}{2}\rho v^2 - u - \rho\phi_g + \varphi\left(\nabla \cdot (\rho \boldsymbol{v}) + \frac{\partial \rho}{\partial t}\right), \tag{4.139}$$

where the quantity in parentheses is the desired equation of continuity. The units of φ are area per time, so that $\dot{\varphi}$ has units of velocity squared. The scalar field $\varphi(\boldsymbol{r}, t)$ is varied and the scalar field Euler–Lagrange equation (3.21) yields

$$\frac{\partial \mathcal{L}}{\partial \varphi} - \partial_\mu \frac{\partial \mathcal{L}}{\partial(\partial_\mu \varphi)} - \frac{\partial \mathcal{L}}{\partial \varphi} = \nabla \cdot (\rho \boldsymbol{v}) + \frac{\partial \rho}{\partial t} = 0, \tag{4.140}$$

which is the desired constraint on the fluid motion.

However, the Euler–Lagrange equations for both ρ and $\boldsymbol{v}$ are now modified by the presence of the constraint term. Since the Lagrangian density is the integrand of the action (4.139), it is possible to perform integrations by parts on the Lagrangian density. This generates *surface terms* that will be discarded, giving

$$\int_V \mathrm{d}^3r \int_0^T \mathrm{d}t\, \varphi\dot{\rho} = \int_V \mathrm{d}^3r \int_0^T \mathrm{d}t \left(\frac{\partial}{\partial t}(\varphi\rho) - \dot{\varphi}\rho\right) = -\int_V \mathrm{d}^3r \int_0^T \mathrm{d}t\, \dot{\varphi}\rho. \tag{4.141}$$

Ignoring the surface terms is equivalent to assuming boundary conditions on φ such that the volume integral of $\rho(\boldsymbol{r}, T)\varphi(\boldsymbol{r}, T) = \rho(\boldsymbol{r}, 0)\varphi(\boldsymbol{r}, 0) = 0$ vanishes. This is achieved by imposing the boundary conditions $\varphi(\boldsymbol{r}, T) = \varphi(\boldsymbol{r}, 0) = 0$. Similarly, the other term in the constraint can be rewritten using the divergence theorem,

$$\begin{aligned}\int_V \mathrm{d}^3r \int_0^T \mathrm{d}t\, \varphi\nabla \cdot (\rho \boldsymbol{v}) &= \int_V \mathrm{d}^3r \int_0^T \mathrm{d}t\, [\nabla \cdot (\varphi\rho \boldsymbol{v}) - \rho \boldsymbol{v} \cdot \nabla\varphi] \\ &= -\int_V \mathrm{d}^3r \int_0^T \mathrm{d}t\, \rho \boldsymbol{v} \cdot \nabla\varphi.\end{aligned} \tag{4.142}$$

The divergence theorem allows the first term to be discarded, since it yields the surface integral

$$\int_0^T \mathrm{d}t \int_V \mathrm{d}^3r\, \nabla \cdot (\varphi\rho \boldsymbol{v}) = \int_0^T \oint_{S(V)} \mathrm{d}\boldsymbol{S} \cdot \boldsymbol{v}\rho\varphi = 0, \tag{4.143}$$

which vanishes if the spatial boundary conditions on φ are chosen such that $\varphi(\boldsymbol{r}, t) = 0$ if $\boldsymbol{r} \in S(V)$. For such boundary conditions the Lagrangian density (4.139) can be written

$$\mathcal{L} = \frac{1}{2}\rho v^2 - u(\rho) - \rho\, \phi_g - \rho\, \dot{\varphi} - \rho\, \boldsymbol{v} \cdot \nabla\varphi, \tag{4.144}$$

where $\boldsymbol{v}$, ρ, and φ will be treated as independent variables during the variation of the action. It is useful to note that only the variable φ has a time derivative appearing in the Lagrangian density (4.144).

Using (4.144) the equation of motion for $\boldsymbol{v}$ gives

$$\frac{\partial\mathcal{L}}{\partial v^j} - \partial_\mu \frac{\partial\mathcal{L}}{\partial(\partial_\mu v^j)} = \frac{\partial\mathcal{L}}{\partial v^j} = \rho\, v_j - \rho\, \partial_j\varphi = 0. \tag{4.145}$$

The equation for φ gives

$$\frac{\partial\mathcal{L}}{\partial\varphi} - \partial_\mu \frac{\partial\mathcal{L}}{\partial(\partial_\mu\varphi)} = -\partial_0 \frac{\partial\mathcal{L}}{\partial\dot{\varphi}} - \partial_j \frac{\partial\mathcal{L}}{\partial(\partial_j\varphi)} = \dot{\rho} + \partial_j(\rho v^j) = \dot{\rho} + \nabla\cdot(\rho\boldsymbol{v}) = 0, \tag{4.146}$$

which is the desired equation of continuity (3.71). Finally, the equation for ρ gives

$$\frac{\partial\mathcal{L}}{\partial\rho} - \partial_\mu \frac{\partial\mathcal{L}}{\partial(\partial_\mu\rho)} = \frac{\partial\mathcal{L}}{\partial\rho} = \frac{1}{2}v^2 - \frac{\partial u}{\partial\rho} - \phi_g - \dot{\varphi} - \boldsymbol{v}\cdot\nabla\varphi = 0. \tag{4.147}$$

All three equations, (4.145), (4.146), and (4.147), must be solved simultaneously.

The solution to (4.145) is simply

$$\boldsymbol{v} = \nabla\varphi. \tag{4.148}$$

This form for $\boldsymbol{v}$ is referred to as *potential flow*, since $\nabla\varphi$ is similar to the gradient of a potential energy giving the force on a particle. It means that the vorticity vanishes, since $\boldsymbol{\Omega} = \nabla\times\boldsymbol{v} = \nabla\times\nabla\varphi = 0$, and so the flow is irrotational. Substituting $\nabla\varphi = \boldsymbol{v}$ into (4.147) gives

$$\frac{1}{2}v^2 + \frac{\partial u}{\partial\rho} + \phi_g + \dot{\varphi} = 0. \tag{4.149}$$

Multiplying (4.149) by ρ and using (4.134), $\rho\, \partial u/\partial\rho = u + P$, allows this equation of motion to be written

$$\frac{1}{2}\rho v^2 + P + u + \rho\phi_g + \rho\dot{\varphi} = 0. \tag{4.150}$$

Alternatively, taking the gradient of result (4.149), multiplying by ρ, and using result (4.136), $\rho\nabla(\partial u/\partial\rho) = \nabla P$, yields a variant of the equation of motion

$$\frac{1}{2}\rho\nabla v^2 + \rho\nabla\left(\frac{\partial u}{\partial\rho}\right) + \rho\nabla\phi_g + \rho\nabla\dot{\varphi} = \frac{1}{2}\rho\nabla v^2 + \nabla P + \rho\nabla\phi_g + \rho\dot{\boldsymbol{v}} = 0, \tag{4.151}$$

where $\nabla\dot{\varphi} = \dot{v}$ was used in the second step. Result (4.151) is the Navier–Stokes equation (4.54) for the special case that $\mathbf{\Omega} = 0$ and no viscosity or additional forces are present.

4.8 The homentropic stress–energy tensor

Now that the Lagrangian density for this case of fluid flow has been identified, the stress–energy tensor for homentropic flow can be found by generalizing the basic definition (3.82). The fluid stress–energy tensor requires summing over all possible degrees of freedom in the Lagrangian, which are v^k, ρ, and φ. Denoting these five degrees of freedom as Φ^α, where $\alpha \in \{1, \ldots, 5\}$, the stress–energy tensor is written

$$T^\mu{}_\nu = \frac{\partial \mathcal{L}(\rho, v, \varphi)}{\partial(\partial_\mu \Phi^\alpha)} \partial_\nu \Phi^\alpha - \delta^\mu{}_\nu \mathcal{L}, \tag{4.152}$$

where the summation convention applies to the index α and it is understood that (4.152) is evaluated using solutions to the equations of motion.

> **Exercise 4.13.** Using the fact that all five degrees of freedom satisfy the Euler–Lagrange equation, show that the stress–energy tensor of (4.152) satisfies $\partial_\mu T^\mu{}_\nu = 0$.

Before explicitly evaluating (4.152) it is necessary to evaluate the Lagrangian density $\mathcal{L}$ for a solution to the three equations of motion. Equation (4.149) then gives

$$\frac{1}{2}\rho v^2 = -\rho \frac{\partial u(\rho)}{\partial \rho} - \rho(\dot{\varphi} + \phi_g), \tag{4.153}$$

so that

$$\dot{\varphi} = -\frac{\partial u(\rho)}{\partial \rho} - \frac{1}{2}v^2 - \phi_g. \tag{4.154}$$

Substituting both (4.154) and $\nabla\varphi = v$ into the Lagrangian density (4.144) gives

$$\mathcal{L} = \left(\frac{1}{2}\rho v^2 - \rho\, v \cdot \nabla\varphi\right) - u(\rho) - \rho\,(\dot{\varphi} + \phi_g) = \rho \frac{\partial u(\rho)}{\partial \rho} - u(\rho). \tag{4.155}$$

Using relation (4.131) for homentropic flow, $\rho\, \partial u / \partial \rho = u + P$, shows that the Lagrangian density (4.144) is therefore identical to the pressure,

$$\mathcal{L} = P, \tag{4.156}$$

when evaluated using the solutions to the equations of motion.

Evaluating the stress–energy tensor (4.152) begins by using the original form for the Lagrangian density (4.144) *prior* to substituting the solutions to the equations of motion. This is necessary since the equations of motion were themselves determined by varying that form for the Lagrangian density. Because φ is the only variable with a time derivative in (4.144), the component T_0^0 is given by

$$\begin{aligned} T_0^0 &= \frac{\partial \mathcal{L}}{\partial(\partial_0\varphi)}\partial_0\varphi - \delta_0^0\mathcal{L} = -\rho\dot{\varphi} - \mathcal{L} = -\frac{1}{2}\rho v^2 + u + \rho\phi_g + \rho \boldsymbol{v}\cdot\nabla\varphi \\ &= \frac{1}{2}\rho v^2 + u + \rho\phi_g \equiv \mathcal{H}, \end{aligned} \tag{4.157}$$

where both relation (4.154) and the solution $\boldsymbol{v} = \nabla\varphi$ were used. Result (4.157) is the sum of the kinetic and potential energy densities, justifying its identification as the Hamiltonian density $\mathcal{H}$. This is similar to (3.85), where T_0^0 was the Hamiltonian density for waves in the stretched string. Similarly, $T^0{}_j$ is given by

$$T_j^0 = \frac{\partial \mathcal{L}}{\partial(\partial_0\varphi)}\partial_j\varphi = -\rho\,\partial_j\varphi = -\rho v_j = -\mathcal{P}_j, \tag{4.158}$$

where $\mathcal{P}_j$ is the jth component of the momentum density, as it was in result (4.60). This shows that T_j^0 is the jth component of the negative momentum density, just as the component T_1^0 was the negative momentum density (3.85) of the stretched string wave. Using (4.152) results in T_0^j being given by

$$T_0^j = \frac{\partial \mathcal{L}}{\partial(\partial_j\varphi)}\partial_0\varphi = -\rho v^j\dot{\varphi}. \tag{4.159}$$

Recalling that the units of $\dot{\varphi}$ are velocity squared, this is once again consistent with result (3.86) for the stretched string.

Taking the time derivative of (4.157) and assuming that $\partial_0\phi_g = \dot{\phi}_g = 0$ gives

$$\partial_0 T_0^0 = \frac{1}{2}\dot{\rho}v^2 + \rho\boldsymbol{v}\cdot\dot{\boldsymbol{v}} + \dot{\rho}\phi_g + \frac{\partial u}{\partial\rho}\dot{\rho}. \tag{4.160}$$

Combining the equation of continuity and $\dot{\boldsymbol{v}} = \nabla\dot{\varphi}$ with the divergence of (4.159) gives

$$\partial_j T_0^j = -\partial_j(\rho v^j\dot{\varphi}) = -\dot{\varphi}\,\partial_j(\rho v^j) - (\rho v^j)\partial_j\dot{\varphi} = \dot{\varphi}\dot{\rho} - \rho v^j\dot{v}_j = \dot{\varphi}\dot{\rho} - \rho\boldsymbol{v}\cdot\dot{\boldsymbol{v}}. \tag{4.161}$$

The sum of these two results yields

$$\partial_\mu T_0^\mu = \partial_0 T_0^0 + \partial_j T_0^j = \dot{\rho}\left(\dot{\varphi} + \frac{1}{2}v^2 + \phi_g + \frac{\partial u}{\partial\rho}\right) = 0, \tag{4.162}$$

which vanishes by virtue of the equation of motion (4.149). Finally, $T^j{}_k$ is given by

$$T_j^k = \frac{\partial \mathcal{L}}{\partial(\partial_k\varphi)}\partial_j\varphi - \delta_j^k\mathcal{L} = -\rho v^k\partial_j\varphi - \delta_j^k P = -\rho v^k v_j - \delta_j^k P, \tag{4.163}$$

where $\mathcal{L} = P$ and $v_j = \partial_j\varphi$ were used. Once again, this is identical to the result found in (4.60).

The action approach to fluid flow was restricted to the homentropic case so that no conversion of mechanical energy into heat occurred. In that regard, it is instructive to consider how the viscosity tensor (4.52) is related to energy. For simplicity a compressible fluid with the coefficient of bulk viscosity $\mu = 0$ will be considered. A scalar energy density $\epsilon(\boldsymbol{v})$ that can be expressed using the coefficient of shearing viscosity η is given by

$$\epsilon(\boldsymbol{v}) = \eta\nabla \cdot \boldsymbol{v}. \tag{4.164}$$

At first glance, because it involves the coefficient of shearing viscosity, it would appear to represent a dissipative process. However, for the case that $\nabla\eta = 0$, the divergence theorem shows that its volume integral becomes a surface integral,

$$E = \int_V \mathrm{d}^3r\, \epsilon(\boldsymbol{v}) = \int_V \mathrm{d}^3\boldsymbol{r}\, \eta\nabla \cdot \boldsymbol{v} = \eta \oint_{S(V)} \mathrm{d}\boldsymbol{S} \cdot \boldsymbol{v}. \tag{4.165}$$

This vanishes for the case that $\boldsymbol{v} = 0$ at the surface of the system. Adding the term (4.164) to the Lagrangian (4.144) is easily seen to have no effect on the Euler–Lagrange equation (4.145) for $\boldsymbol{v}$, since the term trivially satisfies the Euler–Lagrange equation,

$$\partial_j \frac{\partial\epsilon(\boldsymbol{v})}{\partial(\partial_j v^k)} - \frac{\partial\epsilon(\boldsymbol{v})}{\partial v^k} = \partial_j\!\left(\eta\, \delta^j_k\right) = 0. \tag{4.166}$$

Using (4.164) to obtain the shearing components, the viscosity tensor (4.52) can be obtained from the definition

$$\mathcal{V}^{jk}_\eta = \frac{\partial\epsilon(\boldsymbol{v})}{\partial(\partial_j v^\ell)}\partial^k v^\ell + \frac{\partial\epsilon(\boldsymbol{v})}{\partial(\partial_k v^\ell)}\partial^j v^\ell. \tag{4.167}$$

Exercise 4.14. Using $\nabla \cdot \boldsymbol{v} = \delta^j_\ell \partial_j v^\ell = \delta^k_\ell \partial_k v^\ell$ show that definition (4.167) gives

$$\mathcal{V}^{jk}_\eta = \eta(\partial^j v^k + \partial^k v^j), \tag{4.168}$$

and that the shearing viscosity force component is given by $f^k_\eta = \eta\nabla^2 v^k + \eta\partial^k\nabla \cdot \boldsymbol{v}$.

The definition (4.167) is symmetric, since it has the property $\mathcal{V}^{jk}_\eta = \mathcal{V}^{kj}_\eta$. However, using the homentropic flow result $v^j = \partial^j\varphi$ shows that the net force associated with $\mathcal{V}^{jk}_\eta$ is given by

$$F^j_\eta = \int_V \mathrm{d}^3r\, \partial_k \mathcal{V}^{jk}_\eta = \oint_{S(V)} \mathrm{d}S_k\, \mathcal{V}^{jk}_\eta = 2\eta \oint_{S(V)} \mathrm{d}\boldsymbol{S} \cdot \nabla(\partial^j\varphi) = 0. \tag{4.169}$$

The last step follows since $S(V)$ is the *closed surface* of the volume V, and thus has no boundary. This is similar to the result that the surface integral of $\nabla \times v$ vanishes if the surface is closed. This was fully discussed in chapter 1, where differential forms were used to generalize Stokes' theorem.

4.9 The symmetric fluid stress–energy tensor

The stress–energy tensor for Newtonian homentropic flow obtained from the definition made in (3.82) has two drawbacks. First, as was the case with the stretched string, $T^0_j = -\rho v_j$ is not simply related to $T^j_0 = -\rho v^j \dot{\varphi}$. Second, it is unclear how the components such as T^0_j transform, since they involve the time index.

It is possible to present an ad hoc solution to these problems by defining a related second-rank symmetric *contravariant* tensor $\Theta^{\mu\nu}$. This begins by making the following definitions based on the previous results,

$$\begin{aligned} \Theta^{kj} &= \Theta^{jk} = \rho v^j v^k + g^{jk} P, \\ \Theta^{0j} &= \Theta^{j0} = \rho v^j, \\ \Theta^{00} &= \rho. \end{aligned} \tag{4.170}$$

It is important to define ρ and P in the frame of a *comoving observer*, since pressure P is dependent upon fluid velocity through the Bernoulli effect. It follows that $\partial_\mu \Theta^{\mu j} = \partial_\mu \Theta^{j\mu} = 0$, since these components are identical to those appearing in (4.62). This relation once again gives the stress on the element, since

$$\partial_\mu \Theta^{j\mu} = \frac{\partial}{\partial t}(\rho v^j) + \partial_k \Theta^{kj} \implies f^j = -\partial_k \Theta^{kj}, \tag{4.171}$$

where f^j is the force density. This reproduces the previous result (4.65). The changed components Θ^{j0} yield

$$\partial_\mu \Theta^{\mu 0} = \partial_0 \Theta^{00} + \partial_j \Theta^{j0} = \dot{\rho} + \partial_j(\rho v^j) = 0, \tag{4.172}$$

which vanishes by virtue of the equation of continuity (3.71) rather than the equations of motion involving energy and momentum densities. It should be noted that the elements of the symmetric contravariant tensor differ in sign from the mixed tensor that was found using the canonical stress–energy tensor definition. The choices for signs have been made so that T^{00} is a positive quantity and the remaining components are consistent with the equation of continuity. However, the tensor $\boldsymbol{\theta}$ no longer involves energy. Instead, it merges the perfect fluid force equation (4.65) with the equation of continuity, and so it would more correctly be referred to as the stress-continuity tensor.

An important aspect of $\Theta^{\mu\nu}$ is that it transforms as a true second-rank contravariant tensor when a Galilean boost (2.20) is performed. This requires taking the critical step of associating the zero indices of the tensor components $\Theta^{\mu\nu}$ with time, in effect identifying $x^0 = t$ as a coordinate. Doing so for the Galilean boost gives the following set of coordinate changes,

$$\frac{\partial x'^0}{\partial x^0} = \frac{\partial t'}{\partial t} = 1, \tag{4.173}$$

$$\frac{\partial x'^0}{\partial x^j} = \frac{\partial t'}{\partial x^j} = 0, \tag{4.174}$$

$$\frac{\partial x'^j}{\partial x^k} = \delta^j{}_k, \tag{4.175}$$

$$\frac{\partial x'^j}{\partial x^0} = \frac{\partial x'^j}{\partial t} = -v_o^j. \tag{4.176}$$

Applying the transformation rules of a second-rank contravariant tensor to the spatial components Θ^{jk} gives

$$\begin{aligned}\Theta'^{jk}(\boldsymbol{r}', t') &= \frac{\partial x'^j}{\partial t}\frac{\partial x'^k}{\partial t}\Theta^{00}(\boldsymbol{r}', t') + \frac{\partial x'^j}{\partial t}\frac{\partial x'^k}{\partial x^\ell}\Theta^{0\ell}(\boldsymbol{r}', t') \\ &\quad + \frac{\partial x'^j}{\partial x^\ell}\frac{\partial x'^k}{\partial t}\Theta^{\ell 0}(\boldsymbol{r}', t') + \frac{\partial x'^j}{\partial x^\ell}\frac{\partial x'^k}{\partial x^m}\Theta^{\ell m}(\boldsymbol{r}', t') \\ &= \rho(\boldsymbol{r}', t')v_o^j v_o^k - \rho(\boldsymbol{r}', t')v_o^j v^k - \rho(\boldsymbol{r}', t')v^j v_o^k \\ &\quad + \rho(\boldsymbol{r}', t')v^j v^k + g^{jk}P(\boldsymbol{r}', t') \\ &= \rho(\boldsymbol{r}', t')(v^j - v_o^j)(v^k - v_o^k) + g^{jk}P(\boldsymbol{r}', t').\end{aligned} \tag{4.177}$$

Exercise 4.15. Show that a Galilean boost yields

$$\begin{aligned}\Theta'^{0j}(\boldsymbol{r}', t') &= -v_o^j\Theta^{00}(\boldsymbol{r}', t') + \Theta^{0j}(\boldsymbol{r}', t') = \rho(\boldsymbol{r}', t')(v^j - v_o^j), \\ \Theta'^{00}(\boldsymbol{r}', t') &= \Theta^{00}(\boldsymbol{r}', t') = \rho(\boldsymbol{r}', t'),\end{aligned} \tag{4.178}$$

In Newtonian physics $\rho(\boldsymbol{r}', t')$ is a scalar, since mass and volume are observer independent. It was shown earlier that $\boldsymbol{v}$ transforms as $\boldsymbol{v} \to \boldsymbol{v} - \boldsymbol{v}_o$ under a Galilean boost. It follows that results (4.177) and (4.178) give the correct forms for the components of the tensor in a boosted Galilean inertial frame. It is important to note that $P(\boldsymbol{r}', t')$ is *not* assumed to be a scalar, even in Newtonian physics. An inertial observer traveling with the fluid element at $\boldsymbol{r}$, referred to as a *comoving observer*, has $\boldsymbol{v}_o = \boldsymbol{v}(\boldsymbol{r}, t)$, since that choice places the position of the $\boldsymbol{r}'$ origin, $\boldsymbol{r}' = 0$, at $\boldsymbol{r} = \boldsymbol{v}_o t$ in the other coordinate system. For the comoving observer at $\boldsymbol{r}$ at time t, the local components of $\boldsymbol{\Theta}$ become $\Theta'^{00} = \rho$ and $\Theta'^{jk} = g^{jk}P$, while $\Theta'^{0j} = \Theta'^{j0} = 0$. This is an important result, since it shows that the choice of how inertial observers are related dictates the form of local physical measurements. This will hold even when special relativity replaces Galilean relativity.

As a final point, the Galilean boost (2.20), $\boldsymbol{r}' = \boldsymbol{r} - \boldsymbol{v}_o t$ and $t' = t$, can be applied to the equation of motion given by (4.150). Treating φ as a scalar, so that $\varphi'(\boldsymbol{r}', t') = \varphi(\boldsymbol{r}, t)$ and $\partial\varphi'(\boldsymbol{r}', t')/\partial x'^k = v'^k$, the tensor transformation rules give

$$\begin{aligned}\dot{\varphi}(\boldsymbol{r}, t) &= \frac{\partial\varphi(\boldsymbol{r}, t)}{\partial t}\\ &= \frac{\partial x'^j}{\partial t}\frac{\partial\varphi'(\boldsymbol{r}', t')}{\partial x'^j} + \frac{\partial t'}{\partial t}\frac{\partial\varphi'(\boldsymbol{r}', t')}{\partial t'} = \dot{\varphi}'(\boldsymbol{r}', t') - \boldsymbol{v}_o\cdot\boldsymbol{v}'(\boldsymbol{r}', t').\end{aligned} \tag{4.179}$$

The fluid velocities for the two observers are related by the usual Galilean result,

$$\boldsymbol{v}(\boldsymbol{r}, t) = \boldsymbol{v}'(\boldsymbol{r}', t') + \boldsymbol{v}_o. \tag{4.180}$$

The equation of motion (4.150) in the boosted coordinate system is given by

$$\frac{1}{2}\rho(\boldsymbol{r}', t')v'^2(\boldsymbol{r}', t') + P(\boldsymbol{r}', t') + u(\boldsymbol{r}', t') + \rho(\boldsymbol{r}', t')\dot{\varphi}'(\boldsymbol{r}', t') = 0. \tag{4.181}$$

Treating u and ρ as scalars, so that $u(\boldsymbol{r}', t') = u(\boldsymbol{r}, t)$ and $\rho(\boldsymbol{r}', t') = \rho(\boldsymbol{r}, t)$, and using (4.179) and (4.180) shows that (4.181) can be written as

$$\begin{aligned}0 &= \frac{1}{2}\rho(\boldsymbol{r}, t)v^2(\boldsymbol{r}, t) + P(\boldsymbol{r}', t') - \frac{1}{2}\rho v_o^2 + u(\boldsymbol{r}, t)\\ &\qquad + \rho(\boldsymbol{r}, t)\dot{\varphi}'(\boldsymbol{r}', t') - \rho(\boldsymbol{r}, t)\boldsymbol{v}_o\cdot\boldsymbol{v}'(\boldsymbol{r}', t')\\ &= \frac{1}{2}\rho(\boldsymbol{r}, t)v^2(\boldsymbol{r}, t) + P(\boldsymbol{r}', t') - \frac{1}{2}\rho v_o^2 + u(\boldsymbol{r}, t) + \rho(\boldsymbol{r}, t)\dot{\varphi}(\boldsymbol{r}, t).\end{aligned} \tag{4.182}$$

In order for (4.182) to be the unboosted version of the equation of motion (4.150), it follows that

$$P(\boldsymbol{r}', t') = P(\boldsymbol{r}, t) + \frac{1}{2}\rho(\boldsymbol{r}, t)v_o^2. \tag{4.183}$$

This is the Bernoulli effect as derived from the equations of motion. If observer $\mathcal{O}$ sees a fluid flowing at the constant and uniform velocity v_o, the boosted observer $\mathcal{O}'$ will see the fluid at rest, so that $\boldsymbol{v}'(\boldsymbol{r}', t') = 0$. The pressure observed by the *comoving observer* $\mathcal{O}'$ will be greater than that observed by observer $\mathcal{O}$ by precisely the amount predicted by the Bernoulli relation (4.76).

4.10 Fluctuations around solutions and stability

The stability of solutions to the equations of motion is an important consideration in field theory. The Euler–Lagrange equations give a set of functions that yield an *extremal value* for the action, but whether they correspond to a stable configuration of the action is not clear. The question of stability can be answered by considering initially infinitesimal perturbations or fluctuations around the solutions and determining their behavior. If infinitesimal fluctuations grow arbitrarily large in magnitude either spatially or temporally, or are complex quantities for real fields, then the solution is, by definition, not stable against fluctuations. This is similar to considering the second derivative of a function at an extremum point to determine whether it corresponds to a maximum or minimum.

For the case that there is a collection of fields in the system, the fields are given an additional label, Φ^α, to distinguish them. The index α may represent a vector component of a vector field or an entirely different kind of field. In the case of fluid flow this was done in the definition (4.152) for the stress–energy tensor, where the index α stood for the five degrees of freedom in the homentropic Lagrangian (4.144). Designating Φ_c^α as a solution to the associated Euler–Lagrange equation, a perturbation $\delta\Phi^\alpha$ around the classical solution is written $\Phi^\alpha = \Phi_c^\alpha + \delta\Phi^\alpha$. As in the variation of action (3.20), the Lagrangian density is expanded using a Taylor series in powers of $\delta\Phi^\alpha$ and their derivatives, $\partial_\mu\delta\Phi^\alpha$, which are treated as separate degrees of freedom. It is also assumed that the Lagrangian is a function of Φ^α and $\partial_\mu\Phi^\alpha$, with no higher derivative terms appearing. As in (3.20) it is assumed that the perturbations vanish at the surface of V and at the initial and final times, so that integrations by parts can drop the surface terms. The result is

$$\begin{aligned} I[\Phi] = \int_{VT} \mathrm{d}^3r\,\mathrm{d}t \Bigg\{ &\mathcal{L}(\Phi_c) + \delta\Phi^\alpha\left(\frac{\partial\mathcal{L}(\Phi)}{\partial\Phi^\alpha} - \partial_\mu\frac{\partial\mathcal{L}(\Phi)}{\partial(\partial_\mu\Phi^\alpha)}\right)\Bigg|_{\Phi_c^\alpha} \\ &+ \frac{1}{2}\delta\Phi^\alpha\,\delta\Phi^\beta\,\frac{\partial^2\mathcal{L}(\Phi)}{\partial\Phi^\alpha\,\partial\Phi^\beta}\Bigg|_{\Phi_c^\alpha} \\ &+ \partial_\mu\delta\Phi^\alpha\,\delta\Phi^\beta\,\frac{\partial^2\mathcal{L}(\Phi)}{\partial(\partial_\mu\Phi^\alpha)\,\partial\Phi^\beta}\Bigg|_{\Phi_c^\alpha} \\ &+ \frac{1}{2}\partial_\mu\delta\Phi^\alpha\,\partial_\nu\delta\Phi^\beta\,\frac{\partial^2\mathcal{L}(\Phi)}{\partial(\partial_\mu\Phi^\alpha)\,\partial(\partial_\nu\Phi^\beta)}\Bigg|_{\Phi_c^\alpha} + \dots\Bigg\}, \end{aligned} \tag{4.184}$$

where all repeated indices stand for sums. By virtue of Φ_c^α satisfying the Euler–Lagrange equations, the linear terms in $\delta\Phi^\alpha$ vanish, leaving terms that are quadratic and higher in the fluctuations. For the purposes of determining stability, the process is terminated with the quadratic terms, which serve to define a quadratic field theory for the fluctuations.

This method can be demonstrated for the homentropic Lagrangian (4.144), where ϕ_g will be set to zero for simplicity. A simple set of solutions to the equations of motion is given by a stationary or static fluid, so that $v_c = 0$, and a constant and uniform density ρ_c, so that $\dot{\rho}_c = 0$ and $\nabla\rho_c = 0$. This gives the solution to (4.145) as $\varphi_c(\boldsymbol{r}, t) = \varphi_c(t)$, since this satisfies $\nabla\varphi_c(t) = v_c = 0$. The equation of continuity (3.71) is trivially satisfied, since $\dot{\rho}_c = 0$ and $v_c = 0$. Equation (4.147) becomes

$$\dot{\varphi}_c = -\frac{\partial u(\rho_c)}{\partial\rho_c}, \tag{4.185}$$

which is consistent with $\nabla\varphi_c = 0$, since ρ_c is uniform and therefore

$$\partial_j\dot{\varphi}_c = \frac{\partial}{\partial t}\partial_j\varphi_c = -\partial_j\frac{\partial u(\rho_c)}{\partial\rho_c} = -\frac{\partial^2 u(\rho_c)}{\partial{\rho_c}^2}\partial_j\rho_c = 0. \quad (4.186)$$

Using (4.136) shows that $\nabla P_c = 0$, demonstrating that the pressure P_c is a uniform function of ρ_c, which is given by the enthalpy relation (4.134),

$$P_c = \rho_c\frac{\partial u(\rho_c)}{\partial\rho_c} - u(\rho_c). \quad (4.187)$$

The next step is to introduce fluctuations, $\delta\boldsymbol{v}$, $\delta\rho$, and $\delta\varphi$, around these solutions. These fluctuations are functions of *both* $\mathbf{r}$ and t, whose equations of motion are to be determined and solved. Expanding the Lagrangian (4.144) around the solutions and using their properties shows that the linear terms in the fluctuations vanish, as expected, and that the remaining quadratic action is given by

$$\delta^2 I = \int_{VT} \mathrm{d}^3x\,\mathrm{d}t\left(\frac{1}{2}\rho_c\,\delta\boldsymbol{v}\cdot\delta\boldsymbol{v} - \frac{1}{2}\frac{\partial^2 u(\rho_c)}{\partial{\rho_c}^2}\delta\rho^2 - \delta\rho\,\delta\dot{\varphi} - \rho_c\,\delta\boldsymbol{v}\cdot\nabla\delta\varphi\right). \quad (4.188)$$

Exercise 4.16. Using (4.184) verify that (4.188) is the quadratic fluctuation action.

In effect, (4.188) defines a Lagrangian density $\mathcal{L}_\delta$ for the fluctuations,

$$\mathcal{L}_\delta = \frac{1}{2}\rho_c\,\delta\boldsymbol{v}\cdot\delta\boldsymbol{v} - \frac{1}{2}\frac{\partial^2 u(\rho_c)}{\partial{\rho_c}^2}\delta\rho^2 - \delta\rho\,\delta\dot{\varphi} - \rho_c\,\delta\boldsymbol{v}\cdot\nabla\delta\varphi, \quad (4.189)$$

and the behavior of the solutions determines the stability of the original solutions.

Analyzing the equations of motion generated by (4.189) involves the bulk modulus B of the fluid, which was defined in (4.30) as $B = \rho\,\partial P/\partial\rho$ and determines the compressibility of the fluid. Using this definition and relation (4.187) shows that

$$\frac{\partial^2 u(\rho_c)}{\partial{\rho_c}^2} = \frac{\partial}{\partial\rho_c}\left(\frac{P}{\rho_c} + \frac{u}{\rho_c}\right) = -\frac{1}{\rho_c^2}(P+u) + \frac{1}{\rho_c}\left(\frac{\partial P}{\partial\rho_c} + \frac{\partial u}{\partial\rho_c}\right) = \frac{1}{\rho_c}\frac{\partial P}{\partial\rho_c} = \frac{B}{\rho_c^2}. \quad (4.190)$$

Using (4.190) in (4.188) allows (4.188) to be written

$$\mathcal{L}_\delta = \frac{1}{2}\rho_c\,\delta\boldsymbol{v}\cdot\delta\boldsymbol{v} - \frac{1}{2}\frac{B}{\rho_c^2}\,\delta\rho^2 - \delta\rho\,\delta\dot{\varphi} - \rho_c\,\delta\boldsymbol{v}\cdot\nabla\delta\varphi. \quad (4.191)$$

Treating $\delta\rho$ and $\delta\boldsymbol{v}$ as field variables gives the equations of motion for $\delta\boldsymbol{v}$ and $\delta\rho$,

$$
\begin{aligned}
&\frac{\partial \mathcal{L}_\delta}{\partial \delta v} - \partial_\mu \frac{\partial \mathcal{L}_\delta}{\partial(\partial_\mu \delta v)} = 0 \Longrightarrow \delta v = \nabla \delta\varphi, \\
&\frac{\partial \mathcal{L}_\delta}{\partial \rho} - \partial_\mu \frac{\partial \mathcal{L}_\delta}{\partial(\partial_\mu \rho)} = 0 \Longrightarrow \delta\rho = -\frac{\rho_c^2}{B}\,\delta\dot{\varphi}.
\end{aligned} \tag{4.192}
$$

Using results (4.192) in (4.188) gives the Lagrangian density $\mathcal{L}_f$ for the fluctuations $\delta\varphi$ as

$$
\mathcal{L}_f = \frac{1}{2}\frac{\rho_c^2}{B}\left(\frac{\partial \delta\varphi}{\partial t}\right)^2 - \frac{1}{2}\rho_c \nabla\delta\varphi \cdot \nabla\delta\varphi. \tag{4.193}
$$

Result (4.193) is identical in functional form to the Lagrangian density (3.11) found for the stretched string, with the important exception that the action is now three-dimensional. Applying the Euler–Lagrange equation to (4.193) gives rise to the wave equation for $\delta\varphi$,

$$
\frac{\partial \mathcal{L}_\delta}{\partial \delta\varphi} - \partial_\mu \frac{\partial \mathcal{L}_\delta}{\partial(\partial_\mu \delta\varphi)} = \nabla^2 \delta\varphi - \frac{1}{\tilde{v}^2}\frac{\partial^2 \delta\varphi}{\partial t^2} = \Box_{\tilde{v}}\delta\varphi = 0, \tag{4.194}
$$

where the velocity of propagation is given by $\tilde{v} = \sqrt{B/\rho_c}$ and $\Box_{\tilde{v}}$ is the extension of the one-dimensional string d'Alembertian (3.23) to three spatial dimensions. Therefore, the fluctuations around the static fluid solution correspond to waves traveling through the fluid, and are referred to as *sound waves*. They are stable, and represent the flow of energy and momentum through the fluid. Since $\delta\varphi$ also determines the density and velocity fluctuations, $\delta\rho$ and δv, through the relations of (4.192), it suffices to solve (4.194) to determine the five fluctuation fields.

The solutions to (4.194) have a much richer structure than string waves because of the three-dimensional nature of the Laplacian. The string harmonic waves of (3.31) can be extended to three spatial dimensions by introducing the *wave vector* $\boldsymbol{k}$ and writing the solutions as

$$
\delta\varphi_c(\boldsymbol{r}, t) = \begin{cases} A\sin(\boldsymbol{k}\cdot\boldsymbol{r} \pm \omega t + \phi) \\ A\cos(\boldsymbol{k}\cdot\boldsymbol{r} \pm \omega t + \phi) \end{cases}. \tag{4.195}
$$

These are solutions to (4.194) as long as $\boldsymbol{k}$ and ω satisfy the following *dispersion relation*,

$$
\omega^2 = \tilde{v}^2 \boldsymbol{k}\cdot\boldsymbol{k} = \tilde{v}^2 k^2. \tag{4.196}
$$

Exercise 4.17. Using the Euclidean Cartesian scalar product $\boldsymbol{k}\cdot\boldsymbol{r} = k_x x + k_y y + k_z z$ and the Euclidean Cartesian ∇ given by (1.135), verify that the functions of (4.195) are solutions to (4.194) as long as (4.196) holds.

For the choice of positive sign the sound wave is traveling so that $\boldsymbol{k} \cdot \boldsymbol{r}$ is decreasing, while the negative sign corresponds to a sound wave where $\boldsymbol{k} \cdot \boldsymbol{r}$ is increasing. As with the string waves, A is the amplitude, ω is the angular frequency, and ϕ is the phase of the wave. The solutions (4.195) reduce to the same form as that of the string wave solutions (3.31) for the case that $\boldsymbol{k}$ is chosen so that $\boldsymbol{k} \cdot \boldsymbol{r} = kx$. Once again, the angular frequency is given by $\omega = 2\pi f$, where f is the frequency of the wave and is related to the period T by $T = 1/f = 2\pi/\omega_k$. The wave vector magnitude is given by $k = 2\pi/\lambda$, where λ is the wavelength. This follows by noting that the vector $\boldsymbol{\lambda} = \lambda\, \boldsymbol{k}/k$ has the property that

$$\boldsymbol{k} \cdot \boldsymbol{\lambda} = \lambda\, k = 2\pi, \tag{4.197}$$

so that the solutions of (4.195) are periodic under the translation $\boldsymbol{r} \rightarrow \boldsymbol{r} + n\boldsymbol{\lambda}$, where n is an arbitrary integer. It follows from (4.196) that

$$\frac{\omega}{k} = f\lambda = \tilde{v}. \tag{4.198}$$

If the negative sign is chosen in (4.195), the solutions are propagating in the direction parallel to $\boldsymbol{k}$, while the positive sign corresponds to a wave moving in the opposite direction to that of $\boldsymbol{k}$. This follows from considering the trajectory defined by $\boldsymbol{r} = \boldsymbol{r}_o \mp \boldsymbol{v}_o t$, where $\boldsymbol{v}_o = \tilde{v}\boldsymbol{k}/k$ and $\boldsymbol{r}_o$ is arbitrary. Combining this with (4.198) shows that the argument of solution (4.195) along the trajectory becomes

$$\begin{aligned} \boldsymbol{k} \cdot \boldsymbol{r} \pm \omega t + \phi = \boldsymbol{k} \cdot \boldsymbol{r}_o \mp \boldsymbol{k} \cdot \boldsymbol{v}_o t \pm \omega t + \phi = \boldsymbol{k} \cdot \mathrm{r}_o \\ \mp \tilde{v}kt \pm \omega t + \phi \\ = \boldsymbol{k} \cdot \boldsymbol{r}_o + \phi. \end{aligned} \tag{4.199}$$

Along the trajectory the argument of the solution becomes a constant, corresponding to a disturbance moving along the trajectory. This shows that the velocity vector for the disturbance is given by $\boldsymbol{v}_o$, which is parallel or antiparallel to $\boldsymbol{k}$ depending on the choice of sign.

Choosing the sine version of the simple harmonic wave solutions (4.195) for $\delta\varphi$ and setting $\phi = 0$ gives the associated forms for $\delta\boldsymbol{v}$ and $\delta\rho$ from (4.192). The velocity fluctuation is given by

$$\delta\boldsymbol{v}(\boldsymbol{r}, t) = \nabla\delta\varphi = A\boldsymbol{k} \cos(\boldsymbol{k} \cdot \boldsymbol{r} \pm \omega_k t). \tag{4.200}$$

Result (4.200) shows that at the origin $\boldsymbol{r} = 0$ the velocity disturbance obeys $\delta\boldsymbol{v}(0, t) = A\boldsymbol{k} \cos(\omega_k t)$, so that the fluid velocity is alternately parallel and antiparallel to the vector $\boldsymbol{k}$ and along the same direction as the wave motion. Since the medium for the sound wave is the fluid under consideration, the wave is referred to as *longitudinal*, as opposed to the string wave of chapter 3, where the motion of the string medium was found to be *transverse* as the wave passes. The maximum longitudinal speed is given by

$$v_m = Ak = \omega_k A/\tilde{v} \implies \omega_k A = v_m\tilde{v}. \tag{4.201}$$

Similarly, the density fluctuation is given by

$$\begin{aligned}\delta\rho(\boldsymbol{r}, t) &= -\frac{\rho_c^2}{B}\delta\dot{\varphi}(\boldsymbol{r}, t) = \mp\frac{A\omega\rho_c^2}{B}\cos(\boldsymbol{k}\cdot\boldsymbol{r}\pm\omega_k t) \\ &= \mp\left(\frac{v_m}{\tilde{v}}\right)\rho_c\cos(\boldsymbol{k}\cdot\boldsymbol{r}\pm\omega_k t).\end{aligned} \tag{4.202}$$

The density alternately increases and decreases as the sound wave passes, corresponding to compression and rarefaction. The pressure fluctuation is given by

$$\delta P(\boldsymbol{r}, t) = -B\frac{\delta V(\boldsymbol{r}, t)}{V(\boldsymbol{r}, t)} = \frac{B}{\rho_c}\delta\rho(\boldsymbol{r}, t) = \mp A\omega\rho_c\cos(\boldsymbol{k}\cdot\boldsymbol{r}\pm\omega_k t). \tag{4.203}$$

The momentum density canonically conjugate to $\delta\varphi$ is given by applying (3.61). For the simple harmonic sound wave this is given by

$$\begin{aligned}\pi(\boldsymbol{r}, t) &= \frac{\partial\mathcal{L}_f}{\partial\delta\dot{\varphi}(\boldsymbol{r}, t)} = \frac{\rho_c^2}{B}\delta\dot{\varphi}(\boldsymbol{r}, t) = \pm\frac{\rho_c^2}{B}\omega_k A\cos(\boldsymbol{k}\cdot\boldsymbol{r}\pm\omega_k t) \\ &= \pm\frac{\rho_c v_m}{\tilde{v}}\cos(\boldsymbol{k}\cdot\boldsymbol{r}\pm\omega_k t).\end{aligned} \tag{4.204}$$

The canonical momentum has no direction and its units are not those of momentum density. Understanding the energy and momentum densities of the simple harmonic sound wave is easiest using the stress–energy tensor associated with the Lagrangian density (4.193). Using the sine solution of (4.195) with (4.201) gives

$$\begin{aligned}T_0^0 = \mathcal{H} &= \frac{\partial\mathcal{L}_f}{\partial\delta\dot{\varphi}}\delta\dot{\varphi} - \mathcal{L} = \frac{1}{2}\frac{\rho_c^2}{B}\delta\dot{\varphi}^2 + \frac{1}{2}\rho_c\nabla\delta\varphi\cdot\nabla\delta\varphi \\ &= \rho_c v_m^2\cos^2(\boldsymbol{k}\cdot\boldsymbol{r}\pm\omega t),\end{aligned} \tag{4.205}$$

$$T_j^0 = \mathcal{P}_j = \frac{\partial\mathcal{L}_f}{\partial\delta\dot{\varphi}}\partial_j\delta\varphi = \frac{\rho_c^2}{B}\delta\dot{\varphi}\,\partial_j\delta\varphi = \mp\left(\frac{v_m}{\tilde{v}}\right)\rho_c v_{mj}\cos^2(\boldsymbol{k}\cdot\boldsymbol{r}\pm\omega t). \tag{4.206}$$

This shows that the energy density is always positive while the momentum density is a vector quantity pointing in the direction of the wave's propagation.

A unique solution to (4.194) is dependent on the nature of the boundary conditions on the sound wave. Just as the general solution to the string wave equation (3.22) was given by the Fourier integral (3.44), a general solution to (4.194) can be written in terms of a Fourier integral,

$$\delta\varphi(\boldsymbol{r}, t) = \int\frac{d^3k}{(2\pi)^3}\left(\tilde{f}(\boldsymbol{k})\,e^{i\boldsymbol{k}\cdot\boldsymbol{r}-i\omega_k t} + \tilde{f}^*(-\boldsymbol{k})\,e^{i\boldsymbol{k}\cdot\boldsymbol{r}+i\omega_k t}\right), \tag{4.207}$$

which is a solution to (4.194) as long as $\omega_k = \tilde{v}k$. Although $\tilde{f}(\boldsymbol{k})$ is a complex function, the form for (4.207) guarantees that $\delta\varphi$ is a real function, since $\delta\varphi^* = \delta\varphi$. It should be noted that the sound wave need not obey periodicity as the wave in a string of finite length did. Cauchy boundary conditions for $\delta\varphi$ consist of specifying

$F(\boldsymbol{r}) = \delta\varphi(\boldsymbol{r}, 0)$ and $T(\boldsymbol{r}) = \delta\dot{\varphi}(\boldsymbol{r}, 0)$. The relationship of $\tilde{f}(\boldsymbol{k})$ to $F(\boldsymbol{r})$ and $T(\boldsymbol{r})$ is found by using the wave vector Dirac delta (2.101), so that the Fourier transforms of the initial conditions give

$$\begin{aligned}\tilde{F}(\boldsymbol{p}) = \int \mathrm{d}^3r\; F(\boldsymbol{r})\, e^{-i\boldsymbol{p}\cdot\boldsymbol{r}} = \int \mathrm{d}^3k\; (\tilde{f}(\boldsymbol{k}) + \tilde{f}^*(\boldsymbol{k})) \int \frac{\mathrm{d}^3r}{(2\pi)^3}\, e^{i(\boldsymbol{k}-\boldsymbol{p})\cdot\boldsymbol{r}} \\ = \tilde{f}(\boldsymbol{p}) + \tilde{f}^*(-\boldsymbol{p})\end{aligned} \tag{4.208}$$

and

$$\begin{aligned}\tilde{T}(\boldsymbol{p}) = \int \mathrm{d}^3r\; T(\boldsymbol{r})\, e^{-i\boldsymbol{p}\cdot\boldsymbol{r}} = \int \mathrm{d}^3k\; (-i\omega_k \tilde{f}(\boldsymbol{k}) + i\omega_k \tilde{f}^*(\boldsymbol{k})) \int \frac{\mathrm{d}^3r}{(2\pi)^3}\, e^{i(\boldsymbol{k}-\boldsymbol{p})\cdot\boldsymbol{r}} \\ = -\, i\omega_p \tilde{f}(\boldsymbol{p}) + i\omega_p \tilde{f}^*(-\boldsymbol{p}).\end{aligned} \tag{4.209}$$

The pair of equations (4.208) and (4.209) are easily solved to find

$$\tilde{f}(\boldsymbol{p}) = \frac{1}{2}\left(\tilde{F}(\boldsymbol{p}) + \frac{i\tilde{T}(\boldsymbol{p})}{\omega_p}\right), \quad \tilde{f}^*(-\boldsymbol{p}) = \frac{1}{2}\left(\tilde{F}(\boldsymbol{p}) - \frac{i\tilde{T}(\boldsymbol{p})}{\omega_p}\right). \tag{4.210}$$

Since both $F(\boldsymbol{r})$ and $T(\boldsymbol{r})$ are real-valued functions, it follows that $\tilde{F}(\boldsymbol{p}) = \tilde{F}^*(-\boldsymbol{p})$ and $\tilde{T}(\boldsymbol{p}) = \tilde{T}^*(-\boldsymbol{p})$ and the two results of (4.210) are consistent with each other.

4.11 Spherical sound waves, power, and the Doppler effect

It is possible to find a simple harmonic wave that is *spherically symmetric*, radiating away from the origin. This corresponds to a sound wave of fixed frequency produced by a point source in the fluid. The solution, known as a *spherical wave*, takes the form

$$\delta\varphi(r, t) = \frac{A}{r}\sin(kr - \omega_k t), \tag{4.211}$$

where r is the radial distance from the origin, $\omega_k = k\tilde{v}$, and A is a constant with units of volume per time.

> **Exercise 4.18.** Using the spherical coordinate version of ∇^2, show that $\square_{\tilde{v}}\delta\varphi(r, t) = 0$.

The wave of (4.211) describes *wave fronts*, spherical shells of maximum disturbance, radiating outward from the origin so that their radial position is given by $r = \omega_k t/k = \tilde{v}t$ as time passes. Successive wave fronts are separated by the distance $\lambda = 2\pi/k$.

As the wave fronts radiate and r increases, the amount of disturbance drops. This follows from examining the maximum radial velocity associated with (4.211). Using the gradient in spherical coordinates gives the radial velocity as a function of r,

$$v^r(r, t) = \frac{\partial}{\partial r}\delta\varphi(r, t) = -\frac{A}{r^2}\sin(kr - \omega_k t) + \frac{kA}{r}\cos(kr - \omega_k t). \tag{4.212}$$

For large r the first term becomes negligible compared to the second, so that the radial velocity becomes

$$v^r(r, t) \approx \frac{kA}{r}\cos(kr - \omega_k t) \implies v_m^r(r) \approx \frac{kA}{r} = \frac{\omega_k A}{r\tilde{v}}. \tag{4.213}$$

The level of maximum disturbance drops off inversely with radial distance.

This is a product of energy conservation as the energy and momentum of the wave spread out over ever larger spheres. This follows from finding the energy present in a sphere of radius $R = \tilde{v}T$, where T is the time required for the wave to have expanded to a radius R. Using the spherical symmetry of the solution (4.211), the energy contained in the sphere is given by

$$E(R) = \int_V \mathrm{d}^3r\, \mathcal{H} = \int_V \mathrm{d}^3r\left(\frac{1}{2}\frac{\rho_c^2}{B}\delta\dot{\varphi}^2 + \frac{1}{2}\rho_c\nabla\delta\varphi \cdot \nabla\delta\varphi\right). \tag{4.214}$$

By choosing $R = \tilde{v}T$, it follows that $\sin(kR - \omega_k T) = \sin(k(R - \tilde{v}T)) = \sin(0) = 0$. Integrating by parts and using the divergence theorem (1.165) in combination with the equation of motion (4.194) gives

$$\begin{aligned}\int_V \mathrm{d}^3r\, \frac{1}{2}\rho_c\nabla\delta\varphi \cdot \nabla\delta\varphi &= \int_V \mathrm{d}^3r \left\{\nabla \cdot \left(\frac{1}{2}\rho_c\delta\varphi\nabla\delta\varphi\right) - \frac{1}{2}\rho_c\, \delta\varphi\, \nabla^2\delta\varphi\right\} \\ = \int_{S(V)} \mathrm{d}\boldsymbol{S} \cdot \frac{1}{2}\rho_c\, \delta\varphi\, \nabla\delta\varphi &- \int_V \mathrm{d}^3r\, \frac{1}{2}\rho_c\, \delta\varphi\, \nabla^2\delta\varphi = -\int_V \mathrm{d}^3r\, \frac{1}{2}\frac{\rho_c}{\tilde{v}^2}\, \delta\varphi\, \delta\ddot{\varphi}.\end{aligned} \tag{4.215}$$

Using $\tilde{v}^2 = B/\rho_c$ and (4.215) shows that expression (4.214) becomes

$$\begin{aligned}E(R) &= \int_V \mathrm{d}^3r\, \frac{1}{2}\frac{\rho_c}{\tilde{v}^2}(\delta\dot{\varphi}^2 - \delta\varphi\, \delta\ddot{\varphi}) = \int_0^R \mathrm{d}r\, \frac{2\pi\rho_c\omega_k^2A^2}{\tilde{v}^2} \\ &(\cos^2(kr - \omega_k T) + \sin^2(kr - \omega_k T)) = \int_0^R \mathrm{d}r 2\pi\rho_c\frac{\omega_k^2A^2}{\tilde{v}^2} \\ &= 2\pi\rho_c(v_m^r(R))^2R^3 = 4\pi R^2\tilde{v}T\left(\frac{1}{2}\rho_c(v_m^r(R))^2\right),\end{aligned} \tag{4.216}$$

where it was assumed that R is sufficiently large that the approximation (4.213) is valid. Result (4.216) shows that the source of the sound wave is delivering the average power P_s given by

$$P_s = \frac{E(R)}{T} = 4\pi R^2\tilde{v}\left(\frac{1}{2}\rho_c(v_m^r(R))^2\right). \tag{4.217}$$

This is the three-dimensional extension of the one-dimensional result (3.69) for the string wave. It shows that the energy per unit time flowing outward through the spherical surface of area $S = 4\pi R^2$ is the product of the surface area, the velocity of propagation, and the energy density of the disturbance at the surface.

One of the hallmarks of Newtonian physics is the use of the Galilean transformation (2.27) to relate observations made by observers in relative motion. The solutions (4.195) were found in the *rest frame* of the fluid medium, so that the velocity of propagation is the same in all directions. To find the solution in the frame of an observer moving relative to the medium, the transformation $\boldsymbol{r}' = \boldsymbol{r} - \boldsymbol{v}_o t$ is applied to sine-wave solution (4.195) with the result

$$A\sin(\boldsymbol{k}\cdot\boldsymbol{r} \pm \omega_k t) \;\rightarrow\; A\sin(\boldsymbol{k}\cdot\boldsymbol{r}' \pm (\omega_k \pm \boldsymbol{k}\cdot\boldsymbol{v}_o)t'). \tag{4.218}$$

The wave given by (4.218) is still a simple harmonic wave with the same wavelength λ, but the angular frequency is now given by

$$\omega_k \rightarrow \omega_k' = \omega_k \pm \boldsymbol{k}\cdot\boldsymbol{v}_o = \omega_k \pm k v_o\cos\theta = \omega_k\left(1 \pm \frac{v_o\cos\theta}{\tilde{v}}\right), \tag{4.219}$$

where θ is the angle between $\boldsymbol{k}$ and $\boldsymbol{v}_o$. For simplicity, the case where $\cos\theta = \pm 1$ will be considered. Using $\omega_k = 2\pi f_k$ gives

$$f_k' = f_k\left(1 \pm \frac{v_o}{\tilde{v}}\right). \tag{4.220}$$

Result (4.220) encapsulates the *Doppler effect*, which is the change in the observed frequency of a sound wave caused by the observer's motion. The plus sign occurs for the two combinations $\cos\theta = 1$ and $+\omega_k t$ in (4.218) or $\cos\theta = -1$ and $-\omega_k t$ in (4.218). This corresponds to the observer moving in the opposite direction from the wave motion and encountering more wave fronts per unit time. The minus sign occurs for two combinations: $\cos\theta = -1$ and $+\omega_k t$ in (4.218) or $\cos\theta = 1$ and $-\omega_k t$ in (4.218). This corresponds to the observer moving in the same direction as the wave motion and encountering fewer wave fronts per unit time. Result (4.220) is consistent with Galilean velocity addition, since it gives the velocity of propagation $\tilde{v}'$ according to the observer moving relative to the medium as

$$\tilde{v}' = f_k'\lambda_k = f_k\lambda_k\left(1 \pm \frac{v_o}{\tilde{v}}\right) = \tilde{v}\left(1 \pm \frac{v_o}{\tilde{v}}\right) = \tilde{v} \pm v_o. \tag{4.221}$$

This is identical to the result found in the one-dimensional case from the transformed d'Alembertian (3.49).

The general Doppler effect also includes the possibility of the source of the sound wave moving relative to the medium. This is depicted in figure 4.5. If the source is moving with velocity v_s *directly toward* an observer stationary with respect to the medium, the source will observe the wave traveling toward the stationary observer with a velocity of propagation $\tilde{v} - v_s$. However, the source is still emitting wave fronts separated in time by the period $T = 1/f_k$. In that time, the previous wave front will have traveled the distance of one wavelength according to the stationary observer, which gives

$$\lambda_k' = (\tilde{v} - v_s)T = \frac{\tilde{v} - v_s}{f_k}. \tag{4.222}$$

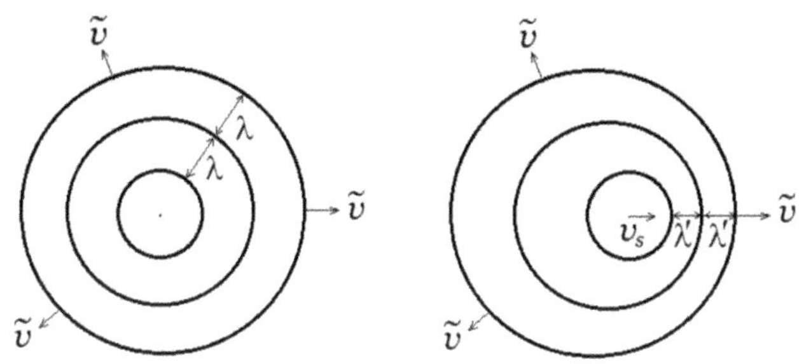

Figure 4.5. The spherical wave emitted by a source stationary with respect to the fluid and the spherical wave emitted by a source moving with the velocity v_s.

The stationary observer will register the frequency f_k' given by the velocity of propagation $\tilde{v}$, so that

$$f_k'\lambda_k' = \tilde{v} = \frac{f_k'}{f_k}(\tilde{v}-v_s) \implies f_k' = f_k \bigg/ \left(1 - \frac{v_s}{\tilde{v}}\right). \qquad (4.223)$$

For the case that the source is moving directly away, an identical argument gives

$$f_k' = f_k \bigg/ \left(1 + \frac{v_s}{\tilde{v}}\right). \qquad (4.224)$$

The two types of Doppler shift in frequency are different, since there is a medium that defines a privileged observer, one who is stationary, $v_o = 0$, with respect to the medium of the fluid.

Exercise 4.19. For the case that both the observer and the source are moving with respect to the medium at the speeds v_o and v_s, show that the observer registers the frequency

$$f_k' = f_k\left(\frac{\tilde{v}+v_o}{\tilde{v}-v_s}\right). \qquad (4.225)$$

In applying this formula v_o is positive if it is toward the source and negative otherwise, while v_s is positive if it is toward the observer and negative otherwise.

References and recommended further reading

Classical field theory texts that cover the topic of fluid mechanics include

- Soper D 2008 *Classical Field Theory* (New York: Dover)
- Nastase H 2019 *Classical Field Theory* (Cambridge: Cambridge University Press)

Physics-oriented monographs dedicated solely to fluid dynamics include

- Batchelor G 1967 *An Introduction to Fluid Dynamics* (Cambridge: Cambridge University Press)
- Landau L and Lifschitz E 1987 *Fluid Dynamics* 2nd edn (Oxford: Pergamon)
- Chorin A and Marsden J 1993 *A Mathematical Introduction to Fluid Mechanics* (Berlin: Springer)
- Bennett A 2006 *Lagrangian Fluid Dynamics* (Cambridge: Cambridge University Press)

Monographs that cover turbulence in fluid flow include

- Feynman R, Leighton R and Sands M 1965 *The Feynman Lectures on Physics* vol II (Reading, MA: Addison-Wesley)
- Tennekes H and Lumley J 1972 *A First Course in Turbulence* (Cambridge, MA: MIT Press)
- Davidson P 2015 *Turbulence: An Introduction for Scientists and Engineers* 2nd edn (Oxford: Oxford University Press)

Fluid dynamics is an important component of astrophysical phenomena. Monographs focusing on astrophysical applications of fluid dynamics include

- Battaner E 1996 *Astrophysical Fluid Dynamics* (Cambridge: Cambridge University Press)
- Choudhuri A 1998 *The Physics of Fluids and Plasmas* (Cambridge: Cambridge University Press)
- Shore S 2007 *Astrophysical Hydrodynamics: An Introduction* 2nd edn (Hoboken, NJ: Wiley)

IOP Publishing

Classical Field Theory and the Stress–Energy Tensor
(Second Edition)

Mark S Swanson

Chapter 5

Galilean covariant complex fields

In the previous two chapters field theories have been developed by applying Newtonian physics to the motion of a medium, either the continuous string of chapter 3 or the fluid of chapter 4. Each medium was shown to support stable wave motion governed by an equation of motion of the form $\Box_{\vec{v}}\varphi = 0$, where φ was a *real* scalar quantity describing the disturbance in the medium. In the cases of string waves and sound waves, the equations of motion were linear, and this allowed the principle of superposition to be applied. In addition, in both cases, the equation of motion changed form under a Galilean transformation. In particular, the one-dimensional string result (3.49) showed that the d'Alembertian $\Box_{\vec{v}}$ factorized into two differential operators associated with differing speeds of propagation when considered in a frame of reference in motion relative to the medium of the string.

In this chapter a theory involving *complex-valued fields* will be developed, based on the requirement that the equation of motion is *Galilean covariant*. This means that the equation of motion for the complex-valued fields will take the same form for all observers related by a Galilean transformation, as opposed to a theory where the Lagrangian and equation of motion change form in different frames of reference. This is accomplished by using complex-valued fields, φ and φ^*, that are *not scalars*. Instead, their behavior under a Galilean transformation allows the equation of motion to take the same form in all Galilean inertial frames. The result is a *field theory version* of the Schrödinger equation without the strict probabilistic interpretation required by quantum mechanics. Instead, it models the behavior of a fluid of identical particles in the classical limit. By including an interaction term, the resulting field theory can be interpreted as a classical model of the much-studied quantum mechanical phenomenon of *Bose–Einstein condensation*. At extremely low temperatures a dilute gas of bosonic particles will condense into the lowest available

doi:10.1088/978-0-7503-3455-6ch5 © IOP Publishing Ltd 2022

energy, forming a macroscopic quantum state known as a Bose–Einstein condensate. The restriction to *bosonic particles* is necessary in order for them to share the same energy. While a detailed explanation of this phenomenon lies in the domain of multiparticle quantum mechanics, its classical field description is referred to as an *effective field theory*, since it encapsulates many of the associated quantum phenomena. It is useful in its own right and offers a reasonably easy description of sound waves and vortices in the condensate without the additional complexity that quantum mechanical analysis requires. However, in order to give an accurate description of Bose–Einstein condensation, it has to borrow some of the quantum mechanical properties of the constituent particles. This is detailed later. In addition, it results in a *nonlinear field theory*, a property of many field theories of relevance to particle physics.

5.1 The complex classical nonrelativistic field

The process of developing a Galilean covariant theory begins by introducing a complex-valued field $\varphi(\boldsymbol{r}, t)$ such that $\varphi^*(\boldsymbol{r}, t) \neq \varphi(\boldsymbol{r}, t)$. Alternatively, the field can be written using the Euler representation (2.80),

$$\varphi(\boldsymbol{r}, t) = R(\boldsymbol{r}, t)e^{i\theta(\boldsymbol{r}, t)} \implies |\varphi(\boldsymbol{r}, t)|^2 = R^2(\boldsymbol{r}, t). \tag{5.1}$$

For the purpose of making a connection to a condensate of particles, the field will be given the following physical meaning. The real quantity $n(\boldsymbol{r}, t) = \varphi^*(\boldsymbol{r}, t)\varphi(\boldsymbol{r}, t) = R^2(\boldsymbol{r}, t)$ is interpreted as the *particle number density* at $\boldsymbol{r}$ at the time t present in the condensate. In a system with an arbitrary number d of spatial dimensions, the field therefore has physical units of $\sqrt{1/V}$, where V is the spatial volume of the system containing the particles. The quantity V may be a length, an area, or a three-dimensional volume, depending on the spatial dimension d. Using the Euler representation (5.1), the number N of particles in the volume is therefore given by

$$N = \int_V \mathrm{d}^d r\, n(\boldsymbol{r}, t) = \int_V \mathrm{d}^d r\, \varphi^*(\boldsymbol{r}, t)\varphi(\boldsymbol{r}, t) = \int_V \mathrm{d}^d r\, R^2(\boldsymbol{r}, t). \tag{5.2}$$

It is assumed that N is the total number of particles in the condensate and is a fixed property of the system. This assumes that no particles are created or destroyed. It is important to note that (5.2) places no constraint on the phase function $\theta(\boldsymbol{r}, t)$. It is assumed that the particles are identical, they have a mass m, and they may interact with each other. If the number N of particles is set to one, then the pure number $R^2(\boldsymbol{r}, t)\, \mathrm{d}^d r$ can be interpreted as the probability of finding that single particle at $\boldsymbol{r}$ at time t in the volume $\mathrm{d}^d r$ around that point. This is precisely the interpretation of the *wave function* in quantum mechanics and it will therefore come as no surprise when the Schrödinger equation emerges from this analysis. However, for the case that $N > 1$, a probabilistic interpretation becomes ambiguous. For example, if there are two particles, then the pure number $R^2(\boldsymbol{r}, t)\, \mathrm{d}^d r$ gives the fraction of two that is present at $\boldsymbol{r}$ and does not yield the probability that one or both of the two particles are present. Instead, for the purposes of this chapter $n(\boldsymbol{r}, t) = |\varphi(\boldsymbol{r}, t)|^2$ is interpreted as the particle number density and all Galilean observers will agree on its value, although they will describe the point using different coordinate systems. In that regard, $m|\varphi|^2$ is identical to the density of a fluid, which is also a Galilean scalar. The fluid analogy will be very useful in interpreting the results of this chapter.

Formulating an action for φ requires finding combinations of φ and φ^* that have units of energy density. Since the fields have only spatial units, this requires the use of the mass m of the particles as well as *some* physical constant with the units of action. Because of the relationship to quantum mechanics, this constant will be denoted by $\hbar$, although it does *not necessarily* have to be Planck's constant. Like the Reynolds number for fluid flow introduced in (4.117), it is often possible to define a constant $\hbar$ from the spatial dimensions and other constants associated with the system being considered. An example is a physical system with a length scale L, a mass scale m, and a velocity of propagation $\tilde{v}$. Such a system allows the definition $\hbar = m\tilde{v}L$, so that $\hbar$ has the desired units of energy ·time. Similarly, if the system is at the Kelvin temperature T_K, it can be associated with the thermal energy $E = k_B T_K$, where k_B is the Boltzmann constant. This gives $\hbar = \sqrt{2mEL^2}$, which does not require a velocity of propagation. It will be seen that both electromagnetism and gravity have naturally occurring values for $\hbar$, while field theories with quantum mechanical origins, such as Bose–Einstein condensation, can simply use Planck's constant $\hbar$.

The terms that make up the Lagrangian density built from these combinations must be real to avoid problems with the stability of fluctuations around the classical solution. This will be discussed later in this chapter. Given the units of φ, there are several possible real *quadratic* terms that involve time derivatives of the field,

$$\mathcal{L}_1(\varphi, \dot{\varphi}, \varphi^*, \dot{\varphi}^*) = \frac{1}{2}i\hbar\left(\varphi^*\frac{\partial\varphi}{\partial t} - \frac{\partial\varphi^*}{\partial t}\varphi\right), \tag{5.3}$$

$$\mathcal{L}_2(\dot{\varphi}, \dot{\varphi}^*) = \frac{\hbar^2}{2m\tilde{v}^2}\frac{\partial\varphi^*}{\partial t}\frac{\partial\varphi}{\partial t}. \tag{5.4}$$

Both $\mathcal{L}_1$ and $\mathcal{L}_2$ are quadratic and real, as well as having the required units of energy density. However, for the purposes of this chapter, $\mathcal{L}_1$ will allow the formulation of a theory that is Galilean covariant, whereas $\mathcal{L}_2$ will not. In addition, $\mathcal{L}_2$ requires a velocity of propagation for its definition, and a condensate does not *a priori* provide such a velocity. A quadratic term involving spatial derivatives is derived from the spatial terms in the string Lagrangian (3.12) and the sound Lagrangian (4.193),

$$\mathcal{L}_s(\varphi, \nabla\varphi) = -\frac{\hbar^2}{2m}\nabla\varphi^* \cdot \nabla\varphi; . \tag{5.5}$$

In addition, there can be a potential $U(\boldsymbol{r}, t)$ which affects the energy of each particle identically. This is modeled by adding a term to the Lagrangian given by

$$-U(\boldsymbol{r}, t)\, n(\boldsymbol{r}, t) = -U(\boldsymbol{r}, t)\varphi^*(\boldsymbol{r}, t)\varphi(\boldsymbol{r}, t). \tag{5.6}$$

The basic quadratic Lagrangian is composed of (5.3), (5.5), and (5.6), while (5.2) is enforced using the method of constraints,

$$\mathcal{L}_Q(\varphi, \partial\varphi, \varphi^*, \partial\varphi^*) = \frac{1}{2}i\hbar\left(\varphi^*\frac{\partial\varphi}{\partial t} - \frac{\partial\varphi^*}{\partial t}\varphi\right) - \frac{\hbar^2}{2m}\nabla\varphi^* \cdot \nabla\varphi - U\varphi^*\varphi + \mu\left(\varphi^*\varphi - \frac{N}{V}\right), \quad (5.7)$$

where V is the spatial volume of the system.

The quantity μ has units of energy and acts as a Lagrange multiplier that enforces (5.2). It is often referred to as the *chemical potential*, since it gives the energy associated with changing the particle number, just as the chemical potential does in thermodynamics. The first law of thermodynamics (4.123) for a volume element at $\boldsymbol{r}$ in a system of identical particles is amended to read

$$\mathrm{d}\mathcal{E}(\boldsymbol{r}) = T(\boldsymbol{r})\,\mathrm{d}S(\boldsymbol{r}) - P(\boldsymbol{r})\,\mathrm{d}V(\boldsymbol{r}) + \mu\,\mathrm{d}N(\boldsymbol{r}), \quad (5.8)$$

where the quantity $\mathrm{d}N(\boldsymbol{r})$ is the change in the number of particles in the volume element. If there are no interactions between the particles, then μ can be set to zero, since there is no additional energy associated with adding or subtracting particles from the volume element. However, if there are interactions among the particles, the term proportional to μ must be kept and its value determined from the energy of interaction. This is a key component of analyzing a Bose–Einstein condensate.

5.2 The Euler–Lagrange equation and its solutions

The first step in analyzing (5.7) is to find the Euler–Lagrange equation of motion by generalizing (3.21) to the case of complex fields. This is done by treating φ and φ^* as *independent degrees of freedom*. The Euler–Lagrange equation for φ is given by

$$\frac{\partial\mathcal{L}}{\partial\varphi^*} - \partial_\mu\frac{\partial\mathcal{L}}{\partial(\partial_\mu\varphi^*)} = \mu\varphi + i\hbar\frac{\partial\varphi}{\partial t} + \frac{\hbar^2}{2m}\nabla^2\varphi - U\varphi = 0, \quad (5.9)$$

along with the complex conjugate of (5.9). Alternatively, the fields may be decomposed as $\varphi(\boldsymbol{r}, t) = X(\boldsymbol{r}, t) + iY(\boldsymbol{r}, t)$, where X and Y are two real fields. Substituting this into the Lagrangian (5.7) gives an alternate expression involving two real fields.

Exercise 5.1. Show that the Euler–Lagrange equations for X and Y are identical to the real and imaginary parts of (5.9).

Equation (5.9) is known as the *Schrödinger equation* and is central to understanding quantum mechanical behavior. Unlike the cases of vibrational waves in the stretched string and sound waves in a fluid, the equation of motion (5.9) does not involve the d'Alembertian. A simple solution to (5.9) is given by

$$\varphi_s(\boldsymbol{r}, t) = \sqrt{\frac{N}{V}}\exp\left(\frac{i}{\hbar}(\boldsymbol{p}\cdot\boldsymbol{r} - \omega_p t)\right), \quad (5.10)$$

where V is the spatial volume of the system, assumed to be very large so that periodic boundary conditions are not needed, and $\boldsymbol{p}$ is a constant vector with units of momentum. The solution (5.10) corresponds to a constant and uniform particle density,

$$n(\boldsymbol{r},\, t) = |\varphi_s(\boldsymbol{r},\, t)|^2 = \frac{N}{V}. \tag{5.11}$$

Constraint (5.2) is satisfied and since there are no interactions among the particles, the chemical potential μ is set to zero. The constant ω_p has the units of energy and is given by the Newtonian expression

$$\omega_p = \frac{p^2}{2m}. \tag{5.12}$$

This does not correspond to a wave with a fixed velocity of propagation. For the case that $\boldsymbol{p} = p\, \hat{\mathbf{e}}_x$, the position of a constant value θ for the phase angle is given by

$$\boldsymbol{p} \cdot \boldsymbol{r} - \omega_p t = \theta \implies x = \frac{\theta + \omega_p t}{p} = \frac{\theta}{p} + \frac{pt}{2m}. \tag{5.13}$$

The speed $\tilde{v}(p)$ at which the fixed phase point moves is dependent upon p,

$$\tilde{v}(p) = \frac{\omega_p}{p} = \frac{p}{2m}. \tag{5.14}$$

Result (5.14) is referred to as the *phase velocity*. The particle number density (5.11) does not have an associated velocity, regardless of the value for p.

In order to understand the physical meaning of solution (5.10), it is useful to examine the stress–energy tensor components. These are defined identically to (3.82), but treat φ and φ^* as two degrees of freedom, and this gives

$$T^\mu_\nu = \frac{\partial \mathcal{L}}{\partial(\partial_\mu \varphi)} \partial_\nu \varphi + \frac{\partial \mathcal{L}}{\partial(\partial_\mu \varphi^*)} \partial_\nu \varphi^* - \delta^\mu_\nu \mathcal{L}. \tag{5.15}$$

Using the Euler–Lagrange equations for both φ and φ^* shows that

$$\partial_\mu T^\mu_\nu = 0, \tag{5.16}$$

as long as the components of (5.15) are evaluated using solution (5.10). The Hamiltonian density $\mathcal{H}$ is given by

$$\mathcal{H} = T^0_0 = \left. \left(\frac{\partial \mathcal{L}}{\partial \dot{\varphi}} \dot{\varphi} + \frac{\partial \mathcal{L}}{\partial \dot{\varphi}^*} \dot{\varphi}^* - \mathcal{L} \right) \right|_{\varphi = \varphi_s} = \frac{\hbar^2}{2m} \nabla \varphi_s^* \cdot \nabla \varphi_s = \frac{N}{V} \frac{p^2}{2m}. \tag{5.17}$$

Integrating (5.17) over the spatial volume V gives the total energy E of the solution (5.10) as $E = Np^2/2m$. This describes the physical situation where each of the N particles in the system has the identical energy $p^2/2m$. Similarly, the spatial components of the stress–energy tensor $T^0{}_j$ give the negative momentum density $-\mathcal{P}_j$, first defined in (3.93) and then used in (4.60),

$$-\mathcal{P}_j = T_j^0 = \left(\frac{\partial \mathcal{L}}{\partial \dot{\varphi}}\partial_j\varphi + \frac{\partial \mathcal{L}}{\partial \dot{\varphi}^*}\partial_j\varphi^*\right)\Bigg|_{\varphi_s} = \frac{1}{2}i\hbar\left(\varphi_s^*\partial_j\varphi_s - \varphi_s\partial_j\varphi_s^*\right) = -\frac{N}{V}p_j. \quad (5.18)$$

The total momentum P_j of the solution is obtained by integrating (5.18) over the spatial volume to find $P_j = Np_j$. This describes the physical situation where each of the N particles in the system has the identical momentum p_j. The Newtonian particle relation $E = Np^2/2m$ has emerged, where $p^2 = \boldsymbol{p} \cdot \boldsymbol{p} = p_j p^j$.

Because the equation of motion (5.9) is linear in φ, its solutions obey the principle of superposition. It is therefore possible to define a *wave packet* solution to (5.9) using the generic solution (5.10). In d spatial dimensions it is given by

$$\varphi(\boldsymbol{r}, t) = \int \frac{\mathrm{d}^d p}{(2\pi\hbar)^d}\, \tilde{\varphi}(\boldsymbol{p}) \exp\left(\frac{i}{\hbar}(\boldsymbol{p} \cdot \boldsymbol{r} - \omega_p t)\right), \quad (5.19)$$

where the integration is over all momentum space. The Fourier transform $\tilde{\varphi}(\boldsymbol{p})$ has units of $\sqrt{V}$, while $\mathrm{d}^d p/(2\pi\hbar)^d$ has units of $1/V$. The form for $\tilde{\varphi}(\boldsymbol{p})$ can be found using the initial condition $\varphi(\boldsymbol{r}, 0) = \varphi_0(\boldsymbol{r})$, using the definition of the Dirac delta,

$$\begin{aligned}\int_V \mathrm{d}^d r\, \varphi_0(\boldsymbol{r}) \exp\left(-\frac{i}{\hbar}\boldsymbol{p} \cdot \boldsymbol{r}\right) &= \int \mathrm{d}^d k \int_V \frac{\mathrm{d}^d r}{(2\pi\hbar)^d}\, \tilde{\varphi}(\boldsymbol{k}) \exp\left(\frac{i}{\hbar}(\boldsymbol{k} - \boldsymbol{p}) \cdot \boldsymbol{r}\right) \\ &= \int \mathrm{d}^d k\, \tilde{\varphi}(\boldsymbol{k})\, \delta^k(\boldsymbol{k} - \boldsymbol{p}) = \tilde{\varphi}(\boldsymbol{p}).\end{aligned} \quad (5.20)$$

It is instructive to consider an example of the behavior of the wave packet solution (5.19). For simplicity, the system will be restricted to one spatial dimension x, and the Fourier transform will be given by

$$\tilde{\varphi}(p) = \sqrt{\sqrt{4\pi}NL}\, \exp\left(-\frac{1}{2}\frac{L^2 p^2}{\hbar^2}\right), \quad (5.21)$$

where L is the size of the one-dimensional system and has units of length. It is understood that L is large such that the average particle density N/L remains constant. This specific form is referred to as a *Gaussian distribution*. Using (5.19), the associated field $\varphi(x, t)$ is given by

$$\begin{aligned}\varphi(x, t) &= \sqrt{\sqrt{4\pi}NL} \int_{-\infty}^{\infty} \frac{\mathrm{d}p}{2\pi\hbar} \exp\left\{-\frac{1}{2}\frac{L^2 p^2}{\hbar^2} + \frac{i}{\hbar}\left(px - \frac{p^2 t}{2m}\right)\right\} \\ &= \sqrt{\sqrt{4\pi}NL} \int_{-\infty}^{\infty} \frac{\mathrm{d}p}{2\pi\hbar} \exp\left\{-\frac{1}{2}\frac{p^2}{\hbar^2}\left(L^2 + \frac{i\hbar t}{m}\right) + \frac{p}{\hbar}(ix)\right\} \\ &= \sqrt{\frac{N}{\sqrt{\pi}\left(L + \dfrac{i\hbar t}{mL}\right)}} \exp\left\{-\frac{x^2}{2\left(L^2 + \dfrac{i\hbar t}{m}\right)}\right\}\end{aligned} \quad (5.22)$$

This field is also a Gaussian, although with a complex coefficient.

Exercise 5.2. Show that the particle density associated with (5.22) is given by

$$n(x,\, t) = \varphi^*(x,\, t)\varphi(x,\, t) = \frac{N}{\sqrt{\pi\left(L^2 + \dfrac{\hbar^2 t^2}{m^2 L^2}\right)}} \exp\left\{ -\frac{x^2}{\left(L^2 + \dfrac{\hbar^2 t^2}{m^2 L^2}\right)} \right\} \tag{5.23}$$

and that the total number of particles is given by

$$\int_{-\infty}^{\infty} \mathrm{d}x\ \varphi^*(x,\, t)\varphi(x,\, t) = N. \tag{5.24}$$

The particle density (5.23) at $t = 0$ peaks at around $x = 0$. However, as time passes, the density at $x = 0$ decreases and the particle density *disperses* uniformly throughout the volume L of the system. For that reason expression $\omega_p = p^2/2m$ is referred to as a *dispersion relation*, since its quadratic nature is the origin of this phenomenon.

If the system is a box with sides of length L, the boundary conditions are $\varphi(\boldsymbol{r},\, t) = 0$ if $\boldsymbol{r}$ corresponds to an edge of the box. The field must obey $\varphi(\boldsymbol{r},\, t) = 0$ if $\boldsymbol{r}$ corresponds to a position outside the box. This guarantees that there are no particles outside the spatial box. For the case that $\boldsymbol{r}$ is a position inside the box, the solutions to (5.9) take the general form

$$\varphi_s(\boldsymbol{r},\, t) = \sqrt{\frac{N}{V}} \exp\left(-\frac{i}{\hbar}\omega_n t\right) \prod_{j=1}^{d} \sin\left(\frac{\pi n_j x^j}{L}\right), \tag{5.25}$$

where n_j are integers and

$$\omega_n = \frac{\hbar^2 \pi^2 n_j n^j}{2mL^2} = \frac{\hbar^2 \pi^2 n^2}{2mL^2}. \tag{5.26}$$

Exercise 5.3. Verify that (5.25) and (5.26) satisfy (5.9) for the case of the box boundary conditions.

5.3 Symmetries of the Lagrangian

An important aspect of the action approach to field theory is the presence of *symmetries*. A symmetry is defined as a set of transformations of the fields and coordinates that leaves the Lagrangian identical in form. In the case of Lagrangian (5.7), an obvious symmetry is simply a change in the field phase by a *constant* quantity α,

$$\varphi \rightarrow e^{i\alpha/\hbar}\varphi, \quad \varphi^* \rightarrow e^{-i\alpha/\hbar}\varphi^* \implies \partial_\mu\varphi \rightarrow e^{i\alpha/\hbar}\partial_\mu\varphi, \quad \partial_\mu\varphi^* \rightarrow e^{-i\alpha/\hbar}\partial_\mu\varphi^*, \tag{5.27}$$

as long as $\nabla\alpha = \dot{\alpha} = 0$.

It is an important consequence that each distinct symmetry of the action gives rise to a *conservation law*. This will be elevated to *Noether's theorem* later in the text, but for now it can be understood for symmetry (5.27) in terms of the Euler–Lagrange equations of motion. Lagrangian (5.7) is unchanged or *invariant* under the action of symmetry (5.27), and this can be expressed as

$$\mathcal{L}(\varphi^*, \partial\varphi^*, \varphi, \partial\varphi) = \mathcal{L}(e^{-i\alpha/\hbar}\varphi^*, e^{-i\alpha/\hbar}\partial\varphi^*, e^{i\alpha/\hbar}\varphi, e^{i\alpha/\hbar}\partial\varphi). \tag{5.28}$$

For the case that α is infinitesimal it follows that $e^{i\alpha/\hbar} \approx 1 + i\alpha/\hbar$. Using this, the right-hand side of relation (5.28) can be expanded in a Taylor series in α that terminates at $O(\alpha)$,

$$\begin{aligned}\mathcal{L}(\varphi^*, \partial\varphi^*, \varphi, \partial\varphi) = \mathcal{L}\left(\varphi^* - \frac{i\alpha}{\hbar}\varphi^*, \partial\varphi^* - \frac{i\alpha}{\hbar}\partial\varphi^*, \varphi + \frac{i\alpha}{\hbar}\varphi, \partial\varphi + \frac{i\alpha}{\hbar}\partial\varphi\right) = \mathcal{L}(\varphi^*, \partial\varphi^*, \varphi, \partial\varphi) \\ + \frac{i\alpha}{\hbar}\left(\varphi\frac{\partial\mathcal{L}}{\partial\varphi} + \partial_\mu\varphi\frac{\partial\mathcal{L}}{\partial(\partial_\mu\varphi)} - \varphi^*\frac{\partial\mathcal{L}}{\partial\varphi^*} - \partial_\mu\varphi^*\frac{\partial\mathcal{L}}{\partial(\partial_\mu\varphi^*)}\right),\end{aligned} \tag{5.29}$$

where there is a sum over μ with $\partial_0 = \partial/\partial t$ and $\partial_j = \partial/\partial x^j$. Since α is an arbitrary infinitesimal factor, the terms in parentheses must sum to zero. If the terms in parentheses are evaluated using the Euler–Lagrange equation solutions φ_s and φ_s^*, they become

$$\begin{aligned}&\frac{i}{\hbar}\left(\varphi_s\frac{\partial\mathcal{L}}{\partial\varphi_s} + \partial_\mu\varphi_s\frac{\partial\mathcal{L}}{\partial(\partial_\mu\varphi_s)} - \varphi_s^*\frac{\partial\mathcal{L}}{\partial\varphi_s^*} - \partial_\mu\varphi_s^*\frac{\partial\mathcal{L}}{\partial(\partial_\mu\varphi_s^*)}\right) \\ &= \frac{i}{\hbar}\left(\varphi_s\partial_\mu\frac{\partial\mathcal{L}}{\partial(\partial_\mu\varphi_s)} + \partial_\mu\varphi_s\frac{\partial\mathcal{L}}{\partial(\partial_\mu\varphi_s)} - \varphi_s^*\partial_\mu\frac{\partial\mathcal{L}}{\partial(\partial_\mu\varphi_s^*)} - \partial_\mu\varphi_s^*\frac{\partial\mathcal{L}}{\partial(\partial_\mu\varphi_s^*)}\right) \\ &= \partial_\mu\left(\frac{i}{\hbar}\varphi_s\frac{\partial\mathcal{L}}{\partial(\partial_\mu\varphi_s)} - \frac{i}{\hbar}\varphi_s^*\frac{\partial\mathcal{L}}{\partial(\partial_\mu\varphi_s^*)}\right) \equiv \partial_\mu J^\mu = 0,\end{aligned} \tag{5.30}$$

where the quantity J^μ, given by

$$J^\mu = \frac{i}{\hbar}\left(\varphi_s\frac{\partial\mathcal{L}}{\partial(\partial_\mu\varphi_s)} - \varphi_s^*\frac{\partial\mathcal{L}}{\partial(\partial_\mu\varphi_s^*)}\right), \tag{5.31}$$

is both real and a *conserved current*.

The statement $\partial_\mu J^\mu = 0$ can be understood in a manner similar to the fluid equation of continuity (4.29). Treating the quantities J^j as the components of the spatial vector $\boldsymbol{J}$, integrating over an arbitrary spatial volume V, and using the divergence theorem gives

$$\int_V \mathrm{d}^3r\ \partial_\mu J^\mu = \frac{\partial}{\partial t}\int_V \mathrm{d}^3r\ J^0 + \int_V \mathrm{d}^3r\ \nabla\cdot\boldsymbol{J} = \frac{\partial}{\partial t}\int_V \mathrm{d}^3r\ J^0 + \oint_{S(V)} \mathrm{d}\boldsymbol{S}\cdot\boldsymbol{J} = 0. \quad (5.32)$$

This shows that the quantity represented by J^0 in the volume V undergoes a rate of change given by

$$\frac{\partial}{\partial t}\int_V \mathrm{d}^3r\ J^0 = -\oint_{S(V)} \mathrm{d}\boldsymbol{S}\cdot\boldsymbol{J}, \quad (5.33)$$

so that it does not disappear, rather flowing out through the surface S enclosing the volume V. In the case of fluid flow, J^0 was the mass density. In the case of Lagrangian (5.7), the components of the current are given by

$$\begin{aligned} J^0 &= \varphi^*\varphi, \\ J^j &= \frac{\hbar}{2mi}(\varphi^*\ \partial^j\varphi - \varphi\ \partial^j\varphi^*). \end{aligned} \quad (5.34)$$

The component J^0 is the particle density defined. In that sense, J^0 plays the same role as the mass density in fluid flow. The analogy to fluid flow is discussed again in the section on vortex solutions in the condensate. Evaluating the components of (5.34) using the basic solution (5.10) yields

$$\begin{aligned} J^0 &= \frac{N}{V}, \\ J^j &= \frac{N}{V}\frac{p^j}{m}, \end{aligned} \quad (5.35)$$

which trivially satisfy $\partial_\mu J^\mu = 0$.

However, a simple version of the wave packet solution (5.19) can yield a nontrivial current. For example, a linear combination of two different momentum solutions of the form (5.10) is given by

$$\varphi(\boldsymbol{r}, t) = \sqrt{\frac{N}{2V}}\left(\exp\left(\frac{i}{\hbar}(\boldsymbol{p}\cdot\boldsymbol{r} - \omega_p t)\right) + \exp\left(\frac{i}{\hbar}(\boldsymbol{k}\cdot\boldsymbol{r} - \omega_k t)\right)\right). \quad (5.36)$$

The particle density is given by

$$J^0(\boldsymbol{r}, t) = \frac{N}{V}\left(1 + \cos\left(\frac{(\boldsymbol{k}-\boldsymbol{p})\cdot\boldsymbol{r}}{\hbar} - \frac{(\omega_k - \omega_p)t}{\hbar}\right)\right). \quad (5.37)$$

Using the limit $V \to \infty$ shows that

$$\begin{aligned} \int_V \mathrm{d}^d r\ J^0(\boldsymbol{r}, t) &= N + \frac{N}{V}\int_V \mathrm{d}^d r\ \cos\left(\frac{(\boldsymbol{k}-\boldsymbol{p})\cdot\boldsymbol{r}}{\hbar} - \frac{(\omega_k - \omega_p)t}{\hbar}\right) \\ &= N + \frac{(2\pi\hbar)^d N}{V}\ \delta^d(\boldsymbol{k}-\boldsymbol{p}). \end{aligned} \quad (5.38)$$

Using the boosted solution (5.47) in the stress–energy tensor yields a total momentum $\boldsymbol{P}^T = N(\boldsymbol{p} - m\boldsymbol{v}_o)$ and a similarly transformed energy, $\omega_p = |\boldsymbol{p} - m\boldsymbol{v}_o|^2/2m$. However, the transformation is such that all observers agree on the number of particles N given by expression (5.2). This is necessary, since N is a true Galilean scalar. It is clear that recipe (5.47) results in an expression that has replaced momentum $\boldsymbol{p}$ with $\boldsymbol{p} - m\boldsymbol{v}_o$, which is consistent with Galilean relativity.

Adding the phase factor to the definition of the boosted solution (5.45) resolves the question of which observer is entitled to use (5.9). Using (5.45) shows that the transformed equation of motion (5.43) becomes

$$\begin{aligned} &\left(i\hbar\frac{\partial}{\partial t'} - i\hbar\, v_0\cdot\nabla' + \frac{\hbar^2}{2m}\nabla'^2\right)\varphi_s(\boldsymbol{r}' + v_o t', t') = 0 \\ \Longrightarrow &\left(i\hbar\frac{\partial}{\partial t'} - i\hbar\, v_0\cdot\nabla' + \frac{\hbar^2}{2m}\nabla'^2\right)\left(e^{i\theta(\boldsymbol{r}',\, t',\, v_o)}\varphi_s'(\boldsymbol{r}', t')\right) = 0. \end{aligned} \tag{5.48}$$

Using the explicit form (5.46) for $\theta(\boldsymbol{r}', t', v_o)$ yields the following relations,

$$\begin{aligned} -i\hbar v_o\cdot\nabla'\left(e^{i\theta(v_o,\, \boldsymbol{r}',\, t')}\varphi_s'(\boldsymbol{r}', t')\right) &= e^{i\theta(v_o,\, \boldsymbol{r}',\, t')}\left(-i\hbar v_o\cdot\nabla' + mv_o^2\right)\varphi_s'(\boldsymbol{r}', t'), \\ \frac{\hbar^2}{2m}\nabla'^2\left(e^{i\theta(v_o,\, \boldsymbol{r}',\, t')}\varphi_s'(\boldsymbol{r}', t')\right) &= e^{i\theta(v_o,\, \boldsymbol{r}',\, t')}\left(\frac{\hbar^2}{2m}\nabla'^2 + i\hbar v_o\cdot\nabla' - \frac{1}{2}mv_o^2\right)\varphi_s'(\boldsymbol{r}', t'), \\ i\hbar\frac{\partial}{\partial t'}\left(e^{i\theta(v_o,\, \boldsymbol{r}',\, t')}\varphi_s'(\boldsymbol{r}', t')\right) &= e^{i\theta(v_o,\, \boldsymbol{r}',\, t')}\left(i\hbar\frac{\partial}{\partial t'} - \frac{1}{2}mv_o^2\right)\varphi_s'(\boldsymbol{r}', t'). \end{aligned} \tag{5.49}$$

Exercise 5.6. Verify the relations listed in (5.49).

Summing the results of (5.49) shows that the second equation in (5.48) is given by

$$\begin{aligned} &\left(i\hbar\frac{\partial}{\partial t'} - i\hbar\, v_0\cdot\nabla' + \frac{\hbar^2}{2m}\nabla'^2\right)\left(e^{i\theta(\boldsymbol{r}',\, t',\, v_o)}\varphi_s'(\boldsymbol{r}', t')\right) \\ &= e^{i\theta(\boldsymbol{r}',\, t',\, v_o)}\left(i\hbar\frac{\partial}{\partial t'} + \frac{\hbar^2}{2m}\nabla'^2\right)\varphi_s'(\boldsymbol{r}', t') = 0. \end{aligned} \tag{5.50}$$

The overall phase factor in (5.50) is irrelevant, and this shows that *both* observers can use the equation of motion (5.9) as long as the derivatives are written in their respective coordinate systems *and* the solutions are transformed according to (5.45). Recipe (5.45) allows the equation of motion (5.9) to be *covariant* in all Galilean coordinate systems, with the understanding that φ and φ^* do *not* transform as true scalars. In that regard, the form of the equation of motion is universally agreed upon by all observers.

5.5 Complex analysis and Cauchy's theorem

Because the fields φ being considered in this chapter are complex-valued functions, the basic results of *complex analysis* often become important. This section assumes familiarity with the definitions and results of subsection 3.4. It then reviews a handful of useful results.

If $f(z)$ is a function of the complex number z, the derivative of $f(z)$ is defined by picking an infinitesimal complex number $\epsilon = \epsilon_R + i\epsilon_I$ and writing

$$\frac{\mathrm{d}f(z)}{\mathrm{d}z} = \lim_{\epsilon \to 0} \frac{f(z + \epsilon) - f(z)}{\epsilon}. \tag{5.51}$$

If (5.51) exists and *is unique*, it must be independent of how the limit $\epsilon \to 0$ is obtained. Starting with $\epsilon_I = 0$ gives

$$\begin{aligned} \frac{\mathrm{d}f(z)}{\mathrm{d}z} &= \lim_{\epsilon_R \to 0} \frac{u(x + \epsilon_R, y) + iv(x + \epsilon_R, y) - u(x, y) - iv(x, y)}{\epsilon_R} \\ &= \frac{\partial u}{\partial x} + i\frac{\partial v}{\partial x}. \end{aligned} \tag{5.52}$$

Starting with $\epsilon_R = 0$ gives

$$\begin{aligned} \frac{\mathrm{d}f(z)}{\mathrm{d}z} &= \lim_{\epsilon_I \to 0} \frac{u(x, y + \epsilon_I) + iv(x, y + \epsilon_I) - u(x, y) - iv(x, y)}{i\epsilon_I} \\ &= -i\frac{\partial u}{\partial y} + \frac{\partial v}{\partial y}. \end{aligned} \tag{5.53}$$

If (5.52) and (5.53) are to be identical, it must be the case that

$$\frac{\partial u}{\partial x} = \frac{\partial v}{\partial y} \quad \text{and} \quad \frac{\partial u}{\partial y} = -\frac{\partial v}{\partial x}, \tag{5.54}$$

which are known as the *Cauchy–Riemann conditions*. If a complex function satisfies the Cauchy–Riemann conditions at a point in the complex plane and at every point in a neighborhood of that point, the function is said to be *analytic* at that point, since it possesses a unique derivative.

Exercise 5.7. Show that the function $f(z) = z^n$ is analytic everywhere for $n \geqslant 0$, while the function $f(z) = z^*$ is not.

A function of a single complex variable $z = x + iy$ can viewed as a mapping from the *complex plane* into itself. The complex plane is visualized as being two-dimensional, and x and y define a point in the plane. It is often important to evaluate the integral of a complex function along a path in the complex plane, which is referred to as a *contour*. This is similar to integrating a vector function along a path in the real plane, as in Stokes' theorem (1.166).

The elements of the contour C are parameterized as $\mathrm{d}z = \mathrm{d}x + i\,\mathrm{d}y$, and this gives $f(z)\,\mathrm{d}z = (u + iv)(\mathrm{d}x + i\,\mathrm{d}y) = (u\,\mathrm{d}x - v\,\mathrm{d}y) + i(u\,\mathrm{d}y + v\,\mathrm{d}x)$. This can be visualized as a pair of separate line integrals in a real two-dimensional plane. In the case of the real term, a two-dimensional vector function $\boldsymbol{A}$ with the components $A_1(x, y) = u(x, y)$ and $A_2(x, y) = -v(x, y)$ is identified. In the imaginary term, a vector function $\boldsymbol{B}$ is defined with the components $B_1(x, y) = v(x, y)$ and $B_2(x, y) = u(x, y)$. A *closed contour* integral can now be analyzed by treating it as two line integrals in the x–y plane. Using S to denote the surface of the disk formed by the closed contour and applying Stokes' theorem (1.166) gives

$$\begin{aligned}\oint_C \mathrm{d}z\, f(z) &= \oint_C (u\,\mathrm{d}x - v\,\mathrm{d}y) + i(u\,\mathrm{d}y + v\,\mathrm{d}x) = \oint_C \boldsymbol{A}\cdot \mathrm{d}\boldsymbol{\ell} + i\oint_C \boldsymbol{B}\cdot \mathrm{d}\boldsymbol{\ell} \\ &= \int_S \mathrm{d}\boldsymbol{S}\cdot\nabla\times(\boldsymbol{A} + i\boldsymbol{B}) = \int_S \left\{-\left(\frac{\partial u}{\partial y} + \frac{\partial v}{\partial x}\right) + i\left(\frac{\partial u}{\partial x} - \frac{\partial v}{\partial y}\right)\right\}\mathrm{d}x\,\mathrm{d}y,\end{aligned} \tag{5.55}$$

where $\nabla\times\boldsymbol{A}$ has the component perpendicular to the disk given by $\partial A_2/\partial x - \partial A_1/\partial y = -(\partial v/\partial x + \partial u/\partial y)$. The result for $\boldsymbol{B}$ gives $\partial B_2/\partial x - \partial B_1/\partial y = \partial v/\partial x - \partial u/\partial y$. Result (5.55) vanishes by virtue of the Cauchy–Riemann conditions (5.54) *if the function is analytic everywhere* in the disk S. As a result, the closed contour integral of an analytic function is zero, regardless of the contour chosen.

If $f(z)$ is analytic everywhere inside a closed contour C, then the function $g(z) = f(z)/(z - z_o)$ is said to have a *simple pole* at $z = z_o$. For a counterclockwise or right-handed circular contour around z_o, it follows that

$$\oint_C \mathrm{d}z\, g(z) = \oint_C \mathrm{d}z\,\frac{f(z)}{(z - z_o)} = 2\pi i f(z_o), \tag{5.56}$$

regardless of the radius of the circular contour. This can be demonstrated by using the fact that $f(z)$ is analytic to expand it in a Taylor series around the singular point. This starts by writing $z = z_o + \delta z$, where the complex number δz is written $\delta z = re^{i\theta}$ using the Euler representation (2.80). The Taylor series for $f(z)$ is then given by

$$f(z) = f(z_o + re^{i\theta}) = \sum_{n=0}^{\infty}\frac{1}{n!}r^n e^{in\theta}\,\frac{\mathrm{d}^n f(z_o)}{\mathrm{d}{z_o}^n}. \tag{5.57}$$

The denominator of the contour integral is given by $1/(z - z_o) = e^{-i\theta}/r$ and, since the contour is a circle of constant radius r, $\mathrm{d}z = ire^{i\theta}\,\mathrm{d}\theta$. This gives $\mathrm{d}z/(z - z_o) = i\,\mathrm{d}\theta$. The circular contour integral around z_o is therefore given by

$$\oint_C \mathrm{d}z\,\frac{f(z)}{z - z_o} = i\sum_{n=0}^{\infty}\frac{1}{n!}r^n\frac{\mathrm{d}^n f(z_o)}{\mathrm{d}{z_o}^n}\int_0^{2\pi}\mathrm{d}\theta\, e^{in\theta} = 2\pi i f(z_o). \tag{5.58}$$

The final step occurs because the integral over θ vanishes unless $n = 0$. Result (5.58) can be extended to any contour C that encloses the simple pole at $z = z_o$ and is known as *Cauchy's integral formula*. Result (5.58) was derived for a right-handed or counterclockwise contour around the pole. If the contour is left-handed or clockwise around the pole, the result changes sign to $-2\pi i f(z_o)$. For the case that there are several different simple poles enclosed by a counterclockwise contour, Cauchy's integral theorem gives

$$\oint_C dz \frac{f(z)}{(z - z_o)(z - z_1)} = 2\pi i \left(\frac{f(z_o) - f(z_1)}{z_o - z_1} \right). \tag{5.59}$$

It is often the case that an integral of a function along the real line can be continued to a contour integral in the complex plane to allow (5.58) to be used. This is especially useful if a closed contour can be chosen for (5.58) such that the integrand vanishes off the real axis. If the integrand $g(z)$ appearing in (5.56) can be written

$$g(z) = e^{iaz} f(z), \tag{5.60}$$

where a is a positive parameter, then *Jordan's lemma* states that a semicircular contour C_R in the upper half plane, parameterized by $z = Re^{i\theta}$ where R is real and $0 \leqslant \theta \leqslant \pi$, is bounded by

$$\left| \int_{C_R} dz\, g(z) \right| \leqslant \frac{\pi}{a} M_R, \tag{5.61}$$

where M_R is the maximum value of $|f(Re^{i\theta})|$ along the semicircular contour. As a result, if

$$\lim_{R \to \infty} f(Re^{i\theta}) = 0, \tag{5.62}$$

then

$$\lim_{R \to \infty} \int_{C_R} dz\, e^{iaz} f(z) = 0. \tag{5.63}$$

For such a case, the integral along the real axis is reduced to identifying the simple poles of the integrand and invoking Cauchy's integral formula (5.58). If a is a negative number, then Jordan's lemma remains valid as long as the semicircular contour is in the *lower half plane*. The Jordan lemma is consistent with the observation that

$$\lim_{z_I \to \infty} e^{iaz} = 0 \tag{5.64}$$

as long as $a > 0$, while

$$\lim_{z_I \to -\infty} e^{-iaz} = 0 \tag{5.65}$$

as long as $a > 0$. This is often used as the reason why the contour is closed, but the Jordan lemma is a more rigorous explanation. The Jordan lemma can be adapted to quarter-circle contours as long as the radius tends to infinity and the argument of the function being integrated tends to zero in that limit.

An example of this is demonstrated by the integral

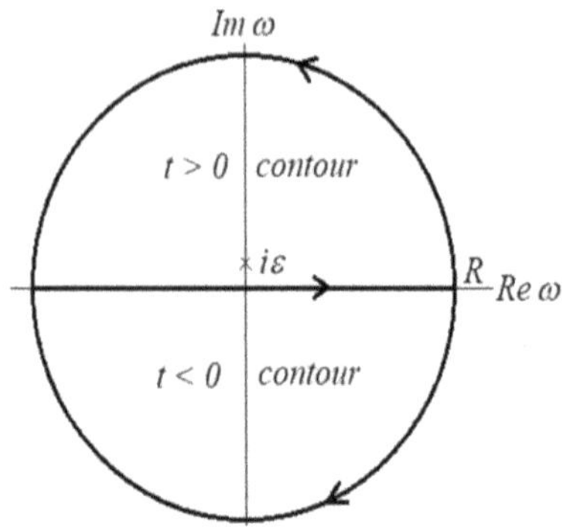

Figure 5.1. The two contours used to evaluate (5.66).

$$\int_{-\infty}^{\infty} \mathrm{d}\omega\, g(\omega) = \lim_{\epsilon \to 0^+} \int_{-\infty}^{\infty} \frac{\mathrm{d}\omega}{2\pi i} \frac{e^{i\omega t}}{\omega - i\epsilon}, \tag{5.66}$$

where t, ω, and ϵ are real and $\epsilon > 0$. If ω is treated as a complex variable, then the function $f(z)$ in (5.60) is identified as

$$f(\omega) = \frac{1}{\omega - i\epsilon} \implies |f(Re^{i\theta} - i\epsilon)| = \frac{1}{\sqrt{R^2 - 2\epsilon R \sin\theta + \epsilon^2}}. \tag{5.67}$$

The maximum value occurs for $\theta = \pi/2$ such that

$$M_R = \frac{1}{\sqrt{R^2 - 2\epsilon R + \epsilon^2}}, \tag{5.68}$$

which vanishes if $R \to \infty$. If $t > 0$ then the closed contour used for Cauchy's theorem includes the semicircular contour in the upper half plane. If $t < 0$ then the closed contour includes a semicircular contour in the lower half plane. This is depicted in figure 5.1, where the limit $R \to \infty$ is understood. However, only for the case $t > 0$ does the closed contour enclose the simple pole at $\omega = i\epsilon$. Cauchy's integral theorem immediately gives

$$\lim_{\epsilon \to 0^+} \int_{-\infty}^{\infty} \frac{\mathrm{d}\omega}{2\pi i} \frac{e^{i\omega t}}{\omega - i\epsilon} = \begin{cases} \lim\limits_{\epsilon \to 0^+} e^{-\epsilon t} = 1 & \text{if } t > 0 \\ 0 & \text{if } t < 0 \end{cases} = \theta(t). \tag{5.69}$$

Result (5.69) shows that (5.66) is the integral representation of the Heaviside step function $\theta(t)$, which was defined initially in (2.45). The relationship between the Heaviside step function and the Dirac delta given by (2.48) shows that result (5.69) is consistent with the integral representation of the Dirac delta given by (2.95). This follows from

$$\frac{\mathrm{d}}{\mathrm{d}t}\theta(t) = \frac{\mathrm{d}}{\mathrm{d}t} \lim_{\epsilon \to 0^+} \int_{-\infty}^{\infty} \frac{\mathrm{d}\omega}{2\pi i} \frac{e^{i\omega t}}{\omega - i\epsilon} = \int_{-\infty}^{\infty} \frac{\mathrm{d}\omega}{2\pi} e^{i\omega t} = \delta(t), \tag{5.70}$$

which is identical to the Fourier representation (2.95) of the Dirac delta.

Exercise 5.8. Show that

$$\lim_{\epsilon \to 0^+} \int_{-\infty}^{\infty} \frac{\mathrm{d}\omega}{2\pi i} \frac{e^{-i\omega t}}{\omega + i\epsilon} = -\theta(t),$$
$$\lim_{\epsilon \to 0^+} \int_{-\infty}^{\infty} \frac{\mathrm{d}\omega}{2\pi i} \frac{e^{-i\omega t}}{\omega - i\epsilon} = \theta(-t). \quad (5.71)$$

5.6 Scattering and the Dirac delta potential

In order to model a Bose–Einstein condensate it is critical to include the effect of interactions among the particles. It will be assumed that the chief form of interaction is through *short-range elastic collisions*. An elastic collision is one in which the energy before and after the collision is the same. The short-range nature of the collision means that before and after the collision the particles are essentially free, so that $\omega_p = p^2/2m$ gives the energy of the particles before and after collision. In what follows it will be shown that a short-range elastic collision can be modeled by using the static potential

$$U(\boldsymbol{r}) = g\,\delta^3(\boldsymbol{r}), \quad (5.72)$$

which represents a *scattering center* located at the origin $\boldsymbol{r} = 0$. The strength of the potential is quantified by g, which has units of energy·volume. Due to the nature of the Dirac delta the potential (5.72) represents a pointlike interaction and is therefore very short-range.

The next goal is to solve the equation of motion (5.9) and then understand how the solution models collisions in the condensate. Since the potential is static the solution for the energy $\omega_p = p^2/2m$ will be written

$$\varphi(\boldsymbol{r}, t) = \varphi_p(\boldsymbol{r})e^{-i\omega_p t/\hbar}. \quad (5.73)$$

Substituting (5.73) into (5.9) and setting $\mu = 0$ gives the equation

$$\left(\frac{\hbar^2}{2m}\nabla^2 + \frac{p^2}{2m}\right)\varphi_p(\boldsymbol{r}) = U(\boldsymbol{r})\varphi_p(\boldsymbol{r}). \quad (5.74)$$

Equation (5.74) will be solved by the method of Green's functions first introduced for the stretched string in (3.109). This begins by finding the function $G_p(\boldsymbol{r} - \boldsymbol{r}')$ that satisfies

$$\left(\frac{\hbar^2}{2m}\nabla^2 + \frac{p^2}{2m}\right)G(\boldsymbol{r} - \boldsymbol{r}') = \delta^3(\boldsymbol{r} - \boldsymbol{r}'). \quad (5.75)$$

The right-hand side of (5.75) is the three-dimensional version of the Green's function equation for the stretched string in (3.109). The Fourier representation of the three-dimensional Dirac delta is given by adapting (2.95), so that

$$\delta^3(\boldsymbol{r} - \boldsymbol{r}') = \int \frac{\mathrm{d}^3k}{(2\pi\hbar)^3}\, e^{i\boldsymbol{k}\cdot(\boldsymbol{r}-\boldsymbol{r}')/\hbar}. \tag{5.76}$$

It follows that the Green's function that satisfies (5.75) is given by the integral expression

$$G_p(\boldsymbol{r} - \boldsymbol{r}') = 2m \lim_{\epsilon \to 0^+} \int \frac{\mathrm{d}^3k}{(2\pi\hbar)^3} \frac{e^{i\boldsymbol{k}\cdot(\boldsymbol{r}-\boldsymbol{r}')/\hbar}}{(p + i\epsilon)^2 - k^2}. \tag{5.77}$$

The *epsilon procedure* used in (5.83) will be shown to match φ_p to an *outgoing* scattering wave.

Exercise 5.9. Verify that (5.77) formally solves (5.75) in the $\epsilon \to 0^+$ limit.

The next step is to express the solution to (5.74) using the Green's function. We start by stipulating the solution to (5.74) for the case that $U = 0$. For such a case it follows that $\varphi(\boldsymbol{r}, t)$ should become the *free* solution (5.10) with the same energy $\omega_p = p^2/2m$, so that

$$\lim_{U \to 0} \varphi(\boldsymbol{r}, t) = \sqrt{n_o}\, e^{i\boldsymbol{p}\cdot\boldsymbol{r}/\hbar} e^{-i\omega_p t/\hbar} \equiv \varphi_0(\boldsymbol{r})\, e^{-i\omega_p t/\hbar}. \tag{5.78}$$

The solution to (5.74) that is consistent with (5.78) can now be formally written using the Green's function as

$$\varphi_p(\boldsymbol{r}) = \varphi_0(\boldsymbol{r}) + \int \mathrm{d}^3r'\, G_p(\boldsymbol{r} - \boldsymbol{r}')\, U(\boldsymbol{r}')\, \varphi_p(\boldsymbol{r}'). \tag{5.79}$$

Because $\varphi_0(\boldsymbol{r})$ satisfies

$$\left(\frac{\hbar^2}{2m}\nabla^2 + \frac{p^2}{2m}\right)\varphi_0(\boldsymbol{r}) = 0, \tag{5.80}$$

the expression (5.79) satisfies (5.74) as well as the limit (5.78).

The solution given by (5.79) is self-referential. It is used to generate a power series in U through iteration, which is simply substituting the expression for φ_p into the right-hand side. After one iteration (5.79) becomes

$$\begin{aligned}\varphi_p(\boldsymbol{r}) = \varphi_0(\boldsymbol{r}) &+ \int \mathrm{d}^3r'\, G_p(\boldsymbol{r} - \boldsymbol{r}')\, U(\boldsymbol{r}')\, \varphi_0(\boldsymbol{r}') \\ &+ \int \mathrm{d}^3r'\, G_p(\boldsymbol{r} - \boldsymbol{r}')\, U(\boldsymbol{r}') \int \mathrm{d}^3r''\, G_p(\boldsymbol{r}' - \boldsymbol{r}'')\, U(\boldsymbol{r}'')\, \varphi_p(\boldsymbol{r}'').\end{aligned} \tag{5.81}$$

If U is a *weak* interaction, it is often the case that only the first iteration is kept and the second integral term is dropped, so that φ_p has the approximate solution

$$\varphi_p(\boldsymbol{r}) \approx \varphi_0(\boldsymbol{r}) + \int \mathrm{d}^3r'\, G_p(\boldsymbol{r} - \boldsymbol{r}')\, U(\boldsymbol{r}')\, \varphi_0(\boldsymbol{r}'). \tag{5.82}$$

Result (5.82) is known as the *Born approximation.*

Using (5.72) and (5.78) in (5.82) and performing the integration over $\boldsymbol{r}'$ gives

$$\varphi_p(\boldsymbol{r}) = \varphi_0(\boldsymbol{r}) + 2mg\sqrt{n_o}\lim_{\epsilon \to 0^+}\int \frac{\mathrm{d}^3k}{(2\pi\hbar)^3}\frac{e^{i\boldsymbol{k}\cdot\boldsymbol{r}/\hbar}}{(p+i\epsilon)^2 - k^2}. \tag{5.83}$$

Evaluating the integral in (5.83) is best done in spherical coordinates in momentum space, given by (k, θ_k, ϕ_k). Choosing k_z to be parallel to $\boldsymbol{r}$ gives $\boldsymbol{k}\cdot\boldsymbol{r} = kr\cos\theta_k$ while $\mathrm{d}^3k = k^2\sin\theta_k\,\mathrm{d}k\,\mathrm{d}\theta_k\,\mathrm{d}\phi_k$. The angular integrations are then readily performed using the result

$$\int_0^\infty \mathrm{d}k\,k^2\int_{-1}^1 \mathrm{d}(\cos\theta_k)\int_0^{2\pi}\mathrm{d}\phi_k\,e^{ikr\cos\theta_k/\hbar} = \frac{\pi\hbar}{ir}\int_{-\infty}^\infty \mathrm{d}k \\ k(e^{ikr/\hbar} - e^{-ikr/\hbar}). \tag{5.84}$$

Expression (5.83) becomes

$$\varphi_p(\boldsymbol{r}) = \varphi_0 + \frac{mg\sqrt{n_o}}{4\pi^2 i\hbar^2 r}\lim_{\epsilon \to 0^+}\int_{-\infty}^\infty \mathrm{d}k\,\frac{k(e^{ikr/\hbar} - e^{-ikr/\hbar})}{(p + i\epsilon - k)(p + i\epsilon + k)}. \tag{5.85}$$

The remaining integral in (5.85) can be evaluated using the Cauchy integral theorem (5.58) in a manner very similar to the Heaviside step function integral (5.66). It is necessary to break (5.85) into the sum of two integrals and choose different contours for them. The two integrals can be written

$$I_\pm(p, r) = \frac{mg\sqrt{n_o}}{4\pi^2 i\hbar^2 r}\lim_{\epsilon \to 0^+}\int_{-\infty}^\infty \mathrm{d}k\,\frac{\pm ke^{\pm ikr/\hbar}}{(p + i\epsilon - k)(p + i\epsilon + k)}, \tag{5.86}$$

so that $\varphi_p(r) = I_+(p, r) + I_-(p, r)$. The integrand has two simple poles, $k = p + i\epsilon$ in the upper half of the complex plane and $k = -p - i\epsilon$ in the lower half of the complex plane. Because $r > 0$ the I_+ integral can be evaluated using the same contour in the upper half plane depicted in figure 5.1. This ensures that $e^{ikr} \to 0$ for the case of an arbitrarily large imaginary value for k. Combining the Cauchy integral theorem with the pole value $k = p + i\epsilon$ gives

$$I_+(p, r) = 2\pi i\left(\frac{mg\sqrt{n_o}}{4\pi^2\hbar^2 ir}\lim_{\epsilon \to 0^+}\frac{1}{2}e^{i(p+i\epsilon)r/\hbar}\right) - \sqrt{n_o}\left(\frac{mg}{4\pi\hbar^2}\right)\frac{e^{ipr/\hbar}}{r}. \tag{5.87}$$

Similarly, the integral $I_-(p, r)$ is analyzed using the same contour in the lower half plane depicted in figure 5.1. This ensures that $e^{-ikr} \to 0$ for the case of an arbitrarily negative imaginary value for k. The result is identical to $I_+(p, r)$. The final result for $\varphi_p(r)$ is

$$\varphi_p(\boldsymbol{r}) = \varphi_0(\boldsymbol{r}) + I_+ + I_- = \sqrt{n_o}\,e^{i\boldsymbol{p}\cdot\boldsymbol{r}/\hbar} + \sqrt{n_o}\left(\frac{mg}{2\pi\hbar^2}\right)\frac{e^{ipr/\hbar}}{r}. \tag{5.88}$$

The term in (5.88) proportional to g becomes arbitrarily large for small r, *regardless* of the strength g of the potential. It is therefore understood that the Born approximation (5.82) breaks down for small r and further iterations are required. Further use of (5.88) will therefore be limited to the case that r is large.

Understanding the correction term proportional to g that has appeared (5.88) begins by noting that the quantity $mg/2\pi\hbar^2$ has units of length and so it defines a quantity a_s, known as the *scattering length*. It is therefore common to write g as a function of a_s,

$$\frac{mg}{2\pi\hbar^2} = a_s \implies g = \frac{2\pi\hbar^2 a_s}{m}. \tag{5.89}$$

As it turns out, this is useful, since the scattering length a_s can be measured experimentally for the specific atoms forming the condensate by using spectrographic methods. Solution (5.88) can then be written

$$\varphi_p(\boldsymbol{r}) = \sqrt{n_o}\left(e^{i\boldsymbol{p}\cdot\boldsymbol{r}/\hbar} + \left(\frac{a_s}{r}\right)e^{ipr/\hbar}\right) \equiv \varphi_0(\boldsymbol{r}) + \varphi_s(r). \tag{5.90}$$

The physical meaning of solution (5.88) can be understood by looking at the spatial component of the density current $\boldsymbol{J}$ defined by (5.34). Using only $\varphi_s(r)$ results in $\boldsymbol{J}$ having solely a radial component and no angular dependence. For the case of large r where the Born approximation is valid, the radial current is given by

$$J^r = \frac{\hbar}{2mi}\left(\varphi_p^* \frac{\partial\varphi_p}{\partial r} - \varphi_p\frac{\partial\varphi_p^*}{\partial r}\right) = n_o\frac{p}{m}\frac{a_s^2}{r^2}. \tag{5.91}$$

Result (5.91) justifies the epsilon procedure used in (5.83), since it has resulted in a positive radial current, representing the process of scattering. This result gives the rate at which particles are exiting a spherical volume with the large radius R around the origin,

$$\frac{\mathrm{d}N}{\mathrm{d}t} = -\oint \mathrm{d}S^r\, J^r = -(4\pi R^2)n_o\frac{p}{m}\frac{a_s^2}{R^2} = -n_o\frac{p}{m}4\pi a_s^2. \tag{5.92}$$

This shares the same characteristics as the rate of energy flow (3.69) in a stretched string. Since $p/m = v$, it follows that (5.92) is the product of the particle flux $n_o v$ and the area $4\pi a_s^2$. This represents the particles in the condensate that are radially scattered by collision with the potential U. The area $4\pi a_s^2$ is known as the *scattering cross-section* associated with the potential (5.72) and gives the *effective size* of the target. The larger the scattering length a_s becomes, the more incident condensate particles will be scattered. The scattering length can therefore be viewed as the effective radius of the atoms comprising the condensate.

5.7 Bose–Einstein condensation

In addition to the terms in (5.7), in order to model the phenomenon of Bose–Einstein condensation, a term representing the energy density associated with the interaction of the particles will be included. This is given by postulating the general potential

$V(\boldsymbol{r}, \boldsymbol{r}', t)$, which gives the energy of interaction between the particle densities at $\boldsymbol{r}$ and $\boldsymbol{r}'$. In *three* spatial dimensions this is written

$$\begin{aligned} V_I(\boldsymbol{r}, t) &= \int_V \mathrm{d}^3r'\, n(\boldsymbol{r}, t)\, V(\boldsymbol{r}, \boldsymbol{r}', t)\, n(\boldsymbol{r}', t) \\ &= \int_V \mathrm{d}^3r'\, \varphi^*(\boldsymbol{r}, t)\varphi(\boldsymbol{r}, t)\, V(\boldsymbol{r}, \boldsymbol{r}', t)\, \varphi^*(\boldsymbol{r}', t)\varphi(\boldsymbol{r}', t). \end{aligned} \tag{5.93}$$

The form for (5.93) is based on the assumption that the energy of interaction between two points in the system is proportional to the particle densities at the two respective points. Including interaction (5.93) in the Lagrangian of (5.7) requires keeping the chemical potential μ, with the result

$$\mathcal{L}(\varphi, \partial\varphi, \varphi^*, \partial\varphi^*) = \frac{1}{2}i\hbar\left(\varphi^*\frac{\partial\varphi}{\partial t} - \frac{\partial\varphi^*}{\partial t}\varphi\right) - \frac{\hbar^2}{2m}\nabla\varphi^* \cdot \nabla\varphi + \mu\left(\varphi^*\varphi - \frac{N}{V}\right) - V_I(\boldsymbol{r}, t). \tag{5.94}$$

In order to make progress in analyzing (5.94), it is assumed that the interaction among the particles is characterized by collisions modeled by the potential (5.72) analyzed in the previous section. For such a case $V(\boldsymbol{r}, \boldsymbol{r}', t)$ is given by the form

$$V(\boldsymbol{r}, \boldsymbol{r}', t) = g\, \delta^3(\boldsymbol{r} - \boldsymbol{r}') = \frac{1}{2}\lambda\, \delta^3(\boldsymbol{r} - \boldsymbol{r}'), \tag{5.95}$$

where the factor of two is for later convenience, so that λ is given by

$$\lambda = 2g = \frac{4\pi\hbar^2 a_s}{m}. \tag{5.96}$$

Using (5.95) in (5.93) gives

$$V_I(\boldsymbol{r}, t) = \int_V \mathrm{d}^3r'\, \varphi^*(\boldsymbol{r}, t)\varphi(\boldsymbol{r}, t)\, V(\boldsymbol{r}, \boldsymbol{r}', t)\, \varphi^*(\boldsymbol{r}', t)\varphi(\boldsymbol{r}', t) = \frac{1}{2}\lambda(\varphi^*(\boldsymbol{r}, t)\varphi(\boldsymbol{r}, t))^2, \tag{5.97}$$

which results in the Lagrangian of (5.94) becoming

$$\begin{aligned} \mathcal{L}(\varphi, \partial\varphi, \varphi^*, \partial\varphi^*) = \frac{1}{2}i\hbar\left(\varphi^*\frac{\partial\varphi}{\partial t} - \frac{\partial\varphi^*}{\partial t}\varphi\right) - \frac{\hbar^2}{2m}\nabla\varphi^* \cdot \nabla\varphi + \mu\left(\varphi^*\varphi - \frac{N}{V}\right) \\ - \frac{1}{2}\lambda(\varphi^*\varphi)^2. \end{aligned} \tag{5.98}$$

This results in a quartic term proportional to λ. Result (5.98) is referred to as the *Gross–Pitaevskii Lagrangian*, although a variant of it was first applied to superconductivity by Ginzburg and Landau. As a model for Bose–Einstein condensation and superconductivity, the Gross–Pitaevskii equation has limitations. It applies only to physical systems at temperatures very close to absolute zero and systems with low particle densities where low-energy collisions predominate.

The Lagrangian density (5.98) gives rise to the Euler–Lagrange equation of motion given by

$$i\hbar\frac{\partial\varphi}{\partial t} + \frac{\hbar^2}{2m}\nabla^2\varphi + \mu\varphi - \lambda|\varphi|^2\varphi = 0. \tag{5.99}$$

Equation (5.99) does not involve the d'Alembertian and, since it is a nonlinear equation, it does not allow the principle of superposition. The constant λ determines the strength of the nonlinearity and is referred to as a *coupling constant*, since it represents the interaction of the field with itself. It has two constant and spatially uniform solutions φ_c such that $\dot{\varphi}_c = 0$ and $\nabla\varphi_c = 0$. The first is the trivial solution $\varphi_c = 0$. Relation (5.2) shows that this solution does not yield the number of particles $N > 0$, so that it does not correspond to the presence of a condensate. The second is the nontrivial solution

$$\varphi_c = \sqrt{\frac{\mu}{\lambda}}. \tag{5.100}$$

It is (5.100) that corresponds to the condensate, since constraint (5.2) gives

$$|\varphi_c|^2 = \frac{\mu}{\lambda} = \frac{N}{V} \equiv n_o \implies \mu = n_o\lambda, \tag{5.101}$$

where $n_o = N/V$ is average particle density. Enforcing constraint (5.2) has led to a solution for the Lagrange multiplier μ that is proportional to the density of the condensate, in a similar way to the Lagrange multipliers that enforced the fixed length of a string and the equation of continuity for a fluid

Comparing the trivial solution $\varphi_c = 0$ with the condensate solution (5.100) is best done from the viewpoint of energy. For constant and uniform solutions, the potential energy density appearing in the Lagrangian (5.98) is given by

$$V(\varphi_c) = -\mu|\varphi_c|^2 + \frac{1}{2}\lambda|\varphi_c|^4 + \mu n_o. \tag{5.102}$$

The trivial solution $\varphi_c = 0$ corresponds to an energy density $V(\varphi_c) = \mu n_o$, while the nontrivial solution (5.100) corresponds to the lower energy density

$$V(\varphi_c) = \mu n_o - \frac{1}{2}\frac{\mu^2}{\lambda} = \frac{1}{2}\mu n_o. \tag{5.103}$$

The presence of the condensate corresponds to a lower energy density, making it energetically preferred.

It is possible to find solutions to (5.99) that correspond to the condensate or its observer in motion. Applying the Galilean transformation recipe (5.45) immediately gives the solution (5.100) according to an observer in motion with velocity v_o,

$$\varphi_c(\boldsymbol{r}, t) = \sqrt{\frac{\mu}{\lambda}}\, \exp\left\{-\frac{i}{\hbar}\left(m\boldsymbol{v}_o \cdot \boldsymbol{r} + \frac{1}{2}mv_o^2 t\right)\right\}. \tag{5.104}$$

Similarly, equation (5.99) is also solved by

$$\varphi_c(\boldsymbol{r}, t) = \sqrt{\frac{\mu}{\lambda}}\, \exp\left\{\frac{i}{\hbar}(\boldsymbol{p} \cdot \boldsymbol{r} - \omega_p t)\right\}, \tag{5.105}$$

where $\omega_p = p^2/2m$. These two solutions correspond to the apparent motion of the individual particles making up the condensate due to the relative motion of the observer and condensate.

5.8 Condensate fluctuations

As in the case of treating sound waves as a density fluctuation in a static fluid, it is important to examine fluctuations in the condensate density for stability. This begins by expanding the Gross–Pitaevskii Lagrangian around the constant solution (5.100). Before doing so, it is important to note that (5.100) can be altered by adding an arbitrary constant phase, since

$$\varphi_c = e^{i\alpha}\sqrt{\frac{\mu}{\lambda}} \tag{5.106}$$

is also a solution of the Euler–Lagrange equation and corresponds to the same particle density. Fluctuations around (5.106) are given by writing the original fields as

$$\varphi(\mathbf{r}, t) = \varphi_c + \eta(\mathbf{r}, t), \quad \varphi^*(\mathbf{r}, t) = \varphi_c^* + \eta^*(\mathbf{r}, t). \tag{5.107}$$

Because the original Lagrangian (5.98) is invariant under a change of phase, $\varphi \to e^{-i\alpha}\varphi$, the phase of φ_c can be removed and absorbed by redefining η according to $\eta \to e^{i\alpha}\eta$. As a result, φ_c can be treated as the real number (5.100). The fluctuation field η represents a deviation in the condensate number density,

$$n = |\varphi|^2 = \varphi_c^2 + \varphi_c(\eta + \eta^*) + |\eta|^2 = n_o + \varphi_c(\eta + \eta^*) + |\eta|^2. \tag{5.108}$$

Invoking the fluid analogy, the fluctuation field therefore represents a density disturbance or sound wave in the condensate.

It is tedious but straightforward to substitute (5.107) into the Gross–Pitaevskii Lagrangian (5.98) and collect terms. Using the specific form for φ_c given by (5.100), the result is

$$\begin{aligned} \mathcal{L}(\eta, \eta^*) &= \frac{1}{2}i\hbar(\eta^*\dot{\eta} - \dot{\eta}^*\eta) - \frac{\hbar^2}{2m}\nabla\eta^* \cdot \nabla\eta \\ &\quad - \frac{\mu N}{V} + \frac{1}{2}\frac{\mu^2}{\lambda} + \frac{1}{2}\mu\,(\eta^* + \eta)^2 - \sqrt{\mu\lambda}\,(\eta^* + \eta)\eta^*\eta - \frac{1}{2}\lambda(\eta^*\eta)^2. \end{aligned} \tag{5.109}$$

> **Exercise 5.10.** Verify result (5.109).

By virtue of using the constant solution (5.100) there are no linear terms in the fluctuation field. Like the Lagrangian (4.193) for sound waves in a fluid, the terms beyond quadratic will be ignored as small compared to the quadratic terms. The quadratic fluctuation Lagrangian is then given by

$$\mathcal{L}_Q(\eta, \eta^*) = \frac{1}{2}i\hbar(\eta^*\dot{\eta} - \dot{\eta}^*\eta) - \frac{\hbar^2}{2m}\nabla\eta^* \cdot \nabla\eta - \mu n_o + \frac{1}{2}\frac{\mu^2}{\lambda} + \frac{1}{2}\mu\,(\eta^* + \eta)^2. \tag{5.110}$$

It is important to note that the constant energy density

$$V_c = \mu n_o - \frac{1}{2}\frac{\mu^2}{\lambda} \tag{5.111}$$

contained in (5.110) yields the same solution (5.100) by extremizing it, since

$$\frac{\partial V_c}{\partial \mu} = 0 \implies \frac{\mu}{\lambda} = n_o, \tag{5.112}$$

so that $\mu = \lambda n_o$.

Applying the Euler–Lagrange equation to (5.110) yields two partial differential equations for the fluctuation fields,

$$i\hbar\frac{\partial \eta}{\partial t} + \frac{\hbar^2}{2m}\nabla^2\eta - \mu(\eta^* + \eta) = 0, \tag{5.113}$$

$$-i\hbar\frac{\partial \eta^*}{\partial t} + \frac{\hbar^2}{2m}\nabla^2\eta^* - \mu(\eta^* + \eta) = 0. \tag{5.114}$$

Solving these two equations simultaneously is expedited by writing them as $\eta = X + iY$ and $\eta^* = X - iY$, where X and Y are both real fields. Substituting these into (5.113) gives

$$i\left(\hbar\frac{\partial X}{\partial t} + \frac{\hbar^2}{2m}\nabla^2 Y\right) + \left(\frac{\hbar^2}{2m}\nabla^2 X - \hbar\frac{\partial Y}{\partial t} - 2\mu X\right) = 0. \tag{5.115}$$

In order for (5.115) to hold, both the real and imaginary parts must vanish, so that

$$\hbar\frac{\partial X}{\partial t} + \frac{\hbar^2}{2m}\nabla^2 Y = 0, \tag{5.116}$$

$$\frac{\hbar^2}{2m}\nabla^2 X - \hbar\frac{\partial Y}{\partial t} - 2\mu X = 0. \tag{5.117}$$

Since (5.114) is the complex conjugate of (5.113), it also yields equations (5.116) and (5.117). Because both fluctuation fields appear in both equations, the two fluctuation fields are said to be *coupled* to each other.

The most effective way to solve (5.116) and (5.117) simultaneously is to make manifestly real Fourier decompositions for X and Y in d spatial dimensions,

$$\begin{aligned} X(\boldsymbol{r}, t) &= \int \frac{\mathrm{d}^d p}{(2\pi\hbar)^d}\left(\tilde{X}(\boldsymbol{p})\, e^{\frac{i}{\hbar}(\boldsymbol{p}\cdot\boldsymbol{r} - \omega_p t)} + \tilde{X}^*(\boldsymbol{p})\, e^{-\frac{i}{\hbar}(\boldsymbol{p}\cdot\boldsymbol{r} - \omega_p t)}\right), \\ Y(\boldsymbol{r}, t) &= \int \frac{\mathrm{d}^d p}{(2\pi\hbar)^d}\left(\tilde{Y}(\boldsymbol{p})\, e^{\frac{i}{\hbar}(\boldsymbol{p}\cdot\boldsymbol{r} - \omega_p t)} + \tilde{Y}^*(\boldsymbol{p})\, e^{-\frac{i}{\hbar}(\boldsymbol{p}\cdot\boldsymbol{r} - \omega_p t)}\right), \end{aligned} \tag{5.118}$$

where the relationship of ω_p to $\boldsymbol{p}$ *is to be determined.* Substituting these expansions into (5.116) gives

$$\hbar\frac{\partial X}{\partial t} + \frac{\hbar^2}{2m}\nabla^2 Y = \int \frac{\mathrm{d}^d p}{(2\pi\hbar)^d}\Bigg\{\left(-i\omega_p\tilde{X}(\boldsymbol{p}) - \frac{p^2}{2m}\tilde{Y}(\boldsymbol{p})\right)e^{\frac{i}{\hbar}(\boldsymbol{p}\cdot\boldsymbol{r} - \omega_p t)} + \left(i\omega_p\tilde{X}^*(\boldsymbol{p}) - \frac{p^2}{2m}\tilde{Y}^*(\boldsymbol{p})\right)e^{-\frac{i}{\hbar}(\boldsymbol{p}\cdot\boldsymbol{r} - \omega_p t)}\Bigg\} = 0. \tag{5.119}$$

Relation (5.119) immediately gives

$$-i\omega_p\tilde{X}(\boldsymbol{p}) - \frac{p^2}{2m}\tilde{Y}(\boldsymbol{p}) = 0 \implies \tilde{Y}(\boldsymbol{p}) = -i\frac{2m\omega_p}{p^2}\tilde{X}(\boldsymbol{p}), \tag{5.120}$$

along with its complex conjugate. Similarly, (5.117) gives

$$\frac{\hbar^2}{2m}\nabla^2 X - \hbar\frac{\partial Y}{\partial t} - 2\mu X = \int \frac{\mathrm{d}^d p}{(2\pi\hbar)^d}\Bigg\{\left(i\omega_p\tilde{Y}(\boldsymbol{p}) - \left(\frac{p^2}{2m} + 2\mu\right)\tilde{X}(\boldsymbol{p})\right)e^{\frac{i}{\hbar}(\boldsymbol{p}\cdot\boldsymbol{r} - \omega_p t)} + \left(-i\omega_p\tilde{Y}^*(\boldsymbol{p}) - \left(\frac{p^2}{2m} + 2\mu\right)\tilde{X}^*(\boldsymbol{p})\right)e^{-\frac{i}{\hbar}(\boldsymbol{p}\cdot\boldsymbol{r} - \omega_p t)}\Bigg\} = 0. \tag{5.121}$$

Relation (5.121) immediately gives

$$i\omega_p\tilde{Y}(\boldsymbol{p}) - \left(\frac{p^2}{2m} + 2\mu\right)\tilde{X}(\boldsymbol{p}) = 0 \implies \tilde{Y}(\boldsymbol{p}) = -\frac{i}{\omega_p}\left(\frac{p^2}{2m} + 2\mu\right)\tilde{X}(\boldsymbol{p}), \tag{5.122}$$

along with its complex conjugate. Equating the results of (5.120) and (5.122) gives

$$-\frac{i}{\omega_p}\left(\frac{p^2}{2m} + 2\mu\right) = -i\frac{2m\omega_p}{p^2} \implies \omega_p = \sqrt{\frac{p^2}{2m}\left(\frac{p^2}{2m} + 2\mu\right)}. \tag{5.123}$$

Result (5.123) is known as the *Bogoliubov dispersion relation.*

The Bogoliubov relation shows that the number density fluctuations of the Bose–Einstein condensate have an energy spectrum quite different from those of standard Newtonian particles, exhibiting the limits

$$\lim_{p\to 0}\omega_p \approx p\sqrt{\frac{\mu}{m}}, \tag{5.124}$$

$$\lim_{p\to\infty}\omega_p \approx \frac{p^2}{2m}. \tag{5.125}$$

The low-energy fluctuations are governed by (5.124) and satisfy the relation $\omega_p = p\sqrt{\mu/m} = p\tilde{v}$, which is identical to the string result (3.81) and the sound result (4.196). By analogy to fluids, $\tilde{v} = \sqrt{\mu/m} = \sqrt{\lambda n_o/m}$ gives the velocity of propagation shared by low-energy sound waves in the condensate. The presence of waves with a constant velocity of propagation can be deduced by combining (5.116) and (5.117). Taking the time derivative of (5.116) gives

$$\frac{\partial^2 X}{\partial t^2} + \frac{\hbar}{2m}\nabla^2\frac{\partial Y}{\partial t} = 0. \tag{5.126}$$

Inserting $\partial Y/\partial t$ from (5.117) into (5.126) gives

$$\frac{\partial^2 X}{\partial t^2} + \frac{\hbar^2}{4m^2}\nabla^2\nabla^2 X - \frac{\mu}{m}\nabla^2 X = 0. \tag{5.127}$$

Result (5.127) can be applied to the Fourier integral (5.118) to determine the Bogoliubov dispersion relation (5.123). The low-momentum limit is given by ignoring the term involving two powers of the Laplacian. For such a limit equation (5.127) becomes

$$\frac{1}{\tilde{v}^2}\frac{\partial^2 X}{\partial t^2} - \nabla^2 X = 0, \tag{5.128}$$

where $\tilde{v}^2 = \mu/m$ is the velocity of propagation for low-energy number density fluctuations found using the Bogoliubov dispersion relation in (5.124). Result (5.128) is the usual wave equation for a disturbance propagating at a constant velocity found for both the stretched string and sound waves in a fluid, and governs low-energy sound waves in the condensate.

It is useful to note that the Bogoliubov spectrum transits from the linear result (5.124) to the quadratic result (5.125) at the value of $p^2/2m = 2\mu$. For that value of $p^2/2m$ the energy of the Bogoliubov dispersion relation can be expressed in two ways,

$$\begin{aligned}\omega_p &= \sqrt{\frac{p^2}{2m}\left(\frac{p^2}{2m} + 2\mu\right)} = \sqrt{\frac{p^2}{2m}\left(\frac{p^2}{m}\right)} = \sqrt{2}\left(\frac{p^2}{2m}\right),\\ \omega_p &= \sqrt{\frac{p^2}{2m}\left(\frac{p^2}{2m} + 2\mu\right)} = \sqrt{2\mu\left(\frac{p^2}{m}\right)} = \sqrt{2}\left(p\sqrt{\frac{\mu}{m}}\right),\end{aligned} \tag{5.129}$$

so that both limits of the dispersion relation occur with an equal factor of $\sqrt{2}$. In anticipation of the next section on vortices, the momentum $p^2 = 4\mu m$ for transiting from linear to quadratic forms of energy can be associated with a length ξ defined by $\xi = \sqrt{2}\ \hbar/p$, with the factor of $\sqrt{2}$ borrowed from (5.129). This gives

$$\frac{p^2}{2m} = 2\mu \implies \xi^2 = \frac{\hbar^2}{2m\mu}. \tag{5.130}$$

The length ξ, known as the *healing length*, represents the distance associated with transiting from the constant-velocity behavior of fluctuations to the dispersive quadratic behavior. It plays a major role in determining the size scale of vortices in the condensate.

5.9 Vortices and the healing length

In addition to the constant solution (5.100) or the boosted solution (5.104), the nonlinear Euler–Lagrange equation (5.99) also admits *vortex solutions*. In order to see this analytically it useful to start with the case that μ and λ are set to zero. A static solution that is symmetric around the z-axis is written in cylindrical coordinates (ρ, ϕ, z) as

$$\varphi_\ell(\rho, \phi) = \nu\rho^{|\ell|}e^{i\ell\phi}, \tag{5.131}$$

where ν is a complex constant, ϕ is the azimuthal angle, and ℓ is a nonzero integer. Restricting ℓ to be an integer gives $\varphi(\rho, \phi + 2\pi) = \varphi(\rho, \phi)$, so that the function $\varphi(\rho, \phi)$ is single valued. Substituting this into the equation of motion in cylindrical coordinates gives

$$\frac{\hbar^2}{2m}\nabla^2\varphi(\rho, \phi) = \frac{\hbar^2}{2m}\left(\frac{\partial^2}{\partial\rho^2} + \frac{1}{\rho}\frac{\partial}{\partial\rho} + \frac{1}{\rho^2}\frac{\partial^2}{\partial\phi^2}\right)(\nu\rho^{|\ell|}e^{i\ell\phi}) = 0, \tag{5.132}$$

which holds for all integer values of ℓ, including $\ell = 0$ and $\ell = \pm 1$.

Exercise 5.11. Verify result (5.132).

In order to understand this solution, it is useful to pursue the analogy to fluid motion discussed earlier in regard to the conserved current (5.34). In the case of fluids the equation of continuity (4.29) was given by

$$\frac{\partial\rho(\boldsymbol{r}, t)}{\partial t} + \nabla\cdot(\boldsymbol{v}\rho(\boldsymbol{r}, t)) = 0, \tag{5.133}$$

where $\rho(\boldsymbol{r}, t)$ is the mass density. Since the particle number density n is given by $\varphi^*\varphi$ in the system under consideration in this chapter, equations (5.133) and (5.34) are formally identical. This allows the conservation law (5.34) to be written

$$\frac{\partial n(\boldsymbol{r}, t)}{\partial t} + \nabla\cdot(\boldsymbol{v}(\boldsymbol{r}, t)\, n(\boldsymbol{r}, t)). \tag{5.134}$$

Comparing the two forms of the conservation law gives the identification

$$\begin{aligned} \boldsymbol{v}(\boldsymbol{r}, t)\, n(\boldsymbol{r}, t) &= \boldsymbol{v}(\boldsymbol{r}, t)\, \varphi^*(\boldsymbol{r}, t)\varphi(\boldsymbol{r}, t) \\ &= \frac{\hbar}{2mi}(\varphi^*(\boldsymbol{r}, t)\nabla\varphi(\boldsymbol{r}, t) - \nabla\varphi^*(\boldsymbol{r}, t)\varphi(\boldsymbol{r}, t)), \end{aligned} \tag{5.135}$$

and this yields the velocity present in the condensate as

$$\boldsymbol{v}(\boldsymbol{r}, t) = \frac{\hbar}{2mi}\frac{(\varphi^*(\boldsymbol{r}, t)\nabla\varphi(\boldsymbol{r}, t) - \nabla\varphi^*(\boldsymbol{r}, t)\varphi(\boldsymbol{r}, t))}{\varphi^*(\boldsymbol{r}, t)\varphi(\boldsymbol{r}, t)}. \tag{5.136}$$

Exercise 5.12. Show that applying definition (5.136) to solution (5.105) gives the Newtonian relation $\boldsymbol{v}(\boldsymbol{r}, t) = \boldsymbol{p}/m$.

If $\varphi = \varphi_o e^{i\theta(r,\, t)}$, where φ_o is a constant, then the velocity expression (5.136) yields

$$v(\mathbf{r},\, t) = \frac{\hbar}{m}\nabla\theta(\mathbf{r},\, t). \tag{5.137}$$

Result (5.137) is identical to expression (4.148) for potential fluid flow in the homentropic case.

Applying (5.136) to solution (5.131) gives the azimuthal velocity

$$v_\phi(\rho) = \frac{\hbar\ell}{m\rho}. \tag{5.138}$$

> **Exercise 5.13.** Verify (5.138) and show that the radial velocity of solution (5.131) is zero.

Result (5.138) drops off as $1/\rho$ and is identical to the velocity (4.81) postulated for a fluid vortex. The constant ℓ determines the angular velocity around the z-axis, and therefore characterizes the presence of an angular momentum density in the condensate. The energy density associated with the solution is given by

$$\mathcal{H} = \frac{\hbar^2}{2m}\nabla\varphi_\ell^* \cdot \nabla\varphi_\ell = \frac{\hbar^2}{2m}\left(\frac{\partial\varphi_\ell^*}{\partial\rho}\frac{\partial\varphi_\ell}{\partial\rho} + \frac{1}{\rho^2}\frac{\partial\varphi_\ell^*}{\partial\phi}\frac{\partial\varphi_\ell}{\partial\phi}\right) = \frac{\hbar^2}{m}|\nu|^2\ell^2\rho^{2|\ell|-2}. \tag{5.139}$$

Result (5.139) shows that the energy density of solution (5.131) is at least a constant and diverges for large ρ unless $\ell = \pm 1$. The solution (5.131) is therefore unphysical for $|\ell| > 1$. This difficulty arises from the diverging velocity for $\rho \to 0$, which is the same problem that the fluid vortex motion of (4.81) exhibited. The simple vortex (5.131) does not solve the full Gross–Pitaevskii equation (5.99) and also yields unphysical results. Nevertheless, it gives a guidepost to the general properties of a realistic vortex solution.

A time-independent solution to (5.99) that is symmetric around the z-axis and corresponds to an azimuthal velocity takes the general form

$$\varphi(\rho,\, \phi) = f_\ell(\rho)e^{i\ell\phi}. \tag{5.140}$$

Using (5.136) the ansatz (5.140) once again gives

$$v_\phi(\rho) = \frac{\hbar\ell}{m\rho}, \qquad v_\rho = 0. \tag{5.141}$$

The Gross–Pitaevskii equation (5.99) becomes

$$\frac{\hbar^2}{2m}\left(\frac{\partial^2}{\partial\rho^2} + \frac{1}{\rho}\frac{\partial}{\partial\rho} - \frac{\ell^2}{\rho^2}\right)f_\ell(\rho) + \mu f_\ell(\rho) - \lambda|f_\ell(\rho)|^2 f_\ell(\rho) = 0. \tag{5.142}$$

For large values of ρ the term involving ℓ^2 is suppressed and the function $f_\ell(\rho)$ becomes the constant solution (5.100), so that

$$\lim_{\rho \to \infty} f_\ell(\rho) = \sqrt{\frac{\mu}{\lambda}}. \tag{5.143}$$

For small values of ρ the function $f_\ell(\rho)$ becomes (5.131), since the limit $\rho \to 0$ suppresses the terms proportional to μ and λ, so that

$$\lim_{\rho \to 0} f_\ell(\rho) = 0. \tag{5.144}$$

Numerical analysis of (5.142) shows that solutions exist that satisfy both limits (5.143) and (5.144).

Unfortunately, there is no exact analytic solution to (5.142). In what follows an *approximate solution* which exhibits these limits will be analyzed. This approximation is given by limiting consideration to $\ell = \pm 1$, since the simple analytic solution (5.131) is energetically uniform for that case and provides a solution for $\rho \to 0$. The ansatz for f_1 is given by

$$f_1(\rho) = \sqrt{\frac{\mu}{\lambda}} \frac{\rho}{\sqrt{\rho^2 + \xi^2}}, \tag{5.145}$$

where ξ is assumed to be a constant characteristic length associated with the condensate. It will be seen to match the *healing length* introduced in (5.130). The ansatz (5.145) has the desired limits,

$$\lim_{\rho \to \infty} f_1(\rho) = \sqrt{\frac{\mu}{\lambda}}, \quad \lim_{\rho \to 0} f_1(\rho) = \sqrt{\frac{\mu}{\lambda}} \frac{\rho}{\xi} \to 0. \tag{5.146}$$

The value of ξ determines the transition from the center of the vortex, where the particle density $n = f_1^2(\rho)$ vanishes, to the situation far from the vortex, where n becomes the constant value $n_o = \mu/\lambda$. If ξ is small, the transition begins at distances close to the z-axis. In effect, ξ determines the radial size of the vortex.

The value of ξ can be approximately determined by substituting (5.145) into (5.142) and choosing ξ to solve the resulting version of (5.142). After considerable algebra, equation (5.142) becomes

$$-\frac{\hbar^2}{2m}\left(\frac{\rho^4 + 4\xi^2\rho^2}{\rho(\rho^2 + \xi^2)^{5/2}}\right)\sqrt{\frac{\mu}{\lambda}}\, e^{i\phi} + \frac{\mu\xi^2(\rho^4 + \xi^2\rho^2)}{\rho(\rho^2 + \xi^2)^{5/2}}\sqrt{\frac{\mu}{\lambda}}\, e^{i\phi} = 0, \tag{5.147}$$

which simplifies to

$$\frac{\hbar^2}{2m}(\rho^2 + 4\xi^2) - \mu\xi^2(\rho^2 + \xi^2) = 0. \tag{5.148}$$

Exercise 5.14. Verify (5.147).

For $\xi/\rho \ll 1$ equation (5.148) is solved by

$$\xi^2 = \frac{\hbar^2}{2m\mu}, \tag{5.149}$$

which matches the transition length (5.130) discovered by considering fluctuations in the condensate. For $\rho/\xi \ll 1$ equation (5.148) is solved by

$$\xi^2 = \frac{2\hbar^2}{m\mu}. \tag{5.150}$$

The two results give values for ξ^2 that differ by a factor of four, showing that ξ cannot be treated as a constant in the ansatz of (5.145) and obtain an exact solution to (5.142). However, equation (5.148) can be solved to find ξ^2 in terms of a slowly varying function of ρ,

$$\xi^2 = \frac{\hbar^2}{m\mu} - \frac{1}{2}\rho^2 + \sqrt{\frac{1}{4}\rho^4 - \frac{\hbar^2}{2m\mu}\rho^2 + \frac{\hbar^4}{m^2\mu^2}}. \tag{5.151}$$

Exercise 5.15. Verify (5.151).

Treating ξ^2 as a constant in deriving (5.148) can be justified if ξ^2 is a slowly varying function of ρ, since its derivatives can then be ignored and (5.148) will hold as a good approximation. To analyze this, it is useful to write ρ in terms of a dimensionless parameter α,

$$\rho^2 = \frac{2\hbar^2}{m\mu}\alpha^2. \tag{5.152}$$

Equation (5.151) then becomes

$$\xi^2 = \frac{\hbar^2}{m\mu}\left(1 - \alpha^2 + \sqrt{\alpha^4 - \alpha^2 + 1}\right). \tag{5.153}$$

Result (5.153) has the same limits as before, $\xi^2 = \hbar^2/2m\mu$ for $\alpha \to \infty$ and $\xi^2 = 2\hbar^2/m\mu$ for $\alpha \to 0$. It also shows that

$$\frac{\partial(\xi^2)}{\partial\rho} = \frac{\partial\alpha}{\partial\rho}\frac{\partial(\xi^2)}{\partial\alpha} = \frac{\hbar}{\sqrt{m\mu}}\left(-2\alpha + \frac{2\alpha^3 - \alpha}{\sqrt{\alpha^4 - \alpha^2 + 1}}\right), \tag{5.154}$$

$$\frac{\partial^2(\xi^2)}{\partial\rho^2} = \left(\frac{\partial\alpha}{\partial\rho}\right)^2\frac{\partial^2(\xi^2)}{\partial\alpha^2} = \left(\frac{2\alpha^6 - 3\alpha^4 + 6\alpha^2 - 1}{(\alpha^4 - \alpha^2 + 1)^{3/2}}\right) - 2. \tag{5.155}$$

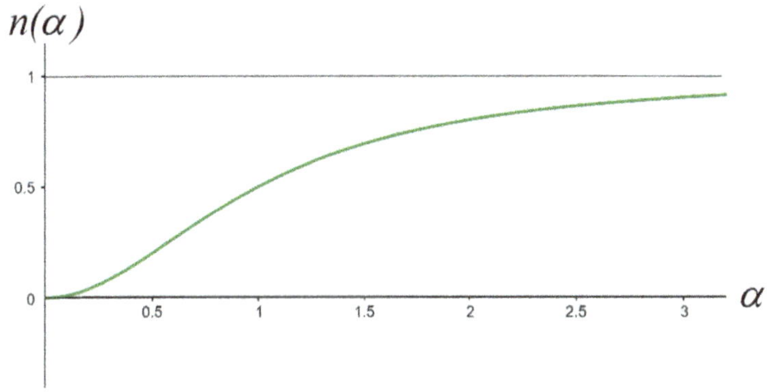

Figure 5.2. The approximate vortex density factor for a Bose–Einstein condensate.

Both (5.154) and (5.155) vanish as $\alpha \to \infty$, becoming approximately zero for $\alpha > 4$. Since $\sqrt{\alpha^4 - \alpha^2 + 1}$ has no zeroes for real values of α^2, both (5.154) and (5.155) remain finite for all α. Numerical analysis shows that (5.154) has a maximum negative value at $\alpha \approx 0.7$, where (5.154) becomes $\approx -1.4\hbar/\sqrt{m\mu}$. Of course, (5.155) vanishes at that point. While (5.154) vanishes at $\alpha = 0$, expression (5.155) becomes -3. Treating the first and second derivatives of ξ^2 as ignorable therefore breaks down for small values of ρ. As a result, it is common in the literature on Bose–Einstein condensates to use result (5.149), since it is the value of ξ^2 for large α where solution (5.145) is most reliable.

The vortex is visualized as a column in the Bose–Einstein condensate. Its size can be inferred in terms of the healing length from (5.145) by finding the value of ρ for which $f_1^2(\rho) > 0.5n_o$. Using the scaling $\rho = \alpha\xi$ in (5.145) gives the particle density

$$f_1^2(\alpha) = \frac{\mu}{\lambda}\frac{\alpha^2}{\alpha^2 + 1} = n(\alpha)\, n_o. \tag{5.156}$$

The function $n(\alpha)$ is displayed in figure 5.2. This function gives $n(\alpha) = 0.5$ when $\alpha = 1$ and $n(\alpha) = 2/3$ when $\alpha = \sqrt{2}$. In effect, the condensate outside the core of the vortex has healed at the radial distance $\rho \approx \sqrt{2}\ \xi$, and this gives the radial size of the vortex.

The healing length ξ can be related to the coupling constant λ found in (5.96). This connection is critical in order that the effective field theory given by the Gross–Pitaevskii model can reflect the properties of the particles of the condensate. Using $\mu = \lambda n_o$ in (5.149) gives

$$\lambda = \frac{\hbar^2}{2mn_o\xi^2}. \tag{5.157}$$

Equating this result to the form for $\lambda = 4\pi\hbar^2 a_s/m$ found in (5.96) gives

$$\xi^2 = \frac{1}{8\pi a_s n_o}. \tag{5.158}$$

Result (5.158) is consistent with the idea that a dilute Bose–Einstein condensate and a small scattering length will have a larger healing length, since it is collisions that damp out the vortex motion. Using (5.158) in (5.149) gives

$$\mu = \frac{\hbar^2}{2m\xi^2} = \frac{4\pi\hbar^2 a_s n_o}{m} = \lambda n_o. \tag{5.159}$$

All the parameters of importance in the effective field theory are now expressed in terms of the condensate density n_o, the scattering length a_s, and the mass m of the constituent particles.

It is possible to use a simple thermodynamic argument to estimate the temperature below which Bose–Einstein condensation can occur. If the average thermal energy available to each particle matches the energy required to change the particle number by one, then thermal effects will evaporate the condensate. The equipartition theorem of thermodynamics states that a typical particle in the condensate has a thermal energy given by $E_K = \frac{3}{2}k_B T$, where k_B is the Boltzmann constant. The change in energy necessary to change the particle number by $\mathrm{d}N = 1$ is obtained from the amended first law of thermodynamics (5.8) by setting $T\,\mathrm{d}S - P\,\mathrm{d}V = 0$ and $\mathrm{d}U = \frac{3}{2}k_B \Delta T = \frac{3}{2}k_B T_c$, so that

$$\mathrm{d}U = \frac{3}{2}k_B T_c = \mu\,\mathrm{d}N = \mu \implies T_c \approx \frac{2\mu}{3k_B}. \tag{5.160}$$

Using expression (5.159) for μ gives the critical temperature in terms of the density,

$$T_c \approx \frac{8\pi\hbar^2 a_s n_o}{3mk_B}. \tag{5.161}$$

The critical temperature computed using quantum statistics is given by

$$T_c = \frac{3.3\,\hbar^2 n_o^{2/3}}{mk_B}. \tag{5.162}$$

While the two results share similar features, the critical temperature obtained from (5.162) is much higher than that obtained from (5.161) and more correctly matches the experimentally observed value. The argument that led to (5.161) suffers from its simplifying assumptions as well as the relatively small number of particles in experimentally observed Bose–Einstein condensates. This number is insufficiently large to create the thermodynamic limit required to make the equipartition theorem valid.

Exercise 5.16. In one of the early experimental observations of Bose–Einstein condensation, the Ketterle group used laser cooling to place $\sim 5 \times 10^5$ sodium atoms into a Bose–Einstein condensate. The number density of the gas was $\sim 1.6 \times 10^{20}$ atoms per cubic meter, approximately a factor of 2×10^5 lower than that of sea-level terrestrial air. Using the mass of a sodium atom $m = 3.8 \times 10^{-26}$ kg and the experimentally measured sodium scattering length $a_s = 2.8 \times 10^{-9}$ m, show that the various parameters associated with the Gross–Pitaevskii Lagrangian are given in SI units by

$$\mu \approx 1.3 \times 10^{-30}\ \text{J}, \quad \lambda \approx 8.4 \times 10^{-51}\ \text{J} \cdot \text{m}^3, \quad \xi \approx 3.3 \times 10^{-7}\ \text{m}. \tag{5.163}$$

Show that (5.161) gives $T_c \approx 6.4 \times 10^{-8}\ {}^o\text{K}$, while (5.162) yields the correct value $T_c \approx 2.0 \times 10^{-6}\ {}^o\text{K}$.

A final point regarding the approximate vortex solution (5.145) is the energy density of (5.140) for the case that $\ell = \pm 1$. This is given by

$$\begin{aligned}\mathcal{H} &= \frac{\hbar^2}{2m}\nabla\varphi^* \cdot \nabla\varphi - \mu(\varphi^*\varphi - n_o) + \frac{1}{2}\lambda(\varphi^*\varphi)^2 \\ &= \frac{\hbar^2\mu}{2m\lambda}\left(\frac{\rho^4 + 2\rho^2\xi^2 + 2\xi^4}{(\rho^2 + \xi^2)^3}\right) - \frac{\mu^2}{\lambda}\left(\frac{\frac{1}{2}\rho^4 + \rho^2\xi^2}{(\rho^2 + \xi^2)^2}\right) + \mu n_o.\end{aligned} \tag{5.164}$$

The large ρ limit of the energy density causes the kinetic energy of the first term to vanish, giving

$$\lim_{\rho \to \infty} \mathcal{H} = -\frac{\mu^2}{2\lambda} + \mu n_o = \frac{\mu^2}{2\lambda}, \tag{5.165}$$

which is identical to the energy density (5.103) of the uniform condensate solution. For small ρ (5.164) gives

$$\lim_{\rho \to 0} \mathcal{H} = \frac{\hbar^2\mu}{m\lambda\xi^2} + \mu n_o. \tag{5.166}$$

Using the healing length (5.150) for small ρ gives

$$\lim_{\rho \to 0} \mathcal{H} = \frac{\mu^2}{2\lambda} + \mu n_o = \frac{3\mu^2}{2\lambda} \tag{5.167}$$

which is three times greater than the energy density at large ρ. The $|\ell| = 1$ vortex is therefore associated with an additional energy density of μ^2/λ in its core.

References and recommended further reading

Bose–Einstein condensation is a quantum mechanical phenomenon. The effective field theory of Bose–Einstein condensation treated in this chapter is developed from quantum mechanics in

- Pitaevskii L and Stringari S 2003 *Bose–Einstein Condensation* (Oxford: Oxford University Press)
- Pethick C and Smith J 2008 *Bose–Einstein Condensation in Dilute Gases* (Cambridge: Cambridge University Press)

Monographs focused on the nonlinear Schrödinger equation of this chapter include

- Sulem C and Sulem P-L 1999 *The Nonlinear Schrödinger Equation* (New York: Springer)

Monographs focused on quantum mechanical systems giving rise to a nonlinear Schrödinger equation include

- Pang X-F and Feng Y-P 2005 *Quantum Mechanics in Nonlinear Systems* (Singapore: World Scientific)

Monographs focusing on complex variables and analysis include

- Gamelin T 2001 *Complex Analysis* (New York: Springer)
- Brown J and Churchill R 2009 *Complex Variables and Applications* 9th edn (New York: McGraw-Hill)

Chapter 6

Basic special relativity

Experimental observations in the latter part of the 19th century forced physicists to abandon Galilean relativity as a universal truth. Although this chapter concentrates on the structure of the special theory of relativity that replaced Galilean relativity, it begins by considering the reasons for the need to alter the rules of relativity. The need for the revision had its origins in electromagnetic phenomena, in particular, the understanding of light as an electromagnetic wave. Although chapter 9 is dedicated to electromagnetic phenomena, it is instructive to review the situation prior to the development of the special theory of relativity in order to understand the conceptual dilemma that physics faced.

6.1 Maxwell's equations

The formulation of the laws governing electric and magnetic phenomena consumed a roughly eighty-year span in physics, starting with the seminal work of Coulomb on electric force in 1785 and ending with Maxwell's analysis of electromagnetic waves in 1865. By Maxwell's time the concepts of electric charge q, the electric field $\mathbf{E}(\boldsymbol{r}, t)$, and the magnetic field $\mathbf{B}(\boldsymbol{r}, t)$ were well established and experimentally quantified. Electric and magnetic fields were understood to be real-valued vector fields, created by the presence and the motion of electric charge. Electric charge was a quantifiable property of *some* inertial matter but, unlike mass, not all inertial matter exhibited this property. The presence and motion of electric charge, much like a fluid, is quantified in terms of an electric charge density scalar $\rho(\boldsymbol{r}, t)$ and an electric current density vector $\mathbf{J}(\boldsymbol{r}, t)$. The relations of the field vectors $\mathbf{E}$ and $\mathbf{B}$ to ρ and $\mathbf{J}$ and to each other are quantified by four fundamental laws, referred to as *Maxwell's equations* due to his codification of them. In the absence of *dielectric materials* and using MKSA units, these four laws are given by

doi:10.1088/978-0-7503-3455-6ch6 © IOP Publishing Ltd 2022

$$
\begin{array}{ll}
\text{Gauss's Law} & \nabla \cdot \mathbf{E} = \rho/\epsilon_0, \\
\text{Ampere's Law} & \nabla \times \mathbf{B} - \mu_0\epsilon_0\, \partial\mathbf{E}/\partial t = \mu_0\mathbf{J}, \\
\text{Faraday's Law} & \nabla \times \mathbf{E} + \partial\mathbf{B}/\partial t = 0, \\
\text{No Magnetic Monopoles} & \nabla \cdot \mathbf{B} = 0,
\end{array}
\tag{6.1}
$$

where ϵ_0 and μ_0 are physical constants referred to as the permittivity and the permeability of free space, respectively. The first two equations have charge and current present and are referred to as the *inhomogeneous equations*. The second two have no reference to electric charge and are referred to as the *homogeneous equations*. The last equation, regarding the absence of pointlike magnetic field sources, can be viewed as an open question, although there has been no experimental evidence to date that magnetic monopoles exist. It is possible to alter Maxwell's equations in the event that such particles are observed, and this is discussed in chapter 9.

In addition, a pointlike particle with an electric charge q moving with velocity $\boldsymbol{v}$ at position $\boldsymbol{r}$ at time t is subjected to the *Lorentz force*,

$$\boldsymbol{F}(\boldsymbol{r}, t) = q\, \mathbf{E}(\boldsymbol{r}, t) + q\, \boldsymbol{v}(\boldsymbol{r}, t) \times \mathbf{B}(\boldsymbol{r}, t). \tag{6.2}$$

It was later experimentally determined that there is a fundamental unit of charge, denoted by e, and all other observable charges are integer multiples of e. The numerous successful applications of Maxwell's equations to physical systems left no one in doubt of their validity.

The two inhomogeneous equations give rise to an important relation between ρ and $\mathbf{J}$. Taking the divergence of Ampere's law and using $\nabla \cdot (\nabla \times \mathbf{B}) = 0$ gives

$$-\epsilon_0 \frac{\partial(\nabla \cdot \mathbf{E})}{\partial t} = \nabla \cdot \mathbf{J}. \tag{6.3}$$

Using Gauss's law in (6.3) immediately gives

$$\frac{\partial \rho}{\partial t} + \nabla \cdot \mathbf{J} = 0, \tag{6.4}$$

which is the familiar form of the conservation law given in (4.29) for the mass of a fluid and in (5.34) for the particle density of a Bose–Einstein condensate. Maxwell's equations are therefore consistent with the experimental observation that electric charge is conserved in all processes.

Exercise 6.1. Using the identity (1.7), show that

$$\nabla \times (\nabla \times \mathbf{E}) = \nabla(\nabla \cdot \mathbf{E}) - \nabla^2\mathbf{E}. \tag{6.5}$$

Maxwell considered the equations of (6.1) in *vacuum*, where ρ and $\mathbf{J}$ vanish. Using the identity (6.5) and Faraday's law gives

$$\nabla \times (\nabla \times \mathbf{E}) + \frac{\partial(\nabla \times \mathbf{B})}{\partial t} = \nabla(\nabla \cdot \mathbf{E}) - \nabla^2\mathbf{E} + \frac{\partial(\nabla \times \mathbf{B})}{\partial t} = 0. \tag{6.6}$$

Using Gauss's law and Ampere's law in (6.6) gives

$$-\nabla^2\mathbf{E} + \epsilon_o\mu_o\frac{\partial^2\mathbf{E}}{\partial t^2} = 0. \tag{6.7}$$

This is the familiar form of the wave equation found for the string in (3.22), sound waves in (4.194), and low-momentum fluctuations in the Bose–Einstein condensate in (5.128).

Exercise 6.2. Show that the magnetic field in vacuum also obeys

$$-\nabla^2\mathbf{B} + \epsilon_o\mu_o\frac{\partial^2\mathbf{B}}{\partial t^2} = 0. \tag{6.8}$$

Results (6.7) and (6.8) show that electric and magnetic fields can travel as a wave whose velocity of propagation is given by

$$c = \frac{1}{\sqrt{\epsilon_o\mu_o}}. \tag{6.9}$$

The experimental values of ϵ_o and μ_o give a value for c that matches the speed of light first measured by Fizeau in 1851 and later measured with ever greater accuracy. As a result, c will stand for the speed of light in vacuum. A simple harmonic wave solution to (6.7), such as the one-dimensional wave (3.31), is given by

$$\mathbf{E}(\boldsymbol{r}, t) = \mathbf{E}_0 \cos(\boldsymbol{k} \cdot \boldsymbol{r} - \omega_k t + \phi), \tag{6.10}$$

where $\boldsymbol{k}$ is the wave vector and satisfies $\boldsymbol{k} \cdot \boldsymbol{k} = k^2 = 4\pi^2/\lambda^2$, while the angular frequency is given by $\omega_k = ck = 2\pi f$. The wavelength λ and the frequency f therefore obey (3.32), so that $f\lambda = c$. The possibility of a phase angle ϕ plays an important role in interference phenomena. A detailed analysis of electromagnetic waves will be given in chapter 9, but for the moment, our attention will be turned to the conceptual problems that arose from treating light as a propagating electromagnetic disturbance in the form given by (6.10).

6.2 The problem with electromagnetic waves

In the three previous chapters the waves that were analyzed represented the propagation of a disturbance in a medium. The string wave's medium was simply the string, the sound wave's medium was the fluid, and the number density fluctuation's medium was the Bose–Einstein condensate. It was natural for 19th century physicists to consider that there might be a medium for the propagation of light. This medium would have to permeate all of space and transparent forms of matter, since light arrives from distant stars as well as passing through transparent materials. Such a medium did not appear to create any discernible impediment to the motion of matter. The medium for light was dubbed the *luminiferous ether* as a result of these ethereal qualities. However, it would then be the case that the speed of light is actually measured *relative to the ether*. It was shown earlier in (3.49) that the wave equation changes form under a Galilean transformation to an observer in motion relative to the medium of the wave disturbance. In the case of light an observer moving with velocity v_o relative to the ether would measure the speed of a light wave moving in the same direction as the observer as $c - v_o$. It was assumed that the Earth's motion around the Sun through the sea of ether would create the situation of an observer moving relative to the medium. For such a case the ratio $v_o/c \approx 10^{-4}$ allowed experiments with light to be devised to observe the effects of motion through the ether.

Before discussing the concepts behind these experiments, it is critical to determine how Galilean relativity and Maxwell's equations interact. The properties of $\mathbf{E}$, $\mathbf{B}$, ρ, and $\mathbf{J}$ under a Galilean boost from the $\mathcal{O}$ frame to the $\mathcal{O}'$ frame can be deduced. This follows from the need for both observers to agree upon the conservation of electric charge (6.4) and from the requirement, first noted in (2.29), that the Lorentz force (6.2) is agreed upon by all Galilean observers in constant relative motion. This starts by treating ρ as a Galilean scalar, so that $\rho = \rho'$. This is consistent with the requirement that both observers agree upon the amount of electric charge contained in a volume of space. Denoting the constant relative velocity of the two observers by v_o, using the transformation rule (5.42) for the partial derivatives shows that the conservation of electric charge equation is agreed upon by both observers only if

$$\frac{\partial \rho}{\partial t} + \nabla \cdot \mathbf{J} = \frac{\partial \rho'}{\partial t'} - v_o \cdot \nabla' \rho + \nabla' \cdot \mathbf{J} = \frac{\partial \rho'}{\partial t'} + \nabla' \cdot \mathbf{J}'. \tag{6.11}$$

It follows that a Galilean boost results in the electric current density becoming

$$\mathbf{J}' = \mathbf{J} - \rho v_o. \tag{6.12}$$

This is consistent with the form of the continuity equation (4.29) for fluids. In that case a fluid moving at the uniform v_o velocity relative to one observer has no current density when observed by the observer boosted to the velocity v_o. Similarly, under a Galilean boost, the velocity appearing in the Lorentz force law becomes $v' = v - v_o$. The requirement that both observers agree upon the force acting on the particle gives

$$\mathbf{F} = q\mathbf{E} + qv \times \mathbf{B} = q\mathbf{E}' + qv' \times \mathbf{B}' = q\mathbf{E}' + q(v - v_o) \times \mathbf{B}' = \mathbf{F}'. \tag{6.13}$$

In order that $\mathbf{F} = \mathbf{F}'$, the Galilean transformed fields must be given by

$$\begin{aligned} &\mathbf{B}' = \mathbf{B}, \\ &\mathbf{E}' = \mathbf{E} + v_o \times \mathbf{B} \implies \mathbf{E} = \mathbf{E}' - v_o \times \mathbf{B}'. \end{aligned} \tag{6.14}$$

While the two Galilean observers will agree upon the magnetic field, they will observe different electric fields. It follows from this that Maxwell's equations change form under a Galilean transformation. For example, Gauss's law becomes

$$\nabla \cdot \mathbf{E}' - v_o \cdot (\nabla \times \mathbf{B}') = \frac{\rho'}{\epsilon_o}. \tag{6.15}$$

The natural question that (6.15) raises is: which observer can assume $v_o = 0$? This raises the specter of a privileged observer who is entitled to the simpler versions of Maxwell's equations. One possible reconciliation was that the equations of (6.1) were not correct and that the correct versions would be invariant or covariant under a Galilean transformation. This idea gained no traction because of the overwhelming success that the equations of (6.1) had demonstrated in explaining numerous phenomena. The second possibility was that the equations of (6.1) were correct in the rest frame of the ether and that the v_o term was necessary when there was motion relative to the ether. There was a third possibility, which was that Galilean relativity was incorrect and needed to be replaced with a new version of relativity that would leave the equations of (6.1) the same for all observers. The third solution endangered the foundations of Newtonian physics, in particular Newtonian mechanics, and called into question the multitude of phenomena that Newton's laws of motion explained. As a result, physicists of the time opted for the ether resolution.

The experimental collaboration of Michelson and Morley that began in 1885 attempted to observe the effects of motion relative to the ether. Their experiment was based on the superposition principle, which states that two light waves traveling via different paths through the ether and then rejoining would exhibit an interference effect due to the difference in the velocity of propagation along the two different paths. This is similar to the interference in (3.35) that gave rise to standing waves in the stretched string. If two waves with no initial phase difference take different amounts of time to reach a point where they overlap or *interfere*, then the principle of superposition and trigonometry applied to the simple harmonic wave (6.10) give

$$\begin{aligned}\mathbf{E} &= \mathbf{E}_0 \cos(\boldsymbol{k} \cdot \boldsymbol{r} - \omega_k t - \omega_k T) + \mathbf{E}_0 \cos(\boldsymbol{k} \cdot \boldsymbol{r} - \omega_k t) \\ &= 2\, \mathbf{E}_0 \cos\left(\frac{1}{2}\omega_k T\right) \cos\left(\boldsymbol{k} \cdot \boldsymbol{r} - \omega_k t + \frac{1}{2}\omega_k T\right),\end{aligned} \tag{6.16}$$

where T is the time difference between the two paths. The difference in paths has caused a phase difference $\phi = \frac{1}{2}\omega_k T$ to appear. If the induced phase $\phi = \frac{1}{2}\omega_k T = \pi c T/\lambda = \pi n$, where n is an integer, the resulting wave will achieve its maximum possible amplitude, corresponding to *constructive interference* or bright light. This will occur if $cT = n\lambda$, which corresponds to the path length difference cT being an integer multiple of the wavelength λ. If $\phi = \pi c T/\lambda = (n + \frac{1}{2})\pi$, then the resulting wave has zero amplitude, corresponding to *destructive interference* or the absence of light. This corresponds to the path length difference cT being a half-integer multiple of the wavelength λ. The enhancement or diminishment of light due to interference is the basis of the experimental device known as an *interferometer*, which can be used to measure the wavelength of light by observing the formation of fringes or bright bands.

Using Galilean relativity shows that the apparent path followed by a light wave would be affected by motion relative to the ether. This is best analyzed in the rest frame of the medium, as depicted in figure 6.1. If the time of travel from source to

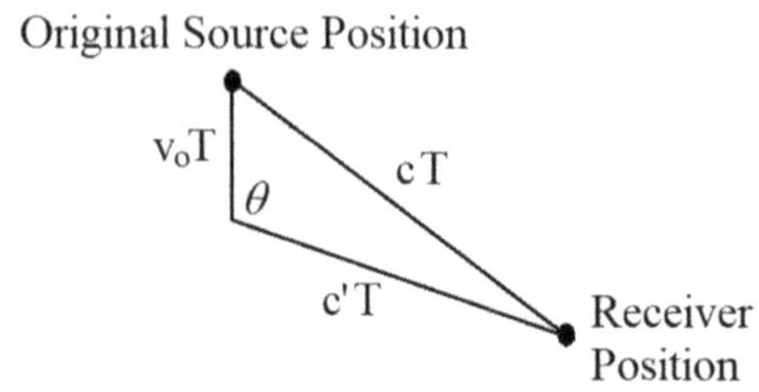

Figure 6.1. The apparent motion of a light pulse emitted by a source in motion relative to the ether.

receiver is T, which all observers agree upon in Galilean relativity, then the observer stationary relative to the ether would see the path length as cT. In the same time T the rest frame observer will see the source move by a distance v_oT. The observer in motion will see the apparent distance between the source and receiver as $c'T$, which both observers will agree upon. The law of cosines gives

$$c' = -v_o \cos\theta + c\sqrt{1 - \frac{v_o^2}{c^2}\sin^2\theta}\,. \tag{6.17}$$

For the case of $\theta = 0$ and $\theta = \pi$ result (6.17) gives $c' = c - v_o$ and $c' = c + v_o$, respectively. This is precisely the same result found in (3.49) by analyzing the Galilean transformation properties of the one-dimensional d'Alembertian. However, it also shows that light propagation perpendicular to the relative velocity results in $c' = c\sqrt{1 - v_o^2/c^2}$. The Michelson and Morley experimental device consisted of a beam splitter that directed equal amounts of light to two mirrors at the same distance L from the beam splitter but oriented at a right angle to each other. The light wave was reflected back to the beam splitter. For the case of the light wave moving parallel and antiparallel to v_o, the time for a round trip is given for small values of v_o/c by

$$T_{\|} = \frac{L}{c + v_o} + \frac{L}{c - v_o} = \frac{2L}{c}\left(1 - \frac{v_o^2}{c^2}\right)^{-1} \approx \frac{2L}{c}\left(1 + \frac{v_o^2}{c^2}\right). \tag{6.18}$$

This exhibits a correction due to the motion of the apparatus relative to the ether. For a round trip in the direction perpendicular to v_o, the time is given by

$$T_{\perp} = \frac{2L}{c\sqrt{1 - v_o^2/c^2}} \approx \frac{2L}{c}\left(1 + \frac{1}{2}\frac{v_o^2}{c^2}\right). \tag{6.19}$$

Using the net difference between (6.18) and (6.19) in result (6.16) gives the phase difference between the two waves due to their different paths and the motion of the apparatus relative to the ether,

$$\phi = \frac{1}{2}\omega_k T = \pi c \frac{(T_\parallel - T_\perp)}{\lambda} \approx \frac{2\pi L}{\lambda} \frac{v_o^2}{c^2}. \tag{6.20}$$

The cosine factor that appears in the interference formula (6.16) therefore has the argument given by

$$\phi = \frac{2\pi L}{\lambda} \frac{v_o^2}{c^2} \equiv \pi n, \tag{6.21}$$

where n gives constructive interference for an integer or destructive interference for a half integer. This would cause the particular wavelength to be shifted when viewed through the interferometer. In addition, the Michelson and Morley apparatus could be rotated, switching the roles of the two paths. Even for the expected small value $v_o/c \approx 10^{-4}$, the short wavelength of visible light, $\lambda \approx 5 \times 10^{-7}$ m, and a long optical path, $L \approx 11$ m, (6.21) gives the expected value of $n \approx 0.44$. This would correspond to the interchange of bright and dark fringes in the apparatus.

Instead, Michelson and Morley conclusively found a *null result*, regardless of how their device was oriented. Experimentally, there was no observable effect caused by moving relative to the hypothetical ether. An attempt to retain the concept of the ether through the idea of ether drag, resulting in $v_o = 0$, suffered a slow experimental death. By the end of the 19th century it had been experimentally established that the observed speed of light c did not depend on motion relative to some medium and the ether concept was abandoned. There was no privileged observer and instead all inertial observers agree upon the form (6.1) for Maxwell's equations. Therefore, the speed of light was the same for all observers, *regardless of relative motion*. It had become clear that Galilean relativity was incorrect. The dilemma facing physics was how to alter Galilean relativity and how to reconcile this alteration with Newtonian physics in a way that retained the successes of Newton's laws of motion. The resolution came with Einstein's publication of the special theory of relativity in 1905.

6.3 Lorentz transformations

The first steps toward the special theory of relativity predated Einstein's publication. In 1904, Lorentz derived a set of transformations which left Maxwell's equations invariant in form and allowed the speed of light to be the same value c for all observers related by these transformations. It was unclear to physicists of the time what the meaning of the Lorentz transformations was. In 1905, Einstein showed that these same transformations could be derived by using simple algebraic manipulations of position x and time t, but only if the concepts of absolute time and length were abandoned. Instead, each individual *inertial observer* is equipped with a clock and meter stick to record the position x and time t of *events*. Events, such as the decay of a radioactive atom or a flash of light, are characterized by a position and a time, just as they were in Newtonian physics. For simplicity, these events occur along the one-dimensional axis of relative motion. Observer $\mathcal{O}$ records the position and time $(x,\ t)$ of the event, while observer $\mathcal{O}'$ records the position and time $(x',\ t')$ of the *same event*. If the two observers are stationary relative to each other and their origins coincide, they will agree that their clocks are ticking at the same rate and that their meter sticks are the same length, so that $x = x'$ and $t = t'$. However, if the two observers are in motion with a constant relative velocity of v, then observer $\mathcal{O}$ sees the origin of the $\mathcal{O}'$ coordinate system located at $x = vt$, while observer $\mathcal{O}'$ sees the origin of the $\mathcal{O}$ coordinate system located at $x' = -vt'$. This assumes that their respective origins coincided at $t = t' = 0$. However, both observers agree that the magnitude of their relative velocity is v.

In the next step both $\mathcal{O}$ and $\mathcal{O}'$ observe the progress of a light pulse that occurs at the moment their origins coincide. Observer $\mathcal{O}$ claims that the two positions of the light pulse along the x-axis, given by $x = \pm ct$, both satisfy $c^2t^2 - x^2 = 0$. Similarly, observer $\mathcal{O}'$ claims that the two positions of the same light pulse along the x'-axis, given by $x' = \pm ct'$, both satisfy $c^2t'^2 - x'^2 = 0$. However, both observers agree that the light wave has the speed of propagation c. In order to reconcile these two observations of the same events, it is assumed that, like the Galilean boost, the measurements are related in a linear way. This assumption can be written as

$$\begin{aligned} ct' &= \Lambda^0{}_0(v)\, ct + \Lambda^0{}_1(v)\, x, \\ x' &= \Lambda^1{}_0(v)\, ct + \Lambda^1{}_1(v)\, x, \end{aligned} \tag{6.22}$$

where the four quantities $\Lambda^\mu{}_\nu$ are independent of x and t and are uniquely determined by the relative velocity v. The relationships of (6.22) are understood as the replacement for the Galilean boost of (2.20) and correctly relate observations made by both *inertial observers*. The form for (6.22) has been chosen to be consistent with the requirement that the two observers are not in relative acceleration, and this requires $\ddot{x}' = 0$. As a result, the relationships of (6.22) must be linear in x and t.

The four quantities of (6.22) can be determined by demanding that both observers agree upon the speed of the two light pulses,

$$c^2t'^2 - x'^2 = (\Lambda^0{}_0(v)\, ct + \Lambda^0{}_1(v)\, x)^2 - (\Lambda^1{}_0(v)\, ct + \Lambda^1{}_1(v)\, x)^2 = c^2t^2 - x^2. \tag{6.23}$$

This yields three equations,

$$\begin{aligned} (\Lambda^0_0(v))^2 - (\Lambda^1_0(v))^2 &= 1, \\ (\Lambda^0_1(v))^2 - (\Lambda^1_1(v))^2 &= -1, \\ \Lambda^0{}_0(v)\, \Lambda^0{}_1(v) - \Lambda^1{}_1(v)\, \Lambda^1{}_0(v) &= 0. \end{aligned} \tag{6.24}$$

A fourth equation is given by the requirement that the origin of the $\mathcal{O}'$ system, $x' = 0$, is located at $x = vt$ according to $\mathcal{O}$. This gives

$$\begin{aligned} &x' = 0 = \Lambda^1{}_0(v)\, ct + \Lambda^1{}_1(v)\, x = \Lambda^1{}_0(v)\, ct + \Lambda^1{}_1(v)\, vt \\ &\implies \Lambda^1{}_0(v) = -\frac{v}{c}\Lambda^1{}_1(v). \end{aligned} \tag{6.25}$$

Solving the system of equations is tedious but straightforward. The result is

$$
\begin{aligned}
\Lambda^0{}_0(v) &= \frac{1}{\sqrt{1-(v/c)^2}} \equiv \gamma_v, \\
\Lambda^0{}_1(v) &= -\frac{v}{c\sqrt{1-(v/c)^2}} = -\frac{v}{c}\gamma_v, \\
\Lambda^1{}_0(v) &= -\frac{v}{c\sqrt{1-(v/c)^2}} = -\frac{v}{c}\gamma_v, \\
\Lambda^1{}_1(v) &= \frac{1}{\sqrt{1-(v/c)^2}} = \gamma_v.
\end{aligned}
\tag{6.26}
$$

Exercise 6.3. Verify that the relations of (6.26) satisfy (6.24) and (6.25).

Returning the results (6.26) to (6.22) gives

$$
\begin{aligned}
t' &= \left(t - \frac{vx}{c^2}\right)\gamma_v, \\
x' &= (x - vt)\gamma_v.
\end{aligned}
\tag{6.27}
$$

The formulas of (6.27) are traditionally known as the *Lorentz transformations* and represent a set of rules that compares the measurement of an event made by $\mathcal{O}$ to the measurement of the *same event* by $\mathcal{O}'$.

Exercise 6.4. Show that the *inverse Lorentz transformations* are given by

$$
\begin{aligned}
t &= \left(t' + \frac{vx'}{c^2}\right)\gamma_v, \\
x &= (x' + vt')\gamma_v.
\end{aligned}
\tag{6.28}
$$

These are the transformations of (6.27) with v replaced by $-v$.

For small v/c, it follows that $\gamma_v \approx 1$ and $v/c^2 \approx 0$, so that the Lorentz transformations (6.27) become the Galilean boost, $x' = x - vt$ and $t' = t$, and the rules of Galilean relativity reappear. However, as $v \to c$, the Lorentz transformations begin to deviate significantly from the Galilean boost and become singular and undefined at $v = c$. In effect, the requirement that all observers agree upon the speed of light prevents two observers from ever having a relative velocity equal to or greater than c, providing the Universe with a speed limit. Because the Lorentz transformations become the Galilean boost for the case that v/c is small, they will likewise be referred to as *Lorentz boosts*.

It is important to note that only the component of the position vector $\boldsymbol{r}$ parallel to $\boldsymbol{v}$, denoted by $\boldsymbol{r}_{\parallel}$, is affected by a Lorentz boost. The component perpendicular to $\boldsymbol{v}$, denoted by $\boldsymbol{r}_{\perp}$, is unaffected. These two components are given by

$$\begin{aligned} \boldsymbol{r}_{\perp} &= \boldsymbol{r} - \frac{\boldsymbol{v}(\boldsymbol{v}\cdot\boldsymbol{r})}{v^2}, \\ \boldsymbol{r}_{\parallel} &= \frac{\boldsymbol{v}(\boldsymbol{v}\cdot\boldsymbol{r})}{v^2}, \end{aligned} \tag{6.29}$$

so that $\boldsymbol{r} = \boldsymbol{r}_{\parallel} + \boldsymbol{r}_{\perp}$, $\boldsymbol{v}\cdot\boldsymbol{r}_{\perp} = 0$, and $\boldsymbol{v}\cdot\boldsymbol{r}_{\parallel} = \boldsymbol{v}\cdot\boldsymbol{r}$. For the case of a relative velocity $\boldsymbol{v}$ that is not directed along a single axis, the Lorentz boost components $\Lambda^{\mu}{}_{\nu}$ are given by

$$\begin{aligned} \Lambda^0{}_0(\boldsymbol{v}) &= \frac{1}{\sqrt{1 - v^2/c^2}} = \gamma_v, \\ \Lambda^0{}_j(\boldsymbol{v}) &= -\frac{v_j}{c\sqrt{1 - v^2/c^2}} = -\frac{v_j}{c}\gamma_v, \\ \Lambda^j{}_0(\boldsymbol{v}) &= -\frac{v^j}{c\sqrt{1 - v^2/c^2}} = -\frac{v^j}{c}\gamma_v, \\ \Lambda^j{}_k(\boldsymbol{v}) &= \delta^j{}_k + (\gamma_v - 1)\frac{v^j v_k}{v^2}. \end{aligned} \tag{6.30}$$

Exercise 6.5. Show that the transformations of (6.30) give the result

$$\begin{aligned} \boldsymbol{r}' &= \boldsymbol{r} + (\gamma_v - 1)\left(\frac{\boldsymbol{v}\cdot\boldsymbol{r}}{v^2}\right)\boldsymbol{v} - \gamma_v \boldsymbol{v} t, \\ t' &= \gamma_v\left(t - \frac{\boldsymbol{v}\cdot\boldsymbol{r}}{c^2}\right). \end{aligned} \tag{6.31}$$

The *inverse transformation* to (6.30) is given by

$$\tilde{\Lambda}^{\mu}_{\nu}(\boldsymbol{v}) = \Lambda^{\mu}_{\nu}(-\boldsymbol{v}), \tag{6.32}$$

which satisfies

$$\tilde{\Lambda}^{\mu}_{\rho}(\boldsymbol{v})\Lambda^{\rho}_{\nu}(\boldsymbol{v}) = \delta^{\mu}_{\nu}. \tag{6.33}$$

Exercise 6.6. Using the explicit forms for $\tilde{\Lambda}^{\mu}_{\rho}$ and $\Lambda^{\rho}_{\nu}(\boldsymbol{v})$, verify (6.33).

Result (6.31) shows that even if the direction of $\boldsymbol{r}$ is orthogonal to $\boldsymbol{v}$, so that $\boldsymbol{v} \cdot \boldsymbol{r} = 0$, $\boldsymbol{r}'$ and $\boldsymbol{r}$ are related by an altered version of the Galilean boost.

The identification of an inertial observer is made by determining that such an observer is moving at a constant velocity relative to *all other inertial observers*. It is the constant velocity requirement that makes this the *special* theory of relativity. It is natural to ask how many inertial observers there are. Both Newtonian physics and Mach's principle treat the bulk matter of the Universe as defining an inertial frame, so that constant motion relative to the distant stars defines an inertial frame. Mach's principle appears in general relativity, which includes arbitrary accelerated frames. Because Newton's laws of motion were developed to reflect Galilean relativity, it becomes critical to understand how special relativity alters Newtonian mechanics into *relativistic mechanics* when accelerated objects are observed. Before proceeding to this goal, it is useful to understand some fundamental differences between special relativity and Galilean relativity.

6.4 Observational effects of special relativity

As mentioned earlier, embracing special relativity requires abandoning the concepts of absolute length and time. Using the Lorentz boosts of (6.30) to compare measurements of events yields results that are quite different from those of Galilean boosts. For simplicity, the effects on different observers' measurements will be presented in the context of the one-spatial-dimension Lorentz boosts of (6.27).

6.4.1 Simultaneity

The first important property of the Lorentz boosts (6.27) occurs when two observers $\mathcal{O}$ and $\mathcal{O}'$ in relative motion compare the time interval between two events. The two events recorded by $\mathcal{O}$ occur *simultaneously* at x_1 and x_2, according to $\mathcal{O}$, so that $t_1 = t_2$. The Lorentz boosts (6.27) show that $\mathcal{O}'$ will claim that these two events occurred at times t_1' and t_2', such that

$$t_2' - t_1' = (t_2 - t_1)\gamma_v - \frac{v}{c^2}(x_2 - x_1)\gamma_v = \frac{v}{c^2}(x_2 - x_1)\gamma_v(v). \tag{6.34}$$

Observer $\mathcal{O}'$ does not agree that the two measurements were made simultaneously by $\mathcal{O}$ unless $x_1 = x_2$. This is in stark contrast to the absolute time of Newton and Galileo, where events that are simultaneous to one observer are simultaneous to all observers, accelerated or inertial.

6.4.2 Length contraction

Another important effect is referred to as *length contraction*. Observer $\mathcal{O}$ places an object along the x-axis whose ends have the positions x_2 and x_1, according to $\mathcal{O}$, at respective times t_1 and t_2. The object is stationary relative to $\mathcal{O}$ and the measurement of the ends are two events recorded by $\mathcal{O}$. Of course, observer $\mathcal{O}$ claims that the length of the object is

$$L = x_2 - x_1. \tag{6.35}$$

Similarly, a second observer $\mathcal{O}'$ measures the length by determining the positions of the ends in the $\mathcal{O}'$ frame. Observer $\mathcal{O}'$ has a relative speed of v parallel to the object and measures its length $x_2' - x_1'$ by *simultaneously* determining the positions x_1' and x_2' of its two ends, so that the two measurements made by $\mathcal{O}'$ have $t_2' = t_1'$. If the positional measurements made by $\mathcal{O}'$ were not simultaneous, $\mathcal{O}'$ would realize that the object's relative motion had resulted in an incorrect measurement of its length. Result (6.27) shows that $\mathcal{O}'$ observes the length

$$L' = x_2' - x_1' = (x_2 - x_1)\gamma_v + v(t_2 - t_1)\gamma_v, \tag{6.36}$$

where t_2 and t_1 are the times of the $\mathcal{O}'$ measurements according to $\mathcal{O}$. The second Lorentz boost formula again shows that the measurements made by $\mathcal{O}'$ are *not simultaneous* according to $\mathcal{O}$,

$$t_2' - t_1' = 0 = (t_2 - t_1)\gamma_v + \frac{v}{c^2}(x_2 - x_1)\gamma_v \implies t_2 - t_1 = -\frac{v}{c^2}(x_2 - x_1). \tag{6.37}$$

Using this result in (6.36) gives

$$L' = x_2' - x_1' = (x_2 - x_1)\gamma_v - \frac{v^2}{c^2}(x_2 - x_1)\gamma_v = (x_2 - x_1)\sqrt{1 - \frac{v^2}{c^2}} = \frac{L}{\gamma_v}. \tag{6.38}$$

This shows that $\mathcal{O}'$ measures a shorter or contracted length for the object in the direction of its motion relative to $\mathcal{O}'$. Observer $\mathcal{O}$ argues that $\mathcal{O}'$ is *not* measuring the positions simultaneously and is therefore allowing the object to move between measurements. Length contraction affects all lengths along the axis of motion, even causing the measurements of areas and volumes to decrease if they are in relative motion.

6.4.3 Time dilation

Another important effect is referred to as *time dilation*. Observer $\mathcal{O}$ has a clock that is stationary at the origin, $x = 0$, and uses it to measure the time interval $t_2 - t_1$ between two events. The Lorentz boosts (6.27) show that the $\mathcal{O}'$ clock at $x' = 0$ will register the time interval for these two events as

$$t_2' - t_1' = (t_2 - t_1)\gamma_v + \frac{v}{c^2}(x_2 - x_1)\gamma_v = (t_2 - t_1)\gamma_v, \tag{6.39}$$

which is larger or dilated compared to the time interval measured by $\mathcal{O}$. This means that $\mathcal{O}'$ will observe that the $\mathcal{O}$ clock ticks slower, although the two clocks behave identically when stationary relative to each other. Using the inverse boost (6.28) shows that $\mathcal{O}$ will observe a time interval measured by the $\mathcal{O}'$ clock located at $x_1' = x_2'$ given by

$$t_2 - t_1 = (t_2' - t_1')\gamma_v + \frac{v}{c^2}(x_2' - x_1')\gamma_v = (t_2' - t_1')\gamma_v, \tag{6.40}$$

which is the same effect. As a result, both observers will claim that the other's clock is ticking slower. This gives rise to what is referred to as the *twin paradox*, asking which of two twin observers will *really* have less time go by according to the local clock as a result of relative motion. The resolution is to place both observers back into a common reference frame and compare their clocks. In order to do this, one or both of the observers must undergo a change in direction or magnitude of velocity, and this will break the symmetry between the two inertial observers and choose one or the other as having had less time elapse. The details of this are presented using the relativistic Doppler effect for light.

6.4.4 The relativistic Doppler effect

Further insight is obtained from the relativistic version of the *Doppler effect*. Observer $\mathcal{O}$ will act as the source of a light wave, which has a frequency f obtained from the light-wave's period T by $f = 1/T$. The light's wavelength is $\lambda = c/f = cT$, according to $\mathcal{O}$. Observer $\mathcal{O}'$ is moving *toward* $\mathcal{O}$ at a relative speed v and is at a distance L', according to $\mathcal{O}'$, when the light source is switched on. The light requires the time $t' = L'/c$ to reach observer $\mathcal{O}'$, so that when the light wave arrives, the distance separating the two observers is given by

$$L' - vt' = L'\left(1 - \frac{v}{c}\right). \tag{6.41}$$

Observer $\mathcal{O}'$ will measure the period of the light T' as larger, due to time dilation, giving $T' = \gamma_v T$. According to $\mathcal{O}'$ during the time interval t' the source emits a number N of cycles given by

$$N = \frac{t'}{T'} = \frac{t'}{\gamma_v T} = \frac{L'}{\gamma_v c\, T} \implies \frac{L'}{N} = c\, T\gamma_v = \lambda\gamma_v. \tag{6.42}$$

Using this result for N, according to $\mathcal{O}'$ the wavelength of each cycle is given by

$$\begin{aligned}\lambda' &= \frac{L'}{N}\left(1 - \frac{v}{c}\right) = \lambda\left(1 - \frac{v}{c}\right)\gamma_v = \lambda\left(1 - \frac{v}{c}\right)\frac{1}{\sqrt{(1 + v/c)(1 - v/c)}} \\ &= \lambda\sqrt{\frac{1 - v/c}{1 + v/c}}.\end{aligned} \tag{6.43}$$

If the source of light is moving toward the receiving observer, then v is positive and the wavelength received by $\mathcal{O}'$ is shorter than the one $\mathcal{O}$ claims to have emitted. This light is referred to as *blueshifted*. If the source of light is receding from the receiving observer, then v is negative and the wavelength is longer or *redshifted*. Using (3.32) allows the Doppler effect for the frequency f to be stated,

$$f' = \frac{c}{\lambda'} = \frac{c}{\lambda}\sqrt{\frac{1 + v/c}{1 - v/c}} = f\sqrt{\frac{1 + v/c}{1 - v/c}}. \tag{6.44}$$

If $v/c = x$ is small then $(1 - x)^{-1} \approx 1 + x$ and (6.44) becomes the approximate relation

$$f' \approx f\sqrt{(1 + v/c)^2} = f\left(1 + \frac{v}{c}\right). \tag{6.45}$$

This is the same as the Galilean Doppler shift for sound found in (4.223) if the relative velocity is small compared to the speed of propagation. However, for $v \approx c$ the relativistic Doppler shift (6.44) differs significantly from the Galilean Doppler shift.

The relativistic Doppler shift resolves the twin paradox of time dilation. One twin is placed on a rocket that leaves the Earth at the relative speed $v = 4c/5$ and travels to a star 20 light-years distant *as measured from Earth*, while the other twin stays on Earth. The twin on board the rocket can communicate the passage of time by sending a pulse of light back to Earth after each year passes on the rocket-based clock. Similarly, the Earth-based twin flashes a signal to the rocket after each year elapses according to the Earth-based clock. Both of these signals can be viewed as having a rest frame frequency of $f = 1$ cycle per year. The Earth-based observer knows the rocket will take $20/(4/5) = 25$ Earth years to reach the star and, if the rocket reverses immediately, it will take another 25 Earth years to return. The Earth-based twin will therefore send a total of 50 signals to the rocket before it returns. On the other hand, the rocket-based twin will claim the distance to the star is contracted to $20\sqrt{1 - (4/5)^2} = 12$ light-years and will measure the time required to reach it as $12/(4/5) = 15$ years. During the trip to the star, the rocket-based twin observes the Earth moving away at the speed $v = -4c/5$ and will observe the light pulses from Earth arriving with the Doppler shifted frequency

$$f' = 1 \cdot \sqrt{\frac{(1 - 4/5)}{(1 + 4/5)}} = 1/3 \text{ pulses per year}. \tag{6.46}$$

During the 15-year trip to the star the rocket twin receives $15 \times 1/3 = 5$ pulses from Earth while sending 15 pulses back to Earth. On the return trip the rocket twin again emits 15 pulses, but sees the Earth sending pulses at the frequency

$$f' = 1 \cdot \sqrt{\frac{(1 + 4/5)}{(1 - 4/5)}} = 3 \text{ pulses per year}. \tag{6.47}$$

Over the 15-year trip back to Earth the rocket twin then receives $15 \times 3 = 45$ pulses, for a total of $5 + 45 = 50$ pulses received from Earth, agreeing with the Earth-based twin that 50 years have passed on Earth while having aged only 30 years.

A similar effect occurs for the Earth-based twin. However, the change in frequency for the signals from the rocket does not immediately occur at the Earth upon the rocket's reversal of direction. Instead, the higher-frequency light pulses will require 20 Earth years to begin arriving at the Earth. This is because the higher-frequency light pulses require 20 years to travel the 20 light-years from the planet to Earth. Thus, for the first $20 + 25 = 45$ Earth years of the rocket twin's trip, the rocket twin's signal frequency appears to be 1/3 pulse per year, so that the Earth-based twin receives $45 \times 1/3 = 15$ pulses from the rocket. For the final five Earth years of the return trip, the rocket twin's signal frequency becomes three pulses per year, so that $3 \times 5 = 15$ pulses arrive from the rocket for a total of 30 pulses received by the Earth-based twin, exactly what the rocket-based twin claims to have emitted. Upon reunion, both agree that the rocket-based twin has aged 30 years, while the Earth-based twin has aged 50 years. It is important to note that the Earth-based twin will observe changes in the rocket-based twin's signal frequency from a redshift to a blueshift, and this identifies the presence of an acceleration. The Earth-based twin is able to determine that no local acceleration effects occurred, which identifies the rocket-based twin as having undergone the acceleration. The rocket-based twin will observe the effects of the local acceleration that accompanied the return trip, in agreement with the Earth-based twin. As a result, the rocket-based twin is distinguished by not always being an inertial observer throughout the trip.

This leads to the concept of *proper time*. If an inertial observer claims that a time interval ΔT has passed, then that inertial observer will see the proper time interval $\Delta T'$ on the clock of another inertial observer given by

$$\Delta T' = \Delta T\sqrt{1 - v^2/c^2}. \tag{6.48}$$

If v is not constant the other observer is non-inertial and cannot make the same claim about the proper time of the inertial observer. In the case just considered the Earth-based inertial observer has an elapsed time of 50 years and sees the accelerated observer with an elapsed time interval of $\Delta T' = 50\sqrt{1 - (4/5)^2} = 30$ years. However, the accelerated twin could not apply (6.48). For the case of an accelerated observer, the infinitesimal proper time interval can still be used. The relationship between force and accelerated motion will be discussed in detail in a later section.

6.4.5 Velocity addition and the concept of space-time

Another effect of special relativity is an altered form of *velocity addition*. This can be stated by considering three observers, $\mathcal{O}$, $\mathcal{O}'$, and $\mathcal{O}''$ in relative motion. According to $\mathcal{O}'$, observer $\mathcal{O}$ is moving along the x^1-axis with a speed $-v_1$, while $\mathcal{O}''$ claims that $\mathcal{O}'$ is moving along the x^1-axis with a speed $-v_2$. The question is: what speed v will $\mathcal{O}''$ claim for $\mathcal{O}$? This is determined from the effect of two consecutive boosts using the velocities $-v_1$ and $-v_2$. Using Galilean boosts gives the familiar Newtonian result $v = -(v_1 + v_2)$. If both $v_1 \approx c$ and $v_2 \approx c$, the consecutive Galilean boosts give the physically impossible velocity $v \approx -2c$. In that regard consecutive Lorentz boosts will maintain c as the maximum possible outcome of velocity addition.

The effect of consecutive Lorentz boosts is best examined by treating the time and position of an event as the contravariant coordinates of a point in what is referred to as *space-time* or *Minkowski space*. Using the space-time approach reveals a wealth of insight into *relativistic behavior*, which has come to mean the difference in measurements made by observers at a relative velocity approaching c. It will also greatly facilitate the development of an action approach for *relativistic field theories*. In order that the coordinates have the same units the *timelike coordinate* is written

$x^0 = ct$, while the spatial coordinates remain the familiar Cartesian coordinates x^j. An event or point in space-time for observer $\mathcal{O}$ is given by $(x^0, \boldsymbol{r})$ and the individual coordinates are given by x^μ. Both the Lorentz boost formulas of (6.27) involving one spatial dimension x^1 can then be written as a single expression by using the summation convention,

$$x'^{\mu} = \Lambda^\mu{}_\nu(v)\, x^\nu, \tag{6.49}$$

where, for the moment, the sum is over $\nu = 0$ and 1. Relation (6.49) allows the $\mathcal{O}''$ space-time coordinates x''^{μ} to be found from the $\mathcal{O}$ space-time coordinates by a double application of (6.49), giving

$$x''^{\mu} = \Lambda^\mu{}_\nu(v_2)\, x'^{\nu} = \Lambda^\mu{}_\nu(v_2)\, \Lambda^\nu{}_\sigma(v_1)\, x^\sigma. \tag{6.50}$$

Using (6.26) gives

$$\begin{aligned}
\Lambda^0{}_\nu(v_2)\, \Lambda^\nu{}_0(v_1) &= \Lambda^0{}_0(v_2)\, \Lambda^0{}_0(v_1) + \Lambda^0{}_1(v_2)\, \Lambda^1{}_0(v_1) \\
&= \frac{1}{\sqrt{(1-(v_1/c)^2)(1-(v_2/c)^2)}} + \frac{v_1 v_2/c^2}{\sqrt{(1-(v_1/c)^2)(1-(v_2/c)^2)}} \\
&= \frac{1 + v_1 v_2/c^2}{\sqrt{(1 + v_1 v_2/c^2)^2 - (v_1 + v_2)^2/c^2}} \\
&= \frac{1}{\sqrt{1 - (v_1 + v_2)^2/c^2(1 + v_1 v_2/c^2)^2}} \\
&= \frac{1}{\sqrt{1 - v^2/c^2}},
\end{aligned} \tag{6.51}$$

where v is given by

$$v = -\frac{v_1 + v_2}{1 + v_1 v_2/c^2}. \tag{6.52}$$

The minus sign is chosen to be consistent with the previous signs of both v_1 and v_2.

Result (6.51) shows that the *composition* of two successive Lorentz boosts has become a single Lorentz boost of $\Lambda^0{}_0(v) = 1/\sqrt{1 - v^2/c^2}$. It is tedious but straightforward to show that similar results hold for the other components,

$$\begin{aligned}
\Lambda^0{}_\nu(v_2)\, \Lambda^\nu{}_1(v_1) &= -\frac{v}{c}\gamma_v = \Lambda^0{}_1(v), \\
\Lambda^1{}_\nu(v_2)\, \Lambda^\nu{}_0(v_1) &= -\frac{v}{c}\gamma_v = \Lambda^1{}_0(v), \\
\Lambda^1{}_\nu(v_2)\, \Lambda^\nu{}_1(v_1) &= \gamma_v = \Lambda^1{}_1(v).
\end{aligned} \tag{6.53}$$

It should be stressed that the same value for v given by (6.52) appears in all the expressions found in (6.53). Formula (6.52) therefore provides the correct method for adding velocities while retaining consistency with the speed of light c remaining constant for all observers. It is clear that the Galilean result appears in the numerator and the denominator will be approximately one for velocities that are small compared to c, resulting in Galilean relativity. However, setting $v_1 = v_2 = \frac{1}{2}c$ in (6.52) gives $v = -4c/5$, rather than the Galilean result $v = -c$.

Exercise 6.7. For three-dimensional space the Lorentz transformations are given by (6.30). Show that the transformations of (6.30) obey $\Lambda^0{}_\nu(\boldsymbol{v}_2)\Lambda^\nu{}_0(\boldsymbol{v}_1) = \Lambda^0{}_0(\boldsymbol{v})$, where the velocity $\boldsymbol{v}$ is given by

$$\boldsymbol{v} = -\left(\frac{\boldsymbol{v}_1 + \boldsymbol{v}_2}{1 + \boldsymbol{v}_1 \cdot \boldsymbol{v}_2/c^2}\right)\sqrt{1 - \frac{v_1^2 v_2^2 - (\boldsymbol{v}_1 \cdot \boldsymbol{v}_2)^2}{c^2(v_1^2 + 2\boldsymbol{v}_1 \cdot \boldsymbol{v}_2 + v_2^2)}}. \tag{6.54}$$

The easiest way to demonstrate this is to show that v^2 satisfies

$$\frac{1}{\sqrt{1 - v^2/c^2}} = \frac{(1 + \boldsymbol{v}_1 \cdot \boldsymbol{v}_2/c^2)}{\sqrt{(1 - v_1^2/c^2)(1 - v_2^2/c^2)}}. \tag{6.55}$$

The concept of space-time and the discovery that Lorentz boosts satisfy the *multiplication rules* of (6.51) and (6.53) opened the door to a profoundly deeper understanding of fundamental phenomena, such as electromagnetism and gravitation. When coupled with the geometric notation of chapter 1, it presents a much more compact and coordinate-free way of describing nature.

6.5 The Minkowski metric and space-time

Rather than the Newtonian treatment of time as a parameter and space as a manifold, the worldview of special relativity is that space and time share equally as the coordinates of an event. The two form a manifold referred to as *space-time.* The four coordinates of an event x^μ can now be understood as the *contravariant components* of a *four-vector in space-time*. The expression (6.49) can be differentiated to find

$$x'^{\mu} = \Lambda^\mu{}_\nu(\boldsymbol{v})\, x^\nu \implies \frac{\partial x'^\mu}{\partial x^\nu} = \Lambda^\mu{}_\nu(\boldsymbol{v}), \tag{6.56}$$

showing that, like (1.21), the Lorentz boost plays the role of a *transformation* of contravariant space-time vectors. The covariant components of an event in space-time are designated as x_μ. There is a Lorentz boost that gives the transformation of the covariant components:

$$x'_\mu = \Lambda_\mu{}^\nu(v)x_\nu, \tag{6.57}$$

which is similar to that of the contravariant components. It is important to note that the two boosts appearing in (6.56) and (6.57) are not the same.

Like the spatial vectors of (1.35), the contravariant space-time components can now be combined with *covariant space-time unit vectors* $\hat{\mathbf{e}}_\mu$ to create the coordinate-free space-time vector,

$$x = x^\mu \, \hat{\mathbf{e}}_\mu = x^0\hat{\mathbf{e}}_0 + x^j\hat{\mathbf{e}}_j. \tag{6.58}$$

In what follows the spatial components will be denoted by x^j or r^j, since they are the components of the spatial vector $\boldsymbol{r}$. The invariant quadratic terms of (6.23) determine the scalar product of such a space-time vector, which is given by

$$x \cdot x = (\hat{\mathbf{e}}_\mu \cdot \hat{\mathbf{e}}_\nu)x^\mu x^\nu = \eta_{\mu\nu}x^\mu x^\nu = x^0x^0 - \boldsymbol{r} \cdot \boldsymbol{r} = c^2t^2 - r^2, \tag{6.59}$$

where $\eta_{\mu\nu} = \hat{\mathbf{e}}_\mu \cdot \hat{\mathbf{e}}_\nu$ are the components of the space-time metric tensor $\boldsymbol{\eta}$. This identification is identical to the manner in which the metric tensor was first introduced for spatial manifolds in (1.65). Relation (6.59) identifies the components of the *Minkowski metric tensor* in Cartesian space-time coordinates as

$$\eta_{00} = -\eta_{11} = -\eta_{22} = -\eta_{33} = 1, \tag{6.60}$$

and all other components vanish. The inverse Minkowski metric has the components $\eta^{\mu\nu}$ such that $\eta^{\mu\rho}\eta_{\rho\nu} = \delta^\mu{}_\nu$. It follows that

$$\eta^{00} = -\eta^{11} = -\eta^{22} = -\eta^{33} = 1, \tag{6.61}$$

and all other components vanish. Many authors use the opposite definition, selecting $\eta_{00} = -\eta_{11} = -\eta_{22} = -\eta_{33} = -1$ for the Minkowski metric. This has the advantage of matching the spatial part of the metric to the Euclidean metric in Cartesian coordinates. There are a number of advantages to choosing the form (6.60), particularly in the definitions of proper time and proper velocity to be made later, where the need to introduce minus signs is avoided. However, doing so requires the Euclidean metric g_{jk} to be matched to $-\eta_{jk}$. Like the covariant unit vectors introduced in (1.45), the space-time vector x can be written using the Cartan notation

$$x = x^\mu(x)e_\mu(x) = x^\mu(x)\partial_\mu. \tag{6.62}$$

Similarly, higher-rank tensors are expressed in the manner of (1.103).

The Minkowski metric tensor transforms as a second-rank covariant tensor under an *arbitrary* coordinate transformation. However, it has a special relationship to the Lorentz boosts of (6.30), since they were derived to give

$$\begin{aligned} c^2t'^2 - \boldsymbol{r}' \cdot \boldsymbol{r}' = c^2t^2 - \boldsymbol{r} \cdot \boldsymbol{r} \implies \eta_{\mu\nu}x'^\mu x'^\nu &= \eta_{\mu\nu}\Lambda^\mu{}_\rho(\boldsymbol{v})\Lambda^\nu{}_\sigma(\boldsymbol{v}) \, x^\rho x^\sigma \\ &= \eta_{\rho\sigma} \, x^\rho x^\sigma. \end{aligned} \tag{6.63}$$

This gives the important result

$$\eta_{\mu\nu}\Lambda^{\mu}{}_{\rho}(v)\Lambda^{\nu}{}_{\sigma}(v) = \eta_{\rho\sigma}. \tag{6.64}$$

> **Exercise 6.8.** Use the specific forms for $\Lambda^{\mu}{}_{\nu}(v)$ given by (6.30) to verify (6.64) explicitly.

This gives the important result that *all inertial observers*, those connected by Lorentz boosts, agree upon the form of the Minkowski metric. This is critical in order for there to be no privileged inertial observer.

Like the spatial metric tensor of (1.71), the Minkowski metric tensor can be used to raise and lower indices, so that the covariant and contravariant components of an event are related according to

$$x^{\mu} = \eta^{\mu\nu}x_{\nu}, \quad x_{\mu} = \eta_{\mu\nu}x^{\nu}. \tag{6.65}$$

This immediately gives the result

$$x^{\mu}x_{\mu} = \eta_{\mu\nu}x^{\mu}x^{\nu} = \eta^{\mu\nu}x_{\mu}x_{\nu}. \tag{6.66}$$

This is the Lorentz invariant quantity identified in (6.63). Using the Lorentz boosts of (6.56) and (6.57) gives

$$x'^{\mu}x'_{\mu} = \Lambda^{\mu}{}_{\alpha}(v)\Lambda_{\mu}{}^{\beta}(v)\, x^{\alpha}x_{\beta}. \tag{6.67}$$

The right-hand side becomes $x^{\alpha}x_{\alpha}$ only if

$$\Lambda_{\mu}{}^{\beta}(v)\Lambda^{\mu}{}_{\alpha}(v) = \delta^{\beta}{}_{\alpha}, \tag{6.68}$$

which reveals that the covariant and contravariant boosts are the inverse of each other. A version of the inverse relationship using contravariant boosts emerges by noting that

$$x'^{\mu}x'_{\mu} = \Lambda^{\mu}{}_{\nu}(v)x^{\nu}\, x'_{\mu} = x^{\nu}x_{\nu} \implies x_{\nu} = x'_{\mu}\Lambda^{\mu}{}_{\nu}(v). \tag{6.69}$$

Similarly, a version of the inverse relationship using covariant boosts emerges by noting that

$$x'^{\mu}x'_{\mu} = x'^{\mu}\Lambda_{\mu}{}^{\nu}(v)x_{\nu} = x^{\nu}x_{\nu} \implies x^{\nu} = x'^{\mu}\Lambda_{\mu}{}^{\nu}(v). \tag{6.70}$$

Contracting (6.69) and (6.70) gives a variant of (6.68),

$$x^{\nu}x_{\nu} = x'_{\rho}x'^{\sigma}\Lambda_{\sigma}{}^{\nu}(v)\Lambda^{\rho}{}_{\nu}(v) = x'^{\rho}x'_{\rho} \implies \Lambda_{\sigma}{}^{\nu}(v)\Lambda^{\rho}{}_{\nu}(v) = \delta^{\rho}{}_{\sigma}. \tag{6.71}$$

Relationships (6.68) through (6.71) will be of great use in building relativistic field theories. It should also be noted that the covariant and contravariant indices of the Lorentz transformation are also raised and lowered using the Minkowski metric,

$$\Lambda^{\mu}{}_{\nu}(v) = \eta^{\mu\sigma}\eta_{\nu\rho}\Lambda_{\sigma}{}^{\rho}(v). \tag{6.72}$$

The nature of the chosen form of the Minkowski metric causes some previous results to be altered. In particular, the covariant spatial components of a four-vector change have the opposite signs to those of the contravariant spatial components. This is written symbolically as

$$x^\mu = (x^0, \boldsymbol{x}) \implies x_\mu = \eta_{\mu\nu}x^\nu = (x^0, -\boldsymbol{x}). \tag{6.73}$$

This applies to all four-vector quantities.

Exercise 6.9. Show that the Levi-Civita symbol in Minkowski space obeys

$$\varepsilon_{\mu\nu\rho\sigma} = -\varepsilon^{\mu\nu\rho\sigma}, \tag{6.74}$$

and the descent equation for the components $W_{\mu\nu\rho}$ of a three-form

$$\varepsilon^{0\mu\nu\rho}W_{\mu\nu\rho} = \varepsilon^{jk\ell}W_{jk\ell}. \tag{6.75}$$

Result (6.74) gives

$$\varepsilon_{\mu\nu\rho\sigma}\varepsilon^{\mu\nu\rho\sigma} = -4! = -24. \tag{6.76}$$

However, when working solely with the spatial components of a vector, the three-dimensional Levi-Civita symbol will still obey

$$\varepsilon_{jk\ell} = \varepsilon^{jk\ell}. \tag{6.77}$$

The Lorentz transformation (6.57) shows that the covariant unit vectors $\hat{\mathbf{e}}_\mu$ transform according to

$$\hat{\mathbf{e}}'_\nu = \Lambda_\nu{}^\mu(v)\hat{\mathbf{e}}_\mu. \tag{6.78}$$

Combining (6.78) with relation (6.68) shows that

$$x' = x'^\nu\hat{\mathbf{e}}'_\nu = \Lambda^\nu{}_\rho(v)\, x^\rho \Lambda_\nu{}^\mu(v)\, \hat{\mathbf{e}}_\mu = \delta^\mu{}_\rho\, x^\rho\hat{\mathbf{e}}_\mu = x^\mu\hat{\mathbf{e}}_\mu = x. \tag{6.79}$$

This gives a coordinate-free or observer-independent expression for events in space-time. In terms of the covariant and contravariant components of an event, the scalar product can be written

$$x \cdot x = \eta_{\mu\nu}x^\mu x^\nu = x_\mu x^\mu = x_0 x^0 + x_j x^j. \tag{6.80}$$

This shows that the contravariant unit vectors are given by $\hat{\mathbf{e}}^\mu = \eta^{\mu\nu}\hat{\mathbf{e}}_\nu$, so that

$$\hat{\mathbf{e}}^\nu \cdot \hat{\mathbf{e}}_\mu = \eta^{\nu\sigma}\hat{\mathbf{e}}_\sigma \cdot \hat{\mathbf{e}}_\mu = \eta^{\nu\sigma}\eta_{\sigma\mu} = \delta^\nu{}_\mu, \tag{6.81}$$

which is identical to the earlier result (1.84) for spatial vectors. It also follows that

$$\hat{\mathbf{e}}^\mu \cdot \hat{\mathbf{e}}^\nu = \eta^{\mu\rho}\eta^{\nu\sigma}\hat{\mathbf{e}}_\rho \cdot \hat{\mathbf{e}}_\sigma = \eta^{\nu\sigma}\eta^{\mu\rho}\eta_{\rho\sigma} = \eta^{\nu\sigma}\eta^{\mu\rho}\eta_{\sigma\rho} = \delta^\nu{}_\rho\eta^{\mu\rho} = \eta^{\mu\nu}. \tag{6.82}$$

These results are identical to the behavior of the spatial metric introduced in chapter 1.

The Minkowski metric is understood to be the metric tensor of *flat Cartesian space-time*. However, interpreting $x \cdot x$ as a length squared shows that space-time has an *indefinite metric*. If $x \cdot x > 0$ then x is referred to as a *timelike vector*, since $c^2t^2 > r^2$. If $x \cdot x < 0$ then x is referred to as a *spacelike vector*, since $r^2 > c^2t^2$. If $x \cdot x = 0$ then x is referred to as a *lightlike vector*, since $r = ct$ corresponds to the motion of a light wave.

Particle motion can be understood as a sequence of events. Therefore, as a particle moves in space-time, it traces out a path of positions and times referred to as its *worldline*. This is depicted in figure 6.2. Because c is the speed limit the length squared is very useful in understanding the *causal relation* of two separate events along the worldline, denoted by x and $\tilde{x}$. For the case that $\tilde{t}>t$, the space-time vector that joins the two events is given by $s = \tilde{x}-x$. The length squared of s is given by

$$s \cdot s = s^2 = c^2(\tilde{t}-t)^2 - |\tilde{\boldsymbol{r}}-\boldsymbol{r}|^2. \tag{6.83}$$

If s^2 is lightlike then the particle can be at both events, since the spatial distance between the two events is less than $c(\tilde{t}-t)$. If s^2 is spacelike then the particle could not travel between those two events, since its velocity would exceed c. The *lightcone*, defined as $(\boldsymbol{r}' - \boldsymbol{r}) \cdot (\boldsymbol{r}' - \boldsymbol{r}) = c^2(t' - t)^2$, separates the two possibilities and represents the path of a light wave in the space-time diagram emanating from the origin. The forward lightcone, defined as $|\boldsymbol{r}' - \boldsymbol{r}| = +c(t' - t)$, contains all future events $(t', \boldsymbol{r}')$ that could be affected by a signal, either light or something slower, from the event $(t, \boldsymbol{r})$.

Using the space-time coordinates it is possible to define the four covariant components of the *space-time gradient*,

$$\partial_\mu = \frac{\partial}{\partial x^\mu} \implies \partial_\mu = \left(\frac{\partial}{\partial(ct)}, \nabla\right). \tag{6.84}$$

Like the spatial gradient components of (1.22), the space-time gradient components (6.84) transform under an *arbitrary* change of coordinates as

$$\partial_\mu^{'} = \frac{\partial x^\nu}{\partial x^{\mu'}}\partial_\nu = \Lambda_\mu{}^\nu(v)\partial_\nu, \tag{6.85}$$

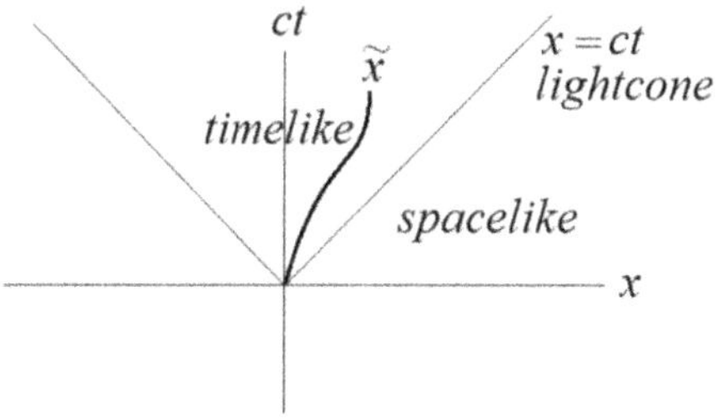

Figure 6.2. The worldline of a particle moving from (0, 0) to $(c\tilde{t}, \tilde{x})$ depicted on a one-dimensional space-time diagram. Spacelike and timelike regions are separated by the lightcone at $x = ct$.

which is the Lorentz boost for a covariant vector. The space-time gradient can be combined with the contravariant components of the Minkowski metric to express the d'Alembertian $\Box_c$ for a light wave in Cartesian coordinates in the $\mathcal{O}$ frame,

$$\Box_c = \eta^{\mu\nu}\frac{\partial}{\partial x^\mu}\frac{\partial}{\partial x^\nu} = \frac{1}{c^2}\frac{\partial^2}{\partial t^2} - \frac{\partial^2}{\partial x^2} - \frac{\partial^2}{\partial y^2} - \frac{\partial^2}{\partial z^2} = \frac{1}{c^2}\frac{\partial^2}{\partial t^2} - \nabla^2. \tag{6.86}$$

The coordinate-free version of the space-time gradient is expressed in terms of the space-time reciprocal basis four-vectors $\hat{\mathbf{e}}^\mu$ as

$$\partial = \hat{\mathbf{e}}^\mu \partial_\mu. \tag{6.87}$$

This definition immediately yields

$$\partial \cdot \partial = \hat{\mathbf{e}}^\mu \cdot (\partial_\mu \hat{\mathbf{e}}^\nu)\partial_\nu + (\hat{\mathbf{e}}^\mu \cdot \hat{\mathbf{e}}^\nu)\partial_\mu\partial_\nu = \eta^{\mu\nu}\partial_\mu\partial_\nu = \frac{1}{c^2}\frac{\partial^2}{\partial t^2} - \nabla^2 = \Box_c, \tag{6.88}$$

where it was assumed that Cartesian coordinates are in use, so that $\partial_\mu \hat{\mathbf{e}}^\nu = 0$.

In order for the speed of light to be invariant the d'Alembertian of (6.86) must be the same for *all inertial observers*. It would appear that (6.86) is a contraction and is therefore invariant under an *arbitrary* change of coordinates. However, this is incorrect. Substituting (6.85) into (6.86) gives

$$\partial \cdot \partial = \Box_c = \eta^{\mu\nu}\partial_\mu\partial_\nu = \eta^{\mu\nu}\partial_\mu\left(\frac{\partial x'^\rho}{\partial x^\nu}\partial'_\rho\right) = \eta^{\mu\nu}\frac{\partial^2 x'^\rho}{\partial x^\mu\,\partial x^\nu}\partial'_\rho + \eta^{\mu\nu}\frac{\partial x'^\rho}{\partial x^\nu}\frac{\partial x'^\sigma}{\partial x^\mu}\partial'_\rho\partial'_\sigma. \tag{6.89}$$

Preventing the appearance of the linear term in ∂'_ρ and ensuring $\Box'_c = \Box_c$ requires the change of coordinates to satisfy two conditions,

$$\begin{aligned}\frac{\partial^2 x'^\rho}{\partial x^\mu\,\partial x^\nu} &= 0,\\ \eta^{\mu\nu}\frac{\partial x'^\rho}{\partial x^\nu}\frac{\partial x'^\sigma}{\partial x^\mu} &= \eta^{\rho\sigma}.\end{aligned} \tag{6.90}$$

These are precisely the conditions that define the Lorentz boost (6.30). The first condition prevents transformations to accelerated or non-inertial observers, while the second expresses the invariance of the Minkowski metric established earlier in (6.64). This does not prevent the transformation of the Minkowski metric to an accelerated coordinate system. However, such a transformation does not necessarily leave the components of the Minkowski metric tensor unchanged; instead, it transforms them to

$$\eta'_{\mu\nu} = \frac{\partial x^\rho}{\partial x'^\mu}\frac{\partial x^\sigma}{\partial x'^\nu}\eta_{\rho\sigma} \implies \frac{\partial}{\partial x'^\alpha}\eta'_{\mu\nu} = \left(\frac{\partial^2 x^\rho}{\partial x'^\mu \partial x'^\alpha}\frac{\partial x^\sigma}{\partial x'^\nu} + \frac{\partial x^\rho}{\partial x'^\mu}\frac{\partial^2 x^\sigma}{\partial x'^\nu \partial x'^\alpha}\right)\eta_{\rho\sigma} \neq 0. \tag{6.91}$$

Such a coordinate system is distinguished by a metric tensor that is *not* the Minkowski metric. It follows that a metric tensor that is not the Minkowski metric must include the effects of forces and their associated accelerations. This forms the basis of Einstein's general theory of relativity presented in chapter 10.

These results have been applied so far to space-time coordinate four-vectors and the gradient four-vector. However, other four-vectors can be identified. For example, the cosine appearing in the electromagnetic wave of (6.10) can be given a relativistically invariant argument by writing it as

$$\begin{aligned}\mathbf{E}(\boldsymbol{r}, t) &= \mathbf{E}_0 \cos(\boldsymbol{k}\cdot\boldsymbol{r} - \omega_k t) = \mathbf{E}_0 \cos\left(\left(\frac{\omega_k}{c}\right)ct - \boldsymbol{k}\cdot\boldsymbol{r}\right)\\ &= \mathbf{E}_0 \cos(\eta_{\mu\nu}k^\mu x^\nu) = \mathbf{E}_0 \cos(k_\mu x^\mu),\end{aligned} \tag{6.92}$$

where the *contravariant four-vector* k^μ components have been identified as

$$k^\mu = \left(\frac{\omega_k}{c}, \boldsymbol{k}\right) \implies k_\mu = \eta_{\mu\nu}k^\nu = \left(\frac{\omega_k}{c}, -\boldsymbol{k}\right). \tag{6.93}$$

As long as the four-vector k^μ transforms as a contravariant vector under a Lorentz boost, the argument of the cosine function is invariant under Lorentz boosts, and is therefore a *Lorentz scalar*. This is necessary since all inertial observers must agree on the form and value of the argument of the cosine function, since it forms a sequence of events in the progression of a light wave. In relativistic notation, it is common to employ the term $k \cdot x$, which stands for

$$k \cdot x = k_\mu x^\mu = \eta_{\mu\nu}k^\nu x^\mu = k^0 x^0 - \boldsymbol{k}\cdot\boldsymbol{x}, \tag{6.94}$$

so that the sign of $\boldsymbol{k}$ determines the direction of spatial motion for the wave.

The function (6.92) solves the wave equation if $\omega_k = c\sqrt{\boldsymbol{k}\cdot\boldsymbol{k}}$, and this gives

$$k_\mu k^\mu = k_0 k^0 + k_j k^j = \eta_{00}k^0 k^0 + \eta_{j\ell}k^j k^\ell = \frac{\omega_k^2}{c^2} - \boldsymbol{k}\cdot\boldsymbol{k} = 0, \tag{6.95}$$

so that k^μ is correctly identified as a *lightlike vector*. A boost to a frame moving in the direction opposite to $\boldsymbol{k}$, so that $v_i k^i = \boldsymbol{v}\cdot\boldsymbol{k} = -vk = -v\omega_k/c$, gives the angular frequency in the boosted frame,

$$\begin{aligned}\frac{\omega_k'}{c} &= \Lambda^0{}_\mu(v)k^\mu = \Lambda^0{}_0(v)\frac{\omega_k}{c} + \Lambda^0{}_j(v)k^j = \gamma_v\frac{\omega_k}{c} + \gamma_v\frac{v\omega_k}{c^2}\\ &= \frac{\omega_k}{c}\frac{(1 + v/c)}{\sqrt{1 - v^2/c^2}}\\ &= \frac{\omega_k}{c}\sqrt{\frac{1 + v/c}{1 - v/c}},\end{aligned} \tag{6.96}$$

where $\Lambda^0{}_0 = \gamma_v$ and $\Lambda^0{}_j = -v_j\gamma_v/c$ were used. Result (6.96) is identical to the relativistic Doppler shift (6.44) for the frequency of an approaching light source that was found using different methods.

6.6 Relativistic energy and momentum

Just as the Galilean boost played a key role in determining the form of Newton's laws of motion, the Lorentz boost determines the structure of relativistic mechanics. All measurable quantities must have well-defined transformation properties, so that once they are known in one inertial frame, they can be found in all inertial frames by applying an appropriate Lorentz boost. In other words, all inertial observers agree on the measurable quantities and the equations of physics that govern them as well as how these quantities are related in different inertial frames. For velocities such that v/c is small, Newtonian quantities must reappear in order to preserve the successes of Newtonian mechanics.

The goal is to identify the relativistic generalizations of momentum and energy and eventually their relationship to force. This begins by deriving these generalizations for a *free point mass* using an action formulation. Motivation can be found in the earlier result (2.117), which showed that the Newtonian action for a *free point mass* was form invariant under a Galilean boost. By analogy, the relativistic action for a *free point mass* should be invariant under Lorentz transformations, since all inertial observers will agree that there is zero force acting on the particle. This begins by noting that each inertial observer will parameterize the motion of a free point particle differently, since both $\boldsymbol{r}$ and t are affected by a Lorentz boost. If observer $\mathcal{O}$ records the infinitesimal change in position $\mathrm{d}\boldsymbol{r}$ during the infinitesimal time interval $\mathrm{d}t$, then observer $\mathcal{O}'$ will record the infinitesimal change in position $\mathrm{d}\boldsymbol{r}'$ during the infinitesimal time interval $\mathrm{d}t'$. It is important to note that both $c\,\mathrm{d}t$ and $\mathrm{d}\boldsymbol{r}$ transform as contravariant differentials, since the Lorentz boost gives

$$\mathrm{d}x'^{\mu} = \mathrm{d}(\Lambda^{\mu}{}_{\nu}(v)x^{\nu}) = (\mathrm{d}\Lambda^{\mu}{}_{\nu}(v))x^{\nu} + \Lambda^{\mu}{}_{\nu}(v)\mathrm{d}x^{\nu} = \Lambda^{\mu}{}_{\nu}(v)\mathrm{d}x^{\nu}. \tag{6.97}$$

The differential $\mathrm{d}\Lambda^{\mu}{}_{\nu}(v)$ vanishes for each inertial observer, since $\mathrm{d}v = 0$ for a free particle. As a result, both inertial observers will agree on the value of the Lorentz invariant expression $\mathrm{d}s^2 = \eta_{\mu\nu}\mathrm{d}x^{\mu}\mathrm{d}x^{\nu}$, so that

$$c^2\mathrm{d}t^2 - \mathrm{d}\boldsymbol{r}\cdot\mathrm{d}\boldsymbol{r} = c^2\mathrm{d}t'^2 - \mathrm{d}\boldsymbol{r}'\cdot\mathrm{d}\boldsymbol{r}'. \tag{6.98}$$

The two observers claim that $\mathrm{d}\boldsymbol{r} = \boldsymbol{v}\,\mathrm{d}t$ and $\mathrm{d}\boldsymbol{r}' = \boldsymbol{v}'\,\mathrm{d}t'$, where $\boldsymbol{v}$ and $\boldsymbol{v}'$ are the velocity of the particle relative to $\mathcal{O}$ and $\mathcal{O}'$, respectively. Using this result in (6.98) gives the equality between the infinitesimal values of the proper time interval first defined in (6.48),

$$\mathrm{d}\tau = \frac{\mathrm{d}t}{\gamma_v} = \mathrm{d}t\sqrt{1 - \frac{v^2}{c^2}} = \mathrm{d}t'\sqrt{1 - \frac{v'^2}{c^2}} = \frac{\mathrm{d}t'}{\gamma_{v'}} = \mathrm{d}\tau. \tag{6.99}$$

Picking a specific representation for $\mathrm{d}\tau$ is equivalent to picking a specific inertial reference frame. Because all inertial observers of the particle's motion will agree on the value of $\mathrm{d}\tau$, it is the infinitesimal version of (6.48) and is therefore referred to as the *infinitesimal proper time interval* for the particle. The infinitesimal proper time interval is the same for all inertial observers, since it is the dilated time interval of the particle due to its relative motion.

As the particle moves the sum of all the proper time intervals will therefore be a Lorentz invariant quantity. A relativistically invariant action for a free particle is therefore given by

$$S[\boldsymbol{r}, \dot{\boldsymbol{r}}] = \int_0^\tau \mathrm{d}\tau\, \mathcal{E} = \int_0^T \mathrm{d}t\, \mathcal{E}\sqrt{1 - \frac{v^2}{c^2}} = \int_0^T \mathrm{d}t\, \mathcal{E}\sqrt{1 - \frac{1}{c^2}\dot{\boldsymbol{r}}\cdot\dot{\boldsymbol{r}}} \\ = \int_0^T \mathrm{d}t\, \mathcal{L}, \tag{6.100}$$

where $\tau = T\sqrt{1 - (v^2(T)/c^2)}$ is the boundary value of the proper time and $\mathcal{E}$ is a constant expressed using units of energy. The constant $\mathcal{E}$ will be determined by the requirement that the action reproduces Newtonian free-particle dynamics in the limit $v/c \to 0$. It is instructive to note that the argument of the action for $\dot{\boldsymbol{r}}/c \ll 1$ is given by the Taylor series expansion,

$$\mathcal{L} = \mathcal{E}\sqrt{1 - \frac{1}{c^2}\dot{\boldsymbol{r}}\cdot\dot{\boldsymbol{r}}} \approx \mathcal{E} - \frac{\mathcal{E}}{2c^2}v^2. \tag{6.101}$$

This includes the usual Newtonian Lagrangian for a free particle, $\mathcal{L} = \frac{1}{2}mv^2$, if

$$\mathcal{E} = -mc^2. \tag{6.102}$$

The presence of the constant $\mathcal{E}$ in the Lagrangian will require further interpretation.

This process begins by noting that

$$\frac{\partial\mathcal{L}}{\partial\dot{x}^j} = \mathcal{E}\frac{\partial}{\partial\dot{x}^j}\sqrt{1 - v^2/c^2} = \mathcal{E}\frac{\partial}{\partial\dot{x}^j}\sqrt{1 - \dot{x}^j\dot{x}^j/c^2} = \frac{-\mathcal{E}\dot{x}^j}{\sqrt{1 - v^2/c^2}}, \tag{6.103}$$

where it is important to note that the *contravariant component* of the velocity v^j appears in the numerator. This leads to the identification of the contravariant component of the canonical relativistic momentum,

$$p^j = \frac{\partial\mathcal{L}}{\partial\dot{x}^j} = -\frac{\mathcal{E}v^j}{c^2\sqrt{1 - v^2/c^2}}. \tag{6.104}$$

Contact with the Newtonian point particle expression $p^j = mv^j$ is found in the $v^2/c^2 \approx 0$ limit, which yields

$$p^j \approx -\frac{\mathcal{E}}{c^2}v^j = mv^j \implies \mathcal{E} = -mc^2, \tag{6.105}$$

which is in agreement with the value predicted by (6.102). Using (6.105) gives the relativistic point particle momentum,

$$p^j = \frac{mv^j}{\sqrt{1 - v^2/c^2}}. \tag{6.106}$$

Expression (6.106) can be solved for v^j in terms of the relativistic momentum p^j, giving

$$v^j = \dot{x}^j = \frac{p^j}{\sqrt{m^2 + \boldsymbol{p} \cdot \boldsymbol{p}/c^2}} \implies \sqrt{1 - v^2/c^2} = \frac{mc}{\sqrt{m^2c^2 + p^2}}, \tag{6.107}$$

where $p^2 = \boldsymbol{p} \cdot \boldsymbol{p} = p^j p^j$. Result (6.107) shows that as the relativistic momentum p^j tends to ∞, the velocity v^j becomes c, while for $p^j \approx 0$ the Newtonian result is obtained,

$$\lim_{p^j \to 0} v^j = \frac{p^j}{m}. \tag{6.108}$$

Treating the components x^j and $\dot{x}^j = \mathrm{d}x^j/\mathrm{d}t$ as the coordinates to be varied and using result (6.104), the Euler–Lagrange equation for (6.100) gives

$$\frac{\partial \mathcal{L}}{\partial x^j} - \frac{\mathrm{d}}{\mathrm{d}t}\frac{\partial \mathcal{L}}{\partial \dot{x}^j} = -\frac{\mathrm{d}}{\mathrm{d}t}\frac{\partial \mathcal{L}}{\partial \dot{x}^j} = -\frac{\mathrm{d}}{\mathrm{d}t}\left(\frac{m\dot{x}_j}{\sqrt{1 - v^2/c^2}}\right) = -\frac{\mathrm{d}p^j}{\mathrm{d}t} = 0, \tag{6.109}$$

which is the generalization of Newton's first and second laws for a free point mass. The energy of the free particle is denoted by K and is found from $\mathcal{L}$ using the Legendre transformation (2.108) and the relations of (6.107),

$$\begin{aligned} K &= \dot{x}^j \frac{\partial \mathcal{L}}{\partial \dot{x}^j} - \mathcal{L} = v^j p^j - \mathcal{L} = \frac{p^2}{\sqrt{m^2 + p^2/c^2}} + mc^2\sqrt{1 - v^2/c^2} \\ &= c\sqrt{m^2c^2 + p^2}. \end{aligned} \tag{6.110}$$

Exercise 6.10. Show that in the small $\boldsymbol{p}$ limit the Taylor series for expression (6.110) gives

$$K \approx mc^2 + \frac{p^2}{2m}. \tag{6.111}$$

Substituting (6.106) into (6.110) gives the more commonly used expression for K,

$$K = \frac{mc^2}{\sqrt{1 - v^2/c^2}}. \tag{6.112}$$

In the limit that $v^2/c^2 \to 0$, expression (6.112) has a first-order Taylor series expansion given by

$$K \approx mc^2 + \frac{1}{2}mv^2. \tag{6.113}$$

In both (6.111) and (6.113), the familiar Newtonian form for the kinetic energy of a free particle, $E_k = p^2/2m = \frac{1}{2}mv^2$, has reappeared, but with the important addition of the *rest mass energy* given by $E_0 = mc^2$. Result (6.109) shows $\mathrm{d}\boldsymbol{v}/\mathrm{d}t = 0$, so that all inertial observers will agree that a free particle obeys

$$\frac{\mathrm{d}K}{\mathrm{d}t} = 0. \tag{6.114}$$

Because there is zero force and the expansions (6.111) and (6.113) contain the Newtonian kinetic energy, the energy K found in (6.112) is understood to be the relativistic version of the *kinetic energy*. In some presentations of special relativity the kinetic energy is defined as $T = K - mc^2$, but this distinction does not affect the results presented here. It is important to note that expression (6.110) gives

$$\lim_{m\to 0} K = cp, \tag{6.115}$$

a relationship that yields an expression for the energy of a massless particle in terms of its momentum. This expression is central to treating a light wave as consisting of massless photons.

These new forms for the free-particle momentum and energy are consistent with Lorentz boosts applied to a four-vector. This is revealed by combining the relativistic momentum (6.106) and the relativistic kinetic energy (6.112) into the energy-momentum four-vector p^μ, where

$$p^0 = \frac{K}{c}. \tag{6.116}$$

Such an identification gives

$$p^2 = \eta^{\mu\nu}p_\mu p_\nu = \frac{K^2}{c^2} - \boldsymbol{p}\cdot\boldsymbol{p} = \frac{m^2c^2}{1 - v^2/c^2} - \frac{m^2v^2}{1 - v^2/c^2} = m^2c^2. \tag{6.117}$$

Since both c and m are Lorentz invariants, it follows that p^2 is the same for all inertial observers of the particle's motion and p_μ is a *timelike vector*. The identification (6.116) will be shown to be consistent with both Lorentz boosts and the relativistic version of Newton's second law.

It is useful to note that a Lorentz invariant quantity can often be evaluated most easily in the rest frame, since all inertial observers will agree upon that value. For example, an observer $\mathcal{O}$ in the rest frame of a particle finds $p^0 = mc$ and $p^j = 0$, which immediately gives $p_\mu p^\mu = m^2c^2$. Using the Lorentz boost of (6.30) shows that a second observer $\mathcal{O}'$ with the velocity $\boldsymbol{v}$ relative to both observer $\mathcal{O}$ and the particle will claim that the particle's kinetic energy is given by

$$\frac{K'}{c} = p'^{\,0} = \Lambda^0{}_\mu(\boldsymbol{v})\,p^\mu = \Lambda^0{}_0(\boldsymbol{v})\,p^0 + \Lambda^0{}_j(\boldsymbol{v})\,p^j = \Lambda^0{}_0(\boldsymbol{v})\,p^0 = \frac{mc}{\sqrt{1 - v^2/c^2}}. \tag{6.118}$$

Similarly, $\mathcal{O}'$ will claim that the spatial momentum components are given by

$$p'^j = \Lambda^j{}_\mu(v)\, p^\mu = \Lambda^j{}_0(v)\, p^0 + \Lambda^j{}_k(v)\, p^k = \Lambda^j{}_0(v)\, p^0 = -\frac{mv^j}{\sqrt{1 - v^2/c^2}}. \quad (6.119)$$

The same relativistic expressions for the energy-momentum four-vector of a particle moving with a relative velocity $-v$ have appeared, showing that the energy-momentum four-vector obtained from definition (6.116) transforms as a four-vector under a Lorentz boost. To complete the demonstration, if another inertial observer $\mathcal{O}''$ moving with a velocity v' relative to $\mathcal{O}'$ measures the energy-momentum four-vector of the particle, the second observer will obtain

$$p''^\mu = \Lambda^\mu{}_\nu(v')\, p'^\nu = \Lambda^\mu{}_\nu(v')\Lambda^\nu{}_0(v)\, p^0 = \Lambda^\mu{}_0(\bar{v})\, p^0, \quad (6.120)$$

where $\bar{v}$ is obtained by applying the relativistic velocity addition formula (6.52) to v and v'. Stated equivalently, $\mathcal{O}''$ observes the energy-momentum four-vector associated with the relative velocity $\bar{v}$.

Exercise 6.11. Use the properties (6.51) and (6.53) of Lorentz boosts to demonstrate (6.120).

6.7 Proper velocity and accelerated motion

The case where a force is present requires that the relative velocity between the particle and the inertial observer must be allowed to change. In order to state this in a way that is consistent with special relativity requires the use of the *proper time derivative*,

$$\frac{\mathrm{d}}{\mathrm{d}\tau} = \gamma_v \frac{\mathrm{d}}{\mathrm{d}t}. \quad (6.121)$$

This derivative compensates for the effect of time dilation found in (6.39) by using the infinitesimal proper time interval $\mathrm{d}\tau$ defined in (6.99) to correct the time interval $\mathrm{d}t$ on the inertial observer's clock. All inertial observers will agree on the rate $\mathrm{d}/\mathrm{d}\tau$ for the particle's clock, regardless of their motion relative to the particle. Using the proper time derivative the contravariant components of the four-momentum can be found from the particle's worldline components $x^\mu(\tau)$,

$$p^\mu = m\frac{\mathrm{d}x^\mu(\tau)}{\mathrm{d}\tau}. \quad (6.122)$$

In expression (6.122) the particle's worldline is parameterized in terms of the proper time τ of the particle, so that $x^\mu = x^\mu(\tau)$. Because τ and m are invariant under a Lorentz boost, two inertial observers will agree that p^μ transforms identically to the four-vector x^μ under a Lorentz boost. This follows from $\mathrm{d}\tau' = \mathrm{d}\tau$ and the contravariant transformation rule

$$p'^{\mu}(\tau) = m\frac{\mathrm{d}x'^{\mu}(\tau)}{\mathrm{d}\tau} = m\frac{\mathrm{d}}{\mathrm{d}\tau}(\Lambda^{\mu}{}_{\nu}(v)\, x^{\nu}(\tau)) = \Lambda^{\mu}{}_{\nu}(v)\, m\frac{\mathrm{d}x^{\nu}(\tau)}{\mathrm{d}\tau} = \Lambda^{\mu}{}_{\nu}(v)\, p^{\nu}(\tau). \tag{6.123}$$

It also relies upon the property of Lorentz boosts between *inertial observers*,

$$\frac{\mathrm{d}}{\mathrm{d}\tau}\Lambda^{\mu}{}_{\nu}(v) = 0, \tag{6.124}$$

which vanishes due to condition (6.90) requiring inertial observers to maintain a constant relative velocity.

By analogy to Newtonian physics, the *four-velocity* of the trajectory $x^{\mu}(\tau)$ is defined as

$$V^{\mu} \equiv \frac{\mathrm{d}x^{\mu}(\tau)}{\mathrm{d}\tau}, \tag{6.125}$$

and is typically referred to as the *proper velocity*. From this point forward the expression V^{μ} will exclusively be used to represent the proper velocity, although it is common in the literature to use U^{μ}. The proper velocity transforms as a Lorentz vector and the spatial components are the velocity of the particle corrected for time dilation,

$$V^{j} = \frac{\mathrm{d}x^{j}(\tau)}{\mathrm{d}\tau} = \gamma_{v}\frac{\mathrm{d}x^{j}(\tau)}{\mathrm{d}t} = \frac{v^{j}(t)}{\sqrt{1 - v^{2}/c^{2}}}, \tag{6.126}$$

where $v^{j}(t)$ is the instantaneous velocity according to the inertial observer. The timelike component of the proper velocity is given by

$$V^{0} = \frac{\mathrm{d}x^{0}}{\mathrm{d}\tau} = \gamma_{v}\frac{\mathrm{d}(ct)}{\mathrm{d}t} = c\gamma_{v} = \frac{c}{\sqrt{1 - v^{2}/c^{2}}}. \tag{6.127}$$

Combining (6.126) and (6.127) gives

$$V^{\mu}V_{\mu} = (V^{0})^{2} - V^{j}V^{j} = \frac{c^{2}}{1 - v^{2}/c^{2}} - \frac{v^{2}}{1 - v^{2}/c^{2}} = c^{2}. \tag{6.128}$$

Using the property velocity the four-momentum of a particle can be written

$$p^{\mu} = m\frac{\mathrm{d}x^{\mu}(\tau)}{\mathrm{d}\tau} = mV^{\mu}(\tau), \tag{6.129}$$

which is similar in form to the Newtonian expression. The relativistic trajectory of the particle gives the differential

$$\mathrm{d}x^{\mu} = \frac{\mathrm{d}x^{\mu}}{\mathrm{d}\tau}\,\mathrm{d}\tau = V^{\mu}\,\mathrm{d}\tau = \frac{\mathrm{d}x^{\mu}}{\mathrm{d}t}\,\mathrm{d}t \implies \mathrm{d}x_{\mu} = \frac{\mathrm{d}x_{\mu}}{\mathrm{d}\tau}\,\mathrm{d}\tau = V_{\mu}\,\mathrm{d}\tau = \frac{\mathrm{d}x_{\mu}}{\mathrm{d}t}\,\mathrm{d}t. \tag{6.130}$$

Combining this with the definition of the proper time interval gives

$$c^{2}\mathrm{d}\tau^{2} = \mathrm{d}x^{\mu}\,\mathrm{d}x_{\mu} = \frac{\mathrm{d}x^{\mu}}{\mathrm{d}\tau}\frac{\mathrm{d}x_{\mu}}{\mathrm{d}\tau}\,\mathrm{d}\tau^{2} = V^{\mu}V_{\mu}\,\mathrm{d}\tau^{2} \implies V^{\mu}V_{\mu} = c^{2}, \tag{6.131}$$

which reproduces (6.128). This also reproduces result (6.117), since (6.129) gives

$$p^\mu p_\mu = m^2 V^\mu V_\mu = m^2c^2. \tag{6.132}$$

In the rest frame of an observer the proper velocity is given by $\bar{V}^0(0) = c$ and all other components vanish. Another inertial observer with a velocity v relative to this observer will then claim that the proper velocity of the particle is given by

$$\begin{aligned} V^\mu(v) &= \Lambda^\mu{}_\nu(v)\bar{V}^\nu(0) = \Lambda^\mu{}_0(v)\bar{V}^0(0) = c\Lambda^\mu{}_0(v) \\ &= \gamma_v(c, v) \implies \Lambda^\mu{}_0(v) = \frac{1}{c}V^\mu(v). \end{aligned} \tag{6.133}$$

The latter is a useful relation for finding relativistic expressions from rest frame results.

The proper velocity allows two inertial observers to have a single Lorentz invariant expression,

$$E = p_\mu(v_1)V^\mu(-v_2), \tag{6.134}$$

to determine the energy E of a particle in their respective inertial frames. In this expression the relativistic momentum $p_\mu(v_1)$ is that of a particle moving with a velocity v_1 relative to the first observer, while $V^\mu(-v_2)$ is the proper velocity of the second inertial observer moving with a velocity $-v_2$ relative to the first inertial observer. The Lorentz invariant quantity (6.134) then gives the energy of the particle according to the second observer. This follows by writing $p_\mu(v_1)$ and $V^\mu(-v_2)$ as boosted rest frame results,

$$\begin{aligned} p_\mu(v_1) &= \Lambda_\mu{}^\nu(v_1)\,\bar{p}_\nu(0), \\ V^\mu(-v_2) &= \Lambda^\mu{}_\rho(-v_2)\,\bar{V}^\rho(0). \end{aligned} \tag{6.135}$$

The rest frame expressions are given by $\bar{p}_0(0) = mc$ and $\bar{V}^0(0) = c$, and all others vanish. Combining the Lorentz boost property (6.51) with these shows that

$$\begin{aligned} p_\mu(v_1)V^\mu(v_2) &= \bar{p}_0\,\bar{V}^0\Lambda_\mu{}^0(v_1)\Lambda^\mu{}_0(-v_2) = mc^2\Lambda^0{}_\mu(-v_1)\Lambda^\mu{}_0(-v_2) \\ &= \frac{mc^2}{\sqrt{1 - v^2/c^2}}, \end{aligned} \tag{6.136}$$

where v is the result of the relativistic velocity addition formula (6.54).

By definition, the presence of a force creates an acceleration for the massive particle it acts upon. A more careful statement of this is that if an object is undergoing an acceleration relative to an inertial observer, then there is a force acting upon the object. In order to state this relationship in a manner that is consistent with special relativity, the proper time derivative and the four-momentum are combined into a tentative relativistic version of Newton's second law,

$$f^\mu = \frac{\mathrm{d}p^\mu(\tau)}{\mathrm{d}\tau} = m\frac{\mathrm{d}^2x^\mu(\tau)}{\mathrm{d}\tau^2}, \tag{6.137}$$

where f^μ is a contravariant four-vector that gives the relativistic version of force and is some function of $x^\mu(\tau)$ and $\mathrm{d}x^\mu/\mathrm{d}\tau$. In order for this to be the correct relativistic replacement for Newton's second law, *all inertial observers* must agree that f^μ transforms as a four-vector. For such a case, statement (6.137) transforms *covariantly*,

$$f'^\mu(\tau) - \frac{\mathrm{d}p'^\mu(\tau)}{\mathrm{d}\tau} = \Lambda^\mu{}_\nu(v)\left(f^\nu(\tau) - \frac{\mathrm{d}p^\nu(\tau)}{\mathrm{d}\tau}\right) = 0, \tag{6.138}$$

and all inertial observers agree that the particle's motion obeys (6.137) when expressed in the coordinates of their inertial frames. Like Newton's second law, the relativistic statement (6.137) is a second-order differential equation, so that the relativistic trajectory of the particle can be found once the four-force f^μ and the boundary conditions on $x^\mu(\tau)$ have been specified. In the case that v^2/c^2 and its time derivatives are ignorable, referred to as the *nonrelativistic limit*, the spatial component of the right-hand side of (6.137) becomes $m\,\mathrm{d}^2\boldsymbol{r}/\mathrm{d}t^2$, which is the Newtonian definition of force. However, no assumptions should be made about how the f^μ appearing in (6.137) are related to the familiar forces of Newtonian physics, in particular, how they are related to their associated Newtonian potential energies.

Demonstrating the validity of (6.137) begins by invoking the relativistic version of the work-energy theorem (2.17). Observer $\mathcal{O}$ measures the spatial displacement $\mathrm{d}\boldsymbol{r}$ in the time interval $\mathrm{d}t$ for a point particle being acted upon by a force. As a result, the relationship $\mathrm{d}\boldsymbol{r} = \boldsymbol{v}(t)\,\mathrm{d}t$ holds in the $\mathcal{O}$ frame of reference, where $\boldsymbol{v}(t)$ is the *instantaneous relative velocity* of the particle according to $\mathcal{O}$. During the same time interval $\mathcal{O}$ observes a change in the relativistic kinetic energy (6.112) given by $\mathrm{d}K$, which $\mathcal{O}$ attributes to a relativistic spatial force $\boldsymbol{f}$ acting on the point mass during the time interval $\mathrm{d}t$. The form of the force $\boldsymbol{f}$ is found by applying the relativistic version of the Newtonian work-energy theorem (2.17) in differential form. This requires the use of the contracted-length version of $\mathrm{d}\boldsymbol{r}$, which is given by (6.38) as $\mathrm{d}\boldsymbol{r}/\gamma_v$. The relativistic work-energy theorem is then given by

$$\mathrm{d}K = \boldsymbol{f}\cdot\frac{\mathrm{d}\boldsymbol{r}}{\gamma_v} = \frac{1}{\gamma_v}\boldsymbol{f}\cdot\frac{\mathrm{d}\boldsymbol{r}}{\mathrm{d}t}\,\mathrm{d}t = \frac{1}{\gamma_v}\boldsymbol{f}\cdot\boldsymbol{v}\,\mathrm{d}t. \tag{6.139}$$

Using the relativistic version of the kinetic energy (6.112) gives

$$\begin{aligned}\mathrm{d}K &= \mathrm{d}\left(\frac{mc^2}{\sqrt{1 - v^2/c^2}}\right) = \frac{m\,\mathrm{d}\boldsymbol{v}\cdot\boldsymbol{v}}{(1 - v^2/c^2)^{3/2}} = \frac{m\,\mathrm{d}\boldsymbol{v}\cdot\boldsymbol{v}}{\sqrt{1 - v^2/c^2}} + \frac{mv^2\,\mathrm{d}\boldsymbol{v}\cdot\boldsymbol{v}}{c^2(1 - v^2/c^2)^{3/2}}\\ &= \frac{m\dot{\boldsymbol{v}}}{\sqrt{1 - v^2/c^2}}\cdot\boldsymbol{v}\,\mathrm{d}t + \left(\frac{m(\dot{\boldsymbol{v}}\cdot\boldsymbol{v})\boldsymbol{v}}{c^2(1 - v^2/c^2)^{3/2}}\right)\cdot\boldsymbol{v}\,\mathrm{d}t = \frac{\mathrm{d}}{\mathrm{d}t}\left(\frac{m\boldsymbol{v}}{\sqrt{1 - v^2/c^2}}\right)\cdot\boldsymbol{v}\,\mathrm{d}t.\end{aligned} \tag{6.140}$$

Comparing (6.140) to (6.139) and using definition (6.121) gives

$$\boldsymbol{f} = \gamma_v \frac{\mathrm{d}}{\mathrm{d}t}\left(\frac{m\boldsymbol{v}}{\sqrt{1 - v^2/c^2}}\right) = \frac{\mathrm{d}\boldsymbol{p}}{\mathrm{d}\tau}, \tag{6.141}$$

where $\boldsymbol{p}$ is the relativistic momentum (6.106). Result (6.141) is precisely the spatial component of the relativistic version of Newton's second law as postulated in (6.137). This shows (6.137) is consistent with the relativistic work-energy theorem and is therefore the correct relativistic version of Newton's second law.

Using (6.137) and result (6.112) gives f^0 as

$$f^0 = \frac{\mathrm{d}p^0}{\mathrm{d}\tau} = \frac{1}{c}\frac{\mathrm{d}K}{\mathrm{d}\tau} = \frac{\gamma_v}{c}\frac{\mathrm{d}K}{\mathrm{d}t} = \gamma_v \frac{\mathrm{d}\boldsymbol{p}}{\mathrm{d}t} \cdot \frac{\boldsymbol{v}}{c} = \frac{\mathrm{d}\boldsymbol{p}}{\mathrm{d}\tau} \cdot \frac{\boldsymbol{v}}{c} = \boldsymbol{f} \cdot \frac{\boldsymbol{v}}{c}. \tag{6.142}$$

Result (6.142) is the relativistic version of the familiar Newtonian result

$$\frac{\mathrm{d}}{\mathrm{d}t}\left(\frac{1}{2}mv^2\right) = m\frac{d\boldsymbol{v}}{\mathrm{d}t} \cdot \boldsymbol{v} = \boldsymbol{F} \cdot \boldsymbol{v}, \tag{6.143}$$

and like (6.143), the result (6.142) yields the generalization of conservation of total energy. Using (6.131) gives

$$0 = \frac{\mathrm{d}}{\mathrm{d}\tau}\left(\frac{1}{2}c^2\right) = \frac{\mathrm{d}}{\mathrm{d}\tau}\left(\frac{1}{2}V^\mu(\tau)V_\mu(\tau)\right) = V_\mu(\tau)\frac{\mathrm{d}V^\mu(\tau)}{\mathrm{d}\tau} = V_\mu(\tau)\mathcal{A}^\mu(\tau), \tag{6.144}$$

where the *relativistic acceleration* is given by

$$\mathcal{A}^\mu(\tau) = \frac{\mathrm{d}^2 x^\mu(\tau)}{\mathrm{d}\tau^2}. \tag{6.145}$$

The relativistic generalization of Newton's second law is given by

$$f^\mu = m\mathcal{A}^\mu, \tag{6.146}$$

which shows that the relativistic force satisfies

$$f^\mu(\tau)V_\mu(\tau) = 0. \tag{6.147}$$

This reproduces result (6.142), since

$$f^\mu(\tau)V_\mu(\tau) = f^0 V_0 - \boldsymbol{f} \cdot \boldsymbol{V} = c\gamma_v f^0 - \gamma_v \boldsymbol{f} \cdot \boldsymbol{v} = 0 \implies f^0 = \boldsymbol{f} \cdot \frac{\boldsymbol{v}}{c}. \tag{6.148}$$

Another useful result follows from

$$\frac{\partial V^\nu(\tau)}{\partial x^\mu(\tau)} = \frac{\partial}{\partial x^\mu(\tau)}\frac{\mathrm{d}}{\mathrm{d}\tau}x^\nu(\tau) = \frac{\mathrm{d}}{\mathrm{d}\tau}\frac{\partial x^\nu(\tau)}{\partial x^\mu(\tau)} = \frac{\mathrm{d}}{\mathrm{d}\tau}\delta^\nu{}_\mu = 0, \tag{6.149}$$

which shows that relativistic position and proper velocity are independent variables.

Having established the consistency of the relativistic version (6.137) of Newton's second law, the next step is to identify the relativistic force f^μ. This is not a trivial effort, since it requires finding a covariant expression for the particular force that reduces in the nonrelativistic limit to the Newtonian expression for the force as the gradient of a potential. It is further complicated by the fact that the Newtonian expression for the force may be incomplete or inconsistent with relativistic physics. A case in point is the gravitational force created by a moving mass. The change in the gravitational force it exerts on another body as a result of a change in its position is transmitted instantaneously throughout all of space *if the Newtonian expression for the force is used.* If the Newtonian expression were correct, then changes in the gravitational potential would travel faster than light. This is clearly inconsistent with the basic tenet of special relativity that the speed of light is the limit for the propagation of signals.

In order to relate the relativistic force to the Newtonian force, it is useful to relate the trajectory $x^\mu(\tau)$ parameterized by the proper time to the trajectory $\tilde{x}^\mu(t)$ of the particle parameterized in terms of the rest frame time t of an observer. This is accomplished by setting

$$\begin{aligned} x^0(\tau) &= \tilde{x}^0(t) = ct, \\ \boldsymbol{x}(\tau) &= \tilde{\boldsymbol{x}}(t), \end{aligned} \tag{6.150}$$

where $\mathrm{d}\tilde{\boldsymbol{x}}(t)/\mathrm{d}t$ is the instantaneous velocity $\boldsymbol{v}(t)$ of the point mass *relative to the observer*,

$$\boldsymbol{v}(t) = \frac{\mathrm{d}\tilde{\boldsymbol{x}}(t)}{\mathrm{d}t}. \tag{6.151}$$

Using the chain rule gives the proper velocity components,

$$\begin{aligned} V^0 &= \frac{\mathrm{d}x^0(\tau)}{\mathrm{d}\tau} = \frac{\mathrm{d}t}{\mathrm{d}\tau}\frac{\mathrm{d}\tilde{x}^0(t)}{\mathrm{d}t} = \gamma_v c, \\ \boldsymbol{V} &= \frac{\mathrm{d}\boldsymbol{x}(\tau)}{\mathrm{d}\tau} = \frac{\mathrm{d}t}{\mathrm{d}\tau}\frac{\mathrm{d}\tilde{\boldsymbol{x}}(t)}{\mathrm{d}t} = \gamma_v \boldsymbol{v}. \end{aligned} \tag{6.152}$$

The components found in (6.152) satisfy the earlier result (6.128),

$$V_\mu V^\mu = V_0^2 - \boldsymbol{V}\cdot\boldsymbol{V} = \gamma_v^2 c^2 - \gamma_v^2 v^2 = \gamma_v^2 c^2 (1 - v^2/c^2) = c^2. \tag{6.153}$$

It is important to note that $\boldsymbol{v}(t)$ is not constant, since, by definition, the particle is not in an inertial frame. Combining these definitions with the relativistic second law (6.137) gives

$$\boldsymbol{p}(\tau) = m\frac{\mathrm{d}\boldsymbol{x}(\tau)}{\mathrm{d}\tau} = m\frac{\mathrm{d}\tilde{\boldsymbol{x}}(t)}{\mathrm{d}\tau} = m\frac{\mathrm{d}t}{\mathrm{d}\tau}\frac{\mathrm{d}\tilde{\boldsymbol{x}}(t)}{\mathrm{d}t} = m\gamma_v\frac{\mathrm{d}\tilde{\boldsymbol{x}}(t)}{\mathrm{d}t} = \frac{m\boldsymbol{v}(t)}{\sqrt{1 - v^2/c^2}}, \tag{6.154}$$

which reproduces the earlier definition of the relativistic momentum (6.106) for the case that $\boldsymbol{v}(t)$ is not constant. Similarly, setting $x^0(\tau) = \tilde{x}^0(t) = ct$, it follows that $p^0(\tau)$ is given by

$$p^0(\tau) = m\frac{\mathrm{d}x^0(\tau)}{\mathrm{d}\tau} = m\frac{\mathrm{d}\tilde{x}^0(t)}{\mathrm{d}\tau} = m\frac{\mathrm{d}t}{\mathrm{d}\tau}\frac{\mathrm{d}\tilde{x}^0(t)}{\mathrm{d}t} = mc\gamma_v = \frac{mc}{\sqrt{1 - v^2/c^2}}, \tag{6.155}$$

which reproduces the relativistic kinetic energy relationship (6.116), $p^0 = K/c$, for the case of an instantaneous velocity. This shows that definition (6.116) is consistent with the relativistic second law (6.137).

The next step uses (6.137) to determine how the relativistic force f^μ is related to the relative velocity $v(t)$ and the relative acceleration $\boldsymbol{a}(t)$. The relative acceleration in the observer's frame is defined as

$$\boldsymbol{a}(t) = \frac{\mathrm{d}\boldsymbol{v}(t)}{\mathrm{d}t}. \tag{6.156}$$

Applying (6.137) to (6.154) and (6.155) yields the relationship of the relativistic force to the relative velocity and acceleration,

$$f^0 = \frac{\mathrm{d}p^0(\tau)}{\mathrm{d}\tau} = \frac{\mathrm{d}t}{\mathrm{d}\tau}\frac{\mathrm{d}p^0(\tau)}{\mathrm{d}t} = \gamma_v \frac{\mathrm{d}}{\mathrm{d}t}\left(\frac{mc}{\sqrt{1 - v^2/c^2}}\right) = \frac{m\boldsymbol{v}\cdot\boldsymbol{a}}{c(1 - v^2/c^2)^2}, \tag{6.157}$$

$$\begin{aligned} \boldsymbol{f} &= \frac{\mathrm{d}\boldsymbol{p}(\tau)}{\mathrm{d}\tau} = \frac{\mathrm{d}t}{\mathrm{d}\tau}\frac{\mathrm{d}\boldsymbol{p}(\tau)}{\mathrm{d}t} = \gamma_v \frac{\mathrm{d}}{\mathrm{d}t}\left(\frac{m\boldsymbol{v}}{\sqrt{1 - v^2/c^2}}\right) \\ &= \frac{m\boldsymbol{a}}{(1 - v^2/c^2)} + \frac{m(\boldsymbol{v}\cdot\boldsymbol{a})\boldsymbol{v}}{c^2(1 - v^2/c^2)^2}. \end{aligned} \tag{6.158}$$

> **Exercise 6.12.** Show that (6.157) and (6.158) satisfy relationship (6.142), which states that $f^0 = \boldsymbol{f}\cdot\boldsymbol{v}/c$.

Results (6.157) and (6.158) are consistent with (6.146), so that the relativistic acceleration $\tilde{a}^\mu$ is given by

$$\begin{aligned} \tilde{a}^0 &= \frac{\boldsymbol{v}\cdot\boldsymbol{a}}{c(1 - v^2/c^2)^2}, \\ \tilde{\boldsymbol{a}} &= \frac{\boldsymbol{a}}{(1 - v^2/c^2)} + \frac{(\boldsymbol{v}\cdot\boldsymbol{a})\boldsymbol{v}}{c^2(1 - v^2/c^2)^2}. \end{aligned} \tag{6.159}$$

Results (6.157) and (6.158) also give a system of equations relating $\boldsymbol{v}$ and $\boldsymbol{a} = \mathrm{d}\boldsymbol{v}/\mathrm{d}t$ to the relativistic force. It is useful to note that for $v^2/c^2 \approx 0$, the two equations (6.157) and (6.158) reduce to $f^0 = 0$ and $\boldsymbol{f} = m\boldsymbol{a}$, the latter of which reproduces Newton's version of the second law.

The two equations simplify considerably for two special cases. The first is the case that $\boldsymbol{v}\cdot\boldsymbol{a} = 0$, which gives $f^0 = 0$ and

$$\boldsymbol{f} = \frac{m\boldsymbol{a}}{(1 - v^2/c^2)}. \tag{6.160}$$

equation (6.160) will hold if the force $\boldsymbol{f}$ acts tangentially to the motion, as is the case with a centripetal force. The second is the case of one-dimensional motion, so that $\boldsymbol{v} = v$ and $\boldsymbol{v} \cdot \boldsymbol{a} = va$. For this case the force equations become

$$f^0 = \frac{vma}{c(1 - v^2/c^2)^2}$$
$$f^1 = \frac{ma}{(1 - v^2/c^2)^2}. \tag{6.161}$$

However, neither (6.157) nor (6.158) identify the components of the relativistic force f^μ. In order to make progress, it will be assumed that *in the rest frame* of the particle there is a spatial force $\boldsymbol{F}$, referred to as the *rest frame force*. The position of the particle does not define an inertial frame, since, by definition, the particle is accelerated and therefore does not have a constant velocity relative to all inertial observers. Therefore, an inertial observer can measure the *instantaneous velocity* $\boldsymbol{v}(t)$ of the particle and apply an *instantaneous Lorentz boost* to the particle's rest frame force. This boost uses a time-dependent $\boldsymbol{v}$ in the Lorentz boost and will yield the force acting on the particle according to the inertial observer. The earlier results (6.118) and (6.119) show that this yields the correct time-dependent four-momentum appearing in (6.137). Such an instantaneous Lorentz boost with a time-dependent velocity no longer connects one inertial observer to another, since it does not obey condition (6.124). Instead, it connects the non-inertial frame of the particle to the frame of an inertial observer.

It is a natural question as to why applying an instantaneous Lorentz boost is the correct method to obtain the force on the particle. The answer is that the form of the Lorentz boost, *even an instantaneous one*, maintains the speed of light as a constant. As a result, the associated relativistic four-force in the inertial observer's frame will be consistent with the basic postulate of special relativity. The relativistic version of the second law (6.137) shows that the rest frame four-force is given by $\bar{f}^0 = 0$ and $\bar{\boldsymbol{f}} = \boldsymbol{F}$, where the rest frame force $\boldsymbol{F}$ is to be identified. In the case of gravitation this does *not* mean that the rest frame force $\boldsymbol{F}$ is given by the Newtonian expression (2.32), for reasons to be discussed. In the case of electrodynamics the Lorentz force (6.2) will be shown be consistent with special relativity in chapter 9.

When setting the boost velocity to $\boldsymbol{v}(t)$ it is very useful to write the rest frame force $\boldsymbol{F}$ as the sum of two uniquely defined components, $\boldsymbol{F} = \boldsymbol{F}_\parallel + \boldsymbol{F}_\perp$, where

$$\boldsymbol{F}_\perp = \boldsymbol{F} - \left(\frac{\boldsymbol{F} \cdot \boldsymbol{v}}{v^2}\right)\boldsymbol{v}, \tag{6.162}$$

$$\boldsymbol{F}_\parallel = \left(\frac{\boldsymbol{F} \cdot \boldsymbol{v}}{v^2}\right)\boldsymbol{v}. \tag{6.163}$$

These two components satisfy $\boldsymbol{v} \cdot \boldsymbol{F}_\perp = 0$ and $\boldsymbol{v} \cdot \boldsymbol{F}_\parallel = \boldsymbol{v} \cdot \boldsymbol{F}$, so that they represent the components of the rest frame force perpendicular and parallel to the velocity. Since $\bar{f}^0 = 0$ the Lorentz boost gives

$$f^0 = \Lambda^0{}_j(v)\bar{f}^j = \Lambda^0{}_j(v)F^j = \gamma_v\, \boldsymbol{F} \cdot \frac{\boldsymbol{v}}{c} = \gamma_v \boldsymbol{F}_\parallel \cdot \frac{\boldsymbol{v}}{c}, \tag{6.164}$$

$$f^j = \Lambda^j{}_k(v)\bar{f}^k = \Lambda^j{}_k(v)F^k = F^j + (\gamma_v - 1)\frac{v^j}{v^2}\boldsymbol{F} \cdot \boldsymbol{v} = F_\perp^j + \gamma_v F_\parallel^j. \tag{6.165}$$

Exercise 6.13. Using (6.162) and (6.163) show that the last steps of (6.164) and (6.165) are correct.

Result (6.164) for f^0 clearly satisfies (6.142) with the identification $\boldsymbol{f}_\parallel = \gamma_v \boldsymbol{F}_\parallel$. Result (6.165) is consistent with the earlier result (6.30) that showed a Lorentz boost affects only the directions parallel to the boost velocity. However, it does not have an obvious connection to the result (6.158) for $\boldsymbol{f}$ in terms of $\boldsymbol{a}$ and $\boldsymbol{v}$. The spatial component of the relativistic version of Newton's second law (6.137) can now be expressed in terms of (6.165) as

$$\frac{\mathrm{d}}{\mathrm{d}\tau}\left(\frac{m\boldsymbol{v}}{\sqrt{1 - v^2/c^2}}\right) = \boldsymbol{F}_\perp + \gamma_v \boldsymbol{F}_\parallel = \boldsymbol{f}, \tag{6.166}$$

where $\boldsymbol{F}_\perp$ and $\boldsymbol{F}_\parallel$ are obtained from the rest frame force $\boldsymbol{F}$ using (6.162) and (6.163). The factor γ_v is found from the instantaneous velocity of the particle. The asymmetric presence of γ_v will limit the use of static versions of force, such as Hooke's law and Newton's law of gravitation.

6.8 Relativistic action in the presence of force

Having found the spatial components of the relativistic force in (6.166), it is important to determine whether the Newtonian action approach to particle motion given by (2.104) can be generalized to the fully relativistic case. This requires the use of a *manifestly Lorentz invariant action* that incorporates force in terms of a potential $U(x, \dot{x})$ that is also a *Lorentz scalar.* This requires the potential to be a scalar function of x^μ and potentially $\mathrm{d}x^\mu/\mathrm{d}\tau$ as well. There are a number of possible ways to proceed.

The first approach is to generalize the free action (6.100) to read

$$S[x, \dot{x}] = \int \mathrm{d}\tau\, (-mc^2 - U(x, \dot{x})) = \int \mathrm{d}t \sqrt{1 - v^2/c^2}(-mc^2 - U(x, \dot{x})). \tag{6.167}$$

For the case that $U(x, \dot{x})$ is a Lorentz scalar, the action (6.167) is manifestly Lorentz invariant and is built on the free-particle action (6.100). It has the advantage that a Taylor series for $v/c \ll 1$ gives

$$\mathcal{L} = \sqrt{1 - v^2/c^2}(-mc^2 - U(x, \dot{x})) \approx -mc^2 + \frac{1}{2}mv^2 - U(x, \dot{x}), \qquad (6.168)$$

which reproduces the nonrelativistic Newtonian expression. However, it displays one of the difficulties in the relativistic action approach, which is that the contribution from the potential term to the Euler–Lagrange equation involves derivatives with respect to the velocity $v^j = \mathrm{d}x^j/\mathrm{d}t$, even if U itself is independent of $\mathrm{d}x^\mu/\mathrm{d}\tau$. Applying the Euler–Lagrange equation to (6.167) gives

$$\frac{\mathrm{d}}{\mathrm{d}t}\left(\frac{m\dot{x}^j}{\sqrt{1 - v^2/c^2}} - \frac{\partial}{\partial \dot{x}^j}\left(\frac{1}{\gamma_v}U(x, \dot{x})\right)\right) + \frac{1}{\gamma_v}\partial_j U(x, \dot{x}) = 0, \qquad (6.169)$$

where $\dot{x}^j = \mathrm{d}x^j/\mathrm{d}t$. Multiplying both sides by γ_v and using $\gamma_v \mathrm{d}/\mathrm{d}t = \mathrm{d}/\mathrm{d}\tau$ gives the contravariant spatial components of the relativistic force,

$$f^j = m\frac{\mathrm{d}^2 x^j}{\mathrm{d}\tau^2} = \frac{\mathrm{d}}{\mathrm{d}\tau}\left(\frac{\partial}{\partial \dot{x}^j}\left(\frac{1}{\gamma_v}U(x, \dot{x})\right)\right) - \partial_j U(x, \dot{x}). \qquad (6.170)$$

A second approach that gives a manifestly covariant Euler–Lagrange equation starts with the action

$$S[x, \dot{x}] = \int_0^{\tau_f} \mathrm{d}\tau \left(-mc\sqrt{\frac{\mathrm{d}x^\mu}{\mathrm{d}\tau}\frac{\mathrm{d}x_\mu}{\mathrm{d}\tau}} - U(x, \dot{x})\right). \qquad (6.171)$$

The free part of the action (6.171) coincides with the free action (6.100), since the proper velocity of the particle's trajectory will obey

$$\sqrt{\frac{\mathrm{d}x^\mu}{\mathrm{d}\tau}\frac{\mathrm{d}x_\mu}{\mathrm{d}\tau}} = \sqrt{V^\mu V_\mu} = \sqrt{c^2} = c. \qquad (6.172)$$

In this approach the action is varied using x^μ and $\dot{x}^\mu = \mathrm{d}x^\mu/\mathrm{d}\tau$ as the independent variables. The previous result (6.149) showed that the relativistic position x^μ and the proper velocity $\mathrm{d}x^\mu/\mathrm{d}\tau$ can be treated as independent degrees of freedom, and so this approach is possible. It is straightforward to repeat the variation of the action that led to the Newtonian Euler–Lagrange equation (2.40) by simply replacing t with τ. Assuming the variation $\delta x^\mu(\tau)$ vanishes at the proper time endpoints and noting that $\delta\dot{x}^\mu = \mathrm{d}(\delta x^\mu)/\mathrm{d}\tau$, an integration by parts yields

$$\delta S[x, \dot{x}] = \int_0^{\tau_f} \mathrm{d}\tau \left(\frac{\partial \mathcal{L}}{\partial x^\mu} - \frac{\mathrm{d}}{\mathrm{d}\tau}\frac{\partial \mathcal{L}}{\partial \dot{x}^\mu}\right)\delta x^\mu(\tau) = 0. \qquad (6.173)$$

This immediately yields the relativistic Euler–Lagrange equation,

$$\frac{\partial \mathcal{L}}{\partial x^\mu} - \frac{\mathrm{d}}{\mathrm{d}\tau}\frac{\partial \mathcal{L}}{\partial \dot{x}^\mu} = 0, \qquad (6.174)$$

where $\dot{x}^\mu = \mathrm{d}x^\mu/\mathrm{d}\tau$. Equation (6.174) transforms like ∂_μ under a Lorentz boost, and so all inertial observers will agree upon its form. Applying (6.174) to the Lagrangian defined by (6.171) gives

$$mc\frac{\mathrm{d}}{\mathrm{d}\tau}\left(\frac{\mathrm{d}x_\mu}{\mathrm{d}\tau}\left(\frac{\mathrm{d}x^\nu}{\mathrm{d}\tau}\frac{\mathrm{d}x_\nu}{\mathrm{d}\tau}\right)^{-\frac{1}{2}}\right) - \frac{\partial U(x, \dot{x})}{\partial x^\mu} + \frac{\mathrm{d}}{\mathrm{d}\tau}\frac{\partial U(x, \dot{x})}{\partial \dot{x}^\mu} = 0. \tag{6.175}$$

The first term in the equation of motion (6.175) reduces to the relativistic force (6.137) for the case that the square root can be replaced by c. However, it is not *a priori* obvious that such a replacement is consistent with (6.175). Using $V_\mu = \mathrm{d}x_\mu/\mathrm{d}\tau$ and the definition (6.137) allows (6.175) to be written

$$\begin{aligned} f_\mu &= \frac{V_\mu}{V^\alpha V_\alpha}V^\nu f_\nu + \frac{1}{c}\sqrt{V^\nu V_\nu}\left(\frac{\partial U(x, \dot{x})}{\partial x^\mu} - \frac{\mathrm{d}}{\mathrm{d}\tau}\frac{\partial U(x, \dot{x})}{\partial \dot{x}^\mu}\right) \\ &= \frac{1}{c^2}(V^\nu f_\nu)V_\mu + \frac{\partial U(x, \dot{x})}{\partial x^\mu} - \frac{\mathrm{d}}{\mathrm{d}\tau}\frac{\partial U(x, \dot{x})}{\partial \dot{x}^\mu}, \end{aligned} \tag{6.176}$$

where (6.128) was used. Contracting both sides with V^μ and again using $V^\mu V_\mu = c^2$ gives

$$V^\mu\left(\frac{\partial U(x, \dot{x})}{\partial x^\mu} - \frac{\mathrm{d}}{\mathrm{d}\tau}\frac{\partial U(x, \dot{x})}{\partial \dot{x}^\mu}\right) = 0. \tag{6.177}$$

This does not uniquely identify f_μ as the quantity in the parentheses, although the quantity in parentheses is orthogonal to the proper velocity V^μ in agreement with the requirement (6.147). Using action (6.171) involves an ambiguity which is typically overlooked.

Exercise 6.14. Using $V^\mu = \mathrm{d}x^\mu/\mathrm{d}\tau$ show that (6.177) can be written as the requirement

$$\frac{\mathrm{d}}{\mathrm{d}\tau}\left(U(x, \dot{x}) - \frac{\mathrm{d}x^\mu}{\mathrm{d}\tau}\frac{\partial U(x, \dot{x})}{\partial \dot{x}^\mu}\right) = 0. \tag{6.178}$$

A third approach is to start with the free action,

$$S_0[x, \dot{x}] = \int_0^T \mathrm{d}\tau\, \frac{1}{2}m\frac{\mathrm{d}x^\mu}{\mathrm{d}\tau}\frac{\mathrm{d}x_\mu}{\mathrm{d}\tau} = \int_0^T \mathrm{d}\tau\, \mathcal{L}(x^\mu, \dot{x}^\mu), \tag{6.179}$$

where $\dot{x}^\mu = \mathrm{d}x^\mu/\mathrm{d}\tau$ is the proper velocity. Like (6.167) and (6.171), the action of (6.179) is also manifestly Lorentz invariant. Applying the Euler–Lagrange equation (6.174) to the Lagrangian defined by (6.179) gives

$$\frac{\partial\mathcal{L}}{\partial x^\mu} - \frac{\mathrm{d}}{\mathrm{d}\tau}\frac{\partial\mathcal{L}}{\partial\dot{x}^\mu} = -\frac{\mathrm{d}}{\mathrm{d}\tau}\left(\eta_{\mu\nu}m\frac{\mathrm{d}x^\nu}{\mathrm{d}\tau}\right) = -\frac{\mathrm{d}}{\mathrm{d}\tau}\left(m\frac{\mathrm{d}x_\mu}{\mathrm{d}\tau}\right) = -\frac{\mathrm{d}p_\mu}{\mathrm{d}\tau} = 0. \tag{6.180}$$

Since there is no force this is the result for a free particle. However, result (6.180) now includes conservation of energy, p_0. There is an important subtlety involved with using (6.179) as the free action, which is that the canonical momentum is given by

$$\frac{\partial \mathcal{L}}{\partial \dot{x}^{\mu}} = m\frac{\mathrm{d}x_{\mu}}{\mathrm{d}\tau} \implies \frac{\partial \mathcal{L}}{\partial \dot{x}^{j}} = m\frac{\mathrm{d}x_{j}}{\mathrm{d}\tau} = -m\frac{\mathrm{d}x^{j}}{\mathrm{d}\tau}. \tag{6.181}$$

This has the opposite sign from those of the canonical momentum spatial components that were obtained from the action (6.100). The source of the minus sign is revealed by writing out the Lagrangian for the case $v^j/c \ll 1$,

$$\frac{1}{2}m\frac{\mathrm{d}x^{\mu}}{\mathrm{d}\tau}\frac{\mathrm{d}x_{\mu}}{\mathrm{d}\tau} = \frac{1}{2}m\frac{\mathrm{d}x^{0}}{\mathrm{d}\tau}\frac{\mathrm{d}x^{0}}{\mathrm{d}\tau} - \frac{1}{2}m\frac{\mathrm{d}x^{j}}{\mathrm{d}\tau}\frac{\mathrm{d}x^{j}}{\mathrm{d}\tau} \approx \frac{1}{2}mc^2 - \frac{1}{2}mv^2, \tag{6.182}$$

which has the nonrelativistic Newtonian Lagrangian $\frac{1}{2}mv^2$ but with the opposite sign. As a result, extending this basis action to the case where force is present requires the use of the opposite sign for the potential.

Using (6.179) as the basis Lagrangian for the case of force requires that the Lagrangian is given by

$$\mathcal{L} = \frac{1}{2}m\frac{\mathrm{d}x^{\mu}}{\mathrm{d}\tau}\frac{\mathrm{d}x_{\mu}}{\mathrm{d}\tau} + U(x, \dot{x}). \tag{6.183}$$

Applying the relativistic Euler–Lagrange equation (6.174) to the Lagrangian (6.183) gives

$$\begin{aligned} -m\frac{\mathrm{d}}{\mathrm{d}\tau}\frac{\mathrm{d}x_{\mu}}{\mathrm{d}\tau} + \frac{\partial U(x, \dot{x})}{\partial x^{\mu}} - \frac{d}{\mathrm{d}\tau}\frac{\partial U(x, \dot{x})}{\partial \dot{x}^{\mu}} = 0 \implies \\ f_{\mu} = \frac{\partial U(x, \dot{x})}{\partial x^{\mu}} - \frac{\mathrm{d}}{\mathrm{d}\tau}\frac{\partial U(x, \dot{x})}{\partial \dot{x}^{\mu}}, \end{aligned} \tag{6.184}$$

which unambiguously identifies the relativistic force as the result inferred from (6.176).

At first glance, the form of the force in (6.184) appears to have the wrong sign. The necessity for this sign difference is demonstrated by considering the spatial components. Using $x^j = -x_j$ gives

$$-f_j = -m\frac{\mathrm{d}^2x_j}{\mathrm{d}\tau^2} = m\frac{\mathrm{d}^2x^j}{\mathrm{d}\tau^2} = f^j = -\partial_j U(x, \dot{x}) + \frac{\mathrm{d}}{\mathrm{d}\tau}\frac{\partial U}{\partial \dot{x}^j}, \tag{6.185}$$

where $-f_j$ is correctly identified as f^j and is given by the Newtonian relationship $f = -\nabla U$ for a static potential. This result is identical to (6.170) for the case that

$$\frac{\mathrm{d}}{\mathrm{d}\tau}\left(\frac{\partial}{\partial(\mathrm{d}x^j/\mathrm{d}t)}\left(\frac{1}{\gamma_v}U(x, \dot{x})\right)\right) = \frac{\mathrm{d}}{\mathrm{d}\tau}\frac{\partial U(x, \dot{x})}{\partial(\mathrm{d}x^j/\mathrm{d}\tau)}. \tag{6.186}$$

Exercise 6.15. Show that requiring the force identified in (6.184) to satisfy the earlier result $f_\mu V^\mu = 0$ found in (6.147) yields the same requirement as (6.178).

It will be seen in chapter 9 that the electromagnetic interaction meets the requirements of both (6.186) and (6.178). For the purposes of this text, action (6.183) will be used for the case that a relativistic force is present. It allows both x^μ and $dx^\mu/d\tau$ to be varied while maintaining $V^\mu V_\mu = c^2$ along the trajectory without the square root term. The difference between the two actions (6.100) and (6.183) will be discussed when the Lorentz force for electromagnetism is derived in chapter 9. The reader should be aware that other authors use the action (6.171) to avoid the minus sign for the potential. All three forms of the relativistic point particle action yield identical results.

So far these results are a tautology, since they simply reproduce the form of the relativistic second law (6.137) for the case of force represented in terms of a Lorentz invariant potential. The latter point is critical, since the success of the manifestly covariant methods requires a *Lorentz invariant* potential to be found such that the force,

$$\frac{\partial U(x, \dot{x})}{\partial x^\mu} - \frac{d}{d\tau}\frac{\partial U(x, \dot{x})}{\partial \dot{x}^\mu} = f_\mu(x, \dot{x}), \tag{6.187}$$

transforms as the covariant components of a four-vector. By construction, $f_\mu(x, \dot{x})$ will transform as the covariant components of a four-vector as long as U is Lorentz invariant, $U(x', \dot{x}') = U(x, \dot{x})$, since then

$$\begin{aligned} f'_\mu(x', \dot{x}') &= \frac{\partial U(x', \dot{x}')}{\partial x'^\mu} - \frac{d}{d\tau}\frac{\partial U(x', \dot{x}')}{\partial \dot{x}'^\mu} \\ &= \Lambda_\mu{}^\nu\left(\frac{\partial U(x, \dot{x})}{\partial x^\nu} - \frac{d}{d\tau}\frac{\partial U(x, \dot{x})}{\partial \dot{x}^\nu}\right) = \Lambda_\mu{}^\nu f_\nu(x, \dot{x}). \end{aligned} \tag{6.188}$$

The full equation of motion (6.185) is then given by

$$\frac{dp_\mu}{d\tau} = \frac{\partial U(x, \dot{x})}{\partial x^\mu} - \frac{d}{d\tau}\frac{\partial U(x, \dot{x})}{\partial \dot{x}^\mu}. \tag{6.189}$$

The zeroth component identifies f^0 as

$$\frac{dp_0}{d\tau} = \frac{dp^0}{d\tau} = f^0 = \frac{\partial U(x, \dot{x})}{\partial x^0} - \frac{d}{d\tau}\frac{\partial U(x, \dot{x})}{\partial \dot{x}^0}, \tag{6.190}$$

while the contravariant components of the force are consistent with (6.185) and given by

$$f^j = \frac{dp^j}{d\tau} = \gamma_v\frac{dp^j}{dt} = -\frac{dp_j}{d\tau} = -\gamma_v\frac{dp_j}{dt} = -\frac{\partial U(x, \dot{x})}{\partial x^j} + \frac{d}{d\tau}\frac{\partial U(x, \dot{x})}{\partial \dot{x}^j}. \tag{6.191}$$

The sign differences for the spatial components are brought about by the representation (6.60) chosen for the Minkowski metric.

The relativistic version of the Legendre transformation uses the covariant form of the Lagrangian (6.183) to define a conserved quantity similar to the Hamiltonian, so that

$$H(p, x) = \dot{x}^{\mu}\frac{\partial \mathcal{L}}{\partial \dot{x}^{\mu}} - \mathcal{L} = \dot{x}^{\mu}p_{\mu} - \mathcal{L}, \tag{6.192}$$

where it is understood that $\dot{x}^{\mu}$ is expressed in terms of p^{μ} and x^{μ} by solving

$$p_{\mu} = \frac{\partial \mathcal{L}}{\partial \dot{x}^{\mu}}. \tag{6.193}$$

When evaluated using a solution x_c of (6.185), the relativistic Hamiltonian obeys the conservation law

$$\left.\frac{\mathrm{d}H}{\mathrm{d}\tau}\right|_{x=x_c} = 0. \tag{6.194}$$

Exercise 6.16. Verify (6.194)

It should be stressed that the relativistically invariant Hamiltonian found by applying (6.192) to action (6.183) may differ from the Hamiltonian found using action (6.100). This difference will be clarified in chapter 9 when the Lorentz equation for charged particles is derived from the two different action approaches. In addition, the relativistic versions of Hamilton's equations (2.112) are given by

$$\begin{aligned}\frac{\mathrm{d}x^{\mu}}{\mathrm{d}\tau} &= \frac{\partial H(p, x)}{\partial p_{\mu}}, \\ \frac{\mathrm{d}p_{\mu}}{\mathrm{d}\tau} &= -\frac{\partial H(p, x)}{\partial x^{\mu}}.\end{aligned} \tag{6.195}$$

Exercise 6.17. Using the relativistic Lagrangian obtained from relation (6.192),

$$\mathcal{L} = p_{\mu}\frac{x^{\mu}}{\mathrm{d}\tau} - H(p, x), \tag{6.196}$$

show that the equations of (6.195) are obtained by variational techniques.

The contravariant components of the relativistic force f^j found earlier in (6.165) can be used to find the potential $U(x, \dot{x})$, assuming that the *correct* rest frame force $\boldsymbol{F}$

is known. Using the f^j obtained from (6.165) in expression (6.187) gives the spatial component relationship,

$$f^j = -\frac{\partial U(x, \dot{x})}{\partial x^j} + \frac{\mathrm{d}}{\mathrm{d}\tau}\frac{\partial U(x, \dot{x})}{\partial \dot{x}^j} = F_\perp^j + \gamma_v F_\parallel^j. \tag{6.197}$$

Result (6.197) exposes the difficulty in adapting a *static* Newtonian potential energy to special relativity, which is that the presence of γ_v makes a velocity-dependent potential $U(x, \dot{x})$ unavoidable. *Assuming* that the rest frame force is given by the Newtonian expression $F^j = -\partial_j U_N(\boldsymbol{r})$, where $U_N(\boldsymbol{r})$ is the *time-independent* Newtonian potential energy, it can be combined with the definitions of (6.162) and (6.163) in (6.197) to find

$$\begin{aligned} &-\frac{\partial U(x, \dot{x})}{\partial x^j} + \frac{\mathrm{d}}{\mathrm{d}\tau}\frac{\partial U(x, \dot{x})}{\partial \dot{x}^j} \\ &= \left(-\partial_j U_N(\boldsymbol{r}) + \frac{v_j}{v^2}\boldsymbol{v}\cdot\nabla U_N(\boldsymbol{r})\right) - \gamma_v \frac{v_j}{v^2}\boldsymbol{v}\cdot\nabla U_N(\boldsymbol{r}). \end{aligned} \tag{6.198}$$

The Lorentz invariant form of U that matches the right-hand side of (6.198) is not obvious. Each Newtonian potential must be treated on a case-by-case basis.

However, there are two generic cases where relation (6.198) simplifies enough to yield an expression for the relativistic force. The first is motion along one spatial dimension. Picking the x-direction for the one-dimensional motion gives $v^j = v^x$, $\boldsymbol{v}\cdot\nabla U_N(\boldsymbol{r}) = v^x\partial_x U_N(x)$, and $v^2 = (v^x)^2$, so that the first two terms on the right-hand side of (6.198) cancel, yielding

$$-\frac{\partial U(x, \dot{x})}{\partial x} + \frac{\mathrm{d}}{\mathrm{d}\tau}\frac{\partial U(x, \dot{x})}{\partial \dot{x}} = -\gamma_v\frac{\partial U_N(x)}{\partial x} \implies f^x = -\gamma_v\frac{\partial U_N(x)}{\partial x}. \tag{6.199}$$

Transporting (6.199) back to the Euler–Lagrange equation (6.191) and canceling the common factor of γ_v gives the one-dimensional relativistic equation of motion for the case of a static Newtonian potential,

$$\frac{\mathrm{d}p^x}{\mathrm{d}t} + \frac{\partial U_N(x)}{\partial x} = 0, \tag{6.200}$$

where p^x is the relativistic momentum (6.106) in the x-direction. It is important to note that it is the rest frame time t that appears in (6.200), rather than the proper time τ. This will parameterize the solution in terms of t. This was discussed earlier where (6.150) introduced the method for doing this.

The second general case occurs if the motion is such that

$$\boldsymbol{v}\cdot\boldsymbol{F} = -\boldsymbol{v}\cdot\nabla U_N = 0. \tag{6.201}$$

This was discussed earlier in (6.160) and is the case for centripetal forces. For this case, (6.198) yields

$$f^j = -\frac{\partial U(x, \dot{x})}{\partial x^j} + \frac{\mathrm{d}}{\mathrm{d}\tau}\frac{\partial U(x, \dot{x})}{\partial \dot{x}^j} = -\partial_j U_N \implies \boldsymbol{f} = -\nabla U_N. \tag{6.202}$$

Transporting (6.202) back to the Euler–Lagrange equation (6.191) yields the relativistic equation of motion for this case,

$$\frac{\mathrm{d}\boldsymbol{p}}{\mathrm{d}\tau} + \nabla U_N = 0 \implies \frac{\mathrm{d}\boldsymbol{p}}{\mathrm{d}t} + \frac{1}{\gamma_v}\nabla U_N = 0, \tag{6.203}$$

where $\boldsymbol{p}$ is the relativistic spatial momentum (6.106). Result (6.203) differs from (6.200) by the factor of γ_v.

The limitation to one-dimensional or circular motion severely hampers the usefulness of static Newtonian potentials in a relativistic context. The fundamental forces of nature, such as gravity and electromagnetism, should yield expressions for motion that are covariant, so that all inertial observers agree upon them regardless of their relative velocity. It will be seen in chapter 9 that there is a Lorentz invariant velocity-dependent potential that reproduces the Lorentz force of electromagnetic phenomena for a relativistic point charge. This should not be a surprise, since Maxwell's equations are *a priori* consistent with special relativity. However, the solution for gravity requires abandoning the Newtonian gravitational potential and introducing a space-time-dependent metric tensor. Chapter 10 examines the relativistic motion of a point particle in a gravitational field using the general relativistic extension of action (6.179). Fully describing the exchange of energy and momentum between matter and electromagnetic and gravitational fields requires the use of field theory to represent the matter, rather than relying on the mechanics of a point particle moving in a force field. In developing such a field theory the stress–energy tensor plays a key role in understanding the nature of the forces. This process began with the stretched string stress–energy tensor results (3.87) and (3.88), which were shown to be identical to the force and energy statements (3.71) and (3.76). The next step is to understand the stress–energy tensor as a relativistic statement. However, before doing so, a few simple examples of accelerated relativistic motion will be presented.

6.8.1 Conservation of relativistic energy

If the case of one-dimensional motion is considered, then the relativistic energy formula (6.164) provides a conservation law for motion by using the solution to (6.200), denoted by $x_c(t)$. Assigning $v_c = \dot{x}_c$, it follows that

$$\begin{aligned}
\frac{\mathrm{d}p^0}{\mathrm{d}\tau} &= \frac{\mathrm{d}}{\mathrm{d}\tau}\left(\frac{mc}{\sqrt{1 - v_c^2/c^2}}\right) = f\frac{v_c}{c} = \gamma_v F(x_c)\frac{v_c}{c},\\
&\implies \frac{\mathrm{d}}{\mathrm{d}t}\left(\frac{mc^2}{\sqrt{1 - v_c^2/c^2}}\right) = -v_c\frac{\partial U_N(x_c)}{\partial x} = -\dot{x}_c\frac{\partial U_N(x_c)}{\partial x} = -\frac{\mathrm{d}}{\mathrm{d}t}U_N(x_c),\\
&\implies \frac{\mathrm{d}}{\mathrm{d}t}\left(\frac{mc^2}{\sqrt{1 - v_c^2/c^2}} + U_N(x_c)\right) = 0,\\
&\implies \frac{mc^2}{\sqrt{1 - v_c^2/c^2}} + U_N(x_c) = E,
\end{aligned} \tag{6.204}$$

where E is a constant energy chosen to be consistent with the boundary conditions of motion. Result (6.204) shows that the sum of the relativistic kinetic energy and the Newtonian potential energy is a constant of motion in the one-dimensional case.

Relation (6.204) is easily solved for v_c^2, finding

$$v_c^2 = c^2 - \frac{m^2c^6}{(E - U_N(x_c))^2}. \tag{6.205}$$

Like the nonrelativistic case of energy conservation, finding the relativistic trajectory has been reduced to solving a first-order differential equation involving $v_c = \dot{x}_c$ and x_c, which is given by

$$\frac{\mathrm{d}x_c}{\mathrm{d}t} = \pm\sqrt{c^2 - \frac{m^2c^6}{(E - U_N(x_c))^2}}, \tag{6.206}$$

where the choice of sign is dictated by the boundary conditions.

Exercise 6.18. For a given choice of E, show that the motion of a particle must satisfy

$$(E - U_N(x_c))^2 > m^2c^4. \tag{6.207}$$

6.8.2 One-dimensional constant-force motion

A very illustrative relativistic example is one-dimensional motion in the presence of a constant rest frame force. In this case a mass m spaceship is undergoing a constant rest frame force $F = mg$ according to the crew in the ship, where for their comfort g is chosen to be the acceleration of gravity on Earth. Using the assumption that v is directed radially away from Earth, the relativistic version of Newton's second law states that earthbound observers will describe the spaceship's motion using the equation,

$$\frac{\mathrm{d}}{\mathrm{d}t}\left(\frac{mv}{\sqrt{1 - v^2/c^2}}\right) = mg. \tag{6.208}$$

For the initial condition $v = 0$ at $t = 0$, this has the solution

$$\frac{v}{\sqrt{1 - v^2/c^2}} = gt. \tag{6.209}$$

Solving (6.209) for v gives

$$v = \frac{\mathrm{d}x}{\mathrm{d}t} = \frac{cg\,t}{\sqrt{g^2t^2 + c^2}}, \tag{6.210}$$

which shows that the Earth-based observer will claim $v \to c$ as t becomes large, so that the spaceship velocity never exceeds c. Result (6.210) gives the relative acceleration as

$$a(t) = \frac{dv}{dt} = \frac{gc^3}{(c^2 + g^2t^2)^{3/2}}, \tag{6.211}$$

so that the Earth-based observer claims $a(t) \to 0$ as t becomes large. The relativistic correction factor γ_v is given by

$$\gamma_v = \frac{1}{\sqrt{1 - v^2/c^2}} = \sqrt{1 + \frac{g^2t^2}{c^2}}, \tag{6.212}$$

so that time dilation and length contraction become profound for large t. According to the Earth-based observer the spaceship travels a distance x in the Earth-based time interval t given by

$$x = \int_0^x dx = \int_0^t dt'\, v(t') = \int_0^t dt' \frac{gt'c}{\sqrt{g^2t'^2 + c^2}} = \frac{c}{g}\left(\sqrt{c^2 + g^2t^2} - c\right). \tag{6.213}$$

For large values of t this gives the result $x \approx ct$, consistent with the spaceship traveling at the speed c.

The next step is to compute the time that has elapsed for the crew in the spaceship. This is computed using the proper time interval $d\tau$, since this gives the time interval elapsed on the clock in the spaceship. Using the expression for v gives the proper time interval

$$d\tau = dt\sqrt{1 - v^2/c^2} = \frac{c\, dt}{\sqrt{c^2 + g^2t^2}}. \tag{6.214}$$

The total proper time τ elapsing according to the spaceship's clock during the Earth-based time interval t is given by

$$\tau = \int_0^\tau d\tau = \int_0^t dt' \frac{c}{\sqrt{c^2 + g^2t'^2}} = \frac{c}{g}\ln\left(\frac{gt}{c} + \sqrt{1 + \frac{g^2t^2}{c^2}}\right) \tag{6.215}$$

If the Earth-based observer claims that $t \approx 7.1 \times 10^{17}$ seconds or 15 billion years has elapsed, then for $g = 10\ \mathrm{m/s^2}$, the proper time for the spaceship's crew is

$$\tau \approx \frac{c}{g}\ln\left(\frac{2gT}{c}\right) = (3.0 \times 10^7\ \mathrm{s})\ln(4.7 \times 10^{10}) = 7.4 \times 10^8\ \mathrm{s} \approx 23\ \text{years}. \tag{6.216}$$

In that proper time interval the spaceship will have traveled 15 billion light-years according to the Earth-based observer. This is an outgrowth of the profound length contraction observed from the accelerated spaceship. Similarly, the kinetic energy of the ship is given at time t by

$$K = \frac{mc^2}{\sqrt{1 - v^2/c^2}} = mc\sqrt{c^2 + g^2t^2}. \tag{6.217}$$

Using (6.213) shows that (6.217) can be written

$$K = mc^2 + mgx, \tag{6.218}$$

where mgx is the work done by the force mg powering the spaceship's acceleration. The energy required to power such a trip exceeds the mass energy of the ship by a factor of $\approx 1.5 \times 10^{10}$ and so such a trip is impossible.

Result (6.215) is easily inverted to find t in terms of the proper time τ,

$$t = \frac{c}{g}\sinh\left(\frac{g\tau}{c}\right). \tag{6.219}$$

This allows the velocity (6.210) to be written

$$v = \frac{cgt}{\sqrt{c^2 + g^2t^2}} = c\tanh\left(\frac{g\tau}{c}\right), \tag{6.220}$$

which in turn gives

$$\gamma_v = \frac{1}{\sqrt{1 - v^2/c^2}} = \cosh\left(\frac{g\tau}{c}\right). \tag{6.221}$$

Using (6.219) and (6.213) gives the trajectory $x^\mu(\tau)$ parameterized by the proper time of the spaceship,

$$\begin{aligned} x^0(\tau) &= ct = \frac{c^2}{g}\sinh\left(\frac{g\tau}{c}\right), \\ x^1(\tau) &= \frac{c^2}{g}\left(\sqrt{c^2 + g^2t^2} - c\right) = \frac{c^2}{g}\left(\cosh\left(\frac{g\tau}{c}\right) - 1\right). \end{aligned} \tag{6.222}$$

These results are specific to the case of a constant rest frame force and give the relativistic force,

$$\begin{aligned} f^0(\tau) &= m\frac{\mathrm{d}^2x^0(\tau)}{\mathrm{d}\tau^2} = mg\sinh\left(\frac{g\tau}{c}\right), \\ f^1(\tau) &= m\frac{\mathrm{d}^2x^1(\tau)}{\mathrm{d}\tau^2} = mg\cosh\left(\frac{g\tau}{c}\right). \end{aligned} \tag{6.223}$$

These results are consistent with the rest frame force $F = mg$, since

$$F = \frac{f^1}{\gamma_v} = mg. \tag{6.224}$$

Exercise 6.19. Find the proper velocity V^μ associated with trajectory (6.222) and show that it and the f^μ of (6.223) obey the identities (6.131) and (6.148).

Exercise 6.20. Show that using $U_N = -mgx$ and $E = mc^2$ in the relativistic conservation law (6.204) yields the same result as (6.210) for one-dimensional constant rest frame acceleration.

6.8.3 Relativistic gravitational infall

Conservation of energy (6.204) can applied to the case of a mass m particle initially at rest at infinity falling radially inward under the Newtonian gravity force created by a point mass M. Such an assumption satisfies the restriction to one-dimensional motion. Using the Newtonian form for the gravitational potential is *a priori* inconsistent with special relativity. This follows from the fact that any change in the position of a gravitating object is communicated instantaneously throughout space if the Newtonian form for the potential is used. In what follows it will be assumed that a *static* gravitating object is present, allowing the gravitational force it creates on a small test mass m to be proportional to $1/r^2$ at every point in space.

For the case that Newtonian gravitation is used, it is straightforward to calculate the time required to fall from rest at the initial position R to the final position r along a radial line. This uses the energy relationship

$$\frac{1}{2}m\left(\frac{\mathrm{d}r}{\mathrm{d}t}\right)^2 - \frac{GMm}{r} = -\frac{GMm}{R}. \tag{6.225}$$

Result (6.204) gives

$$\frac{mc^2}{\sqrt{1 - v_r^2/c^2}} - \frac{GMm}{r} = E, \tag{6.226}$$

where v_r is the radial speed. The total energy E at infinity is given by mc^2, since $v_r = 0$. Using (6.205) gives

$$v_r = \sqrt{\frac{2GM}{r}}\left(\frac{\sqrt{1 + GM/2rc^2}}{1 + GM/rc^2}\right). \tag{6.227}$$

For the case that $GM/rc^2 \ll 1$, result (6.227) gives the familiar Newtonian result $v_r = \sqrt{2GM/r}$. It is instructive to note that the relativistic correction term GM/rc^2 has the value of $\approx 7 \times 10^{-10}$ at the surface of the Earth and $\approx 2 \times 10^{-6}$ at the surface of the Sun.

However, there is a significant deviation from the Newtonian result as r grows small. Using Newtonian dynamics and the Newtonian gravitational potential predicts that the infall velocity will reach c when $r = 2GM/c^2 = r_s$, which is known as the *Schwarzschild radius*. Using relativistic dynamics and the static Newtonian gravitational potential predicts that the velocity at the Schwarzschild radius is $v_r = \sqrt{5}\ c/3$. For the case that $r \to 0$, the relativistic corrections become large and expression (6.227) becomes

$$\lim_{r\to 0} v_r = \sqrt{\frac{2GM}{r}}\sqrt{\frac{rc^2}{2GM}} = c, \tag{6.228}$$

so that only at the position of the gravitating point mass does the infall velocity reach c. Of course, treating the gravitating body as a point particle is unphysical. Instead, the point particle could be envisioned as a spherical object of mass M with the size of its associated Schwarzschild radius. For such a case its density would be given by

$$\rho = \frac{M}{V} = \frac{3M}{4\pi r_s^3} = \frac{3Mc^6}{32\pi G^3}. \tag{6.229}$$

In SI units this gives the density of the mass M object in kg/m^3 as $7.2 \times 10^{79} M$.

However, in the gravitational model provided by general relativity (to be developed in chapter 10), the Schwarzschild radius is once again found to correspond to the velocity c for an infalling particle, in contradiction to result (6.228). The source of this contradiction is that result (6.227) combined relativistic dynamics with the static Newtonian gravitational potential. This can be addressed in an ad hoc manner by modifying the Newtonian gravitational potential energy of the test mass m to

$$\tilde{U}(r) = mc^2\left(1 - \frac{1}{\sqrt{1 - 2GM/rc^2}}\right). \tag{6.230}$$

For the case that $2GM/rc^2$ is small the Taylor series expansion of (6.230) gives

$$\tilde{U}(r) \approx -\frac{GMm}{r}, \tag{6.231}$$

reducing to the usual Newtonian gravitational potential. However, using (6.205) gives the relativistic result

$$v^2 = \frac{2GM}{r}, \tag{6.232}$$

so that once again v $= c$ at the Schwarzschild radius.

For such a case an object with the density (6.229) would have an escape velocity matching the speed of light. Such an object is known as a *black hole*. This is *not* to say that (6.230) is the correct static Newtonian gravitational potential, but rather it is to say that the static Newtonian potential is accurate up to small corrections of the order of GM/rc^2 for all but the most extreme circumstances of a black hole. For that reason, in developing a relativistic version of gravitation, the Newtonian gravitational potential is matched to the nonrelativistic limit of the theory to ensure its accuracy.

6.8.4 Relativistic centripetal force and circular orbits

The relativistic force expression (6.165) results in treating the components $\boldsymbol{F}_\perp$ and $\boldsymbol{F}_\parallel$ differently when used in the relativistic version of Newton's second law (6.137). Solving (6.137) for the case that $\boldsymbol{F}_\parallel = 0$ leads to a relativistic version of centripetal force. This can be applied to the situation of a circular orbit for a point mass resulting from a central force $\boldsymbol{F}(\boldsymbol{r})$. Using spherical coordinates it is assumed that the orbiting point mass has a velocity $\boldsymbol{v} = v^\phi \hat{\mathbf{e}}_\phi$, where v^ϕ is constant. For the case that $\boldsymbol{F}_\parallel = 0$ and spherical coordinates are chosen, Newton's second law (6.137) gives

$$\frac{\mathrm{d}\boldsymbol{p}}{\mathrm{d}\tau} = \frac{1}{\sqrt{1 - v^2/c^2}} \frac{\mathrm{d}}{\mathrm{d}t}\left(\frac{mv^\phi \hat{\mathbf{e}}_\phi}{\sqrt{1 - v^2/c^2}}\right) = \frac{mv^\phi}{1 - v^2/c^2} \frac{\mathrm{d}\hat{\mathbf{e}}_\phi}{\mathrm{d}t} = \boldsymbol{F}_\perp. \tag{6.233}$$

Result (2.6) showed that for a circular orbit $\dot{\hat{\mathbf{e}}}_\phi = -\dot{\phi}\hat{\mathbf{e}}_r = -v_\phi \hat{\mathbf{e}}_r/r$, so that (6.233) becomes

$$-\frac{mv^2}{r(1 - v^2/c^2)} \hat{\mathbf{e}}_r = -\gamma_v^2 \frac{mv^2}{r} \hat{\mathbf{e}}_r = \boldsymbol{F}_\perp. \tag{6.234}$$

It follows that the relativistic centripetal force f_c is given in terms of the Newtonian centripetal force F_c by

$$f_c = \gamma_v^2 \frac{mv^2}{r} = \frac{mv^2}{r(1 - v^2/c^2)} = \frac{ma_c}{(1 - v^2/c^2)}. \tag{6.235}$$

The relativistic centripetal force (6.235) exactly matches the form (6.160) found for the case $\boldsymbol{v} \cdot \boldsymbol{a} = 0$.

This result can be coupled with the form of the force acting on the particle. Assuming $\boldsymbol{F}_\perp$ is an attractive force and writing it $\boldsymbol{F}_\perp = -F(r)\,\hat{\mathbf{e}}_r$ allows (6.235) to be solved for the tangential velocity squared, giving

$$v^2 = \frac{rF(r)/m}{1 + rF(r)/mc^2}. \tag{6.236}$$

The denominator of (6.236) gives the relativistic correction to the orbital speed. If the Newtonian form for the gravitational force is used in (6.236), the tangential velocity is given by

$$v^2 = \frac{GM}{r}\left(\frac{1}{1 + GM/rc^2}\right). \tag{6.237}$$

For large r, result (6.237) matches the Kepler speed. However, as $r \to 0$, the right-hand side becomes c^2. As with (6.228) it will be seen that the small r result is inconsistent with the results of general relativity because the Newtonian form for gravitation was employed.

Exercise 6.21. Using the modified Newtonian potential (6.230), show that the orbital speed v is given by

$$v^2 = \frac{GM}{r}\left(\left(1 - \frac{2GM}{rc^2}\right)^{3/2} + \frac{GM}{rc^2}\right)^{-1}. \tag{6.238}$$

It should be noted that at the Schwarzschild radius the modified potential gives $v = c$, and for r less than the Schwarzschild radius, v becomes complex valued.

6.9 Relativistic quantities

In order to develop field theories that are consistent with special relativity, it is critical to identify the mathematical objects that have clearly defined Lorentz transformation properties. To do this properly requires the reader to be familiar with the basic aspects of group theory. This is presented in the next chapter after reviewing the requisite linear algebra and matrix manipulation. However, before developing those mathematical tools, it is still possible to identify some important relativistic quantities and relations—in particular, the relativistic form of the fluid stress–energy tensor.

The conservation of electric charge is encapsulated in (6.4). The four-vector electric current J^μ can be identified as $J^0 = \rho c$, with $\boldsymbol{J}$ coinciding with the current density. The conservation of electric charge (6.4) is then written in a relativistically invariant expression as

$$\partial_\mu J^\mu = 0. \tag{6.239}$$

All inertial observers agree on the form of electric charge conservation. It also means that the charge density ρc is not a Lorentz scalar quantity. This is consistent with the fact that a charge that is stationary relative to one observer will appear to be in motion according to other inertial observers. As a result there will be a charge current density for other inertial observers.

The relativistic version of the equation of continuity (4.29) is written

$$\partial_\mu(\rho V^\mu) = 0, \tag{6.240}$$

where ρ is the rest frame *baryon density* of the fluid. Result (6.240) holds, since the rest frame version reduces to

$$\lim_{v\to 0}\partial_\mu(\rho V^\mu) = \frac{\partial\rho}{\partial t} = 0, \tag{6.241}$$

which simply states that the rest frame baryon density $\bar{J}^0 = \rho$ is independent of time. Since it is true in the rest frame, it is true in all inertial frames. Result (6.240) is identical in form to the electric charge conservation law (6.239) if the baryon current is identified as $J^\mu = \rho V^\mu$. Baryons include the nuclear building blocks neutrons and protons, which constitute the overwhelming bulk of terrestrial and stellar matter. Although neutrons decay into a proton, an electron, and a neutrino, the proton is stable within experimental limits. However, limiting consideration to the baryon density bypasses the mass conversion processes that take place in stellar interiors where light elements are fused into heavier elements. Although astronomical observations indicate the existence of dark nonbaryonic matter, baryonic matter accounts for the overwhelming bulk of the mass present in dense objects, such as planets and stars.

The *canonical stress–energy tensor* for a collection of fields φ^α with the Lagrangian $\mathcal{L}$ is defined identically to (3.82), taking the form

$$T^\mu{}_\nu = \frac{\partial \mathcal{L}}{\partial(\partial_\mu \varphi^\alpha)} \partial_\nu \varphi^\alpha - \delta^\mu{}_\nu \mathcal{L}, \tag{6.242}$$

where ∂_0 was identified as $\partial/\partial t$ in the symmetric fluid stress–energy tensor (4.170). From this point forward $\partial_0 = \partial/\partial x^0$, where $x^0 = ct$. This requires a relativistic Lagrangian to replace $\partial/\partial t$ with $c(\partial/\partial x^0)$ in order to be consistent with the use of relativistic notation. If the Lagrangian is Lorentz invariant, then the canonical stress–energy tensor of (6.242) has a well-defined Lorentz transformation property as a mixed second-rank tensor,

$$T'^\mu{}_\nu(x') = \Lambda^\mu{}_\rho(v)\Lambda_\nu{}^\sigma(v)\, T^\rho{}_\sigma(x). \tag{6.243}$$

It is now possible to find $T^{\mu\nu}$ or $T_{\mu\nu}$ using the Minkowski metric, so that

$$T^{\mu\nu} = \eta^{\nu\sigma} T^\mu{}_\sigma = \frac{\partial \mathcal{L}}{\partial(\partial_\mu \varphi^\alpha)} \eta^{\nu\sigma} \partial_\sigma \varphi^\alpha - \eta^{\nu\sigma} \delta^\mu{}_\sigma \mathcal{L} = \frac{\partial \mathcal{L}}{\partial(\partial_\mu \varphi^\alpha)} \partial^\nu \varphi^\alpha - \eta^{\mu\nu} \mathcal{L}, \tag{6.244}$$

where $\eta^{\nu\mu} = \eta^{\mu\nu}$ was used.

The Euler–Lagrange equations for the fields φ^α are given by

$$\frac{\partial \mathcal{L}}{\partial \varphi^\alpha} - \partial_\mu \frac{\partial \mathcal{L}}{\partial(\partial_\mu \varphi^\alpha)} = 0. \tag{6.245}$$

Evaluating $T^{\mu\nu}$ using a solution of (6.245) ensures that $\partial_\mu T^{\mu\nu} = 0$, just as it did in (3.84). The demonstration is worth repeating to reiterate the assumptions that go into the result. The Lagrangian is assumed to be a function of φ^α and $\partial_\mu \varphi^\alpha$. Higher derivative theories exhibit pathologies in their quantized versions, leading to nonphysical quantum processes and a failure of the probabilistic interpretation required by quantum theory. As these issues lie outside the purview of this text, such theories will therefore not be considered. The conservation of $T^\mu{}_\nu$ depends upon this assumption, since the Euler–Lagrange equation (6.245) was also derived using the same assumption. It follows that

$$\partial_\mu T^{\mu\nu} = \partial_\mu \frac{\partial \mathcal{L}}{\partial(\partial_\mu \varphi^\alpha)}\, \partial^\nu \varphi^\alpha + \frac{\partial \mathcal{L}}{\partial(\partial_\mu \varphi^\alpha)} \partial_\mu \partial^\nu \varphi^\alpha - \partial^\nu \mathcal{L}(\varphi^\alpha, \partial\varphi^\alpha) \tag{6.246}$$

$$=\frac{\partial\mathcal{L}}{\partial\varphi^\alpha}\partial^\nu\varphi^\alpha+\frac{\partial\mathcal{L}}{\partial(\partial_\mu\varphi^\alpha)}\partial_\mu\partial^\nu\varphi^\alpha-\frac{\partial\mathcal{L}}{\partial\varphi^\alpha}\partial^\nu\varphi^\alpha-\frac{\partial\mathcal{L}}{\partial(\partial_\mu\varphi^\alpha)}\partial_\mu\partial^\nu\varphi^\alpha=0. \tag{6.247}$$

However, it is not clear that $\partial_\nu T^{\mu\nu}=0$, since this is given by

$$\partial_\nu T^{\mu\nu}=\partial_\nu\left(\frac{\partial\mathcal{L}}{\partial(\partial_\mu\varphi^\alpha)}\partial^\nu\varphi^\alpha\right)-\partial^\mu\mathcal{L}. \tag{6.248}$$

It is not apparent how or even whether the Euler–Lagrange equation enters (6.248) to ensure that it is conserved. The question disappears if

$$\frac{\partial\mathcal{L}}{\partial(\partial_\mu\varphi^\alpha)}\partial^\nu\varphi^\alpha=\frac{\partial\mathcal{L}}{\partial(\partial_\nu\varphi^\alpha)}\partial^\mu\varphi^\alpha, \tag{6.249}$$

since substituting (6.249) into (6.248) shows that the right-hand side vanishes once again by virtue of the Euler–Lagrange equation. It follows that if condition (6.249) holds then $T^{\mu\nu}=T^{\nu\mu}$ and the stress–energy tensor is symmetric. However, if condition (6.249) does not hold then it is not possible to prove that $\partial_\nu T^{\mu\nu}=0$, since $T^{\mu\nu}\neq T^{\nu\mu}$. If that is the case then integrations by parts require great care. For example, in the case of a *vector field* A_ρ, two very similar contractions have very different meanings,

$$T^{\mu\rho}\partial_\mu A_\rho=\partial_\mu\left(T^{\mu\rho}A_\rho\right), \tag{6.250}$$

$$T^{\rho\mu}\partial_\mu A_\rho=\partial_\mu\left(T^{\rho\mu}A_\rho\right)-\partial_\mu T^{\rho\mu}A_\rho. \tag{6.251}$$

A symmetric stress–energy tensor is therefore critical to calculational consistency.

In that regard, it was previously found that the canonical stress–energy tensor was not symmetric for either the stretched string or the Newtonian homentropic fluid flow, since $T^0{}_j$ differed from T_0^j, where $T_j^0=-\mathcal{P}_j$ is the momentum density of the wave in the string or the flowing fluid. One advantage of using the Minkowski metric as defined in (6.60) is that the contravariant version of this is given by

$$T^{0k}=\eta^{kj}T_j^0=-T_k^0=\mathcal{P}^k, \tag{6.252}$$

so that T^{0k} is the momentum density of the system. This is consistent with the components of the contravariant symmetric perfect fluid stress–energy tensor (4.170), which was developed by demanding consistency with Galilean relativity. Most of the problems with the canonical stress–energy tensor of (6.242) arise because it is an ad hoc construction, having loosely evolved from the Euler–Lagrange equations of motion to represent a collection of conservation laws relating momentum and energy. It is important to develop the stress–energy tensor in a more systematic way. In the next chapter the stress–energy tensor and its associated conservation laws will be placed on a firmer basis through *Noether's theorem*, where they are seen to originate from symmetries of the Lagrangian and the equations of motion obtained from it. Eventually a symmetric stress–energy tensor will be defined in chapter 10 by combining functional methods with general relativity.

Beginning with Belinfante in 1939, a number of proposals were made to amend the definition (6.242) to deal with its various flaws in a way that did not alter the energy and momentum interpretation of the canonical stress–energy tensor. One option is to define an alternate stress–energy tensor as

$$\Theta^{\mu}{}_{\nu} = T^{\mu}_{\nu} - \partial_{\rho} B^{\mu\rho}_{\nu}, \tag{6.253}$$

where T^{μ}_{ν} is the canonical stress–energy tensor. The *Belinfante tensor* $B^{\mu\rho}{}_{\nu}$ is chosen to have the property

$$B^{\mu\rho}_{\nu} = -B^{\rho\mu}_{\nu}. \tag{6.254}$$

Using the antisymmetry property of (6.254) and manipulating dummy indices gives

$$\partial_{\mu}\partial_{\rho}B^{\mu\rho}_{\nu} = -\partial_{\mu}\partial_{\rho}B^{\rho\mu}_{\nu} = -\partial_{\rho}\partial_{\mu}B^{\mu\rho}_{\nu} = -\partial_{\mu}\partial_{\rho}B^{\mu\rho}_{\nu} \implies \partial_{\mu}\partial_{\rho}B^{\mu\rho}_{\nu} = 0. \tag{6.255}$$

As a result, the tensor of (6.253) also obeys $\partial_{\mu}\,\Theta^{\mu}_{\nu} = 0$. In addition, because Θ^{μ}_{ν} and T^{μ}_{ν} differ only by the exact divergence of the Belinfante tensor, it is possible to choose the Belinfante tensor so that spatial integrals of the important quantities of energy and momentum densities do not differ from the canonical stress–energy tensor. The Belinfante tensor approach offers a means to remedy flaws that may appear in the canonical stress–energy tensor, but once again in a somewhat ad hoc manner. It will be used in chapter 9 to find the stress–energy tensor for the case of electromagnetic fields.

The relativistic version of the symmetric fluid stress–energy tensor (4.170) is important in cosmology and astrophysical processes. The method for finding it is identical to the method used to find the relativistic force of (6.164) and (6.165). The process starts by defining an observer in the rest frame of a fluid element, referred to as the *comoving frame*. The comoving observer measures the rest frame baryon density ρ and the pressure P in the fluid element. Setting the velocity to zero in (4.170) gives the nonzero rest frame components,

$$\begin{aligned} \bar{\Theta}^{00} &= \rho c^2, \\ \bar{\Theta}^{jk} &= -\,\eta^{jk} P, \end{aligned} \tag{6.256}$$

where the spatial metric g^{jk} has been replaced with $-\eta^{jk}$ and the factor of c^2 converts $\bar{\Theta}^{00}$ into an energy density to match the units of $\bar{\Theta}^{jk}$. As a result, $\Theta^{\mu\nu}$ will once again be a stress–energy tensor. Using (6.133), the perfect fluid stress–energy tensor in a boosted frame is then given by

$$\begin{aligned} \Theta'^{\,00} &= \Lambda^{0}{}_{\rho}(v)\Lambda^{0}{}_{\sigma}(v)\,\bar{\Theta}^{\rho\sigma} = \frac{1}{c^2}V_{\rho}V_{\sigma}\,\bar{\Theta}^{\rho\sigma} = \frac{1}{c^2}\left(V_0^2\bar{\Theta}^{00} + V_jV_k\bar{\Theta}^{jk}\right) \\ &= \gamma_v^2\left(\rho c^2 + \frac{v^2}{c^2}P\right) = \left(\rho + \frac{P}{c^2}\right)V^0V^0 - \eta^{00}P. \end{aligned} \tag{6.257}$$

Similarly, the other components are given by

$$\begin{aligned}\Theta'^{j0} &= \Lambda^j{}_\rho(v)\Lambda^0{}_\sigma(v)\,\bar{\Theta}^{\rho\sigma} = \Lambda^j{}_0(v)\Lambda^0{}_0(v)\bar{\Theta}^{00} + \Lambda^j{}_i(v)\Lambda^0{}_\ell(v)\bar{\Theta}^{i\ell} \\ &= V^j V^0\left(\rho + \frac{P}{c^2}\right) - \eta^{j0}P,\end{aligned} \tag{6.258}$$

$$\begin{aligned}\Theta'^{jk} &= \Lambda^j{}_\rho(v)\Lambda^k{}_\sigma(v)\,\bar{\Theta}^{\rho\sigma} = \Lambda^j{}_0(v)\Lambda^k{}_0(v)\bar{\Theta}^{00} + \Lambda^j{}_i(v)\Lambda^k{}_\ell(v)\bar{\Theta}^{i\ell} \\ &= V^j V^k\left(\rho + \frac{P}{c^2}\right) - \eta^{jk}P.\end{aligned} \tag{6.259}$$

Exercise 6.22. Verify the last steps of (6.257), (6.258), and (6.259).

Assembling results (6.257), (6.258), and (6.259) gives the relativistic perfect fluid stress–energy tensor,

$$\Theta^{\mu\nu} = \left(\rho + \frac{P}{c^2}\right)V^\mu V^\nu - \eta^{\mu\nu}P, \tag{6.260}$$

where V^μ is the proper velocity of the fluid at the point under consideration.

The equation that follows from $\partial_\mu\Theta^{\mu\nu} = 0$ can now be shown to yield the relativistic version of the Euler equation (4.18). Assigning $\varepsilon = \rho + P/c^2$ and using (6.260) gives

$$\partial_\mu\Theta^{\mu\nu} = (\partial_\mu\varepsilon)V^\mu V^\nu + \varepsilon(\partial_\mu V^\mu)V^\nu + \varepsilon V^\mu(\partial_\mu V^\nu) - \partial^\nu P = 0. \tag{6.261}$$

Using $V_\nu V^\nu = c^2$ and $\partial_\mu c^2 = \partial_\mu(V_\nu V^\nu) = 2V_\nu\partial_\mu V^\nu = 0$, it follows that

$$V_\rho\partial_\mu\Theta^{\mu\rho} = c^2(\partial_\mu\varepsilon)V^\mu + c^2\varepsilon(\partial_\mu V^\mu) - V_\mu\partial^\mu P = 0. \tag{6.262}$$

Multiplying (6.262) by V^ν/c^2 gives

$$\frac{1}{c^2}V^\nu V_\rho\partial_\mu\Theta^{\mu\rho} = (\partial_\mu\varepsilon)V^\mu V^\nu + \varepsilon(\partial_\mu V^\mu)V^\nu - \frac{1}{c^2}V^\nu V_\mu\partial^\mu P = 0. \tag{6.263}$$

Subtracting (6.263) from (6.261) gives the relativistic form of the Euler equation (4.18) in the absence of gravity and viscosity,

$$E^\nu = \left(\rho + \frac{P}{c^2}\right)V^\mu\partial_\mu V^\nu - \partial^\nu P + \frac{1}{c^2}V^\nu V_\mu\partial^\mu P = 0. \tag{6.264}$$

Exercise 6.23. As a way to check that (6.264) is correct, show that $E^\nu V_\nu = 0$.

It is possible to include the relativistic version of the viscosity tensor (4.52) as well as thermal gradients to model an imperfect fluid that is not in thermal equilibrium. Such extensions will not be considered here, and so the interested reader should consult the references at the end of this chapter.

In the case of field theory the stress–energy tensor is composed of components representing the energy and momentum densities of the participating fields. For example, $\Theta^{\mu 0}$ is the four-momentum density $\mathcal{P}^{\mu}$. The total energy-momentum in the field is then defined as

$$P^{\mu} = \int_V \mathrm{d}^3x \, \Theta^{\mu 0}. \tag{6.265}$$

At first glance, (6.265) does not appear to be a Lorentz covariant statement. However, this is easily remedied by writing (6.265) in terms of the three-volume Lorentz vector $\mathrm{d}S_{\mu}$ that is found using the wedge product in four-dimensions,

$$\mathrm{d}x^{\mu} \wedge \mathrm{d}x^{\nu} \wedge \mathrm{d}x^{\rho} = \varepsilon^{\mu\nu\rho\sigma}\mathrm{d}S_{\sigma}. \tag{6.266}$$

This shows that $\mathrm{d}S_0$ is given by

$$\mathrm{d}S_0 = \mathrm{d}x^1 \wedge \mathrm{d}x^2 \wedge \mathrm{d}x^3 = \varepsilon^{123}\mathrm{d}^3x = \mathrm{d}^3x. \tag{6.267}$$

The expression for the total four-momentum can therefore be written

$$P^{\mu} = \int_V \mathrm{d}S_{\nu} \, \Theta^{\mu\nu}, \tag{6.268}$$

which is manifestly covariant and reduces to (6.265) in the rest frame. Result (6.265) is consistent with the second law (6.137), since it yields the manifestly covariant expression

$$\frac{\mathrm{d}P^{\mu}}{\mathrm{d}\tau} = \int_V \mathrm{d}S_{\nu} \frac{\mathrm{d}}{\mathrm{d}\tau}\Theta^{\mu\nu} = f^{\mu}. \tag{6.269}$$

The integral of (6.269) therefore transforms as the contravariant components of the four-vector force f^{μ}. This is understood to be the net force acting on the field, since it represents the rate at which the four-momentum is changing. This will be presented in more detail when the electromagnetic field and its interaction with currents is analyzed. It is possible to define angular momentum using the stress–energy tensor, but this will require further developments.

References and recommended further reading

The foundational papers of special relativity are available in

- Einstein A *et al* 1952 *The Principle of Relativity* (New York: Dover)

Monographs focusing on special relativity include

- Rindler W 1991 *Introduction to Special Relativity* 2nd edn (Oxford: Oxford University Press)
- Woodhouse N 1992 *Special Relativity* (Berlin: Springer)
- French A 2009 *Special Relativity* (Oxfordshire: Taylor and Francis)

Monographs on classical field theory that cover special relativity include

- Landau L and Lifshitz E 1975 *The Classical Theory of Fields* 4th edn (Amsterdam: Elsevier)
- Scheck F 2012 *Classical Field Theory* (Berlin: Springer)

Monographs on classical mechanics that cover special relativity include

- Goldstein H, Poole C, and Safko J 2000 *Classical Mechanics* 3rd edn (San Francisco, CA: Addison-Wesley)
- Thornton S and Marion J 2004 *Classical Dynamics of Particles and Systems* (Boston, MA: Cengage)

Monographs on modern physics that cover special relativity include

- Bernstein J, Fishbane P and Gasiorowicz S 2000 *Modern Physics* (Englewood Cliffs, NJ: Prentice-Hall)
- Tipler P and Llewellyn R 2008 *Modern Physics* 5th edn (New York: Freeman)

Monographs on classical electrodynamics that cover special relativity include

- Jackson J 1999 *Classical Electrodynamics* 3rd edn (Hoboken, NJ: Wiley)
- Rohrlich F 2007 *Classical Charged Particles* 3rd edn (Singapore: World Scientific)

Monographs on general relativity that cover special relativity include

- Hartle J 2003 *Gravity: An Introduction to Einstein's General Relativity* (San Francisco, CA: Addison-Wesley)
- Misner C, Thorne K, and Wheeler J 2017 *Gravitation* (Princeton, NJ: Princeton University Press)

Mark S Swanson

Chapter 7

Linear algebra and group theory

Identifying and classifying fundamental field theories is greatly facilitated by the mathematical tools of basic linear algebra and group theory. The reader familiar with these topics may wish to skip to the end of this chapter, where the properties of the Lorentz group are discussed and their representations are developed. Readers unfamiliar with these techniques can use the brief review of these methods presented in the first two sections to identify where further study is required.

7.1 Linear algebra and matrices

A matrix $\mathbf{A}$ is a rectangular array of *elements*—mathematical objects—possessing n rows and m columns. Such an array is referred to as an $n \times m$*-dimensional matrix* and its elements are denoted by their row i and column j, so that A_{ij} is the element of $\mathbf{A}$ at row i and column j. In the case of a matrix these indices do not necessarily correspond to contravariant or covariant components, so that there is no need to distinguish superscript and subscript indices in that case. If the elements of a matrix are identified as being related to position or velocity vectors, then superscript and subscript indices are used. Although matrices can have any dimensions, this book exclusively deals with square matrices $\mathbf{A}$, where $n = m$. Such a matrix can be displayed as

$$\mathbf{A} = \begin{pmatrix} A_{11} & A_{12} & \dots & A_{1n} \\ A_{21} & A_{22} & \dots & A_{2n} \\ \dots & \dots & \dots & \dots \\ A_{n1} & A_{n2} & \dots & A_{nn} \end{pmatrix}, \tag{7.1}$$

where the ellipses '…' refer to all the rest of the sequentially numbered elements. Two other important types of matrix are the row $\mathbf{R}$ and column $\mathbf{C}$ matrices, which are $1 \times n$- and $n \times 1$-dimensional, respectively. These are displayed as

doi:10.1088/978-0-7503-3455-6ch7

© IOP Publishing Ltd 2022

$$\mathbf{R} = (R_{11}\ R_{12}\ \dots\ R_{1n}),\ \ \mathbf{C} = \begin{pmatrix} C_{11} \\ C_{21} \\ \dots \\ C_{n1} \end{pmatrix}. \tag{7.2}$$

Row and column matrices are referred to as *row and column vectors* because of their resemblance to the components of a vector in an n-dimensional coordinate system. Two matrices, $\mathbf{A}$ and $\mathbf{B}$, are equal, written $\mathbf{A} = \mathbf{B}$, if and only if $A_{ij} = B_{ij}$ for all i and j. This requires two equal matrices to have the same dimensions and the same elements. For a square matrix the *diagonal elements* A_{ij} are those where $i = j$.

The addition of two matrices, denoted by $\mathbf{A} + \mathbf{B}$, is defined by specifying the elements of their sum,

$$(\mathbf{A} + \mathbf{B})_{ij} = A_{ij} + B_{ij}, \tag{7.3}$$

which requires both matrices to have identical dimensions. A matrix $\mathbf{A}$ can be multiplied by a real or complex number a to obtain a new matrix, $\mathbf{C} = a\mathbf{A}$, whose elements are given by $C_{ij} = aA_{ij}$. The additive inverse of a matrix is defined by $\mathbf{A} + (-\mathbf{A}) \equiv \mathbf{A} - \mathbf{A} = 0$, where 0 is the *zero matrix* with the same dimensions as $\mathbf{A}$ with the elements $(0)_{ij} = 0$. The zero matrix has the usual property that $\mathbf{A} + 0 = \mathbf{A}$. The additive inverse $-\mathbf{A}$ has the elements $-A_{ij}$, so that $\mathbf{A} - \mathbf{A} = 0$. Similarly, the product $\mathbf{A} \cdot \mathbf{B}$ of two matrices is obtained by defining the elements of the product according to

$$(\mathbf{A} \cdot \mathbf{B})_{ik} = \sum_{j=1}^{n} A_{ij} B_{jk} \equiv A_{ij} B_{jk}, \tag{7.4}$$

where the second version uses the familiar summation convention. It is assumed that the dimension n is contextually understood. As with vector products the index that is summed over is an arbitrary label in the sense that $A_{ij}B_{jk} = A_{i\ell}B_{\ell k}$ and is again referred to as a dummy index. For consistency in matrix multiplication, it must be the case that $\mathbf{A}$ has n columns and $\mathbf{B}$ has n rows. The product $\mathbf{A} \cdot \mathbf{B}$ results in another matrix in which the number of rows is inherited from $\mathbf{A}$ and the number of columns is inherited from $\mathbf{B}$, so that multiplying an $n \times 1$ column vector matrix on the left by an $n \times n$ matrix results in another column vector matrix with dimensions of $n \times 1$. It also follows that $\mathbf{A} \cdot 0 = 0$, where 0 is the previously defined zero matrix.

In the case of row and column matrices the '1' in the element is dropped, and the $1 \times n$ matrices are simply referred to as n-dimensional *row vectors*, while the $n \times 1$ matrices are referred to as n-dimensional *column vectors*. A connection to the position vectors of basic physics occurs by multiplying an $1 \times n$ row vector $\mathbf{R}$ by an $n \times 1$ column vector $\mathbf{C}$. If the n elements of $\mathbf{R}$ are x_j and the n elements of $\mathbf{C}$ are y^j, then the matrix product can be expressed using the summation convention as

$$\mathbf{R} \cdot \mathbf{C} = x_j y_j. \tag{7.5}$$

The product of these two matrices yields a 1×1 matrix, which is simply a number. Result (7.5) is identical in form to the scalar product of two vectors, $\boldsymbol{x}$ and $\boldsymbol{y}$, expressed in Cartesian components and given by $\boldsymbol{x} \cdot \boldsymbol{y} = x_j y_j$. Linear algebra uses the term *inner product* to designate the matrix product (7.5), and two vectors are said to be *orthogonal* if their inner product vanishes.

If $\mathbf{A}$ is an $n \times n$ square matrix and $\mathbf{C}$ is an n-dimensional column vector, the product $\mathbf{A} \cdot \mathbf{C} = \mathbf{C}'$ yields another n-dimensional column vector denoted by $\mathbf{C}'$. The elements of $\mathbf{C}'$ are given by

$$C'_j = A_{jk} C_k, \tag{7.6}$$

which resembles a coordinate transformation without the distinction of covariant or contravariant indices. Similarly, if $\mathbf{R}$ is an n-dimensional row vector, the product $\mathbf{R} \cdot \mathbf{A} = \mathbf{R}'$ yields another n-dimensional row vector denoted by $\mathbf{R}'$. In this sense, a matrix can represent an action, such as a rotation, on a spatial vector through matrix multiplication. This is developed in the sections on group theory. It follows that this action is *linear*, since

$$(\mathbf{A} + \mathbf{B}) \cdot \mathbf{C} = \mathbf{A} \cdot \mathbf{C} + \mathbf{B} \cdot \mathbf{C}. \tag{7.7}$$

Because of their similarity to spatial vectors, column vectors will be written $\mathbf{x}$.

The definition of matrix multiplication creates the possibility that $\mathbf{A} \cdot \mathbf{B}$ is not necessarily the same as $\mathbf{B} \cdot \mathbf{A}$, even when both matrices are square and both products are defined. In other words, it is possible that two matrices do not *commute* when they are multiplied. Given two square matrices, $\mathbf{A}$ and $\mathbf{B}$, with the same dimensions, the quantity defined as

$$\mathbf{A} \cdot \mathbf{B} - \mathbf{B} \cdot \mathbf{A} \equiv [\mathbf{A}, \mathbf{B}], \tag{7.8}$$

is referred to as the *commutator* of $\mathbf{A}$ and $\mathbf{B}$. The commutator of two square matrices is also a square matrix of the same dimension, and its definition satisfies $[\mathbf{A}, \mathbf{B}] = -[\mathbf{B}, \mathbf{A}]$. Two matrices are said to commute if $[\mathbf{A}, \mathbf{B}] = 0$. The commutator satisfies an important property known as the *Jacobi identity*. For *any* three square matrices, $\mathbf{A}$, $\mathbf{B}$, and $\mathbf{C}$, all of which have the same dimension, the definition of the commutator gives

$$[\mathbf{A}, [\mathbf{B}, \mathbf{C}]] + [\mathbf{C}, [\mathbf{A}, \mathbf{B}]] + [\mathbf{B}, [\mathbf{C}, \mathbf{A}]] = 0. \tag{7.9}$$

The order of the three matrices in each term is a *cyclic permutation* of the previous term, where the last matrix in the previous term becomes the first matrix in the next term. The proof of (7.9) is straightforward, simply amounting to substituting the definition of each commutator into the appropriate expression and noticing that the twelve terms all cancel. The Jacobi identity is unique in the sense that there are no similar identities involving four or more matrices. This is because for n matrices there are $n!$ possible permutations of products involving all n matrices, while each set of commutators involving $n - 1$ matrices generates 2^{n-1} terms and there are n of them for cyclic permutations. In order to guarantee that all possible permutations would cancel, it is necessary for the commutators to generate $2n!$ terms, one positive and one negative for each permutation. It follows that $2n! = 2^{n-1}n$ only if n is two or three, completing the assertion.

The *anticommutator* of **A** and **B** is another n-dimensional square matrix $\bar{\mathbf{C}}$ defined as

$$\bar{\mathbf{C}} = \{\mathbf{A}, \mathbf{B}\} \equiv \mathbf{A} \cdot \mathbf{B} + \mathbf{B} \cdot \mathbf{A}. \tag{7.10}$$

The anticommutator clearly satisfies $\{\mathbf{A}, \mathbf{B}\} = \{\mathbf{B}, \mathbf{A}\}$. Two matrices are said to *anticommute* if $\{\mathbf{A}, \mathbf{B}\} = 0$. There are no Jacobi-like identities for the anticommutator as long as the elements of the matrices are standard real or complex numbers. The possibility of other types of element, such as anticommuting Grassmann numbers, will not be considered in the case of classical field theory. There are a number of useful identities that involve commutators and anticommutators for the product of matrices. For example,

$$\begin{aligned}[\mathbf{A}, \mathbf{B} \cdot \mathbf{C}] &= \mathbf{A} \cdot \mathbf{B} \cdot \mathbf{C} - \mathbf{B} \cdot \mathbf{C} \cdot \mathbf{A} = \mathbf{A} \cdot \mathbf{B} \cdot \mathbf{C} - \mathbf{B} \cdot \mathbf{A} \cdot \mathbf{C} \\ &\quad + \mathbf{B} \cdot \mathbf{A} \cdot \mathbf{C} - \mathbf{B} \cdot \mathbf{C} \cdot \mathbf{A} \\ &= [\mathbf{A}, \mathbf{B}] \cdot \mathbf{C} + \mathbf{B} \cdot [\mathbf{A}, \mathbf{C}].\end{aligned} \tag{7.11}$$

Exercise 7.1. Show that

$$[\mathbf{A}, \mathbf{B} \cdot \mathbf{C}] = \{\mathbf{A}, \mathbf{B}\} \cdot \mathbf{C} - \mathbf{B} \cdot \{\mathbf{A}, \mathbf{C}\}. \tag{7.12}$$

Restricting our attention to square matrices allows the definition of an n-dimensional *unit or identity matrix* **I**, such that its product with any other n-dimensional square matrix **A** gives $\mathbf{I} \cdot \mathbf{A} = \mathbf{A} \cdot \mathbf{I} = \mathbf{A}$. The elements of the unit matrix are given by the Kronecker delta δ_{ij}, which is verified by using (7.4) since it gives $(\mathbf{A} \cdot \mathbf{I})_{ik} = A_{ij}\delta_{jk} = A_{ik}$ or $\mathbf{A} \cdot \mathbf{I} = \mathbf{A}$. The unit matrix **I** can therefore be visualized as a square array which has one for each of its diagonal elements and zero for all others. The unit matrix **I** plays the role of 'one' in all square matrix manipulations.

Using the unit matrix allows the inverse of a square matrix **A**, denoted by $\mathbf{A}^{-1}$, to be defined as the matrix such that $\mathbf{A}^{-1} \cdot \mathbf{A} = \mathbf{A} \cdot \mathbf{A}^{-1} = \mathbf{I}$. Clearly, the inverse $\mathbf{A}^{-1}$ must have the same dimensions as **A**. Since the unit matrix **I** leaves square matrices unchanged by multiplication from either the left or the right, it follows that the inverse **L** from the left, $\mathbf{L} \cdot \mathbf{A} = \mathbf{I}$, is the same as the inverse **R** from the right, $\mathbf{A} \cdot \mathbf{R} = 1$, since

$$\mathbf{L} \cdot \mathbf{A} = \mathbf{I} \Longrightarrow \mathbf{L} \cdot \mathbf{A} \cdot \mathbf{R} = \mathbf{I} \cdot \mathbf{R} \Longrightarrow \mathbf{L} \cdot \mathbf{I} = \mathbf{I} \cdot \mathbf{R} \Longrightarrow \mathbf{L} = \mathbf{R}. \tag{7.13}$$

An example of a square matrix inverse is given in the 2×2 case by the two matrices

$$\mathbf{A} = \begin{pmatrix} 0 & 1 \\ -1 & 0 \end{pmatrix}, \quad \mathbf{B} = \begin{pmatrix} 0 & -1 \\ 1 & 0 \end{pmatrix},$$

which have the property that

$$\mathbf{A}\cdot\mathbf{B}=\mathbf{B}\cdot\mathbf{A}=\begin{pmatrix}1 & 0\\ 0 & 1\end{pmatrix}=\mathbf{I}\implies\mathbf{B}=\mathbf{A}^{-1}. \tag{7.14}$$

However, *not every matrix possesses an inverse.* The zero matrix 0 defined earlier is a trivial example of a matrix with no inverse. However, even non-trivial matrices may not be invertible. For example, the matrix

$$\mathbf{A}=\begin{pmatrix}0 & 1\\ 0 & 1\end{pmatrix} \tag{7.15}$$

has no inverse. A general criterion for the existence of an inverse is presented later.

Given a matrix **A**, other matrices can be formed from it. The transpose of **A**, denoted by $\mathbf{A}^T$, is the matrix whose elements are given by $(\mathbf{A}^T)_{ij}=A_{ji}$. In effect, the transpose of a square matrix is formed by simply interchanging rows and columns, so that the ith row of a matrix becomes the ith column of its transpose. A matrix is *symmetric* if it is equal to its transpose, $\mathbf{A}=\mathbf{A}^T$, since, for such a case, $A_{ij}=A_{ji}$ and the matrix is symmetric about its diagonal elements. For example, any matrix with only diagonal elements is symmetric. A matrix is *antisymmetric* if it is the additive inverse of its transpose, $\mathbf{A}^T=-\mathbf{A}$, so that $A_{ij}=-A_{ji}$. The diagonal elements of an antisymmetric matrix must vanish. The transpose of the product of two matrices **A** and **B** satisfies

$$(\mathbf{A}\cdot\mathbf{B})^{\mathrm{T}}_{ij}=(\mathbf{A}\cdot\mathbf{B})_{ji}=A_{jk}B_{ki}=(\mathbf{B}^T)_{ik}(\mathbf{A}^T)_{kj}=(\mathbf{B}^T\cdot\mathbf{A}^T)_{ij}, \tag{7.16}$$

so that

$$(\mathbf{A}\cdot\mathbf{B})^{\mathrm{T}}=\mathbf{B}^T\cdot\mathbf{A}^T. \tag{7.17}$$

The act of transposing a column vector **x** gives a row vector and vice-versa, since a column vector has the elements x_{j1}, so that its transpose must have the elements $(\mathbf{x}^T)_{1j}=x_{j1}$, making it a row vector. The inner product of two column vectors **x** and **y** is denoted by (**x**, **y**) and, *for the case of real elements*, is defined using matrix multiplication as

$$(\mathbf{x},\mathbf{y})\equiv\mathbf{x}^T\cdot\mathbf{y}, \tag{7.18}$$

which is consistent with the earlier result (7.5). Definition (7.18) yields a linear function of a column vector, since matrix operations give

$$(a\mathbf{x}+b\mathbf{y},\mathbf{z})=a\mathbf{x}^T\cdot\mathbf{z}+b\mathbf{y}^T\cdot\mathbf{z}=a(\mathbf{x},\mathbf{z})+b(\mathbf{y},\mathbf{z}), \tag{7.19}$$

where a and b are two arbitrary numbers. If **y** is the column vector found by acting on the column vector **x** with the matrix **A**, so that $\mathbf{y}=\mathbf{A}\cdot\mathbf{x}$, then it follows that

$$\begin{aligned}(\mathbf{y}^T)_{1i}&=(\mathbf{A}\cdot\mathbf{x})^{\mathrm{T}}_{1i}=(\mathbf{A}\cdot\mathbf{x})_{i1}=A_{ij}x_{j1}=(\mathbf{A}^T)_{ji}(\mathbf{x}^T)_{1j}\\ &=(\mathbf{x}^T)_{1j}(\mathbf{A}^T)_{ji}=(\mathbf{x}^T\cdot\mathbf{A}^T)_{1i},\end{aligned} \tag{7.20}$$

so that

$$\mathbf{y} = \mathbf{A} \cdot \mathbf{x} \Longrightarrow \mathbf{y}^T = \mathbf{x}^T \cdot \mathbf{A}^T. \tag{7.21}$$

A matrix **A** is called *orthogonal* if its transpose is also its inverse, so that

$$\mathbf{A}^T \cdot \mathbf{A} = \mathbf{I} \Longrightarrow \mathbf{A}^T = \mathbf{A}^{-1}. \tag{7.22}$$

An example of a simple orthogonal matrix is given by the **A** appearing in (7.14). If $\mathbf{x}' = \mathbf{A} \cdot \mathbf{x}$ and $\mathbf{y}' = \mathbf{A} \cdot \mathbf{y}$ are the two column vectors created by multiplying two column vectors **x** and **y** by an orthogonal matrix **A**, then the previous results give

$$\begin{aligned}(\mathbf{y}', \mathbf{x}') = (\mathbf{y}')^T \cdot \mathbf{x}' = (\mathbf{A} \cdot y)^T \cdot \mathbf{A} \cdot \mathbf{x} = \mathbf{y}^T \cdot \mathbf{A}^T \cdot \mathbf{A} \cdot \mathbf{x} \\ = \mathbf{y}^T \cdot \mathbf{1} \cdot \mathbf{x} = \mathbf{y}^T \cdot \mathbf{x} = (\mathbf{y}, \mathbf{x}).\end{aligned} \tag{7.23}$$

The action of an orthogonal matrix therefore preserves the inner product of two real-valued vectors. This is identical to the action of a rotation on two three-dimensional vectors, and so orthogonal matrices are used to represent rotations in developing group theory.

If a matrix **A** has complex elements, then it is useful to define its *Hermitian adjoint* or *complex transpose*, denoted by $\mathbf{A}^\dagger$, as the complex conjugate of its transpose. The elements of the Hermitian adjoint are therefore given by $(\mathbf{A}^\dagger)_{ij} = A^*_{ji}$. A matrix **A** is called *Hermitian* if it satisfies $\mathbf{A} = \mathbf{A}^\dagger$. Clearly, the identity matrix **I** is Hermitian. If the elements of a matrix are all real numbers, the definition of the Hermitian adjoint and the transpose are identical. A Hermitian matrix must have real numbers along the diagonal, since the complex conjugate of a diagonal element must not change the element. A matrix is *antihermitian* if it satisfies $\mathbf{A}^\dagger = -\mathbf{A}$. Similarly, an antihermitian matrix must have imaginary numbers along its diagonal, since the diagonal elements must change sign under complex conjugation. The action of the Hermitian adjoint operation on products of matrices is found identically to (7.20) and (7.16); the transposition operation is replaced by the Hermitian adjoint, yielding

$$(\mathbf{A} \cdot \mathbf{B})^\dagger = \mathbf{B}^\dagger \cdot \mathbf{A}^\dagger. \tag{7.24}$$

In the event that the elements of the matrices and vectors are complex numbers, the inner product of two vectors presented earlier is generalized using the Hermitian adjoint. For the case that the column vectors **x** and **y** have complex elements, the inner product is defined to be identical to the earlier definition (2.84),

$$(\mathbf{y}, \mathbf{x}) = \mathbf{y}^\dagger \cdot \mathbf{x} = y_j^* x_j. \tag{7.25}$$

In general, this will be a complex number, since $y_j^* x_j \neq y_j x_j^*$. However, for the case of the inner product of a complex vector with itself, $(\mathbf{x}, \mathbf{x}) = x_j^* x_j$, the result is a real and positive number by virtue of each term in the sum being real and positive. Therefore, $(\mathbf{x}, \mathbf{x})$ can still be viewed as the magnitude squared of the vector **x**, despite its complex values. This extension of the inner product reduces to the previous definition (7.18) in the event that the vectors are real valued. Although outside the scope of this book, this form of the inner product is central to the definition of a *Hilbert space*, which is critical to the formulation of functional analysis and quantum mechanics.

A matrix is *unitary* if its Hermitian adjoint is also its inverse, so that $\mathbf{A}^{\dagger} \cdot \mathbf{A} = \mathbf{I}$. The unitary matrix is the extension of the orthogonal matrix to the complex case, as seen from its action on the inner product for complex vectors. If $\mathbf{x}' = \mathbf{A} \cdot \mathbf{x}$ and $\mathbf{y}' = \mathbf{A} \cdot \mathbf{y}$ are the vectors created by multiplying two vectors $\mathbf{x}$ and $\mathbf{y}$ by a unitary matrix $\mathbf{A}$, then it follows from the previous results that

$$\begin{aligned}(\mathbf{y}', \mathbf{x}') = (\mathbf{y}')^{\dagger} \cdot \mathbf{x}' = (\mathbf{A} \cdot y)^{\dagger} \cdot \mathbf{A} \cdot \mathbf{x} = \mathbf{y}^{\dagger} \cdot \mathbf{A}^{\dagger} \cdot \mathbf{A} \cdot \mathbf{x} = \mathbf{y}^{\dagger} \cdot \mathbf{I} \cdot \mathbf{x} = \mathbf{y}^{\dagger} \cdot \mathbf{x} \\ = (\mathbf{y}, \mathbf{x}).\end{aligned} \tag{7.26}$$

The action of a unitary matrix therefore preserves the generalized inner product of two complex vectors.

Similarly, functions of a matrix act as a mapping of a matrix to another matrix. Using the substitution $1 \to \mathbf{I}$, the power series representation of a real or complex function can be adapted to the corresponding function of a square matrix. For example, given a number x, the exponential function e^x has the power series representation $e^x = \sum_{n=0}^{\infty} x^n/n!$. It follows that the exponential of the square matrix $\mathbf{A}$ multiplied by the number θ is obtained by the replacement $x \to \theta\mathbf{A}$ in the power series to obtain

$$\exp(\theta\mathbf{A}) = e^{(\theta\mathbf{A})} = \sum_{n=0}^{\infty} \frac{1}{n!}\theta^n \mathbf{A}^n, \tag{7.27}$$

where $\mathbf{A}^n$ is simply the product of n copies of the matrix $\mathbf{A}$ with the understanding that $\mathbf{A}^0 = \mathbf{I}$ and $\mathbf{A}^{-n} = (\mathbf{A}^{-1})^n$. The result of (7.27) is a square matrix with the same size as $\mathbf{A}$. It is critical to note that $\exp \mathbf{A}$ is *not* the matrix with the exponential of the original elements of $\mathbf{A}$, so that $(\exp \mathbf{A})_{ij} \neq \exp A_{ij}$.

As an example, the matrix

$$\mathbf{A} = \begin{pmatrix} 0 & -1 \\ -1 & 0 \end{pmatrix} \tag{7.28}$$

has the properties that $\mathbf{A}^{2n} = \mathbf{I}$ and $\mathbf{A}^{2n+1} = \mathbf{A}$, where n is an integer. The exponential power series (7.27) gives

$$e^{\theta\mathbf{A}} = \begin{pmatrix} \cosh\theta & -\sinh\theta \\ -\sinh\theta & \cosh\theta \end{pmatrix}. \tag{7.29}$$

This follows from the two Taylor series representations given by

$$\sinh\theta = \sum_{n=0}^{\infty} \frac{\theta^{2n+1}}{(2n+1)!}, \qquad \cosh\theta = \sum_{n=0}^{\infty} \frac{\theta^{2n}}{(2n)!}. \tag{7.30}$$

In a similar manner, the Taylor series $(1 + x)^{-1} = 1 - x + x^2 - x^3 + \ldots$ and the properties of the same matrix **A** show that

$$(\mathbf{1} + \alpha\mathbf{A})^{-1} = \begin{pmatrix} 1 & \alpha \\ \alpha & 1 \end{pmatrix}^{-1} = \frac{1}{1 - \alpha^2}\begin{pmatrix} 1 & -\alpha \\ -\alpha & 1 \end{pmatrix}, \tag{7.31}$$

when coupled with the two identities

$$\frac{1}{1 - \alpha^2} = \sum_{n=0}^{\infty} \alpha^{2n}, \quad -\frac{\alpha}{1 - \alpha^2} = -\sum_{n=0}^{\infty} \alpha^{2n+1}, \tag{7.32}$$

along with the assumption that $\alpha^2 < 1$.

Exercise 7.2. Show that the right-hand side of (7.31) is the matrix inverse of $\mathbf{1} + \alpha\mathbf{A}$.

Result (7.29) demonstrates an important property. This starts by noting that

$$e^{-\theta\mathbf{A}} = \begin{pmatrix} \cosh(-\theta) & -\sinh(-\theta) \\ -\sinh(-\theta) & \cosh(-\theta) \end{pmatrix} = \begin{pmatrix} \cosh\theta & \sinh\theta \\ \sinh\theta & \cosh\theta \end{pmatrix}. \tag{7.33}$$

Using matrix multiplication it follows that

$$e^{-\theta\mathbf{A}}e^{\theta\mathbf{A}} = \begin{pmatrix} \cosh^2\theta - \sinh^2\theta & 0 \\ 0 & \cosh^2\theta - \sinh^2\theta \end{pmatrix} = \begin{pmatrix} 1 & 0 \\ 0 & 1 \end{pmatrix} = \mathbf{I}. \tag{7.34}$$

Although result (7.34) was found for the specific matrix **A** of (7.28), it can be shown to hold for an arbitrary square matrix, so that $\exp(-\theta\mathbf{A})$ is the inverse of $\exp(\theta\mathbf{A})$.

Using the Levi-Civita symbol (1.5) with n indices allows the *determinant* of an n-dimensional square matrix **A**, denoted $\det\mathbf{A}$, to be defined as

$$\det\mathbf{A} = \varepsilon^{abc\ldots}A_{1a}A_{2b}A_{3c}\ldots, \tag{7.35}$$

where the number of superscripts n on the Levi-Civita symbol matches the dimension n of the matrix. For example, the determinant of a 2×2 square matrix **A** is

$$\det\mathbf{A} = \varepsilon^{ab}A_{1a}A_{2b} = A_{11}A_{22} - A_{12}A_{21}. \tag{7.36}$$

Result (7.36) is identical to the Jacobian (1.12) that occurs in a change of two-dimensional coordinates when the matrix elements are identified with $\partial x^a/\partial y^b$. Using the definition of the determinant given by (7.35), the Jacobian of (1.12) for an arbitrary change of coordinates can be written

$$J = \det\left(\frac{\partial y}{\partial x}\right), \tag{7.37}$$

where $\partial y/\partial x$ stands for the matrix $\mathbf{M}$, which has the elements

$$M^{\mu}_{\nu} = \frac{\partial y^{\mu}}{\partial x^{\nu}}. \tag{7.38}$$

Exercise 7.3. Show that the Jacobian of (1.12) is identical to $\det \mathbf{M}$.

All terms in the determinant of an $n \times n$ matrix are products of n matrix elements. If the matrix is diagonal then the determinant is simply the product of all the diagonal elements, since all other terms in (7.35) vanish. This gives $\det \mathbf{I} = 1$ and $\det 0 = 0$. The determinant has a number of important properties that are uncovered by first noting the matrix version of the Jacobian result (1.14), which is given by

$$\varepsilon_{abc\ldots} \det \mathbf{A} = \varepsilon^{a'b'c'\ldots} A_{aa'} A_{bb'} A_{cc'} \ldots. \tag{7.39}$$

It is important to note that (1.11) assumes that $\varepsilon_{abc\ldots} = \varepsilon^{abc\ldots}$ is valid. As with (1.14), the proof of (1.11) hinges on noting that if any of the indices in the set $abc\ldots$ are equal then the right-hand side of (1.11) vanishes, just as the left-hand side does, since

$$\varepsilon^{a'b'c'\ldots} A_{aa'} A_{ab'} \ldots = \varepsilon^{b'a'c'\ldots} A_{ab'} A_{aa'} \ldots = -\varepsilon^{a'b'c'\ldots} A_{aa'} A_{ab'} \ldots, \tag{7.40}$$

where the freedom to relabel dummy indices was combined with the antisymmetry of the Levi-Civita symbol. Result (7.40) shows that both sides vanish unless $abc\ldots$ is an odd or even permutation. The final step follows by noting that the right-hand side of (1.11) is $\det \mathbf{A}$ if $abc\ldots$ is an even permutation and is $-\det \mathbf{A}$ for an odd permutation, which matches the value of the left-hand side's Levi-Civita symbol. Result (7.39) can be rewritten by using identity (1.11) to obtain

$$\det \mathbf{A} = \frac{1}{n!} \varepsilon^{abc\ldots} \varepsilon^{a'b'c'\ldots} A_{aa'} A_{bb'} A_{cc'} \ldots. \tag{7.41}$$

Letting B_{ij} denote the elements of a second n-dimensional square matrix and using (1.11) gives

$$\begin{aligned} \varepsilon^{abc\ldots} B_{1a} B_{2b} B_{3c} \ldots \det \mathbf{A} &= \varepsilon^{a'b'c'\ldots} (B_{1a} A_{aa'})(B_{2b} A_{bb'})(B_{3c} A_{cc'}) \ldots \\ &= \varepsilon^{a'b'c'\ldots} (\mathbf{B} \cdot \mathbf{A})_{1a'} (\mathbf{B} \cdot \mathbf{A})_{2b'} (\mathbf{B} \cdot \mathbf{A})_{3c'} \ldots. \end{aligned} \tag{7.42}$$

Using (7.35) shows that (7.42) gives a very useful result,

$$\det \mathbf{B} \det \mathbf{A} = \det(\mathbf{B} \cdot \mathbf{A}). \tag{7.43}$$

Combining (7.43) with $\det(\mathbf{A} \cdot \mathbf{A}^{-1}) = \det \mathbf{I} = 1$ shows that

$$\det(\mathbf{A}^{-1}) = \frac{1}{\det \mathbf{A}}. \tag{7.44}$$

If $\det \mathbf{A} = 0$ then $\det \mathbf{A}^{-1}$ does not exist, since the number zero has no inverse, and for that case, matrix $\mathbf{A}$ has no inverse. An example of this is the matrix appearing in (7.15), which has a determinant of zero and therefore has no inverse. Result (7.43) is easily extended to show that

$$\det(\mathbf{A} \cdot \mathbf{B} \cdots \mathbf{C}) = \det \mathbf{A} \det \mathbf{B} \cdots \det \mathbf{C}. \tag{7.45}$$

Using this result and assuming $\mathbf{A}^{-1}$ exists, replacing $\mathbf{B}$ with $\mathbf{A}^{-1} \cdot \mathbf{B}$ in (7.43) gives

$$\det(\mathbf{A}^{-1} \cdot \mathbf{B} \cdot \mathbf{A}) = \det \mathbf{B}. \tag{7.46}$$

Form (7.41) also gives

$$\det \mathbf{A} = \det \mathbf{A}^T. \tag{7.47}$$

This follows by simply substituting $A_{aa'} = A^T_{a'a}$ for all the factors in (7.41) and then relabeling the dummy indices $a \leftrightarrow a'$.

An important characteristic of a matrix is its set of *eigenvalues*. If $\mathbf{A}$ is a square matrix of dimension n, the number λ is an eigenvalue if there is a nonzero N-dimensional column vector $\mathbf{x}_\lambda$ such that

$$\mathbf{A} \cdot \mathbf{x}_\lambda = \lambda\, \mathbf{x}_\lambda. \tag{7.48}$$

The vector $\mathbf{x}_\lambda$ is called the *eigenvector* associated with the eigenvalue λ. The eigenvalues of a matrix can be found by noting that for any given eigenvalue λ, there is a related matrix $\mathbf{B} = \mathbf{A} - \lambda \mathbf{I}$. The action of $\mathbf{B}$ on the associated eigenvector gives

$$\mathbf{B} \cdot \mathbf{x}_\lambda = (\mathbf{A} - \lambda \mathbf{I}) \cdot \mathbf{x}_\lambda = \mathbf{A} \cdot \mathbf{x}_\lambda - \lambda \mathbf{x}_\lambda = \mathbf{x}_0, \tag{7.49}$$

where $\mathbf{x}_0$ is the null vector, since it has $x_j = 0$ by virtue of (7.48). As a result, $\mathbf{B}$ has no inverse, since otherwise

$$\mathbf{x}_0 = \mathbf{B}^{-1} \cdot \mathbf{x}_0 = \mathbf{B}^{-1} \cdot \mathbf{B} \cdot \mathbf{x}_\lambda = \mathbf{I} \cdot \mathbf{x}_\lambda = \mathbf{x}_\lambda, \tag{7.50}$$

in contradiction to result (7.49). Since $\mathbf{B}$ has no inverse it must be that $\det \mathbf{B} = 0$, and this gives the *characteristic polynomial* for the eigenvalues,

$$\det(\mathbf{A} - \lambda \mathbf{I}) = 0. \tag{7.51}$$

For an n-dimensional square matrix, (7.51) gives an nth-order polynomial in λ whose solutions are the eigenvalues of the matrix. The fundamental theorem of algebra then states that an n-dimensional square matrix has at most n distinct eigenvalues. Once the eigenvalues are determined the general form of the eigenvectors can be determined by solving (7.48). The eigenvectors determined in this way are not unique. Any multiple of an eigenvector, $\alpha \mathbf{x}_\lambda$, is still an eigenvector, since the α factor drops out of (7.48). The eigenvectors are made unique by imposing a *normalization* condition,

$$(\mathbf{x}_\lambda, \mathbf{x}_\lambda) = \mathbf{x}^\dagger_\lambda \cdot \mathbf{x}_\lambda = 1, \tag{7.52}$$

which fixes the overall multiplicative factor. It will be assumed that all eigenvectors $\mathbf{x}_\lambda$ obey (7.52).

Square Hermitian matrices play an important role in physics. The primary reason is that a Hermitian matrix **A** has real eigenvalues. This is demonstrated by starting with

$$\mathbf{A} \cdot \mathbf{x}_\lambda = \lambda \mathbf{x}_\lambda \implies \mathbf{x}_\lambda^\dagger \cdot \mathbf{A} \cdot \mathbf{x}_\lambda = \lambda \mathbf{x}_\lambda^\dagger \cdot \mathbf{x}_\lambda = \lambda. \tag{7.53}$$

Taking the Hermitian adjoint and using $\mathbf{A}^\dagger = \mathbf{A}$ gives

$$\mathbf{x}_\lambda^\dagger \cdot \mathbf{A}^\dagger = \lambda^* \mathbf{x}_\lambda^\dagger \implies \mathbf{x}_\lambda^\dagger \cdot \mathbf{A} \cdot \mathbf{x}_\lambda = \lambda^* \mathbf{x}_\lambda^\dagger \cdot \mathbf{x}_\lambda = \lambda^*. \tag{7.54}$$

Since the two results must be the same, it follows that λ is real, since $\lambda^* = \lambda$. A second important property of Hermitian matrices is the orthogonality of their eigenvectors. In other words, if $\mathbf{x}_\lambda$ and $\mathbf{x}_{\lambda'}$ are eigenvectors of the Hermitian matrix **A** associated with *different eigenvalues*, $\lambda \neq \lambda'$, then

$$(\mathbf{x}_\lambda, \mathbf{x}_{\lambda'}) = \mathbf{x}_\lambda^\dagger \cdot \mathbf{x}_{\lambda'} = 0. \tag{7.55}$$

Demonstrating this starts by noting that

$$\mathbf{A} \cdot \mathbf{x}_{\lambda'} = \lambda' \mathbf{x}_{\lambda'} \implies \mathbf{x}_\lambda^\dagger \cdot \mathbf{A} \cdot \mathbf{x}_{\lambda'} = \lambda' \mathbf{x}_\lambda^\dagger \cdot \mathbf{x}_{\lambda'} = \lambda'(\mathbf{x}_\lambda, \mathbf{x}_{\lambda'}). \tag{7.56}$$

Because $\mathbf{A}^\dagger = \mathbf{A}$ and the eigenvalues are real it follows that

$$\mathbf{x}_\lambda^\dagger \cdot \mathbf{A} = \lambda \mathbf{x}_\lambda^\dagger \implies \mathbf{x}_\lambda^\dagger \cdot \mathbf{A} \cdot \mathbf{x}_{\lambda'} = \lambda \mathbf{x}_\lambda^\dagger \cdot \mathbf{x}_{\lambda'} = \lambda(\mathbf{x}_\lambda, \mathbf{x}_{\lambda'}). \tag{7.57}$$

Setting (7.56) equal to (7.57) gives

$$(\lambda' - \lambda)(\mathbf{x}_\lambda, \mathbf{x}_{\lambda'}) = 0. \tag{7.58}$$

Result (7.58) shows that the inner product must vanish if $\lambda' - \lambda \neq 0$. The eigenvectors of a Hermitian matrix are therefore orthogonal.

An n-dimensional Hermitian matrix **A** has n eigenvectors that are both orthogonal and normalized, which are referred to as *orthonormal* eigenvectors. Denoting these eigenvectors by $\mathbf{x}_{\lambda_i}$, they can be used to form an n-dimensional square matrix **M**, given by

$$\mathbf{M} = (\mathbf{x}_{\lambda_1} \;\; \mathbf{x}_{\lambda_2} \;\; \cdots \;\; \mathbf{x}_{\lambda_n}), \tag{7.59}$$

where the n eigenvectors $\mathbf{x}_\lambda$ form the n columns of the matrix **M**. It follows that the Hermitian adjoint $\mathbf{M}^\dagger$ is given by

$$\mathbf{M}^\dagger = \begin{pmatrix} \mathbf{x}_{\lambda_1}^\dagger \\ \mathbf{x}_{\lambda_2}^\dagger \\ \vdots \\ \mathbf{x}_{\lambda_n}^\dagger \end{pmatrix}, \tag{7.60}$$

where the Hermitian adjoint of the n eigenvectors forms the n rows of $\mathbf{M}^\dagger$. For simplicity, it is assumed that the n eigenvalues of **A**, denoted by λ_i, are distinct. It follows that the product $\mathbf{M}^\dagger \cdot \mathbf{M}$ has the elements

$$(\mathbf{M}^\dagger \cdot \mathbf{M})_{ij} = \mathbf{x}_{\lambda_i}^\dagger \cdot \mathbf{x}_{\lambda_j} = \delta_{ij} \implies \mathbf{M}^\dagger \cdot \mathbf{M} = \mathbf{I}. \tag{7.61}$$

where (7.66) was used, thereby demonstrating (7.70).

Result (7.70) can be used to demonstrate a useful result regarding the product of two matrices, **A** and **B**. It follows that (7.43) and (7.70) give

$$\det \mathbf{A} \det \mathbf{B} = \det(\mathbf{A} \cdot \mathbf{B}) = \exp \operatorname{Tr} \ln(\mathbf{A} \cdot \mathbf{B}). \tag{7.78}$$

Taking the logarithm of both sides and using (7.70) gives

$$\ln \det \mathbf{A} + \ln \det \mathbf{B} = \operatorname{Tr} \ln \mathbf{A} + \operatorname{Tr} \ln \mathbf{B} = \operatorname{Tr}(\ln \mathbf{A} + \ln \mathbf{B}) = \operatorname{Tr} \ln(\mathbf{A} \cdot \mathbf{B}), \tag{7.79}$$

which also proves (7.73).

7.2 Basic group theory

Group theory has its origins in the classification of transformations and their relation to differential equations. This section provides a very brief and incomplete introduction to this rich subject, which is motivated by a study of finite matrix transformations. Consideration will be limited to what are called *continuous Lie groups*, since they form the basis of special relativity as well as nonabelian gauge field theories. Loosely speaking, a group is a collection of mathematical objects such that the product or composition of any two members of the collection yields another member of the collection. It also possesses an identity element so that its product with any member of the collection yields that member.

7.2.1 Spatial rotations and continuous groups

A simple and useful example of a group is the matrix representation of a transformation corresponding to the rotation of a coordinate system in a three-dimensional Euclidean space. The three Cartesian coordinates of a point can be viewed as the components of a contravariant position vector from the origin to the point. Since rotation is a type of coordinate transformation, it will be accompanied by a change in the components of the vector. The effect of rotating the x–y plane around the z-axis by the angle θ is depicted in figure 7.1.

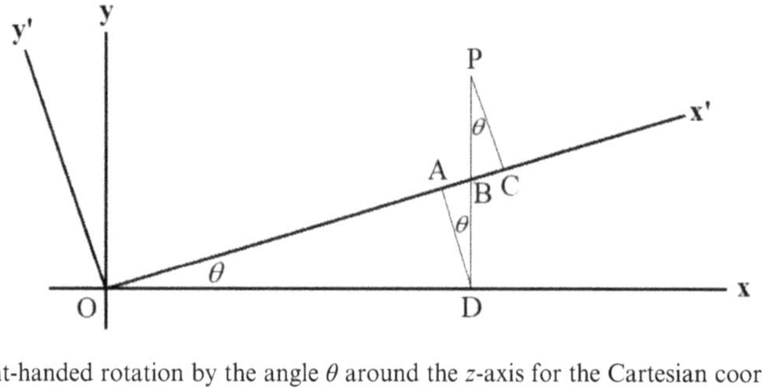

Figure 7.1. A right-handed rotation by the angle θ around the z-axis for the Cartesian coordinate system in a three-dimensional Euclidean space. The right triangles OAD, ABD, and PBC are all similar.

Actively rotating the coordinate system through an angle θ counterclockwise around the z-axis causes the x and y coordinates of the point or equivalently the components of the position vector to change.

Exercise 7.4. If the previous coordinates of the point P were (x, y, z), where $x = \bar{\mathrm{OD}}$ and $y = \bar{\mathrm{PD}}$, show that the new coordinates of point P in figure 7.1 are

$$\bar{\mathrm{OC}} = x' = x\cos\theta + y\sin\theta, \tag{7.80}$$

$$\bar{\mathrm{PC}} = y' = y\cos\theta - x\sin\theta, \tag{7.81}$$

$$z' = z. \tag{7.82}$$

It is important to note that this rotation does not change the magnitude of the position vector, since

$$x'^2 + y'^2 + z'^2 = x^2 + y^2 + z^2. \tag{7.83}$$

These results are consistent with the transformation property of contravariant vectors (1.21). For example, the relation (7.80) can be written as

$$x'^1 = x^1\cos\theta + x^2\sin\theta = \frac{\partial x'}{\partial x}x + \frac{\partial x'}{\partial y}y + \frac{\partial x'}{\partial z}z = \frac{\partial x'^1}{\partial x^j}x^j. \tag{7.84}$$

The rotated coordinates can be expressed in matrix notation by writing the position vector $\boldsymbol{x}$ as a column vector using the three coordinates x, y, and z of the point as its elements,

$$\boldsymbol{x} = \begin{pmatrix} x \\ y \\ z \end{pmatrix} \equiv \begin{pmatrix} x^1 \\ x^2 \\ x^3 \end{pmatrix}. \tag{7.85}$$

The effect of the rotation around the z-axis given by (7.80), (7.81), and (7.82) can be written in the form of a matrix $\mathbf{R}_z(\theta)$ acting on the column vector $\boldsymbol{x}$,

$$\boldsymbol{x}' = \begin{pmatrix} x'^1 \\ x'^2 \\ x'^3 \end{pmatrix} = \begin{pmatrix} \cos\theta & \sin\theta & 0 \\ -\sin\theta & \cos\theta & 0 \\ 0 & 0 & 1 \end{pmatrix} \cdot \begin{pmatrix} x^1 \\ x^2 \\ x^3 \end{pmatrix} \equiv \mathbf{R}_z(\theta)\cdot\boldsymbol{x}. \tag{7.86}$$

Using the rules of matrix multiplication, (7.4) shows that (7.86) reproduces the results of the rotation found in (7.80) and (7.81).

The matrix $\mathbf{R}_z(\theta)$ has a number of important properties. The first is that it is parameterized by a *continuous* variable θ in such a way that $\mathbf{R}_z(0)$ is the 3×3 identity matrix $\mathbf{I}$. Second, using the basic trigonometric identities $\cos(\theta_1 + \theta_2) = \cos\theta_1\cos\theta_2 - \sin\theta_1\sin\theta_2$ and $\sin(\theta_1 + \theta_2) = \sin\theta_1\cos\theta_2 + \cos\theta_1\sin\theta_2$ shows that the product of $\mathbf{R}_z$ for two different angles gives

$$\mathbf{R}_z(\theta_1)\cdot\mathbf{R}_z(\theta_2)=\begin{pmatrix}\cos\theta_1 & \sin\theta_1 & 0\\ -\sin\theta_1 & \cos\theta_2 & 0\\ 0 & 0 & 1\end{pmatrix}\cdot\begin{pmatrix}\cos\theta_2 & \sin\theta_2 & 0\\ -\sin\theta_2 & \cos\theta_2 & 0\\ 0 & 0 & 1\end{pmatrix}$$
$$=\begin{pmatrix}\cos(\theta_1+\theta_2) & \sin(\theta_1+\theta_2) & 0\\ -\sin(\theta_1+\theta_2) & \cos(\theta_1+\theta_2) & 0\\ 0 & 0 & 1\end{pmatrix}=\mathbf{R}_z(\theta_1+\theta_2). \tag{7.87}$$

The product is a matrix *identical* in form to the two matrices forming the product. This is a fundamental property of a *continuous group*, which is a set of mathematical objects, in this case matrices, characterized by continuous parameters and *closed* under multiplication or *composition*. In other words, multiplying any two matrices in this group gives a matrix which is also a member of the group in that it shares the same properties as all the other members, regardless of the values of the parameter θ. The second basic property is the existence of an inverse for each transformation. Result (7.87) shows that

$$\mathbf{R}_z(-\theta)\cdot\mathbf{R}_z(\theta)=\mathbf{I}\implies\mathbf{R}_z(-\theta)=\mathbf{R}^{-1}(\theta). \tag{7.88}$$

In this particular case the two matrices also satisfy

$$\mathbf{R}_z(\theta_1)\cdot\mathbf{R}_z(\theta_2)=\mathbf{R}_z(\theta_2)\cdot\mathbf{R}_z(\theta_1)\implies[\mathbf{R}_z(\theta_1),\mathbf{R}_z(\theta_2)]=0, \tag{7.89}$$

so that the two matrices commute. This is reflected in the fact that their product only depends on the sum of the two parameter angles. Because they commute, the matrices of the form $\mathbf{R}_z(\theta)$ form an *abelian group*, since the order of their composition does not matter. Most continuous or *Lie groups* do not satisfy this property and are referred to as *nonabelian*.

It is important to identify the properties that the matrices $\mathbf{R}_z(\theta)$ share, regardless of the value of θ. First, they are orthogonal since

$$(\mathbf{R}_z)^T(\theta)=\mathbf{R}_z(-\theta)=(\mathbf{R}_z)^{-1}(\theta). \tag{7.90}$$

Result (7.17) shows that the product of two orthogonal matrices is also an orthogonal matrix, since

$$(\mathbf{A}\cdot\mathbf{B})^T(\mathbf{A}\cdot\mathbf{B})=\mathbf{B}^T\cdot\mathbf{A}^T\cdot\mathbf{A}\cdot\mathbf{B}=\mathbf{B}^T\cdot\mathbf{B}=\mathbf{I}, \tag{7.91}$$

so that $(\mathbf{A}\cdot\mathbf{B})^T$ is the inverse of $\mathbf{A}\cdot\mathbf{B}$. Second, they have the *special* property that their determinant is unity,

$$\det(\mathbf{R}_z(\theta))=\cos^2\theta+\sin^2\theta=1. \tag{7.92}$$

The product of two matrices, each with determinant one, is also a matrix with determinant one, since (7.43) gives

$$\det(\mathbf{A}\cdot\mathbf{B})=\det\mathbf{A}\det\mathbf{B}=1. \tag{7.93}$$

As a result, grouping together all 3×3 orthogonal matrices with determinant one has the property that the product of any two of these matrices will also be a member of that same group. This group of matrices is therefore referred to as SO(3, R), which stands for the set of all *special orthogonal 3 × 3 matrices with real elements.*

However, $\mathbf{R}_z(\theta)$ does not exhaust all possible 3 × 3 orthogonal matrices with a determinant of +1, since it only represents rotations around the z-axis, and rotations around the x- and y-axes are also possible. The matrices of the form $\mathbf{R}_z(\theta)$ therefore constitute an abelian *subgroup* of SO(3, R), since they are closed under multiplication but do not exhaust all possible members of SO(3, R). Further investigation into SO(3, R) is required to understand its full range of parameters and its structure. Because these matrices represent the action of a rotation on the coordinates of a point, the column vector $\boldsymbol{x}$ is referred to as the *representation space*, and the matrices $\mathbf{R}_z$ give a *representation* of the subgroup for that choice of representation space. While all this terminology may seem unnecessary for such a simple case, it will be useful when these concepts are generalized to *nonabelian* groups as well as to other representation spaces.

7.2.2 Generators and the exponential map

The first step in understanding the full nature of SO(3, R) is to examine the behavior of the subgroup $\mathbf{R}_z(\theta)$ in the neighborhood of $\mathbf{I}$. The continuous nature of the parameter θ allows an expansion of $\mathbf{R}_z(\theta)$ in a Taylor series around $\mathbf{I}$, and to $O(\theta)$ this gives

$$\mathbf{R}_z(\theta) \approx \mathbf{I} + \theta \frac{\partial \mathbf{R}_z}{\partial \theta}\bigg|_{\theta=0} + \dots. \tag{7.94}$$

Using (7.86) gives the constant matrix

$$\mathbf{T}_z \equiv \frac{\partial \mathbf{R}_z}{\partial \theta}\bigg|_{\theta=0} = \begin{pmatrix} 0 & 1 & 0 \\ -1 & 0 & 0 \\ 0 & 0 & 0 \end{pmatrix}. \tag{7.95}$$

The matrix $\mathbf{T}_z$ is referred to as a *generator*, since it generates $\mathbf{R}_z(\theta)$ for *finite values* of θ. The demonstration of this starts by creating an infinitesimal parameter θ/N, where N is an arbitrarily large integer. For such a case, the infinitesimal transformation becomes $\mathbf{I} + (\theta/N)\mathbf{T}_z$, where the limit $N \to \infty$ is understood. The form of the transformation for the finite value θ is generated through N applications of the infinitesimal expression, giving

$$\mathbf{R}_z(\theta) = \lim_{N\to\infty}\left(\mathbf{I} + \frac{\theta}{N}\mathbf{T}_z\right)^N = \exp(\theta \mathbf{T}_z). \tag{7.96}$$

The emergence of the exponential follows by taking the logarithm of both sides and using the power series representation $\ln(1 + x) = x - x^2/2 + \dots$ to find

$$\ln \mathbf{R}_z(\theta) = \lim_{N\to\infty} N \ln\left(\mathbf{I} + \frac{\theta}{N}\mathbf{T}_z\right) = \lim_{N\to\infty}\left(\theta \mathbf{T}_z - \frac{\theta^2}{2N}(\mathbf{T}_z)^2 + \dots\right) = \theta \mathbf{T}_z, \tag{7.97}$$

which is the property of an exponential. This can be explicitly demonstrated for the special case under consideration by combining the exponential power series (7.27) with the result that even and odd powers of $\mathbf{T}_z$ are given by

$$(\mathbf{T}_z)^{2n} = (-1)^n \begin{pmatrix} 1 & 0 & 0 \\ 0 & 1 & 0 \\ 0 & 0 & 0 \end{pmatrix} \quad \text{and} \quad (\mathbf{T}_z)^{2n+1} = (-1)^n \begin{pmatrix} 0 & 1 & 0 \\ -1 & 0 & 0 \\ 0 & 0 & 0 \end{pmatrix}. \tag{7.98}$$

The next step is to use the power series representations

$$\cos\theta = \sum_{n=0}^{\infty} \frac{(-1)^n}{(2n)!}\theta^{2n} \quad \text{and} \quad \sin\theta = \sum_{n=0}^{\infty} \frac{(-1)^n}{(2n+1)!}\theta^{2n+1}, \tag{7.99}$$

and notice that they appear as the elements of the exponential series for (7.96).

Exercise 7.5. Show that (7.98) and (7.99) combine with the exponential power series (7.27) to yield the matrix $\mathbf{R}_z(\theta)$ appearing in (7.86).

Result (7.96) occurs for all matrix groups and is referred to as the *exponential map.*

Arbitrary rotations about the other two axes are given by $\mathbf{R}_x(\alpha)$ and $\mathbf{R}_y(\beta)$, with the similar results that

$$\mathbf{R}_x(\alpha) = \begin{pmatrix} 1 & 0 & 0 \\ 0 & \cos\alpha & \sin\alpha \\ 0 & -\sin\alpha & \cos\alpha \end{pmatrix}, \quad \mathbf{R}_y(\beta) = \begin{pmatrix} \cos\beta & 0 & -\sin\beta \\ 0 & 1 & 0 \\ \sin\beta & 0 & \cos\beta \end{pmatrix}. \tag{7.100}$$

Following the same procedure as (7.94) yields two new generators, which are given by

$$\mathbf{T}_x = \begin{pmatrix} 0 & 0 & 0 \\ 0 & 0 & 1 \\ 0 & -1 & 0 \end{pmatrix}, \quad \mathbf{T}_y = \begin{pmatrix} 0 & 0 & -1 \\ 0 & 0 & 0 \\ 1 & 0 & 0 \end{pmatrix}. \tag{7.101}$$

Denoting as 1, 2, and 3 by x, y, and z, the elements of the three matrices can be written using the Levi-Civita symbol,

$$(\mathbf{T}_a)_{bc} = \varepsilon_{abc}. \tag{7.102}$$

A full three-dimensional rotation can then be viewed as a combination of rotations around all three axes. Denoting three angles by θ^a, (7.96) is generalized to a general three-dimensional rotation in terms of the three generators of (7.95) and (7.101) and the three angles as

$$\mathbf{R}(\theta) = \exp(\theta^a \mathbf{T}_a). \tag{7.103}$$

The 3×3 matrix defined by (7.103) is understood to act on the representation space of vectors, $\boldsymbol{x}$, and gives a representation of the *rotation group.*

For infinitesimal rotations (7.103) becomes $\mathbf{R}(\theta) \approx \mathbf{I} + \theta^a \mathbf{T}_a$, and its action on the representation space is

$$x^{i\prime} = x^i + \theta^k (\mathbf{T}_k)_{ij} x^j = x^i + \varepsilon_{kij} \theta^k x^j. \tag{7.104}$$

This result will be of importance in understanding the angular momentum content of field theories. For infinitesimal rotations about the x- and y-axes, the commutator $[\mathbf{R}_x(\alpha), \mathbf{R}_y(\beta)]$ is given by

$$[\mathbf{I} + \alpha \mathbf{T}_x, \mathbf{I} + \beta \mathbf{T}_y] = \alpha\beta [\mathbf{T}_x, \mathbf{T}_y] = -\alpha\beta\, \mathbf{T}_z, \tag{7.105}$$

where straightforward matrix multiplication results in the appearance of $\mathbf{T}_z$. This result shows two things: first, the order in which rotations are performed produces different results; and second, the generators themselves form a closed set of mathematical objects under multiplication. Because the generators $\mathbf{T}_a$ do not commute, they are the generators of *a nonabelian group.*

These results emerged by considering rotations around all three possible axes. It is important to confirm that these are the specific properties of SO(3, R). The act of rotating a coordinate system does not change the length of a position vector, and this means that a general rotation matrix $\mathbf{R}$, characterized by three angles, cannot change the inner product of a vector with itself. Using $\boldsymbol{x}' = \mathbf{R} \cdot \boldsymbol{x}$, this requirement is stated mathematically as

$$\boldsymbol{x}'^T \cdot \boldsymbol{x}' = \boldsymbol{x}^T \cdot \mathbf{R}^T \cdot \mathbf{R} \cdot \boldsymbol{x} = \boldsymbol{x}^T \cdot \boldsymbol{x}. \tag{7.106}$$

As anticipated with (7.23), this requires that $\mathbf{R}$ is an orthogonal matrix, since it must satisfy $\mathbf{R}^T \cdot \mathbf{R} = \mathbf{I}$. In three dimensions and for real values, this means that the elements of $\mathbf{R}$ must obey the set of equations given by

$$(\mathbf{R}^T \cdot \mathbf{R})_{ij} = R_{ki} R_{kj} = \delta_{ij}. \tag{7.107}$$

Three of the requirements of (7.107) are symmetric for an interchange of i and j, e.g. $i = 2$ and $j = 3$ gives the same equation as $i = 3$ and $j = 2$. This means that (7.107) gives a total of $3 \times 3 - 3 = 6$ *unique* equations. In turn, this means that the nine elements of the matrix $\mathbf{R}$ are determined entirely by $9 - 6 = 3$ free parameters. Once any three elements are specified, all nine elements of $\mathbf{R}$ are uniquely determined through the six equations required for orthogonality. The three angles of rotation around the three possible rotational axes are therefore sufficient to determine the matrix $\mathbf{R}$. Each of the three generators $\mathbf{T}_a$ has the transpose property that $(\mathbf{T}_a)^T = -\mathbf{T}_a$. It follows from the exponential map of (7.103) that its transpose is given by

$$\mathbf{R}^T = \exp(\theta^a (\mathbf{T}_a)^T) = \exp(-\theta^a \mathbf{T}_a) = \mathbf{R}^{-1}, \tag{7.108}$$

where the property of the matrix exponential, $(e^{\mathbf{A}})^{-1} = e^{-\mathbf{A}}$, was used. As a result, (7.103) is an orthogonal 3×3 matrix with three parameters, as it must be. The generators also have the property that $\mathrm{Tr}(\mathbf{T}_a) = 0$. Therefore, theorem (7.77) shows that (7.103) gives

$$\det \mathbf{R} = \exp\mathrm{Tr}(\ln \mathbf{R}) = \exp\mathrm{Tr}(\ln e^{(\theta^a \mathbf{T}_a)}) = \exp\mathrm{Tr}(\theta^a \mathbf{T}_a) = \exp(0) = 1. \quad (7.109)$$

Because it is an orthogonal 3×3 matrix with three free parameters and a determinant of one, matrix (7.103) represents a general member of SO(3, R).

Result (7.47) established that $\det \mathbf{R}^T = \det \mathbf{R}$, and so the orthogonality requirement for a real matrix gives

$$(\det \mathbf{R})^2 = \det \mathbf{R}^T \det \mathbf{R} = \det(\mathbf{R}^T \cdot \mathbf{R}) = \det \mathbf{I} = 1. \quad (7.110)$$

As a result, an orthogonal matrix $\mathbf{R}$ must have $\det \mathbf{R} = \pm 1$. An example of a 3×3 orthogonal matrix with $\det \mathbf{R} = -1$ is $\mathbf{R} = -\mathbf{I}$. This matrix corresponds to a reflection or a change in the handedness of the coordinate system, which was discussed in relation to pseudovectors following result (1.187) in chapter 1.

Exercise 7.6. Show that there is no unitary transformation that connects (7.103) to $-\mathbf{I}$.

As a result, $\mathbf{R} = -\mathbf{I}$ is not a true rotation. Letting O(3, R) denote the set of *orthogonal* 3×3 real matrices, it follows that O(3, R) breaks into two disjoint sets, one labeled $\mathrm{O}_{\downarrow}(3, R)$ that corresponds to *improper rotations* with $\det \mathbf{R} = -1$, and one labeled SO(3, R) that corresponds to *proper rotations* with $\det \mathbf{R} = 1$. The latter can be written as (7.103). The product of an even number of members of $\mathrm{O}_{\downarrow}(3)$ will be a member of SO(3, R), while the product of any number of members of SO(3, R) will be a member of SO(3, R). This is identical to the behavior of negative and positive numbers.

7.2.3 Lie algebras and structure constants

While these ideas have come from considering rotations in three dimensions, they form the prototype for the behavior of all continuous groups. The classification of all possible continuous groups and their properties is referred to as *Lie group theory*. An essential aspect of Lie groups is revealed by examining the commutators of their generators. In the case of SO(3, R) the generators satisfy the commutation relation

$$[\mathbf{T}_a, \mathbf{T}_b] = -\varepsilon_{abc} \mathbf{T}_c, \quad (7.111)$$

where the Levi-Civita symbol (1.5) in three dimensions has appeared.

Exercise 7.7. Verify result (7.111).

It is more useful to deal with Hermitian matrices when performing matrix manipulations. This is accomplished by defining the Hermitian matrices $\tau_a = -i\mathbf{T}_a$. Replacing $\mathbf{T}_a$ by $i\tau_a$ in (7.111) gives

$$[\boldsymbol{\tau}_a, \boldsymbol{\tau}_b] = i\varepsilon_{abc}\boldsymbol{\tau}_c. \tag{7.112}$$

The Levi-Civita symbol is said to provide the *structure constants* of the *rotation group* SO(3, R).

While the matrix generators derived from three-dimensional rotations satisfy (7.111), there are other mathematical objects that also satisfy (7.111). An important example is the three differential operators defined by the three-dimensional Euclidean cross product in Cartesian coordinates,

$$\mathbf{L} = \boldsymbol{r} \times \nabla \Longrightarrow L_x = y\partial_z - z\partial_y, \;\; L_y = z\partial_x - x\partial_z, \;\; L_z = x\partial_y - y\partial_x, \tag{7.113}$$

where the manifold is assumed to be Euclidean. When acting on a function $f(\boldsymbol{r})$ the three operators of (7.113) do not commute, instead satisfying the rotation algebra (7.111). For example,

$$\begin{aligned} L_x L_y f(\boldsymbol{r}) &= (yz\partial_x\partial_y - yx\partial_z^2 - z^2\partial_y\partial_x + zx\partial_y\partial_z + y\partial_x)f(\boldsymbol{r}), \\ L_y L_x f(\boldsymbol{r}) &= (yz\partial_y\partial_x - yx\partial_z^2 - z^2\partial_x\partial_y + zx\partial_z\partial_y + x\partial_y)f(\boldsymbol{r}), \end{aligned} \tag{7.114}$$

which immediately yields

$$[L_x, L_y]f(\boldsymbol{r}) = (L_x L_y - L_y L_x)f(\boldsymbol{r}) = -(x\partial_y - y\partial_x)f(\boldsymbol{r}) = -L_z f(\boldsymbol{r}). \tag{7.115}$$

Exercise 7.8. Complete the proof that the differential operators of (7.113) satisfy

$$[L_a, L_b] = -\varepsilon_{abc}L_c. \tag{7.116}$$

The differential operators of (7.113) therefore satisfy the same SO(3, R) relations of (7.111) when acting on functions $f(\boldsymbol{r})$. The argument of a scalar function can therefore be rotated by applying the exponential map,

$$f(\boldsymbol{r}') = e^{\boldsymbol{\theta}\cdot\boldsymbol{L}}f(\boldsymbol{r}) = e^{i\boldsymbol{\theta}\cdot(-i\boldsymbol{L})}f(\boldsymbol{r}), \tag{7.117}$$

where $-i\boldsymbol{L}$ is the Hermitian or *self-adjoint version* of the differential operator, just as $\tau_a = -iT_a$ provided the Hermitian matrix form of the generator in (7.112).

Because there are different sets of objects that satisfy (7.111), it should be viewed as the definition of a *Lie algebra*. The algebra associated with SO(3, R) is denoted by $\mathfrak{so}(3)$, and does not require the objects obeying that algebra to be matrices. The matrices used to derive the rotation algebra $\mathfrak{so}(3)$ are a specific *representation* of the Lie algebra for the choice of three-dimensional vectors $\boldsymbol{r}$ as the representation space. A general Lie algebra consists of a set of mathematical objects τ_a that obey commutation relations written

$$[\boldsymbol{\tau}_a, \boldsymbol{\tau}_b] = if_{abc}\,\boldsymbol{\tau}_c, \tag{7.118}$$

where f_{abc} are called the *structure constants* of the algebra. Because of the antisymmetry of the commutator, the structure constants must obey $f_{abc} = -f_{bac}$. Relation (7.112) shows that the structure constants for SO(3, R) are $f_{abc} = \varepsilon_{abc}$, which clearly exhibit this property. While the Jacobi identity (7.9) was derived in the context of matrices, it can be expressed using other non-commuting mathematical objects. Relation (7.118) can be used to reduce the double commutators of (7.9) to a single factor of $\boldsymbol{\tau}_e$. For example,

$$[\boldsymbol{\tau}_a, [\boldsymbol{\tau}_b, \boldsymbol{\tau}_c]] = if_{bcd}[\boldsymbol{\tau}_a, \boldsymbol{\tau}_d] = -f_{bcd}f_{ade}\boldsymbol{\tau}_e. \tag{7.119}$$

Repeating (7.119) for the other two double commutators in (7.9) and using the idea that the $\boldsymbol{\tau}_e$ are independent quantities gives the Jacobi identity for the structure constants,

$$f_{bcd}f_{ade} + f_{abd}f_{cde} + f_{cad}f_{bde} = 0. \tag{7.120}$$

The Jacobi identity (7.120) can be converted into a matrix statement by identifying the ab matrix element of the generator $\boldsymbol{\tau}_c$ as $(\boldsymbol{\tau}_c)_{ab} = -if_{cab}$. Using this identification and the property $f_{abc} = -f_{bac}$ shows that (7.120) can be written in terms of matrix multiplication as

$$\boldsymbol{\tau}_a \cdot \boldsymbol{\tau}_b - \boldsymbol{\tau}_b \cdot \boldsymbol{\tau}_a = if_{abd}\,\boldsymbol{\tau}_d. \tag{7.121}$$

Exercise 7.9. Use (7.120) to demonstrate that $(\boldsymbol{\tau}_c)_{ab} = -if_{cab}$ satisfies (7.121). It is useful to note that $-if_{cad} = if_{acd} = -(\boldsymbol{\tau}_a)_{cd}$.

The structure constants themselves therefore give a matrix representation of the Lie algebra (7.118) known as the *adjoint representation* of the algebra. Using the structure constants $f_{abc} = \varepsilon_{abc}$ of the group SO(3, R) gives $(\boldsymbol{\tau}_a)_{bc} = -i\varepsilon_{abc}$, which are the 3×3 Hermitian matrix versions of the generators (7.95) and (7.101) given by $-i\mathbf{T}_a$.

Lie algebras are an important tool in theoretical physics, and many of them have their origin in matrix groups. The group SO(3, R) has been defined in terms of real-valued orthogonal matrices with determinant one that generate rotations in a three-dimensional space. This result can be generalized by identifying SO(n, R) as the $n \times n$ orthogonal matrices with real elements and determinant one. Such matrices form a group under multiplication and leave the magnitude of an n-dimensional real vector unchanged, just as SO(3, R) left three-dimensional vectors unchanged in length. There are n^2 elements in a real $n \times n$ matrix. Orthogonality gives a set of equations that the elements must satisfy through the requirement that $\mathbf{A}^T \cdot \mathbf{A} = \mathbf{I}$. However, some of the equations given by this requirement are redundant. This is easily seen by noting that the matrix $\mathbf{B} = \mathbf{A}^T \cdot \mathbf{A}$ is symmetric, since $B_{ij} = A_{ik}A_{jk} = A_{jk}A_{ik} = B_{ji}$. This means that only half of the off-diagonal elements of $\mathbf{B}$ give a unique equation. In an $n \times n$ matrix there are $n^2 - n = n(n - 1)$

off-diagonal elements and only half of them provide a unique equation, so that the off-diagonal elements provide $\frac{1}{2}n(n-1)$ unique equations. There are n diagonal elements in $\mathbf{B}$ and these provide n unique equations. There are therefore a total of $\frac{1}{2}n(n-1)+n=\frac{1}{2}n(n+1)$ unique requirements. This reduces the number of free parameters in SO(n) for a real matrix from n^2 to

$$n^2-\frac{1}{2}n(n+1)=\frac{1}{2}n(n-1). \tag{7.122}$$

The restriction to having a determinant of one does not further reduce the number of free parameters, since the orthogonality of matrix $\mathbf{A}$ gives

$$1=\det(\mathbf{A}^T\cdot\mathbf{A})=\det\mathbf{A}^T\det\mathbf{A}=(\det\mathbf{A})^2 \implies \det\mathbf{A}=\pm 1. \tag{7.123}$$

If matrix $\mathbf{A}$ has $\det\mathbf{A}=-1$, then the matrix $-\mathbf{A}=-\mathbf{I}\cdot\mathbf{A}$ has a determinant of $+1$ for odd values of n. If n is even, then there is always a real and diagonal matrix $\mathbf{D}$ such that $\det\mathbf{D}=-1$, so that $\mathbf{D}\cdot\mathbf{A}$ has determinant of $+1$. For SO(3, R) there are three arbitrary angles, while for SO(4, R) there are six arbitrary parameters. This result plays a key role in understanding special relativity and the Lorentz group.

7.2.4 SU(n) and spinor rotations

Another important group in physics is known as the *special unitary group of order n*, SU(n), which are the complex-valued $n\times n$ matrices that are both unitary and have a determinant of one. The previous matrix analysis shows that the product of two such matrices is another unitary matrix with determinant one, and therefore this set of matrices forms a group. An arbitrary $n\times n$ complex-valued matrix has $2n^2$ parameters, since the n^2 elements can be complex, corresponding to both a real and an imaginary component. For complex elements the unitarity requirement,

$$(\mathbf{A}^\dagger\cdot\mathbf{A})_{\ell j}=A^*_{k\ell}A_{kj}=\delta_{\ell j}, \tag{7.124}$$

gives n^2 distinct equations. This follows from the fact that the n diagonal cases, where $\ell=j$ in the unitarity requirement, give a real number on the left-hand side and therefore correspond to only n constraints that must be satisfied. For the non-diagonal cases, where $\ell\neq j$, the unitarity requirement will result in a complex number,

$$A^*_{k\ell}A_{kj}=\alpha+\beta i=0 \implies \alpha=\beta=0, \tag{7.125}$$

reducing the free parameters by two. However, interchanging ℓ and j results in

$$A^*_{kj}A_{k\ell}=(A^*_{k\ell}A_{kj})^*=(\alpha+\beta i)^*=\alpha-\beta i=0 \implies \alpha=\beta=0, \tag{7.126}$$

which is the same pair of constraints. As a result, $2\times\frac{1}{2}n(n-1)=n(n-1)$ unique constraints arise from the $n(n-1)$ off-diagonal elements of the unitarity requirement. The total number of unique equations obtained from the unitarity requirement is $n+n(n-1)=n^2$. The unitarity requirement gives

$$\det(\mathbf{A}^\dagger \cdot \mathbf{A}) = \det \mathbf{I} = 1. \tag{7.127}$$

Writing $\det \mathbf{A} = \gamma + \delta i$, the definition of the determinant (7.35) and the Hermitian adjoint $(\mathbf{A}^\dagger)_{ij} = A^*_{ji}$ show that

$$\det \mathbf{A}^\dagger = (\det \mathbf{A})^* = \gamma - \delta i \implies \det \mathbf{A} \det \mathbf{A}^\dagger = (\gamma + \delta i)(\gamma - \delta i) = \gamma^2 + \delta^2 = 1, \tag{7.128}$$

so that the unitarity requirement already includes the constraint $\gamma^2 + \delta^2 = 1$. The demand that the determinant is unity therefore gives only one unique constraint, since it requires the imaginary part of the determinant, δ, to be zero. The unitarity requirement then reduces to the single constraint $\gamma^2 = 1$. The matrices of SU(n) therefore have the total number of free parameters given by

$$2n^2 - n^2 - 1 = n^2 - 1. \tag{7.129}$$

For example, SU(2) has three free parameters, just as SO(3) has three parameters. It will now be demonstrated that SU(2) has a representation that reproduces the three-dimensional rotation algebra (7.111). Using two complex numbers, α and β, a general member of SU(2) can be written as the 2×2 matrix

$$\mathbf{U} = \begin{pmatrix} \alpha & \frac{1}{2}\beta \\ -\frac{1}{2}\beta^* & \alpha^* \end{pmatrix}, \tag{7.130}$$

where the factor of one-half is for later convenience.

Exercise 7.10. Verify that $\mathbf{U}^\dagger \cdot \mathbf{U} = \mathbf{I}$ and $\det \mathbf{U} = 1$ if

$$\alpha^*\alpha + \frac{1}{4}\beta^*\beta = 1. \tag{7.131}$$

Constraint (7.131) is a real-valued equation, and so it removes only one of the four parameters of α and β, reducing the number of free parameters to three, which is consistent with the $2^2 - 1 = 3$ free parameters predicted for SU(2) by (7.129). Using the polar form $\alpha = ae^{i\theta/2}$, this constraint becomes $a^2 = 1 - \frac{1}{4}(\beta_R^2 + \beta_I^2)$, where β_R and β_I are the real and imaginary parts of $\beta = \beta_R + i\beta_I$. For the infinitesimal case the quadratic terms are ignorable, so that $a = 1$ solves equation (7.131). The generators for SU(2) can now be found by treating θ, β_R, and β_I as the three infinitesimal parameters of the matrix (7.130). The infinitesimal transformation is then given by

$$\mathbf{U} \approx \begin{pmatrix} 1 + \frac{1}{2}i\theta & \frac{1}{2}\beta_R + \frac{1}{2}i\beta_I \\ -\frac{1}{2}\beta_R + \frac{1}{2}i\beta_I & 1 - \frac{1}{2}i\theta \end{pmatrix} = \mathbf{I} + \frac{1}{2}\begin{pmatrix} i\theta & \beta_R + i\beta_I \\ -\beta_R + i\beta_I & -i\theta \end{pmatrix}. \tag{7.132}$$

Employing the procedure (7.95) applied to SO(3, R) gives the generators for SU(2),

$$\begin{aligned} \mathbf{T}_1 &= \frac{\partial}{\partial \beta_I}\mathbf{U}\bigg|_{\theta,\beta=0} = \frac{1}{2}i\begin{pmatrix} 0 & 1 \\ 1 & 0 \end{pmatrix} \equiv \frac{1}{2}i\boldsymbol{\sigma}_1, \\ \mathbf{T}_2 &= \frac{\partial}{\partial \beta_R}\mathbf{U}\bigg|_{\theta,\beta=0} = \frac{1}{2}i\begin{pmatrix} 0 & -i \\ i & 0 \end{pmatrix} \equiv \frac{1}{2}i\boldsymbol{\sigma}_2, \\ \mathbf{T}_3 &= \frac{\partial}{\partial \theta}\mathbf{U}\bigg|_{\theta,\beta=0} = \frac{1}{2}i\begin{pmatrix} 1 & 0 \\ 0 & -1 \end{pmatrix} \equiv \frac{1}{2}i\boldsymbol{\sigma}_3, \end{aligned} \tag{7.133}$$

where $\boldsymbol{\sigma}_a$ are known as the *Pauli spin matrices*. The $\mathbf{T}_a$ generators are antihermitian, satisfying $\mathbf{T}_a^\dagger = -\mathbf{T}_a$, while the Pauli spin matrices are Hermitian, satisfying $\boldsymbol{\sigma}_a^\dagger = \boldsymbol{\sigma}_a$. Explicit calculation shows that the $\mathbf{T}_a$ generators satisfy

$$[\mathbf{T}_a, \mathbf{T}_b] = -\varepsilon_{abc}\mathbf{T}_c, \tag{7.134}$$

which is the same Lie algebra found in (7.111) for the generators of the rotation group SO(3, R). In addition, $\mathbf{T}_a$ and $\boldsymbol{\sigma}_a$ satisfy the anticommutators

$$-\{\mathbf{T}_a, \mathbf{T}_b\} = \frac{1}{4}\{\boldsymbol{\sigma}_a, \boldsymbol{\sigma}_b\} = \frac{1}{2}\delta_{ab}\,\mathbf{I}, \tag{7.135}$$

a property not shared by the SO(3, R) generators. The Pauli matrices satisfy the algebra given by

$$[\boldsymbol{\sigma}_a, \boldsymbol{\sigma}_b] = 2i\varepsilon_{abc}\boldsymbol{\sigma}_c, \tag{7.136}$$

which is consistent with both (7.134) and their Hermitian nature. In effect, the identification $\frac{1}{2}\boldsymbol{\sigma}_a = -i\mathbf{T}_a$ has been made, with the factor of $\frac{1}{2}$ for later convenience.

Because of its relationship to the rotation algebra $\mathfrak{so}(3)$, the group SU(2) can be used to give what is known as a *spinor representation* of rotations. The representation space is the set of two-dimensional complex vectors, known as *spinors*, which the 2 × 2 matrices of SU(2) rotate. If ξ is a spinor, then it can be written

$$\xi = \begin{pmatrix} \xi_1 \\ \xi_2 \end{pmatrix} \implies \xi^\dagger = \begin{pmatrix} \xi_1^* & \xi_2^* \end{pmatrix}, \tag{7.137}$$

where the ξ_a are the spinor components. The rotation of a spinor through the angles θ^a is given by

$$\xi' = \mathbf{U}(\theta) \cdot \xi \equiv \mathbf{U}\xi, \tag{7.138}$$

where the indication of matrix multiplication will be dropped in future notation. In complete analogy to SO(3), the rotation matrix $\mathbf{U}$ is written using the exponential map,

$$\mathbf{U} = \exp(\theta^a\mathbf{T}_a) = \exp\left(\frac{1}{2}i\theta^a\boldsymbol{\sigma}_a\right). \tag{7.139}$$

The transformation matrix $\mathbf{U}$ is unitary, since $\mathbf{T}_a^\dagger = -\mathbf{T}_a$ and therefore

$$\mathbf{U}^\dagger = \exp(\theta^a \mathbf{T}_{a)}^\dagger) = \exp(-\theta^a \mathbf{T}_{a)}) = \mathbf{U}^{-1}. \tag{7.140}$$

Both the generators $\mathbf{T}_a$ and the Pauli matrices have the property $\mathrm{Tr}(\mathbf{T}_{a)}) = \mathrm{Tr}(\boldsymbol{\sigma}_{a)}) = 0$, and so theorem (7.77) again shows that

$$\det \mathbf{U} = \exp(\mathrm{Tr}\ln \mathbf{U}) = \exp(\mathrm{Tr}(\theta^a \mathbf{T}_{a)})) = \exp(0) = 1, \tag{7.141}$$

verifying that $\mathbf{U}$ is a member of SU(2). By virtue of unitarity the transformed spinor satisfies

$$\xi_1^{*\prime}\xi_1^\prime + \xi_2^{*\prime}\xi_2^\prime = \xi^{\prime\dagger}\xi^\prime = \xi^\dagger \mathbf{U}^\dagger \mathbf{U}\xi = \xi^\dagger\xi = \xi_1^*\xi_1 + \xi_2^*\xi_2, \tag{7.142}$$

so that the magnitude of the spinor is maintained. Although ξ can have complex elements, the bilinear $\xi^\dagger\xi$ is a real number and can therefore correspond to a physical quantity. Physical systems that are modeled using spinors typically interpret $\xi^\dagger \cdot \xi$ as a particle density, which is similar to the complex field in the Gross–Pitaevskii model of Bose–Einstein condensation analyzed in chapter 5. This will be discussed in the next chapter, where relativistic spinor field theories are constructed.

To demonstrate the spinor version of rotation it is necessary to combine a vector $\mathbf{A}$ in three-dimensional space with the Pauli matrices to form the matrix

$$\not{A} = A^j \boldsymbol{\sigma}_j. \tag{7.143}$$

This matrix has the property that its determinant is given by

$$\det \not{A} = \det\begin{pmatrix} A^3 & A^1 - iA^2 \\ A^1 + iA^2 & -A^3 \end{pmatrix} = -A^j A^j = -\mathbf{A}\cdot\mathbf{A}. \tag{7.144}$$

Therefore, $\det \not{A}$ depends only on the coordinate-independent magnitude of the vector. Because the Pauli matrices are Hermitian, it follows that

$$\not{A}^\dagger = \not{A}. \tag{7.145}$$

Using $\not{A}$ the quantity s is defined as

$$s = \xi^\dagger \cdot \not{A} \cdot \xi. \tag{7.146}$$

Exercise 7.11. Use (7.145) to show that (7.146) is a real number.

Using the unitarity of $\mathbf{U}$ the expression for s can be written as

$$s = \xi^\dagger \not{A}\xi = \xi^\dagger \mathbf{U}^\dagger \mathbf{U}\not{A}\mathbf{U}^\dagger \mathbf{U}\xi = \xi^{\prime\dagger}\mathbf{U}\not{A}\mathbf{U}^\dagger\xi^\prime. \tag{7.147}$$

It follows that the rotated version of matrix $\not{A}$ is given by

$$\not{A}^\prime = \mathbf{U}\cdot\not{A}\cdot\mathbf{U}^\dagger. \tag{7.148}$$

Transformation (7.148) represents a rotation, since (7.43) and the unitarity of **U** show that

$$\begin{aligned} -\mathbf{A}' \cdot \mathbf{A}' &= \det \not{A}' = \det(\mathbf{U} \cdot \not{A} \cdot \mathbf{U}^\dagger) \\ &= \det(\mathbf{U}^\dagger \cdot \mathbf{U}) \det \not{A} = \det \mathbf{I} \det \not{A} = \det \not{A} = -\mathbf{A} \cdot \mathbf{A}. \end{aligned} \tag{7.149}$$

The transformation preserves the magnitude of the vector **A**, which is a property of rotation.

To see this in detail a spinor rotation around the z-axis through the angle θ is performed using

$$\mathbf{U}_z(\theta) = \exp\left(\frac{1}{2} i\theta\sigma_3\right) = \sum_{n=0}^{\infty} \frac{1}{n!}\left(\frac{1}{2} i\theta\sigma_3\right)^n. \tag{7.150}$$

If n is an even integer, then

$$\left(\frac{1}{2} i\theta\sigma_3\right)^n = \left(\frac{1}{2} i\theta\right)^n \mathbf{I} = \begin{pmatrix} \left(\frac{1}{2} i\theta\right)^n & 0 \\ 0 & \left(\frac{1}{2} i\theta\right)^n \end{pmatrix}, \tag{7.151}$$

while for an odd integer n,

$$\left(\frac{1}{2} i\theta\sigma_3\right)^n = \left(\frac{1}{2} i\theta\right)^n \sigma_3 = \begin{pmatrix} \left(\frac{1}{2} i\theta\right)^n & 0 \\ 0 & -\left(\frac{1}{2} i\theta\right)^n \end{pmatrix}. \tag{7.152}$$

Using (7.151) and (7.152) in (7.150) gives

$$\mathbf{U}_z(\theta) = \begin{pmatrix} e^{\frac{1}{2} i\theta} & 0 \\ 0 & e^{-\frac{1}{2} i\theta} \end{pmatrix} \Longrightarrow \mathbf{U}_z^\dagger(\theta) = \begin{pmatrix} e^{-\frac{1}{2} i\theta} & 0 \\ 0 & e^{\frac{1}{2} i\theta} \end{pmatrix}. \tag{7.153}$$

Performing the spinor transformation (7.148) gives

$$\not{A}' = \mathbf{U}_z(\theta) \not{A} \mathbf{U}_z^\dagger(\theta) = \begin{pmatrix} A^3 & (A^1 - iA^2)e^{i\theta} \\ (A^1 + iA^2)e^{-i\theta} & -A^3 \end{pmatrix}. \tag{7.154}$$

Using $e^{i\theta} = \cos\theta + i\sin\theta$ shows that the elements of the transformed matrix $\not{A}'$ are given by

$$\begin{aligned} A'^1 + A'^2 &= (A^1\cos\theta + A^2\sin\theta) + i(A^2\cos\theta - A^1\sin\theta), \\ A'^3 &= A^3. \end{aligned} \tag{7.155}$$

The real and imaginary parts of the elements must be equal separately, so that

$$\begin{aligned} A'^1 &= A^1 \cos\theta + A^2 \sin\theta, \\ A'^2 &= A^2 \cos\theta - A^1 \sin\theta, \\ A'^3 &= A^3. \end{aligned} \tag{7.156}$$

These are identical to the previous results (7.80) and (7.81) found for the rotation of a vector around the z-axis, justifying the claim that the transformation $\mathbf{U}_z(\theta)$ represents a spinor rotation around the z-axis.

There are several unusual aspects of the spinor version of rotation. The first follows from noting that that the generators found in (7.133) are not unique. The Pauli matrices may undergo an arbitrary unitary transformation V and still obey the rotation algebra. Defining $\bar{\boldsymbol{\sigma}}_a$ as

$$\bar{\boldsymbol{\sigma}}_a = V\boldsymbol{\sigma}_a V^\dagger, \tag{7.157}$$

it follows that

$$[\bar{\boldsymbol{\sigma}}_a, \bar{\boldsymbol{\sigma}}_b] = V[\boldsymbol{\sigma}_a, \boldsymbol{\sigma}_b]V^\dagger = 2i\varepsilon_{abc}V\boldsymbol{\sigma}_c V^\dagger = 2i\varepsilon_{abc}\bar{\boldsymbol{\sigma}}_c, \tag{7.158}$$

which is identical to the original algebra (7.136). As a result, there is a manifold of possible representations, in much the same way that there is a manifold of possible three-dimensional coordinate systems. By convention, the representation given in (7.133) is used.

The second unusual aspect of the spinor version of rotation is apparent when the angle of rotation $\theta = 2\pi$ is considered. In that case, the transformation matrix (7.153) is given by

$$\mathbf{U}_z(2\pi) = -\mathbf{I} \implies \mathbf{U}_z(2\pi) \cdot \boldsymbol{\xi} = -\boldsymbol{\xi}. \tag{7.159}$$

No directly observable quantity can change sign under a rotation of 2π, since it would be required to vanish. As a result, a spinor does not correspond to a directly observable physical quantity. However, the bilinear $\boldsymbol{\xi}^\dagger \cdot \boldsymbol{\xi}$ is a real number and does not change sign under a rotation of 2π, so it can correspond to a physical quantity such as the previously mentioned particle density. The spinor version of a vector $\not{A}$ does not change sign under a rotation of 2π, and therefore does not face this dilemma of interpretation. Although rotation can be given a representation using SU(2), its representation space of two-dimensional complex spinors is fundamentally different from the representation space of three-dimensional real vectors used for SO(3, R). It is not possible to connect the SO(3, R) transformation of (7.103) to an improper rotation $-\mathbf{I}$ through a unitary transformation, since such a transformation has a determinant of -1 in the 3×3 case. However, $-\mathbf{I}$ is a member of SU(2), since it has a determinant of $+1$ in the 2×2 case. In effect, the transformations of SU(2) map the interval $[0, 4\pi]$ into two copies of the same rotation for the vector $\not{A}$. Such a representation of rotations is referred to as *unfaithful*, since it is two-to-one. However, because SU(2) has the same Lie algebra as the rotation group, SU(2) is said to be the *universal covering group* for rotation in three dimensions and is often referred to as Spin(3) due to its relation to spin angular momentum. This will be discussed in the next chapter.

7.3 SO (3,1) and the Lorentz group

The basic concepts of special relativity introduced in chapter 6 concentrated on the *Lorentz boost*. This was a transformation of spatial and time coordinates that preserved the generalized inner product for space-time events, $x_\mu x^\mu = \eta_{\mu\nu} x^\mu x^\nu = c^2t^2 - \boldsymbol{x} \cdot \boldsymbol{x}$, leaving it invariant for all inertial observers. It was parameterized by the three components of the relative spatial velocity between two inertial observers. It was shown earlier in (6.53) that successive Lorentz boosts obey the composition rule

$$\Lambda^\mu{}_\rho(v_1)\Lambda^\rho{}_\nu(v_2) = \Lambda^\mu{}_\nu(v), \tag{7.160}$$

where v was given by the velocity addition rule (6.54). This is the property of a group, so that the set of all Lorentz boosts is closed under the matrix multiplication of (7.160). However, this does not exhaust the coordinate transformations that leave $x_\mu x^\mu$ invariant, since rotation of the spatial coordinates $\boldsymbol{x}$ leaves the length squared $\boldsymbol{x} \cdot \boldsymbol{x}$ invariant. The rotation group SO(3, R) is parameterized by three angles. The Lorentz group unites both boosts and rotations into a general invariance of the inner product. The union of the two operations must therefore be parameterized by six continuous variables representing both the three boost-velocity components and the three rotation angles. In this case the representation space is comprised of the four-vectors of space-time. The Lorentz group is therefore identified as SO(4, R) adapted to the indefinite metric of Minkowski space, so that it is written SO(3, 1).

The starting point for understanding SO(3, 1) is the previously derived property (6.68) of the Lorentz boost,

$$\Lambda^\mu_\nu(v)\Lambda^\nu_\sigma(v) = \Lambda^{\rho\mu}(v)\Lambda_{\rho\sigma}(v) = \delta^\mu_\sigma. \tag{7.161}$$

Exercise 7.12. Show that the second expression follows from the first by using the Minkowski metric relation (6.72) to convert the indices.

Although the presence of both covariant and contravariant indices complicates viewing this as a matrix equation, (7.161) is essentially a statement of orthogonality adapted to the space-time metric. In effect, Lorentz boosts correspond to rotations using one space axis and the time axis. In order to clarify this it is useful to treat the boost parameters $\Lambda^\mu{}_\nu$ as the components of a 4 × 4 matrix $\boldsymbol{\Lambda}$. For simplicity, the boost will be along the z-axis, so that the boost matrix $\boldsymbol{\Lambda}$ is given explicitly by

$$\boldsymbol{\Lambda}(v) = \begin{pmatrix} \gamma_v & 0 & 0 & -\gamma_v v/c \\ 0 & 1 & 0 & 0 \\ 0 & 0 & 1 & 0 \\ -\gamma_v v/c & 0 & 0 & \gamma_v \end{pmatrix}, \tag{7.162}$$

where v is the boost velocity. Treating space-time events as a column vector x, where

$$x = \begin{pmatrix} x^0 \\ x^1 \\ x^2 \\ x^3 \end{pmatrix}, \tag{7.163}$$

the Lorentz boost (6.27) in the z-direction can be written as the matrix equation

$$x' = \mathbf{\Lambda}(v)\, x, \tag{7.164}$$

where the matrix $\mathbf{\Lambda}$ acts on the column vector x.

In order to understand the meaning of (7.161) it is useful to define the matrix $\bar{\mathbf{\Lambda}}$ as having the elements $\Lambda_{\mu\nu}$ and the matrix $\tilde{\mathbf{\Lambda}}$ as having the elements $\Lambda^{\mu\nu}$. Using these matrices the relationship (7.161) can be expressed as a matrix relation,

$$\tilde{\mathbf{\Lambda}}^T(v)\bar{\mathbf{\Lambda}}(v) = \mathbf{I}. \tag{7.165}$$

In order to relate $\mathbf{\Lambda}$ to these two matrices, the matrix version of the metric tensor is required, which is given by

$$\boldsymbol{\eta} = \begin{pmatrix} 1 & 0 & 0 & 0 \\ 0 & -1 & 0 & 0 \\ 0 & 0 & -1 & 0 \\ 0 & 0 & 0 & -1 \end{pmatrix}. \tag{7.166}$$

The metric matrix (7.166) gives the usual Minkowski inner product by combining (7.163) with the definition

$$x^2 = x \cdot x = x^T\boldsymbol{\eta}\, x. \tag{7.167}$$

It follows from

$$\Lambda_{\mu\nu} = \eta_{\mu\rho}\Lambda^{\rho}_{\nu}, \quad \Lambda^{\mu\nu} = \eta^{\nu\rho}\Lambda^{\mu}_{\rho}, \tag{7.168}$$

that the two matrices $\tilde{\mathbf{\Lambda}}$ and $\bar{\mathbf{\Lambda}}$ are related to $\mathbf{\Lambda}$ by

$$\tilde{\mathbf{\Lambda}} = \mathbf{\Lambda}\boldsymbol{\eta}^T, \quad \bar{\mathbf{\Lambda}} = \boldsymbol{\eta}\mathbf{\Lambda}. \tag{7.169}$$

The matrix condition (7.165) becomes

$$\tilde{\mathbf{\Lambda}}^T\bar{\mathbf{\Lambda}} = (\mathbf{\Lambda}\boldsymbol{\eta}^T)^T\boldsymbol{\eta}\mathbf{\Lambda} = \boldsymbol{\eta}\mathbf{\Lambda}^T\boldsymbol{\eta}\mathbf{\Lambda} = \mathbf{I}. \tag{7.170}$$

Exercise 7.13. Using the boost matrix (7.162), verify that (7.170) holds.

Result (7.170) is the expression of orthogonality consistent with the Minkowski metric.

Using $\det\boldsymbol{\eta} = -1$, it follows from (7.170) that

$$\det(\eta\mathbf{\Lambda}^T\eta\mathbf{\Lambda}) = \det\eta \det\mathbf{\Lambda}^T \det\eta \det\mathbf{\Lambda} = \det\mathbf{\Lambda}^T \det\mathbf{\Lambda} = (\det\mathbf{\Lambda})^2 = \det\mathbf{I} = 1. \tag{7.171}$$

Like the orthogonal matrices the Lorentz boost satisfies

$$\det\mathbf{\Lambda} = \pm 1. \tag{7.172}$$

A *proper Lorentz transformation* satisfies the condition

$$\det\mathbf{\Lambda} = 1. \tag{7.173}$$

For this choice it follows that the product of two distinct Lorentz boosts satisfies

$$\det(\mathbf{\Lambda}(v')\mathbf{\Lambda}(v)) = \det\mathbf{\Lambda}(v')\det\mathbf{\Lambda}(v) = 1, \tag{7.174}$$

and therefore forms a group that is consistent with the Minkowski metric. This property extends to the full set of Lorentz transformations that include both boosts and rotations.

Exercise 7.14. Verify that the $\mathbf{\Lambda}$ given by (7.162) satisfies $\det\mathbf{\Lambda} = 1$.

Condition (7.173) excludes Lorentz transformations with the determinant -1. Examples of such transformations are given by combining $\Lambda^0{}_0 = 1$ with $\Lambda^j{}_k = -\delta^j{}_k$ or combining $\Lambda^0{}_0 = -1$ with $\Lambda^j{}_k = \delta^j{}_k$, so that both combinations have $\det\mathbf{\Lambda} = -1$. The first corresponds to a parity transformation of the spatial coordinates, $\boldsymbol{x} \to -\boldsymbol{x}$, while the second corresponds to time reversal, $t \to -t$. Neither of these *discrete transformations* can be smoothly connected back to the identity transformation $\mathbf{I}$ by an orthogonal transformation. The set of *proper Lorentz transformations* excludes both parity and time reversal transformations and forms the group SO(3, 1), which satisfies (7.174). The inclusion of parity or time reversal as a symmetry is dependent upon the physical phenomenon being modeled.

Since relativistic physics is intertwined with the properties of the Lorentz group, it is very useful to understand the basic properties of this group. The group SO(3, 1) has the same number of free parameters as SO(4), so that a proper Lorentz transformation has $\frac{1}{2}\cdot 4\cdot 3 = 6$ free parameters. The form of the parameters for SO(3, 1) can be deduced by looking at an infinitesimal Lorentz transformation,

$$\Lambda^\rho{}_\mu \approx \delta^\rho{}_\mu + \lambda^\rho{}_\mu. \tag{7.175}$$

Inserting this into the fundamental requirement for Lorentz boosts (6.64) gives, to $O(\lambda)$,

$$\eta_{\mu\nu} = \Lambda^\rho_\mu\Lambda^\sigma_\nu\eta_{\rho\sigma} \approx \delta^\rho_\mu\delta^\sigma_\nu\eta_{\rho\sigma} + \delta^\rho_\mu\lambda^\sigma_\nu\eta_{\rho\sigma} + \lambda^\rho_\mu\delta^\sigma_\nu\eta_{\rho\sigma} = \eta_{\mu\nu} + \lambda_{\nu\mu} + \lambda_{\mu\nu}, \tag{7.176}$$

which shows that the infinitesimal parameters $\lambda_{\mu\nu}$ must obey

$$\lambda_{\mu\nu} = -\lambda_{\nu\mu}. \tag{7.177}$$

The six infinitesimal parameters of SO(3, 1) can therefore be associated with an antisymmetric 4×4 matrix which has the components $\lambda_{\mu\nu}$.

For the case that the representation space consists of the four-vectors $x = (x^0, \boldsymbol{x})$ of space-time, where $\boldsymbol{x}$ is a spatial vector in a Cartesian manifold, three of the six parameters correspond to three angles θ^k found for the rotation of the spatial vector $\boldsymbol{x}$ in (7.104). The remaining three dimensionless parameters correspond to the three components of the boost velocity v/c. These are redefined in terms of the parameter ϕ^j,

$$\tanh \phi^j = \frac{v^j}{c}, \tag{7.178}$$

where the three ϕ^j are referred to as the components of the *rapidity* $\boldsymbol{\phi}$. The rapidity encapsulates the *hyperbolic geometry* of space-time. In one spatial dimension $v/c = \tanh \phi$, and combining this with the hyperbolic identities

$$\cosh^2 \phi - \sinh^2 \phi = 1, \quad \tanh \phi = \frac{\sinh \phi}{\cosh \phi}, \tag{7.179}$$

shows that

$$\begin{aligned} \gamma_v &= \frac{1}{\sqrt{1 - v^2/c^2}} = \cosh \phi, \\ \gamma_v \frac{v}{c} &= \cosh \phi \tanh \phi = \sinh \phi. \end{aligned} \tag{7.180}$$

This shows that the Lorentz boost factors have simple expressions in terms of the rapidity.

Exercise 7.15. Verify the results of (7.180).

Using the rapidity results of (7.180) allows the Lorentz boosts in one spatial dimension first found in (6.27) to be written as

$$\begin{aligned} x'^0 = ct' &= \gamma_v\left(ct - \frac{vx}{c}\right) = \cosh \phi(ct - x \tanh \phi) \\ &= x^0 \cosh \phi - x^1 \sinh \phi, \end{aligned} \tag{7.181}$$

$$\begin{aligned} x'^1 = x' &= \gamma_v(x - vt) = \cosh \phi(x - ct \tanh \phi) \\ &= x^1 \cosh \phi - x^0 \sinh \phi. \end{aligned} \tag{7.182}$$

These can be combined into a 2 × 2 matrix expression,

$$\begin{pmatrix} x'^0 \\ x'^1 \end{pmatrix} = \begin{pmatrix} \cosh\phi & -\sinh\phi \\ -\sinh\phi & \cosh\phi \end{pmatrix} \cdot \begin{pmatrix} x^0 \\ x^1 \end{pmatrix}. \tag{7.183}$$

Expression (7.183) is easily inverted to find

$$\begin{pmatrix} x^0 \\ x^1 \end{pmatrix} = \begin{pmatrix} \cosh\phi & \sinh\phi \\ \sinh\phi & \cosh\phi \end{pmatrix} \cdot \begin{pmatrix} x'^0 \\ x'^1 \end{pmatrix}. \tag{7.184}$$

The matrix form of the boost found in (7.183) can be used to find the relativistic velocity addition formula. This follows from the product of two boosts with different rapidities,

$$\begin{aligned} \mathbf{U}(\phi')\mathbf{U}(\phi) &= \begin{pmatrix} (\cosh\phi'\cosh\phi + \sinh\phi'\sinh\phi) & -(\cosh\phi'\sinh\phi + \sinh\phi'\cosh\phi) \\ -(\cosh\phi'\sinh\phi + \sinh\phi'\cosh\phi) & (\cosh\phi'\cosh\phi + \sinh\phi'\sinh\phi) \end{pmatrix} \\ &= \begin{pmatrix} \cosh(\phi'+\phi) & -\sinh(\phi'+\phi) \\ -\sinh(\phi'+\phi) & \cosh(\phi'+\phi) \end{pmatrix} = \mathbf{U}(\phi'+\phi). \end{aligned} \tag{7.185}$$

The result v_s of relativistic velocity addition is given by the property of the hyperbolic tangent,

$$v_s = c\tanh(\phi' + \phi) = \frac{c\tanh\phi' + c\tanh\phi}{1 + \tanh\phi'\tanh\phi} = \frac{v' + v}{1 + v'v/c^2}, \tag{7.186}$$

which is identical to result (6.52). This demonstrates the *hyperbolic nature* of Minkowski space-time.

The previous matrix result found in (7.29) shows that the matrix appearing in (7.183) can be written

$$\begin{pmatrix} \cosh\phi & -\sinh\phi \\ -\sinh\phi & \cosh\phi \end{pmatrix} = \exp\begin{pmatrix} 0 & -\phi \\ -\phi & 0 \end{pmatrix} = \exp\left\{\phi\begin{pmatrix} 0 & -1 \\ -1 & 0 \end{pmatrix}\right\}. \tag{7.187}$$

Result (7.187) shows that a Lorentz boost in the x^1 direction has the 4 × 4 generator

$$\mathbf{K}_x = \begin{pmatrix} 0 & -1 & 0 & 0 \\ -1 & 0 & 0 & 0 \\ 0 & 0 & 0 & 0 \\ 0 & 0 & 0 & 0 \end{pmatrix}. \tag{7.188}$$

Similarly, the two generators for boosts in the y and z directions are given by

$$\mathbf{K}_y = \begin{pmatrix} 0 & 0 & -1 & 0 \\ 0 & 0 & 0 & 0 \\ -1 & 0 & 0 & 0 \\ 0 & 0 & 0 & 0 \end{pmatrix}, \quad \mathbf{K}_z = \begin{pmatrix} 0 & 0 & 0 & -1 \\ 0 & 0 & 0 & 0 \\ 0 & 0 & 0 & 0 \\ -1 & 0 & 0 & 0 \end{pmatrix}. \tag{7.189}$$

The three remaining generators for SO(3, 1) correspond to spatial rotations, which were found earlier in (9.234). Adapting the generators given in (9.234) to act on the spatial part of a four-vector gives

$$\mathbf{R}_x = \begin{pmatrix} 0 & 0 & 0 & 0 \\ 0 & 0 & 0 & 0 \\ 0 & 0 & 0 & 1 \\ 0 & 0 & -1 & 0 \end{pmatrix}, \quad \mathbf{R}_y = \begin{pmatrix} 0 & 0 & 0 & 0 \\ 0 & 0 & 0 & -1 \\ 0 & 0 & 0 & 0 \\ 0 & 1 & 0 & 0 \end{pmatrix}, \quad \mathbf{R}_z = \begin{pmatrix} 0 & 0 & 0 & 0 \\ 0 & 0 & 1 & 0 \\ 0 & -1 & 0 & 0 \\ 0 & 0 & 0 & 0 \end{pmatrix}. \tag{7.190}$$

The exponential map for SO(3, 1) gives the proper Lorentz transformation in terms of the rapidity $\boldsymbol{\phi}$ and the rotation angles $\boldsymbol{\theta}$,

$$\mathrm{U}(\boldsymbol{\phi}, \boldsymbol{\theta}) = \exp\left(\phi^j \mathbf{K}_j + \theta^j \mathbf{R}_j\right) \equiv \exp\left(\boldsymbol{\phi} \cdot \mathbf{K} + i\boldsymbol{\theta} \cdot \mathbf{L}\right), \tag{7.191}$$

where the definition $\mathbf{L}_a = -i\mathbf{R}_a$ gives a set of Hermitian matrices that satisfy the rotation algebra (7.112). It is tedious but not difficult to show that the six matrices $\mathbf{K}_a$ and $\mathbf{R}_a$ satisfy the fifteen commutation relations given by

$$[\mathbf{K}_a, \mathbf{K}_b] = \varepsilon_{abc}\mathbf{R}_c, \quad [\mathbf{K}_a, \mathbf{R}_b] = -\varepsilon_{abc}\mathbf{K}_c, \quad [\mathbf{R}_a, \mathbf{R}_b] = -\varepsilon_{abc}\mathbf{R}_c. \tag{7.192}$$

Exercise 7.16. Verify the commutation relations of (7.192).

The commutation relations of (7.192) define the *Lorentz algebra* in four space-time dimensions. Any set of mathematical objects $\mathbf{K}_a$ and $\mathbf{R}_a$ that satisfy the algebra (7.192) can be used to obtain the associated Lorentz transformation $\mathrm{U}(\boldsymbol{\phi}, \boldsymbol{\theta})$ using the expression (7.191). The representation space defines the nature of the generators that are used. It is instructive to note that the commutator of two boost generators yields a rotation, referred to as a *Wigner* or *Thomas rotation*. This gives rise to the physical effect known as *Thomas precession*, which is important in atomic physics but lies outside the scope of this text.

A different form for the Lorentz algebra can be derived by exploiting the analogy to SO(3, R). The algebra associated with SO(3, 1) can be found from the space-time differential operators, which are similar to the differential operators (7.113) discussed for the spatial rotation algebra of (7.111). These differential operators are found from the infinitesimal Lorentz transformation of a scalar function $\varphi(x)$ by using a Taylor series expansion, which gives

$$\begin{aligned} \varphi'(x') &= \varphi(\Lambda \cdot x) = \varphi(x + \lambda \cdot x) \\ &\approx \varphi(x) + \lambda^\mu_\nu x^\nu \partial_\mu \varphi(x) = \varphi(x) - \frac{1}{2}\lambda_{\mu\nu} J^{\mu\nu} \varphi(x), \end{aligned} \tag{7.193}$$

where $\mathbf{J}^{\mu\nu}$ are six antisymmetric differential operators given by

$$\mathbf{J}^{\mu\nu} = x^{\mu}\partial^{\nu} - x^{\nu}\partial^{\mu}. \tag{7.194}$$

Expression (7.193) is the infinitesimal version of the exponential map,

$$\varphi'(x') = \exp\left(-\frac{1}{2}\lambda_{\mu\nu}\mathbf{J}^{\mu\nu}\right)\varphi(x). \tag{7.195}$$

It should be noted that some authors attach a factor of i to $\mathbf{J}^{\mu\nu}$ and $\lambda_{\mu\nu}$ to compensate for the minus sign in (7.195). This was done earlier with the differential operators (7.113) for the case of SO(3, R).

The differential operators $\mathbf{J}^{\mu\nu}$ define a useful variant of the Lorentz algebra when they act on a scalar function. Using the result

$$\partial^{\nu}x^{\rho} = \eta^{\nu\sigma}\partial_{\sigma}x^{\rho} = \eta^{\nu\sigma}\delta_{\sigma}^{\rho} = \eta^{\nu\rho}, \tag{7.196}$$

it follows that

$$x^{\mu}\partial^{\nu}(x^{\rho}\partial^{\sigma}) = \eta^{\nu\rho}x^{\mu}\partial^{\sigma} + x^{\mu}x^{\rho}\partial^{\nu}\partial^{\sigma}. \tag{7.197}$$

Using (7.197) shows that the six $J^{\mu\nu}$ differential operators obey the Lorentz algebra given by their fifteen distinct commutators,

$$[\mathbf{J}^{\mu\nu}, \mathbf{J}^{\rho\sigma}] = \eta^{\nu\rho}\mathbf{J}^{\mu\sigma} - \eta^{\mu\rho}\mathbf{J}^{\nu\sigma} + \eta^{\mu\sigma}\mathbf{J}^{\nu\rho} - \eta^{\nu\sigma}\mathbf{J}^{\mu\rho}. \tag{7.198}$$

Exercise 7.17. Verify (7.198).

Like the $\mathbf{K}_a$ and $\mathbf{R}_a$ of (7.192), any set of mathematical objects $\mathbf{J}^{\mu\nu}$ that satisfies the algebra (7.198) gives a representation of the Lorentz group in four dimensions. As a result, the exponential map for these operators,

$$U(\boldsymbol{\phi}, \boldsymbol{\theta}) = \exp\left(-\frac{1}{2}\lambda_{\mu\nu}(\boldsymbol{\phi}, \boldsymbol{\theta})\mathbf{J}^{\mu\nu}\right), \tag{7.199}$$

is an SO(3, 1) transformation *once a representation space is chosen*. For example, the differential operators of (7.194) are the generators when a scalar function is chosen as the representation space. For the case of a four-vector representation space it should come as no surprise that the matrix generators for the two versions of the Lorentz algebra are related to each other,

$$\begin{aligned} \mathbf{J}^{0j} &= -\,\mathbf{J}^{j0} = -\mathbf{K}^{j} = \mathbf{K}_{j}, \\ \mathbf{J}^{jk} &= -\,\varepsilon^{jk\ell}\mathbf{R}_{\ell}. \end{aligned} \tag{7.200}$$

These generators are coupled with the parameters given by

$$\lambda_{0j} = -\lambda_{0j} = \phi_j = -\phi^j, \quad \lambda_{jk} = \varepsilon_{jk\ell}\theta^\ell. \tag{7.201}$$

Exercise 7.18. Verify that the matrix identifications of (7.200) obey the version of the Lorentz algebra given by (7.198) and that combining them with the parameters of (7.201) in the exponential map (7.199) yields a transformation identical to (7.191).

7.4 Spinor representations of the Lorentz group

Like the related spatial rotations, the Lorentz group also has spinor representations. The proper Lorentz group has six free parameters corresponding to the three angles θ^k of a rotation associated with the generators $\mathbf{R}_j$ and the three components ϕ^k of the rapidity associated with the generators $\mathbf{K}_k$. These can be related to the elements of a 2×2 complex matrix $\mathbf{U}$. Since $\mathbf{U}$ is a 2×2 complex-valued matrix, it has four complex elements yielding a total of eight free parameters. Therefore, two additional conditions are required to reduce the number to six parameters, and these two conditions are given by demanding that $\det \mathbf{U} = 1$. This requires the imaginary part of $\det \mathbf{U}$ to vanish, while the real part must be one. It also guarantees that $\mathbf{U}$ has an inverse. If $\mathbf{U}_1$ and $\mathbf{U}_2$ both have determinant unity, then so does $\mathbf{U} = \mathbf{U}_1\mathbf{U}_2$, since

$$\det \mathbf{U} = \det(\mathbf{U}_1\mathbf{U}_2) = \det \mathbf{U}_1 \det \mathbf{U}_2 = 1. \tag{7.202}$$

These matrices therefore form a group known as the *special linear group* SL(2, C), which consists of all 2×2 complex matrices with determinant one.

7.4.1 The Weyl spinor

The *Weyl spinor* is a two-component complex vector $\xi(x)$, so that

$$\xi(x) = \begin{pmatrix} \xi^1(x) \\ \xi^2(x) \end{pmatrix}, \tag{7.203}$$

where $\xi^a(x)$ are complex-valued functions. The Lorentz transformation of a Weyl spinor takes the form

$$\xi'(x') = \mathbf{U}\xi(x), \tag{7.204}$$

where $\mathbf{U}$ is member of SL(2, C) that will be constructed to implement a Lorentz transformation. The notation of (7.204) stands for the action of the 2×2 matrix $\mathbf{U}$ on the 2×1 spinor ξ, resulting in another 2×1 spinor. The notation $\xi^\dagger\xi$ stands for the spinor inner product,

$$\xi^\dagger\xi = \xi_1^*\xi_1 + \xi_2^*\xi_2. \tag{7.205}$$

Other expressions are defined contextually as either matrix multiplication or the inner product. Expression (7.204) means that a second inertial observer will not only use the

Lorentz transformed event x', but will also have to change the components of the first observer's spinor to match what the second observer will claim for the spinor.

Finding a representation of the Lorentz group for the Weyl spinor begins by augmenting the Pauli spin matrices $\boldsymbol{\sigma}_j$ defined in (7.133) with the 2×2 identity matrix $\boldsymbol{\sigma}_0 = \mathbf{I}$. The Pauli spin matrices defined in (7.133) obey the algebra (7.136),

$$[\boldsymbol{\sigma}_j, \boldsymbol{\sigma}_k] = 2i\varepsilon_{jk\ell}\boldsymbol{\sigma}_\ell, \tag{7.206}$$

where $\varepsilon_{123} = 1$. The set of four matrices can be used to define two distinct four-matrices,

$$\begin{aligned} \boldsymbol{\sigma}_\mu &= (\boldsymbol{\sigma}^0, \boldsymbol{\sigma}), \\ \tilde{\boldsymbol{\sigma}}_\mu &= (\boldsymbol{\sigma}^0, -\boldsymbol{\sigma}). \end{aligned} \tag{7.207}$$

Both of these four-matrices are Hermitian, since both $\boldsymbol{\sigma}_0$ and the Pauli spin matrices are Hermitian,

$$\begin{aligned} \boldsymbol{\sigma}_\mu^\dagger &= \boldsymbol{\sigma}_\mu, \\ \tilde{\boldsymbol{\sigma}}_\mu^\dagger &= \tilde{\boldsymbol{\sigma}}_\mu. \end{aligned} \tag{7.208}$$

These two four-spinors provide two distinct representations of the Lorentz algebra. The two four-matrices of (7.207) satisfy the useful anticommutator identity,

$$\frac{1}{2}\{\tilde{\boldsymbol{\sigma}}_\mu, \boldsymbol{\sigma}_\nu\} = \eta_{\mu\nu}\, \mathbf{I}. \tag{7.209}$$

Exercise 7.19. Use the specific forms for the $\boldsymbol{\sigma}_j$ to verify (7.209).

Either of the two four-matrices of (7.207) can be contracted with a covariant four-vector x_μ to define two distinct 2×2 matrices,

$$\begin{aligned} x^\mu \boldsymbol{\sigma}_\mu &= x^0\boldsymbol{\sigma}_0 + x^j\boldsymbol{\sigma}_j, \\ x^\mu \tilde{\boldsymbol{\sigma}}_\mu &= x^0\boldsymbol{\sigma}_0 - x^j\boldsymbol{\sigma}_j. \end{aligned} \tag{7.210}$$

Using the identity

$$\boldsymbol{\sigma}_\mu\tilde{\boldsymbol{\sigma}}_\nu = \frac{1}{2}[\boldsymbol{\sigma}_\mu, \tilde{\boldsymbol{\sigma}}_\nu] + \frac{1}{2}\{\boldsymbol{\sigma}_\mu, \tilde{\boldsymbol{\sigma}}_\nu\}, \tag{7.211}$$

along with the anticommutator (7.209), shows that the product $x^\mu\boldsymbol{\sigma}_\mu x^\nu\tilde{\boldsymbol{\sigma}}_\nu$ is given by

$$\begin{aligned} x^\mu\boldsymbol{\sigma}_\mu x^\nu\tilde{\boldsymbol{\sigma}}_\nu &= x^\mu x^\nu\frac{1}{2}[\boldsymbol{\sigma}_\mu, \tilde{\boldsymbol{\sigma}}_\nu] + x^\mu x^\nu\frac{1}{2}\{\boldsymbol{\sigma}_\mu, \tilde{\boldsymbol{\sigma}}_\nu\} = x^\mu x^\nu\frac{1}{2}\{\boldsymbol{\sigma}_\mu, \tilde{\boldsymbol{\sigma}}_\nu\} \\ &= x^\mu x^\nu\eta_{\mu\nu}\, \mathbf{I} = x_\mu x^\mu\, \mathbf{I}. \end{aligned} \tag{7.212}$$

Using the explicit forms for the Pauli spin matrices found in (7.133) gives the important result

$$\det(x^\mu \boldsymbol{\sigma}_\mu) = \det\begin{pmatrix} x^0 + x^3 & x^1 - ix^2 \\ x^1 + ix^2 & x^0 - x^3 \end{pmatrix} = (x^0)^2 - (x^1)^2 - (x^2)^2 - (x^3)^2 \\ = x_\mu x^\mu, \tag{7.213}$$

along with the identical result

$$\det(x^\mu \tilde{\boldsymbol{\sigma}}_\mu) = \det\begin{pmatrix} x^0 - x^3 & -x^1 + ix^2 \\ -x^1 - ix^2 & x^0 + x^3 \end{pmatrix} = (x^0)^2 - (x^1)^2 - (x^2)^2 - (x^3)^2 \\ = x_\mu x^\mu. \tag{7.214}$$

Both determinants therefore yield expressions that are invariant under Lorentz transformations.

Expressions such as $x^\mu \boldsymbol{\sigma}_\mu$ have the appearance of a Lorentz invariant. However, the matrices $\boldsymbol{\sigma}_\mu$, like the Kronecker delta, do *not* change under a Lorentz transformation. Instead, all inertial observers can agree upon their form before comparing observations, in much the same way that they synchronize their clocks and meter sticks. It is possible to use the Minkowski metric to define $\boldsymbol{\sigma}^\mu$ and $\tilde{\boldsymbol{\sigma}}^\mu$. These are given by

$$\boldsymbol{\sigma}^\mu = \eta^{\mu\nu}\boldsymbol{\sigma}_\nu = (\boldsymbol{\sigma}_0, -\boldsymbol{\sigma}) = \tilde{\boldsymbol{\sigma}}_\mu, \\ \tilde{\boldsymbol{\sigma}}^\mu = \eta^{\mu\nu}\tilde{\boldsymbol{\sigma}}_\nu = (\boldsymbol{\sigma}_0, \boldsymbol{\sigma}) = \boldsymbol{\sigma}_\mu, \tag{7.215}$$

so that the two choices are transformed into each other

Using the rapidity $\tanh\phi = v/c$ in the direction of the boost, the *spinor Lorentz boost* $\mathbf{U}(\phi)$ corresponding to the *inverse* space-time Lorentz boost $\Lambda^\nu_\mu(-\phi)$ is defined as the member of SL(2, C) that satisfies the relation

$$\mathbf{U}^\dagger(\phi)\boldsymbol{\sigma}_\mu \mathbf{U}(\phi) = \Lambda_\mu{}^\nu(\phi)\,\boldsymbol{\sigma}_\nu, \tag{7.216}$$

where the Lorentz boost satisfies the inverse relation of (6.70), so that

$$x^\nu = x'^\mu \Lambda^\mu_\nu(\phi). \tag{7.217}$$

This definition of $\mathbf{U}(\phi)$ has the property that

$$\mathbf{U}^\dagger(\phi)x'^\mu\boldsymbol{\sigma}_\mu\mathbf{U}(\phi) = x'^\mu\mathbf{U}^\dagger(\phi)\boldsymbol{\sigma}_\mu\mathbf{U}(\phi) = x'^\mu\Lambda^{\mu\nu}_\nu(\phi)\boldsymbol{\sigma}_\nu = x^\nu\boldsymbol{\sigma}_\nu. \tag{7.218}$$

Combining (7.214) with the property (7.43) of determinants shows that (7.218) gives

$$x_\mu x^\mu = \det(x^\mu\boldsymbol{\sigma}_\mu) = \det(\mathbf{U}^\dagger x'^\mu\boldsymbol{\sigma}_\mu\mathbf{U}) \\ = \det\mathbf{U}^\dagger\det(x'^\mu\boldsymbol{\sigma}_\mu)\det\mathbf{U} = \det(x'^\mu\boldsymbol{\sigma}_\mu) = x'_\mu x'^\mu. \tag{7.219}$$

Since the unitary transformation $\mathbf{U}$ leaves $x^\mu x_\mu$ invariant, it corresponds to a Lorentz transformation. The final step is to prove that there is a $\mathbf{U}$ that satisfies (7.216). This starts by using result (7.209) and the property of the trace to show that

$$\begin{aligned}\eta_{\mu\nu} &= \frac{1}{2}\,\mathrm{Tr}(\eta_{\mu\nu}\mathbf{I}) = \frac{1}{4}\,\mathrm{Tr}(\{\tilde{\boldsymbol{\sigma}}_\mu, \boldsymbol{\sigma}_\nu\}) \\ &= \frac{1}{4}\,\mathrm{Tr}(\tilde{\boldsymbol{\sigma}}_\mu\boldsymbol{\sigma}_\nu) + \frac{1}{4}\,\mathrm{Tr}(\boldsymbol{\sigma}_\nu\tilde{\boldsymbol{\sigma}}_\mu) = \frac{1}{2}\,\mathrm{Tr}(\tilde{\boldsymbol{\sigma}}_\mu\boldsymbol{\sigma}_\nu).\end{aligned} \tag{7.220}$$

For a given **U**, the associated inverse Lorentz transformation $\Lambda^\nu_\mu(\phi)$ can be found from the definition (7.216) by using (7.220), which gives

$$\Lambda^\nu_\mu(\phi) = \frac{1}{2}\eta^{\nu\rho}\,\mathrm{Tr}(\tilde{\boldsymbol{\sigma}}_\rho\mathbf{U}^\dagger\boldsymbol{\sigma}_\mu\mathbf{U}). \tag{7.221}$$

Result (7.221) constitutes a *mapping* from the members of the group SL(2, C) to the Lorentz group.

Exercise 7.20. Use (7.216) to derive relation (7.221).

The nature of the spinor Lorentz transformation **U** can be clarified by examining its explicit form. Rather than limiting consideration to spinor boosts, the full form for **U** is first determined by identifying the spinor equivalents for $\mathbf{K}_a$ and $\mathbf{R}_a$ that satisfy the Lorentz algebra given by (7.192) and then using them in the exponential map for $\mathbf{U}(\boldsymbol{\phi}, \boldsymbol{\theta})$ given by (7.191). In the spinor case the six Lorentz algebra generators are defined in terms of the 2 × 2 Pauli matrices,

$$\mathbf{K}_j = \frac{1}{2}\boldsymbol{\sigma}_j, \quad \mathbf{R}_j = \frac{1}{2}i\boldsymbol{\sigma}_j. \tag{7.222}$$

Using the commutation relation (7.136) for the Pauli matrices, it follows that

$$\begin{aligned}[\mathbf{K}_j, \mathbf{K}_k] &= \frac{1}{4}[\boldsymbol{\sigma}_j, \boldsymbol{\sigma}_k] = \frac{1}{2}i\varepsilon_{jk\ell}\boldsymbol{\sigma}_\ell = \varepsilon_{jk\ell}\mathbf{R}_\ell, \\ [\mathbf{K}_j, \mathbf{R}_k] &= \frac{1}{4}i[\boldsymbol{\sigma}_j, \boldsymbol{\sigma}_k] = -\frac{1}{2}\varepsilon_{jk\ell}\boldsymbol{\sigma}_\ell = -\varepsilon_{jk\ell}\mathbf{K}_\ell, \\ [\mathbf{R}_j, \mathbf{R}_k] &= -\frac{1}{4}[\boldsymbol{\sigma}_j, \boldsymbol{\sigma}_k] = -\frac{1}{2}i\varepsilon_{jk\ell}\boldsymbol{\sigma}_\ell = -\varepsilon_{jk\ell}\mathbf{R}_\ell,\end{aligned} \tag{7.223}$$

which match the form of the Lorentz algebra given by (7.192). At first glance this appears to be sleight of hand, since there are only three Pauli matrices making up six generators. The explanation lies in the fact that $\boldsymbol{\sigma}_j$ is Hermitian while $i\boldsymbol{\sigma}_j$ is not, and so the two generators are distinct. However, there is a subtlety in definition (7.222), which is that the alternate definition

$$\tilde{\mathbf{K}}_j = -\frac{1}{2}\boldsymbol{\sigma}_j, \quad \mathbf{R}_j = \frac{1}{2}i\boldsymbol{\sigma}_j, \tag{7.224}$$

also satisfies the algebra (7.223).

Exercise 7.21. Show that the generators (7.224) also satisfy the Lorentz algebra (7.223).

The resolution to the dilemma posed by the two possible representations of the Lorentz algebra lies in recalling that two four-spinors $\boldsymbol{\sigma}_\mu$ and $\tilde{\boldsymbol{\sigma}}_\mu$ were defined in (7.207). In order to satisfy the defining equation (7.218) it will now be demonstrated that the representation of (7.222) must be associated with $\boldsymbol{\sigma}_\mu$, while the representation of (7.224) must be associated with $\tilde{\boldsymbol{\sigma}}_\mu$. The two possibilities yield *inequivalent representations* of the spinor Lorentz group.

Using representation (7.222) in exponential map (7.199) gives the first spinor version of the Lorentz transformation,

$$\mathbf{U}(\boldsymbol{\phi}, \boldsymbol{\theta}) = \exp(\phi^j \mathbf{K}_j + \theta^j \mathbf{R}_j) = \exp\left\{\frac{1}{2}(i\theta^j + \phi^j)\boldsymbol{\sigma}_j\right\}. \tag{7.225}$$

The three complex parameters of SL(2, C) are given by $\phi^j + i\theta^j$ in terms of rapidity and angle of rotation. Since $\boldsymbol{\sigma}_j$ are traceless, this form for $\mathbf{U}$ yields

$$\det \mathbf{U} = \exp \operatorname{Tr} \ln \mathbf{U} = \exp \operatorname{Tr}\left\{\frac{1}{2}\boldsymbol{\sigma}_j(i\theta^j + \phi^j)\right\} = \exp(0) = 1, \tag{7.226}$$

showing that $\mathbf{U}$ belongs to SL(2, C). However, in the case of a boost, the matrix $\mathbf{U}$ is *not* unitary. Setting $\theta^k = 0$ and $\phi^j = \phi\delta^j{}_3$, the form for $\boldsymbol{\sigma}_3$ gives

$$\mathbf{U}(\phi) = \exp\left(\frac{1}{2}\phi\boldsymbol{\sigma}_3\right) = \begin{pmatrix} e^{\frac{1}{2}\phi} & 0 \\ 0 & e^{-\frac{1}{2}\phi} \end{pmatrix} \Longrightarrow \mathbf{U}^\dagger(\phi) = \begin{pmatrix} e^{\frac{1}{2}\phi} & 0 \\ 0 & e^{-\frac{1}{2}\phi} \end{pmatrix} \neq \mathbf{U}^{-1}(\phi). \tag{7.227}$$

Using this transformation shows that relation (7.216) gives

$$\begin{aligned} \mathbf{U}^\dagger x'^\mu \boldsymbol{\sigma}_\mu \mathbf{U} &= \begin{pmatrix} e^{\frac{1}{2}\phi} & 0 \\ 0 & e^{-\frac{1}{2}\phi} \end{pmatrix} \begin{pmatrix} (x'^0 + x'^3) & (x'^1 - ix'^2) \\ (x'^1 + ix'^2) & (x'^0 - x'^3) \end{pmatrix} \begin{pmatrix} e^{\frac{1}{2}\phi} & 0 \\ 0 & e^{-\frac{1}{2}\phi} \end{pmatrix} \\ &= \begin{pmatrix} e^{\phi}(x'^0 + x'^3) & (x'^1 - ix'^2) \\ (x'^1 + ix'^2) & e^{-\phi}(x'^0 - x'^3) \end{pmatrix} = \begin{pmatrix} (x^0 + x^3) & (x^1 - ix^2) \\ (x^1 + ix^2) & (x^0 - x^3) \end{pmatrix} = x^\mu \boldsymbol{\sigma}_\mu. \end{aligned} \tag{7.228}$$

The last two steps require adapting the boost equations (7.181) to the x^3 direction and inverting them, which gives

$$\begin{aligned} x^0 &= x'^0 \cosh\phi + x'^3 \sinh\phi, \\ x^1 &= x'^1, \\ x^2 &= x'^2, \\ x^3 &= x'^3 \cosh\phi + x'^0 \sinh\phi. \end{aligned} \tag{7.229}$$

Using these results shows that

$$\begin{aligned}x^0 + x^3 &= x'^0 \cosh\phi + x'^3 \sinh\phi + x'^3 \cosh\phi + x'^0 \sinh\phi \\ &= e^{\phi}(x'^0 + x'^3), \\ x^0 - x^3 &= x'^0 \cosh\phi + x'^3 \sinh\phi - x'^3 \cosh\phi - x'^0 \sinh\phi \\ &= e^{-\phi}(x'^0 - x'^3), \\ x^1 - ix^2 &= x'^1 - ix'^2.\end{aligned} \tag{7.230}$$

As a result, the required relation (7.218) holds for the choice of the Lorentz algebra representation (7.222) and $x^\mu \boldsymbol{\sigma}_\mu$. This means that the quantity

$$\xi'^\dagger x'^\mu \boldsymbol{\sigma}_\mu \xi' = \xi^\dagger \mathbf{U}^\dagger x'^\mu \boldsymbol{\sigma}_\mu \mathbf{U} \xi = \xi^\dagger x^\mu \boldsymbol{\sigma}_\mu \xi \tag{7.231}$$

is invariant under spinor Lorentz transformations.

Exercise 7.22. Use (7.227) in (7.221) to show that the mapping to Lorentz transformations gives

$$\begin{aligned}\Lambda_0{}^0(\phi) &= \frac{1}{2}\eta^{00}\,\mathrm{Tr}(\boldsymbol{\sigma}_0 \mathrm{U}^\dagger \boldsymbol{\sigma}_0 \mathrm{U}) = \cosh\phi, \\ \Lambda_3{}^0(\phi) &= \frac{1}{2}\eta^{00}\,\mathrm{Tr}(\boldsymbol{\sigma}_0 \mathrm{U}^\dagger \boldsymbol{\sigma}_3 \mathrm{U}) = \sinh\phi,\end{aligned} \tag{7.232}$$

which matches the inverse Lorentz boost of (7.229).

Relation (7.216) can be adapted to the matrices of (7.215). It follows that

$$\mathbf{U}^\dagger \boldsymbol{\sigma}^\mu \mathbf{U} = \eta^{\mu\nu} \mathbf{U}^\dagger \boldsymbol{\sigma}_\nu \mathbf{U} = \eta^{\mu\nu} \Lambda_\nu^\rho(\phi)\, \boldsymbol{\sigma}_\rho = \eta^{\mu\nu} \Lambda_\nu^\rho(\phi)\, \eta_{\rho\beta} \boldsymbol{\sigma}^\beta = \Lambda_\beta^\mu(\phi)\, \boldsymbol{\sigma}^\beta. \tag{7.233}$$

This shows that $\xi^\dagger \boldsymbol{\sigma}^\mu \partial_\mu \xi$ is Lorentz invariant, since

$$\mathbf{U}^\dagger \boldsymbol{\sigma}^\mu \partial'_\mu \mathbf{U} = \Lambda_\beta^\mu(\phi) \boldsymbol{\sigma}^\beta \partial'_\mu = \Lambda_\beta^\mu(\phi) \boldsymbol{\sigma}^\beta \Lambda_\mu^\nu(\phi) \partial_\nu = \delta_\beta^\nu\, \boldsymbol{\sigma}^\beta \partial_\nu = \boldsymbol{\sigma}^\nu \partial_\nu. \tag{7.234}$$

Similarly, the second representation (7.224) can be used to find the spinor Lorentz transformation $\tilde{\mathbf{U}}$ given by

$$\begin{aligned}\tilde{\mathbf{U}} &= \exp(\phi^k \tilde{\mathbf{K}}_k + \theta^k \mathbf{R}_k) = \exp\!\left(-\frac{1}{2}\phi^k \boldsymbol{\sigma}_k + \frac{1}{2} i\theta^k \boldsymbol{\sigma}_k\right) \\ &= \exp\left\{\frac{1}{2}(i\theta^k - \phi^k)\boldsymbol{\sigma}_k\right\}.\end{aligned} \tag{7.235}$$

The spinors transformed by $\tilde{\mathbf{U}}$ are denoted by $\tilde{\xi}$, so that

$$\tilde{\xi}'(x') = \tilde{\mathbf{U}}\tilde{\xi}(x). \tag{7.236}$$

It is left as an exercise to show that

$$\tilde{\xi}'^{\dagger} x'^{\mu} \tilde{\boldsymbol{\sigma}}_{\mu} \tilde{\xi}' = \tilde{\xi}^{\dagger} \tilde{\mathbf{U}}^{\dagger} x'^{\mu} \tilde{\boldsymbol{\sigma}}_{\mu} \tilde{\mathbf{U}} \tilde{\xi} = \tilde{\xi}^{\dagger} x^{\mu} \tilde{\boldsymbol{\sigma}}_{\mu} \tilde{\xi}, \tag{7.237}$$

which provides a second Lorentz scalar.

Exercise 7.23. Verify (7.237).

In the spinor representation the Lorentz group is mapped into two distinct spinor versions. It is impossible to find a unitary transformation between $\mathbf{U}$ and $\tilde{\mathbf{U}}$. This is similar to the double covering of rotation that was found in the spinor version (7.159). In that regard, it is important to note that $\mathbf{U}$ and $\tilde{\mathbf{U}}$ are *not* unitary. This follows from examining the representation for $\mathbf{U}$ given by (7.225), which shows that

$$\mathbf{U}^{\dagger} = \exp\left(\phi^{j}\boldsymbol{\sigma}_{j} - i\theta^{j}\boldsymbol{\sigma}_{j}\right) \neq \mathbf{U}^{-1}. \tag{7.238}$$

A similar result follows for the representation of $\tilde{\mathbf{U}}$ given by (7.235). Because neither $\mathbf{U}$ nor $\tilde{\mathbf{U}}$ is unitary, it follows that

$$\xi'^{\dagger}\xi' = \xi^{\dagger}\mathbf{U}^{\dagger}\mathbf{U}\xi \neq \xi^{\dagger}\xi, \tag{7.239}$$

which shows that $\xi^{\dagger}\xi$ is *not* Lorentz invariant. A similar result follows for $\tilde{\xi}^{\dagger}\tilde{\xi}$. However, using the result

$$\tilde{\mathbf{U}}^{\dagger} = \exp\left\{-\frac{1}{2}(\phi^{k} + i\theta^{k})\boldsymbol{\sigma}_{k}\right\}, \tag{7.240}$$

a simple comparison to (7.225) shows that

$$\tilde{\mathbf{U}}^{\dagger} = \mathbf{U}^{-1} \implies (\tilde{\mathbf{U}}^{\dagger}\mathbf{U})^{\dagger} = \mathbf{U}^{\dagger}\tilde{\mathbf{U}} = \mathbf{I} \implies \mathbf{U}^{\dagger} = \tilde{\mathbf{U}}^{-1}. \tag{7.241}$$

As a result, both $\tilde{\xi}^{\dagger}\xi$ and $\xi^{\dagger}\tilde{\xi}$ are Lorentz invariant. This follows from using (7.241) to obtain

$$\xi'^{\dagger}\tilde{\xi}' = \xi^{\dagger}\mathbf{U}^{\dagger}\tilde{\mathbf{U}}\tilde{\xi} = \xi^{\dagger}\tilde{\xi}, \tag{7.242}$$

with a similar result for $\tilde{\xi}^{\dagger}\xi$.

7.4.2 The Dirac spinor

Result (7.242) allows the construction of the *bispinor* or *Dirac spinor* ψ using both representations of the Weyl spinor,

$$\psi = \begin{pmatrix} \xi \\ \bar{\xi} \end{pmatrix}. \tag{7.243}$$

The Dirac spinor has four components corresponding to the two possible spinor representations of the Lorentz algebra. The 4×4 matrix γ_0 is defined as

$$\gamma_0 = \begin{pmatrix} 0 & \mathbf{I} \\ \mathbf{I} & 0 \end{pmatrix} = \begin{pmatrix} 0 & \boldsymbol{\sigma}_0 \\ \boldsymbol{\sigma}_0 & 0 \end{pmatrix}, \tag{7.244}$$

where the entries in (7.244) are 2×2 matrices. Using γ_0 the *adjoint Dirac spinor* $\bar{\psi}$ is defined as

$$\bar{\psi} = \psi^{\dagger}\gamma_0. \tag{7.245}$$

Using $\bar{\psi}$ and ψ it follows that

$$\bar{\psi}\psi = \psi^{\dagger}\gamma_0\psi = \begin{pmatrix} \xi^{\dagger} & \bar{\xi}^{\dagger} \end{pmatrix}\begin{pmatrix} 0 & \mathbf{I} \\ \mathbf{I} & 0 \end{pmatrix}\begin{pmatrix} \xi \\ \bar{\xi} \end{pmatrix} = \xi^{\dagger}\bar{\xi} + \bar{\xi}^{\dagger}\xi. \tag{7.246}$$

Result (7.242) shows that both of the final two terms in (7.246) are Lorentz invariant, and so $\bar{\psi}\psi$ is a Lorentz invariant.

Defining four 4×4 γ_μ matrices is achieved by using the earlier definitions (7.207), so that

$$\gamma_\mu = \begin{pmatrix} 0 & \bar{\boldsymbol{\sigma}}_\mu \\ \boldsymbol{\sigma}_\mu & 0 \end{pmatrix}. \tag{7.247}$$

The matrices of (7.247) are known as the *Weyl or chiral representation* of the *Dirac gamma matrices*. The definition (7.247) gives

$$\gamma_0 = \begin{pmatrix} 0 & \boldsymbol{\sigma}_0 \\ \boldsymbol{\sigma}_0 & 0 \end{pmatrix}, \quad \gamma_j = \begin{pmatrix} 0 & -\boldsymbol{\sigma}_j \\ \boldsymbol{\sigma}_j & 0 \end{pmatrix}. \tag{7.248}$$

The matrices of (7.248) have the property that their anticommutator satisfies

$$\frac{1}{2}\{\gamma_\mu, \gamma_\nu\} = \eta_{\mu\nu}\mathbf{I}. \tag{7.249}$$

For example,

$$\begin{aligned} \{\gamma_j, \gamma_k\} &= \begin{pmatrix} 0 & -\boldsymbol{\sigma}_j \\ \boldsymbol{\sigma}_j & 0 \end{pmatrix}\begin{pmatrix} 0 & -\boldsymbol{\sigma}_k \\ \boldsymbol{\sigma}_k & 0 \end{pmatrix} + \begin{pmatrix} 0 & -\boldsymbol{\sigma}_k \\ \boldsymbol{\sigma}_k & 0 \end{pmatrix}\begin{pmatrix} 0 & -\boldsymbol{\sigma}_j \\ \boldsymbol{\sigma}_j & 0 \end{pmatrix} \\ &= \begin{pmatrix} -\{\boldsymbol{\sigma}_j, \boldsymbol{\sigma}_k\} & 0 \\ 0 & -\{\boldsymbol{\sigma}_j, \boldsymbol{\sigma}_k\} \end{pmatrix} = 2\,\eta_{jk}\begin{pmatrix} \mathbf{I} & 0 \\ 0 & \mathbf{I} \end{pmatrix} = 2\,\eta_{jk}\,\mathbf{I}, \end{aligned} \tag{7.250}$$

which follows from the property of the Pauli matrices given by (7.135). It is useful to note that γ_0 is Hermitian, $\gamma_0^{\dagger} = \gamma_0$, while the γ_j are antihermitian, $\gamma_j^{\dagger} = -\gamma_j$. The representation (7.247) is not unique and can be unitarily transformed, which preserves the algebra (7.249) of the gamma matrices.

It is possible to define the set of four matrices γ^μ in the usual manner,

$$\gamma^\mu = \eta^{\mu\rho}\gamma_\rho = \begin{pmatrix} 0 & \eta^{\mu\rho}\tilde{\boldsymbol{\sigma}}_\rho \\ \eta^{\mu\rho}\boldsymbol{\sigma}_\rho & 0 \end{pmatrix} = \begin{pmatrix} 0 & \tilde{\boldsymbol{\sigma}}^\mu \\ \boldsymbol{\sigma}^\mu & 0 \end{pmatrix} = \begin{pmatrix} 0 & \boldsymbol{\sigma}_\mu \\ \tilde{\boldsymbol{\sigma}}_\mu & 0 \end{pmatrix}. \tag{7.251}$$

It follows that γ^μ and γ_μ are related by

$$\gamma^0 = \gamma_0, \quad \gamma^j = -\gamma_j. \tag{7.252}$$

The γ^μ matrices satisfy the relation similar to (7.249),

$$\frac{1}{2}\{\gamma^\mu, \gamma^\nu\} = \frac{1}{2}\eta^{\mu\rho}\{\gamma_\rho, \gamma_\sigma\}\eta^{\sigma\nu} = \eta^{\mu\rho}\eta_{\rho\sigma}\eta^{\sigma\nu}\mathbf{I} = \eta^{\mu\nu}\mathbf{I}. \tag{7.253}$$

The contravariant components of a four-vector A can be combined with the Dirac matrices, similarly to the case of rotation, by defining

$$\not{A} = A^\mu\gamma_\mu. \tag{7.254}$$

Combining this with the Dirac adjoint shows that

$$\begin{aligned} \bar{\psi}\not{A}\psi &= \begin{pmatrix} \xi^\dagger & \tilde{\xi}^\dagger \end{pmatrix}\begin{pmatrix} 0 & \mathbf{I} \\ \mathbf{I} & 0 \end{pmatrix}\begin{pmatrix} 0 & A^\mu\tilde{\boldsymbol{\sigma}}_\mu \\ A^\mu\boldsymbol{\sigma}_\mu & 0 \end{pmatrix}\begin{pmatrix} \xi \\ \tilde{\xi} \end{pmatrix} \\ &= \begin{pmatrix} \xi^\dagger & \tilde{\xi}^\dagger \end{pmatrix}\begin{pmatrix} A^\mu\boldsymbol{\sigma}_\mu & 0 \\ 0 & A^\mu\tilde{\boldsymbol{\sigma}}_\mu \end{pmatrix}\begin{pmatrix} \xi \\ \tilde{\xi} \end{pmatrix} = \xi^\dagger A^\mu\boldsymbol{\sigma}_\mu\xi + \tilde{\xi}^\dagger A^\mu\tilde{\boldsymbol{\sigma}}_\mu\tilde{\xi}. \end{aligned} \tag{7.255}$$

The two Weyl spinor Lorentz invariants found in (7.231) and (7.237) have reappeared. As a result, the Lorentz transformation matrix for the Dirac spinor ψ is found by combining the two Weyl transformations $\mathbf{U}$ and $\tilde{\mathbf{U}}$ of (7.225) and (7.235) into the 4×4 *block diagonal matrix* $\mathbf{S}$ given by

$$\mathbf{S} = \begin{pmatrix} \mathbf{U} & 0 \\ 0 & \tilde{\mathbf{U}} \end{pmatrix}, \tag{7.256}$$

since this gives the desired result

$$\mathbf{S}\psi = \begin{pmatrix} \mathbf{U} & 0 \\ 0 & \tilde{\mathbf{U}} \end{pmatrix}\begin{pmatrix} \xi \\ \tilde{\xi} \end{pmatrix} = \begin{pmatrix} \mathbf{U}\xi \\ \tilde{\mathbf{U}}\tilde{\xi} \end{pmatrix} = \begin{pmatrix} \xi' \\ \tilde{\xi}' \end{pmatrix} = \psi'. \tag{7.257}$$

Using $\gamma_0\gamma_0 = \mathbf{I}$, the Lorentz transformation of the adjoint Dirac spinor is given by

$$\bar{\psi}' = \psi^\dagger\mathbf{S}^\dagger\gamma_0 = \psi^\dagger\gamma_0\gamma_0\mathbf{S}^\dagger\gamma_0 = \bar{\psi}\mathbf{S}^{-1}, \tag{7.258}$$

which follows from (7.241),

$$\gamma_0\mathbf{S}^\dagger\gamma_0 = \begin{pmatrix} 0 & \mathbf{I} \\ \mathbf{I} & 0 \end{pmatrix}\begin{pmatrix} \mathbf{U}^\dagger & 0 \\ 0 & \tilde{\mathbf{U}}^\dagger \end{pmatrix}\begin{pmatrix} 0 & \mathbf{I} \\ \mathbf{I} & 0 \end{pmatrix} = \begin{pmatrix} \tilde{\mathbf{U}}^\dagger & 0 \\ 0 & \mathbf{U}^\dagger \end{pmatrix} = \begin{pmatrix} \mathbf{U}^{-1} & 0 \\ 0 & \tilde{\mathbf{U}}^{-1} \end{pmatrix} = \mathbf{S}^{-1}. \tag{7.259}$$

Both of the Dirac invariants (7.246) and (7.255) will be used to develop the Dirac equation in the next chapter.

It will also be useful for future applications to show how the Lorentz algebra of (7.198) can be expressed in terms of the Dirac matrices. The generators $\mathbf{J}^{\mu\nu}$ of (7.198) are defined as

$$\mathbf{J}^{\mu\nu} = \frac{1}{4}[\gamma^{\mu}, \gamma^{\nu}], \tag{7.260}$$

which has the required antisymmetry. Demonstrating that (7.260) satisfies the Lorentz algebra (7.198) regardless of the representation for γ^{μ} begins by using (7.12) and (7.253) to show that

$$\begin{aligned}[\mathbf{J}^{\mu\nu}, \gamma^{\rho}] &= \frac{1}{4}[\gamma^{\rho}, \gamma^{\nu}\gamma^{\mu}] - \frac{1}{4}[\gamma^{\rho}, \gamma^{\mu}\gamma^{\nu}] \\ &= \frac{1}{4}\{\gamma^{\rho}, \gamma^{\nu}\}\gamma^{\mu} - \frac{1}{4}\gamma^{\nu}\{\gamma^{\rho}, \gamma^{\mu}\} - \frac{1}{4}\{\gamma^{\rho}, \gamma^{\mu}\}\gamma^{\nu} + \frac{1}{4}\gamma^{\mu}\{\gamma^{\rho}, \gamma^{\nu}\} \\ &= \eta^{\nu\rho}\gamma^{\mu} - \eta^{\mu\rho}\gamma^{\nu}.\end{aligned} \tag{7.261}$$

The next step is to use (7.11) and (7.261) to obtain

$$\begin{aligned}[\mathbf{J}^{\mu\nu}, \mathbf{J}^{\rho\sigma}] &= \frac{1}{4}[\mathbf{J}^{\mu\nu}, \gamma^{\rho}\gamma^{\sigma}] - \frac{1}{4}[\mathbf{J}^{\mu\nu}, \gamma^{\sigma}\gamma^{\rho}] \\ &= \frac{1}{4}[\mathbf{J}^{\mu\nu}, \gamma^{\rho}]\gamma^{\sigma} + \frac{1}{4}\gamma^{\rho}[\mathbf{J}^{\mu\nu}, \gamma^{\sigma}] - \frac{1}{4}[\mathbf{J}^{\mu\nu}, \gamma^{\sigma}]\gamma^{\rho} - \frac{1}{4}\gamma^{\sigma}[\mathbf{J}^{\mu\nu}, \gamma^{\rho}] \\ &= \frac{1}{4}\eta^{\nu\rho}[\gamma^{\mu}, \gamma^{\sigma}] + \frac{1}{4}\eta^{\nu\sigma}[\gamma^{\rho}, \gamma^{\mu}] + \frac{1}{4}\eta^{\mu\sigma}[\gamma^{\nu}, \gamma^{\rho}] + \frac{1}{4}\eta^{\mu\rho}[\gamma^{\sigma}, \gamma^{\nu}] \\ &= \eta^{\nu\rho}\mathbf{J}^{\mu\sigma} + \eta^{\nu\sigma}\mathbf{J}^{\rho\mu} + \eta^{\mu\sigma}\mathbf{J}^{\nu\rho} + \eta^{\mu\rho}\mathbf{J}^{\sigma\nu} \\ &= \eta^{\nu\rho}\mathbf{J}^{\mu\sigma} - \eta^{\nu\sigma}\mathbf{J}^{\mu\rho} + \eta^{\mu\sigma}\mathbf{J}^{\nu\rho} - \eta^{\mu\rho}\mathbf{J}^{\nu\sigma},\end{aligned} \tag{7.262}$$

which is the Lorentz algebra of (7.198).

For the specific case of the Weyl or chiral representation of the γ matrices given by (7.251), the generators of (7.260) are given by

$$\begin{aligned}\mathbf{J}^{0j} &= \frac{1}{4}[\gamma^{0}, \gamma^{j}] = \frac{1}{2}\begin{pmatrix} -\boldsymbol{\sigma}^{j} & 0 \\ 0 & \boldsymbol{\sigma}^{j}\end{pmatrix}, \\ \mathbf{J}^{jk} &= \frac{1}{4}[\gamma^{j}, \gamma^{k}] = -\frac{1}{2}i\varepsilon^{jk\ell}\begin{pmatrix} \boldsymbol{\sigma}_{\ell} & 0 \\ 0 & \boldsymbol{\sigma}_{\ell}\end{pmatrix}.\end{aligned} \tag{7.263}$$

Using these in the exponential map (7.199) along with the parameters (7.201) gives the bispinor Lorentz transformation matrix

$$\mathbf{S} = \exp(-\frac{1}{2}\lambda_{\mu\nu}\mathbf{J}^{\mu\nu}) = \begin{pmatrix} \frac{1}{2}(i\boldsymbol{\theta} + \boldsymbol{\phi})\cdot\boldsymbol{\sigma} & 0 \\ 0 & \frac{1}{2}(i\boldsymbol{\theta} - \boldsymbol{\phi})\cdot\boldsymbol{\sigma}\end{pmatrix} = \begin{pmatrix} \mathbf{U} & 0 \\ 0 & \tilde{\mathbf{U}}\end{pmatrix}, \tag{7.264}$$

which matches (7.256), since $\mathbf{U}$ and $\tilde{\mathbf{U}}$ are given by (7.225) and (7.235), respectively.

In order to understand the physical meaning of the two Weyl representations used in the Dirac spinor, it is useful to examine the behavior of the Dirac spinor under the discrete transformations of time reversal and parity change. Because $x^\mu\gamma_\mu$ is central to the definition of the proper Lorentz transformation of the Dirac spinor, in the case of time reversal, this requires $\gamma_0 \to -\gamma_0$ and $\gamma_j \to \gamma_j$, while a parity change requires $\gamma_0 \to \gamma_0$ and $\gamma_j \to -\gamma_j$. These two changes correspond to two distinct unitary transformations and are accomplished by using the two 4×4 unitary matrices

$$\mathbf{T} = \begin{pmatrix} 0 & -i\mathbf{I} \\ i\mathbf{I} & 0 \end{pmatrix}, \quad \mathbf{P} = \begin{pmatrix} 0 & \mathbf{I} \\ \mathbf{I} & 0 \end{pmatrix}. \tag{7.265}$$

It is straightforward to show that they are unitary,

$$\mathbf{T}^\dagger\mathbf{T} = \mathbf{P}^\dagger\mathbf{P} = \mathbf{I}, \tag{7.266}$$

and that they satisfy the two required behaviors,

$$\begin{aligned} \mathbf{T}^\dagger\gamma_0\mathbf{T} &= -\gamma_0, \quad \mathbf{T}^\dagger\gamma_j\mathbf{T} = \gamma_j, \\ \mathbf{P}^\dagger\gamma_j\mathbf{P} &= -\gamma_j, \quad \mathbf{P}^\dagger\gamma_0\mathbf{P} = \gamma_0. \end{aligned} \tag{7.267}$$

The time-reversed Dirac spinor is given by

$$\psi_T = \mathbf{T}\psi, \tag{7.268}$$

while the parity-changed Dirac spinor is given by

$$\psi_P = \mathbf{P}\psi = \begin{pmatrix} 0 & \mathbf{I} \\ \mathbf{I} & 0 \end{pmatrix}\begin{pmatrix} \xi \\ \tilde{\xi} \end{pmatrix} = \begin{pmatrix} \tilde{\xi} \\ \xi \end{pmatrix}. \tag{7.269}$$

In the case of (7.269) the change of parity exchanges ξ and $\tilde{\xi}$, revealing that the two Weyl spinors correspond to different *handednesses*. For that reason, ξ is referred to as a left-handed Weyl spinor and $\tilde{\xi}$ is referred to as a right-handed Weyl spinor. The justification for these choices will require the solutions to the associated Dirac equation, which are discussed in the next chapter. However, it is consistent with the result

$$\mathbf{P}^\dagger\gamma_\mu\mathbf{P} = \begin{pmatrix} 0 & \mathbf{I} \\ \mathbf{I} & 0 \end{pmatrix}\begin{pmatrix} 0 & \tilde{\boldsymbol{\sigma}}_\mu \\ \boldsymbol{\sigma}_\mu & 0 \end{pmatrix}\begin{pmatrix} 0 & \mathbf{I} \\ \mathbf{I} & 0 \end{pmatrix} = \begin{pmatrix} 0 & \boldsymbol{\sigma}_\mu \\ \tilde{\boldsymbol{\sigma}}_\mu & 0 \end{pmatrix}, \tag{7.270}$$

which corresponds to the exchange $\boldsymbol{\sigma}_j \leftrightarrow -\boldsymbol{\sigma}_j$ among the spatial components of $\boldsymbol{\sigma}_\mu$. This is discussed again in the coming chapter, where the *helicity* of the spinors is defined and analyzed. It also shows that

$$\bar{\psi}_P\psi_P = \bar{\psi}\mathbf{P}^\dagger\mathbf{P}\psi = \bar{\psi}\psi, \tag{7.271}$$

with a similar statement for $\mathbf{T}$. The two transformations do not commute, instead satisfying

$$\mathbf{PT} + \mathbf{TP} = 0, \tag{7.272}$$

which is another example of the anticommutation relations that characterize spinor representations.

The gamma matrices have many properties. The first is that they obey the bispinor extension of relation (7.216),

$$\mathbf{S}^{-1}\gamma_\mu\mathbf{S} = \begin{pmatrix} \tilde{\mathbf{U}}^\dagger & 0 \\ 0 & \mathbf{U}^\dagger \end{pmatrix}\begin{pmatrix} 0 & \tilde{\boldsymbol{\sigma}}_\mu \\ \boldsymbol{\sigma}_\mu & 0 \end{pmatrix}\begin{pmatrix} \mathbf{U} & 0 \\ 0 & \tilde{\mathbf{U}} \end{pmatrix}$$
$$= \begin{pmatrix} 0 & \tilde{\mathbf{U}}^\dagger\tilde{\boldsymbol{\sigma}}_\mu\tilde{\mathbf{U}} \\ \mathbf{U}^\dagger\boldsymbol{\sigma}_\mu\mathbf{U} & 0 \end{pmatrix} = \Lambda^\nu_\mu(\phi)\begin{pmatrix} 0 & \tilde{\boldsymbol{\sigma}}_\nu \\ \boldsymbol{\sigma}_\nu & 0 \end{pmatrix} = \Lambda^\nu_\mu(\phi)\,\gamma_\nu. \tag{7.273}$$

Relation (7.273) shows that

$$\mathbf{S}^{-1}\gamma^\mu\mathbf{S} = \eta^{\mu\nu}\mathbf{S}^{-1}\gamma_\nu\mathbf{S} = \eta^{\mu\nu}\Lambda^\beta_\nu(\phi)\,\gamma_\beta = \eta^{\mu\nu}\Lambda^\beta_\nu(\phi)\,\eta_{\beta\rho}\gamma^\rho = \Lambda^\mu_\rho(\phi)\gamma^\rho. \tag{7.274}$$

Exercise 7.24. Use result (7.274) to show that

$$\mathbf{S}^{-1}\mathbf{J}^{\mu\nu}\mathbf{S} = \frac{1}{4}\mathbf{S}^{-1}[\gamma^\mu, \gamma^\nu]\mathbf{S} = \Lambda^\mu_\alpha(\phi)\Lambda^\nu_\beta(\phi)\mathbf{J}^{\alpha\beta}. \tag{7.275}$$

Result (7.274) shows that the γ^μ matrices can be combined with the gradient ∂_μ to give

$$\mathbf{S}^{-1}\gamma^\mu\partial'_\mu\mathbf{S} = \Lambda^\mu_\nu(\phi)\,\gamma^\nu\Lambda^\rho_\mu(\phi)\partial_\rho = \delta^\rho_\nu\gamma^\nu\partial_\rho = \gamma^\nu\partial_\nu, \tag{7.276}$$

so that $\bar{\psi}\gamma^\mu\partial_\mu\psi$ is a Lorentz invariant.

Using all four of the gamma matrices defined in (7.248), it is possible to define a fifth matrix γ_5,

$$\gamma_5 = -i\gamma_0\gamma_1\gamma_2\gamma_3 = i\gamma^0\gamma^1\gamma^2\gamma^3 = \begin{pmatrix} -\mathbf{I} & 0 \\ 0 & \mathbf{I} \end{pmatrix}. \tag{7.277}$$

The matrix γ_5 is invariant under a proper Lorentz transformation, since

$$\mathbf{S}^{-1}\gamma_5\mathbf{S} = \begin{pmatrix} \mathbf{U}^{-1} & 0 \\ 0 & \tilde{\mathbf{U}}^{-1} \end{pmatrix}\begin{pmatrix} -\mathbf{I} & 0 \\ 0 & \mathbf{I} \end{pmatrix}\begin{pmatrix} \mathbf{U} & 0 \\ 0 & \tilde{\mathbf{U}} \end{pmatrix} = \gamma_5, \tag{7.278}$$

and so the Dirac spinor expression $\bar{\psi}\gamma_5\psi$ is invariant under a proper Lorentz transformation. However, it is not invariant under a parity change, since

$$\mathbf{P}^\dagger\gamma_5\mathbf{P} = -\gamma_5. \tag{7.279}$$

As a result, $\bar{\psi}\gamma_5\psi$ is a Lorentz *pseudoscalar*, since it changes sign under a change in parity. Another useful property of the γ matrices is given by

$$\gamma^\dagger_\mu\gamma^\dagger_0 = \gamma_0\gamma_\mu. \tag{7.280}$$

Exercise 7.25. Demonstrate (7.280) using the Weyl representation for the γ matrices given by (7.247).

References and recommended further reading

Monographs that present linear algebra include

- Bamberg P and Sternberg S 1988 *A Course in Mathematics for Students of Physics* vol I (Cambridge: Cambridge University Press)
- Courant R and Hilbert D 1989 *Methods of Mathematical Physics* vol II (New York: Wiley)
- Byron F and Fuller R 1992 *Mathematics of Classical and Quantum Physics* (New York: Dover)
- Arfken G, Weber H and Harris F 2013 *Mathematical Methods for Physicists* 7th edn (Amsterdam: Elsevier)
- Hassani S 2013 *Mathematical Physics* 2nd edn (Berlin: Springer)

Monographs that focus on Lie groups and Lie algebras include

- Wybourne B 1974 *Classical Groups for Physicists* (New York: Wiley)
- Sternberg S 1994 *Group Theory and Physics* (Cambridge: Cambridge University Press)

Classical field theory monographs that cover Lie group theory include

- Nastase H 2019 *Classical Field Theory* (Cambridge: Cambridge University Press)

Monographs on particle physics and quantum field theory that cover Lie groups and Lie algebras include

- Kaku M 1993 *Quantum Field Theory: A Modern Introduction* (Oxford: Oxford University Press)
- Georgi H 1999 *Lie Algebras in Particle Physics* 2nd edn (Boca Raton, FL: CRC Press)

IOP Publishing

Classical Field Theory and the Stress–Energy Tensor (Second Edition)

Mark S Swanson

Chapter 8

Scalar and spinor field theories

In this chapter Lorentz invariant field theories will be developed for the case that fields are scalar or spinor quantities. The action method used in previous chapters will be developed so that the action is invariant under the action of the Lorentz group. In so doing, the action of the Lorentz group will become a *symmetry* of the action, representing an action on the fields and their arguments that leaves the action *invariant*. As a result, all inertial observers will agree on the form of the action as well as the associated Euler–Lagrange equations of motion. Lorentz invariance is now understood to be a requirement for *all fundamental field theories*. 'Fundamental field theory' means a field theory representing a fundamental force present in the world. Although treating the world as built from point particles interacting with fields has proven to be an incomplete approach to fundamental interactions, it still yields important insights. As a result, this chapter will begin by defining the conserved currents and stress–energy tensor associated with charged and massive point particles. The next step is to understand what a Lorentz invariant action requires and then develop it for the case of scalar and spinor fields. The chapter concludes with the derivation of *Noether's theorem*, which yields the conserved quantities associated with a symmetry or invariance of the action. In the case of scalar and spinor fields this allows the definition of conserved orbital angular momentum, and in the case of spinor fields it allows the definition of the intrinsic angular momentum, known as *helicity*.

8.1 Classical point particles

An important component of classical field theory is the ability to model the interactions between relativistic point particles and fields. A point particle is characterized by its four-trajectory $x_c^\mu(\tau)$, which is parameterized by the proper time τ and is a solution of the relativistic version of the second law (6.137). This is identical in spirit to (6.222), which was used to solve the case of a constant rest frame force. In some cases there is a conserved quantity associated with the particle, such

doi:10.1088/978-0-7503-3455-6ch8

© IOP Publishing Ltd 2022

as its electric charge q or its mass m. As the particle moves, it carries this conserved quantity with it, creating a conserved current density $J^{\mu}(x)$. Using the trajectory's proper velocity,

$$V_c^{\mu}(\tau) = \frac{\mathrm{d}x_c^{\mu}(\tau)}{\mathrm{d}\tau}, \tag{8.1}$$

the current density of a point charge can be written

$$\begin{aligned} J^{\mu}(\boldsymbol{x}, t) &= \int_{-\infty}^{\infty} \mathrm{d}\tau' \; q \; V_c^{\mu}(\tau') \, \delta^4(x - x_c(\tau')) \\ &= \int_{-\infty}^{\infty} \mathrm{d}\tau' \; q \; \frac{\mathrm{d}x_c^{\mu}(\tau')}{\mathrm{d}\tau'} \, \delta^4(x - x_c(\tau')). \end{aligned} \tag{8.2}$$

In expression (8.2) $\mathrm{d}\tau' = \mathrm{d}t'/\gamma_v$ and the Dirac delta is given by

$$\delta^4(x - x_c(\tau')) = \delta(c(t - t'))\delta^3(\boldsymbol{x} - \tilde{\boldsymbol{x}}_c(t')). \tag{8.3}$$

The form for (8.3) uses results (6.150) and (6.154) that show the two relations

$$\begin{aligned} x_c^0(\tau') &= \tilde{x}_c^0(t') = ct', \\ \boldsymbol{x}_c(\tau') &= \tilde{\boldsymbol{x}}_c(t'), \end{aligned} \tag{8.4}$$

correctly parameterize noninertial motion as long as the instantaneous relative velocity of the particle $\boldsymbol{v}_c(t)$ is given by

$$\boldsymbol{v}_c(t) = \frac{\mathrm{d}\tilde{\boldsymbol{x}}_c(t)}{\mathrm{d}t}. \tag{8.5}$$

Justifying expression (8.2) for the electric current density of a point charge begins by noting that it transforms as a contravariant four-vector under a proper Lorentz transformation $\boldsymbol{\Lambda}$, since both τ' and the Dirac delta are invariant. The invariance of the Dirac delta follows from property (2.61) of the Dirac delta. Combining (2.61) with result (7.37) for the Jacobian J of the Lorentz transformation gives

$$\begin{aligned} \delta^4(x' - x_c') &= \delta^4(\Lambda \cdot (x - x_c)) = \frac{1}{J}\delta^4(x - x_c) \\ &= \frac{1}{\det \Lambda}\delta^4(x - x_c) = \delta^4(x - x_c), \end{aligned} \tag{8.6}$$

since $\det \Lambda = 1$ for a proper Lorentz transformation. Using the relation

$$\mathrm{d}\tau' \; V_c^0(\tau') = \mathrm{d}\tau' \; \frac{\mathrm{d}x_c^0(\tau')}{\mathrm{d}\tau'} = \mathrm{d}x_c^0(\tau') = \mathrm{d}\tilde{x}_c^0(t') = \mathrm{d}(ct') = c \, \mathrm{d}t', \tag{8.7}$$

the expression for $J^0(\boldsymbol{x}, t)$ becomes

$$J^0(\boldsymbol{x}, t) = q \int_{-\infty}^{\infty} \mathrm{d}t' \; \delta(t - t') \, \delta^3(\boldsymbol{x} - \tilde{\boldsymbol{x}}_c(t')) = q \, \delta^3(\boldsymbol{x} - \tilde{\boldsymbol{x}}_c(t)). \tag{8.8}$$

Result (8.8) has units of charge per spatial volume and so it is the charge density of the point charge. The integral of $J^0(\boldsymbol{x}, t)$ over a spatial volume that includes the current location of the charge, $\tilde{\boldsymbol{x}}_c(t)$, yields q. The Dirac delta is singular at the location $\tilde{\boldsymbol{x}}_c(t)$ and zero everywhere else, providing an expression for charge density that is consistent with the pointlike nature of the charge. Expression (8.2) also satisfies the required electric charge conservation law $\partial_\mu J^\mu = 0$. This follows from the property of Dirac delta derivatives given by (2.53),

$$\begin{aligned}\partial_\mu J^\mu(\boldsymbol{x}, t) &= \int_{-\infty}^{\infty} \mathrm{d}\tau\, q\, V_c^\mu(\tau) \frac{\partial}{\partial x^\mu} \delta^4(x - x_c(\tau)) \\ &= -q \int_{-\infty}^{\infty} \mathrm{d}\tau \frac{\mathrm{d}x_c^\mu(\tau)}{\mathrm{d}\tau} \frac{\partial}{\partial x_c^\mu(\tau)} \delta^4(x - x_c(\tau)) \\ &= -q \int_{-\infty}^{\infty} \mathrm{d}\tau \frac{\mathrm{d}}{\mathrm{d}\tau} \delta^4(x - x_c(\tau)) \\ &= -q\delta^4(x - x_c(\tau))|_{-\infty}^{\infty} = 0.\end{aligned} \tag{8.9}$$

The Dirac delta expression resulting from the end point terms in (8.9) vanishes everywhere unless $t \to \pm\infty$, and is therefore zero for finite t. As a result $\partial_\mu J^\mu = 0$ at all finite times, and J^μ is a conserved current. Expression (8.2) generalizes to a collection of N point charges, so that the current density is given by

$$J^\mu(x) = \sum_{n=1}^{N} \int_{-\infty}^{\infty} \mathrm{d}\tau\, q^{(n)}\, V_c^{(n)\mu}(\tau)\, \delta^4(x - x_c^{(n)}(\tau)), \tag{8.10}$$

where $x_c^{(n)}(\tau)$ is the trajectory of the nth particle with electric charge $q^{(n)}$.

Similarly, a symmetric stress–energy tensor can be constructed using the classical trajectory for a massive point particle,

$$T^{\mu\nu}(\boldsymbol{x}, t) = \int_{-\infty}^{\infty} \mathrm{d}\tau'\, mc\, V_c^\mu(\tau') V_c^\nu(\tau')\, \delta^4(x - x_c(\tau')), \tag{8.11}$$

where the factor of mc is present to give $T^{\mu\nu}$ the correct units of energy per three-dimensional volume. Expression (8.11) transforms as a second-rank symmetric tensor. However, it is not necessarily conserved. Using (2.53) and (6.137) gives

$$\begin{aligned}\partial_\mu T^{\mu\nu}(x) &= \int_{-\infty}^{\infty} \mathrm{d}\tau'\, mc\, V_c^\mu(\tau') V_c^\nu(\tau')\, \partial_\mu \delta^4(x - x_c(\tau')) \\ &= = -\int_{-\infty}^{\infty} \mathrm{d}\tau'\, mc\, V_c^\nu(\tau') \frac{\mathrm{d}x_c^\mu(\tau')}{\mathrm{d}\tau'} \frac{\partial}{\partial x_c^\mu} \delta^4(x - x_c(\tau')) \\ &= -\int_{-\infty}^{\infty} \mathrm{d}\tau'\, mc\, V_c^\nu(\tau') \frac{\mathrm{d}}{\mathrm{d}\tau'} \delta^4(x - x_c(\tau')) \\ &= -\int_{-\infty}^{\infty} \mathrm{d}\tau' \frac{\mathrm{d}}{\mathrm{d}\tau'} (mc\, V_c^\nu(\tau')\, \delta^4(x - x_c(\tau'))) \\ &\quad + \int_{-\infty}^{\infty} \mathrm{d}\tau'\, mc \frac{\mathrm{d}V_c^\nu(\tau')}{\mathrm{d}\tau'} \delta^4(x - x_c(\tau')) \\ &= \int_{-\infty}^{\infty} \frac{\mathrm{d}t'}{\gamma_v} c f_c^\nu(\tau')\, \delta(c(t - t'))\delta^3(\boldsymbol{x} - \tilde{\boldsymbol{x}}_c(t')) \\ &= \frac{1}{\gamma_v} \tilde{f}_c^\nu(t)\, \delta^3(\boldsymbol{x} - \tilde{\boldsymbol{x}}_c(t)),\end{aligned} \tag{8.12}$$

where $\tilde{f}^{\nu}(t) = f^{\nu}(\tau)$ and the end point terms were discarded as they were for (8.9). The remaining term in (8.12) is a nonzero *force density* if there is an external agent creating a force on the particle. The stress–energy tensor of the solitary particle is therefore not conserved unless the particle is free. In order to formulate a conserved stress–energy tensor, the energy and momentum changes for the external agent creating the force must also be included. For a closed system the total stress–energy tensor, that of both the particle and the agent of the force, will be conserved as an expression of Newton's third law.

8.2 Lorentz invariant actions

Field theories that describe the fundamental behavior of nature at arbitrary energies and velocities approaching c must be consistent with the rules of special relativity. The physical quantities modeled by a field theory must therefore have well-defined transformation properties for *space-time* coordinate changes. The second critical requirement of a fundamental field theory is that the equations of motion for the field theory take the same form *for all inertial observers, as defined by special relativity*. This means that a field theory's equations of motion must also transform as a representation of the Lorentz group. This is often referred to as *Lorentz covariance*, but it is not directly related to covariant tensors. The first step in defining a relativistic field theory requires identifying the Lorentz invariant quantities to be used as the building blocks of the theory. This requires identifying the representation space to be used to define the Lorentz invariant quantities. The second step is to determine the equations of motion for these quantities that are both covariant and reflect experimentally observed qualities. The equations of motion are not Lorentz invariant, but they take the same form and hold in all inertial frames. The second step will hold if the equations of motion are derived from an action that is *invariant* under Lorentz transformations, so that *the action is a Lorentz scalar*. Lorentz invariance of the action determines a set of conservation laws via Noether's theorem, which is to be discussed in a subsequent section.

The first step in this process is to note that the field theory actions developed in previous chapters were given by integrations over both space and time. In a four-dimensional world this means that the action will be given by

$$S[\varphi^{\alpha}, \partial\varphi^{\alpha}] = \int_V \mathrm{d}^4x \; \mathcal{L}(\varphi^{\alpha}(x), \partial\varphi^{\alpha}(x)), \tag{8.13}$$

where $\varphi^{\alpha}(x)$ constitute a set of fields with well-defined transformation properties. In the language of group theory the fields are a representation space where the Lorentz algebra (7.198) is realized by a specific form for $J^{\mu\nu}$. However, before undertaking the process of identifying the fields, it is important to note that the Lagrangian is assumed to be a *local expression*, dependent only on x. The alternative, allowing nonlocal interactions, often leads to a breakdown in *causality* in the relativistic case, a situation where events separated by space-like distances still influence one another.

As a result, such theories will not be considered in the relativistic case. However, in the case where there are interacting fields, it is possible to create an *effective field theory* with nonlocal interactions, like the one considered in (5.93) describing the Bose–Einstein condensate. This will be discussed later when the effect of interactions is considered.

Before proceeding to the identification of the possible fields, it is important to note that d^4x is *not invariant* under an *arbitrary* change of space-time coordinates, instead picking up a factor of the Jacobian (1.12). For the case of a Lorentz transformation the elements of the matrix $\mathbf{M}$ are given by

$$y^\mu = \Lambda^\mu{}_\nu x^\nu \implies \frac{\partial y^\mu}{\partial x^\nu} = \Lambda^\mu{}_\nu, \tag{8.14}$$

which holds since $\Lambda^\mu{}_\nu$ is independent of x. Result (7.174) shows that the Jacobian of a proper Lorentz transformation is given by

$$J = \det \mathbf{\Lambda} = 1, \tag{8.15}$$

and so

$$d^4x = J\, d^4y = d^4y. \tag{8.16}$$

The measure of the integral is a scalar for a proper Lorentz transformation. For rotation this occurs because the lengths of spatial vectors are invariant. In the case of a Lorentz boost this is the result of simultaneous length contraction and time dilation effects that offset each other. It is worth noting that improper Lorentz transformations of parity and time reversal give $d^4x = -d^4y$.

In addition to Lorentz invariance, the action must have units of energy × time. Therefore, the units of the fields must combine with the units of the gradient, the units of the integration, and any *relevant* constants to yield an action with those units. For example, in chapter 3, the action for a non-relativistic stretched string was developed by combining a scalar field φ, which had units of length, with derivatives and the relevant constants of tension F and mass per unit length μ. In a relativistic theory the speed of light c will always be available and used to define $x^0 = ct$ for the purposes of Lorentz transformations. Other constants might include a relevant mass m, as well as universal constants, such as the fundamental unit of electric charge e, Coulomb's constant K, Newton's gravitational constant G, as well as a constant $\hbar$ with units of energy × time. While $\hbar$ is most commonly associated with Planck's constant and quantum effects, it is sometimes possible to use classical constants to define a factor with these units. In the case of electromagnetism the combination of Ke^2/c has units of energy × time. Combining $\hbar$ and c with a mass m allows the definition of an associated length, $\ell = \hbar/mc$, or an associated time, $T = \hbar/mc^2$. It is standard practice to *scale* the fields using a combination of constants so that the fields have units measured as an inverse power of length. An example is the electric potential ϕ, which has units of energy per unit charge. Its scaled version, $\varphi = e\phi/\hbar c$, has units of inverse length.

8.3 Relativistic scalar field theory

A relativistic field theory can be formulated using a real or complex Lorentz scalar field $\varphi(x)$ that is assumed to have units of inverse length. If $\varphi(x)$ is a Lorentz scalar, it transforms as

$$\varphi'(x') = \varphi(x). \tag{8.17}$$

This means that another inertial observer, denoted by the prime, will use the same function φ but will transform the event x into their inertial frame using a Lorentz transformation, $x' = \Lambda \cdot x$. At first glance it would appear that any power of φ could appear in the action, as long as the necessary constants are available to create the correct units for the action. There are results from quantum field theory that place limits on the possible powers of φ in order to correspond to a well-behaved theory. While any power of φ may appear in a *classical theory*, quantum field theories are limited to no higher than quartic powers. An action that is quadratic in the field variable will give an Euler–Lagrange equation that is linear in the field. Because such an equation satisfies the principle of superposition, its solutions can be used as a basis for more complicated field theories where terms higher than quadratic appear. For that reason, a quadratic action is the best starting point for constructing a field theory.

8.3.1 The real Klein–Gordon field

To begin with the scalar field will be treated as real valued, so that $\varphi^* = \varphi$. It is useful to examine how the scalar field satisfies the SO(3, 1) transformation law (7.193), which requires the scalar field to obey

$$\varphi'(x') = \varphi(x) \implies \lambda_{\mu\nu} x^\mu \partial^\nu \varphi(x) = 0, \tag{8.18}$$

where $\lambda_{\mu\nu} = -\lambda_{\nu\mu}$ are the six infinitesimal Lorentz group parameters. Requirement (8.18) is satisfied if $\varphi(x)$ is a function of the Lorentz invariant $x = x_\mu x^\mu$. The chain rule then gives

$$\begin{aligned} \lambda_{\mu\nu} x^\mu \partial^\nu \varphi(x) &= \frac{\partial \varphi(x)}{\partial x} \lambda_{\mu\nu} x^\mu (\partial^\nu x) \\ &= 2 \frac{\partial \varphi(x)}{\partial x} \lambda_{\mu\nu} x^\mu x^\nu = 0, \end{aligned} \tag{8.19}$$

which vanishes due to the antisymmetry of $\lambda_{\mu\nu}$ and the symmetry of $x^\mu x^\nu$. The second possibility is that $\varphi = \varphi(k_\mu x^\mu)$, where k_μ are the covariant components of a four-vector such as the wave vector. For such a case, the differential operators associated with the SO(3, 1) algebra do not appear in (7.193), since $k'_\mu x'^\mu = k_\mu x^\mu$ is Lorentz invariant and $\varphi'(k'_\mu x'^\mu) = \varphi(k_\mu x^\mu)$. A scalar function can therefore be a function of either $x_\mu x^\mu$ or $k_\mu x^\mu$. The equation of motion will result in the latter choice for the argument of φ.

A particularly simple nontrivial Lorentz invariant quadratic action involving a scalar field is given by

$$S = \hbar \int \mathrm{d}^4x\, \mathcal{L}(\varphi, \partial\varphi) = \hbar \int \mathrm{d}^4x\, \frac{1}{2}\, \partial_\mu\varphi(x)\, \partial^\mu\varphi(x). \tag{8.20}$$

The inverse units of length associated with φ and ∂_μ cancel the quartic units of length associated with d^4x, leaving the action with the correct physical units of $\hbar$. In the absence of a mass associated with the scalar field, there is no naturally occurring set of constants that can be used to define $\hbar$. As a result, Planck's constant $\hbar$ must be employed in defining the action of (8.20). Because $\hbar$ is simply an overall factor in (8.20), it is common practice to suppress it by working in *natural units*, where $\hbar$ is unity. Natural units will be developed as the analysis of this chapter proceeds. Using the orthogonality of Lorentz transformations found in (7.161) shows that the Lagrangian $\mathcal{L}$ is a Lorentz invariant,

$$\begin{aligned}\partial'_\mu\varphi'(x')\, \partial^{\mu'}\varphi'(x') &= \Lambda_{\mu\sigma}\, \partial^\sigma\varphi(x)\, \Lambda^{\mu\rho}\, \partial_\rho\varphi(x) \\ &= \delta^\rho{}_\sigma\, \partial_\rho\varphi(x)\, \partial^\sigma\varphi(x) = \partial_\rho\varphi(x)\, \partial^\rho\varphi(x).\end{aligned} \tag{8.21}$$

Result (8.16) has already shown that d^4x is invariant for a proper Lorentz transformation. Action (8.20) is therefore a Lorentz invariant. Applying the Euler–Lagrange equation (6.245) to the Lagrangian of (8.20) gives

$$\partial_\mu \frac{\partial\mathcal{L}}{\partial(\partial_\mu\varphi)} - \frac{\partial\mathcal{L}}{\partial\varphi} = \partial_\mu\partial^\mu\varphi(x) = \left(\frac{1}{c^2}\frac{\partial^2}{\partial t^2} - \nabla^2\right)\varphi(x) = \Box_c\varphi(x) = 0, \tag{8.22}$$

which is the wave equation with a velocity of propagation c. From this point forward $\Box$ will be understood to be the Lorentz invariant $\Box_c$. This equation of motion is *covariant*, since a Lorentz transformation gives

$$\Box'\varphi'(x') = \Box\varphi(x) = 0, \tag{8.23}$$

so that all inertial observers agree on its form.

If the field φ is associated with the inertial mass M, then it is possible to define a constant m with units of inverse length given by

$$m = \frac{Mc}{\hbar}. \tag{8.24}$$

Exercise 8.1: Verify that the m of (8.24) has units of inverse length.

Using m it is possible to add a Lorentz invariant term to the Lagrangian that is quadratic in the field φ and has the required units of inverse length to the fourth power,

$$\mathcal{L} = \frac{1}{2}\partial_\mu\varphi\, \partial^\mu\varphi - \frac{1}{2}m^2\varphi^2, \tag{8.25}$$

where the term proportional to m^2 plays the role of a quadratic potential energy. The Euler–Lagrange equation of motion gives

$$\partial_\mu \frac{\partial \mathcal{L}}{\partial(\partial_\mu \varphi)} - \frac{\partial \mathcal{L}}{\partial \varphi} = (\Box + m^2)\varphi(x) = 0, \tag{8.26}$$

which is known as the *Klein–Gordon equation*. At the time of its development the Klein–Gordon equation was a candidate for the relativistic quantum mechanical wave-function describing the electron. It was found experimentally that a scalar field was inadequate to describe the electron. This is discussed in a coming section where the Weyl and Dirac equations are developed. Since there was no experimentally known fundamental scalar field at the time, the Klein–Gordon equation appeared to have no immediate application. Subsequently, it was discovered that certain subatomic particles such as the pi meson are *approximately scalar*, in the sense that they are tightly bound pairs of quarks behaving like a scalar field, but complicated by the fact that they carry electric charge. Later, the Higgs field was proposed theoretically and then found to exist experimentally. As of this writing, the Higgs field is the only experimentally verified *fundamental* scalar field present in nature, although it is not the single real scalar field φ of (8.25). This is discussed in the final chapter of this book. Although the Higgs field can be treated as a classical field theory, ultimately it is quantum mechanical in nature and the $\hbar$ appearing in the definition (8.24) is Planck's constant.

8.3.2 Solving the Klein–Gordon equation

The Klein–Gordon equation (8.26) has wave-like solutions similar to those of the one-dimensional simple harmonic waves (3.31) for the stretched string,

$$\varphi(x) = A \sin(\boldsymbol{k} \cdot \boldsymbol{x} - \omega_k t), \tag{8.27}$$

as long as the following *dispersion relation* is obeyed,

$$\omega_k^2 = c^2(\boldsymbol{k} \cdot \boldsymbol{k} + m^2). \tag{8.28}$$

However, (8.28) has both positive and negative solutions for ω_k. The presence of the negative frequency solution is problematic *at the quantum level*, corresponding to negative energy particles. In developing the solutions to (8.26) the negative frequency solution of (8.28) will be suppressed. Following (6.96) the first step is to identify the wave vector $\boldsymbol{k}$ and the frequency ω_k as the covariant components of a four-vector k_μ, where $k_0 = \omega_k/c$. This gives

$$k \cdot x = k_\mu x^\mu = k_0 x^0 - \boldsymbol{k} \cdot \boldsymbol{x} = \omega_k t - \boldsymbol{k} \cdot \boldsymbol{x}, \tag{8.29}$$

which is manifestly Lorentz invariant.

Since the Klein–Gordon equation is linear the principle of superposition can be invoked to write a general solution to (8.26) in terms of a manifestly Lorentz invariant Fourier transform. For the case of a *real scalar field*, where $\varphi^* = \varphi$, this Fourier transform is given by

$$\varphi(\boldsymbol{x}, t) = \int \frac{d^4k}{(2\pi)^3} \delta(k^2 - m^2)\, \theta(k_0)(\tilde{\varphi}(k)e^{-ik\cdot x} + \tilde{\varphi}^*(k)e^{ik\cdot x}). \tag{8.30}$$

The presence of the Heaviside step function $\theta(k_0)$ suppresses the negative frequency solution to (8.28), since only $k_0 > 0$ contributes to (8.30). Three of the constituent factors, d^4k, $\delta(k^2 - m^2)$, and $e^{\pm ik\cdot x}$ are all obviously Lorentz invariant, since k transforms as a four-vector. The function $\tilde{\varphi}(k)$ is an arbitrary complex *Lorentz scalar function* of k with units of length. The argument of (8.19) can be adapted to show that $\tilde{\varphi}(k)$ can be an arbitrary function of $k^2 = k_\mu k^\mu$ and m and still satisfy the requirement to be a Lorentz scalar. If other four-vectors not involving x are available, the function may include them as well. The Lorentz invariance of the step function $\theta(k_0)$ is less obvious, but follows from the time-like nature of k, which gives the following inequality for the positive frequency,

$$k_\mu k^\mu = k_0^2 - \boldsymbol{k} \cdot \boldsymbol{k} = m^2c^2 > 0 \implies k_0 > |\boldsymbol{k}|. \tag{8.31}$$

After a *proper* Lorentz boost k_0 is given by

$$k_0' = \gamma_v\left(k_0 - \frac{\boldsymbol{v}}{c} \cdot \boldsymbol{k}\right). \tag{8.32}$$

Since $v/c < 1$ and $k_0 > |\boldsymbol{k}|$, it follows that $k_0' > 0$. As a result, the positivity of k_0 is invariant under the proper Lorentz group. If time reversal transformations are allowed, then it is not Lorentz invariant.

It is important to note that the k_0 appearing in (8.30) is *not* ω_k/c. Instead, the dispersion relation (8.28) is enforced by the Dirac delta, while the step function $\theta(k_0)$ ensures that only the positive frequency is selected. This follows from property (2.56) of the Dirac delta, which shows that

$$\begin{aligned}\delta(k^2 - m^2) &= \delta(k_0^2 - \boldsymbol{k} \cdot \boldsymbol{k} - m^2) = \delta(k_0^2 - \omega_k^2/c^2)\\ &= \delta((k_0 - \omega_k/c)(k_0 + \omega_k/c))\\ &= \frac{c}{2|\omega_k|}(\delta(k_0 - |\omega_k|/c) + \delta(k_0 + |\omega_k|/c)).\end{aligned} \tag{8.33}$$

As a result,

$$\delta(k^2 - m^2)\, \theta(k_0) = \frac{c}{2|\omega_k|}\, \delta(k_0 - |\omega_k|/c), \tag{8.34}$$

and this allows the k_0 integral to be performed, resulting in

$$\varphi(\boldsymbol{x}, t) = \int \frac{d^3k}{(2\pi)^3} \frac{c}{2|\omega_k|} \left(\tilde{\varphi}(\boldsymbol{k})e^{-i(\omega_k t - \boldsymbol{k}\cdot\boldsymbol{x})} + \tilde{\varphi}^*(\boldsymbol{k})e^{i(\omega_k t - \boldsymbol{k}\cdot\boldsymbol{x})}\right). \tag{8.35}$$

Result (8.35) clearly satisfies the Klein–Gordon equation (8.26) by virtue of the dispersion relation (8.28). It is also invariant under the Lorentz group. It is clearly invariant under rotations, since $\boldsymbol{k} \cdot \boldsymbol{k}$ is invariant, as is d^3k. For the case of a boost in the z direction, it follows that

$$
\begin{aligned}
k_3' &= \cosh\phi\, k_3 + \sinh\phi\, \omega_k \implies \mathrm{d}k_3' = \left(\cosh\phi + \sinh\phi\, \frac{k_3}{\omega_k}\right)\mathrm{d}k_3, \\
\omega_k' &= \cosh\phi\, \omega_k + \sinh\phi\, k_3 = \omega_k\left(\cosh\phi ++ \sinh\phi\, \frac{k_3}{\omega_k}\right).
\end{aligned}
\tag{8.36}
$$

The first expression uses $\mathrm{d}\omega_k$ as the change in ω_k under an infinitesimal Lorentz boost in the z direction. It is clear from (8.36) that $\mathrm{d}k_3/\omega_k$ is Lorentz invariant and therefore (8.35) is Lorentz invariant.

It is useful to express $\tilde{\varphi}(\boldsymbol{k})$ in terms of a function a_k that has units of the square root of a spatial volume. This can be accomplished in a Lorentz invariant way by combining $\tilde{\varphi}(k)$ with ω_k in the same manner as that used for the stress–energy tensor in (6.265). In that case, k_μ and the volume three-form $\mathrm{d}S^\mu$ were contracted into the Lorentz invariant quantity $k_\mu \mathrm{d}S^\mu$. In this case, k_μ is contracted with S^μ to give $k_\mu S^\mu$. In the rest frame this takes the form $k_0 S^0$, where S^0 is the spatial volume. Since it is defined in the rest frame, it can be boosted into all inertial frames. Therefore,

$$
\sqrt{k_\mu S^\mu} = \sqrt{k_0 S^0} = \sqrt{\frac{\omega_k}{c}}\sqrt{S^0}
\tag{8.37}
$$

defines a Lorentz scalar. It follows that the scalar $\tilde{\varphi}(\boldsymbol{k})$ can be written

$$
\tilde{\varphi}(\boldsymbol{k}) = \sqrt{\frac{2\omega_k}{c}}\, a_k,
\tag{8.38}
$$

where a_k now has units of $\sqrt{S^0}$, the square root of a spatial volume. Expression (8.35) becomes

$$
\begin{aligned}
\varphi(\boldsymbol{x}, t) &= \int \frac{\mathrm{d}^3k}{(2\pi)^3}\sqrt{\frac{c}{2\omega_k}}\left(a_k e^{-i(\omega_k t - \boldsymbol{k}\cdot\boldsymbol{x})} + a_k^* e^{i(\omega_k t - \boldsymbol{k}\cdot\boldsymbol{x})}\right) \\
&= \int \frac{\mathrm{d}^3k}{(2\pi)^3}\left(a_k f_k(x) + a_k^* f_k^*(x)\right),
\end{aligned}
\tag{8.39}
$$

where the $\boldsymbol{k}$ th *mode* is given by

$$
f_k(x) = \sqrt{\frac{c}{2\omega_k}}\, e^{-i(\omega_k t - \boldsymbol{k}\cdot\boldsymbol{x})} = \sqrt{\frac{1}{2k_o}}\, e^{-ik_\mu x^\mu},
\tag{8.40}
$$

and $k_0 = \omega_k/c$. Expression (8.39) still satisfies the Klein–Gordon equation (8.26) by virtue of the dispersion relation (8.28) appearing in the definition of the modes (8.40).

The version of the Fourier expansion given by (8.39) is often referred to as a *mode expansion* and can be viewed as a linear superposition of all possible modes. The modes satisfy a three-dimensional variant of the orthogonality condition (2.99), where the inner product is given by

$$
\begin{aligned}
\left(f_k(x), f_p(x)\right) &\equiv i \int \frac{d^3x}{(2\pi)^3}\left(f_k^*(x)\frac{\partial f_p(x)}{\partial x^0} - \frac{\partial f_k^*(x)}{\partial x^0} f_p(x)\right) \\
&= \int \frac{d^3x}{(2\pi)^3}\frac{(p_0 + k_0)}{2\sqrt{k_0 p_0}} e^{-i(\boldsymbol{k}-\boldsymbol{p})\cdot \boldsymbol{x}} e^{i(k_0-p_0)x^0} \\
&= \frac{(\omega_p + \omega_k)}{2\sqrt{\omega_k \omega_p}} \delta^3(\boldsymbol{k} - \boldsymbol{p}) e^{i(\omega_k - \omega_p)t} = \delta^3(\boldsymbol{k} - \boldsymbol{p}),
\end{aligned} \tag{8.41}
$$

where the property of the Dirac delta $(x + y)\,\delta(x - y) = 2x\,\delta(x - y)$ was used.

Exercise 8.2: Show that the form of the inner product (8.41) also gives

$$
\begin{aligned}
\left(f_k^*(x), f_p^*(x)\right) &= \delta^3(\boldsymbol{k} - \boldsymbol{p}), \\
\left(f_k(x), f_p^*(x)\right) &= \left(f_k^*(x), f_p(x)\right) = 0.
\end{aligned} \tag{8.42}
$$

The orthogonal nature of the modes immediately yields

$$
a_k = i \int d^3x \left(f_k^*(x)\, \partial_0 \varphi(x) - \partial_0 f_k^*(x)\, \varphi(x)\right). \tag{8.43}
$$

Matching the boundary conditions placed on $\varphi(x)$ requires using the method of (3.45),

$$
\begin{aligned}
\int d^3x\, e^{-i\boldsymbol{p}\cdot\boldsymbol{x}} \varphi(\boldsymbol{x}, t) &= \sqrt{\frac{c}{2\,\omega_p}}\left(a_p e^{-i\omega_p t} + a_{-p}^* e^{i\omega_p t}\right), \\
\int d^3x\, e^{-i\boldsymbol{p}\cdot\boldsymbol{x}} \dot{\varphi}(\boldsymbol{x}, t) &= -\, i\sqrt{\frac{c\,\omega_p}{2}}\left(a_p e^{-i\omega_p t} - a_{-p}^* e^{i\omega_p t}\right),
\end{aligned} \tag{8.44}
$$

where $\omega_{-p} = \omega_p$ was used. Result (8.44) provides a set of equations that can be solved for a_p to match the boundary conditions provided by $\varphi(\boldsymbol{x}, t)$ and $\dot{\varphi}(\boldsymbol{x}, t)$.

Exercise 8.3: Solve (8.44) for the case that $\varphi(\boldsymbol{x}, 0)$ and $\varphi(\boldsymbol{x}, T)$ are given as the boundary conditions.

An important property of the *wave packet* (8.39) is revealed for the case that $\tilde{f}_k$ is sharply peaked around the value $\boldsymbol{k}_p$. Using a Taylor series shows that the principal contribution to (8.39) comes from the frequencies

$$\omega_k \approx \omega_{k_p} + (\boldsymbol{k} - \boldsymbol{k}_p) \cdot \frac{\partial \omega_k}{\partial \boldsymbol{k}}\bigg|_{k = k_p} \equiv \omega_{k_p} + (\boldsymbol{k} - \boldsymbol{k}_p) \cdot \boldsymbol{v}_g, \tag{8.45}$$

where $\boldsymbol{v}_g$ is referred to as the *group velocity* and is given by

$$\boldsymbol{v}_g = \frac{\partial \omega_k}{\partial \boldsymbol{k}}\bigg|_{k = k_p} = \frac{c\, \boldsymbol{k}_p}{\sqrt{\boldsymbol{k}_p \cdot \boldsymbol{k}_p + m^2}}. \tag{8.46}$$

Substituting (8.45) into the integral (8.39) and using Ω_p to denote $\omega_{k_p} - \boldsymbol{k}_p \cdot \boldsymbol{v}_g$ gives the general result

$$\begin{aligned} \varphi(\boldsymbol{x}, t) &\approx \int \frac{\mathrm{d}^3 k}{(2\pi)^3 \sqrt{2\omega_k / c}} \left(\tilde{f}_k \, e^{ik\cdot(x - v_g t)} e^{-i\Omega t} + \tilde{f}_k^{\,*} e^{-ik\cdot(x - v_g t)} e^{i\Omega t} \right) \\ &= e^{-i\Omega_p t} \int \frac{\mathrm{d}^3 k}{(2\pi)^3 \sqrt{2\omega_k / c}} \tilde{f}_k \, e^{ik\cdot(x - v_g t)} \\ &\quad + e^{i\Omega_p t} \int \frac{\mathrm{d}^3 k}{(2\pi)^3 \sqrt{2\omega_k / c}} \tilde{f}_k^{*} \, e^{-ik\cdot(x - v_g t)} \\ &= e^{-i\Omega_p t} f(\boldsymbol{x} - \boldsymbol{v}_g t) + e^{i\Omega_p t} f^{*}(\boldsymbol{x} - \boldsymbol{v}_g t). \end{aligned} \tag{8.47}$$

For such a case the Fourier solution to the Klein–Gordon equation (8.39) describes a *wave packet* with a spatial shape given by

$$f(\boldsymbol{x}) = \int \frac{\mathrm{d}^3 k}{(2\pi)^3 \sqrt{2\omega_k / c}} \tilde{f}_k \, e^{ik\cdot x}, \tag{8.48}$$

traveling at the *group velocity* $\boldsymbol{v}_g$ while oscillating at the frequency

$$\Omega = \omega_o - \boldsymbol{k}_o \cdot \boldsymbol{v}_g = \frac{m^2 c}{\sqrt{k_o^2 + m^2}}, \tag{8.49}$$

which has a maximum value $\Omega_{\max} = mc = Mc^2/\hbar$.

This result is consistent with the relativistic limitation for massive particles, since the group velocity (8.46) approaches c only if $\boldsymbol{k}_p \to \infty$. Inverting the definition (8.46) of $\boldsymbol{v}_g$ gives

$$c\, \boldsymbol{k}_p = \frac{m\boldsymbol{v}_g}{\sqrt{1 - v_g^2/c^2}}. \tag{8.50}$$

Since $m = Mc/\hbar$, it follows that

$$\hbar\, \boldsymbol{k}_p = \frac{M\boldsymbol{v}_g}{\sqrt{1 - v_g^2/c^2}}, \tag{8.51}$$

which is the relativistic particle momentum associated with the inertial mass M used in the Klein–Gordon wave equation. This is identical to the result found by applying a Lorentz boost to the rest frame frequency $\bar{\omega} = mc$, which gives

$$\begin{aligned} k' &= \Lambda^j{}_0\, k^0 = \Lambda^j{}_0\left(\frac{\bar{\omega}}{c}\right) = \Lambda^j{}_0 m \\ &= \gamma_v \frac{mv^j}{c} = \gamma_v \frac{Mv^j}{\hbar} \implies \hbar k'^j = \frac{Mv^j}{\sqrt{1 - v^2/c^2}}. \end{aligned} \tag{8.52}$$

This is an important aspect of field theories and defines a core property of such fields that is carried over into *quantum field theory*. A relativistic scalar field with a quadratic mass term allows the identification

$$\hbar \boldsymbol{k} = \frac{M\boldsymbol{v}}{\sqrt{1 - v^2/c^2}} = \boldsymbol{p} \tag{8.53}$$

to be made, which is consistent with the relativistic mechanics of massive particles.

8.3.3 Energy and momentum in the Klein–Gordon field

The Hamiltonian for the relativistic massive scalar field is constructed by the method of (3.62) with the minor modification that the units of $\mathcal{L}$, which are energy × time over (length)^4, require an additional overall factor of c so that $c\mathcal{L}$ has the units of a spatial energy density. The canonical momentum is given by

$$\pi(x) = \frac{\partial \mathcal{L}}{\partial \dot{\varphi}(x)} = \frac{\hbar}{c^2}\frac{\partial \varphi(x)}{\partial t} = \frac{\hbar}{c}\partial_0 \varphi(x) \implies \partial_0 \varphi(x) = \frac{c}{\hbar}\pi(x). \tag{8.54}$$

The total energy of the field φ is then given by

$$\begin{aligned} H[\varphi, \pi](t) &= \int \mathrm{d}^3x\; c\,(\pi\dot{\varphi} - \mathcal{L}) \\ &= \int \mathrm{d}^3x\; \frac{1}{2}\hbar c\left(\frac{c^2}{\hbar^2}\pi^2 + \partial_i\varphi\,\partial_i\varphi + m^2\varphi^2\right) \\ &= \int \mathrm{d}^3x\; \mathcal{H}(\pi, \varphi). \end{aligned} \tag{8.55}$$

Because of the proliferation of factors of $\hbar$ and c it will be useful introduce natural units, where $\hbar = c = 1$. For the moment, the existing units will be kept for the purpose of making contact with SI units. Although the Hamiltonian is a spatial functional, the Hamiltonian (8.55) itself is time independent if it is evaluated using a solution φ_s to (8.26) since

$$\begin{aligned} \partial_0 H[\varphi, \pi](t) &= \int \mathrm{d}^3x\; \hbar c(\partial_0\varphi_s\, \partial_0^2\varphi_s + \nabla\partial_0\varphi_s \cdot \nabla\varphi_s + m^2\partial_0\varphi_s\, \varphi_s) \\ &= \int \mathrm{d}^3x\; \hbar c\; \partial_0\varphi_s(\Box + m^2)\varphi_s = 0, \end{aligned} \tag{8.56}$$

where a spatial integration by parts on the gradient was used under the assumption the fields vanish at the boundary of the spatial volume. This means that the energy can be evaluated at $t = 0$ using the general form (8.39) and the definition of the canonical momentum, written in the forms

$$\begin{aligned}\varphi(\boldsymbol{x}, 0) &= \int \frac{\mathrm{d}^3k}{(2\pi)^3}\sqrt{\frac{c}{2\omega_k}}\, e^{i\boldsymbol{k}\cdot\boldsymbol{x}}(a_k + a_{-k}^*) \\ \pi(\boldsymbol{x}, 0) &= \frac{\hbar}{c^2}\frac{\partial\varphi(\boldsymbol{x}, t)}{\partial t}\bigg|_{t=0} = -i\frac{\hbar}{c}\int \frac{\mathrm{d}^3p}{(2\pi)^3}\sqrt{\frac{\omega_p}{2c}}\, e^{i\boldsymbol{p}\cdot\boldsymbol{x}}\left(a_p - a_{-p}^*\right).\end{aligned} \tag{8.57}$$

Using these in (8.55) and canceling numerous factors of c and $\hbar$ gives

$$\begin{aligned}H = &-\hbar\int \mathrm{d}^3x\,\frac{\mathrm{d}^3k}{(2\pi)^3}\frac{\mathrm{d}^3p}{(2\pi)^3}\, e^{i(\boldsymbol{p}+\boldsymbol{k})\cdot\boldsymbol{x}}\left\{\frac{\sqrt{\omega_k\,\omega_p}}{4}(a_k - a_{-k}^*)(a_p - a_{-p}^*)\right. \\ &\left.+\frac{c^2(\boldsymbol{k}\cdot\boldsymbol{p} - m^2)}{4\sqrt{\omega_k\,\omega_p}}(a_k + a_{-k}^*)(a_p + a_{-p}^*)\right\} \\ = &-\hbar\int \frac{\mathrm{d}^3k}{(2\pi)^3}\frac{1}{4}\omega_k\left\{(a_k - a_{-k}^*)(a_{-k} - a_k^*) - (a_k + a_{-k}^*)(a_{-k} + a_k^*)\right\} \\ = &\int \frac{\mathrm{d}^3k}{(2\pi)^3}\,\hbar\omega_k\, a_k^* a_k,\end{aligned} \tag{8.58}$$

where the dispersion relation, $\boldsymbol{k}\cdot\boldsymbol{k} + m^2 = \omega_k^2/c^2$, was used, as well as the result

$$\int \frac{\mathrm{d}^3k}{(2\pi)^3}\,\hbar\omega_k\, a_{-k}^* a_{-k} = \int \frac{\mathrm{d}^3k}{(2\pi)^3}\,\hbar\omega_{-k}\, a_k^* a_k = \int \frac{\mathrm{d}^3k}{(2\pi)^3}\,\hbar\omega_k\, a_k^* a_k. \tag{8.59}$$

While (8.58) is a classical result, quantum field theory also finds that $\hbar\omega_k$ is the energy of the Klein–Gordon modes, which are associated with particle excitations.

The canonical stress–energy tensor for the Klein–Gordon field is given by the basic definition (6.252) and the Klein–Gordon Lagrangian (8.25),

$$T^{\mu\nu}(x) = \eta^{\nu\rho}\frac{\partial\mathcal{L}}{\partial_\mu\varphi}\partial_\rho\varphi - \eta^{\mu\nu}\mathcal{L} = \eta^{\mu\sigma}\eta^{\nu\rho}\partial_\sigma\varphi\,\partial_\rho\varphi - \frac{1}{2}\eta^{\mu\nu}(\partial_\rho\varphi\,\partial^\rho\varphi - m^2\varphi^2). \tag{8.60}$$

It follows that the Hamiltonian density $\mathcal{H}$ appearing in (8.55) is identical to cT^{00}, which gives the correct units for $\mathcal{P}^0$. It follows from (6.265) that the spatial part of the four-momentum density $\mathcal{P}^\mu$ of the Klein–Gordon field is given by

$$\mathcal{P}^j = T^{j0} = \hbar\,\partial^j\varphi\,\partial^0\varphi = c\,\partial^j\varphi\,\pi \implies \mathcal{P}(\boldsymbol{x}, t) = -c\,\pi(\boldsymbol{x}, t)\,\nabla\varphi(\boldsymbol{x}, t). \tag{8.61}$$

Using (8.57) shows that integrating (8.61) gives

$$\boldsymbol{P}(t) = \int \mathrm{d}^3x\,\mathcal{P}(\boldsymbol{x}, t) = -\int \mathrm{d}^3x\, c\,\pi\,\nabla\varphi = \int \frac{\mathrm{d}^3k}{(2\pi)^3}\,\hbar\boldsymbol{k}\, a_k^* a_k. \tag{8.62}$$

Exercise 8.4: Using the expansions (8.57), show that (8.62) is the result of the spatial integration. In that regard, it is useful to start by proving the following two results,

$$\int d^3k\ \hbar\boldsymbol{k}\ a^*_{-k}a_{-k} = -\int d^3k\ \hbar\boldsymbol{k}\ a^*_k a_k, \qquad \int d^3k\ \hbar\boldsymbol{k}\ a_k a_{-k} = 0. \tag{8.63}$$

Results (8.58) and (8.62) show that the momentum four-vector associated with the Klein–Gordon mode is

$$p^\mu(\boldsymbol{k}) = (\hbar\omega_k/c,\ \hbar\boldsymbol{k}). \tag{8.64}$$

Using this with (8.24) gives

$$p_\mu(\boldsymbol{k})p^\mu(\boldsymbol{k}) = \hbar^2\left(\frac{\omega_k^2}{c^2} - \boldsymbol{k}\cdot\boldsymbol{k}\right) = \hbar^2 m^2 = M^2c^2, \tag{8.65}$$

which is the previous result (6.117) for the relativistic momentum of a mass M particle.

The Poisson bracket formalism introduced in (3.154) can be extended to the Klein–Gordon field. If $A[\varphi,\pi](t)$ and $B[\varphi,\pi](t)$ are two spatial functionals of φ and π, the relativistic Poisson bracket is defined as

$$\begin{aligned}\{A[\varphi,\pi](t), B[\varphi,\pi](t)\} = \frac{1}{c}\int d^3x \\ \left(\frac{\delta A[\varphi,\pi](t)}{\delta\varphi(\boldsymbol{x},t)}\frac{\delta B[\varphi,\pi](t)}{\delta\pi(\boldsymbol{x},t)} - \frac{\delta A[\varphi,\pi](t)}{\delta\pi(\boldsymbol{x},t)}\frac{\delta B[\varphi,\pi](t)}{\delta\varphi(\boldsymbol{x},t)}\right)\Bigg|_{\varphi_s},\end{aligned} \tag{8.66}$$

where the spatial functional derivatives are evaluated using a solution φ_s to the Euler–Lagrange equation. As an example, using $A = \varphi(\boldsymbol{y}, t)$ and $B = H(t)$ gives

$$\begin{aligned}\{\varphi(\boldsymbol{y},t), H(t)\} &= \frac{1}{c}\int d^3x \frac{\delta\varphi(\boldsymbol{y},t)}{\delta\varphi(\boldsymbol{x},t)}\frac{\delta H(t)}{\delta\pi(\boldsymbol{x},t)} \\ &= \frac{c^2}{\hbar}\int d^3x\ \delta^3(\boldsymbol{x}-\boldsymbol{y})\pi(\boldsymbol{x},t) = \frac{\partial\varphi(\boldsymbol{y},t)}{\partial t},\end{aligned} \tag{8.67}$$

where relation (8.54) was used. As another example, the mode coefficients a_k and a_k^* defined in (8.43) can be rewritten as a spatial functional of φ and π,

$$\begin{aligned}a_k[\varphi,\pi](t) &= i\int d^3x \left(f_k^*(\boldsymbol{x},t)\frac{c}{\hbar}\pi(\boldsymbol{x},t) - \partial_0 f_k^*(\boldsymbol{x},t)\,\varphi(\boldsymbol{x},t)\right), \\ a_p^*[\varphi,\pi](t) &= -i\int d^3x \left(f_p(\boldsymbol{x},t)\frac{c}{\hbar}\pi(\boldsymbol{x},t) - \partial_0 f_p(\boldsymbol{x},t)\,\varphi(\boldsymbol{x},t)\right).\end{aligned} \tag{8.68}$$

Evaluating the Poisson bracket for the modes gives

$$\begin{aligned}\left\{a_k, a_p^*\right\} &= -\frac{c}{\hbar}\int \mathrm{d}^3x \left(\partial_0 f_k^*(\boldsymbol{x}, t) f_p(\boldsymbol{x}, t) - f_k^*(\boldsymbol{x}, t)\partial_0 f_p(\boldsymbol{x}, t)\right) \\ &= -\frac{ic}{\hbar}(2\pi)^3\delta^3(\boldsymbol{k} - \boldsymbol{p}).\end{aligned} \tag{8.69}$$

8.3.4 The Klein–Gordon field and point particle interactions

It is possible to add a simple interaction term to the Klein–Gordon Lagrangian (8.25), modifying it to be

$$\mathcal{L} = \frac{1}{2}\partial_\mu\varphi\partial^\mu\varphi - \frac{1}{2}m^2\varphi^2 - J\varphi, \tag{8.70}$$

where the scalar function $J(x)$ is referred to as a *source*. The Euler–Lagrange equation (8.26) is now given by

$$(\square + m^2)\varphi(x) = -J(x). \tag{8.71}$$

Solving (8.71) can be done in a general manner using Green's function techniques, similar to those used for (3.109) and (5.75).

The Green's function $G(x - x')$ for the Klein–Gordon equation is defined as the function that satisfies

$$(\square + m^2)G(x - y) = -\delta^4(x - y). \tag{8.72}$$

It should be noted that some authors do not have the minus sign in definition (8.72). The solution to (8.71) is then given by

$$\varphi(x) = \varphi_0(x) + \int \mathrm{d}^4y\; G(x - y)J(y), \tag{8.73}$$

where $\varphi_0(x)$ satisfies the original Klein–Gordon equation (8.26),

$$(\square + m^2)\varphi_0(x) = 0, \tag{8.74}$$

but is otherwise arbitrary.

The tentative form of the Green's function is readily found using the Fourier representation of the delta function,

$$\delta^4(x - y) = \delta(x_0 - y_0)\delta^3(\boldsymbol{x} - \boldsymbol{y}) = \int \frac{\mathrm{d}^4k}{(2\pi)^4}\, e^{-ik_\mu(x^\mu - y^\mu)}, \tag{8.75}$$

which is manifestly Lorentz invariant. The Green's function can then be written

$$G(x - y) = \int \frac{\mathrm{d}^4k}{(2\pi)^4}\frac{e^{-ik_\mu(x^\mu - y^\mu)}}{k_\mu k^\mu - m^2} = \int \frac{\mathrm{d}^4k}{(2\pi)^4}\frac{e^{-ik\cdot(x-y)}}{k^2 - m^2}. \tag{8.76}$$

Applying $(\square + m^2)$ cancels the denominator, leaving $-\delta^4(x - y)$. In addition, expression (8.76) is manifestly Lorentz invariant. However, it is undefined due to the highly singular nature of the integrand. The integrand is singular at those values of k_0 that satisfy

$$k_0 = \pm\sqrt{\boldsymbol{k} \cdot \boldsymbol{k} + m^2} = \pm\omega_k/c, \tag{8.77}$$

which occurs for *all values* of $\boldsymbol{k}$. This continuous line of poles is referred to as a *branch cut*, and renders the integral in (8.76) undefined.

As a result, the tentative Green's function of (8.76) must be redefined in such a way as to make it integrable. There are a number of ways to do this, generically referred to as *epsilon prescriptions*, similar to the procedure used in (5.77). A commonly used version is given by

$$G_+(x - y) = \lim_{\epsilon\to 0^+} \int \frac{\mathrm{d}^4k}{(2\pi)^4} \frac{e^{-ik\cdot(x-y)}}{(k_0 + i\epsilon)^2 - \boldsymbol{k} \cdot \boldsymbol{k} - m^2}, \tag{8.78}$$

where $\epsilon > 0$ as the limit $\epsilon \to 0$ is taken. The integrand no longer has singularities on the real k_0 axis. Instead the singularities have become simple poles in the complex k_0 plane. To see this explicitly the function $G_+(x - y)$ is written

$$\begin{aligned} G_+(x - y) = &\lim_{\epsilon\to 0^+} \int \frac{\mathrm{d}^3k}{(2\pi)^3} e^{i\boldsymbol{k}\cdot(\boldsymbol{x}-\boldsymbol{y})} \\ &\int_{-\infty}^{\infty} \frac{\mathrm{d}k_0}{2\pi} \frac{c}{2\omega_k}\left(\frac{1}{k_0 - \omega_k/c + i\epsilon} - \frac{1}{k_0 + \omega_k/c + i\epsilon}\right)e^{-ik_0(x_0-y_0)}. \end{aligned} \tag{8.79}$$

Since $\epsilon > 0$ the integrand in (8.79) has two simple poles in the lower half of the complex k_0 plane. The integral over k_0 can therefore be evaluated using the Cauchy residue theorem (5.59), similarly to the evaluation of the related step function representations (5.69) and (5.71). The result is

$$\begin{aligned} &\lim_{\epsilon\to 0^+} \int_{-\infty}^{\infty} \frac{\mathrm{d}k_0}{2\pi} \frac{c}{2\omega_k}\left(\frac{1}{k_0 - \omega_k/c + i\epsilon} - \frac{1}{k_0 + \omega_k/c + i\epsilon}\right)e^{-ik_0(x_0-y_0)} \\ &= -\lim_{\epsilon\to 0^+} \theta(x_0 - y_0)\frac{c}{\omega_k} \sin\left(\frac{\omega_k}{c}(x_0 - y_0)\right)e^{-\epsilon(x_0-y_0)}. \end{aligned} \tag{8.80}$$

Exercise 8.5: Verify (8.80).

Returning (8.80) to (8.79) gives

$$G_+(x - y) = -\theta(x_0 - y_0)\lim_{\epsilon\to 0^+} \int \frac{\mathrm{d}^3k}{(2\pi)^3} \frac{c}{\omega_k} e^{i\boldsymbol{k}\cdot(\boldsymbol{x}-\boldsymbol{y})} \sin\left(\frac{\omega_k}{c}(x_0 - y_0)\right)e^{-\epsilon(x_0-y_0)}. \tag{8.81}$$

The function (8.79) is known as the *retarded Green's function*, since it is nonzero only if $x_0 > y_0$.

Using G_+ in (8.73) shows that the current form of φ can be found from the *past events* of the source. It is instructive to evaluate (8.73) for the case that the source $J(x)$ corresponds to a *static point particle* at the origin $\boldsymbol{x} = 0$, so that

$$J(x) = g\,\delta^3(\boldsymbol{x}), \tag{8.82}$$

where g measures the strength of the coupling. For simplicity, φ_0 will be set to zero in (8.73). It is critical to pass to the limit $\epsilon \to 0^+$ only after performing the y_0 integration, since retaining the factor $\exp(-\epsilon(x_0 - y_0))$ gives

$$\begin{aligned} &\lim_{\epsilon\to 0^+} \int_{-\infty}^{x_0} \mathrm{d}y_0 \sin\left(\frac{\omega_{\boldsymbol{k}}}{c}(x_0 - y_0)\right) e^{-\epsilon(x_0-y_0)} \\ &= -\frac{1}{2i}\lim_{\epsilon\to 0^+}\left(\frac{1}{i\omega_{\boldsymbol{k}}/c - \epsilon} + \frac{1}{i\omega_{\boldsymbol{k}}/c + \epsilon}\right) = \frac{c}{\omega_{\boldsymbol{k}}}. \end{aligned} \tag{8.83}$$

Using this result and performing the integral over $\boldsymbol{y}$ using $J(\boldsymbol{y}) = g\delta^3(\boldsymbol{y})$ gives the static field configuration created by the static point source,

$$\varphi(\boldsymbol{x}) = -g\int \frac{\mathrm{d}^3k}{(2\pi)^3}\frac{c^2}{\omega_{\boldsymbol{k}}^2} e^{i\boldsymbol{k}\cdot\boldsymbol{x}} = -g\int \frac{\mathrm{d}^3k}{(2\pi)^3}\frac{e^{i\boldsymbol{k}\cdot\boldsymbol{x}}}{\boldsymbol{k}\cdot\boldsymbol{k} + m^2}. \tag{8.84}$$

Evaluating the remaining integration is expedited by using spherical coordinates for $\boldsymbol{k}$. This uses

$$\mathrm{d}^3k = k^2 \sin\theta_k\, \mathrm{d}k\, \mathrm{d}\theta_k\, \mathrm{d}\phi_k, \tag{8.85}$$

where θ_k and ϕ_k are the polar and azimuthal angles in $\boldsymbol{k}$-space. The direction of the k_z axis can be chosen to match the z-axis of $\boldsymbol{x}$, so that $\boldsymbol{k}\cdot\boldsymbol{x} = kr\cos\theta_k$, where r is the magnitude of $\boldsymbol{x}$. Performing the $\mathrm{d}\phi_k$ and $\mathrm{d}\theta_k$ integrations gives

$$\begin{aligned} \varphi(\boldsymbol{x}) &= -\frac{g}{2\pi^2 r}\int_0^\infty \mathrm{d}k\,\frac{k\sin kr}{k^2 + m^2} \\ &= -\frac{g}{4\pi^2 r}\int_{-\infty}^\infty \mathrm{d}k\,\frac{k\sin kr}{k^2 + m^2} \\ &= -\frac{g}{8\pi^2 ir}\int_{-\infty}^\infty \mathrm{d}k\,\frac{k(e^{ikr} - e^{-ikr})}{k^2 + m^2}. \end{aligned} \tag{8.86}$$

The remaining integral over k can be evaluated using complex analysis techniques, which gives

$$\varphi(\boldsymbol{x}) = -\frac{g}{8\pi^2 ir}\int_{-\infty}^\infty \mathrm{d}k\,\frac{k(e^{ikr} - e^{-ikr})}{(k + mi)(k - mi)} = -\frac{g}{4\pi r}e^{-mr}. \tag{8.87}$$

The form of the field in (8.87) is known as the Yukawa potential. The point particle at the origin creates a short-range field through its coupling to the Klein–Gordon field.

It is important to note that the field configuration is exponentially damped by the factor $\exp(-mr) = \exp(-Mcr/\hbar)$. For example, the pi meson field around the nucleon is responsible for the transfer of force between nucleons. It can be described using a scalar field with an SI value of $Mc/\hbar \approx 7 \times 10^{14}$ m^{-1}. Due to the exponential damping factor, the pion field around a nucleon becomes virtually negligible for distances greater than 2×10^{-15} meters, which explains the short-range nature of the nuclear force and the compact size of the nucleus.

Another commonly used epsilon prescription defines the Green's function as

$$G_F(x - y) = \lim_{\epsilon \to 0^+} \int \frac{\mathrm{d}^4 k}{(2\pi)^4} \frac{e^{-ik_\mu(x^\mu - y^\mu)}}{k^2 - m^2 + i\epsilon}, \tag{8.88}$$

which is known from quantum field theory as the *Feynman propagator*. For simplicity of notation, the quantity ε_k is defined as

$$\varepsilon_k = \frac{\omega_k}{c} = \sqrt{\boldsymbol{k} \cdot \boldsymbol{k} + m^2}. \tag{8.89}$$

The pole structure of the integrand is revealed by writing it as

$$\frac{1}{k^2 - m^2 + i\epsilon} = \frac{1}{(k_0 - (\varepsilon_k - i\epsilon/2\varepsilon_k))(k_0 + (\varepsilon_k - i\epsilon/2\varepsilon_k)) - \epsilon^2/4\varepsilon_k^2}. \tag{8.90}$$

In the limit $\epsilon \to 0$ the term involving ϵ^2 can be dropped. This shows that there is a simple pole at $\varepsilon_k - i\epsilon/\varepsilon_k$, which is in the lower half of the complex k_0 plane, and a simple pole at $-\varepsilon_k + i\epsilon/\varepsilon_k$, which is in the upper half of the complex k_0 plane. Cauchy's theorem can now be used to perform the integration over k_0 by choosing a closed contour in the lower half plane if $x^0 - y^0 > 0$ and a closed contour in the upper half plane if $x^0 - y^0 < 0$. The sum of the residues for the two cases gives, in the $\epsilon \to 0$ limit,

$$\begin{aligned} G_F(x - y) = i \int \frac{\mathrm{d}^3 k}{(2\pi)^3} \frac{e^{i\boldsymbol{k}\cdot(\boldsymbol{x}-\boldsymbol{y})}}{2\varepsilon_k} \\ \left(\theta(x^0 - y^0)e^{-i\varepsilon_k(x^0 - y^0)} + \theta(y^0 - x^0)e^{i\varepsilon_k(x^0 - y^0)}\right). \end{aligned} \tag{8.91}$$

This includes the retarded Green's function found earlier as well as the second term, which is known as the *advanced Green's function*, since it includes contributions from the future of the event.

The Feynman propagator can be explicitly evaluated using a technique often referred to as the *Schwinger parameterization*. This exponentiates the integrand of the Green's function by writing it as

$$G_F(x) = -i \lim_{\epsilon \to 0^+} \int \frac{\mathrm{d}^4 k}{(2\pi)^4} \int_0^\infty \mathrm{d}\alpha \; e^{i\alpha(k^2 - m^2 + i\epsilon)} e^{-ik\cdot x}. \tag{8.92}$$

It is important to note that this expression is undefined without the presence of the $i\epsilon$ term, which serves to damp the $\alpha \to \infty$ limit of the integration. The exponential integrals over k can be performed by a method known as *completing the square*. For example, the integral over k_0 contained in (8.92) can be written

$$\begin{aligned}\int_{-\infty}^{\infty}\frac{\mathrm{d}k_0}{2\pi}\exp\left(i\alpha k_0^2 - ik_0x_0\right) &= \int_{-\infty}^{\infty}\frac{\mathrm{d}k_0}{2\pi}\exp\left(i\alpha\left(k_0 - \frac{x_0}{2\alpha}\right)^2\right)\exp\left(-i\frac{x_0^2}{4\alpha}\right)\\ &= \sqrt{\frac{i}{4\pi\alpha}}\exp\left(-i\frac{x_0^2}{4\alpha}\right),\end{aligned} \tag{8.93}$$

where changing the variable to $k_0' = k_0 - x_0/2\alpha$ allows (8.93) to be evaluated using a variant of the *Fresnel integral*,

$$\int_{-\infty}^{\infty}\mathrm{d}k_0'\, e^{-i\alpha k_0'^2} = \sqrt{\frac{\pi}{i\alpha}}. \tag{8.94}$$

The remaining integrations over the spatial parts of k are evaluated identically, but result in the opposite sign. For example,

$$\begin{aligned}\int_{-\infty}^{\infty}\frac{\mathrm{d}k_1}{2\pi}\exp\left(-i\alpha k_1^2 + ik_1x_1\right) &= \int_{-\infty}^{\infty}\frac{\mathrm{d}k_1}{2\pi}\exp\left(-i\alpha\left(k_1 - \frac{x_1}{2\alpha}\right)^2\right)\exp\left(i\frac{x_1^2}{4\alpha}\right)\\ &= \sqrt{\frac{1}{4\pi i\alpha}}\exp\left(i\frac{x_1^2}{4\alpha}\right).\end{aligned} \tag{8.95}$$

After all four k integrations are complete the integral (8.92) is given by

$$G_F(x) = \frac{i}{16\pi^2}\lim_{\epsilon\to 0^+}\int_0^{\infty}\frac{\mathrm{d}\alpha}{\alpha^2}e^{-i\alpha(m^2-i\epsilon)}e^{-ix\cdot x/4\alpha}, \tag{8.96}$$

where $x\cdot x = x_0^2 - \boldsymbol{x}\cdot\boldsymbol{x}$ is the manifestly Lorentz invariant inner product. The remaining integral in (8.96) is well-known and tabulated in many standard references on special functions. It is crucially dependent upon the sign of $x\cdot x$. If x^μ is space-like, then $x\cdot x < 0$, while if x^μ is time-like, then $x\cdot x > 0$. In the space-like case the real quantity $s = \sqrt{-x\cdot x}$ is used and the final result is given by

$$G_F(x) = -\frac{m}{4\pi^2 s}K_1(ms), \tag{8.97}$$

where K_1 is the modified Bessel function of the second kind and index one. It is singular as $s\to 0$, $K_1(ms)\approx 1/ms$, and exponentially damped as $s\to\infty$, just as the Yukawa potential is (8.87). This result points to an underlying property of field theory, even at the classical level, which is that events at a space-like separation can still affect one another, even though this property can be highly suppressed. For the time-like case $x\cdot x > 0$, the real quantity $s' = \sqrt{x\cdot x}$ is used and the result of the integral is given by

$$G_F(x) = -\frac{im}{8\pi s'}H_1^{(2)}(ms'), \tag{8.98}$$

where $H_1^{(2)}$ is the Hankel function of the second kind and index one. For $s' \to 0$ the Hankel function is singular, $H_1^{(2)} \approx 2i/\pi m s'$. The two expressions, (8.97) and (8.98), result in the same small limit, so that for light-like vectors the result is singular,

$$\lim_{x\cdot x\to 0} G_F(x) = \lim_{x\cdot x\to 0} \frac{1}{4\pi x \cdot x}. \tag{8.99}$$

Another widely used approach to defining the Green's function is known as *Wick rotation*, which consists of simultaneously continuing both x_0 and k_0 to imaginary values,

$$x^0 \to -i\tau, \quad k_0 \to +i\kappa. \tag{8.100}$$

The Wick rotation gives

$$\begin{aligned} k \cdot k &= k_0^2 - \boldsymbol{k} \cdot \boldsymbol{k} \to -(\kappa^2 + \boldsymbol{k} \cdot \boldsymbol{k}), \\ x \cdot x &= x_0^2 - \boldsymbol{x} \cdot \boldsymbol{x} \to -(\tau^2 + \boldsymbol{x} \cdot \boldsymbol{x}), \\ k \cdot x &= k_0 x_0 - \boldsymbol{k} \cdot \boldsymbol{x} \to \kappa\tau - \boldsymbol{k} \cdot \boldsymbol{x}. \end{aligned} \tag{8.101}$$

The Wick rotation results in negative definite expressions for $x \cdot x$ and $k \cdot k$, but preserves the direction of wave motion. Integrals over k_0 become integrals over κ with the accompanying replacement

$$\mathrm{d}k_0 \to i\, \mathrm{d}\kappa. \tag{8.102}$$

The τ and κ variables are both real. At the end of the calculation, the inverse Wick rotation $\tau \to ix_0$ is performed to regain the Minkowski metric. In the interim calculations the Wick rotation alleviates the need to distinguish between space-like and time-like vectors and bypasses the need to shift poles in expressions in order to apply Cauchy's theorem. In effect, the integrals over k_0 along the real axis are rotated 90° to lie along the imaginary axis parameterized by κ. This is depicted in figure 8.1 for the case of the Feynman propagator (8.88).

The Jordan lemma (5.63) can be adapted to show that there is no contribution from either of the quarter-circle contours. Since neither of the original poles of the Feynman propagator, marked with an x, are enclosed by the contour, it follows that integration over k_0 and integration over $i\kappa$ yield identical results, with the understanding that the limits on the κ integration are reversed from ∞ to $-\infty$, which yields a minus sign that offsets the minus sign from the denominator resulting from the Wick rotation. Because there are no poles along the κ-axis, ϵ can be set to zero. After the Wick rotation the Feynman propagator (8.88) is given by

$$G_W(x_W) = i \int \frac{\mathrm{d}^3 k}{(2\pi)^3} \int_{-\infty}^{\infty} \frac{\mathrm{d}\kappa}{(2\pi)} \frac{e^{-i(\kappa\tau - \boldsymbol{k}\cdot\boldsymbol{x})}}{\kappa^2 + \boldsymbol{k} \cdot \boldsymbol{k} + m^2}, \tag{8.103}$$

so that the integrand has no singularities.

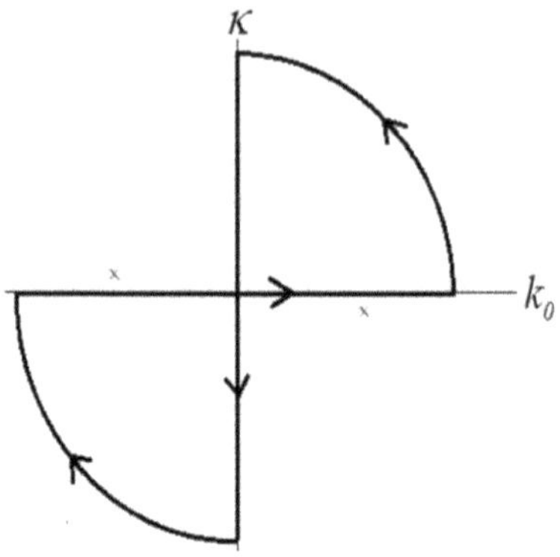

Figure 8.1. The Wick rotation of the Feynman propagator.

The Schwinger parameterization is now given by

$$G_W(x_W) = -i\int\frac{\mathrm{d}^3k}{(2\pi)^3}\frac{\mathrm{d}\kappa}{(2\pi)}\int_0^\infty \mathrm{d}\alpha\; e^{-i(\kappa\tau-\boldsymbol{k}\cdot\boldsymbol{x})}e^{-\alpha(\kappa^2+\boldsymbol{k}\cdot\boldsymbol{k}+m^2)}, \tag{8.104}$$

which is well-defined without an epsilon prescription. The integration over κ gives

$$\int_{-\infty}^{\infty}\frac{\mathrm{d}\kappa}{2\pi}\,e^{-\alpha\kappa^2}e^{i\kappa\tau} = \sqrt{\frac{1}{4\pi\alpha}}\,e^{-\tau^2/4\alpha}, \tag{8.105}$$

with identical results for the integrations over k_j. The Wick-rotated Feynman propagator is now given by

$$G_W(x_W) = -\frac{i}{16\pi^2}\int_0^\infty\frac{\mathrm{d}\alpha}{\alpha^2}\,e^{-\alpha m^2-(\tau^2+\boldsymbol{x}\cdot\boldsymbol{x})/4\alpha}. \tag{8.106}$$

Using s to denote the real variable $\sqrt{\tau^2+\boldsymbol{x}\cdot\boldsymbol{x}}$, the integral (8.106) gives

$$G_W(x_W) = -\frac{im}{4\pi^2 s}K_1(ms), \tag{8.107}$$

where K_1 is the modified Bessel function of the second kind with index one. The inverse Wick rotation gives the Feynman propagator,

$$G_F(x) = -i\;G_W(x_W)|_{\tau\,=\,ix_0} = -\frac{m}{4\pi^2 s}K_1(ms), \tag{8.108}$$

where $s=\sqrt{-x_0^2+\boldsymbol{x}\cdot\boldsymbol{x}}$. The factor of $-i$ is required to return the Wick-rotated contour to the original form of the Feynman propagator (8.88) via $\kappa\to -ik_0$. For the case of a space-like vector, (8.108) is precisely the result (8.97) found earlier using the epsilon prescription. However, if x^μ is a time-like vector, then inverse Wick rotation results in an imaginary value for s given by

$$s=\sqrt{-x_0^2+\boldsymbol{x}\cdot\boldsymbol{x}} = i\sqrt{x_0^2-\boldsymbol{x}\cdot\boldsymbol{x}} = is'. \tag{8.109}$$

The modified Bessel function K_1 has the property that for imaginary arguments it is proportional to the Hankel function $H_1^{(2)}$,

$$K_1(ims') = -\frac{\pi}{2} H_1^{(2)}(ms'). \tag{8.110}$$

For time-like Minkowski vectors, (8.107) therefore becomes

$$G_F(x) = \frac{m}{4\pi^2 is'} K_1(ims') = -\frac{im}{8\pi s'} H_1^{(2)}(ms'), \tag{8.111}$$

which is identical to the previous result (8.98) for time-like vectors. As a result, Wick rotation yields results identical to the epsilon prescription employed in evaluating (8.88). Although beyond the scope of this text, the Wick rotation also connects quantum processes to statistical mechanics and is used throughout the technique of path integrals to regularize them.

8.4 Classical scalar solutions and broken symmetry

So far, our attention has been limited to the quadratic action (8.25) that defines the Klein–Gordon equation. It is possible to add a nonlinear interaction term to the Lagrangian and rewrite it in the form

$$\mathcal{L} = \frac{1}{2}\partial_\mu\varphi\partial^\mu\varphi + \frac{1}{2}\mu^2\varphi^2 - \frac{1}{4}\lambda\varphi^4, \tag{8.112}$$

where μ has units of mass and λ is a dimensionless positive constant. The quadratic mass term has entered with the opposite sign from that of (8.25), and so the quartic term is often combined with it to define

$$V(\varphi) = -\frac{1}{2}\mu^2\varphi^2 + \frac{1}{4}\lambda\varphi^4, \tag{8.113}$$

which is known as the *Higgs potential* and is depicted in figure 8.2. The Higgs potential has the *discrete symmetry* $\varphi \to -\varphi$, since this operation leaves $\mathcal{H}_I$ unchanged.

The Euler–Lagrange equation for (8.112) is given by

$$\partial_\mu \frac{\partial\mathcal{L}}{\partial(\partial_\mu)} + \frac{\partial V(\varphi)}{\partial\varphi} = \left(\Box + \frac{\partial V}{\partial\varphi}\right)\varphi = \Box\varphi - \mu^2\varphi + \lambda\varphi^3 = 0. \tag{8.114}$$

The Euler–Lagrange equation for this system has two constant solutions given by

$$\nu = \pm\sqrt{\frac{\mu^2}{\lambda}}, \tag{8.115}$$

which correspond to the two minima of the Higgs potential, so that

$$\left.\frac{\partial V(\varphi)}{\partial\varphi}\right|_{\varphi=\nu} = 0. \tag{8.116}$$

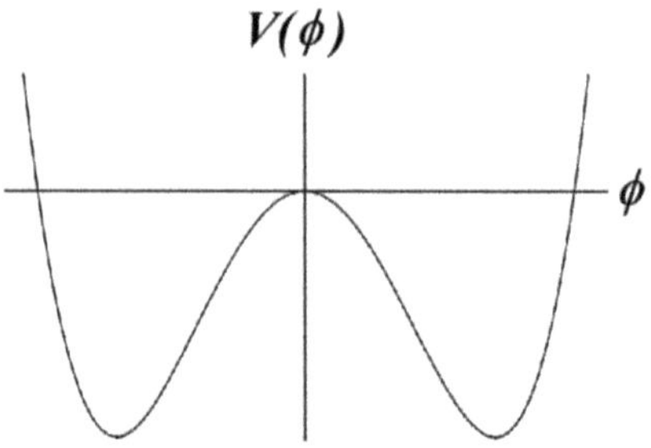

Figure 8.2. The Higgs potential.

As in the case of sound waves, the stability of fluctuations for the field φ around one of these values is revealed by writing the field as

$$\varphi(x) = \xi(x) + \nu, \tag{8.117}$$

where ξ is the fluctuation around the constant value ν. The Lagrangian (8.112) becomes

$$\mathcal{L}(\xi + \nu) = \frac{1}{2}\partial_\mu \xi \partial^\mu \xi - \frac{\mu^4}{4\lambda} - \mu^2\xi^2 - \lambda\nu\xi^3 - \frac{1}{4}\lambda\xi^4. \tag{8.118}$$

The quadratic terms in (8.118) give rise to a Klein–Gordon equation with a positive mass term,

$$(\Box + 2\mu^2)\xi = 0, \tag{8.119}$$

which corresponds to well-defined fluctuations of the form

$$\xi = A\cos(kx - \omega_k t), \quad \omega_k^2 = k^2 + 2\mu^2. \tag{8.120}$$

However, the remaining nonlinear terms, given by

$$\mathcal{H}_I = \lambda\nu\xi^3 + \frac{1}{4}\lambda\xi^4, \tag{8.121}$$

no longer possesses the symmetry $\xi \to -\xi$. The symmetry has been *broken* by fluctuating around one of the two minima.

The Higgs equation of motion (8.114) also possesses a spatially dependent *static* solution. For simplicity, consideration will focus on one spatial dimension, denoted by x, so that the equation of motion becomes

$$-\frac{\partial^2\varphi}{\partial x^2} + \frac{\partial V(\varphi)}{\partial\varphi} = -\frac{\partial^2\varphi(x)}{\partial x^2} - \mu^2\varphi(x) + \lambda\varphi^3(x) = 0. \tag{8.122}$$

This equation has the solution

$$\varphi_c(x) = \sqrt{\frac{\mu^2}{\lambda}} \tanh\left(\frac{\mu}{\sqrt{2}}(x - x_o)\right), \tag{8.123}$$

where x_o is arbitrary.

> **Exercise 8.6:** Verify that (8.123) is a solution of (8.114).

Solution (8.123) has the property that

$$\lim_{x \to \pm\infty} \varphi_c(x) = \pm\sqrt{\frac{\mu^2}{\lambda}}, \tag{8.124}$$

so that the solution interpolates between the two constant solutions of (8.115). For that reason it is referred to as the *kink solution.*

Like (8.119) the fluctuations of the field around the kink solution are found by expanding the Lagrangian (8.112) around $\varphi_c(x)$,

$$\varphi = \xi + \varphi_c, \tag{8.125}$$

and retaining the quadratic terms $\mathcal{L}_Q$. Setting $x_o = 0$ for simplicity the result is

$$\begin{aligned} \mathcal{L}_Q &= \frac{1}{2}\partial_\mu \xi \partial^\mu \xi - \frac{1}{2}\frac{\partial^2 V(\varphi)}{\partial \varphi^2}\bigg|_{\varphi = \varphi_c} \xi^2 \\ &= \frac{1}{2}\partial_\mu \xi \partial^\mu \xi - \frac{1}{2}\mu^2\left(3\tanh^2\left(\frac{\mu}{\sqrt{2}}x\right) - 1\right)\xi^2, \end{aligned} \tag{8.126}$$

where the constant terms have been discarded and the linear terms in ξ vanish, since φ_c satisfies the equation of motion (8.122). This gives the basic fluctuation equation of motion

$$\left(\Box + \frac{\partial^2 V(\varphi)}{\partial \varphi^2}\bigg|_{\varphi = \varphi_c}\right)\xi = \Box\xi + \mu^2\left(3\tanh^2\left(\frac{\mu}{\sqrt{2}}x\right) - 1\right)\xi = 0. \tag{8.127}$$

Investigating the stability of the theory requires finding the time-dependent solutions to (8.127). These are found by *separation of variables*, so that ξ is written

$$\xi(x, t) = u(x)\cos(\omega t), \tag{8.128}$$

which take the form of standing waves. After using (8.128) and changing to the new variable $z = \mu x/\sqrt{2}$, the quadratic equation of motion (8.127) becomes

$$\left(-\frac{1}{2}\frac{\mathrm{d}^2}{\mathrm{d}z^2} + 3\tanh^2(z) - 1\right)u(z) = \frac{\omega^2}{\mu^2}u(z). \tag{8.129}$$

Equation (8.129) provides an *eigenvalue equation* which determines both the *eigenvalue* ω^2 and the associated *eigenfunction* $u(z)$.

In order for the fluctuations to be stable the eigenvalue ω^2 cannot be negative. There is no standard method for solving eigenvalue equations. Typical methods include Fourier transforms as well as power series expansions. In the case of (8.129) the solutions include two discrete eigenvalue and eigenfunction pairs,

$$\begin{aligned} \omega_0^2 &= 0, \quad u_0(z) = \mathrm{sech}^2(z), \\ \omega_1^2 &= \frac{3}{2}\mu^2, \quad u_1(z) = \tanh(z)\,\mathrm{sech}(z), \end{aligned} \tag{8.130}$$

and a set of continuous eigenvalue and eigenfunction pairs,

$$\begin{aligned} \omega_k^2 &= k^2 + 2\mu^2, \\ u_k(z) &= \left(3\tanh^2(z) - 1 - \frac{2k^2}{\mu}\right)\cos\left(\frac{\sqrt{2}kz}{\mu}\right) + \frac{3\sqrt{2}k}{\mu}\sin\left(\frac{\sqrt{2}kz}{\mu}\right). \end{aligned} \tag{8.131}$$

It should be noted that the frequency ω_k for the continuous eigenfunctions matches the frequency of oscillations around the constant solutions given by (8.120). These correspond to the fluctuation field *scattering* off the position of the kink through nonlinear self-interactions. However, the second of the two modes given by (8.130) corresponds to an eigenfunction localized around $x = 0$ with a frequency *or energy* $\omega_1 = \sqrt{3/2}\ \mu$ that is less than $\sqrt{2}\mu$. This corresponds to a *bound state* of the fluctuation field that has localized on the position of the kink.

The first of the two modes given by (8.130) is referred to as the *translation mode*. Such a zero-frequency mode will always occur in the presence of a spatially dependent classical solution. This follows from differentiating the static equation of motion (8.122) evaluated at φ_c, which gives

$$\frac{\partial}{\partial x}\left(-\frac{\partial^2\varphi_c}{\partial x^2} + \frac{\partial V(\varphi_c)}{\partial \varphi_c}\right) = \left(-\frac{\partial^2}{\partial x^2} + \frac{\partial^2 V(\varphi)}{\partial \varphi_c^2}\right)\frac{\partial \varphi_c}{\partial x} = 0. \tag{8.132}$$

This shows that $\partial_x\varphi_c$ corresponds to an eigenfunction of the fluctuation equation (8.127) with zero frequency. The zero-frequency eigenfunction $\mathrm{sech}^2(z)$ of (8.130) is precisely $\partial_z \tanh(z)$. It is known as the translation mode, since the solution $\varphi_c(x)$ has the property that

$$\varphi_c(x + \epsilon) = \varphi_c(x) + \epsilon\ \partial_x\varphi_c(x), \tag{8.133}$$

for the case that $\epsilon \approx 0$. The position of the kink can be boosted into motion or translated, and this gives rise to the zero-energy *translation mode*.

8.5 Relativistic spinor fields and quadratic actions

In this section relativistically invariant actions for Weyl spinors and Dirac bispinors will be constructed using the results of the previous chapter, where spinor representations of the Lorentz group were developed. The key results necessary to introduce a Lorentz invariant action are (7.216) in the case of the Weyl spinor and (7.273) in the case of Dirac bispinors. These two results associate an inverse Lorentz transformation with the spinor transformation and allow the construction of Lorentz invariant actions using the gradient ∂_μ.

8.5.1 The Weyl spinor and its solutions

The construction of a quadratic action in the two-component Weyl spinor field is limited, since a term of the form $\xi^\dagger\xi$ is *not* Lorentz invariant. This is because the Weyl spinor transformation $\mathbf{U}$ given by (7.225) is not unitary, so that $\mathbf{U}^\dagger \neq \mathbf{U}^{-1}$. The same problem also prevents a term of the form $\partial_\mu\xi^\dagger\partial^\mu\xi$ from being Lorentz invariant. A Lorentz invariant action for Weyl spinors requires using relation (7.234), which gives

$$\begin{aligned}\xi^{\dagger\prime}(x')\boldsymbol{\sigma}^\mu\partial'_\mu\xi'(x') &= \xi^\dagger(x)\mathbf{U}^\dagger\boldsymbol{\sigma}^\mu\mathbf{U}\partial'_\mu\xi(x)\\ &= \xi^\dagger(x)\boldsymbol{\sigma}^\nu\Lambda_\nu{}^\mu(-\phi)\partial'_\mu\xi(x) = \xi^\dagger\boldsymbol{\sigma}^\nu\partial_\nu\xi.\end{aligned} \tag{8.134}$$

The bilinear form $\xi^\dagger\sigma^\mu\partial_\mu\xi$ is a Lorentz invariant, but it is not real valued,

$$(\xi^\dagger\boldsymbol{\sigma}^\mu\partial_\mu\xi)^* = (\partial_\mu\xi^\dagger)\boldsymbol{\sigma}^\mu\xi \neq \xi^\dagger\boldsymbol{\sigma}^\mu\partial_\mu\xi. \tag{8.135}$$

If the action is not real, there are *typically* difficulties in identifying real-valued quantities, such as energy, or in finding stable solutions at the classical level. The solution is to use the Weyl spinor Lagrangian,

$$\mathcal{L}_\xi = \frac{1}{2}i\hbar\,\xi^\dagger\boldsymbol{\sigma}^\mu\partial_\mu\xi - \frac{1}{2}i\hbar\,(\partial_\mu\xi^\dagger)\boldsymbol{\sigma}^\mu\xi, \tag{8.136}$$

where the factor of $\hbar$ will be retained in what follows. The Lagrangian of (8.136) is both Lorentz invariant and real, since the two terms transform into each other under complex conjugation.

The resulting Weyl Lagrangian is of the first order in derivatives and involves a complex spinor field that must have units of inverse length to the 3/2 power in order for the Lagrangian of (8.136) to have units of $\hbar$ per inverse length to the fourth power. This is identical to the units of the complex scalar field φ considered in chapter 5, where $\varphi^*\varphi$ was interpreted as the particle density of the Bose–Einstein condensate. At the classical level $\xi^\dagger\xi$ can also be interpreted as a particle density, although doing so is not necessary in order to discuss the mathematical properties of the Weyl field. Making such an identification is not Lorentz invariant, since neither $\xi^\dagger\xi$ nor the particle density is Lorentz invariant. In that regard it is instructive to perform a spinor boost along the z-axis using the transformation found in (7.227). This gives the boosted version of $\xi^\dagger\xi$ as

$$
\begin{aligned}
\xi'^{\dagger}\xi' = \xi^{\dagger}\mathbf{U}^{\dagger}\mathbf{U}\xi = \xi^{\dagger}\begin{pmatrix} e^{\phi} & 0 \\ 0 & e^{-\phi} \end{pmatrix}\xi &= e^{\phi}\xi_1^*\xi_1 + e^{-\phi}\xi_2^*\xi_2 \\
&= \cosh\phi\,(\xi_1^*\xi_1 + \xi_2^*\xi_2) + \sinh\phi\,(\xi_1^*\xi_1 - \xi_2^*\xi_2) \\
&= \gamma_v(\xi_1^*\xi_1 + \xi_2^*\xi_2) + \gamma_v\frac{v}{c}(\xi_1^*\xi_1 - \xi_2^*\xi_2) \\
&= \gamma_v\,\xi^{\dagger}\xi + \gamma_v\frac{v}{c}\xi^{\dagger}\boldsymbol{\sigma}_3\xi,
\end{aligned} \tag{8.137}
$$

which shows that $\xi^{\dagger}\xi$ transforms like a spatial density. The second term will be discussed in more detail in the upcoming section on Noether's theorem, where it is shown to represent the appearance of a current density in the boosted frame.

The two quantities ξ and $\xi^{\dagger}$ are independent degrees of freedom, and so there are two Euler–Lagrange equations that follow from the Weyl Lagrangian. These are given in matrix form by

$$
\begin{aligned}
\frac{\partial\mathcal{L}_{\xi}}{\partial\xi} - \partial_{\mu}\frac{\partial\mathcal{L}_{\xi}}{\partial(\partial_{\mu}\xi)} = 0 &\implies -i(\partial_{\mu}\xi^{\dagger})\boldsymbol{\sigma}^{\mu} = -i(\partial_0\xi^{\dagger}\boldsymbol{\sigma}^0 - \nabla\xi^{\dagger}\cdot\boldsymbol{\sigma}) = 0, \\
\frac{\partial\mathcal{L}_{\xi}}{\partial\xi^{\dagger}} - \partial_{\mu}\frac{\partial\mathcal{L}_{\xi}}{\partial(\partial_{\mu}\xi^{\dagger})} = 0 &\implies i\boldsymbol{\sigma}^{\mu}\partial_{\mu}\xi = i(\boldsymbol{\sigma}^0\partial_0 - \boldsymbol{\sigma}\cdot\nabla)\xi = 0,
\end{aligned} \tag{8.138}
$$

which are collectively referred to as the *Weyl equation*, since they are the Hermitian adjoints of each other. If ξ is a solution to the second equation, then $\xi^{\dagger}$ automatically solves the first equation.

The wave content of (8.138) is revealed by relation (7.212), which can be adapted to show

$$
\tilde{\boldsymbol{\sigma}}^{\mu}\partial_{\mu}\boldsymbol{\sigma}^{\nu}\partial_{\nu} = \mathbf{I}\,\partial_{\mu}\partial^{\mu} = \mathbf{I}\,\Box, \tag{8.139}
$$

where $\tilde{\boldsymbol{\sigma}}^{\mu} = (\boldsymbol{\sigma}^0, \boldsymbol{\sigma})$. Dropping the factor of i, the equation of motion then gives

$$
\boldsymbol{\sigma}^{\mu}\partial_{\mu}\xi = 0 \implies \tilde{\boldsymbol{\sigma}}^{\nu}\partial_{\nu}\boldsymbol{\sigma}^{\mu}\partial_{\mu}\xi = \mathbf{I}\,\Box\xi = \Box\xi = 0, \tag{8.140}
$$

which shows that both components of the Weyl spinor field satisfy the wave equation with the velocity of propagation c. The solution to the Weyl equation therefore has the general form

$$
\xi(x) = e^{\pm ip\cdot x}u(p), \tag{8.141}
$$

where

$$
\begin{aligned}
p\cdot p = p_{\mu}p^{\mu} &= p_0^2 - \boldsymbol{p}\cdot\boldsymbol{p} = 0, \\
p\cdot x = p_{\mu}x^{\mu} &= p_0x^0 - \boldsymbol{p}\cdot\boldsymbol{r} = 0,
\end{aligned} \tag{8.142}
$$

while $u(p)$ is a two-component spinor function of $\boldsymbol{p}$, since $p_0 = +\sqrt{\boldsymbol{p}\cdot\boldsymbol{p}}$. The Weyl spinor field therefore corresponds to a massless field that propagates at the speed of light. The equation for the two-component momentum spinor $u(\boldsymbol{p})$ is determined by inserting the form for $\xi(x)$ into the original equation of motion (8.138), which shows that $u(\boldsymbol{p})$ must satisfy the matrix equation

$$p_\mu \sigma^\mu u(\boldsymbol{p}) = (p_0 \sigma^0 + p_j \sigma^j) u(\boldsymbol{p}) = (p_0 \sigma^0 + \boldsymbol{p} \cdot \boldsymbol{\sigma}) u(\boldsymbol{p}) = 0, \tag{8.143}$$

where $\boldsymbol{\sigma}$ are the Pauli spin matrices. Before finding specific solutions to (8.143) it is important to note that all solutions will have the property

$$\left(\frac{1}{2p_0} \boldsymbol{p} \cdot \boldsymbol{\sigma} \right) u(\boldsymbol{p}) = -\frac{1}{2} u(\boldsymbol{p}), \tag{8.144}$$

so that $u(\boldsymbol{p})$ is an *eigenspinor* of the matrix

$$\boldsymbol{h} = \frac{1}{2} \boldsymbol{p} \cdot \boldsymbol{\sigma} / p_0 \tag{8.145}$$

with the eigenvalue $-\frac{1}{2}$. The matrix $\boldsymbol{h}$ is known as the *helicity operator* and is related to the intrinsic angular momentum of the spinor field referred to as *spin*. This will be clarified when Noether's theorem is discussed in a later section.

Finding the solutions to this matrix equation is greatly simplified by noting the relationship similar to (8.139),

$$\tilde{\boldsymbol{\sigma}}^\nu p_\nu \boldsymbol{\sigma}^\mu p_\mu = \mathbf{I}\, \eta^{\mu\nu} p_\mu p_\nu = \mathbf{I}\, p^2 = 0. \tag{8.146}$$

As a result, the momentum spinor that solves (8.143) is given by

$$u(\boldsymbol{p}) = \tilde{\boldsymbol{\sigma}}^\nu p_\nu u(0) = (p_0 \sigma^0 - \boldsymbol{p} \cdot \boldsymbol{\sigma}) u(0), \tag{8.147}$$

where $u(0)$ is a constant nonzero spinor.

There are two independent constant spinors that can be used to construct specific solutions and, as a result, there are *two solutions* to the Weyl equation (8.138). Inserting normalization factors $N_\pm$ for the moment, these solutions are given by using the Pauli matrices in the form of (7.133):

$$\begin{aligned} N_+ u_+(\boldsymbol{p}) &= \tilde{\boldsymbol{\sigma}}^\mu p_\mu \begin{pmatrix} 1 \\ 0 \end{pmatrix} \\ &= \begin{pmatrix} p^0 - p^3 & -p^1 + ip^2 \\ -p^1 - ip^2 & p^0 + p^3 \end{pmatrix} \begin{pmatrix} 1 \\ 0 \end{pmatrix} = \begin{pmatrix} p^0 - p^3 \\ -p^1 - ip^2 \end{pmatrix}, \\ N_- u_-(\boldsymbol{p}) &= \tilde{\boldsymbol{\sigma}}^\mu p_\mu \begin{pmatrix} 0 \\ 1 \end{pmatrix} = \begin{pmatrix} p^0 - p^3 & -p^1 + ip^2 \\ -p^1 - ip^2 & p^0 + p^3 \end{pmatrix} \begin{pmatrix} 0 \\ 1 \end{pmatrix} \\ &= \begin{pmatrix} -p^1 + ip^2 \\ p^0 + p^3 \end{pmatrix}, \end{aligned} \tag{8.148}$$

where $p^0 = +\sqrt{\boldsymbol{p} \cdot \boldsymbol{p}}$. It is important to note that the Pauli matrices can be unitarily transformed, and the explicit forms (8.148) will change, as will the transformation $\mathbf{U}$

defined in (7.204). The two momentum spinors of (8.148) are orthogonal in the following way,

$$N_+^* N_- u_+^\dagger(\boldsymbol{p}) u_-(-\boldsymbol{p}) = N_-^* N_+ u_-^\dagger(\boldsymbol{p}) u_+(-\boldsymbol{p}) = 0. \tag{8.149}$$

The normalized spinors, denoted by $u_+(\boldsymbol{p})$ and $u_-(\boldsymbol{p})$, define $N_\pm$ so that the inner product is given by

$$u_+^\dagger(\boldsymbol{p}) u_+(\boldsymbol{p}) = u_-^\dagger(\boldsymbol{p}) u_-(\boldsymbol{p}) = 1. \tag{8.150}$$

Exercise 8.7: Show that the normalized spinors are given by

$$\begin{aligned} u_+(\boldsymbol{p}) &= \frac{1}{\sqrt{2p^0(p^0 - p^3)}} \begin{pmatrix} p^0 - p^3 \\ -p^1 - ip^2 \end{pmatrix}, \\ u_-(\boldsymbol{p}) &= \frac{1}{\sqrt{2p^0(p^0 + p^3)}} \begin{pmatrix} -p^1 + ip^2 \\ p^0 + p^3 \end{pmatrix}. \end{aligned} \tag{8.151}$$

Exercise 8.8: Show that both spinors of (8.151) explicitly satisfy the helicity relation

$$\boldsymbol{h} u_\pm(\boldsymbol{p}) = \frac{1}{2p_0} p^j \sigma_j u_\pm(\boldsymbol{p}) = \frac{1}{2p_0} \begin{pmatrix} p^3 & p^1 - ip^2 \\ p^1 + ip^2 & -p^3 \end{pmatrix} u_\pm(\boldsymbol{p}) = -\frac{1}{2} u_\pm(\boldsymbol{p}). \tag{8.152}$$

The two independent solutions to the Weyl equation (8.138) are written in terms of the normalized momentum spinors as

$$\xi_+(\boldsymbol{p}, x) = e^{-ip\cdot x}\, u_+(\boldsymbol{p}), \tag{8.153}$$

$$\xi_-(\boldsymbol{p}, x) = e^{ip\cdot x}\, u_-(\boldsymbol{p}), \tag{8.154}$$

where $p_0 = +\sqrt{\boldsymbol{p} \cdot \boldsymbol{p}}$. These spinors are orthogonal, since they obey the relation

$$\begin{aligned} (\xi_+, \xi_-) &\equiv \int \frac{\mathrm{d}^3 x}{(2\pi)^3}\, \xi_+^\dagger(\boldsymbol{p}, x) \xi_-(\boldsymbol{k}, x) \\ &= \delta^3(\boldsymbol{k} + \boldsymbol{p})\, u_+^\dagger(\boldsymbol{p}) u_-(\boldsymbol{k}) e^{-i(k_0 + p_0)x_0} \\ &= \delta^3(\boldsymbol{k} + \boldsymbol{p})\, u_+^\dagger(\boldsymbol{p}) u_-(-\boldsymbol{p}) e^{-2ip_0 x_0} = 0, \end{aligned} \tag{8.155}$$

where (8.149) was used. The solutions (8.153) and (8.154) are therefore independent and comprise the modes of the Weyl equation. A general solution to the equation of motion (8.138) can therefore be written as a linear superposition of the two independent modes (8.153) and (8.154),

$$\xi(x) = \int \frac{d^3p}{(2\pi)^3} \left(\alpha_p\, \xi_+(\boldsymbol{p}, x) + \beta_p^*\, \xi_-(\boldsymbol{p}, x)\right), \tag{8.156}$$

where α_p and β_p are arbitrary complex coefficients to be discussed. The units of α_p and β_p are length to the 3/2 power, so that ξ has units of length to the $-3/2$ power.

> **Exercise 8.9:** Combining relation (7.216) with the general form of the solutions given by (8.148), show that (8.156) transforms according to (7.204) under a Lorentz transformation.

The nature of the general solution can be clarified by examining the Hamiltonian, which is constructed according to the standard procedure of (3.62). The canonical momenta are given by

$$\begin{aligned} \pi &= \frac{\partial \mathcal{L}_\xi}{\partial \dot{\xi}} = \frac{1}{2} i \frac{\hbar}{c}\, \xi^\dagger \boldsymbol{\sigma}^0 = \frac{1}{2} i \frac{\hbar}{c}\, \xi^\dagger, \\ \pi^\dagger &= \frac{\partial \mathcal{L}_\xi}{\partial \dot{\xi}^\dagger} = -\frac{1}{2} i \frac{\hbar}{c}\, \boldsymbol{\sigma}^0 \xi = -\frac{1}{2} i \frac{\hbar}{c}\, \xi, \end{aligned} \tag{8.157}$$

where the factors of $\hbar$ and c will be kept. Even though it is not possible to solve for $\dot{\xi}$ solely in terms of π and ξ, the Hamiltonian density is found to be

$$\begin{aligned} \mathcal{H} &= c(\pi\dot{\xi} + \dot{\xi}^\dagger \pi^\dagger - \mathcal{L}_\xi) = -\frac{1}{2} i\hbar\, \xi^\dagger \boldsymbol{\sigma}^j \partial_j \xi + \frac{1}{2} i\hbar\, \partial_j \xi^\dagger \boldsymbol{\sigma}^j \xi \\ &= -\, \pi \boldsymbol{\sigma}^j \partial_j \xi - \partial_j \xi^\dagger \boldsymbol{\sigma}^j \pi^\dagger. \end{aligned} \tag{8.158}$$

Evaluating (8.158) using the equations of motion for ξ and $\xi^\dagger$,

$$\boldsymbol{\sigma}^j \partial_j \xi = -\boldsymbol{\sigma}^0 \partial_0 \xi = -\frac{1}{c}\dot{\xi}, \quad \partial_j \xi^\dagger \boldsymbol{\sigma}^j = -\partial_0 \xi^\dagger \boldsymbol{\sigma}^0 = -\frac{1}{c}\dot{\xi}^\dagger, \tag{8.159}$$

shows that the Hamiltonian can be written

$$\mathcal{H} = \frac{1}{2} i\hbar\, \xi^\dagger \dot{\xi} - \frac{1}{2} i\hbar\, \dot{\xi}^\dagger \xi. \tag{8.160}$$

Inserting the general solution (8.156) into (8.160) gives the Hamiltonian in terms of the α_p and β_p coefficients,

$$H = \frac{1}{2} i\hbar \int d^3x\, (\xi^\dagger \dot{\xi} - \dot{\xi}^\dagger \xi) = \int \frac{d^3p}{(2\pi)^3}\, \hbar\omega_p \left(\alpha_p^* \alpha_p - \beta_p \beta_p^*\right), \tag{8.161}$$

where the orthogonality relations of (8.155) were used and $\omega_p = +c\sqrt{\boldsymbol{p}\cdot\boldsymbol{p}}$.

Exercise 8.10: Verify (8.161).

It is instructive to note that the order of the coefficients has been preserved in (8.161). If β_p is a standard complex number then

$$\beta_p\beta_p^* = \beta_p^*\beta_p, \tag{8.162}$$

and the product is a real positive quantity. For such a case $\xi_-(\boldsymbol{p}, x)$ represents a negative energy mode and the energy of the classical Weyl spinor is unbounded from below. This is an unphysical situation, since the Weyl spinor would serve as an unbounded source of positive energy as more and more negative energy modes were created through its interaction with other fields. The origin of the negative energy is the use of $\exp(ip\cdot x)$ in solution (8.154), and it is worth considering changing the sign of p to be the same as that of the positive energy solution (8.153). However, doing so would cause the orthogonality relation (8.155) to break down and the two solutions would no longer be independent. The appearance of negative energy solutions is a general property of relativistic spinor fields, and remedying it requires the techniques of quantum field theory. It is worth noting that if $\beta_p\beta_p^* = -\beta_p^*\beta_p$, then (8.161) takes a step toward becoming positive. Such a possibility can only occur if β_p and β_p^* are *Grassmann variables* or if β_p and β_p^* become *anticommuting operators.* Both topics lie far from classical field theory, since they require the introduction of the *Pauli exclusion principle* as well as the definition of *antimatter states.*

It is also important to note that there is a second version of the Weyl spinor field, denoted by $\tilde{\xi}$, found by using the Lorentz algebra of (7.224). The Lagrangian is given by

$$\mathcal{L}_{\tilde{\xi}} = \frac{1}{2}i\hbar\,\tilde{\xi}^\dagger\tilde{\boldsymbol{\sigma}}^\mu\partial_\mu\tilde{\xi} - \frac{1}{2}i\hbar\,(\partial_\mu\tilde{\xi}^\dagger)\tilde{\boldsymbol{\sigma}}^\mu\tilde{\xi}, \tag{8.163}$$

where $\tilde{\boldsymbol{\sigma}}^\mu = (\sigma_0, \boldsymbol{\sigma})$ in terms of the Pauli spin matrices. The equation of motion for (8.163) is given by

$$i\tilde{\boldsymbol{\sigma}}^\mu\partial_\mu\tilde{\xi} = i(\tilde{\boldsymbol{\sigma}}^0\partial_0 + \tilde{\boldsymbol{\sigma}}^j\partial_j)\tilde{\xi} = i(\boldsymbol{\sigma}^0\partial_0 + \boldsymbol{\sigma}\cdot\nabla)\tilde{\xi} = 0. \tag{8.164}$$

The Lagrangian of (8.163) and the associated equation of motion (8.164) are related to those of (8.136) and (8.138) through a change of parity, $\nabla \to -\nabla$. As mentioned in the previous chapter, there is no unitary transformation that changes $\boldsymbol{\sigma}_j \to -\boldsymbol{\sigma}_j$ and so the two representations are inequivalent. The analysis of (8.163) is literally the mirror image of the analysis of (8.136) and the details are left as an exercise. Using the form

$$\tilde{\xi}_{\pm} = e^{\mp ip\cdot x}\tilde{u}(\boldsymbol{p}) \tag{8.165}$$

reduces the second Weyl equation (8.164) to

$$(p^0\sigma^0 - \boldsymbol{p}\cdot\boldsymbol{\sigma})\tilde{u}(\boldsymbol{p}) = 0, \tag{8.166}$$

so that all solutions to (8.166) have the property

$$\left(\frac{1}{2p^0}\boldsymbol{p}\cdot\boldsymbol{\sigma}\right)\tilde{u}(\boldsymbol{p}) = h\tilde{u}(\boldsymbol{p}) = +\frac{1}{2}\tilde{u}(\boldsymbol{p}). \tag{8.167}$$

This shows that $\tilde{u}(\boldsymbol{p})$ is an eigenspinor of the helicity with the eigenvalue $+\frac{1}{2}$. This will be discussed further in the section on Noether's theorem.

> **Exercise 8.11:** Determine the modes for $\tilde{\xi}$ associated with the equation of motion given by (8.163) and evaluate the Hamiltonian associated with a general solution to show the appearance of negative energy modes.

8.5.2 The Dirac bispinor and its solutions

Beyond the appearance of negative energy modes, the Weyl spinor has several additional drawbacks that limit its utility. There is no parity operation and it is impossible to add an inertial mass term to either of the Lagrangians for the two-component Weyl spinor. This was remedied by using both types of Weyl spinor to define the Dirac bispinor ψ in (7.243) and the γ matrices in (7.247),

$$\psi = \begin{pmatrix}\xi\\ \tilde{\xi}\end{pmatrix}, \quad \gamma^\mu = \begin{pmatrix}0 & \tilde{\boldsymbol{\sigma}}^\mu\\ \boldsymbol{\sigma}^\mu & 0\end{pmatrix}. \tag{8.168}$$

The reader is reminded that this specific choice for the γ matrices is referred to as the *Weyl or chiral representation*. It is not unique, since the Dirac matrices can be redefined using a unitary transformation. The two Lorentz invariant bispinor expressions (7.246) and (7.255) can be used to define a Lorentz invariant Lagrangian density for the Dirac bispinor,

$$\mathcal{L}_\psi = \frac{1}{2}i\hbar\left(\bar{\psi}\gamma^\mu\partial_\mu\psi - \partial_\mu\bar{\psi}\gamma^\mu\psi\right) - \hbar m\,\bar{\psi}\psi, \tag{8.169}$$

where $\bar{\psi} = \psi^\dagger\gamma_0$ and $m = Mc/\hbar$ is an inverse length associated with the inertial mass M. The Lagrangian (8.169) is real, since

$$(i\bar{\psi}\gamma^\mu\partial_\mu\psi)^* = -i\partial_\mu\psi^\dagger\gamma^{\mu\dagger}\gamma_0^\dagger\psi = -i\partial_\mu\psi^\dagger\gamma_0\gamma^\mu\psi = -i\partial_\mu\bar{\psi}\gamma^\mu\psi, \tag{8.170}$$

where the identity (7.280) was used. The Dirac spinor has the same units as the Weyl spinor, so that $\bar{\psi}\psi$ has units of inverse spatial volume. The *Dirac equation* for ψ and $\bar{\psi}$ follows from the Euler–Lagrange equation,

$$\frac{\partial \mathcal{L}_\psi}{\partial \bar{\psi}} - \partial_\mu \frac{\partial \mathcal{L}_\psi}{\partial(\partial_\mu \bar{\psi})} = i\hbar(\gamma^\mu \partial_\mu + im)\psi = 0, \tag{8.171}$$

along with the conjugate equation

$$\frac{\partial \mathcal{L}_\psi}{\partial \psi} - \partial_\mu \frac{\partial \mathcal{L}_\psi}{\partial(\partial_\mu \psi)} = -i\hbar(\partial_\mu \bar{\psi}\, \gamma^\mu - im\bar{\psi}) = 0. \tag{8.172}$$

The conjugate equation can also be obtained by taking the Hermitian adjoint of (8.171) and using the identity (7.280).

Before discussing the specific solutions to (8.171) it is useful to find their general properties. Using the chiral version (8.168) for the γ matrices, the Dirac equation can be written as

$$i\hbar(\gamma^\mu \partial_\mu + im)\psi = \hbar \begin{pmatrix} -m & i\tilde{\boldsymbol{\sigma}}^\mu \partial_\mu \\ i\boldsymbol{\sigma}^\mu \partial_\mu & -m \end{pmatrix} \begin{pmatrix} \xi \\ \tilde{\xi} \end{pmatrix} = 0. \tag{8.173}$$

Using matrix multiplication, this can be viewed as two equations,

$$i\boldsymbol{\sigma}^\mu \partial_\mu \xi - m\tilde{\xi} = 0, \tag{8.174}$$

$$i\tilde{\boldsymbol{\sigma}}^\mu \partial_\mu \tilde{\xi} - m\xi = 0, \tag{8.175}$$

so that the two two-component spinor fields ξ and $\tilde{\xi}$ are coupled to each other if $m \neq 0$. For the case that $m = 0$ the two equations decouple and become the two inequivalent Weyl spinor equations of motion (8.138) and (8.164). The Dirac equation therefore involves both types of Weyl spinor, and this allows a parity transformation.

Analyzing the Dirac equation is facilitated by the useful product

$$\begin{aligned}(\gamma^\nu \partial_\nu - im\mathbf{I})(\gamma^\mu \partial_\mu + im\mathbf{I}) &= \frac{1}{2}\{\gamma^\mu, \gamma^\nu\}\partial_\mu \partial_\nu + m^2 \mathbf{I} \\ &= \mathbf{I}(\eta^{\mu\nu}\partial_\mu \partial_\nu + m^2) = \mathbf{I}(\Box + m^2),\end{aligned} \tag{8.176}$$

where the anticommutation relation (7.249) was used. Combining this with (8.171) shows that the Dirac bispinor satisfies

$$(\gamma^\nu \partial_\nu - im\mathbf{I})(\gamma^\mu \partial_\mu + im\mathbf{I})\psi = (\Box + m^2)\psi = 0. \tag{8.177}$$

As a result, each component of the Dirac bispinor satisfies the massive wave equation familiar from the scalar Klein–Gordon field. Result (8.177) shows that the solutions to (8.171) take the mode form

$$\psi_{\boldsymbol{k}}^{+}(x) = e^{-ik\cdot x} u(\boldsymbol{k}), \tag{8.178}$$

$$\psi_{\boldsymbol{k}}^{-}(x) = e^{ik\cdot x} v(\boldsymbol{k}), \tag{8.179}$$

where $k^2 - m^2 = 0$ and $u(\boldsymbol{k})$ and $v(\boldsymbol{k})$ are the momentum bispinors associated with the respective possible choices of sign for $\exp(\pm ik \cdot x)$. The apparent backwards naming of $\psi_k^{\pm}(x)$ will be justified when the energies of the modes are calculated. Inserting the forms (8.178) and (8.179) into the Dirac equation (8.171) gives the equations which the momentum bispinors must solve,

$$\begin{aligned}(k_\mu\gamma^\mu - m\mathbf{I})u(\boldsymbol{k}) &= 0,\\ (k_\mu\gamma^\mu + m\mathbf{I})v(\boldsymbol{k}) &= 0.\end{aligned} \tag{8.180}$$

Because the field obeys the massive relation $k^2 - m^2 = k_0^2 - \boldsymbol{k}\cdot\boldsymbol{k} - m^2 = 0$, the momentum spinor equations of (8.180) can be transformed to the rest frame where $\boldsymbol{k} = 0$ and $k_0 = +m$. In the rest frame the equations of (8.180) become

$$\begin{aligned}\gamma^0 u(0) &= u(0),\\ \gamma^0 v(0) &= -\,v(0).\end{aligned} \tag{8.181}$$

The relations of (8.181) can be understood as eigenvalue equations. The rest frame bispinors are *eigenspinors* of the matrix γ^0, with $u(0)$ corresponding to the eigenvalue $+1$ and $v(0)$ corresponding to the eigenvalue -1. Demonstrating that these eigenspinors exist follows from finding the eigenvalues λ of γ^0 from the characteristic equation (7.51), which gives

$$\det(\gamma^0 - \lambda\mathbf{I}) = \lambda^4 - 2\lambda^2 + 1 = (\lambda^2 - 1)^2 = 0. \tag{8.182}$$

Exercise 8.12: Using the chiral representation of γ_0 given by (7.244), verify (8.182).

The fourth-order equation (8.182) has four solutions; two of them are given by $\lambda = +1$ and two of them are given by $\lambda = -1$. There are therefore two eigenspinors $u^{(s)}(0)$ associated with the two positive eigenvalues and two eigenspinors $v^{(s)}(0)$ associated with the two negative eigenvalues, where the eigenspinors are indexed with $s = 1, 2$. The eigenspinors are explicitly given by

$$u^{(1)}(0) = \frac{1}{\sqrt{2}}\begin{pmatrix}1\\0\\1\\0\end{pmatrix},\quad u^{(2)}(0) = \frac{1}{\sqrt{2}}\begin{pmatrix}0\\1\\0\\1\end{pmatrix}, \tag{8.183}$$

$$v^{(1)}(0) = \frac{1}{\sqrt{2}}\begin{pmatrix}1\\0\\-1\\0\end{pmatrix},\quad v^{(2)}(0) = \frac{1}{\sqrt{2}}\begin{pmatrix}0\\1\\0\\-1\end{pmatrix}. \tag{8.184}$$

Exercise 8.13: Using the chiral representation of γ_0 given by (7.244), verify that (8.183) and (8.184) are the four eigenspinors of γ_0. Verify that the four eigenspinors are independent degrees of freedom that satisfy

$$u^{(s)\dagger}(0)u^{(s')}(0) = v^{(s)\dagger}(0)v^{(s')}(0) = \delta^{ss'}, \quad u^{(s)\dagger}(0)v^{(s')}(0) = 0. \tag{8.185}$$

The solutions to the Dirac momentum spinor equation (8.180) can now be found from (8.183) and (8.184) by using a method identical to that used in (8.148) for the Weyl spinor. Utilizing the γ matrix property (7.249) shows that

$$\begin{aligned}(k^\mu\gamma_\mu + m\mathbf{I})(k^\nu\gamma_\nu - m\mathbf{I}) &= \left(\frac{1}{2}k^\mu k^\nu\{\gamma_\mu, \gamma_\nu\} - m^2\mathbf{I}\right) \\ &= \mathbf{I}\,(\eta_{\mu\nu}k^\mu k^\nu - m^2) = \mathbf{I}\,(k^2 - m^2) = 0.\end{aligned} \tag{8.186}$$

Using (8.186) enables the construction of the solution to (8.180) for arbitrary k from the four eigenspinors $w_\pm^{(s)}(0)$. Letting $N_+(\boldsymbol{k})$ denote the normalization factor, the two momentum eigenspinors for the positive eigenvalue are given by

$$u^{(s)}(\boldsymbol{k}) = \frac{1}{N_+(\boldsymbol{k})}(\gamma^\nu k_\nu + m)u^{(s)}(0). \tag{8.187}$$

Similarly, those for the negative eigenvalues are given by

$$v^{(s)}(\boldsymbol{k}) = \frac{1}{N_-(\boldsymbol{k})}(\gamma^\nu k_\nu - m)v^{(s)}(0), \tag{8.188}$$

where $k_0 = +\sqrt{\boldsymbol{k}\cdot\boldsymbol{k} + m^2}$. The general solution to the Dirac equation is given by a mode expansion,

$$\psi(x) = \int \frac{d^3k}{(2\pi)^3}\sum_{s=1}^{2}\left(\alpha_k^{(s)}u^{(s)}(\boldsymbol{k})e^{-ik\cdot x} + \beta_k^{(s)*}v^{(s)}(\boldsymbol{k})e^{ik\cdot x}\right), \tag{8.189}$$

where $\alpha_k^{(s)}$ and $\beta_k^{(s)}$ are four arbitrary complex coefficients. A particularly simple set of solutions is given by the rest frame forms of (8.178) and (8.179),

$$\psi_+^{(s)}(x^0) = \frac{1}{\sqrt{V}}\, e^{-imx^0}u^{(s)}(0) \tag{8.190}$$

$$\psi_-^{(s)}(x^0) = \frac{1}{\sqrt{V}}\, e^{imx^0}v^{(s)}(0), \tag{8.191}$$

where V is the spatial volume of the system.

The physical meanings of the rest solutions (8.190) and (8.191) are once again revealed by the Hamiltonian of the Dirac bispinor field. The momentum canonically conjugate to $\psi(x)$ is given by

$$\pi(x) = \frac{\partial \mathcal{L}_\psi}{\partial \dot{\psi}(x)} = i\frac{\hbar}{c}\bar{\psi}(x)\gamma_0 = i\frac{\hbar}{c}\psi^\dagger(x). \tag{8.192}$$

Using this in the Legendre transformation gives the Hamiltonian density

$$\mathcal{H}_\psi = c(\pi\dot{\psi} - \mathcal{L}_\psi) = -i\hbar c\, \bar{\psi}(\gamma_j\partial^j + im)\psi. \tag{8.193}$$

Evaluating (8.193) using a solution to the Dirac equation gives

$$\mathcal{H}_\psi = i\hbar c\, \bar{\psi}\gamma_0\partial^0\psi = i\hbar\, \psi^\dagger\dot{\psi}. \tag{8.194}$$

For the rest frame solution (8.190), the Hamiltonian is given by

$$H = \int \mathrm{d}^3x\, \mathcal{H}_\psi = \frac{\hbar c\, m}{V}\int \mathrm{d}^3x\, u^{(s)\dagger}(0)u^{(s)}(0) = \hbar c\, m = Mc^2, \tag{8.195}$$

where the orthonormality relation (8.185) was used. Solution (8.190) can therefore represent the particle density for a single massive particle at rest in the volume V. However, negative energy solutions are present, since the rest frame solution (8.191) gives

$$H = \int \mathrm{d}^3x\, \mathcal{H}_\psi = -\frac{\hbar c\, m}{V}\int \mathrm{d}^3x\, v^{(s)\dagger}(0)v^{(s)}(0) = -\hbar c\, m = -Mc^2. \tag{8.196}$$

The Klein–Gordon Feynman propagator $G_F(x - y)$ defined in (8.88) can be adapted to find the Green's function for the Dirac equation. The Green's function for the Dirac equation is given by the matrix

$$\boldsymbol{S}_F(x - y) = \left(i\gamma_\mu\partial^\mu + m\mathbf{I}\right)G_F(x - y). \tag{8.197}$$

Using (8.176) shows that

$$\begin{aligned}(i\gamma_\nu\partial^\nu - m\mathbf{I})\boldsymbol{S}_F(x - y) &= (i\gamma_\nu\partial^\nu - m\mathbf{I})\left(i\gamma_\mu\partial^\mu + m\mathbf{I}\right)G_F(x - y)\\ &= -\mathbf{I}\,(\Box + m^2)G_F(x - y) = \mathbf{I}\,\delta^4(x - y).\end{aligned} \tag{8.198}$$

Using results (8.97) and (8.98) allows the evaluation of the Dirac equation Green's function in terms of the space-time vector $x - y$.

8.6 Symmetry and conservation laws

So far many of the important concepts in classical field theory have been developed, but much of the presentation has lacked a unifying principle. In particular, the identification of conserved quantities, which are critical to the understanding of physical processes, has been ad hoc and primarily based on use of the Euler–Lagrange equations of motion to justify them. This is remedied by *Noether's theorem*, which provides a systematic means of identifying conserved quantities in field theory. The key concept is that of *symmetry*. In the context of field theory, a symmetry means that the action defining the theory is invariant under a *symmetry operation*. For example, the action in a relativistic field theory is constructed to be *a priori* invariant under Lorentz transformations, so that Lorentz invariance is automatically a symmetry of the action. Typically, a symmetry of the action in a field theory can involve both a change in the *form or components* of the fields as well as in the *arguments* of the fields, as in the case of a spinor field Lorentz transformation, $\xi'(x') = \mathbf{U}\xi(x)$. It should be stressed that this statement means that a second inertial observer will use the event x' found from x using special relativity, but will also perform a transformation of the components of the field. Noether's theorem connects each symmetry of the action to a conservation law.

The action S for a general field theory is assumed to be a functional of a set of fields, $\varphi_\alpha(x)$, and their first derivatives, $\partial_\mu\varphi_\alpha$, where

$$S[\varphi] = \int \mathrm{d}^4x \ \mathcal{L}(\varphi_\alpha(x), \ \partial_\mu\varphi_\alpha(x)). \tag{8.199}$$

The subscript α appearing on the field represents a set of indices labeling *all* possible independent degrees of freedom for the fields appearing in the action. The indices can include tensor indices, spinor indices, and *isospin* indices. Isospin indices label a collection of fields that share a set of common properties. The first use of isospin grouped the neutron and proton together into a doublet and referred to the doublet as the *nucleon*. This was useful for building a theory of the nucleus, since the neutron and proton share many similar properties, such as an almost identical mass as well as similar interactions with other particles. The concept of isospin symmetry underlies the current formulation of many subatomic processes using field theory.

8.6.1 Noether's theorem

The goal is to characterize all actions on the fields and their arguments that leave the action invariant. To this end a general symmetry operation on a field $\varphi_\alpha(x)$ is written

$$\varphi_\alpha(x) \to \varphi'_\alpha(x'), \tag{8.200}$$

and is assumed to be a *continuous symmetry* that depends on a set of continuous parameters such as the rapidity and rotation angles of the Lorentz group. Not all symmetries are continuous. For example, an action might be invariant under the operation $\varphi_\alpha \to -\varphi_\alpha$, which cannot be expressed in an infinitesimal form. Such a symmetry is typically referred to as a *discrete symmetry*. If the symmetry operation is *continuous*, it is characterized by a set of parameters and can be written in infinitesimal form,

$$\varphi'_\alpha(x') = \varphi_\alpha(x) + \delta\varphi_\alpha(x), \tag{8.201}$$

where $\delta\varphi_\alpha$ is the infinitesimal version of the symmetry operation. For example, the Lorentz transformation of a scalar field obeys $\varphi'(x') = \varphi(x)$, and so the infinitesimal Lorentz transformation of a scalar field is characterized by $\delta\varphi(x) = 0$. The form of the scalar field function does not change, while $x' = \Lambda \cdot x$ is the original point in the new coordinate system. The Lorentz transformation of the spinor field is given by

$$\xi'(x') = \mathbf{U} \cdot \xi(x) \implies \delta\xi(x) = (\mathbf{U} - \mathbf{I}) \cdot \xi(x) = \delta\mathbf{U} \cdot \xi(x), \tag{8.202}$$

where $\delta \mathbf{U}$ is understood to be the *infinitesimal version* of the spinor transformation. The infinitesimal nature of the transformation allows terms higher than $O(\delta\varphi_\alpha)$ to be dropped. The term $\delta\varphi_\alpha$ is not to be confused with the variation of the field used to derive the Euler–Lagrange equation in chapter 2. In that case the variation vanished at the boundaries of integration so that total derivatives could be dropped. This will *not* be the case with the symmetry operation, and total derivatives involving $\delta\varphi_\alpha$ must be retained.

In addition to a change in the form of the field, the argument of the field may also undergo a continuous coordinate transformation, such as the Lorentz transformation. In its infinitesimal form this is written

$$x^\mu \to x^{\mu\,\prime} = x^\mu + \delta x^\mu. \tag{8.203}$$

All terms higher than first order in δx^μ, *including* cross terms with $\delta\varphi_\alpha$, can be dropped. It is clear that (8.203) will affect the measure of the action (8.199). A symmetry of the action therefore requires that

$$\begin{aligned} S \to S' &= \int \mathrm{d}^4x'\ \mathcal{L}(\varphi'_\alpha(x'),\ \partial'_\mu\varphi'_\alpha(x')) \\ &= \int \mathrm{d}^4x\ \mathcal{L}(\varphi_\alpha(x),\ \partial_\mu\varphi_\alpha(x)) = S. \end{aligned} \tag{8.204}$$

Because there may be a simultaneous transformation in the form of the field and its argument, the action of the symmetry operation on derivatives of the fields is more complicated. If ∂_μ undergoes a transformation, it will be the case that

$$\delta(\partial_\mu\varphi_\alpha) \neq \partial_\mu(\delta\varphi_\alpha). \tag{8.205}$$

It should be stressed that this is different from the variational approach of chapter 2, where the variation of the derivative of the field was the derivative of the variation of the field. In order to distinguish this aspect of symmetry operations, a Taylor series expansion of the symmetry operation is used to obtain

$$\varphi'_\alpha(x') = \varphi'_\alpha(x + \delta x) = \varphi'_\alpha(x) + \partial_\mu\varphi'_\alpha(x)\,\delta x^\mu = \varphi'_\alpha(x) + \partial_\mu\varphi_\alpha(x)\,\delta x^\mu, \tag{8.206}$$

where the last step occurs since $\delta\varphi_\alpha\delta x^\mu = 0$. This is rewritten to obtain

$$\varphi'_\alpha(x) = \varphi'_\alpha(x') - \partial_\mu\varphi_\alpha(x)\,\delta x^\mu. \tag{8.207}$$

Expression (8.207) allows the definition of a second kind of variation of the field, denoted by $\bar{\delta}$ and given by

$$\bar{\delta}\varphi_\alpha(x) = \varphi'_\alpha(x) - \varphi_\alpha(x). \tag{8.208}$$

Because $\bar{\delta}$ does not involve x^μ, it has the critical property that

$$\bar{\delta}(\partial_\mu\varphi_\alpha(x)) = \partial_\mu\varphi'_\alpha(x) - \partial_\mu\varphi_\alpha(x) = \partial_\mu(\varphi'_\alpha(x) - \varphi_\alpha(x)) = \partial_\mu\bar{\delta}\varphi_\alpha(x). \tag{8.209}$$

Variation (8.208) is related to the variation of (8.201) through (8.207) by

$$\bar{\delta}\varphi_\alpha(x) = \varphi'_\alpha(x') - \partial_\mu\varphi_\alpha(x)\,\delta x^\mu - \varphi_\alpha(x) = \delta\varphi_\alpha(x) - \partial_\mu\varphi_\alpha(x)\,\delta x^\mu. \tag{8.210}$$

Since it is infinitesimal, the variation $\bar{\delta}$ acts linearly on field products, so that

$$\frac{1}{2}\bar{\delta}(\partial_\mu\varphi_\alpha\partial^\mu\varphi_\alpha) = \partial^\mu\varphi_\alpha\partial_\mu\bar{\delta}\varphi_\alpha. \tag{8.211}$$

Under the action of the symmetry operation the change in the Lagrangian density is given by

$$\begin{aligned}\delta\mathcal{L}(x,\,\varphi_\alpha(x)) &= \mathcal{L}'(x',\,\varphi'_\alpha(x')) - \mathcal{L}(x,\,\varphi_\alpha(x))\\ &= \mathcal{L}'(x,\,\varphi_\alpha(x)) + \partial_\mu\mathcal{L}(x,\,\varphi_\alpha(x))\delta x^\mu - \mathcal{L}(x,\,\varphi_\alpha(x))\\ &= \bar{\delta}\mathcal{L}(x,\,\varphi_\alpha(x)) + \partial_\mu\mathcal{L}(x,\,\varphi_\alpha(x))\,\delta x^\mu.\end{aligned} \tag{8.212}$$

Using (8.211) and then (8.210) gives

$$\begin{aligned}\bar{\delta}\mathcal{L}(x,\,\varphi_\alpha(x)) &= \frac{\partial\mathcal{L}}{\partial\varphi_\alpha}\bar{\delta}\varphi_\alpha + \frac{\partial\mathcal{L}}{\partial(\partial_\mu\varphi_\alpha)}\partial_\mu\bar{\delta}\varphi_\alpha\\ &= \left(\frac{\partial\mathcal{L}}{\partial\varphi_\alpha} - \partial_\mu\frac{\partial\mathcal{L}}{\partial(\partial_\mu\varphi_\alpha)}\right)\bar{\delta}\varphi_\alpha + \partial_\mu\left(\frac{\partial\mathcal{L}}{\partial(\partial_\mu\varphi_\alpha)}\bar{\delta}\varphi_\alpha\right)\\ &= \partial_\mu\left(\frac{\partial\mathcal{L}}{\partial(\partial_\mu\varphi_\alpha)}\bar{\delta}\varphi_\alpha\right) = \partial_\mu\left(\frac{\partial\mathcal{L}}{\partial(\partial_\mu\varphi_\alpha)}(\delta\varphi_\alpha(x) - \partial_\nu\varphi_\alpha(x)\,\delta x^\nu)\right)\\ &= \partial_\mu\left(\frac{\partial\mathcal{L}}{\partial(\partial_\mu\varphi_\alpha)}\delta\varphi_\alpha - \frac{\partial\mathcal{L}}{\partial(\partial_\mu\varphi_\alpha)}\partial_\nu\varphi_\alpha\delta x^\nu\right),\end{aligned} \tag{8.213}$$

where it is assumed that φ_α is a solution of the Euler–Lagrange equation. Returning this result to (8.212) gives the variation of the Lagrangian density under the symmetry operation,

$$\begin{aligned}\delta\mathcal{L}(x,\,\varphi_\alpha(x)) &= \partial_\mu\left(\frac{\partial\mathcal{L}}{\partial(\partial_\mu\varphi_\alpha)}\delta\varphi_\alpha - \frac{\partial\mathcal{L}}{\partial(\partial_\mu\varphi_\alpha)}\partial_\nu\varphi_\alpha\delta x^\nu\right)\\ &\quad + \partial_\mu\mathcal{L}(x,\,\varphi_\alpha(x))\,\delta x^\mu\\ &= \partial_\mu\left[\left(\mathcal{L}\,\delta^\mu{}_\nu - \frac{\partial\mathcal{L}}{\partial(\partial_\mu\varphi_\alpha)}\partial_\nu\varphi_\alpha\right)\delta x^\nu + \frac{\partial\mathcal{L}}{\partial(\partial_\mu\varphi_\alpha)}\delta\varphi_\alpha\right]\\ &\quad - \mathcal{L}\,\partial_\mu\delta x^\mu.\end{aligned} \tag{8.214}$$

If there is a nonzero change δx^μ in the coordinates, then the integral of the Lagrangian that determines the action must include the Jacobian J of the coordinate change, so that

$$d^4x' = d^4x\,J. \tag{8.215}$$

The Jacobian can be analyzed for the infinitesimal case by using

$$\frac{\partial x'^{\mu}}{\partial x^{\nu}} = \delta^{\mu}_{\nu} + \frac{\partial \delta x^{\mu}}{\partial x^{\nu}}, \tag{8.216}$$

in the expression (7.41) for the determinant, so that the Jacobian is given by

$$J = \det\left(\frac{\partial x'}{\partial x}\right) = \frac{1}{4!}\varepsilon^{\mu\nu\rho\sigma}\varepsilon_{\alpha\beta\gamma\epsilon}\frac{\partial x'^{\alpha}}{\partial x^{\mu}}\frac{\partial x'^{\beta}}{\partial x^{\nu}}\frac{\partial x'^{\gamma}}{\partial x^{\rho}}\frac{\partial x'^{\epsilon}}{\partial x^{\sigma}}. \tag{8.217}$$

Since δx^{μ} is infinitesimal, only its zero or first power may appear in the Jacobian. For the zero-power case, (8.217) becomes

$$J = \frac{1}{4!}\varepsilon^{\mu\nu\rho\sigma}\varepsilon_{\alpha\beta\gamma\epsilon}\,\delta^{\alpha}_{\mu}\delta^{\beta}_{\nu}\delta^{\gamma}_{\rho}\delta^{\epsilon}_{\sigma} = \frac{1}{4!}\varepsilon^{\mu\nu\rho\sigma}\varepsilon_{\mu\nu\rho\sigma} = 1. \tag{8.218}$$

For the case of one power there are four permutations of δx^{μ}, so the Jacobian for the infinitesimal transformation becomes $J = 1 + \delta J$, where

$$\delta J = \frac{1}{3!}\varepsilon^{\mu\nu\rho\sigma}\varepsilon_{\alpha\beta\gamma\epsilon}\delta^{\alpha}_{\mu}\delta^{\beta}_{\nu}\delta^{\gamma}_{\rho}\frac{\partial \delta x^{\epsilon}}{\partial x^{\sigma}} = \frac{1}{3!}\varepsilon^{\mu\nu\rho\sigma}\varepsilon_{\mu\nu\rho\epsilon}\frac{\partial \delta x^{\epsilon}}{\partial x^{\sigma}} = \delta^{\sigma}_{\epsilon}\frac{\partial \delta x^{\epsilon}}{\partial x^{\sigma}} = \partial_{\mu}\delta x^{\mu}. \tag{8.219}$$

Under the infinitesimal transformation the action S therefore changes to

$$S' = \int \mathrm{d}^1x\,(1 + \delta J)(\mathcal{L} + \delta\mathcal{L}) = \int \mathrm{d}^4x\,(\mathcal{L} + \mathcal{L}\,\delta J + \delta\mathcal{L}). \tag{8.220}$$

It follows that

$$\delta S = S' - S = \int \mathrm{d}^4x\,(\mathcal{L}\,\delta J + \delta\mathcal{L}), \tag{8.221}$$

where (8.212) and (8.219) give the respective quantities $\delta\mathcal{L}$ and δJ in terms of the specific transformation and the Lagrangian density. Using (8.219) shows that the term $\mathcal{L}\,\delta J$ is canceled by an identical term in (8.212). The remaining terms in (8.221) form a total divergence,

$$\delta S = \int \mathrm{d}^4x\,\partial_{\mu}J^{\mu}, \tag{8.222}$$

where J^{μ} is given by

$$J^{\mu} = \left(\delta^{\mu}_{\nu}\mathcal{L} - \frac{\partial\mathcal{L}}{\partial(\partial_{\mu}\varphi_{\alpha})}\partial_{\nu}\varphi_{\alpha}\right)\delta x^{\nu} + \frac{\partial\mathcal{L}}{\partial(\partial_{\mu}\varphi_{\alpha})}\delta\varphi_{\alpha}. \tag{8.223}$$

If the action is invariant under the transformation then $\delta S = 0$, and this gives

$$\delta S = \int \mathrm{d}^4x\,\partial_{\mu}J^{\mu} = 0. \tag{8.224}$$

In the case of a symmetry, δx^{μ} and $\delta\varphi_{\alpha}$ do not vanish at the boundaries of the space-time volume, and this requires that

$$\partial_{\mu}J^{\mu} = 0, \tag{8.225}$$

which must hold everywhere throughout the space-time volume. The quantity J^μ given by (8.223) is referred to as a *Noether current*, and (8.225) shows that *each* symmetry of the action is associated with a *conserved current*. This is the essence of Noether's theorem.

Conserved quantities were first discussed in the context of the stress–energy tensor in (3.94) and later with electric charge conservation in (6.239). The component J^0 is a *volume density* associated with a physical quantity, and its integral over a spatial volume is the amount Q of that quantity contained in the volume. Using $\partial_\mu J^\mu = 0$ and Gauss's theorem gives

$$\frac{\mathrm{d}Q}{\mathrm{d}t} = \frac{\mathrm{d}}{\mathrm{d}t}\int_V \mathrm{d}^3x\ J_0 = \int_V \mathrm{d}^3x\ \partial_0 J^0 = -\int_V \mathrm{d}^3x\ \partial_i J^i = -\oint_{S(V)} \mathrm{d}\mathbf{S}\cdot \boldsymbol{J}. \tag{8.226}$$

This shows that the change in the charge contained in a volume is identical to the flux of charge into or out of the volume through its surface, so that the charge is conserved. If the volume of integration is infinite or the fields vanish at the boundary, then Q is constant in time. The definition of Q can manifestly be made Lorentz invariant through the method applied to the total four-momentum (6.265) in (6.268) by defining

$$Q = \int \mathrm{d}S_\mu\ J^\mu, \tag{8.227}$$

which reduces to the definition in (8.226) in the rest frame.

8.6.2 Matter currents and the Weyl spinor

The Weyl spinor Lagrangian $\mathcal{L}_\xi$ given by (8.136) possesses a simple symmetry, which is the simultaneous changes

$$\xi(x) \to e^{-i\alpha}\xi(x), \quad \xi^\dagger(x) \to e^{i\alpha}\xi^\dagger(x), \tag{8.228}$$

where α is an arbitrary real constant. A symmetry with constant parameters is referred to as a *global symmetry*, since it is the same at every space-time point. The infinitesimal versions of this symmetry operation are given by

$$\delta\xi = -i\alpha\xi, \quad \delta\xi^\dagger = i\alpha\xi^\dagger. \tag{8.229}$$

For this symmetry $\delta x^\mu = 0$ and the conserved Noether current (8.223) is given by

$$\begin{aligned} J^\mu &= \left(\frac{\partial \mathcal{L}_\xi}{\partial(\partial_\mu \xi)}\right)\delta\xi + \delta\xi^\dagger\left(\frac{\partial \mathcal{L}_\xi}{\partial(\partial_\mu \xi^\dagger)}\right) \\ &= \left(\frac{1}{2}i\hbar\xi^\dagger\boldsymbol{\sigma}^\mu\right)(-i\alpha\xi) + (i\alpha\xi^\dagger)\left(-\frac{1}{2}i\hbar\boldsymbol{\sigma}^\mu\xi\right) = \alpha\hbar\ \xi^\dagger\boldsymbol{\sigma}^\mu\xi. \end{aligned} \tag{8.230}$$

Since $\alpha\hbar$ is an arbitrary constant, the conservation law $\partial_\mu J^\mu = 0$ can be written

$$\partial_\mu(\xi^\dagger\boldsymbol{\sigma}^\mu\xi) = 0, \tag{8.231}$$

where it is understood that (8.231) is evaluated using solutions to the equations of motion.

Exercise 8.14: Show that (8.231) is consistent with the solutions to the equation of motion.

Result (8.231) gives $J^0 = \xi^\dagger\xi$ and $J^k = \xi^\dagger\sigma^k\xi$. The earlier result (8.137) is now seen to be an expression of (8.231), since boosting the rest frame version of $\xi^\dagger\xi$ in the z direction produced $J^3 = \gamma_v\xi^\dagger\sigma^3\xi$, a matter current in the z direction.

8.6.3 Poincaré invariance and the stress–energy tensor

The stress–energy tensor can now be understood in terms of a symmetry. If the space-time volume of the action is infinite *and* there are no functions of the form $V_\alpha(x - x_o)$ appearing in the action, then invariance under Lorentz transformations can be extended to invariance under *Poincaré transformations*, which are defined as

$$x^\mu \to \Lambda^\mu{}_\nu x^\nu - b^\mu, \tag{8.232}$$

where $\Lambda^\mu{}_\nu$ is a proper Lorentz transformation and b^μ is an arbitrary constant four-vector. This means that all the field arguments can be replaced by $x^\mu - b^\mu$ and a subsequent change of variables in the integration returns the action to its original form. This action on the field is referred to as a *translation*, and if it is a symmetry then the action is said to be *translationally invariant*. In the infinitesimal case the field changes according to

$$\varphi_\alpha(x') = \varphi_\alpha(x - b) \approx \varphi_\alpha(x) - b^\mu\partial_\mu\varphi_\alpha(x). \tag{8.233}$$

Result (8.233) identifies the generator of translations as $P_\mu = \partial_\mu$, and it follows from the Taylor expansion that a finite translation is given by the exponential map,

$$\varphi_\alpha(x - b) = \exp(-b_\mu P^\mu)\varphi_\alpha(x). \tag{8.234}$$

The generator P^μ can be combined with the set of differential Lorentz generators given by (7.194) to define the *Poincaré algebra*. For the case of continuous functions φ_α, it follows that

$$[P^\mu, P^\nu]\varphi_\alpha = (\partial^\mu\partial^\nu - \partial^\nu\partial^\mu)\varphi_\alpha = 0 \implies [P^\mu, P^\nu] = 0. \tag{8.235}$$

For the Lorentz generators, it follows from the definition of $J^{\mu\nu}$ given by (7.194) that

$$\begin{aligned}[J^{\mu\nu}, P^\sigma]\varphi_\alpha &= ((x^\mu\partial^\nu - x^\nu\partial^\mu)\partial^\sigma - \partial^\sigma(x^\mu\partial^\nu - x^\nu\partial^\mu))\\ &\varphi_\alpha = (\eta^{\sigma\nu}\partial^\mu - \eta^{\sigma\mu}\partial^\nu)\varphi_\alpha,\\ &\implies [J^{\mu\nu}, P^\sigma] = \eta^{\sigma\nu}P^\mu - \eta^{\sigma\mu}P^\nu.\end{aligned} \tag{8.236}$$

The set of commutators is expressed entirely in terms of the Lorentz generators $J^{\mu\nu}$ and the translation generators P^{σ}, and so the Poincaré algebra defines a group. The Poincaré group in four dimensions involves the six parameters of the proper Lorentz group and the four translation parameters b^{μ} for a total of ten continuous parameters. It also shows that translations and the Lorentz transformations do not commute. Instead, a difference in their order results in a translation.

The case of an infinitesimal translation corresponds to

$$\delta x^{\mu} = -b^{\mu}, \tag{8.237}$$

while the action on the field is given by

$$\varphi_{\alpha}'(x') = \varphi_{\alpha}(x') \implies \delta\varphi_{\alpha} = 0. \tag{8.238}$$

This states that the components of the field φ_{α}' at $x^{\mu'} = x^{\mu} - b^{\mu}$ are unchanged. For the case that the theory is translationally invariant, the associated conserved Noether current is found by setting $\delta x^{\mu} = -b^{\mu}$ and $\delta\varphi_{\alpha} = 0$ in (8.223), giving

$$J^{\mu} = \left(\frac{\partial\mathcal{L}}{\partial(\partial_{\mu}\varphi_{\alpha})}\partial_{\nu}\varphi_{\alpha} - \delta^{\mu}{}_{\nu}\mathcal{L}\right)b^{\nu} = T^{\mu}_{\nu}\, b^{\nu}, \tag{8.239}$$

where the familiar canonical stress–energy tensor T^{μ}_{ν} defined in (6.242) has appeared. Because b^{μ} is an arbitrary infinitesimal constant, it follows that the conservation of the Noether current gives

$$\partial_{\mu}J^{\mu} = (\partial_{\mu}T^{\mu}_{\nu})b^{\nu} = 0 \implies \partial_{\mu}T^{\mu}_{\nu} = 0. \tag{8.240}$$

The conserved charge density T^{μ}_{0} has previously been identified as the four-momentum density $\mathcal{P}^{\mu}$ of the fields. The conservation of four-momentum is now seen as the result of the translational invariance of the action.

The canonical stress–energy tensor defined by Noether's theorem has the same drawbacks as those discussed in (6.248). It is not manifestly symmetric, since it does not necessarily satisfy $T^{\mu\nu} = T^{\nu\mu}$, where $T^{\mu\nu} = \eta^{\nu\rho}T^{\mu}_{\rho}$. In addition, it is also *not unique*. If there is a third-rank tensor $B^{\mu\rho}_{\nu}$ that satisfies $\partial_{\mu}\partial_{\rho}B^{\mu\rho}_{\nu} = 0$, then the term

$$\partial_{\mu}\partial_{\rho}B^{\mu\rho}_{\nu}\delta x^{\nu} = 0 \tag{8.241}$$

can be subtracted from the second line of (8.214) in the process of deriving Noether's theorem. For the case of translational invariance this results in a modified Noether current, where the term proportional to b^{ν} is then given by

$$\Theta^{\mu}_{\nu}b^{\nu} = (T^{\mu}_{\nu} - \partial_{\rho}B^{\mu\rho}_{\nu})b^{\nu}. \tag{8.242}$$

The new tensor Θ^{μ}_{ν} also satisfies the conservation law $\partial_{\mu}\Theta^{\mu}_{\nu} = 0$. This is precisely the Belinfante approach to modifying the stress–energy tensor discussed earlier in (6.254), but it is now placed in the context of an ambiguity in Noether's theorem. The Belinfante approach to remedying these drawbacks will be implemented in chapter 9 and will be replaced in chapter 10 by a redefinition of the stress–energy tensor.

8.6.4 Angular momentum

Lorentz invariance is an *a priori* symmetry of relativistic field theories and is therefore associated with conserved currents. The proper Lorentz group includes rotations, and an infinitesimal rotation is parameterized by three angles of rotation θ_k around the three spatial axes. An infinitesimal spatial rotation is given by (7.104), so that it can be written

$$x'^j = x^j + \delta x^j = x^j + \theta_k \varepsilon^{kji} x^i \implies \delta x^j = \varepsilon^{kji} \theta^k x^i, \tag{8.243}$$

where θ^k are arbitrary infinitesimal angles of rotation around the respective spatial axes. For the case of a single scalar field φ it follows that there are no components that change, so that

$$\delta\varphi = \varphi'(x') - \varphi(x) = 0. \tag{8.244}$$

The Noether current (8.223) associated with the rotational symmetry of the scalar field action is given by

$$J^\mu = \left(\delta^\mu_j \mathcal{L} - \frac{\partial \mathcal{L}}{\partial(\partial_\mu \varphi)} \partial_j \varphi \right) \varepsilon^{kji} \theta^k x^i = -\theta^k \varepsilon^{kji} x^i T^\mu_j, \tag{8.245}$$

where the stress–energy tensor of (8.239) has appeared. Because the three θ^k parameters are arbitrary, (8.245) gives three distinct conserved currents, denoted by the index k,

$$(J^k)^\mu = -\varepsilon^{kji} x^i T^\mu_j = \varepsilon^{kij} x^i T^\mu_j. \tag{8.246}$$

The densities associated with these three currents are chosen to be

$$\ell^k = -J^{k\,0} = -\varepsilon^{kij} x^i T^0_j = \varepsilon^{kij} x^i T^{0j}, \tag{8.247}$$

where $T^0_j = \eta_{jk} T^{0k} = -T^{0j}$ was used. Previous analysis has identified $T^{0j} = \mathcal{P}^j$ as the spatial momentum density of the field, so that

$$\ell^k = \varepsilon^{kij} x^i \mathcal{P}^j. \tag{8.248}$$

These are recognized as the components of the *orbital angular momentum density* of the field, since this result is identical in form to the Newtonian result for the orbital angular momentum of a particle,

$$L^k = \varepsilon^{k\ell j} x^\ell p^j. \tag{8.249}$$

The symmetry of the action under rotation implies the conservation of orbital angular momentum for the case of a scalar field. It is important to note that the orbital angular momentum density (8.248) is not translationally invariant, just as the particle orbital angular momentum (8.249) is not translationally invariant

As opposed to a scalar field, a field with components, such as a spinor field, will receive an additional contribution to the angular momentum from the action of the rotation transformation on the field present in (8.223). For simplicity, attention is limited to a theory with a single Weyl spinor field with the Lagrangian density $\mathcal{L}_\xi$ given by (8.136). The stress–energy tensor is found from (8.223) as

$$T^{\mu}_{\nu} = \frac{\partial \mathcal{L}_{\xi}}{\partial(\partial_{\mu}\xi)}\partial_{\nu}\xi - \delta^{\mu}_{\nu}\mathcal{L}_{\xi} = i\hbar\, \xi^{\dagger}(\boldsymbol{\sigma}^{\mu}\partial_{\nu} - \delta^{\mu}_{\nu}\, \boldsymbol{\sigma}^{\rho}\partial_{\rho})\xi. \tag{8.250}$$

Result (8.250) shows that the Weyl spinor's momentum density is

$$\mathcal{P}_j = T^0_j = i\hbar\, \xi^{\dagger}\partial_j\xi. \tag{8.251}$$

For a positive energy solution of the form

$$\xi(\boldsymbol{p}, x) = e^{-ip\cdot x}\xi(p), \tag{8.252}$$

the spinor's momentum density is given by

$$\mathcal{P}_j = -p_j\xi^{\dagger}(p)\xi(p). \tag{8.253}$$

The orbital angular momentum density is found in the same manner as the scalar field, so that

$$\ell^k = -J^{k\,0} = -\varepsilon^{kij}x^iT^0_j = \varepsilon^{kij}x^iT^{0j} = \varepsilon^{kij}x^i\mathcal{P}^j. \tag{8.254}$$

However, the effect of a rotation for a Weyl spinor field must also include the effect of the infinitesimal rotation obtained from the Lorentz transformation matrix $\mathbf{U}$ by setting $\phi^k = 0$ and treating θ^k as infinitesimal. This gives

$$\mathbf{U} \approx \mathbf{I} + i\frac{1}{2}\theta^k\boldsymbol{\sigma}_k, \tag{8.255}$$

so that

$$\delta\xi = i\frac{1}{2}\theta^k\boldsymbol{\sigma}_k\xi. \tag{8.256}$$

The total conserved current density therefore receives a contribution from the additional term proportional to $\delta\varphi_\alpha$ in (8.223), so that

$$\begin{aligned} J^0 &= -\,\theta^k\ell^k + \frac{\partial \mathcal{L}_{\xi}}{\partial(\partial_0\xi)}\delta\xi = -\theta^k\ell^k + (i\hbar\xi^{\dagger})\left(\frac{1}{2}i\theta^k\boldsymbol{\sigma}^k\xi\right) \\ &= -\,\theta^k\left(\ell^k + \frac{1}{2}\hbar\, \xi^{\dagger}\boldsymbol{\sigma}^k\xi\right). \end{aligned} \tag{8.257}$$

Once again, (8.257) yields three contributions, one for each of the three angles θ^k. The final result for the Weyl spinor is the conserved angular momentum density $\ell^k + s^k$, where

$$s^k \equiv \frac{1}{2}\hbar\, \xi^{\dagger}\boldsymbol{\sigma}^k\xi. \tag{8.258}$$

Rotational symmetry implies conservation of *total* angular momentum density, which includes a contribution s^k identified as the *intrinsic* or *spin angular momentum density* of the spinor field.

The definition of the helicity (8.145) of the Weyl spinor can now be related to the spin (8.258)

$$\xi^{\dagger}\boldsymbol{h}\xi = \frac{1}{2}\frac{\boldsymbol{p}}{p^0}\cdot\xi^{\dagger}\boldsymbol{\sigma}\xi = \frac{\boldsymbol{p}}{\hbar p^0}\cdot\boldsymbol{s}. \tag{8.259}$$

In the previous section it was shown that the two possible Weyl spinors have two distinct helicities, given by (8.144) for ξ and (8.167) for $\bar{\xi}$. For the case that the spinor ξ is the normalized positive energy solution corresponding to a *single particle* in the spatial volume V, the spinor is written

$$\xi(x) = e^{-ip\cdot x}u(\boldsymbol{p}), \tag{8.260}$$

where

$$u^{\dagger}(\boldsymbol{p})u(\boldsymbol{p}) = \frac{1}{V}. \tag{8.261}$$

Using the helicity result (8.144) it follows that

$$-\frac{1}{2V} = \xi^{\dagger}\boldsymbol{h}\xi = \frac{\boldsymbol{p}}{\hbar p_0}\cdot\boldsymbol{s} = \frac{\boldsymbol{p}}{\hbar p_0}\cdot\frac{\boldsymbol{\Sigma}}{V}, \tag{8.262}$$

where $\boldsymbol{\Sigma}$ is defined as the spin angular momentum *of the individual particle*, so that the spin angular momentum density $\boldsymbol{s}$ can be written $\boldsymbol{\Sigma}/V$. For the case of Weyl spinor ξ, it follows that

$$\frac{\boldsymbol{p}}{p_0}\cdot\boldsymbol{\Sigma} = -\frac{1}{2}\hbar. \tag{8.263}$$

The inner product

$$\frac{\boldsymbol{p}}{p_0}\cdot\frac{\boldsymbol{p}}{p_0} = \frac{\boldsymbol{p}\cdot\boldsymbol{p}}{p_0^2} = \frac{\boldsymbol{p}\cdot\boldsymbol{p}}{\boldsymbol{p}\cdot\boldsymbol{p}} = 1, \tag{8.264}$$

shows that the vector $\boldsymbol{p}/p_0 = \hat{\boldsymbol{p}}$ is a unit vector in the direction of motion. Result (8.263) reveals that the Weyl spinor ξ has a spin projection of $\boldsymbol{\Sigma}\cdot\hat{\boldsymbol{p}} = \frac{1}{2}\hbar$ along the direction of motion. For that reason ξ is known as the left-handed Weyl spinor, since its intrinsic angular momentum is in the opposite direction to that of its motion.

> **Exercise 8.15:** Show that the Weyl spinor $\bar{\xi}$ has a positive spin projection along the direction of motion and is therefore the right-handed Weyl spinor.

Because the spin contents of the two possible Weyl spinor fields are $\pm\frac{1}{2}\hbar$, the Weyl field is referred to as a *spin-1/2 field. Chirality* or handedness, as related to helicity, is an important aspect of the phenomena that can be modeled by a field theory. In the case of the Weyl equation the initial choice of chirality for the theory, either left- or right-handed, cannot be changed by a *parity transformation*, since there is no unitary transformation that represents a parity change. This follows from the impossibility of boosting into the frame of the Weyl spinor. A Weyl spinor field therefore has a definite chirality that coincides with its helicity. The neutrino has a helicity identical to the Weyl spinor ξ, since it is only left-handed. If it is massless, the neutrino field is an ideal candidate for being modeled by the Weyl spinor ξ. However, current experimental evidence indirectly indicates a tiny but non-vanishing mass for the neutrino. The reasons for this tiny mass are a subject of much current research.

The electron has values for helicity like that of a Weyl spinor. However, unlike the Weyl spinor, it is massive and it is therefore possible to boost into a frame where the momentum of the electron switches sign. The electron must therefore be modeled with both chiralities, and *this requires the use of a Dirac bispinor*. For rotation only, the Lorentz transformation for a bispinor (7.264) becomes

$$\mathbf{S} \approx \mathbf{I} + \frac{1}{2} i\theta^k \mathbf{R}_k, \tag{8.265}$$

where $\mathbf{R}_k$ is given by the 4×4 matrix

$$\mathbf{R}_k = \begin{pmatrix} \boldsymbol{\sigma}_k & 0 \\ 0 & \boldsymbol{\sigma}_k \end{pmatrix}. \tag{8.266}$$

Since this is two copies of (8.255), it identifies the spin angular momentum density for the Dirac bispinor field as

$$s^k = \frac{1}{2}\hbar\psi^\dagger \mathbf{R}^k \psi = \frac{1}{2}\hbar(\xi^\dagger \boldsymbol{\sigma}^k \xi + \tilde{\xi}^\dagger \boldsymbol{\sigma}^k \tilde{\xi}). \tag{8.267}$$

This is the spin angular momentum density of both possible Weyl spinors, the left-handed ξ and the right-handed $\tilde{\xi}$, so that the Dirac bispinor is a spin-1/2 field with both possible helicities, which is consistent with the experimentally observed electron. The parity operation for the Dirac bispinor is given by (7.269), which results in the exchange $\xi \leftrightarrow \tilde{\xi}$. This result is consistent with the helicity, since a right-handed solution becomes a left-handed solution and vice versa under a parity transformation.

The demonstration of the chiral components of the Dirac bispinor appears to depend on the nature of the chiral representation γ matrices, and these can be altered by a unitary transformation. In order to define the chiral components of the Dirac bispinor in a representation-independent manner, the matrix γ_5 initially defined using the chiral representation in (7.277) will be employed. In order to achieve a representation-independent version of γ_5, it is written

$$\gamma_5 = i\gamma^0\gamma^1\gamma^2\gamma^3 = \frac{i}{4!}\varepsilon_{\mu\nu\rho\sigma}\gamma^\mu\gamma^\nu\gamma^\rho\gamma^\sigma. \tag{8.268}$$

The anticommuting algebra of the γ matrices given by (7.249), which holds regardless of representation, allows this equality to hold. This shows that a coordinate change will result in a factor of the Jacobian appearing, since $\varepsilon_{\mu\nu\rho\sigma}$ is a pseudotensor. This generates a change in sign for γ_5 under a parity transformation, since the Jacobian is $J = -1$ for $\boldsymbol{x} \to -\boldsymbol{x}$, and so γ_5 remains a *pseudoscalar* quantity, regardless of the representation used for the γ matrices. From its definition

$$\gamma_5^\dagger = -i\gamma^{3\dagger}\gamma^{2\dagger}\gamma^{1\dagger}\gamma^{0\dagger} = i\gamma^3\gamma^2\gamma^1\gamma^0 = i\gamma^0\gamma^1\gamma^2\gamma^3 = \gamma_5, \tag{8.269}$$

where the last step uses the anticommutativity of the γ matrices. Because each γ^μ commutes with itself and anticommutes with the other three, the anticommutator with γ_5 gives

$$\{\gamma^5, \gamma^\mu\} = 0. \tag{8.270}$$

Combining the two equivalent expressions for γ_5 appearing in (8.269) with the properties $(\gamma^j)^2 = -\mathbf{I}$ and $(\gamma^0)^2 = \mathbf{I}$ shows that

$$\gamma_5^2 = (i\gamma^3\gamma^2\gamma^1\gamma^0)(i\gamma^0\gamma^1\gamma^2\gamma^3) = \mathbf{I}. \tag{8.271}$$

Property (8.271) allows the eigenvalues of γ_5 to be determined. If ψ_5 is an eigenspinor of γ_5 with the eigenvalue λ, it follows that

$$\begin{aligned} \psi_5 &= \gamma_5^2\psi_5 = \gamma_5(\lambda\psi_5) \\ &= \lambda^2\psi_5 \implies \lambda = \pm 1. \end{aligned} \tag{8.272}$$

Two *projection operators* $P_\pm$ are introduced by the definitions

$$P_\pm = \frac{1}{2}(\mathbf{I} \pm \gamma_5). \tag{8.273}$$

These projection operators allow an *arbitrary* Dirac bispinor ψ to be written

$$\begin{aligned} \psi &= \frac{1}{2}(\mathbf{I} + \gamma_5)\psi + \frac{1}{2}(\mathbf{I} - \gamma_5)\psi \\ &= P_+\psi + P_-\psi \equiv \psi_R + \psi_L. \end{aligned} \tag{8.274}$$

The term $\psi_R = P_+\psi$ is an eigenspinor of γ_5 with eigenvalue +1, since

$$\gamma_5\psi_R = \gamma_5 P_+\psi = \frac{1}{2}\gamma_5(\mathbf{I} + \gamma_5)\psi = \frac{1}{2}(\gamma_5 + \gamma_5^2)\psi = \frac{1}{2}(\mathbf{I} + \gamma_5)\psi = P_+\psi = \psi_R. \tag{8.275}$$

Similarly, $\psi_L = P_-\psi$ is an eigenspinor of γ_5 with eigenvalue −1, since

$$\gamma_5\psi_L = \gamma_5 P_-\psi = \frac{1}{2}\gamma_5(\mathbf{I} - \gamma_5)\psi = \frac{1}{2}(\gamma_5 - \gamma_5^2)\psi = -\frac{1}{2}(\mathbf{I} - \gamma_5)\psi = -P_-\psi = -\psi_L. \tag{8.276}$$

The parity operator $\mathbf{P}$, defined in (7.269) using the chiral representation of the γ matrices, anticommutes with γ_5,

$$\mathbf{P}\gamma_5\mathbf{P}^\dagger = -\mathbf{P}\mathbf{P}^\dagger\gamma_5 = -\gamma_5. \tag{8.277}$$

As a result, the parity operator has the effect

$$\mathbf{P}P_\pm\mathbf{P}^\dagger = \frac{1}{2}\mathbf{P}(\mathbf{I} \pm \gamma_5)\mathbf{P}^\dagger = \frac{1}{2}(\mathbf{I} \mp \gamma_5) = P_\mp, \tag{8.278}$$

so that the two projection operators change into one another under a parity transformation. In the chiral representation of (8.168) γ_5 is found to take the explicit form

$$\gamma_5 = \begin{pmatrix} -\boldsymbol{\sigma}^0 & 0 \\ 0 & \boldsymbol{\sigma}^0 \end{pmatrix} \Longrightarrow \frac{1}{2}(\mathbf{I} + \gamma_5) = \begin{pmatrix} 0 & 0 \\ 0 & \boldsymbol{\sigma}^0 \end{pmatrix}, \ \frac{1}{2}(\mathbf{I} - \gamma_5) = \begin{pmatrix} \boldsymbol{\sigma}^0 & 0 \\ 0 & 0 \end{pmatrix}. \tag{8.279}$$

The Dirac bispinor ψ in the chiral representation of (8.168) has the projections

$$\begin{aligned} P_+\psi &= \frac{1}{2}(\mathbf{I} + \gamma_5)\psi = \begin{pmatrix} 0 & 0 \\ 0 & \boldsymbol{\sigma}^0 \end{pmatrix}\begin{pmatrix} \xi \\ \tilde{\xi} \end{pmatrix} = \begin{pmatrix} 0 \\ \tilde{\xi} \end{pmatrix}, \\ P_-\psi &= \frac{1}{2}(\mathbf{I} - \gamma_5)\psi = \begin{pmatrix} \boldsymbol{\sigma}^0 & 0 \\ 0 & 0 \end{pmatrix}\begin{pmatrix} \xi \\ \tilde{\xi} \end{pmatrix} = \begin{pmatrix} \xi \\ 0 \end{pmatrix}, \end{aligned} \tag{8.280}$$

which are the right-handed and left-handed components, respectively, of the Dirac bispinor in the chiral representation. The action of γ_5 given by (8.279) then gives +1 for a right-handed spinor $\tilde{\xi}$ and −1 for a left-handed spinor ξ, and so γ_5 gives the *chirality* of the Dirac bispinor. This is true regardless of the representation chosen for the γ matrices, since the definition of γ_5 is made using all four of them.

References and recommended further reading

Monographs on classical field theory covering the topics of chapter 8 include

- Nastase H 2019 *Classical Field Theory* (Cambridge: Cambridge University Press)

Monographs on quantum field theory that introduce the topics of chapter 8 in classical terms include

- Kaku M 1993 *Quantum Field Theory: A Modern Introduction* (Oxford: Oxford University Press)
- Greiner W and Reinhardt J 1996 *Field Quantization* (Berlin: Springer)
- Srednicki M 2007 *Quantum Field Theory* (Cambridge: Cambridge University Press)

Mathematical monographs covering Green's functions include

- Arfken G, Weber H and Harris F 2013 *Mathematical Methods for Physicists* 7th edn (Amsterdam: Elsevier)
- Hassani S 2013 *Mathematical Physics* 2nd edn (Berlin: Springer)

Physics applications of Noether's theorem are reviewed in

- Hill E 1951 *Rev. Mod. Phys.* **23 253**

IOP Publishing

Classical Field Theory and the Stress-Energy Tensor (Second Edition)

Mark S Swanson

Chapter 9

Classical relativistic electrodynamics

The development of classical electrodynamics in the 19th Century revolutionized physics. Not only did it provide a set of mathematical tools that have explained a vast array of physical phenomena, but also, as shown in chapter 6, Maxwell's equations and their predictions also forced the development of special relativity, which underpins the conceptual basis of modern physics. The primary purpose of this chapter is to present the structure of electrodynamics as a relativistic field theory, the underlying symmetry of *gauge invariance*, and the nature of electrodynamic force and energy. Many of the important physical phenomena that classical electrodynamics explains will unfortunately be omitted, and the reader is referred to the bibliography.

9.1 Aspects of Maxwell's equations

The traditional versions of Maxwell's equations in vacuum were presented in (6.1). The first step in analyzing them as a relativistic theory is to scale the fields and currents that appear in (6.1) so that the fields and currents are measured in powers of length. This is achieved by using Coulomb's constant K, the speed of light c, and the fundamental unit of charge e to redefine $\mathbf{E}$, $\mathbf{B}$, ρ, and $\mathbf{J}$ according to

$$\begin{aligned} \mathbf{E} &\to Ke\,\mathbf{E}, \\ \mathbf{B} &\to \frac{Ke}{c}\,\mathbf{B}, \\ \rho &\to \frac{e}{4\pi}\,\rho, \\ \mathbf{J} &\to \frac{ec}{4\pi}\,\mathbf{J}. \end{aligned} \tag{9.1}$$

doi:10.1088/978-0-7503-3455-6ch9

© IOP Publishing Ltd 2022

Using $c = 1/\sqrt{\epsilon_o \mu_o}$ and $K = 1/4\pi\epsilon_0$ shows that the versions of Maxwell's equations given by (6.1) become

$$\begin{aligned} &\text{Gauss's Law} && \nabla \cdot \mathbf{E} = \rho, \\ &\text{Ampere's Law} && \nabla \times \mathbf{B} - \frac{\partial \mathbf{E}}{\partial (ct)} = \mathbf{J}, \\ &\text{Faraday's Law} && \nabla \times \mathbf{E} + \frac{\partial \mathbf{B}}{\partial (ct)} = 0, \\ &\text{No Magnetic Monopoles} && \nabla \cdot \mathbf{B} = 0. \end{aligned} \tag{9.2}$$

Exercise 9.1: Verify (9.2) and show that **E** and **B** now have units of inverse length squared.

The redefined charge density ρ and spatial current **J** have units of inverse length cubed, and the statement of electric charge conservation (6.4) becomes

$$\frac{\partial \rho}{\partial (ct)} + \nabla \cdot \mathbf{J} = \partial_0 J^0 + \partial_j J^j = \partial_\mu J^\mu = 0, \tag{9.3}$$

which is the relativistically invariant statement of charge conservation and agreed upon by all inertial observers.

9.2 The Helmholtz decomposition and the Coulomb potential

In order to develop the relativistic version of Maxwell's equations, it is critical to understand the degrees of freedom present in the electric and magnetic fields and how these solve the four Maxwell equations. This is facilitated by applying the *Helmholtz decomposition* to **E** and **B**, which breaks each of the two spatial vector fields into the sum of two unique spatial vector fields, the *solenoidal or transverse components* $\mathbf{E}^T$ and $\mathbf{B}^T$ and the *irrotational or curl-free components* $\mathbf{E}^L$ and $\mathbf{B}^L$, so that

$$\begin{aligned} \mathbf{E} &= \mathbf{E}^T + \mathbf{E}^L, \\ \mathbf{B} &= \mathbf{B}^T + \mathbf{B}^L. \end{aligned} \tag{9.4}$$

These components obey the respective equations

$$\begin{aligned} \nabla \cdot \mathbf{E}^T &= \nabla \cdot \mathbf{B}^T = 0, \\ \nabla \times \mathbf{E}^L &= \nabla \times \mathbf{B}^L = 0. \end{aligned} \tag{9.5}$$

A variant of Helmholtz's theorem can be demonstrated by using the *Coulomb Green's function* $G(\boldsymbol{x} - \boldsymbol{x}')$, which satisfies the static equation

$$\nabla^2 G(\boldsymbol{x} - \boldsymbol{x}') = \delta^3(\boldsymbol{x} - \boldsymbol{x}'). \tag{9.6}$$

The transverse and irrotational components of **E** are then defined as

$$\begin{aligned}\mathbf{E}^T(\boldsymbol{x}, t) &= \mathbf{E}(\boldsymbol{x}, t) - \nabla \int \mathrm{d}^3x'\, G(\boldsymbol{x} - \boldsymbol{x}')\nabla' \cdot \mathbf{E}(\boldsymbol{x}', t), \\ \mathbf{E}^L(\boldsymbol{x}, t) &= \nabla \int \mathrm{d}^3x'\, G(\boldsymbol{x} - \boldsymbol{x}')\nabla' \cdot \mathbf{E}(\boldsymbol{x}', t),\end{aligned} \tag{9.7}$$

where ∇' is the gradient for the $\boldsymbol{x}'$ spatial coordinates. An identical expression holds for **B**. Using (9.6) and (9.7) gives

$$\begin{aligned}\nabla \cdot \mathbf{E}^T(\boldsymbol{x}, t) &= \nabla \cdot \mathbf{E}(\boldsymbol{x}, t) - \nabla^2 \int \mathrm{d}^3x'\, G(\boldsymbol{x} - \boldsymbol{x}')\nabla' \cdot \mathbf{E}(\boldsymbol{x}', t) \\ &= \nabla \cdot \mathbf{E}(\boldsymbol{x}, t) - \int \mathrm{d}^3x'\, \delta^3(\boldsymbol{x} - \boldsymbol{x}')\nabla' \cdot \mathbf{E}(\boldsymbol{x}', t) \\ &= \nabla \cdot \mathbf{E}(\boldsymbol{x}, t) - \nabla \cdot \mathbf{E}(\boldsymbol{x}, t) = 0.\end{aligned} \tag{9.8}$$

Using result (1.162) shows that

$$\nabla \times \mathbf{E}^L(\boldsymbol{x}, t) = \nabla \times \nabla \int \mathrm{d}^3x'\, G(\boldsymbol{x} - \boldsymbol{x}')\nabla' \cdot \mathbf{E}(\boldsymbol{x}', t) = 0. \tag{9.9}$$

It follows by inspection of (9.7) that the decomposition (9.4) holds. It should be apparent that the Helmholtz decomposition is not invariant under boosts.

The Coulomb Green's function is given by

$$G(\boldsymbol{x} - \boldsymbol{x}') = -\frac{1}{4\pi|\boldsymbol{x} - \boldsymbol{x}'|}. \tag{9.10}$$

There are several ways to demonstrate this. The simplest is to set $\boldsymbol{x}' = 0$ and apply the divergence theorem in spherical coordinates to the integral

$$\int_V \mathrm{d}^3x\, \nabla^2\left(\frac{1}{4\pi|\boldsymbol{x}|}\right) = \int_V \mathrm{d}^3x\, \nabla \cdot \nabla\left(\frac{1}{4\pi|\boldsymbol{x}|}\right) = \oint_S \mathrm{d}\boldsymbol{S} \cdot \nabla\left(\frac{1}{4\pi r}\right), \tag{9.11}$$

where $\mathrm{d}\boldsymbol{S} = r^2\, \mathrm{d}\Omega\, \hat{\mathbf{e}}_r$. Using the radial component of the gradient $\hat{\mathbf{e}}_r\, \partial/\partial r$ gives

$$\int_V \mathrm{d}^3x\, \nabla^2\left(\frac{1}{4\pi|\boldsymbol{x}|}\right) = \oint_S \mathrm{d}\Omega\, r^2 \frac{\partial}{\partial r}\left(\frac{1}{4\pi r}\right) = -\frac{1}{4\pi}\oint_S \mathrm{d}\Omega = -1, \tag{9.12}$$

which holds for an *arbitrarily* small volume V around the origin. This is the property of a Dirac delta function, and so it follows that

$$\nabla^2\left(\frac{1}{4\pi|\boldsymbol{x}|}\right) = -\delta^3(\boldsymbol{x}). \tag{9.13}$$

This result can also be derived by evaluating the Fourier representation of the Coulomb Green's function,

$$G(\boldsymbol{x} - \boldsymbol{x}') = -\lim_{\epsilon \to 0} \int \frac{\mathrm{d}^3k}{(2\pi)^3} \frac{e^{i\boldsymbol{k}\cdot(\boldsymbol{x} - \boldsymbol{x}')}}{k^2 + \epsilon^2}. \tag{9.14}$$

The Fourier integral (9.14) is identical to the static Green's function used to derive the Yukawa potential (8.87). Setting $g = 1$ and $m = 0$ in (8.87) immediately gives (9.10).

The Helmholtz decomposition breaks each of the two spatial vectors **E** and **B** into two transverse degrees of freedom and one irrotational degree of freedom. This follows from a consideration of the Fourier integral representation for a static **E**,

$$\mathbf{E}(\boldsymbol{x}) = \int \frac{\mathrm{d}^3k}{(2\pi)^3}\, \tilde{\boldsymbol{E}}(\boldsymbol{k}) e^{i\boldsymbol{k}\cdot\boldsymbol{x}}, \tag{9.15}$$

where the real nature of **E** is ensured if

$$\tilde{\boldsymbol{E}}^*(\boldsymbol{k}) = \tilde{\boldsymbol{E}}(-\boldsymbol{k}). \tag{9.16}$$

Combining (9.15) with (9.7) and (9.14) gives

$$\mathbf{E}^T(\boldsymbol{x}) = \int \frac{\mathrm{d}^3k}{(2\pi)^3} \left(\tilde{\boldsymbol{E}}(\boldsymbol{k}) - \frac{\boldsymbol{k}\cdot\tilde{\boldsymbol{E}}(\boldsymbol{k})}{k^2}\boldsymbol{k} \right) e^{i\boldsymbol{k}\cdot\boldsymbol{x}}. \tag{9.17}$$

It follows that **E** and $\mathbf{E}^T$ coincide if

$$\boldsymbol{k}\cdot\tilde{\boldsymbol{E}}(\boldsymbol{k}) = 0. \tag{9.18}$$

For each value of $\boldsymbol{k}$ there is a two-dimensional plane perpendicular to $\boldsymbol{k}$ formed by $\tilde{\boldsymbol{E}}(\boldsymbol{k})$ that satisfies this condition. This shows that there are two independent components $\tilde{\boldsymbol{E}}(\boldsymbol{k})$ that define $\mathbf{E}^T$. Similarly, the irrotational component satisfies

$$\nabla\times\mathbf{E} = i\int \frac{\mathrm{d}^3k}{(2\pi)^3} (\boldsymbol{k}\times\tilde{\boldsymbol{E}}(\boldsymbol{k})e^{i\boldsymbol{k}\cdot\boldsymbol{x}}) = 0, \tag{9.19}$$

and this requires that $\boldsymbol{k}\times\tilde{\boldsymbol{E}}(\boldsymbol{k}) = 0$. Only the projection of $\tilde{\boldsymbol{E}}(\boldsymbol{k})$ parallel or antiparallel to $\boldsymbol{k}$ satisfies this requirement, demonstrating that there is one component of $\tilde{\boldsymbol{E}}(\boldsymbol{k})$ that defines $\mathbf{E}^L$. Identical statements follow for **B**.

There are six total degrees of freedom in **E** and **B** in the form of their components. There are three total degrees of freedom in electric current, since it must satisfy (9.3). The six components of **E** and **B** satisfy four Maxwell equations and it is useful to see how this occurs. The single irrotational components of **E** and **B** satisfy

$$\begin{aligned} \nabla\cdot\mathbf{E} &= \nabla\cdot(\mathbf{E}^L + \mathbf{E}^T) = \nabla\cdot\mathbf{E}^L = J^0, \\ \nabla\cdot\mathbf{B} &= \nabla\cdot(\mathbf{B}^T + \mathbf{B}^L) = \nabla\cdot\mathbf{B}^L = 0, \end{aligned} \tag{9.20}$$

and so the two degrees of freedom $\mathbf{E}^L$ and $\mathbf{B}^L$ are fixed by two of Maxwell's equations. The transverse components satisfy

$$\begin{aligned} \nabla\times\mathbf{B} &= \nabla\times(\mathbf{B}^T + \mathbf{B}^L) = \nabla\times\mathbf{B}^T = \boldsymbol{J} + \frac{\partial\mathbf{E}^L}{\partial(ct)} + \frac{\partial\mathbf{E}^T}{\partial(ct)}, \\ \nabla\times\mathbf{E} &= \nabla\times(\mathbf{E}^T + \mathbf{E}^L) = \nabla\times\mathbf{E}^T = -\frac{\partial\mathbf{B}^L}{\partial(ct)} - \frac{\partial\mathbf{B}^T}{\partial(ct)}. \end{aligned} \tag{9.21}$$

Since $\mathbf{E}^L$ and $\mathbf{B}^L$ are already fixed, these two equations fix two of the four transverse degrees of freedom. Out of the initial six degrees of freedom, Maxwell's equations fix both of the irrotational degrees of freedom and two of the four transverse degrees of freedom. This leaves **E** and **B** with *two transverse degrees of freedom* to match the boundary conditions.

9.3 The field strength tensor

The natural emergence of $x^0 = ct$ in (9.2) and (9.3) opens the way to expressing Maxwell's equations in relativistic terms. To start, identifying $J^0 = \rho$ allows the statement of charge conservation (9.3) to be written in the familiar Lorentz invariant form,

$$\partial_\mu J^\mu = 0, \tag{9.22}$$

so that $J^\mu = (\rho, \mathbf{J})$ is a conserved contravariant four-vector. Consistency between Gauss's law and this identification requires that $\nabla \cdot \mathbf{E}$ transforms as the time component of a four-vector. In addition, the combination of **E** and **B** has a total of six degrees of freedom, while a four-vector has four independent degrees of freedom.

The solution to both these problems is to introduce an *antisymmetric* second-rank contravariant *field strength tensor* $F^{\mu\nu}$, so that

$$F^{\nu\mu} = -F^{\mu\nu}. \tag{9.23}$$

A 4×4 antisymmetric tensor has $(4 \times 4 - 4)/2 = 6$ degrees of freedom, and this matches the number of **E** and **B** degrees of freedom. Gauss's law will transform as the time component of a four-vector if it can be written in the covariant form

$$\partial_\mu F^{\mu 0} = J^0. \tag{9.24}$$

Making the identifications

$$F^{j0} = E_j, \quad F^{0j} = -E_j, \tag{9.25}$$

and using antisymmetry, which fixes F^{00} to be zero, immediately gives

$$\partial_\mu F^{\mu 0} = \partial_0 F^{00} + \partial_j F^{j0} = \partial_j F^{j0} = \partial_j E_j = \nabla \cdot \mathbf{E} = J^0. \tag{9.26}$$

It should be noted that E_j stands for the components of the vector **E** in a specific frame and it is the case that **E** is neither a covariant or contravariant vector. Instead, **E** and **B** are the components of the second-rank tensor $F^{\mu\nu}$.

The next step is to identify F^{jk} so that the statement

$$\partial_\mu F^{\mu j} = J^j \tag{9.27}$$

reproduces Ampere's law. The previous identification gives

$$\partial_0 F^{0j} = -\frac{\partial E_j}{\partial (ct)}, \tag{9.28}$$

which is one of the two terms present on the left side of Ampere's law. The other term on the left side of Ampere's law is included by requiring that F^{kj} satisfy

$$(\nabla \times \mathbf{B})_j = \varepsilon^{jk\ell}\partial_k B_\ell = \partial_k \varepsilon^{jk\ell} B_\ell = -\partial_k \varepsilon^{kj\ell} B_\ell = \partial_k F^{kj}. \tag{9.29}$$

This is achieved by identifying F^{kj} as

$$F^{kj} = -F^{jk} = -\varepsilon^{kj\ell} B_\ell. \tag{9.30}$$

Using (9.28) and (9.30) together reproduces Ampere's law,

$$\partial_\mu F^{\mu j} = \partial_0 F^{0j} + \partial_k F^{kj} = -\frac{\partial E_j}{\partial(ct)} + (\nabla \times \mathbf{B})_j = J^j. \tag{9.31}$$

Relation (9.30) can be inverted using the identity (1.8), which gives

$$B_k = -\frac{1}{2}\varepsilon^{kj\ell} F^{j\ell}. \tag{9.32}$$

Exercise 9.2: Verify (9.32).

With these identifications Gauss's law and Ampere's law combine into a single relativistically covariant expression,

$$\partial_\nu F^{\nu\mu} = J^\mu, \tag{9.33}$$

so that both sides transform as a contravariant four-vector. The conservation of current, $\partial_\mu J^\mu = 0$, follows from the antisymmetry of $F^{\mu\nu}$, since relabeling dummy indices and using $[\partial_\mu, \partial_\nu] = 0$ gives

$$\partial_\mu J^\mu = \partial_\mu \partial_\nu F^{\nu\mu} = \partial_\nu \partial_\mu F^{\mu\nu} = -\partial_\nu \partial_\mu F^{\nu\mu} = -\partial_\mu \partial_\nu F^{\nu\mu} = -\partial_\mu J^\mu \implies \partial_\mu J^\mu = 0. \tag{9.34}$$

Using (9.25) and (9.30) allows the second-rank tensor $F^{\mu\nu}$ to be viewed as the components of a 4×4 antisymmetric matrix $\mathbf{F}$ obtained from the Cartesian components of $\mathbf{E}$ and $\mathbf{B}$,

$$\mathbf{F} = \begin{pmatrix} 0 & -E_x & -E_y & -E_z \\ E_x & 0 & -B_z & B_y \\ E_y & B_z & 0 & -B_x \\ E_z & -B_y & B_x & 0 \end{pmatrix}, \tag{9.35}$$

where the rows and columns are labeled (0, 1, 2, 3), which is consistent with relativistic notation. Rather than the spatial vectors $\mathbf{E}$ and $\mathbf{B}$ directly, it is $F^{\mu\nu}$ that obeys the Lorentz transformation rules for a second-rank tensor,

$$F'^{\mu\nu}(x') = \Lambda^\mu{}_\rho \Lambda^\nu{}_\sigma F^{\rho\sigma}(x). \tag{9.36}$$

As a result, the fields **E** and **B** transform in a different manner than either a four-vector or two spatial vectors. Labeling **E** and **B** as contravariant or covariant is not possible, since it is the transformation properties of $F^{\mu\nu}$ that determine the transformation properties of **E** and **B**.

For example, if an observer claims there is solely a uniform electric field E_z in the z-direction, the field strength tensor components are given by

$$F^{03} = -F^{30} = E_z, \tag{9.37}$$

and all other components vanish. The Lorentz transformed version of this physical system is given by

$$F'^{\mu\nu}(x') = \Lambda^{\mu}_{\sigma}\Lambda^{\nu}_{\sigma}F^{\rho\sigma} = \Lambda^{\mu}_{0}\Lambda^{\nu}_{3}F^{03} + \Lambda^{\mu}_{3}\Lambda^{\nu}_{0}F^{30} = E_z(\Lambda^{\mu}_{0}\Lambda^{\nu}_{3} - \Lambda^{\mu}_{3}\Lambda^{\nu}_{0}). \tag{9.38}$$

For the case of a boost in the y-direction corresponding to the velocity v^y, the Lorentz boosts of (6.30) give

$$\begin{aligned} E_z' &= F'^{03} = E_z\Lambda^0_0\Lambda^3_3 = \gamma_v E_z, \\ B_x' &= -F'^{23} = -E_z\Lambda^2{}_0\Lambda^3_3 = -\gamma_v\frac{v^y}{c}E_z. \end{aligned} \tag{9.39}$$

A magnetic field component is present in the boosted frame where there was none in the original frame. As a result, the individual electric and magnetic fields are *observer dependent.*

It is possible to find the covariant version $F_{\mu\nu}$ from the Minkowski metric by the usual procedure,

$$F_{\mu\nu} = \eta_{\mu\rho}\eta_{\nu\sigma}F^{\rho\sigma}. \tag{9.40}$$

It is left as an exercise to show that the matrix associated with $F_{\mu\nu}$ is given by

$$\bar{\mathbf{F}} = \begin{pmatrix} 0 & E_x & E_y & E_z \\ -E_x & 0 & -B_z & B_y \\ -E_y & B_z & 0 & -B_x \\ -E_z & -B_y & B_x & 0 \end{pmatrix}, \tag{9.41}$$

so that only the sign of **E** is changed.

Exercise 9.3: Verify (9.41).

With the identifications made so far, the two homogeneous Maxwell equations can also be combined into one relativistic equation. This starts by noting that (9.32) gives

$$\nabla \cdot \mathbf{B} = -\frac{1}{2}\partial_j\varepsilon^{jk\ell}F^{k\ell} = -\frac{1}{2}\partial_j\varepsilon^{jk\ell}F_{k\ell} = 0. \tag{9.42}$$

Property (6.75) of the four-dimensional Levi-Civita symbol $\varepsilon^{\mu\nu\rho\sigma}$ shows that this expression can be written

$$-\frac{1}{2}\partial_j\varepsilon^{jk\ell}F_{k\ell} = -\frac{1}{2}\varepsilon^{0\nu\rho\sigma}\partial_\nu F_{\rho\sigma} = 0. \tag{9.43}$$

Adapting the definition of (9.29) to **E** and $F_{\mu\nu}$,

$$(\nabla \times \mathbf{E})_j = \varepsilon^{jk\ell}\partial_k E_\ell = \partial_k\varepsilon^{jk\ell}E_\ell = -\partial_k\varepsilon^{kj\ell}E_\ell = \partial_k\varepsilon^{kj\ell}F_{\ell 0}, \tag{9.44}$$

and using

$$\varepsilon^{0j\rho\sigma} = -\varepsilon^{j0\rho\sigma} = \varepsilon^{j\rho 0\sigma} = -\varepsilon^{j\rho\sigma 0}, \tag{9.45}$$

the spatial components of (9.43) give

$$\begin{aligned} -\frac{1}{2}\varepsilon^{j\nu\rho\sigma}\partial_\mu F_{\rho\sigma} &= -\frac{1}{2}(\varepsilon^{j0\rho\sigma}\partial_0 F_{\rho\sigma} + \varepsilon^{j\nu 0\sigma}\partial_\nu F_{0\sigma} + \varepsilon^{j\nu\rho 0}\partial_\nu F_{\rho 0}) = -\frac{1}{2}\varepsilon^{jk\ell}(-\partial_0 F_{k\ell} + \partial_k F_{0\ell} \\ &- \partial_k F_{\ell 0}) = \frac{1}{2}\varepsilon^{jk\ell}\varepsilon_{k\ell r}\partial_0 B_r - \varepsilon^{jk\ell}\partial_k E_\ell = -\partial_0 B_j - (\nabla \times \mathbf{E})_j = 0, \end{aligned} \tag{9.46}$$

which is Faraday's law written in terms of the components of **E** and **B**. Combining (9.43) and (9.46) and dropping the minus sign shows that the two inhomogenous Maxwell's equations can be written in relativistic format as

$$\frac{1}{2}\varepsilon^{\mu\nu\rho\sigma}\partial_\nu F_{\rho\sigma} = 0. \tag{9.47}$$

This expression is often written using the second-rank pseudotensor defined by

$${}^*F^{\mu\nu} \equiv \frac{1}{2}\varepsilon^{\mu\nu\rho\sigma}F_{\rho\sigma}, \tag{9.48}$$

which is referred to as the *dual field strength pseudotensor*. It is an antisymmetric second-rank pseudotensor, since $\varepsilon^{\mu\nu\rho\sigma}$ picks a factor of the Jacobian when transformed. It corresponds to the replacements $\mathbf{E} \to \mathbf{B}$ and $\mathbf{B} \to -\mathbf{E}$ in the field strength tensor $F^{\mu\nu}$, referred to as a *duality transformation*, so that its matrix form is

$${}^*\mathbf{F} = \begin{pmatrix} 0 & -B_x & -B_y & -B_z \\ B_x & 0 & E_z & -E_y \\ B_y & -E_z & 0 & E_x \\ B_z & E_y & -E_x & 0 \end{pmatrix}. \tag{9.49}$$

The dual field strength obeys the equation of motion

$$\partial_\nu {}^*F^{\nu\mu} = \frac{1}{2}\varepsilon^{\nu\mu\rho\sigma}\partial_\nu F_{\rho\sigma} = -\frac{1}{2}\varepsilon^{\mu\nu\rho\sigma}\partial_\nu F_{\rho\sigma} = 0, \tag{9.50}$$

which is similar in form to the inhomogeneous Maxwell equations but without a charge current. If there were point sources of magnetic fields, referred to as *magnetic monopoles*, then the associated current would be present on the right-hand side of

(9.50). For such a case the two pairs of Maxwell's equations would be identical in structure. However, despite intensive experimental searches over many years, no magnetic monopoles have ever been reliably detected.

9.4 Electromagnetic fields and the gauge field

The covariant forms for Maxwell's equations derived in the previous section lead directly to the introduction of a Lorentz contravariant vector field $A^\mu(x)$ with units of inverse length. The contravariant four-vector A^μ is referred to as a *gauge field* for reasons to be discussed. The covariant version of the gauge field is given by the usual expression

$$A_\mu(x) = \eta_{\mu\nu} A^\nu(x), \tag{9.51}$$

so that A^μ and A_μ are related according to

$$A^\mu = (A^0, \boldsymbol{A}) \implies A_\mu = (A_0, -\boldsymbol{A}). \tag{9.52}$$

The two homogeneous Maxwell equations can be satisfied if the antisymmetric field strength tensor $F_{\mu\nu}$ is written in terms of A_μ as

$$F_{\mu\nu} = \partial_\mu A_\nu - \partial_\nu A_\mu. \tag{9.53}$$

Definition (9.53) satisfies the required antisymmetry $F_{\mu\nu} = -F_{\nu\mu}$. It follows from the relabeling of dummy indices and from the properties of the Levi-Civita symbol that

$$\varepsilon^{\mu\nu\rho\sigma}\partial_\nu\partial_\sigma A_\rho = \varepsilon^{\mu\rho\nu\sigma}\partial_\rho\partial_\sigma A_\nu = \varepsilon^{\mu\rho\sigma\nu}\partial_\rho\partial_\nu A_\sigma = \varepsilon^{\mu\nu\rho\sigma}\partial_\rho\partial_\nu A_\sigma, \tag{9.54}$$

so that the inhomogeneous equation (9.50) becomes

$$\begin{aligned} \partial_\nu {}^*F^{\nu\mu} &= \frac{1}{2}\varepsilon^{\mu\nu\rho\sigma}\partial_\nu\partial_\rho A_\sigma - \frac{1}{2}\varepsilon^{\mu\nu\rho\sigma}\partial_\nu\partial_\sigma A_\rho \\ &= \frac{1}{2}\varepsilon^{\mu\nu\rho\sigma}(\partial_\nu\partial_\rho - \partial_\rho\partial_\nu)A_\sigma = \frac{1}{2}\varepsilon^{\mu\nu\rho\sigma}[\partial_\nu, \partial_\rho]A_\sigma. \end{aligned} \tag{9.55}$$

Expression (9.55) will therefore vanish as long as

$$(\partial_\nu\partial_\rho - \partial_\rho\partial_\nu)A_\sigma = [\partial_\nu, \partial_\rho]A_\sigma = 0, \tag{9.56}$$

and this will be the case as long as the vector field A_μ is *not a singular function*. If singular fields are excluded, then (9.53) will automatically satisfy the two homogeneous Maxwell equations. *If singular fields are considered*, then the homogeneous Maxwell equations will not be satisfied. This would need to be the case if magnetic charges or monopoles were present, since the homogeneous Maxwell equations would no longer be true. If magnetic monopoles were to be discovered, it is still possible that (9.53) could hold, but this would require lifting the restriction on singular fields. This will be discussed in the section on magnetic monopoles.

As long as the fields are not singular, the field strength tensor (9.53) automatically satisfies the homogeneous Maxwell equations as well as what is known as the *Bianchi identity*,

$$\partial_\rho F_{\mu\nu} + \partial_\nu F_{\rho\mu} + \partial_\mu F_{\nu\rho} = 0. \tag{9.57}$$

It is important to note the cyclic permutation of the three indices that occurs in (9.57). The Bianchi identity holds trivially if any two of the indices are identical. For example, setting $\mu = \nu$ gives

$$\partial_\rho F_{\mu\mu} + \partial_\mu F_{\rho\mu} + \partial_\mu F_{\mu\rho} = \partial_\mu(F_{\rho\mu} + F_{\mu\rho}) = 0, \tag{9.58}$$

which vanishes due to the antisymmetry of $F_{\mu\rho}$. Since there are three four-vector indices characterizing the Bianchi identity, there are $4!/3! = 4$ distinct versions. These four distinct Bianchi identities are demonstrated by direct substitution of (9.53) into (9.57) and noting that the result can be written

$$[\partial_\rho, \partial_\mu]A_\nu + [\partial_\nu, \partial_\rho]A_\nu + [\partial_\mu, \partial_\nu]A_\rho = 0, \tag{9.59}$$

which vanishes for nonsingular fields. As a result, the Bianchi identity is identical in content to the homogeneous equation (9.50). For example, picking $\rho = 1$, $\mu = 2$, and $\nu = 3$ in (9.57) and reading the field strength entries from (9.41) gives

$$\partial_1 F_{23} + \partial_3 F_{12} + \partial_2 F_{31} = -\partial_x B_x - \partial_z B_z - \partial_y B_y = -\nabla \cdot \mathbf{B} = 0, \tag{9.60}$$

and the three components of Faraday's law can be demonstrated similarly.

The relation of A_μ to **E** and **B** follows from (9.53). It is important to note that the spatial indices of **E** and **B** are not covariant or contravariant. Instead, they represent the frame-dependent observations of the magnitude and direction of the electric and magnetic fields. For example, (9.53) gives

$$E_j = -F_{j0} = -\partial_j A_0 + \partial_0 A_j = -\partial_j A^0 - \partial_0 A^j. \tag{9.61}$$

This expression gives the spatial vector statement

$$\mathbf{E} = -\nabla A^0 - \frac{1}{c}\frac{\partial \boldsymbol{A}}{\partial t}, \tag{9.62}$$

which is the usual expression for the electric field in terms of the *scalar potential* A^0 and the *vector potential* $\boldsymbol{A}$. The terms scalar potential and vector potential predate the understanding that they are the components of a single four-vector. Similarly, (9.53) gives

$$B_j = -\frac{1}{2}\varepsilon^{jk\ell}F_{k\ell} = -\frac{1}{2}\varepsilon^{jk\ell}(\partial_k A_\ell - \partial_\ell A_k) = -\varepsilon^{jk\ell}\partial_k A_\ell = \varepsilon^{jk\ell}\partial_k A^\ell. \tag{9.63}$$

This expression gives the vector statement

$$\mathbf{B} = \nabla \times \boldsymbol{A}, \tag{9.64}$$

which is the usual relationship of **B** to the vector potential.

The contravariant version of (9.53) is given by

$$F^{\mu\nu} = \partial^\mu A^\nu - \partial^\nu A^\mu, \tag{9.65}$$

where

$$\partial^\mu = (\partial^0, -\nabla). \tag{9.66}$$

Using (9.65) in the *inhomogeneous* Maxwell equations of (9.33) gives

$$\partial_\nu F^{\nu\mu} = \partial_\nu \partial^\nu A^\mu - \partial_\nu \partial^\mu A^\nu = \Box A^\mu - \partial^\mu (\partial_\nu A^\nu) = J^\mu, \tag{9.67}$$

where it is assumed that condition (9.56) holds. This shows that the four Maxwell equations involving six degrees of freedom have been reduced to four equations of motion, one for each component of A^μ. It will be seen in the next section that these equations are not independent. These four equations couple the four-vector A^μ to the electric current four-vector J^μ. Like the equations of motion for previous field theories, the d'Alembertian is present, but (9.67) has an additional term, $\partial^\mu(\partial_\nu A^\nu)$, involving the gauge field. Understanding this term requires examining the structure of electrodynamics further.

9.5 Gauge transformations and gauge conditions

A *gauge transformation* on the gauge field A_μ is defined as

$$A_\mu \to A_\mu - \partial_\mu \Lambda, \tag{9.68}$$

where Λ is an arbitrary dimensionless scalar function referred to as a *gauge function*. The term *gauge transformation* has its origin in the idea of changing the scale or *gauge* of a field. The gauge transformation causes the field strength tensor to change according to

$$F_{\mu\nu} = \partial_\mu A_\nu - \partial_\nu A_\mu \to \partial_\mu A_\nu - \partial_\nu A_\mu - \partial_\mu \partial_\nu \Lambda + \partial_\nu \partial_\mu \Lambda = F_{\mu\nu} - [\partial_\mu, \partial_\nu]\Lambda. \tag{9.69}$$

The field strength tensor is unaffected by the gauge transformation as long as the scalar function Λ satisfies

$$[\partial_\mu, \partial_\nu]\Lambda = 0. \tag{9.70}$$

For the case that Λ is not singular, (9.70) will hold and $F_{\mu\nu}$ is said to be *gauge invariant*.

The covariant version of the equation of motion (9.67) is easily seen to be invariant under a nonsingular gauge transformation,

$$\Box A_\mu - \partial_\mu(\partial_\nu A^\nu) \to \Box A_\mu - \partial_\mu(\partial_\nu A^\nu) - \Box \partial_\mu \Lambda + \partial_\mu \Box \Lambda = \Box A_\mu - \partial_\mu(\partial_\nu A^\nu), \tag{9.71}$$

as long as

$$\Box \partial_\mu \Lambda = \partial_\mu \Box \Lambda. \tag{9.72}$$

It is important to note that the current J_μ has also been assumed to be invariant under a gauge transformation. This is revisited later in this chapter, when *minimal coupling* is defined and examined. The gauge invariance of the equation of motion (9.71) reveals that A_μ are *not all* independent degrees of freedom. This follows since the solutions of (9.71) will *not be unique*, even if sufficient and consistent boundary conditions are placed on **E** and **B**, since the solution can always undergo a gauge transformation. As a result, all four components of the gauge field A_μ do not correspond to physically observable degrees of freedom.

The freedom that gauge transformations represent can be used to remove the additional term $\partial_\mu(\partial_\nu A^\nu)$ from (9.71) as well to render the solutions for A_μ unique. In order to solve (9.67) it is necessary to make A_μ unique, and the freedom to gauge transform A_μ must be removed by choosing a *subsidiary condition*, often referred to as a *gauge condition*. Because Λ is a scalar, it represents a single degree of freedom, while A_μ has four degrees of freedom. It is therefore not possible to remove more than one degree of freedom from A_μ using a gauge transformation. Since $\partial_\mu A^\mu$ is a scalar quantity, and therefore a single degree of freedom, it can be set to zero using a gauge transformation,

$$\partial_\mu A^\mu \to \partial_\mu A^\mu - \Box\Lambda = 0. \tag{9.73}$$

Choosing Λ to solve

$$\Box\Lambda = \partial_\mu A^\mu \tag{9.74}$$

fixes A_μ to satisfy

$$\partial_\mu A^\mu = 0. \tag{9.75}$$

This choice for A_μ is known as the *Lorenz gauge condition*, although it is often mislabeled as the Lorentz condition. The process of choosing a subsidiary or gauge condition for A_μ is referred to as *gauge fixing*. The equation of motion in the Lorenz gauge is given by

$$\Box A^\mu = J^\mu. \tag{9.76}$$

Using conservation of electric current (9.34) shows that

$$\Box\partial_\mu A^\mu = \partial_\mu J^\mu = 0, \tag{9.77}$$

which reveals that $\partial_\mu A^\mu$ is completely decoupled from the electric current and propagates as a *free quantity*. The solution to the d'Alembertian equation has been discussed in chapter 3. Using (3.45) shows that Dirichlet boundary conditions, where $\partial_\mu A^\mu$ is zero everywhere at an initial and final time, will result in the solution to (9.77) *vanishing everywhere at all other times*. The Lorenz gauge condition is therefore consistent. Since it is also a relativistically invariant gauge condition, the Lorenz gauge is a common choice for removing the extra degree of freedom from A_μ.

Since Λ is arbitrary, there are other common choices for a gauge condition. For example, choosing $\nabla \cdot \boldsymbol{A} = 0$ is known as the *Coulomb gauge*, while choosing $A_0 = 0$ is known as the *temporal gauge*. Both of these gauge conditions can be useful in various physical applications. Even though they are not relativistically invariant, they can still be used by applying the correct boundary conditions, and this will be discussed in an upcoming section. However, even after gauge fixing there is often a residual gauge symmetry that places a restriction on the kinds of gauge function that can be used. For example, a gauge transformation of the Lorenz condition gives

$$\partial_\mu A^\mu \to \partial_\mu A^\mu - \Box\Lambda, \tag{9.78}$$

and so the Lorenz condition is invariant under gauge transformations that satisfy

$$\Box\Lambda = 0. \tag{9.79}$$

Similarly, a gauge transformation of the temporal gauge condition gives

$$A_0 \to A_0 - \partial_0 \Lambda. \tag{9.80}$$

For the case that Λ is time-independent the temporal gauge condition is unchanged.

The presence of gauge invariance arises from a mismatch between the number of degrees of freedom in the **E** and **B** field formulation, given by (9.2), and the formulation in terms of A_μ, given by (9.67). It was shown earlier in (9.20) and (9.21) that there are two independent transverse degrees of freedom in the **E** and **B** formulation. There are four degrees of freedom in A^μ, and it obeys just one equation, leaving three degrees of freedom. Gauge invariance reflects the ability to adjust the extra degree of freedom present in the A^μ formulation in an arbitrary manner, and gauge fixing is a *constraint* that removes the extra degree of freedom in the A^μ formulation. As a result, the two formulations then match in the number of independent degrees of freedom. Because of these varied and different choices for a subsidiary condition, the choice of gauge condition or subsequent gauge transformations *can have no measurable effect on the* **E** *and* **B** *fields found by solving* (9.67).

9.6 Natural units

There are a number of natural constants that occur regularly in relativistic field theories. Chief among these are $\hbar$ and c. Although these constants are useful they can become clutter in the formulas that occur. For that reason it is common to set them to unity and measure all physical quantities in either energy or length. This is referred to as the use of *natural units*. For the case of electrodynamics the scaling of (9.1) shows how the **E** and **B** fields can be measured in inverse length squared. Similarly, the current J^μ is measured in inverse length cubed, while the gauge field A^μ is measured in inverse length.

For the four-dimensional relativistic fields discussed so far, the scalar and vector fields have units of inverse length, while the spinor fields have units of inverse length to the 3/2 power. However, these fields and their derivatives and integrals are often used to calculate energy, momentum, and angular momentum, and it is important to understand how the units of length are converted into the relevant physical units of a particular system, such as SI or the centimetre–gram–second (cgs) system. This requires the appropriate use of factors of $\hbar$ and c to convert the results found into the relevant physical quantities.

For example, in natural units, the energy ε of a system can be measured in inverse length. This inverse length is converted back into units of energy by using the factor $\hbar c$,

$$\varepsilon \to E = \hbar c \varepsilon. \tag{9.81}$$

In natural units the energy density ε, which is the energy per unit spatial volume, is measured in inverse length to the fourth power, so that it is also converted into units of energy density $\mathcal{E}$ by multiplying it by $\hbar c$,

$$\varepsilon \to \mathcal{E} = \hbar c \varepsilon. \tag{9.82}$$

Similarly, if an inertial mass is measured in inverse length m, it is converted back into the units of inertial mass–energy by the same factor of $\hbar c$,

$$m \rightarrow Mc^2 = \hbar cm. \tag{9.83}$$

For the case that momentum k is measured in inverse length, the inverse length is converted into units of momentum by multiplying it by $\hbar$,

$$k \rightarrow p = \hbar k. \tag{9.84}$$

The angular momentum s in natural units is a dimensionless quantity, so that it is also converted into units of angular momentum by multiplying it by $\hbar$,

$$s \rightarrow L = \hbar s. \tag{9.85}$$

In natural units the energy flux Φ, which is the energy per unit time per unit area, is measured in inverse length to the fourth power, so that it is converted into units of momentum density $\mathcal{F}$ by multiplying it by $\hbar$,

$$\Phi \rightarrow \mathcal{F} = \hbar\Phi. \tag{9.86}$$

In natural units the energy per unit time w is measured in inverse length squared, so that it is converted into units of energy per unit time W by multiplying it by $\hbar c^2$,

$$w \rightarrow W = \hbar c^2 w. \tag{9.87}$$

Electrodynamics allows the natural definition of $\hbar$, given by the fundamental charge e, Coulomb's constant K, and the speed of light c, which is

$$\hbar = \frac{Ke^2}{c}, \tag{9.88}$$

which has the value of 7.7×10^{-37} J · s in meter, kilogram, second, ampere (MKSA) units. Natural units can be converted into MKSA units using the $\hbar$ found from the constants associated with electrodynamics. If $\mathbf{E}_M$ is the electric field in MKSA units of force per unit charge, and $\mathbf{E}_n$ is the electric field in natural units in inverse length squared, then $\hbar$ follows from the requirement that

$$\frac{1}{2}\epsilon_0 E_M^2 = \frac{1}{2}\hbar c E_n^2, \tag{9.89}$$

where the left-hand side is the energy density of the electric field in MKSA units and the right-hand side is the conversion from natural units for energy density. The derivation of this expression follows in a later section. Using the scaling of (9.1), $E_M = KeE_n$ as well as $\epsilon_0 = 1/4\pi K$ in (9.89) yields

$$\hbar = \frac{Ke^2}{4\pi c}. \tag{9.90}$$

As a result, it is possible to simply treat $\hbar$ and c as factors of unity throughout a calculation. At the end of the calculation all that is required is to add the correct factors of $\hbar$ and c to restore the correct physical units of whatever was being calculated. The advantage is that during interim calculations, ∂_0 becomes

$$\frac{\partial}{\partial x^0} = \frac{1}{c}\frac{\partial}{\partial t} \to \frac{\partial}{\partial t}, \tag{9.91}$$

while all factors of $\hbar$ and c in the Lagrangian density are simply suppressed. For such a case the four-dimensional action, which is the space-time integral of the Lagrangian density, becomes dimensionless as long as the Lagrangian density has units of inverse length to the fourth power. From this point forward natural units will be employed.

9.7 The gauge field action and minimal coupling

The action for electrodynamics is almost always written in terms of the gauge field A^μ. Using A^μ rather than **E** and **B** enables the *principle of gauge invariance* to define the *minimal coupling* of A^μ to other fields. In effect, gauge invariance is elevated to a required symmetry of the action involving A^μ, and doing so results in terms higher than quadratic in the fields appearing in the action. As a result, the fields become *coupled*, and the other fields to which A^μ is coupled create the charged current J^μ which acts as the *source* of the electromagnetic field. This was done in an ad hoc manner in chapter 3 for the case of the stretched string, which was coupled to some point masses by their gravitational potential. Minimal coupling uses gauge invariance to determine the types of coupling that can occur.

To start, the quadratic Lagrangian density for the electromagnetic field in the absence of a current is written

$$\mathcal{L}_A = -\frac{1}{4}F_{\mu\nu}F^{\mu\nu} = -\frac{1}{2}\left(\partial_\mu A_\nu\,\partial^\mu A^\nu - \partial_\mu A_\nu\,\partial^\nu A^\mu\right), \tag{9.92}$$

where (9.53) was used. The Lagrangian is manifestly Lorentz invariant and has units of inverse length to the fourth power, so that its space-time integral is dimensionless. Because $F_{\mu\nu}$ is gauge invariant, the Lagrangian density (9.92) is gauge invariant. Treating A_ν as the field degree of freedom and noting that

$$\partial_\mu A_\nu \partial^\mu A^\nu = \eta^{\mu\rho}\eta^{\nu\sigma}\partial_\mu A_\nu\,\partial_\rho A_\sigma, \tag{9.93}$$

the Euler–Lagrange equation yields

$$\frac{\partial \mathcal{L}_A}{\partial A_\mu} - \partial_\nu\frac{\partial \mathcal{L}_A}{\partial(\partial_\nu A_\mu)} = \partial_\nu(\partial^\nu A^\mu - \partial^\mu A^\nu) = \Box A^\mu - \partial^\mu(\partial_\nu A^\nu) = 0, \tag{9.94}$$

which is the equation of motion (9.67) with $J^\mu = 0$. The momentum that is canonically conjugate to A^μ is given by

$$\pi_\mu = \frac{\partial \mathcal{L}_A}{\partial(\partial_0 A^\mu)} = -F_{0\mu} = F_{\mu 0} = \partial_\mu A_0 - \partial_0 A_\mu. \tag{9.95}$$

However, this gives

$$\pi_0 = 0. \tag{9.96}$$

This is a result of gauge invariance. Defining a momentum that is canonically conjugate to A^0 can be resolved by picking a subsidiary or gauge condition. For example, choosing the Lorenz condition (9.75) allows the Lagrangian density to be altered to

$$\mathcal{L}_A = -\frac{1}{4}F_{\mu\nu}F^{\mu\nu} - \frac{1}{2}\lambda(\partial_\mu A^\mu)^2, \tag{9.97}$$

where λ plays the role of a Lagrange multiplier. The Lorenz condition then becomes a *constraint*. The term proportional to λ is referred to as a *gauge-fixing term*, and the Lagrangian of (9.97) is said to be *gauge fixed*. It is no longer invariant under an arbitrary continuous gauge transformation. Instead, it is invariant under gauge transformations that satisfy (9.79), since they leave the gauge-fixing term invariant. Such a choice of gauge-fixing term yields

$$\pi_0 = \frac{\partial \mathcal{L}_A}{\partial(\partial_0 A^0)} = -\lambda\, \partial_\mu A^\mu. \tag{9.98}$$

This will be revisited when the stress–energy tensor is analyzed.

Combining (9.25) and (9.30) with the identity (1.8) allows the Lagrangian density (9.92) to be written in terms of **E** and **B**:

$$\begin{aligned}\mathcal{L}_A &= -\frac{1}{4}F_{\mu\nu}F^{\mu\nu} = -\frac{1}{4}F_{j0}F^{j0} - \frac{1}{4}F_{0j}F^{0j} - \frac{1}{4}F_{jk}F^{jk}\\ &= \frac{1}{2}E_jE_j - \frac{1}{4}\varepsilon^{jk\ell}\varepsilon_{jkr}B_\ell B_r = \frac{1}{2}(\mathbf{E}^2 - \mathbf{B}^2).\end{aligned} \tag{9.99}$$

The next step is to introduce a field to which the gauge field will be coupled to provide the conserved electric current J^μ appearing in (9.67). In order to model electrons as the particles that will couple to the gauge field, it was determined in the previous chapter that it is necessary to use a massive Dirac bispinor, ψ. The quadratic Lagrangian for such a field is given by (8.169); in natural units, it is

$$\mathcal{L}_\psi = \frac{1}{2}i\left(\bar{\psi}\gamma^\mu\partial_\mu\psi - \partial_\mu\bar{\psi}\gamma^\mu\psi\right) - m\bar{\psi}\psi. \tag{9.100}$$

The key idea is to use Noether's theorem to define a conserved electric current by enforcing the symmetry of gauge invariance. The first step is to notice that the Dirac Lagrangian is invariant under the simultaneous transformations

$$\psi \to e^{i\theta}\psi, \quad \bar{\psi} \to \bar{\psi}e^{-i\theta}, \tag{9.101}$$

where θ is a constant. This is the same symmetry operation as (8.228), which gave rise to the conserved Weyl spinor particle current (8.230). Applying Noether's theorem gives the conserved bispinor current

$$J^\mu = \bar{\psi}\gamma^\mu\psi. \tag{9.102}$$

Exercise 9.4: Adapt the derivation of the Weyl current (8.230) to derive the conserved bispinor current (9.102) using Noether's theorem.

The global symmetry of (9.101) is now elevated to a *local symmetry*,

$$\psi(x) \to e^{i\alpha\Lambda(x)}\psi(x), \quad \bar{\psi}(x) \to \bar{\psi}(x)e^{-i\alpha\Lambda(x)}, \tag{9.103}$$

where α is the dimensionless version of the fundamental electron charge. The transformation of (9.103) has the property that two consecutive transformations commute,

$$\psi(x) \to e^{i\alpha\Lambda(x)}\psi(x) \to e^{i\alpha\Lambda(x)}e^{i\alpha\Lambda'(x)}\psi(x) = e^{i\alpha\Lambda'(x)}e^{i\alpha\Lambda(x)}\psi(x). \tag{9.104}$$

This is similar to the consecutive rotations around the z-axis that were presented in (7.89). The z-rotations are an abelian subgroup of all possible rotations, since they commute. The action of (9.103) is therefore referred to as an *abelian gauge transformation*. This transformation of the Dirac field leaves $m\bar{\psi}\psi$ and the current of (9.102) invariant. However, elevating this to a *gauge symmetry* of the total Dirac Lagrangian requires an additional term to compensate for the action of the derivatives in the bispinor Lagrangian (9.100). These are not invariant under (9.103), instead giving

$$\partial_\mu\psi(x) \to \partial_\mu(e^{i\alpha\Lambda(x)}\psi(x)) = e^{i\alpha\Lambda(x)}\partial_\mu\psi(x) + i\alpha e^{i\alpha\Lambda(x)}\partial_\mu\Lambda(x)\,\psi(x), \tag{9.105}$$

with a similar term for $\partial_\mu\bar{\psi}$. As a result, the transformation of (9.103) causes the Dirac Lagrangian $\mathcal{L}_\psi$ to become

$$\mathcal{L}_\psi \to \mathcal{L}_\psi - \alpha\,\bar{\psi}\gamma^\mu\psi\,\partial_\mu\Lambda, \tag{9.106}$$

where the conserved current (9.102) has appeared.

This additional term is canceled by a simultaneous gauge transformation of the gauge field if the Lagrangian also contains the *interaction term*

$$\mathcal{L}_I = -\alpha A_\mu\bar{\psi}\gamma^\mu\psi \tag{9.107}$$

The action of the gauge transformation on both ψ and A_μ causes the interaction term of (9.107) to become

$$\mathcal{L}_I \to \mathcal{L}_I + \alpha\,\bar{\psi}\gamma^\mu\psi\,\partial_\mu\Lambda, \tag{9.108}$$

which cancels the term that has appeared in (9.106). Using the Lorentz transformation property for ψ, given by (7.274), and the covariant four-vector transformation rule for A_μ gives

$$\begin{aligned} gA_\mu\bar{\psi}\gamma^\mu\psi \to \alpha\Lambda^\nu_\mu(\phi)A_\nu\bar{\psi}\mathbf{U}^{-1}\gamma^\mu\mathbf{U}\psi &= \alpha A_\nu\Lambda^\nu_\mu(\phi)\Lambda^\mu_\rho(-\phi)\,\bar{\psi}\gamma^\rho\psi \\ &= \alpha\delta^\nu_\rho A_\nu\bar{\psi}\gamma^\rho\psi = \alpha A_\nu\bar{\psi}\gamma^\nu\psi, \end{aligned} \tag{9.109}$$

which shows that $\mathcal{L}_I$ is Lorentz invariant. Ignoring gauge fixing for the moment, adding this term to the two quadratic Lagrangians gives the *minimally coupled* gauge and Lorentz invariant Lagrangian for *bispinor electrodynamics*,

$$\begin{aligned}\mathcal{L} &= i\frac{1}{2}(\bar{\psi}\gamma^{\mu}\partial_{\mu}\psi - \partial_{\mu}\bar{\psi}\gamma^{\mu}\psi) - m\bar{\psi}\psi - \frac{1}{4}F_{\mu\nu}F^{\mu\nu} - \alpha A_{\mu}\bar{\psi}\gamma^{\mu}\psi \\ &= i\frac{1}{2}\left(\bar{\psi}\gamma^{\mu}D_{\mu}\psi - D_{\mu}^{\dagger}\bar{\psi}\,\gamma^{\mu}\psi\right) - m\bar{\psi}\psi - \frac{1}{4}F_{\mu\nu}F^{\mu\nu},\end{aligned} \tag{9.110}$$

where the *gauge covariant derivative* is defined as

$$D_{\mu} \equiv \partial_{\mu} + i\alpha A_{\mu}. \tag{9.111}$$

The gauge covariant derivative has the property that a gauge transformation gives

$$D_{\mu}\psi \to (\partial_{\mu} + i\alpha A_{\mu} - i\alpha\partial_{\mu}\Lambda)(e^{i\alpha\Lambda}\psi) = e^{i\alpha\Lambda}(\partial_{\mu} + i\alpha A_{\mu})\psi = e^{i\alpha\Lambda}D_{\mu}\psi, \tag{9.112}$$

with a similar result for $\bar{\psi}$,

$$D_{\mu}^{\dagger}(e^{-i\alpha\Lambda}\bar{\psi}) = e^{-i\alpha\Lambda}D_{\mu}^{\dagger}\bar{\psi}. \tag{9.113}$$

Applying the Euler–Lagrange equation to (9.110) gives the equations of motion for A^{μ}, ψ, and $\bar{\psi}$,

$$\begin{aligned}\partial_{\nu}F^{\nu\mu} &= \Box A^{\mu} - \partial^{\mu}(\partial_{\nu}A^{\nu}) = \alpha\bar{\psi}\gamma^{\mu}\psi \equiv J^{\mu}, \\ i\gamma^{\mu}\partial_{\mu}\psi - m\psi &= \alpha A_{\mu}\gamma^{\mu}\psi \implies i(\gamma^{\mu}D_{\mu} + im)\psi = 0, \\ -i\partial_{\mu}\bar{\psi}\,\gamma^{\mu} - m\bar{\psi} &= \alpha A_{\mu}\bar{\psi}\gamma^{\mu} \implies -i(D_{\mu}^{\dagger}\bar{\psi}\gamma^{\mu} - im\bar{\psi}) = 0,\end{aligned} \tag{9.114}$$

and these are supplemented by the Lorenz condition and conservation of charge,

$$\begin{aligned}\partial_{\mu}A^{\mu} &= 0, \\ \partial_{\mu}J^{\mu} &= 0.\end{aligned} \tag{9.115}$$

The real constant g is referred to as the *coupling constant*, since its value determines the strength of the interaction between the two fields. The charge density $J^0 = \alpha\bar{\psi}\gamma^0\psi = \alpha\psi^{\dagger}\psi$ is real and can therefore represent the presence of electric charge, so that α can be chosen to model the presence of electric charge in the form of fundamental charged particles. The behaviors of the two fields, A_{μ} and ψ, are *coupled* by the *nonlinear differential equations* of (9.114), and these equations must be solved simultaneously. In the limit $\alpha \to 0$, the gauge field and the bispinor field decouple and are described by the *free* solutions found earlier. The coupled field theory of (9.114) is referred to as *spinor electrodynamics*.

The Noether current associated with gauge invariance can now be found. The infinitesimal versions of these simultaneous gauge transformations are given by

$$\begin{aligned}\delta A_{\mu}(x) &= -\partial_{\mu}\Lambda(x), \\ \delta\psi(x) &= e^{i\alpha\Lambda(x)}\psi(x) - \psi(x) = i\alpha\Lambda(x)\,\psi(x), \\ \delta\bar{\psi}(x) &= e^{-i\alpha\Lambda(x)}\bar{\psi}(x) - \bar{\psi}(x) = -i\alpha\Lambda(x)\,\bar{\psi}(x),\end{aligned} \tag{9.116}$$

while $\delta x^\mu = 0$. Using (9.116) and the Lagrangian (9.110) in (8.223) gives the conserved Noether current $\mathcal{J}^\mu$,

$$\mathcal{J}^\mu = \frac{\partial \mathcal{L}}{\partial(\partial_\mu A_\nu)} \delta A_\nu + \frac{\partial \mathcal{L}}{\partial(\partial_\mu \psi)} \delta\psi + \delta\bar{\psi} \frac{\partial \mathcal{L}}{\partial(\partial_\mu \bar{\psi})} = F^{\mu\nu}\, \partial_\nu \Lambda - J^\mu \Lambda. \tag{9.117}$$

The conservation of $\mathcal{J}^\mu$ implies the *conservation of the electric current* J^μ,

$$\partial_\mu \mathcal{J}^\mu = \partial_\mu F^{\mu\nu}\, \partial_\nu \Lambda + F^{\mu\nu}\, \partial_\mu \partial_\nu \Lambda - \partial_\mu J^\mu \Lambda - J^\mu \partial_\mu \Lambda = -\partial_\mu J^\mu \Lambda = 0, \tag{9.118}$$

where

$$\begin{aligned} \partial_\mu F^{\mu\nu} &= J^\nu, \\ F^{\mu\nu} \partial_\mu \partial_\nu \Lambda &= 0, \end{aligned} \tag{9.119}$$

were both used. Since Λ is both nonzero and arbitrary, conservation of J^μ immediately follows. The conservation of J^μ also follows from the equations of motion (9.114), which give

$$\begin{aligned} \partial_\mu J^\mu &= \alpha(\partial_\mu \bar{\psi} \gamma^\mu)\psi + \alpha\bar{\psi}(\gamma^\mu \partial_\mu \psi) \\ &= i\alpha(\bar{\psi}(m + A_\mu \gamma^\mu))\psi - i\alpha\bar{\psi}((m + A_\mu \gamma^\mu)\psi) = 0. \end{aligned} \tag{9.120}$$

Solving the nonlinear equations of (9.114) for a general situation is not possible. Instead, approximation techniques are employed to analyze situations where the behavior of the charge current is specified or the electromagnetic field is specified. This is discussed throughout the coming sections.

9.8 Relativistic point charges and electromagnetic interactions

As an example of analyzing the equations of motion, it is instructive to adapt the interaction term (9.107) to provide the potential for use in the relativistic equation of motion (6.189) for a point charge. This ignores the possibility that the point charge may lose mechanical energy through radiation and assumes that the gauge field A_μ is unaffected by the charge's motion.

The first step is to evaluate the contribution of (9.107) to the point particle's action. This is accomplished by using the expression for J^μ given by the point particle expression (8.2) in the integral of the interaction term (9.107). Designating the trajectory of the point charge q as $x_c^\mu(\tau)$, the term $A_\mu J^\mu$ can be integrated using the expression (8.2) for a point particle, which is given by

$$J^\mu(x) = q \int_{-\infty}^{\infty} \mathrm{d}\tau' \, \frac{\mathrm{d}x^\mu(\tau')}{\mathrm{d}\tau'} \, \delta^4(x - x_c(\tau')). \tag{9.121}$$

The energy of the interaction term (9.107) for a positive charge density is ρA_0, which gives the relativistic energy for a point charge as

$$\begin{aligned}\int_{-\infty}^{\infty} \mathrm{d}\tau\; U(x_c(\tau), \dot{x}_c(\tau)) &= \int \mathrm{d}^4x\; A_\mu(x) J^\mu(x) \\ &= \int_{-\infty}^{\infty} \mathrm{d}\tau' \int \mathrm{d}^4x\; qA_\mu(x) \frac{\mathrm{d}x^\mu(\tau')}{\mathrm{d}\tau'}\, \delta^4(x - x_c(\tau')) \\ &= \int_{-\infty}^{\infty} \mathrm{d}\tau'\; qA_\mu(x_c(\tau')) \frac{\mathrm{d}x_c^\mu(\tau')}{\mathrm{d}\tau'},\end{aligned} \tag{9.122}$$

from which the Lorentz invariant potential for the point charge is found to be

$$U(x_c, \dot{x}_c) = qA_\mu(x_c(\tau)) \frac{\mathrm{d}x_c^\mu(\tau)}{\mathrm{d}\tau}. \tag{9.123}$$

It is important to determine whether the contribution of this potential to the action is gauge invariant. A gauge transformation gives

$$\begin{aligned}\int_{-\infty}^{\infty} \mathrm{d}\tau\; U(x_c(\tau), \dot{x}_c(\tau)) &\to \int_{-\infty}^{\infty} \mathrm{d}\tau\; U(x_c(\tau), \dot{x}_c(\tau)) - \int_{-\infty}^{\infty} \mathrm{d}\tau\; q \frac{\partial \Lambda(x_c)}{\partial x_c^\mu} \frac{\mathrm{d}x_c^\mu(\tau)}{\mathrm{d}\tau} \\ &= \int_{-\infty}^{\infty} \mathrm{d}\tau\; U(x_c(\tau), \dot{x}_c(\tau)) - \int_{-\infty}^{\infty} \mathrm{d}\tau\; q \frac{\mathrm{d}\Lambda(x_c(\tau))}{\mathrm{d}\tau} \\ &= -\, q\Lambda(x_c(\tau))\big|_{-\infty}^{\infty} + \int_{-\infty}^{\infty} \mathrm{d}\tau\; U(x_c(\tau), \dot{x}_c(\tau)).\end{aligned} \tag{9.124}$$

Because it is nonzero only at the limits $\pm\infty$, the first term will have no effect on the motion of the charged particle.

Combining (9.123) with the relativistic Lagrangian (6.183) gives the action for a massive point charge,

$$\mathcal{L}_q = \frac{1}{2} m \frac{\mathrm{d}x_\mu}{\mathrm{d}\tau} \frac{\mathrm{d}x^\mu}{\mathrm{d}\tau} + U(x, \dot{x}) = \frac{1}{2} m \frac{\mathrm{d}x_\mu}{\mathrm{d}\tau} \frac{\mathrm{d}x^\mu}{\mathrm{d}\tau} + qA_\mu(x(\tau)) \frac{\mathrm{d}x^\mu(\tau)}{\mathrm{d}\tau}. \tag{9.125}$$

This gives the relativistic momentum for the point charge,

$$p_\mu = \frac{\partial \mathcal{L}_q}{\partial \dot{x}^\mu} = m \frac{\mathrm{d}x_\mu}{\mathrm{d}\tau} + qA_\mu(x) \implies \frac{\mathrm{d}x_\mu}{\mathrm{d}\tau} = \dot{x}_\mu = \frac{1}{m}\big(p_\mu - qA_\mu(x)\big). \tag{9.126}$$

Using (9.126) in the relativistic Hamiltonian definition (6.192) shows that the Hamiltonian for this system is given by

$$\begin{aligned}H(p, x) &= \dot{x}^\mu \frac{\partial \mathcal{L}_q}{\partial \dot{x}^\mu} - \mathcal{L}_q = \dot{x}^\mu p_\mu - \frac{1}{2} m \dot{x}^\mu \dot{x}_\mu - qA_\mu(x)\dot{x}^\mu \\ &= \frac{1}{2m}(p_\mu - qA_\mu)(p^\mu - qA^\mu).\end{aligned} \tag{9.127}$$

It is now possible to find the equation of motion by either applying the relativistic Euler–Lagrange equation (6.185) to (9.125) or by using the relativistic version of Hamilton's equations (6.195) with (9.127), both of which give

$$f_\mu = m \frac{\mathrm{d}^2 x_\mu}{\mathrm{d}\tau^2} = q\partial_\mu A_\nu \frac{\mathrm{d}x^\nu}{\mathrm{d}\tau} - q \frac{\mathrm{d}A_\mu}{\mathrm{d}\tau} = q(\partial_\mu A_\nu - \partial_\nu A_\mu) \frac{\mathrm{d}x^\nu}{\mathrm{d}\tau} = qF_{\mu\nu} V^\nu, \tag{9.128}$$

where $V^\mu(\tau) = \mathrm{d}x^\mu/\mathrm{d}\tau$ is the proper velocity of the charge. This yields the relativistic version of the Lorentz force law. For example, selecting $\mu = 1$ and using the entries from matrix (9.41) gives the relativistic force in the x-direction on the charge as it moves along the trajectory,

$$f^x(x_c) = -f_x(x_c) = -q(F_{10}V^0 + F_{12}V^2 + F_{13}V^3) = q\gamma_v(E_x + v^y B_z - v^z B_y), \quad (9.129)$$

where $V^\mu = \gamma_v v^\mu$ was used. This generalizes to all three components, so that the relativistic spatial force on the charge is

$$f = q\gamma_v(\mathbf{E} + v \times \mathbf{B}), \quad (9.130)$$

which reduces to the standard version of the Lorentz force law in the nonrelativistic limit. Similarly, relation (9.128) gives f^0,

$$f^0 = f_0 = qF_{0\nu}V^\nu = qF_{0j}V^j = qE_jV^j = q\gamma_v\mathbf{E} \cdot v = f_E \cdot v = \frac{\mathrm{d}p^0}{\mathrm{d}\tau}. \quad (9.131)$$

Result (9.131) is the electrodynamic version of the result (6.142) found earlier, but is now expressed in natural units, where $c = 1$. Result (9.131) relates the change in the relativistic kinetic energy p^0 of the charge to its motion through the electromagnetic field. It shows that the magnetic field does no work on the charge. Combining the two results gives the relativistic four-force on the electric point charge expressed in terms of **E** and **B**,

$$\begin{aligned} f^\mu &= (f \cdot v, f) = (q\gamma_v(\mathbf{E} + v \times \mathbf{B}) \cdot v,\ q\gamma_v(\mathbf{E} + v \times \mathbf{B})) \\ &= (q\gamma_v\mathbf{E} \cdot v,\ q\gamma_v(\mathbf{E} + v \times \mathbf{B})). \end{aligned} \quad (9.132)$$

These results have been derived by starting with the relativistic action (9.125) written in terms of $\mathrm{d}x^\mu/\mathrm{d}\tau$. The difference between this approach and adapting the free-point particle action (6.100) can now be clarified. The contribution of the potential (9.123) to the action can be written in natural units as

$$\int \mathrm{d}\tau\ U(x, \dot{x}) = \int \mathrm{d}\tau\ qA_\mu \frac{\mathrm{d}x^\mu}{\mathrm{d}\tau} = \int \mathrm{d}t\ qA_\mu \frac{\mathrm{d}x^\mu}{\mathrm{d}t} = \int \mathrm{d}t \left(qA_0 + qA_j \frac{\mathrm{d}x^j}{\mathrm{d}t} \right). \quad (9.133)$$

Using this in the free-particle action (6.167) gives the alternate action for the charged particle,

$$S(x, \dot{x}) = \int \mathrm{d}\tau\ (-mc^2 - U(x, \dot{x})) = \int \mathrm{d}t \left(-mc^2\sqrt{1 - \frac{v^2}{c^2}} - qA_0 - qA_j \frac{\mathrm{d}x^j}{\mathrm{d}t} \right). \quad (9.134)$$

This action uses x^j and $\dot{x}^j = \mathrm{d}x^j/\mathrm{d}t$ as the variables, so that the Euler–Lagrange equation is given by

$$-\frac{\mathrm{d}}{\mathrm{d}t}\frac{\partial \mathcal{L}}{\partial \dot{x}^j} + \frac{\partial \mathcal{L}}{\partial x^j} = -\frac{\mathrm{d}}{\mathrm{d}t}\left(\frac{m\dot{x}^j}{\sqrt{1 - v^2/c^2}} - qA_j \right) - q\partial_j A_0 - q(\partial_j A_k)\dot{x}^k = 0. \quad (9.135)$$

This can be written

$$\frac{\mathrm{d}}{\mathrm{d}t}\left(\frac{m\dot{x}^j}{\sqrt{1-\dot{x}^k\dot{x}^k/c^2}}\right) = q(\partial_0 A_j - \partial_j A_0) - q(\partial_j A_k - \partial_k A_j)\dot{x}^k = qF_{0j} - qF_{jk}\dot{x}^k. \quad (9.136)$$

Multiplying by γ_v and using the components of $F_{\mu\nu}$ found in (9.41) shows that this expression yields the same *contravariant spatial components* of the relativistic Lorentz force law that were found in (9.130). The reason that the two versions of the action give the same equation of motion is that the potential $U(x, \dot{x})$ satisfies the required identity (6.186),

$$\frac{\mathrm{d}}{\mathrm{d}\tau}\left(\frac{\partial}{\partial(\mathrm{d}x^j/\mathrm{d}t)}\left(\frac{1}{\gamma_v}qA_\mu\frac{\mathrm{d}x^\mu}{\mathrm{d}\tau}\right)\right) = \frac{\mathrm{d}}{\mathrm{d}\tau}(qA_\mu) = \frac{\mathrm{d}}{\mathrm{d}\tau}\left(\frac{\partial}{\partial(\mathrm{d}x^j/\mathrm{d}\tau)}\left(qA_\mu\frac{\mathrm{d}x^\mu}{\mathrm{d}\tau}\right)\right), \quad (9.137)$$

which holds because the potential is linear in $\mathrm{d}x^\mu/\mathrm{d}\tau$.

However, the two actions differ in their result for the Hamiltonian. For the action (9.125) the Hamiltonian was found in (9.127). Using the expression for p^μ given by (9.126), the Hamiltonian can be written in coordinate representation as

$$H = \frac{1}{2}m\frac{\mathrm{d}x^\mu}{\mathrm{d}\tau}\frac{\mathrm{d}x_\mu}{\mathrm{d}\tau}. \quad (9.138)$$

The equation of motion (9.128) gives

$$\frac{\mathrm{d}H}{\mathrm{d}\tau} = m\frac{\mathrm{d}x^\mu}{\mathrm{d}\tau}\frac{\mathrm{d}^2x_\mu}{\mathrm{d}\tau^2} = qm\frac{\mathrm{d}x^\mu}{\mathrm{d}\tau}F_{\mu\nu}\frac{\mathrm{d}x^\nu}{\mathrm{d}\tau} = 0, \quad (9.139)$$

which vanishes trivially because of the antisymmetry of $F_{\mu\nu}$. For the action (9.134) the momentum is given by

$$p_j = \frac{\partial\mathcal{L}}{\partial\dot{x}^j} = \frac{m\dot{x}^j}{\sqrt{1-v^2/c^2}} - qA_j, \quad (9.140)$$

so that the Hamiltonian for the action (9.134) is given in coordinate representation by

$$\tilde{H} = \dot{x}^j p_j - \mathcal{L} = \frac{mv^2}{\sqrt{1-v^2/c^2}} + mc^2\sqrt{1-v^2/c^2} + qA_0 = \frac{mc^2}{\sqrt{1-v^2/c^2}} + qA_0, \quad (9.141)$$

which shows that the total energy of the system is the combination of the mechanical energy p_0 and the electric field energy qA_0 and is independent of the magnetic field.

Demonstrating that $\tilde{H}$ is conserved begins by noting that in coordinate representation, the time derivative of $\tilde{H}$ is given by

$$\frac{\mathrm{d}\tilde{H}}{\mathrm{d}t} = \frac{m\boldsymbol{v}\cdot\boldsymbol{a}}{\sqrt{(1-v^2/c^2)^3}} + q\frac{\mathrm{d}A_0}{\mathrm{d}t} = \frac{m\boldsymbol{v}\cdot\boldsymbol{a}}{\sqrt{(1-v^2/c^2)^3}} + q(\partial_0 A_0 + \nabla A_0\cdot\boldsymbol{v}). \quad (9.142)$$

The equation of motion gives

$$\frac{\mathrm{d}p_j}{\mathrm{d}t} = \frac{\mathrm{d}}{\mathrm{d}t}\frac{mv^j}{\sqrt{1 - v^2/c^2}} = \frac{ma^j}{\sqrt{1 - v^2/c^2}} + \frac{mv^j(v \cdot a)}{c^2\sqrt{(1 - v^2/c^2)^3}} = qE_j + qF_{jk}v^k. \tag{9.143}$$

Contracting (9.143) with v^j gives

$$\frac{mv \cdot a}{\sqrt{(1 - v^2/c^2)^3}} = q\mathbf{E} \cdot v. \tag{9.144}$$

It is important to note that the Hamiltonian $\tilde{H}$ is *not gauge invariant*, although the action (9.134) from which it was derived is gauge invariant. It is therefore possible to perform a gauge transformation $A_\mu \to A_\mu - \partial\Lambda$ on the last two terms in (9.142), where the Lorenz gauge condition $\partial_\mu A^\mu$ is being used. The last two terms become

$$q(\partial_0 A_0 + \nabla A_0 \cdot v) \to q(\partial_0 A_0 + \nabla A_0 \cdot v) - q\ddot{\Lambda} - q\nabla\dot{\Lambda} \cdot v. \tag{9.145}$$

The gauge function Λ is chosen so that

$$\ddot{\Lambda} = \partial_0 A^0. \tag{9.146}$$

The two terms can then be written

$$q(\partial_0 A_0 + \nabla A_0 \cdot v) = -q(\partial_0(\nabla\Lambda) - \nabla A_0) \cdot v = -q\mathbf{E} \cdot v, \tag{9.147}$$

where $\partial_0\nabla\Lambda$ is the *pure gauge* version of $-\dot{A}$. This is consistent with the Lorenz gauge condition, since $\Box\Lambda = 0$ gives

$$\ddot{\Lambda} = \nabla^2\Lambda, \tag{9.148}$$

which in turn gives

$$\partial_\mu A^\mu \to \partial_\mu A^\mu - \Box\Lambda = \partial_j A^j + \nabla^2\Lambda = 0 \implies \partial_j(\dot{A}^j + \partial_0\partial_j\Lambda) = 0. \tag{9.149}$$

Returning (9.144) and (9.147) to (9.142) shows that the Hamiltonian $\tilde{H}$ is conserved.

This result is only half of the picture; the other half shows how the electric charge and its motion alter the electromagnetic field through *radiative processes*. This requires an evaluation of the stress–energy tensor.

9.9 The stress–energy tensor and electrodynamics

Using the Lagrangian density (9.110) and result (8.239) gives the canonical stress–energy tensor for a gauge field coupled to a Dirac spinor,

$$T^\nu_\mu = \frac{\partial\mathcal{L}}{\partial(\partial_\mu A_\rho)}\partial_\nu A_\rho + \frac{\partial\mathcal{L}}{\partial(\partial_\mu\psi)}\partial_\nu\psi + \partial_\nu\bar{\psi}\frac{\partial\mathcal{L}}{\partial(\partial_\mu\bar{\psi})} - \delta^\mu_\nu\mathcal{L} \equiv T^\mu_{A_\nu} + T^\mu_{\psi_\nu}, \tag{9.150}$$

where

$$T^\mu_{A_\nu} = -F^{\mu\rho}\partial_\nu A_\rho + \frac{1}{4}\delta^\mu_\nu F^{\alpha\beta}F_{\alpha\beta} \tag{9.151}$$

and

$$T^{\mu}_{\psi\,\nu} = \frac{1}{2}i(\bar{\psi}\gamma^{\mu}\partial_{\nu}\psi - \partial_{\nu}\bar{\psi}\gamma^{\mu}\psi) - \delta^{\mu}_{\nu}\left\{\frac{1}{2}i\bar{\psi}(\gamma^{\alpha}D_{\alpha} + im)\psi - \frac{1}{2}i(D^{\dagger}_{\alpha}\bar{\psi}\gamma^{\alpha} - im\bar{\psi})\psi\right\}. \quad (9.152)$$

The stress–energy tensor is evaluated using the solutions to the equations of motion. Using the spinor equations of motion (9.114), the spinor contribution (9.152) reduces to

$$T^{\mu}_{\psi\,\nu} = \frac{1}{2}i(\bar{\psi}\gamma^{\mu}\partial_{\nu}\psi - \partial_{\nu}\bar{\psi}\gamma^{\mu}\psi). \quad (9.153)$$

As defined, the stress–energy tensor of (9.150) is *not* gauge invariant, since neither (9.151) nor (9.153) is gauge invariant. As a result, (9.150) is not consistent with the Lagrangian and the equations of motion. In addition, neither contribution is symmetric under the interchange $\mu \leftrightarrow \nu$. However, it was shown in (6.253) as well as in the derivation of Noether's theorem (8.241) that the canonical stress–energy tensor is not unique and can be modified by adding terms using the Belinfante tensor,

$$\Theta^{\mu}{}_{\nu} = T^{\mu}_{\nu} + \partial_{\rho}B^{\mu\rho}_{\nu}, \quad (9.154)$$

as long as the Belinfante tensor satisfies

$$B^{\mu\rho}_{\nu} = -B^{\rho\mu}_{\nu}. \quad (9.155)$$

Choosing the Belinfante tensor to be

$$B^{\mu\rho}_{\nu} = F^{\mu\rho}A_{\nu} \quad (9.156)$$

clearly satisfies condition (9.155). Using the equations of motion (9.114) and the form for J^{μ} gives

$$\partial_{\rho}B^{\mu\rho}_{\nu} = \partial_{\rho}F^{\mu\rho}A_{\nu} + F^{\mu\rho}\partial_{\rho}A_{\nu} = -J^{\mu}A_{\nu} + F^{\mu\rho}\partial_{\rho}A_{\nu} = -\alpha\bar{\psi}\gamma^{\mu}\psi A_{\nu} + F^{\mu\rho}\partial_{\rho}A_{\nu}. \quad (9.157)$$

The two terms of (9.157) can therefore be added to the canonical stress–energy tensor (9.150) to obtain the two modified forms,

$$\Theta^{\mu}_{A\,\nu} = F^{\mu\rho}F_{\rho\nu} + \frac{1}{4}\delta^{\mu}_{\nu}\, F^{\alpha\beta}F_{\alpha\beta}, \quad (9.158)$$

$$\Theta^{\mu}_{\psi\,\nu} = \frac{1}{2}i(\bar{\psi}\gamma^{\mu}D_{\nu}\psi - D^{\dagger}_{\nu}\bar{\psi}\gamma^{\mu}\psi). \quad (9.159)$$

Both (9.158) and (9.159) are now manifestly gauge invariant and the modified gauge field contribution (9.158) is symmetric. The contravariant version of the gauge field stress–energy tensor can be written

$$\Theta_{A}{}^{\mu\nu} = \eta_{\rho\sigma}F^{\mu\rho}F^{\sigma\nu} + \frac{1}{4}\eta^{\mu\nu}\, F^{\alpha\beta}F_{\alpha\beta}, \quad (9.160)$$

which has the obvious property $\Theta_A{}^{\mu\nu} = \Theta_A{}^{\nu\mu}$. However, the modified spinor contribution $\Theta_\psi^{\mu\nu}$ is *not symmetric*. For example, in its contravariant form it has the components

$$\Theta_\psi^{0j} = \frac{1}{2}i(\bar{\psi}\gamma^0 D^j\psi - D^{\dagger j}\bar{\psi}\gamma^0\psi) = \frac{1}{2}i\psi^\dagger(D^j\psi) - \frac{1}{2}i(D^{\dagger j}\psi^\dagger)\psi,\ \Theta_\psi^{j0} = \frac{1}{2}i(\bar{\psi}\gamma^j D^0\psi - D^{\dagger 0}\bar{\psi}\gamma^j\psi) = \frac{1}{2}i\psi^\dagger\gamma^0\gamma^j(D^0\psi) - \frac{1}{2}i(D^{\dagger 0}\psi^\dagger)\gamma^0\gamma^j\psi, \tag{9.161}$$

which are clearly different, even after the Belinfante tensor modification has restored gauge invariance. The spinor stress–energy tensor can be made symmetric by simply defining

$$\bar{\Theta}_\psi^{\mu\nu} = \frac{1}{2}\Theta^{\mu\nu} + \frac{1}{2}\Theta^{\nu\mu}. \tag{9.162}$$

However, at first glance, this appears to be an ad hoc method, and the physical meaning of the symmetrized tensor $\bar{\Theta}_\psi^{\mu\nu}$ is unclear. The recipe for finding the symmetric version of the spinor stress–energy tensor is provided in chapter 10, where the spin connection is introduced. This issue will be revisited in the next section, where the angular momentum of the spinor field is considered.

Demonstrating the conservation of the modified stress–energy tensor,

$$\partial_\mu\Theta^\mu{}_\nu = \partial_\mu(\Theta^\mu_{A_\nu} + \Theta^\mu_{\psi_\nu}) = 0, \tag{9.163}$$

uses the equations of motion (9.114). In particular, the gauge field component gives, after relabeling dummy indices,

$$\begin{aligned}\partial_\mu\Theta^\mu_{A_\nu} &= \partial_\mu F^{\mu\rho}F_{\rho\nu} + F^{\mu\rho}\partial_\mu F_{\rho\nu} + \frac{1}{2}F^{\alpha\beta}\partial_\nu F_{\alpha\beta}\\ &= J^\rho F_{\rho\nu} + F^{\alpha\beta}\left(\partial_\alpha F_{\beta\nu} + \frac{1}{2}\partial_\nu F_{\alpha\beta}\right)\\ &= J^\rho F_{\rho\nu} + \frac{1}{2}F^{\alpha\beta}\left(\partial_\nu F_{\alpha\beta} + \partial_\beta F_{\nu\alpha} + \partial_\alpha F_{\beta\nu}\right)\\ &- J^\rho F_{\rho\nu},\end{aligned} \tag{9.164}$$

which combined the Bianchi identity (9.57) with the identity

$$F^{\alpha\beta}\partial_\alpha F_{\beta\nu} = F^{\beta\alpha}\partial_\beta F_{\alpha\nu} = F^{\alpha\beta}\partial_\beta F_{\nu\alpha} \Longrightarrow F^{\alpha\beta}\partial_\alpha F_{\beta\nu} = \frac{1}{2}F^{\alpha\beta}(\partial_\beta F_{\nu\alpha} + \partial_\alpha F_{\beta\nu}). \tag{9.165}$$

The bispinor field contribution is given by

$$\partial_\mu\Theta^\mu_{\psi_\nu} = \frac{1}{2}i\partial_\mu(\bar{\psi}\gamma^\mu D_\nu\psi) - \frac{1}{2}i\partial_\mu(D_\nu^\dagger\bar{\psi}\gamma^\mu\psi). \tag{9.166}$$

The first term on the right-hand side of (9.166) becomes the sum of two contributions,

$$\frac{1}{2}i\partial_\mu(\bar{\psi}\gamma^\mu D_\nu\psi) = \frac{1}{2}i(\partial_\mu\bar{\psi}\gamma^\mu)D_\nu\psi + \frac{1}{2}i\bar{\psi}\gamma^\mu(\partial_\mu D_\nu\psi). \tag{9.167}$$

Using the equations of motion, these two terms can be written

$$\begin{aligned}\frac{1}{2}i(\partial_\mu\bar{\psi}\gamma^\mu)D_\nu\psi &= -\frac{1}{2}m\bar{\psi}D_\nu\psi - \frac{1}{2}\alpha A_\mu\bar{\psi}\gamma^\mu D_\nu\psi,\\ \frac{1}{2}i\bar{\psi}\gamma^\mu(\partial_\mu D_\nu\psi) &= \frac{1}{2}i\bar{\psi}D_\nu(\gamma^\mu\partial_\mu\psi) - \frac{1}{2}\alpha\partial_\mu A_\nu\bar{\psi}\gamma^\mu\psi\\ &= \frac{1}{2}i\bar{\psi}D_\nu(-im\psi - i\alpha A_\mu\gamma^\mu\psi) - \frac{1}{2}\alpha\partial_\mu A_\nu\bar{\psi}\gamma^\mu\psi\end{aligned} \tag{9.168}$$

$$= \frac{1}{2}m\bar{\psi}D_\nu\psi + \frac{1}{2}\alpha A_\mu\bar{\psi}\gamma^\mu D_\nu\psi - \frac{1}{2}\alpha(\partial_\mu A_\nu - \partial_\nu A_\mu)\bar{\psi}\gamma^\mu\psi. \tag{9.169}$$

The sum of the two terms (9.168) and (9.169) gives

$$\begin{aligned}\frac{1}{2}i\partial_\mu(\bar{\psi}\gamma^\mu D_\nu\psi) &= \frac{1}{2}i(\partial_\mu\bar{\psi}\gamma^\mu)D_\nu\psi + \frac{1}{2}i\bar{\psi}\gamma^\mu\partial_\mu(D_\nu\psi)\\ &= -\frac{1}{2}\alpha(\partial_\mu A_\nu - \partial_\nu A_\mu)\bar{\psi}\gamma^\mu\psi = -\frac{1}{2}J^\mu F_{\mu\nu}.\end{aligned} \tag{9.170}$$

Exercise 9.5: Show that the same result follows for the second term on the right-hand side of (9.166),

$$-\frac{1}{2}i\partial_\mu(D_\nu^\dagger\bar{\psi}\gamma^\mu\psi) = -\frac{1}{2}J^\mu F_{\mu\nu}. \tag{9.171}$$

Combining (9.170) and (9.171) shows that (9.166) is given by

$$\partial_\mu\Theta^{\mu}_{\psi\,\nu} = -J^\mu F_{\mu\nu}, \tag{9.172}$$

which is the exact opposite of the gauge field result (9.164). It follows that (9.163) holds for the total stress–energy tensor. The two results (9.164) and (9.172) can be written in contravariant form as

$$\partial_\mu\Theta_\psi{}^{\mu\nu} = -J_\mu F^{\mu\nu}, \tag{9.173}$$

$$\partial_\mu\Theta_A{}^{\mu\nu} = J_\mu F^{\mu\nu}, \tag{9.174}$$

which clearly shows the conservation of their sum. However, because of lack of symmetry for the spinor contribution, it follows that

$$\partial_\nu\Theta_\psi^{\mu\nu} \neq -J_\nu F^{\nu\mu}. \tag{9.175}$$

Exercise 9.6: Show that

$$\partial_\nu \Theta_\psi^{\mu\nu} = \frac{1}{2i}(\bar{\psi}\gamma^\mu \Box \psi - (\Box\bar{\psi})\gamma^\mu \psi) - A^\nu \partial_\nu J^\mu. \quad (9.176)$$

In the case that the bispinor is not coupled, the derived equation (8.177) shows that (9.176) vanishes. However, in the coupled case, the right-hand side of expression (9.176) does not reduce to $-J_\nu F^{\nu\mu}$.

Because (9.163) generalizes the Newtonian concepts of force, momentum, and work to the field theory setting, the Lagrangian density (9.110) describes a closed system where the *total* energy and momentum are conserved. The two separate parts of $\Theta^{\mu\nu}$ show how energy and momentum are exchanged between the electromagnetic field A^μ, described by $\Theta_A^{\mu\nu}$, and the matter field ψ, described by $\Theta_\psi^{\mu\nu}$. For example, the energy density of the gauge field is given by

$$\mathcal{E}_A = \Theta_A{}^{00} = \Theta^0_{A0}. \quad (9.177)$$

Combining (9.158) with (9.35), (9.41), and (9.99) gives

$$\begin{aligned}\mathcal{E}_A = \Theta_A{}^{00} &= \eta_{\mu\nu} F^{0\mu} F^{\nu 0} + \frac{1}{4} F^{\alpha\beta} F_{\alpha\beta} = -F^{0j} F^{j0} + \frac{1}{4} F^{\alpha\beta} F_{\alpha\beta} \\ &= \mathbf{E}^2 - \frac{1}{2}(\mathbf{E}^2 - \mathbf{B}^2) = \frac{1}{2}(\mathbf{E}^2 + \mathbf{B}^2).\end{aligned} \quad (9.178)$$

Similarly, Θ_A^{j0} is given by

$$\Theta_A{}^{j0} = -F^{jk} F^{k0} = -\varepsilon^{kj\ell} B_\ell E_k = \varepsilon^{jk\ell} E_k B_\ell = (\mathbf{E} \times \mathbf{B})_j. \quad (9.179)$$

The results for Θ_A^{j0} are the components of the *Poynting vector* $\mathbf{S}$, which represents the momentum density $\mathcal{P}^j$ or, since the electromagnetic field is massless, the energy density flux of the electromagnetic field,

$$\mathbf{S} = \mathbf{E} \times \mathbf{B}. \quad (9.180)$$

This identification follows from relation (9.164), which gives

$$\partial_\mu \Theta_A^{\mu 0} = \partial_0 \Theta_A^{00} + \partial_j \Theta_A^{j0} = J_\mu F^{\mu 0} \implies \frac{\partial \mathcal{E}_A}{\partial t} + \nabla \cdot \mathbf{S} = -\mathbf{J} \cdot \mathbf{E}, \quad (9.181)$$

where $J_j = -J^j$ was used. If $\mathbf{J} \cdot \mathbf{E}$ is positive, the minus sign on the right-hand side of (9.181) shows that net energy is being lost from the electromagnetic field in the process of doing positive work on the charge current.

Like the case of the canonical stress–energy tensor first found in (3.91), the components of the momentum density $\boldsymbol{P}_A$ for the electromagnetic field are given by using the velocity of propagation squared (c^2) and Θ^j_{A0},

$$\mathcal{P}^j_A = \frac{1}{c^2} \Theta^j_{A0} \implies \boldsymbol{P}_A = \frac{\mathbf{S}}{c^2} = \frac{1}{c^2}(\mathbf{E} \times \mathbf{B}). \quad (9.182)$$

Of course, in natural units, the factor c^2 is suppressed. When (9.181) is integrated over a spatial volume, the contribution from the Poynting vector becomes a surface integral through Gauss's theorem, and therefore represents the flow of electromagnetic energy in or out of that spatial volume through its surface. If the product of the electric and magnetic fields drops off faster than $1/r^2$, there will be no flow of energy into or out of the volume as its radius tends to infinity. However, if the product of the electric and magnetic fields drops off as $1/r^2$, there will be a net flow of energy out through the surface at infinity, representing the radiation of energy by the charge current. Electromagnetic radiation by accelerated charges is discussed in a later section.

The spatial momentum density $\mathcal{P}^j_\psi$ of the bispinor field is defined using Θ^{0j}_ψ, which gives

$$\mathcal{P}^j_\psi = \Theta^{0j}_\psi = \frac{1}{2}i(\bar{\psi}\gamma^0 D^j\psi - D^{\dagger j}\bar{\psi}\gamma^0\psi) = \frac{1}{2}i\psi^\dagger(D^j\psi) - \frac{1}{2}i(D^{\dagger j}\psi^\dagger)\psi. \tag{9.183}$$

Combining $\partial_\mu\Theta^{\mu\nu}_\psi = -J_\mu F^{\mu\nu}$ with $F^{0j} = -E_j$, $F^{kj} = -\varepsilon^{kj\ell}B_\ell$, and $J_k = -J^k$ gives the force density F^j_ψ on the current,

$$\begin{aligned} F^j_\psi = \frac{\partial}{\partial t}\mathcal{P}^j_\psi &= \partial_0\Theta^{0j}_\psi = -\partial_k\Theta^{kj}_\psi - J_\mu F^{\mu j} = -\,\partial_k{\Theta_\psi}^{kj} - J_0F^{0j} - J_kF^{kj} \\ &= -\,\partial_k\Theta^{kj}_\psi + J^0E_j + \varepsilon^{jk\ell}J^kB_k. \end{aligned} \tag{9.184}$$

When integrated over an infinite volume, the divergence can often be ignored, since it represents the flux of momentum carried by the Dirac field into or out of the surface of the volume. The second and third terms represent the spatial components of the relativistic Lorentz force (9.130), written in terms of the bispinor electric current J^μ,

$$\mathbf{F}_\psi = J_0\mathbf{E} + \mathbf{J}\times\mathbf{B}. \tag{9.185}$$

Similarly, the timelike component of the four-force density for the bispinor is given by

$$F^0_\psi = \frac{\partial}{\partial t}\Theta^{00}_\psi = -\partial_k\Theta^{k0}_\psi - J_\mu F^{\mu 0} = -\partial_k\Theta^{k0}_\psi - J_jF^{j0} = -\,\partial_k\Theta^{k0}_\psi + \mathbf{J}\cdot\mathbf{E}, \tag{9.186}$$

where $F^{j0} = E_j$ and $J_j = -J^j$ were used. The combination of (9.185) and (9.186) gives the Lorentz four-force density acting on the bispinor current,

$$F^\mu_\psi = (\mathbf{J}\cdot\mathbf{E},\ J_0\mathbf{E} + \mathbf{J}\times\mathbf{B}), \tag{9.187}$$

written in terms of the four-vector current density J^μ and the non-covariant **E** and **B** fields. Result (9.187) is the electric current version of the point charge Lorentz force result (9.132), where the role of **J** was played by $q\gamma_v v\ \delta^3(\boldsymbol{x} - \boldsymbol{x}_c(\tau))$.

9.10 Angular momentum for gauge and spinor fields

The conservation of the Noether current associated with invariance under rotations was used in (8.245) to derive the nonrelativistic expression for the orbital angular momentum of the scalar field. In that case, the asymmetric form of the stress–energy tensor appeared. The symmetry of the stress–energy tensor derived in the previous section allows the definition of a relativistic expression for the angular momentum of a field. This begins by noting that the orbital angular momentum **L** of a spinless Newtonian particle is the spatial vector defined as

$$\boldsymbol{L} = \boldsymbol{r} \times \boldsymbol{p} = \boldsymbol{r} \times m\boldsymbol{v}. \tag{9.188}$$

The definition (9.188) is dependent on the coordinate system for its value. For the case of a translation $\boldsymbol{r} \to \boldsymbol{r} + \boldsymbol{a}$, which is included in the Poincaré transformations (8.232) defining the Noether version of the stress–energy tensor, the orbital angular momentum changes according to

$$\boldsymbol{L} \to \boldsymbol{L} + \boldsymbol{a} \times \boldsymbol{p}. \tag{9.189}$$

For the case that $\boldsymbol{p}$ is constant it is possible to choose $\boldsymbol{a}$ so that the orbital angular momentum is zero. If there is no force acting on the particle the orbital angular momentum is conserved in all coordinate systems,

$$\frac{\mathrm{d}\boldsymbol{L}}{\mathrm{d}t} = \dot{\boldsymbol{r}} \times m\boldsymbol{v} + \boldsymbol{r} \times \dot{\boldsymbol{p}} = m\boldsymbol{v} \times \boldsymbol{v} + \boldsymbol{r} \times \boldsymbol{F} = 0. \tag{9.190}$$

As a first attempt to define the angular momentum for electromagnetic fields, it would be reasonable to write the spatial angular momentum $\boldsymbol{J}$, not to be confused with the spatial components of the charge current **J**, in terms of the momentum density $\mathcal{P}^j$ obtained from the *conserved symmetric* stress–energy tensor $\mathcal{P}^j = T^{0j} = T^{j0}$ as

$$\boldsymbol{J} = \int \mathrm{d}^3x\, \boldsymbol{r} \times \boldsymbol{\mathcal{P}}, \tag{9.191}$$

so that the spatial components of the *angular momentum density*, denoted by $\mathcal{J}$, are given by

$$\mathcal{J}_j = \varepsilon_{jk\ell} x^k T^{0\ell}. \tag{9.192}$$

However, this expression does not transform as a spatial vector. The relation of Poincaré transformations to angular momentum is clarified by defining the space-time tensor $M^{\alpha\beta\gamma}$, given by

$$M^{\alpha\beta\gamma} = T^{\alpha\beta}x^\gamma - T^{\alpha\gamma}x^\beta, \tag{9.193}$$

which is antisymmetric in β and γ, since $M^{\alpha\beta\gamma} = -M^{\alpha\gamma\beta}$. This tensor contains an antisymmetric form of the spatial angular momentum density (9.192) for the case $\alpha = 0$,

$$M^{0jk} = T^{0j}x^k - T^{0k}x^j = x^k\mathcal{P}^j - x^j\mathcal{P}^k = \frac{1}{2}\varepsilon^{kjm}\mathcal{J}_m, \tag{9.194}$$

where $\mathcal{J}$ is the angular momentum density of (9.192). This identification follows from identity (1.8), which gives

$$\frac{1}{2}\varepsilon_{kj\ell}\varepsilon^{kjm}\mathcal{J}_m = \delta^m{}_\ell\mathcal{J}_m = \mathcal{J}_\ell = \frac{1}{2}\varepsilon_{kj\ell}(x^k\mathcal{P}^j - x^j\mathcal{P}^k) = \varepsilon_{\ell kj}x^k\mathcal{P}^j. \tag{9.195}$$

This shows that the angular momentum density of the fields has the same coordinate dependence as the mechanical angular momentum (9.188). This also demonstrates that it is necessary to include the interaction of the matter fields with the other fields present, since the total stress–energy tensor must include the contribution of both in order to be conserved. This allows the matter fields and the other fields to exchange angular momentum in a conserved manner.

The tensor of (9.193) has a number of important properties. The first is that its divergence is given by

$$\partial_\alpha M^{\alpha\beta\gamma} = (\partial_\alpha T^{\alpha\beta})x^\gamma + T^{\alpha\beta}\delta^\gamma_\alpha - (\partial_\alpha T^{\alpha\gamma})x^\beta - T^{\alpha\gamma}\delta^\beta_\alpha = T^{\gamma\beta} - T^{\beta\gamma}, \tag{9.196}$$

which vanishes if the stress–energy tensor is both conserved *and symmetric*. In the case of the gauge field interacting with a spinor field, the Belinfante stress–energy tensor found as the sum of (9.158) and (9.159) is conserved, but is *not* symmetric. Using the stress–energy tensor defined in this manner would not yield a conserved version of the tensor (9.193) for the interacting case, and therefore would not yield the total angular momentum of the interacting fields. Rather than relying upon the Belinfante procedure or another ad hoc procedure to piece together a symmetric version of the stress–energy tensor, a universal definition of the symmetric stress–energy tensor will be given in chapter 10.

9.10.1 Gauge field angular momentum

In what follows attention will therefore be restricted to the case of a free electromagnetic field, so that the stress–energy tensor is given by the modified electromagnetic component (9.158) for the case that the matter J_μ vanishes. For such a restriction relation, (9.194) shows that the *relativistic angular momentum tensor* $J_A^{\mu\nu}$ for the gauge field is defined as

$$J_A^{\mu\nu} = M_A^{0\mu\nu} = \Theta_A^{0\mu}x^\nu - \Theta_A^{0\nu}x^\mu. \tag{9.197}$$

This tensor is clearly antisymmetric, $J_A^{\mu\nu} = -J_A^{\nu\mu}$, and since $\partial_\alpha M_A^{\alpha\mu\nu} = 0$, its spatial integral satisfies

$$\frac{\partial}{\partial t}\int \mathrm{d}^3x\, J^{\mu\nu} = \int \mathrm{d}^3x\, \partial_0 M^{0\mu\nu} = -\int \mathrm{d}^3x\, \partial_j M^{j\mu\nu} = -\oint_{S(V)} \mathrm{d}S_j M^{j\mu\nu} = 0, \tag{9.198}$$

where it is assumed that the field configurations vanish at the surface of the spatial volume. Combining (9.198) with (9.194) shows that this definition of angular momentum yields a conserved quantity, since the required conditions of conservation and symmetry of the stress–energy tensor are met. Under a coordinate translation $x^\mu \to x^\mu + a^\mu$, the angular momentum tensor changes according to

$$\begin{aligned} J_A^{\mu\nu} &= M_A^{0\mu\nu} \to M_A^{0\mu\nu} + \Theta_A^{0\mu} a^\nu - \Theta_A^{0\nu} a^\mu = J_A^{\mu\nu} + \Theta_A^{0\mu} a^\nu - \Theta_A^{0\nu} a^\mu \\ &= J_A^{\mu\nu} + \mathcal{P}_A^\mu a^\nu - \mathcal{P}_A^\nu a^\mu, \end{aligned} \tag{9.199}$$

which is the property of the particle angular momentum (9.188). While (9.194) shows that J_A^{jk} is related to the spatial angular momentum of the gauge field, the other components are given by

$$J_A^{0j} = M_A^{00j} = \Theta_A^{00} x^j - \Theta_A^{0j} x^0 = \mathcal{E}_A x^j - \mathcal{P}_A^j t. \tag{9.200}$$

Like the angular momentum, the spatial integral of (9.200) for field configurations that vanish at the surface of the spatial volume obeys the conservation law

$$\begin{aligned} \partial_0 \int \mathrm{d}^3x\, J_A^{0j} &= \int \mathrm{d}^3x\, \partial_0 M_A^{00j} = -\int \mathrm{d}^3x\, \partial_k M_A^{k0j} \\ &= -\oint_{S(V)} \mathrm{d}S_k\, M_A^{k0j} = 0, \end{aligned} \tag{9.201}$$

which is known as the *center of mass theorem*. Since the integral is spatial it can be written

$$\int \mathrm{d}^3x J_A^{0j} = \int \mathrm{d}^3x\, \mathcal{E}_A x^j - t \int \mathrm{d}^3x \mathcal{P}_A^j = \int \mathrm{d}^3x\, \mathcal{E}_A x^j - t P_A^j, \tag{9.202}$$

where P_A^j is the jth component of the gauge field momentum. The quantity $\langle x^j \rangle$ is defined by

$$\int \mathrm{d}^3x\, x^j \mathcal{E}_A = \langle x^j \rangle \int \mathrm{d}^3x\, \mathcal{E}_A = \langle x^j \rangle\, U, \tag{9.203}$$

where U is the total gauge field energy. Since the integral (9.202) is a constant vector, denoted by $\boldsymbol{X}_c$, it follows that (9.202) becomes

$$\langle x^j \rangle = \frac{X_c^j}{U} + \frac{\mathcal{P}_A^j}{U} t, \tag{9.204}$$

which shows that there is a point around which the energy is centered, like the center of mass of an object, and that this point moves at a constant velocity of $v^j = \mathcal{P}_A^j / U$. While interesting, this is of marginal value in calculations.

It is useful to write the angular momentum (9.195) as the sum of the orbital angular momentum and an intrinsic angular momentum, the latter of which is referred to as the *spin angular momentum* or simply the *spin*. This was previously introduced for the spinor field in (8.258) and its conservation followed from Noether's theorem. Noether's theorem will be applied to the electromagnetic case in the next section to derive results similar to the spinor field for the electromagnetic field. However, the intrinsic angular momentum for the electromagnetic field can also be extracted from the expression (9.195) for the spatial angular momentum by restricting attention to the electromagnetic part of the stress–energy tensor. This allows the angular momentum to be conserved, since the free electromagnetic stress–energy tensor is both conserved and symmetric, so that (9.193) is a conserved tensor.

This starts by using the electromagnetic momentum density (9.182), the relation $\mathbf{B} = \nabla \times \boldsymbol{A}$, and the identity (1.7) to find

$$x^k \mathcal{P}_A^j = x^k \varepsilon^{j\ell m} E_\ell B_m = x^k \varepsilon^{j\ell m} E_\ell \, \varepsilon_{mrp} \, \partial^r A^p = x^k E_\ell \, \partial^j A^\ell - x^k E_\ell \partial^\ell A^j. \quad (9.205)$$

Using this result in (9.195) shows that the electromagnetic angular momentum is given by the spatial integral

$$J_{Ai} = \int \mathrm{d}^3 x \, \varepsilon_{ikj} x^k \mathcal{P}_A^j = \int \mathrm{d}^3 x \, \varepsilon_{ikj} (x^k E_m \partial^j A^m - x^k E_m \partial^m A^j). \quad (9.206)$$

The total integrand in (9.206) is gauge invariant although the two individual terms are not. As a result, separating the two terms results in two quantities that are individually *not gauge invariant*.

Performing a coordinate translation, $x^k \to x^k + a^k$, causes the first integral, denoted by L_{Ai}, to change according to

$$L_{Ai} = \int \mathrm{d}^3 x \, \varepsilon_{ikj} x^k E_m \partial^j A^m \;\to\; L_{Ai} + \varepsilon_{ikj} a^k \int \mathrm{d}^3 x \, E_m \partial^j A^m. \quad (9.207)$$

In order to understand the spatial integral that is generated by the translation, it is noted that the electromagnetic field momentum is given by

$$\begin{aligned} P_A^j &= \int \mathrm{d}^3 x \, (\mathbf{E} \times \mathbf{B})^j = \int \mathrm{d}^3 x (\varepsilon^{jk\ell} E_k \, \varepsilon_{\ell mn} \partial^m A^n) \\ &= \int \mathrm{d}^3 x \, (E_k \partial^j A^k - E_k \partial^k A^j) \\ &= \int \mathrm{d}^3 x \, E_k \partial^j A^k - \int \mathrm{d}^3 x \, \partial^k (E_k A^j) + \int \mathrm{d}^3 x \, A^j \nabla \cdot \mathbf{E} \\ &= \int \mathrm{d}^3 x \, E_k \partial^j A^k, \end{aligned} \quad (9.208)$$

where Gauss's law $\nabla \cdot \mathbf{E} = 0$ for the free field was used and it was assumed that the field configurations vanish at the boundary of the spatial volume so that the divergence can be discarded. Expression (9.208) for P_A^j is gauge invariant, since integrating by parts and dropping surface terms gives

$$\begin{aligned} \int \mathrm{d}^3 x \, E_k \partial^j (A^k - \partial^k \Lambda) &= \int \mathrm{d}^3 x \, E_k \partial^j A^k + \int \mathrm{d}^3 x \, \partial^j \Lambda \nabla \cdot \mathbf{E} \\ &= \int \mathrm{d}^3 x \, E_k \partial^j A^k. \end{aligned} \quad (9.209)$$

Using result (9.208) shows that the translation of (9.207) can be written as the vector relationship,

$$\boldsymbol{L}_A \to \boldsymbol{L}_A + \boldsymbol{a} \times \boldsymbol{P}_A, \quad (9.210)$$

which contains the familiar transformation property (9.189) of the orbital angular momentum of a point particle. Expression (9.207) for $\boldsymbol{L}_A$ is therefore identified as the total *orbital angular momentum* present in the electromagnetic field configuration.

The second term in (9.206) does not transform like the orbital angular momentum and is therefore identified as the total *intrinsic angular momentum* or the spin Σ_{Ai} of the gauge field. The intrinsic nature of the second term follows from writing it as

$$\begin{aligned}\Sigma_{Ai} &= -\int d^3x\ \varepsilon_{ikj} x^k E_\ell \partial_\ell A^j = -\int d^3x\ \varepsilon_{ikj}(\partial_\ell(x^k E_\ell A^j) - E^k A^j - x^k A^j \partial_\ell E_\ell) \\ &= \int d^3x\ \varepsilon_{ikj} E^k A^j,\end{aligned} \tag{9.211}$$

where Gauss's law $\nabla \cdot \mathbf{E} = 0$ was used and the field configuration was assumed to vanish at the boundary of the spatial volume, allowing the total divergence to be discarded. The intrinsic angular momentum has the property that a translation results in

$$\boldsymbol{\Sigma}_A \to \boldsymbol{\Sigma}_A, \tag{9.212}$$

so that the sum $\boldsymbol{J}_A = \boldsymbol{L}_A + \boldsymbol{\Sigma}_A$ has the property that a translation has the expected angular momentum translation property,

$$\boldsymbol{J}_A \to \boldsymbol{J}_A + \boldsymbol{a} \times \boldsymbol{P}_A. \tag{9.213}$$

This shows that the total intrinsic angular momentum density $\boldsymbol{s}_A$ is given by

$$\boldsymbol{s}_A = \mathbf{E} \times \boldsymbol{A}, \tag{9.214}$$

which does not transform under a coordinate translation as (9.199) does. The quantity $\boldsymbol{s}_A$ is therefore identified as the *electromagnetic spin density*, so that (9.206) shows that the total angular momentum $\boldsymbol{J}$ is given by the gauge invariant sum,

$$\boldsymbol{J}_A = \boldsymbol{L}_A + \int d^3x\ \boldsymbol{s}_A = \boldsymbol{L}_A + \boldsymbol{\Sigma}_A. \tag{9.215}$$

Result (9.214) will also be obtained from Noether's theorem in the next section.

9.10.2 Spinor field angular momentum

It is natural to consider analyzing the angular momentum of the Dirac spinor field by a similar method. Taking such an approach immediately encounters the problem generated by the asymmetric nature of the spinor stress–energy tensor. The asymmetry of the spinor stress–energy tensor (9.153) results in an expression for the tensor (9.193) that generalizes angular momentum that is *not conserved*. In this section a symmetric spinor stress–energy tensor will be defined by symmetrizing (9.153) for the case of the non-interacting spinor field using the method of (9.162). The result is a symmetric stress–energy tensor for the free Dirac spinor field that is given by

$$\bar{\Theta}^{\mu\nu}_\psi = \frac{1}{4} i\bar{\psi}(\gamma^\mu \partial^\nu + \gamma^\nu \partial^\mu)\psi - \frac{1}{4} i(\partial^\nu \bar{\psi}\gamma^\mu + \partial^\mu \bar{\psi}\gamma^\nu)\psi. \tag{9.216}$$

Exercise 9.7: Show that the stress–energy tensor of (9.216) is conserved by using the Dirac equations, $\partial_\mu \bar{\psi}\gamma^\mu = im\bar{\psi}$ and $\gamma^\mu \partial_\mu \psi = -im\psi$, and the secondary equations of (8.177), $(\Box + m^2)\psi = 0$ and $(\Box + m^2)\bar{\psi} = 0$.

For the purpose of facilitating calculation, it is also useful to note that (9.216) is the sum of two tensors,

$$\begin{aligned} \bar{T}^{1\mu\nu}_{\psi} &= \frac{1}{4} i\bar{\psi}(\gamma^\mu \partial^\nu + \gamma^\nu \partial^\mu)\psi, \\ \bar{T}^{2\mu\nu}_{\psi} &= -\frac{1}{4} i(\partial^\nu \bar{\psi}\gamma^\mu + \partial^\mu \bar{\psi}\gamma^\nu)\psi, \end{aligned} \tag{9.217}$$

which have the property that

$$\left(\bar{T}^{1\mu\nu}_{\psi}\right)^* = \bar{T}^{2\mu\nu}_{\psi}. \tag{9.218}$$

This guarantees that the stress–energy tensor is real valued.

Exercise 9.8: Using result (7.280) show that (9.218) holds.

As a result, calculations of integrals made with one tensor can be found for the other tensor by simply taking the complex conjugate.

The contribution to the spatial angular momentum made by the tensor $\bar{T}^{1\mu\nu}_{\psi}$ is given by

$$J^1_\ell = \varepsilon_{\ell jk} \int d^3x\, x^j \bar{T}^{10k}_{\psi} = \frac{1}{4} i\varepsilon_{\ell jk} \int d^3x\, x^j (\bar{\psi}(\gamma^0 \partial^k + \gamma^k \partial^0)\psi). \tag{9.219}$$

The first term in the integral of (9.219) can be written

$$\frac{1}{4} i\varepsilon_{\ell jk} \int d^3x\, x^j \bar{\psi}\gamma^0 \partial^k \psi = \frac{1}{4} i\varepsilon_{\ell jk} \int d^3x\, x^j \psi^\dagger \partial^k \psi. \tag{9.220}$$

Using the anticommutation relations of the Dirac matrices, the second term in the integral of (9.219) can be evaluated with the equation of motion for ψ,

$$\begin{aligned} \frac{1}{4} i\varepsilon_{\ell jk} \int d^3x\, x^j \bar{\psi}\gamma^k \partial^0 \psi &= \frac{1}{4} i\varepsilon_{\ell jk} \int d^3x\, x^j \psi^\dagger \gamma^0 \gamma^k \partial^0 \psi = -\frac{1}{4} i\varepsilon_{\ell jk} \int d^3x \\ & x^j \psi^\dagger \gamma^k (\gamma^0 \partial^0 \psi) = \frac{1}{4} i\varepsilon_{\ell jk} \int d^3x\, x^j \psi^\dagger \gamma^k \gamma^r \partial^r \psi \\ & - \frac{1}{4} m\, \varepsilon_{\ell jk} \int d^3x\, x^j \psi^\dagger \gamma^k \psi. \end{aligned} \tag{9.221}$$

The product of the gamma matrices can be written

$$\gamma^k\gamma^r = \frac{1}{2}[\gamma^k, \gamma^r] + \frac{1}{2}\{\gamma^k, \gamma^r\} = 2\,\mathbf{S}^{kr} + \eta^{kr}\mathbf{I}, \tag{9.222}$$

where the commutators $\mathbf{S}^{kr}$ are the spatial components of the Lorentz algebra generators found earlier in (7.260) and relabeled here as $\mathbf{S}$ to avoid confusion with the angular momentum. These were shown to generate spatial rotations for the Dirac spinor in chapter 7. Using (9.222) in (9.221) gives

$$\frac{1}{4}i\varepsilon_{\ell jk}\int \mathrm{d}^3x\; x^j\bar{\psi}\gamma^k\partial^0\psi = \frac{1}{4}\varepsilon_{\ell jk}\int \mathrm{d}^3x\; x^j(i\psi^\dagger\partial^k\psi + 2i\psi^\dagger\mathbf{S}^{kr}\partial^r\psi - m\psi^\dagger\gamma^k\psi). \tag{9.223}$$

Adding (9.220) to this result gives the contribution of $\bar{T}^1_\psi$ to the angular momentum,

$$J^1_\ell = \frac{1}{2}i\varepsilon_{\ell jk}\int \mathrm{d}^3x\; x^j\left(\psi^\dagger\partial^k\psi + \psi^\dagger\mathbf{S}^{kr}\partial^r\psi + \frac{1}{2}im\psi^\dagger\gamma^k\psi\right). \tag{9.224}$$

The contribution of the tensor T^{2jk}_ψ to the angular momentum is found by taking the complex conjugate of result (9.224) and adding it to (9.224). The first term gives

$$(i\psi^\dagger\partial^k\psi)^* = -i(\partial^k\psi^\dagger)\psi. \tag{9.225}$$

Using $\mathbf{S}^{kr\dagger} = -\mathbf{S}^{kr}$ shows that the second term gives

$$\left(\frac{1}{2}i\psi^\dagger\mathbf{S}^{kr}\partial^r\psi\right)^* = -\frac{1}{2}i(\partial^r\psi^\dagger)\mathbf{S}^{kr\dagger}\psi = \frac{1}{2}i(\partial^r\psi^\dagger)\mathbf{S}^{kr}\psi. \tag{9.226}$$

The Weyl representation of the Dirac matrices has the property $\gamma^{k\dagger} = -\gamma^k$, so that the third term gives

$$\left(-\frac{1}{4}m\psi^\dagger\gamma^k\psi\right)^* = -\frac{1}{4}\psi^\dagger\gamma^{k\dagger}\psi = \frac{1}{4}m\psi^\dagger\gamma^k\psi. \tag{9.227}$$

Using (9.225), (9.226), and (9.227) shows that the spinor angular momentum is given by

$$J^\psi_\ell = \frac{1}{2}i\int \mathrm{d}^3x\; \varepsilon_{\ell jk}x^j(\psi^\dagger\partial^k\psi - \partial^k\psi^\dagger\psi + \partial^r(\psi^\dagger\mathbf{S}^{kr}\psi)). \tag{9.228}$$

The term involving $\mathbf{S}$ can be integrated by parts and the surface terms discarded, giving the final result for the spinor angular momentum,

$$J^\psi_\ell = \frac{1}{2}i\int \mathrm{d}^3x\; \varepsilon_{\ell jk}(x^j(\psi^\dagger\partial^k\psi - \partial^k\psi^\dagger\psi) + \psi^\dagger\mathbf{S}^{jk}\psi), \tag{9.229}$$

where $\mathbf{S}^{kj} = -\mathbf{S}^{jk}$ was used.

The orbital angular momentum of the Dirac spinor is identified as

$$L^\psi_\ell = \frac{1}{2}i\varepsilon_{\ell jk}\int \mathrm{d}^3x\; x^j(\psi^\dagger\partial^k\psi - \partial^k\psi^\dagger\psi), \tag{9.230}$$

since the translation $x^j \to x^j + a^j$ causes it to transform identically to (9.189). It takes the usual form for orbital angular momentum,

$$L_\ell^\psi = \varepsilon_{\ell jk} \int \mathrm{d}^3x\; x^j \mathcal{P}^k, \tag{9.231}$$

where the momentum density is given by

$$\mathcal{P}^k = \frac{1}{2} i(\psi^\dagger \partial^k \psi - \partial^k \psi^\dagger \psi) = -\frac{1}{2} i(\psi^\dagger \partial_k \psi - \partial_k \psi^\dagger \psi). \tag{9.232}$$

The intrinsic or spin angular momentum for the Dirac spinor is identified from (9.229) as

$$\Sigma_\ell^\psi = \int \mathrm{d}^3x \left(\frac{1}{2} i\varepsilon_{\ell jk} \psi^\dagger \mathbf{S}^{jk} \psi \right). \tag{9.233}$$

In the Weyl representation the rotation generators $\mathbf{S}^{jk}$ are given in terms of the Pauli spin matrices $\boldsymbol{\sigma}_j$ by (7.263), so that

$$\mathbf{S}^{jk} = \frac{1}{4}[\gamma^j, \gamma^k] = -\frac{1}{2} i\varepsilon^{jkr} \begin{pmatrix} \boldsymbol{\sigma}_r & 0 \\ 0 & \boldsymbol{\sigma}_r \end{pmatrix} \equiv -\frac{1}{2} i\varepsilon^{jkr} \boldsymbol{S}_r. \tag{9.234}$$

Using $\varepsilon_{\ell jk} \varepsilon^{jkr} = 2\delta^r{}_\ell$ gives

$$\Sigma_\ell^\psi = \int \mathrm{d}^3x\; \frac{1}{2} \psi^\dagger \boldsymbol{S}_\ell \psi. \tag{9.235}$$

Apart from the factor of $\hbar$ that is absent due to using natural units, result (9.235) for the Dirac spinor is identical to the previous result (8.258) found for the spin density of the Weyl spinor by using the Noether current. This further identifies $\frac{1}{2}\boldsymbol{\sigma}_j$ as the spin matrix for spinor fields.

9.11 Electromagnetic waves and spin

It is important to analyze the vector function $a^\mu(x)$ that satisfies both the equation of motion in the absence of electric charges,

$$\Box a^\mu(x) = 0, \tag{9.236}$$

as well as the Lorenz condition

$$\partial_\mu a^\mu(x) = 0. \tag{9.237}$$

The function $a^\mu(x)$ is referred to as a *vacuum solution* of Maxwell's equations, since no dielectric materials or electric charges are present.

Restricting consideration to real-valued solutions, a particularly simple form is given by

$$a^\mu(x) = \alpha^\mu \sin(k \cdot x) = \alpha^\mu \sin(k^0 x^0 - \boldsymbol{k} \cdot \boldsymbol{x}), \tag{9.238}$$

where α^μ is a constant vector with units of inverse length, not to be confused with the coupling constant α, and

$$k^0 = \omega_k = +\sqrt{\boldsymbol{k} \cdot \boldsymbol{k}}\,. \tag{9.239}$$

Relation (9.239) ensures that $k^\mu = (\omega_k, \boldsymbol{k})$ is a lightlike four-vector,

$$k^2 = k_\mu k^\mu = \omega_k^2 - \boldsymbol{k} \cdot \boldsymbol{k} = 0, \tag{9.240}$$

which in turn solves (9.236),

$$\Box a^\mu(x) = -k^2 a^\mu(x) = 0. \tag{9.241}$$

The constant α^μ is referred to as the *polarization vector* of the solution. The two spatial vectors $\boldsymbol{\alpha}$ and $\boldsymbol{k}$ define a plane, and so the solution (9.238) is known as a *plane wave*. The Lorenz condition requires that

$$\partial_\mu a^\mu(x) = k_\mu \alpha^\mu \cos(k \cdot x) = 0, \tag{9.242}$$

which is satisfied if the following Lorentz invariant condition holds,

$$k_\mu \alpha^\mu = \omega_k \alpha^0 - \boldsymbol{k} \cdot \boldsymbol{\alpha} = 0. \tag{9.243}$$

This is easily solved to find

$$\alpha^0 = \frac{\boldsymbol{k} \cdot \alpha}{\omega_k}. \tag{9.244}$$

The choice of subsidiary condition is arbitrary and it was pointed out in (9.21) that, regardless of the subsidiary condition, Maxwell's equations possess two *transverse* degrees of freedom. This can be demonstrated for the Lorenz condition by using its residual gauge invariance (9.79) under scalar gauge functions $\Lambda(x)$ that satisfy $\Box\Lambda = 0$. By construction, the gauge function

$$\Lambda(x) = -\lambda \cos(k \cdot x), \tag{9.245}$$

satisfies this requirement if λ is a constant. The gauge transformed version of $a^\mu(x)$ is given by

$$\tilde{a}^\mu(x) = a^\mu(x) - \partial^\mu \Lambda(x) = (\alpha^\mu - k^\mu \lambda)\sin(k \cdot x), \tag{9.246}$$

which still satisfies the equation of motion (9.236) and the Lorenz condition (9.237) as long as $k_\mu k^\mu = 0$. By choosing the constant λ to be

$$\lambda = \frac{\alpha^0}{\omega_k} = \frac{\boldsymbol{k} \cdot \boldsymbol{\alpha}}{\omega_k^2}, \tag{9.247}$$

the spatial part of the gauge transformed solution is given by

$$\tilde{\boldsymbol{a}}(x) = \left(\boldsymbol{\alpha} - \frac{\boldsymbol{k} \cdot \boldsymbol{\alpha}}{\omega_k^2}\boldsymbol{k}\right) \sin(k \cdot x) \equiv \tilde{\boldsymbol{\alpha}}\sin(k \cdot x), \tag{9.248}$$

while $\tilde{a}^0(x) = 0$. For the case that $\boldsymbol{k} \cdot \boldsymbol{k} = \omega_k^2$, the gauge transformed field $\tilde{a}^\mu(x)$ satisfies both the Lorenz and Coulomb subsidiary conditions, since $\boldsymbol{k} \cdot \tilde{\boldsymbol{\alpha}} = 0$ gives

$$\nabla \cdot \tilde{\boldsymbol{a}}(x) = \boldsymbol{k} \cdot \tilde{\boldsymbol{\alpha}} \cos(k \cdot x) = 0, \tag{9.249}$$

revealing that the gauge field solution is *transverse* to the direction of propagation given by $\boldsymbol{k}$. This identifies the two transverse spatial degrees of freedom found earlier using the Helmholtz decomposition (9.7).

Using the transverse solution (9.248) reveals many properties of vacuum electromagnetic waves. Since $\tilde{a}^0 = 0$ the electric field is given by

$$\mathbf{E} = -\partial_0 \tilde{\boldsymbol{a}}(x) = -\omega_k \tilde{\boldsymbol{\alpha}} \cos(k \cdot x), \tag{9.250}$$

while the magnetic field is given by

$$\mathbf{B} = \nabla \times \tilde{\boldsymbol{a}}(x) = -(\boldsymbol{k} \times \tilde{\boldsymbol{\alpha}}) \cos(k \cdot x). \tag{9.251}$$

These two results demonstrate the transverse nature of the wave,

$$\boldsymbol{k} \cdot \mathbf{E} = \boldsymbol{k} \cdot \mathbf{B} = 0. \tag{9.252}$$

It follows that $\mathbf{E}$ and $\mathbf{B}$ are perpendicular, since

$$\mathbf{E} \cdot \mathbf{B} = \omega_k \, \tilde{\boldsymbol{\alpha}} \cdot (\boldsymbol{k} \times \tilde{\boldsymbol{\alpha}}) \cos^2(k \cdot x) = 0. \tag{9.253}$$

Using the vector identity

$$\tilde{\boldsymbol{\alpha}} \times (\boldsymbol{k} \times \tilde{\boldsymbol{\alpha}}) = (\tilde{\boldsymbol{\alpha}} \cdot \tilde{\boldsymbol{\alpha}}) \, \boldsymbol{k} - (\tilde{\boldsymbol{\alpha}} \cdot \boldsymbol{k}) \, \tilde{\boldsymbol{\alpha}} = \boldsymbol{k} \, \tilde{\alpha}^2, \tag{9.254}$$

gives the spatial momentum density of the electromagnetic wave from the Poynting vector (9.180),

$$\mathbf{S} = \boldsymbol{P}_A = \mathbf{E} \times \mathbf{B} = \boldsymbol{k} \, \omega_k \tilde{\alpha}^2 \cos^2(k \cdot x), \tag{9.255}$$

showing that the flow is in the direction of propagation. Similarly, using the vector identity

$$(\boldsymbol{k} \times \tilde{\boldsymbol{\alpha}}) \cdot (\boldsymbol{k} \times \tilde{\boldsymbol{\alpha}}) = (\boldsymbol{k} \cdot \boldsymbol{k})(\tilde{\alpha} \cdot \tilde{\alpha}) - (\boldsymbol{k} \cdot \tilde{\alpha})^2 = \omega_k^2 \tilde{\alpha}^2, \tag{9.256}$$

the energy density of the wave is given by

$$\mathcal{E}_A = \frac{1}{2}(\mathbf{E}^2 + \mathbf{B}^2) = \omega_k^2 \tilde{\alpha}^2 \cos^2(k \cdot x). \tag{9.257}$$

Using (9.240), the magnitudes of (9.255) and (9.257) satisfy the relativistic equation

$$\mathcal{E}_A^2 - \boldsymbol{P}_A \cdot \boldsymbol{P}_A = 0, \tag{9.258}$$

which relation (6.117) shows must correspond to a massless quantity.

In quantum field theory this massless quantity is identified as the *photon*, the constituent particle that makes up an electromagnetic wave. The photon has the energy and momentum

$$\begin{aligned} E &= \hbar \omega_k = \hbar c |\boldsymbol{k}|, \\ \boldsymbol{p} &= \hbar \boldsymbol{k}, \end{aligned} \tag{9.259}$$

which become ω_k and $\boldsymbol{k}$ in natural units. This result can be derived from the classical electromagnetic wave, just as the Dirac field for a single particle was defined in (8.190), and this will prove useful in connecting classical results to the particle concept of quantum field theory. This is accomplished by replacing the plane wave over an infinite volume with the standing-wave solution in a finite volume given by a cubical box with sides of $2L$. The solution must vanish at the walls of the box, and so it is given by

$$\tilde{\boldsymbol{a}}(x) = \tilde{\alpha}_n \cos(\omega_k t) \sin\left(\frac{n_x \pi x}{L}\right) \sin\left(\frac{n_y \pi y}{L}\right) \sin\left(\frac{n_z \pi z}{L}\right), \tag{9.260}$$

where n_j are integers with either sign that can be treated as a vector,

$$\boldsymbol{n} = (n_x, n_y, n_z). \tag{9.261}$$

This is a solution of the wave equation if

$$\omega_k^2 = \frac{(n_x^2 + n_y^2 + n_z^2)\pi^2}{L^2} = \boldsymbol{k} \cdot \boldsymbol{k} = \frac{(\boldsymbol{n} \cdot \boldsymbol{n})\pi^2}{L^2} = \frac{n^2 \pi^2}{L^2}, \tag{9.262}$$

where the wave vector is given by

$$\boldsymbol{k} = \left(\frac{n_x \pi}{L}, \frac{n_y \pi}{L}, \frac{n_z \pi}{L}\right) = \frac{\pi \boldsymbol{n}}{L}. \tag{9.263}$$

In the limit that $L \to \infty$ the frequency (9.262) and the wave vector (9.263) become continuous, with the advantage that interim calculations before the limit is taken are well-defined. For example, the $\mathbf{E}$-field is given by

$$\mathbf{E} = -\partial_0 \tilde{\boldsymbol{a}}(x) = -\omega_k \tilde{\boldsymbol{\alpha}} \sin(\omega_k t) \sin\left(\frac{n_x \pi x}{L}\right) \sin\left(\frac{n_y \pi y}{L}\right) \sin\left(\frac{n_z \pi z}{L}\right), \tag{9.264}$$

while the $\mathbf{B}$-field is given by $\nabla \times \tilde{\boldsymbol{a}}$. It is left as an exercise to show that, for large L, the energy density $\mathcal{E}_A$ of the solution can be normalized,

$$\int_{L^3} \mathrm{d}^3x\, \mathcal{E}_A - \int_{L^3} \mathrm{d}^3x\, \frac{1}{2}(\mathbf{E}^2 + \mathbf{B}^2) = \omega_k, \tag{9.265}$$

by choosing the constant vector $\tilde{\boldsymbol{\alpha}}$ to obey

$$\tilde{\boldsymbol{\alpha}} \cdot \tilde{\boldsymbol{\alpha}} = \frac{1}{\omega_k L^3}. \tag{9.266}$$

This choice is referred to as *box normalization*. Demonstrating this result is facilitated by using the identity

$$\mathbf{B} \cdot \mathbf{B} = (\nabla \times \tilde{\boldsymbol{a}}) \cdot (\nabla \times \tilde{\boldsymbol{a}}) = \varepsilon^{ijk} \varepsilon^{i\ell m} \partial_j \tilde{a}_k \partial_\ell \tilde{a}_m = \partial_j \tilde{a}_k \partial_j \tilde{a}_k, \tag{9.267}$$

where $\nabla \cdot \tilde{\boldsymbol{a}} = 0$ was used.

Exercise 9.9: Prove (9.267) and then use it to verify (9.265) holds if the factor $\tilde{a}$ obeys (9.266) in the large L limit.

These results can be extended to a Fourier transform, allowing the construction of arbitrary electromagnetic wave packets. For an infinite volume the *randomly polarized* vector solution can be written similarly to the scalar field case (8.30),

$$a_\mu(x) = \int \frac{d^4k}{(2\pi)^3}\, \delta(k^2)\, \theta(k_0) \sum_{\lambda=0}^{3} \epsilon_\mu^{(\lambda)}(k)(\tilde{a}_\lambda(k)\, e^{-ik\cdot x} + \tilde{a}_\lambda^*(k)\, e^{ik\cdot x}), \tag{9.268}$$

where $\epsilon_\mu^{(\lambda)}(k)$ are four *polarization vectors* indexed by λ, so that there is a sum over λ. The utility of using polarization vectors in (9.268) emerges from writing the first two as spatial vectors,

$$\epsilon_\mu^{(1,\,2)}(k) = (0,\, \boldsymbol{\epsilon}^{(1,2)}(\boldsymbol{k})), \tag{9.269}$$

where the spatial components are chosen to satisfy

$$\boldsymbol{k} \cdot \boldsymbol{\epsilon}^{(1,2)}(\boldsymbol{k}) = 0. \tag{9.270}$$

These two spatial polarization vectors are also chosen to be orthogonal,

$$\boldsymbol{\epsilon}^{(1)}(\boldsymbol{k}) \cdot \boldsymbol{\epsilon}^{(2)}(\boldsymbol{k}) = 0, \tag{9.271}$$

and they are normalized, so that

$$\boldsymbol{\epsilon}^{(1)}(\boldsymbol{k}) \cdot \boldsymbol{\epsilon}^{(1)}(\boldsymbol{k}) = \boldsymbol{\epsilon}^{(2)}(\boldsymbol{k}) \cdot \boldsymbol{\epsilon}^{(2)}(\boldsymbol{k}) = 1. \tag{9.272}$$

These two polarization vectors form the unit vector basis of the plane perpendicular to $\boldsymbol{k}$, since they are both transverse to $\boldsymbol{k}$ and each other and are normalized.

In order to be independent of these two unit vectors, the third spatial polarization vector must be chosen to be perpendicular to the first two and therefore parallel or *longitudinal* to $\boldsymbol{k}$, so that it can be written

$$\epsilon_\mu^{(3)}(k) = (0,\, \hat{\boldsymbol{k}}) = \left(0,\, \frac{\boldsymbol{k}}{\omega_k}\right), \tag{9.273}$$

where $\hat{\boldsymbol{k}}$ is a unit vector in the direction of propagation $\boldsymbol{k}$. Relations (9.270) and (9.273) allow the properties of all three spatial polarization vectors to be written in terms of the Minkowski metric,

$$\boldsymbol{\epsilon}^{(j)}(\boldsymbol{k}) \cdot \boldsymbol{\epsilon}^{(\ell)}(\boldsymbol{k}) = -\eta^{j\ell}. \tag{9.274}$$

These three spatial polarization vectors span the $\boldsymbol{k}$-space. Their *completeness* is written using the tensor product for a Cartesian space given by (1.105),

$$\sum_{\lambda=1}^{3} \epsilon_j^{(\lambda)}(\boldsymbol{k})\, \epsilon_\ell^{(\lambda)}(\boldsymbol{k}) = \sum_{\lambda=1}^{3} (\boldsymbol{\epsilon}^{(\lambda)}(\boldsymbol{k}) \otimes \boldsymbol{\epsilon}^{(\lambda)}(\boldsymbol{k}))_{j\ell} = -\eta_{j\ell}. \tag{9.275}$$

Using the longitudinal vector (9.273) gives the following useful result,

$$\sum_{\lambda=1}^{2} \epsilon_j^{(\lambda)}(\boldsymbol{k})\epsilon_\ell^{(\lambda)}(\boldsymbol{k}) = -\eta_{j\ell} - \epsilon_j^{(3)}(\boldsymbol{k})\epsilon_\ell^{(3)}(\boldsymbol{k}) = -\eta_{j\ell} - \frac{k_j k_\ell}{\omega_{\boldsymbol{k}}^2}. \tag{9.276}$$

Exercise 9.10: Using

$$k^j k_j = -k^j k^j = -\boldsymbol{k} \cdot \boldsymbol{k} = -\omega_{\boldsymbol{k}}^2, \tag{9.277}$$

show that the space spanned by (9.276) is indeed orthogonal to $\boldsymbol{k}$.

The final polarization vector points in the timelike direction, so that

$$\epsilon_\mu^0(k) = (1, 0). \tag{9.278}$$

Grouping all the results together gives

$$\epsilon_\mu^{(\lambda)}\tilde{a}_\lambda(k) = (\tilde{a}_0(k), \tilde{a}_1(k)\boldsymbol{\epsilon}^{(1)}(\boldsymbol{k}) + \tilde{a}_2(k)\boldsymbol{\epsilon}^{(2)}(\boldsymbol{k}) + \tilde{a}_3(k)\hat{\boldsymbol{k}}). \tag{9.279}$$

It follows that $\tilde{a}_0(k)$ is known as the *timelike component* and $\tilde{a}_3(k)$ is known as the *longitudinal component* of (9.268), while $\tilde{a}_1(k)$ and $\tilde{a}_2(k)$ are known as the transverse components. The usual relativistic properties are enforced for the polarization vectors, so that

$$\begin{aligned} \epsilon_{(\lambda)\mu} &= \eta_{\lambda\lambda'}\epsilon_\mu^{(\lambda')}, \\ \epsilon^{(\lambda)\mu} &= \eta^{\mu\nu}\epsilon_\nu^{(\lambda)}, \end{aligned} \tag{9.280}$$

so that $\epsilon_\mu^{(\lambda)}\tilde{a}_\lambda$ transforms as a covariant vector in (9.268). Direct substitution of their definitions shows that the polarization vectors obey

$$\epsilon_{(1)\mu} = -\epsilon_\mu^{(1)}, \quad \epsilon_{(2)\mu} = -\epsilon_\mu^{(2)}, \quad \epsilon_{(3)\mu} = -\epsilon_\mu^{(3)}, \quad \epsilon_{(0)\mu} = \epsilon_\mu^{(0)}. \tag{9.281}$$

Using these gives the second version of (9.276),

$$\sum_{\lambda=1}^{2} \epsilon_j^{(\lambda)}(\boldsymbol{k})\epsilon_{(\lambda)\ell}(\boldsymbol{k}) = \eta_{j\ell} - \epsilon_j^{(3)}(\boldsymbol{k})\epsilon_{(3)\ell}(\boldsymbol{k}) = \eta_{j\ell} + \frac{k_j k_\ell}{\omega_{\boldsymbol{k}}^2}. \tag{9.282}$$

The solution (9.268) must also satisfy the Lorenz condition (9.237). This starts by noting that the polarization vectors obey

$$k^\mu \epsilon_\mu^{(1,2)} = \boldsymbol{k} \cdot \boldsymbol{\epsilon}^{(1,2)}(\boldsymbol{k}) = 0. \tag{9.283}$$

As a result, the Lorenz condition for (9.268) will be satisfied if

$$\begin{aligned} \partial_\mu a^\mu(x) = 0 \implies k_\mu \epsilon^{(\lambda)\mu}\tilde{a}_\lambda &= k^\mu \epsilon_\mu^{(0)}\tilde{a}_0 + k^\mu \epsilon_\mu^{(3)}\tilde{a}_3 \\ &= k^0\tilde{a}_0 - \boldsymbol{k} \cdot \frac{\boldsymbol{k}}{\omega_{\boldsymbol{k}}}\tilde{a}_3 = \omega_{\boldsymbol{k}}(\tilde{a}_0 - \tilde{a}_3) = 0. \end{aligned} \tag{9.284}$$

The Lorenz condition is therefore satisfied if

$$\tilde{a}_0(k) = \tilde{a}_3(k), \tag{9.285}$$

along with its complex conjugate in (9.268).

Using the definition of the polarization vectors along with the Lorenz condition (9.285) allows the contributions from the longitudinal and timelike components in (9.268) to be combined and written

$$\epsilon_\mu^{(0)}\tilde{a}_0(k) + \epsilon_\mu^{(3)}\tilde{a}_3(k) = \tilde{a}_0(k)(\epsilon_\mu^{(0)} + \epsilon_\mu^{(3)}) = \tilde{a}_0(k)\,\frac{k_\mu}{\omega_{\boldsymbol{k}}}, \tag{9.286}$$

where $k_\mu k^\mu = 0$. The residual gauge invariance of the theory allows the remaining contribution of (9.286) to be removed from (9.268) by performing a gauge transformation with the scalar function $\Lambda(x)$ given by

$$\Lambda(x) = \int \frac{\mathrm{d}^4k}{(2\pi)^3}\,\delta(k^2)\,\theta(k_0)\,\frac{i}{k_0}(\tilde{a}_0 e^{-ik\cdot x} - \tilde{a}_0^* e^{ik\cdot x}). \tag{9.287}$$

The Dirac delta function and step function ensure that (9.287) satisfies the Lorenz gauge requirement $\Box\Lambda = 0$ by enforcing $k_0 = \omega_{\boldsymbol{k}}$. This choice for (9.287) gives

$$\partial_\mu\Lambda(x) = \int \frac{\mathrm{d}^4k}{(2\pi)^3}\,\delta(k^2)\,\theta(k_0)\,\frac{k_\mu}{k_0}(\tilde{a}_0 e^{-ik\cdot x} + \tilde{a}_0^* e^{ik\cdot x}). \tag{9.288}$$

The remaining part of the Lorenz gauge solution (9.268),

$$\tilde{a}_\mu(x) = a_\mu(x) - \partial_\mu\Lambda(x), \tag{9.289}$$

has only the first two polarization vectors and these are transverse to $\boldsymbol{k}$. This choice of gauge condition leaves $\tilde{a}_0(x) = 0$, so that the electric and magnetic fields are given by

$$\mathbf{E} = -\partial_0\,\tilde{\boldsymbol{a}}, \quad \mathbf{B} = \nabla \times \tilde{\boldsymbol{a}}. \tag{9.290}$$

By virtue of the transverse nature of $\tilde{\boldsymbol{a}}$ these satisfy the *vacuum* Maxwell equations

$$\begin{aligned}\nabla\cdot\mathbf{E} = -\,\partial_0\nabla\cdot\tilde{\boldsymbol{a}} = 0,\ \nabla\times\mathbf{B} - \partial_0\mathbf{E} = \nabla\times(\nabla\times\tilde{\boldsymbol{a}}) + \partial_0^2\tilde{\boldsymbol{a}} = \nabla(\nabla\cdot\tilde{\boldsymbol{a}})\\ -\,\nabla^2\tilde{\boldsymbol{a}} + \partial_0^2\tilde{\boldsymbol{a}} = \Box\tilde{\boldsymbol{a}} = 0.\end{aligned} \tag{9.291}$$

The remaining transverse degrees of freedom, $\tilde{a}_{(1,2)}(k)$, in (9.268) are available to match the boundary conditions placed on $\tilde{\boldsymbol{a}}(x)$.

As a final step the k_0 integration in (9.268) is performed using

$$\delta(k^2)\,\theta(k_0) = \frac{\delta(k_0 - \omega_{\boldsymbol{k}})}{2\omega_{\boldsymbol{k}}}. \tag{9.292}$$

As in the treatment of (8.39), $\tilde{a}_\lambda(k)$ are scaled according to

$$\tilde{a}_\lambda = \sqrt{2\omega_{\boldsymbol{k}}}\,\tilde{f}_\lambda, \tag{9.293}$$

in order to define a manifestly covariant expression. The result of integrating over k_0 is a commonly used form for a_μ,

$$a_\mu(x) = \int \frac{d^3k}{(2\pi)^3\sqrt{2\omega_k}} \sum_{\lambda=1}^{2} \epsilon_\mu^{(\lambda)}(\boldsymbol{k})\left(\tilde{f}_\lambda(\boldsymbol{k})\, e^{-ik\cdot x} + \tilde{f}_\lambda^*(\boldsymbol{k})\, e^{ik\cdot x}\right), \tag{9.294}$$

where k_0x^0 is replaced with $\omega_k t$.

The vector nature of the gauge field results in a spin contribution to the angular momentum in a manner similar to that found for the spinor field in (8.258). The rotation of a vector $\boldsymbol{A}$ through the infinitesimal angles θ^a is given by

$$\delta A^j = \theta^a \mathbf{T}^a_{jk} A^k = (\theta^a \mathbf{T}^a \cdot \boldsymbol{A})^j, \tag{9.295}$$

where $\mathbf{T}^a$ are the three 3×3 rotation generators found in (7.95) and (7.101). For the case that the Lagrangian is rotationally invariant, Noether's theorem (8.223) gives the spin angular momentum density s^a of the gauge field as the timelike component of the current associated with rotations, so that

$$J^0 = \theta^a s^a = \frac{\partial \mathcal{L}}{\partial(\partial_0 A^j)}\delta A^j = \theta^a \frac{\partial \mathcal{L}}{\partial(\partial_0 A^j)}\mathbf{T}^a_{j\ell} A^\ell = \theta^a E_j \mathbf{T}^a_{j\ell} A^\ell. \tag{9.296}$$

Since θ^a is arbitrary this gives the spin density vector component s^a,

$$s^a = E_j \mathbf{T}^a_{j\ell} A^\ell. \tag{9.297}$$

Noting that the matrix forms for the generators given by (7.95) and (7.101) can be written $\mathbf{T}^a_{j\ell} = \varepsilon^{aj\ell}$ shows that (9.297) is identical to the earlier result (9.214) for the spin density when J_0 vanishes. The simple plane-wave solution (9.248) for A^ℓ along with the associated $\mathbf{E}$ field (9.250) can be used to evaluate (9.296). Using $\mathbf{T}^a_{j\ell} = \varepsilon^{aj\ell}$ results in

$$\theta^a s^a = -\omega_k \cos(k \cdot x)\sin(k \cdot x)(\tilde{\alpha}^x \;\; \tilde{\alpha}^y \;\; \tilde{\alpha}^z) \cdot \begin{pmatrix} 0 & \theta^z & -\theta^y \\ -\theta^z & 0 & \theta^x \\ \theta^y & -\theta^x & 0 \end{pmatrix} \cdot \begin{pmatrix} \tilde{\alpha}^x \\ \tilde{\alpha}^y \\ \tilde{\alpha}^z \end{pmatrix} = 0. \tag{9.298}$$

This shows that the plane polarized solution (9.248) has no spin angular momentum.

In order to identify a nonzero spin angular momentum for electromagnetic waves, it is necessary to introduce *circularly polarized waves*. With no loss of generality the coordinate system can be chosen so that the wave propagates along the $+z$-axis with $\boldsymbol{k} = k\hat{\mathbf{e}}_z$, so that $\omega_k = |k|$. Using the two orthogonal spatial unit vectors, $\hat{\mathbf{e}}_x$ and $\hat{\mathbf{e}}_y$, the two complex polarization vectors $\boldsymbol{\epsilon}_\pm$ are defined as

$$\begin{aligned} \epsilon_+ &= \frac{1}{2}(\hat{\mathbf{e}}_x + i\hat{\mathbf{e}}_y), \\ \epsilon_- &= \frac{1}{2}(\hat{\mathbf{e}}_x - i\hat{\mathbf{e}}_y). \end{aligned} \tag{9.299}$$

The positive circularly polarized electromagnetic wave is given by

$$\begin{aligned} \boldsymbol{a}_+(x) &= \beta\, \boldsymbol{\epsilon}_+ e^{-ik\cdot x} + \beta\, \boldsymbol{\epsilon}_+^* e^{ik\cdot x} = \beta\, \boldsymbol{\epsilon}_+ e^{-i\omega_k t + ikz} + \beta\, \boldsymbol{\epsilon}_+^* e^{i\omega_k t - ikz} \\ &= \beta\, \hat{\mathbf{e}}_x \cos(\omega_k t - kz) + \beta\, \hat{\mathbf{e}}_y \sin(\omega_k t - kz), \end{aligned} \tag{9.300}$$

where the Euler identity (2.83) was used and β is an arbitrary *real* constant with units of inverse length. The wave of (9.300) has a constant magnitude,

$$\boldsymbol{a}_+(x) \cdot \boldsymbol{a}_+(x) = \beta^2 \cos^2(\omega_k t - kz) + \beta^2 \sin^2(\omega_k t - kz) = \beta^2, \tag{9.301}$$

which is visualized as a ray of magnitude β rotating around the z-axis as the wave passes. This visualization follows by considering the origin, $z = 0$, where the wave is given by

$$\boldsymbol{a}_+(0, t) = \beta\, \hat{\mathbf{e}}_x \cos(\omega_k t) + \beta\, \hat{\mathbf{e}}_y \sin(\omega_k t). \tag{9.302}$$

This corresponds to a ray of length β rotating counterclockwise in the x–y plane around the origin at a frequency of $\omega_k = k$. This is referred to as right-handed polarization.

Exercise 9.11: Show that the other circularly polarized wave, given by

$$\boldsymbol{a}_-(x) = \beta\, \boldsymbol{\epsilon}_- e^{-ik\cdot x} + \beta\, \boldsymbol{\epsilon}_-^* e^{ik\cdot x}, \tag{9.303}$$

corresponds to left-handed polarization.

The electric field associated with (9.300) and (9.303) is given by

$$\mathbf{E}_\pm(x) = -\partial_0 \boldsymbol{a}_\pm(x) = \beta\omega_k\big(\hat{\mathbf{e}}_x \sin(\omega_k t - kz) \mp \hat{\mathbf{e}}_y \cos(\omega_k t - kz)\big), \tag{9.304}$$

while the magnetic field is given by

$$\mathbf{B}_\pm(x) = \nabla \times \boldsymbol{a}_\pm(x) = \beta k\big(\pm\hat{\mathbf{e}}_x \cos(\omega_k t - kz) + \hat{\mathbf{e}}_y \sin(\omega_k t - kz)\big). \tag{9.305}$$

These are depicted in figure 9.1.

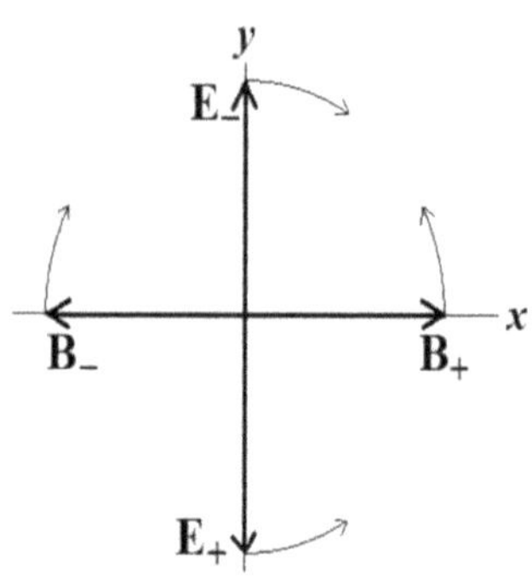

Figure 9.1. The circularly polarized electric and magnetic fields $\mathbf{E}_\pm$ and $\mathbf{B}_\pm$ of (9.304) and (9.305) with their directions of rotation.

The momentum density in the z-direction for the two circularly polarized waves is given by

$$\begin{aligned}\mathcal{P}^z_\pm &= (\mathbf{E}_\pm \times \mathbf{B}_\pm)^z = E^x_\pm B^y_\pm - E^y_\pm B^x_\pm \\ &= k\omega_k\beta^2(\sin^2(\omega_k t - kz) + \cos^2(\omega_k t - kz)) = k\omega_k\beta^2,\end{aligned} \tag{9.306}$$

so that both polarizations correspond to a momentum in the $+z$-direction. This momentum can be normalized to that of a single photon by choosing

$$\omega_k\beta^2 = \frac{1}{V}, \tag{9.307}$$

where V is the volume of the system. For such a choice, the total momentum of the system is given by

$$P^z = \int_V \mathrm{d}^3x\, \mathcal{P}^z = \int_V \mathrm{d}^3x\, k\omega_k\beta^2 = \frac{k}{V}\int_V \mathrm{d}^3x = k, \tag{9.308}$$

which is the momentum of a photon. For the case of circularly polarized waves propagating along the z-axis, the helicity, which is the component of the spin density along the z-axis, is found by using (9.297),

$$\begin{aligned}s_{z\pm} &= E_{\pm j}\mathbf{T}^z_{j\ell}A^\ell_\pm = E_{\pm j}\varepsilon^{3j\ell}A^\ell_\pm = E_{\pm 1}A^2_\pm - E_{\pm 2}A^1_\pm \\ &= \pm\beta^2\omega_k(\sin^2(\omega_k t - kz) + \cos^2(\omega_k t - kz)) = \pm\beta^2\omega_k = \pm\frac{1}{V}.\end{aligned} \tag{9.309}$$

The total spin angular momentum in the z-direction for a single photon is therefore given by

$$\Sigma_{z\pm} = \int_V \mathrm{d}^3x\, s_{z\pm} = \pm\frac{1}{V}\int_V \mathrm{d}^3x = \pm 1, \tag{9.310}$$

when measured in natural units. Reinserting the factor of $\hbar$ shows the helicity of the photon is $\pm\hbar$, either parallel or antiparallel to the momentum of the photon. The absence of zero helicity is due to the massless nature of the electromagnetic wave, which renders it impossible to boost into the rest frame of the wave. If the gauge field were massive, boosting into the rest frame would render the momentum zero and the projection of the spin would be zero.

9.12 The Proca field

It is important to note that including a mass term of the form $m^2A_\mu A^\mu$ in the electrodynamic Lagrangian would result in an action that is *not* fully gauge invariant, since

$$m^2A_\mu A^\mu \to m^2A_\mu A^\mu - 2m^2A^\mu\partial_\mu\Lambda + m^2\partial_\mu\Lambda\,\partial^\mu\Lambda. \tag{9.311}$$

In general, the requirement of gauge invariance precludes an *a priori* mass term for the gauge field. However, there is a loophole in the gauge invariance argument against a mass term. If the two terms generated in (9.311) are integrated in the action and the gauge function Λ vanishes at the boundary of the integral, the two terms can be integrated by parts to obtain

$$\int d^4x \left(-2m^2 A^\mu \partial_\mu \Lambda + m^2 \partial_\mu \Lambda \, \partial^\mu \Lambda\right) = \int d^4x \left(2m^2 \Lambda \, \partial_\mu A^\mu - m^2 \Lambda \, \Box \Lambda\right), \quad (9.312)$$

which will vanish if the Lorenz condition holds and the gauge function satisfies (9.79). In that sense, the mass term is invariant under a restricted set of gauge transformations. However, the method of *spontaneously broken symmetry* provides a mechanism to bypass these restrictions and allow a gauge invariant action to represent a gauge field with an inertial mass. The classical version of spontaneously broken symmetry was discussed in chapter 8 and will be revisited in chapter 11.

However, for the case that a mass term is present, the Lagrangian for the gauge field Z_μ coupled to a current J_μ is given by

$$\mathcal{L}_Z = -\frac{1}{4} F_{\mu\nu} F^{\mu\nu} + \frac{1}{2} m^2 Z_\mu Z^\mu - J^\mu Z_\mu, \quad (9.313)$$

where

$$F_{\mu\nu} = \partial_\mu Z_\nu - \partial_\nu Z_\mu. \quad (9.314)$$

It is worth noting that the quadratic term in Z_μ enters the Lagrangian with the opposite sign from the mass term in the Klein–Gordon Lagrangian (8.25). The reason for this follows from the equation of motion for the gauge field,

$$\begin{aligned} -\partial_\mu \frac{\partial \mathcal{L}_Z}{\partial(\partial_\mu Z_\mu)} + \frac{\partial \mathcal{L}_Z}{\partial Z_\nu} &= \partial_\mu(\partial^\mu Z^\nu - \partial^\nu Z^\mu) + m^2 Z^\nu - J^\nu \\ &= (\Box + m^2) Z^\nu - \partial^\nu(\partial_\mu Z^\mu) - J^\nu = 0, \end{aligned} \quad (9.315)$$

so that the correct massive d'Alembertian has appeared. Such a vector field is referred to as the *Proca field* or often as the *Proca boson*. If *either* the current is zero *or* if it is conserved, so that $\partial_\mu J^\mu = 0$, it follows that the divergence of the equation of motion (9.315) gives

$$(\Box + m^2)\partial_\nu Z^\nu - \Box(\partial_\mu Z^\mu) = \partial_\nu J^\nu = 0 \implies \partial_\mu Z^\mu = 0, \quad (9.316)$$

so that the Lorenz condition is required. For the case that the current is zero, it follows that

$$(\Box + m^2) Z^\mu = 0, \quad (9.317)$$

and the gauge field Z^μ obeys an equation of motion similar to (8.26) for the massive Klein–Gordon field, and this allows the massive momentum-energy equation to hold. The Lorenz condition is invariant under gauge transformations that satisfy $\Box\Lambda = 0$, but the equation of motion (9.317) is not, even if J^μ is gauge invariant. As a result, it is not possible to use a gauge transformation on the Proca field to remove the longitudinal degree of freedom from Z^μ, as it was for the electrodynamic gauge field A^μ.

It is instructive to analyze the solutions to the equation of motion for $J^\mu = 0$ as well as the ramifications for the spin content of the Proca boson. The counterpart to the electromagnetic plane-wave solution (9.238) is given by

$$Z^\mu = \beta^\mu \sin(k \cdot x), \tag{9.318}$$

where k_μ is a spacelike vector that must satisfy

$$k \cdot k = k_\mu k^\mu = k_0^2 - \boldsymbol{k} \cdot \boldsymbol{k} = m^2 \implies k_0 = +\sqrt{\boldsymbol{k} \cdot \boldsymbol{k} + m^2}, \tag{9.319}$$

while the Lorenz condition requires that the polarization constant satisfies

$$k_\mu \beta^\mu = 0. \tag{9.320}$$

The latter requirement gives

$$\beta^0 = \frac{\boldsymbol{\beta} \cdot \boldsymbol{k}}{k_0}, \tag{9.321}$$

which removes *one* of the four degrees of freedom from β^μ, but leaves $\boldsymbol{\beta}$ arbitrary. Since it is not possible to perform a gauge transformation, it is not possible to remove the longitudinal degree of freedom, which is given by

$$\boldsymbol{\beta}_L = \frac{(\boldsymbol{\beta} \cdot \boldsymbol{k})\boldsymbol{k}}{\boldsymbol{k} \cdot \boldsymbol{k}}. \tag{9.322}$$

As a result, the Proca field has three possible polarizations, the two transverse polarizations,

$$\boldsymbol{\beta}_T = \boldsymbol{\beta} - \frac{(\boldsymbol{\beta} \cdot \boldsymbol{k})\boldsymbol{k}}{\boldsymbol{k} \cdot \boldsymbol{k}}, \tag{9.323}$$

and the longitudinal polarization (9.322). It is left as an exercise to show that a general solution for the Proca field is given by

$$Z_\mu(x) = \int \frac{d^4k}{(2\pi)^3}\, \delta(k^2 - m^2)\, \theta(k_0) \sum_{\lambda=1}^{3} \epsilon_\mu^{(\lambda)}(k)(z_\lambda(k)\, e^{-ik\cdot x} + z_\lambda^*(k)\, e^{ik\cdot x}), \tag{9.324}$$

where $\epsilon_\mu^{(1,\,2)}$ are the transverse polarization vectors of (9.270) that satisfy $k^\mu \epsilon_\mu^{(1,\,2)} = 0$. The polarization vector $\epsilon_\mu^{(3)}$ differs from the longitudinal polarization vector (9.273) for the electromagnetic gauge field, and is given explicitly by

$$\epsilon_\mu^{(3)} = \left(\frac{|\boldsymbol{k}|}{k_0}, \frac{\boldsymbol{k}}{|\boldsymbol{k}|} \right). \tag{9.325}$$

It has the properties that $\epsilon^{(3)\mu} \epsilon_\mu^{(1,\,2)} = 0$ and $k^\mu \epsilon_\mu^{(3)} = 0$; the latter ensures that Z^μ satisfies the Lorenz condition.

Exercise 9.12: Verify that (9.324) satisfies the equation of motion (9.317) for $J^\mu = 0$ and satisfies the Lorenz condition for the choice of polarization vector (9.325).

Because it is a vector field like the electromagnetic gauge field, the Lagrangian is invariant under spatial rotations. Repeating the steps that led to (9.300) and (9.303) shows the Proca field has a spin projection of ± 1 for the case that $\tilde{z}_3$ is zero. However, unlike the electromagnetic gauge field, the Proca field has three possible helicities, i.e. two transverse and one longitudinal. It is the longitudinal degree of freedom, $\tilde{z}_3(k)$, that allows the Proca field to have a zero projection of the spin angular momentum. For the general solution (9.324), the spatial divergence gives the longitudinal content of the Proca field,

$$\begin{aligned}\nabla \cdot \boldsymbol{Z} &= i \int \frac{\mathrm{d}^4 k}{(2\pi)^3}\, \delta(k^2 - m^2)\, \theta(k_0) \sum_{\lambda=1}^{3} k_j \epsilon_j^{(\lambda)}(k)(\tilde{z}_\lambda(k)\, e^{-ik\cdot x} - \tilde{z}_\lambda^*(k)\, e^{ik\cdot x}) \\ &= i \int \frac{\mathrm{d}^4 k}{(2\pi)^3}\, \delta(k^2 - m^2)\, \theta(k_0)\, |\boldsymbol{k}|(\tilde{z}_3(k)\, e^{-ik\cdot x} - \tilde{z}_3^*(k)\, e^{ik\cdot x}) \neq 0.\end{aligned} \tag{9.326}$$

Once again, this is because it is not possible to perform a gauge transformation similar to (9.287) to remove the longitudinal degree of freedom $\tilde{z}_3(k)$ from the general Fourier solution (9.324).

9.13 Green's functions and electromagnetic radiation

The Lorenz condition gauge field A^μ interaction with a spinor field ψ is governed by the partial differential equations of (9.114),

$$\Box A^\mu = \alpha \bar{\psi} \gamma^\mu \psi = J^\mu, \tag{9.327}$$

$$(i\gamma^\mu \partial_\mu - m)\psi = \alpha A_\mu \gamma^\mu \psi = \alpha \not{A} \psi. \tag{9.328}$$

The two equations (9.327) and (9.328) are nonlinear and coupled, where the charge α quantifies the strength of the nonlinearity. Clearly, if $\alpha \to 0$, the two equations decouple and the solutions become the solutions to the linear equations of motion, denoted by a^μ and ψ_0. For that reason, α is referred to as a *coupling constant.*

Like many partial differential equations, solving these equations of motion is expedited by using Green's function techniques. The Green's function $\boldsymbol{S}_F(x - y)$ for the Dirac equation was found earlier in (8.197), where it was shown to solve

$$(i\gamma^\mu \partial_\mu - m)\boldsymbol{S}_F(x - y) = \mathbf{I}\, \delta^4(x - y). \tag{9.329}$$

Denoting $\psi_0(x)$ as a solution to the uncoupled Dirac equation,

$$(i\gamma^\mu \partial_\mu - m)\psi_0(x) = 0, \tag{9.330}$$

it is possible to write a *formal solution* to (9.328) as

$$\psi(x) = \psi_0(x) + \alpha \int d^4y\, S_F(x-y)A_\mu(y)\gamma^\mu\psi(y). \tag{9.331}$$

It is important to note that the solutions to both (9.327) and (9.328) appear in the integration, and it is for that reason that (9.331) is classified as formal, since these solutions are unknown.

The Green's function $\Delta(x-y)$ for the gauge field is the solution of

$$\Box\Delta(x-y) = \delta^4(x-y), \tag{9.332}$$

consistent with the choice of boundary conditions. If $a^\mu(x)$ is the solution to the homogeneous wave equation, where $J^\mu = 0$, chosen to satisfy the boundary conditions and the Lorenz condition, then a *formal solution* to (9.327) can be written

$$\begin{aligned} A^\mu(x) &= a^\mu(x) + \int d^4y\, \Delta(x-y)\, J^\mu(y) \\ &= a^\mu(x) + \alpha \int d^4y\, \Delta(x-y)\, \bar{\psi}(y)\gamma^\mu\psi(y). \end{aligned} \tag{9.333}$$

Once again, the solution (9.333) for A^μ requires the solution for ψ, which is unavailable. However, the Lorenz gauge condition is satisfied for A^μ, since integrating by parts and using the conservation of current gives

$$\begin{aligned} \partial_\mu A^\mu(x) &= \int d^4y\, \frac{\partial}{\partial x^\mu}\Delta(x-y)\, J^\mu(y) = -\int d^4y\, J^\mu(y)\frac{\partial}{\partial y^\mu}\Delta(x-y) \\ &= \int d^4y\, \Delta(x-y)\, \frac{\partial}{\partial y^\mu}J^\mu(y) = 0, \end{aligned} \tag{9.334}$$

where $\Delta(x-y)$ is assumed to vanish at the boundaries of the integration.

Like many Green's functions, the easiest method to find $\Delta(x-y)$ is by a Fourier transform. In an infinite volume, $\Delta(x-y)$ can be written in a Lorentz invariant form as

$$\Delta(x-y) = -\int \frac{d^4k}{(2\pi)^4}\frac{1}{k^2}\, e^{-ik\cdot(x-y)}. \tag{9.335}$$

In expression (9.335), k_0 is *not* constrained to be $\sqrt{\boldsymbol{k}\cdot\boldsymbol{k}} \equiv |\boldsymbol{k}| = \omega_k$. Since

$$\Box e^{-ik\cdot(x-y)} = -k^2 e^{-ik\cdot(x-y)}, \tag{9.336}$$

it follows that

$$\Box\Delta(x-y) = \int \frac{d^4k}{(2\pi)^4}\, e^{-ik\cdot(x-y)} = \delta^4(x-y), \tag{9.337}$$

where the Fourier representation (2.95) of the Dirac delta has been used. As written, (9.335) suffers from the same malady as the tentative Green's function (8.76) for the Klein–Gordon equation, which is the presence of a branch cut for the k_0 integration.

This is the continuous line of singularities that occur along the k_0 axis at those points where $k_0 = |\boldsymbol{k}|$.

The solution is to redefine (9.335) using an epsilon procedure that shifts the singularities off the real axis into the complex k_0 plane. There are many prescriptions for doing this, each of which has a different form of usefulness. The *retarded Green's function* is defined as

$$\Delta_+(x-y) = -\lim_{\epsilon\to 0}\int \frac{d^4k}{(2\pi)^4}\frac{e^{-ik\cdot(x-y)}}{(k_0+i\epsilon)^2-|k|^2}. \tag{9.338}$$

The retarded Green's function integrand possesses two simple poles in the lower half of the complex k_0 plane at $k_0 = -i\epsilon \pm |k|$, and so the integral over k_0 can be evaluated using the residue theorem. The integral along the real k_0 axis, which gives the Green's function, can be combined with a contour in the upper or lower half of the complex plane as long as the contribution from this second contour vanishes. This is identical to the method for evaluating the step function used in (5.69) and (5.71) as well as for the retarded Klein–Gordon propagator of (8.80). Using Jordan's lemma (5.63) shows that the sign of $x^0 - y^0$ determines how the contour must be closed, either in the upper or in the lower half of the complex k_0 plane. If $x^0 - y^0 < 0$, then the upper half-plane closure must be chosen and there are no poles enclosed by the total contour, giving zero for the integral. If $x^0 - y^0 > 0$, then the lower half-plane closure must be chosen and both poles are enclosed in the left-handed sense. Cauchy's theorem then gives $-2\pi i$ times the residues at the two poles. After letting $\epsilon \to 0$, the result is

$$\begin{aligned}\Delta_+(x-y) = i\,\theta(x^0-y^0)\int \frac{d^3k}{(2\pi)^3}\frac{1}{2|k|}e^{ik\cdot(x-y)}\\ \left(e^{-i\omega_k(x^0-y^0)} - e^{i\omega_k(x^0-y^0)}\right),\end{aligned} \tag{9.339}$$

where $\theta(x^0 - y^0)$ is the Heaviside step function and $\omega_k = |\boldsymbol{k}|$. Setting $k = |k|$, the argument of the integral can be written in spherical coordinates as

$$d^3k\; e^{ik\cdot(x-y)} = k^2 dk\; d(\cos\theta_k)\; d\phi_k\; e^{ik|x-y|\cos\theta_k}. \tag{9.340}$$

Performing the integrations over $\cos\theta_k$ and ϕ_k results in

$$\begin{aligned}\Delta_+(x-y) = \frac{\theta(x^0-y^0)}{2\pi|\boldsymbol{x}-\boldsymbol{y}|}\int_0^\infty \frac{dk}{2\pi}\\ (\cos(k|\boldsymbol{x}-\boldsymbol{y}| - k(x^0-y^0)) - \cos(k|\boldsymbol{x}-\boldsymbol{y}| + k(x^0-y^0))).\end{aligned} \tag{9.341}$$

The integral over k results in delta functions, since

$$\begin{aligned}\int_0^\infty \frac{dk}{2\pi}\cos(kx) &= \frac{1}{2}\int_0^\infty \frac{dk}{2\pi}(e^{ikx}+e^{-ikx})\\ &= \frac{1}{2}\int_{-\infty}^\infty \frac{dk}{2\pi}e^{ikx} = \frac{1}{2}\delta(x).\end{aligned} \tag{9.342}$$

The result of the integration over k is

$$\Delta_+(x-y) = \frac{\theta(x^0 - y^0)}{4\pi|\boldsymbol{x}-\boldsymbol{y}|}(\delta(|\boldsymbol{x}-\boldsymbol{y}| - (x^0 - y^0)) - \delta(|\boldsymbol{x}-\boldsymbol{y}| + (x^0 - y^0))). \quad (9.343)$$

However, because $|\boldsymbol{x}-\boldsymbol{y}| > 0$ and the step function requires $x^0 - y^0 > 0$ to be nonzero, the second delta function will always be zero, since its argument is never zero. The retarded Green's function is therefore

$$\Delta_+(x-y) = \frac{\theta(x^0 - y^0)}{4\pi|\boldsymbol{x}-\boldsymbol{y}|}\delta(|\boldsymbol{x}-\boldsymbol{y}| - (x^0 - y^0)). \quad (9.344)$$

It is instructive to note that the Coulomb Green's function (9.10) has appeared in the relativistic retarded Green's function. The term *retarded* stems from the fact that only events on the *past* light cone, given by the relation

$$|\boldsymbol{x}-\boldsymbol{y}| = (x^0 - y^0), \quad (9.345)$$

can contribute to the current value of the gauge field. This is referred to as a *causal* restriction, since it is consistent with the constant and finite speed of light.

Result (9.344) can be written in a manifestly covariant form by using the identity

$$\begin{aligned} \theta(t)\,\delta(t^2 - x^2) &= \theta(t)\,\delta[(t + |x|)(t - |x|)] \\ &= \theta(t)\,\frac{\delta(t + |x|)}{(t - |x|)} + \theta(t)\,\frac{\delta(t - |x|)}{(t + |x|)} \\ &= \theta(t)\,\frac{\delta(t - |x|)}{2|x|}, \end{aligned} \quad (9.346)$$

where the step function prevents the argument of the first delta function from vanishing, since $t + |x| > 0$. Extending this identity to

$$\begin{aligned} \theta(x^0 - y^0)\delta[(x-y)^2] &= \theta(x^0 - y^0)\delta((x^0 - y^0)^2 - |\boldsymbol{x}-\boldsymbol{y}|^2) \\ &= \theta(x^0 - y^0)\frac{\delta(x^0 - y^0 - |\boldsymbol{x}-\boldsymbol{y}|)}{2|\boldsymbol{x}-\boldsymbol{y}|}, \end{aligned} \quad (9.347)$$

allows the retarded Green's function (9.344) to be written

$$\Delta_+(x-y) = \frac{\theta(x^0 - y^0)}{2\pi}\,\delta[(x-y)^2]. \quad (9.348)$$

The Heaviside step function $\theta(x^0 - y^0)$ was shown to be invariant under Lorentz transformations in (8.31), since the sign of $x^0 - y^0$ is invariant under a Lorentz transformation.

There are several ways to use the formal solutions (9.333) and (9.331). One method is to pick both free solutions ψ_0 and a^μ based on initial and boundary conditions and then *iterate* the formal solutions to generate a power series in the coupling constant e to find the interacting fields ψ and A^μ. For the case that e is small, this is expected to generate a *perturbation* series whose first few terms constitute the important result of the interaction. This is similar to the method used earlier in (5.82) to find the Born approximation to scattering in a Bose–Einstein condensate. After one iteration and retaining only terms $O(\alpha)$, which is equivalent to the approximations $A^\mu \approx a^\mu$ and $\psi \approx \psi_0$ on the right-hand sides of (9.333) and (9.331), the formal solutions give

$$A^\mu(x) = a^\mu(x) + \alpha \int \mathrm{d}^4y\, \Delta_+(x-y)\, \bar{\psi}_0(y)\gamma^\mu\psi_0(y), \tag{9.349}$$

$$\psi = \psi_0(x) + \alpha \int \mathrm{d}^4y\, \boldsymbol{S}_F(x-y)\, a_\mu(y)\gamma^\mu\psi_0(y). \tag{9.350}$$

In the case of several fields such as this, higher-order powers of an iterative solution quickly become cumbersome and are of questionable validity. The latter is true because the convergence of the power series in α becomes unclear. This can be a problem in any *perturbative* solution to a nonlinear system, even at the classical level.

A second method is to forgo a fully interacting solution for both ψ and A^μ and model one component of the system with a particular behavior. For example, the gauge field solution can be written

$$A^\mu(x) = a^\mu(x) + \int \mathrm{d}^4y\, \Delta_+(x-y)\, J^\mu(y), \tag{9.351}$$

and the electric current J^μ and the external field a^μ can be assigned a particular form. Using (9.351) then gives the nature of the gauge field A^μ associated with that particular current. A particularly simple example is the case of a static point charge q located at $\boldsymbol{a}$ with respect to the origin, so that

$$J^0(\boldsymbol{x}, t) = q\, \delta^3(\boldsymbol{x} - \boldsymbol{a}), \quad \boldsymbol{J} = 0. \tag{9.352}$$

For $a^\mu = 0$ the integral in (9.351) is easily evaluated using the Dirac delta functions, resulting in

$$\begin{aligned} A^0(x) &= q \int \mathrm{d}^4y\, \frac{\theta(x^0 - y^0)}{4\pi|\boldsymbol{x} - \boldsymbol{y}|}\delta(|\boldsymbol{x} - \boldsymbol{y}| - (x^0 - y^0))\delta^3(\boldsymbol{y} - \boldsymbol{a}) \\ &= \frac{q\, \theta(|\boldsymbol{x} - \boldsymbol{a}|)}{4\pi|\boldsymbol{x} - \boldsymbol{a}|} = \frac{q}{4\pi|\boldsymbol{x} - \boldsymbol{a}|}, \end{aligned} \tag{9.353}$$

where the step function is always one, since its argument is always positive.

Result (9.353) is the Coulomb potential familiar from elementary physics. Using the result

$$\frac{\partial}{\partial x^j}|\boldsymbol{x} - \boldsymbol{a}| = \frac{x^j - a^j}{|\boldsymbol{x} - \boldsymbol{a}|}, \tag{9.354}$$

this form for A^0 gives the familiar inverse square electric field

$$\mathbf{E} = -\nabla A^0(\boldsymbol{x}) = \frac{q}{4\pi}\frac{(\boldsymbol{x} - \boldsymbol{a})}{|\boldsymbol{x} - \boldsymbol{a}|^3}, \tag{9.355}$$

which points radially away from positive point charges and radially toward negative point charges. The energy density of (9.355) is

$$\mathcal{E}_A = \frac{1}{2}\mathbf{E} \cdot \mathbf{E} = \frac{1}{32\pi^2}\frac{q^2}{|\boldsymbol{x} - \boldsymbol{a}|^4}. \tag{9.356}$$

The spatial integral of the energy density (9.356) is divergent and requires a *cutoff* r_c on the radial integration in order to be defined. A change of origin, $\boldsymbol{x} \to \boldsymbol{x} + \boldsymbol{a}$, gives the total electrostatic energy

$$E_S = \int \mathrm{d}^3x \, \frac{q^2}{32\pi^2 r^4} = \frac{q^2}{32\pi^2}\int_{r_c}^{\infty}\frac{\mathrm{d}r}{r^2}\int \mathrm{d}\Omega = \frac{q^2}{8\pi r_c}. \tag{9.357}$$

The quantity E_S is referred to as the *self-energy* of the point charge and diverges as $r_c \to 0$. This creates a mathematically undefined quantity and therefore a result with no physical interpretation. Historically, this dilemma arose from treating the electron as a point particle, so that the limit $r_c \to 0$ was believed to reflect the *correct experimental nature* of the electron. In the latter part of the 19th century and well into the 20th century, numerous attempts were made, particularly by Lorentz, to remedy this problem on a classical level by replacing the point charge with finite-sized or *extended* charge distributions. These models encountered additional problems when such extended charges had to incorporate the self-force created by the charge distribution on itself. Special relativity brought a further complication in that the shape of a moving extended charge distribution was dependent on its relative velocity, altering the self-force and self-energy. In addition, a moving charge receives a contribution to its momentum from the effective inertia of its self-energy, $m_s = E_s/c^2$, which is once again divergent in the limit $r_c \to 0$ for a point charge. Such mechanical models were abandoned when it was generally conceded that the limit $r_c \to 0$ entered the realm where classical mechanics is no longer valid and quantum mechanics becomes the rule. Eventually, the development of *quantum electrodynamics* in the mid-20th Century provided a means to deal with the problems of self-interaction and infinite self-energies using the systematic methods of *renormalization*. For the purposes of this text, the classical self-energy problem will be ignored and the interested reader is referred to the references.

In that regard, it is instructive to find the field for two static point charges located at separate positions $\boldsymbol{a}$ and $\boldsymbol{b}$,

$$J^0(x) = Q_1\delta^3(\boldsymbol{x} - \boldsymbol{a}) + Q_2\delta^3(\boldsymbol{x} - \boldsymbol{b}). \tag{9.358}$$

Using this charge distribution in (9.351) gives two copies of result (9.353),

$$A^0(\boldsymbol{x}) = \frac{Q_1}{4\pi|\boldsymbol{x} - \boldsymbol{a}|} + \frac{Q_2}{4\pi|\boldsymbol{x} - \boldsymbol{b}|}. \tag{9.359}$$

The electric field of the two point charges is

$$\mathbf{E} = -\nabla A^0(x) = \frac{Q_1}{4\pi}\frac{(\boldsymbol{x} - \boldsymbol{a})}{|\boldsymbol{x} - \boldsymbol{a}|^3} + \frac{Q_2}{4\pi}\frac{(\boldsymbol{x} - \boldsymbol{b})}{|\boldsymbol{x} - \boldsymbol{b}|^3}. \tag{9.360}$$

The energy density in the field around the two point charges is given by

$$\mathcal{E}_A = \frac{1}{2}\mathbf{E}\cdot\mathbf{E} = \frac{1}{32\pi^2}\left\{\frac{Q_1{}^2}{|\boldsymbol{x} - \boldsymbol{a}|^4} + \frac{Q_2{}^2}{|\boldsymbol{x} - \boldsymbol{b}|^4} + \frac{2Q_1Q_2(\boldsymbol{x} - \boldsymbol{a})\cdot(\boldsymbol{x} - \boldsymbol{b})}{|\boldsymbol{x} - \boldsymbol{a}|^3|\boldsymbol{x} - \boldsymbol{b}|^3}\right\}. \tag{9.361}$$

The integrals of the first two terms reproduce the divergent self-energy of (9.357) for both point charges and will be ignored. The integral of the third term in (9.361) is finite, and can be analyzed by noting that

$$\frac{(\boldsymbol{x} - \boldsymbol{a})}{|\boldsymbol{x} - \boldsymbol{a}|^3} = -\nabla\left(\frac{1}{|\boldsymbol{x} - \boldsymbol{a}|}\right). \tag{9.362}$$

This allows the spatial integral of the third term to be integrated by parts, giving

$$\begin{aligned} E_I &= \frac{Q_1Q_2}{16\pi^2}\int d^3x\ \nabla\left(\frac{1}{|\boldsymbol{x} - \boldsymbol{a}|}\right)\cdot\nabla\left(\frac{1}{|\boldsymbol{x} - \boldsymbol{b}|}\right) \\ &= -\frac{Q_1Q_2}{16\pi^2}\int d^3x\left(\frac{1}{|\boldsymbol{x} - \boldsymbol{a}|}\right)\nabla^2\left(\frac{1}{|\boldsymbol{x} - \boldsymbol{b}|}\right) \\ &= \frac{Q_1Q_2}{16\pi^2}\int d^3x\left(\frac{1}{|\boldsymbol{x} - \boldsymbol{a}|}\right)4\pi\delta^3(\boldsymbol{x} - \boldsymbol{b}) \\ &= \frac{1}{4\pi}\frac{Q_1Q_2}{|\boldsymbol{b} - \boldsymbol{a}|}, \end{aligned} \tag{9.363}$$

where the previous delta function result (9.13) for the Laplacian of the Coulomb potential was used. The interaction energy (9.363) gives the usual Coulomb energy associated with attraction or repulsion between two point charges, depending on the sign of Q_1Q_2. It is both finite and goes to zero at infinite separation.

The field generated by a single point charge q in motion along the trajectory $x_c^\mu(\tau)$ can be found by combining the point charge current (8.2) with expression (9.333) and the retarded Green's function (9.348). This gives

$$\begin{aligned} A^\mu(x) &= \int d^4y\ \Delta_+(x - y)J^\mu(y) \\ &= q\int d\tau\ d^4y\ \frac{\theta(x^0 - y^0)}{2\pi}\delta[(x - y)^2]\frac{dy_c^\mu(\tau')}{d\tau'}\delta^4(y - y_c(\tau)) \\ &= \frac{q}{2\pi}\int d\tau\ \theta(t - t_c(\tau))\ V^\mu(\tau)\ \delta[(x - x_c(\tau))^2], \end{aligned} \tag{9.364}$$

where $V^\mu(\tau)$ is the proper velocity of the particle. The Dirac delta in (9.364) can be visualized as being sharply peaked around the value of x^μ that satisfies

$$f(x,\tau) \equiv (x - x_c(\tau))^2 = (t - t_c(\tau))^2 - |\boldsymbol{x} - \boldsymbol{x}_c(\tau)|^2 = 0. \tag{9.365}$$

Reversing this argument, for a given value of x there is a value of τ, designated τ_o and referred to as the *retarded proper time*, that satisfies the equation

$$f(x,\tau_o) = 0, \tag{9.366}$$

subject to the retardation constraint of the step function,

$$t - t_c(\tau_o) > 0. \tag{9.367}$$

The retarded proper time τ_o identifies those points where the particle's trajectory crosses the past light cone of the space-time point x under consideration. This allows a signal traveling at the speed of light to reach the position $\boldsymbol{x}$ at the time t. This incorporates the finite speed of propagation for light into the transmission of force and energy.

The argument of the Dirac delta appearing in (9.364) can be expanded in the neighborhood of τ_o using

$$f(x,\tau) \approx (\tau - \tau_o) \left.\frac{\mathrm{d}f(x,\tau)}{\mathrm{d}\tau}\right|_{\tau=\tau_o}, \tag{9.368}$$

where (9.365) gives

$$\begin{aligned}\left.\frac{\mathrm{d}f(x,\tau)}{\mathrm{d}\tau}\right|_{\tau=\tau_o} &= -2\,(x - x_c(\tau))_\mu \left.\frac{\mathrm{d}x_c^\mu(\tau)}{\mathrm{d}\tau}\right|_{\tau=\tau_o} \\ &= -2\,(x - x_c(\tau_o))_\mu V^\mu(\tau_o) = -2\,(x - x_c(\tau_o))\cdot V(\tau_o).\end{aligned} \tag{9.369}$$

Using the property of the Dirac delta $\delta(ax) = \delta(x)/|a|$ gives

$$\delta[(x - x_c(\tau))^2] = \frac{\delta(\tau - \tau_o)}{2\,|V(\tau_o)\cdot(x - x_c(\tau_o))|}. \tag{9.370}$$

Substituting (9.370) into (9.364) and performing integration over τ gives the gauge field

$$A^\mu(x) = \frac{q}{4\pi}\left.\frac{V^\mu(\tau)}{|V\cdot(x - x_c(\tau))|}\right|_{\tau=\tau_o}. \tag{9.371}$$

The components of (9.371) are known as the *Liénard–Wiechert potentials.*

For the case of a static charge at position $\boldsymbol{x}_c$, the retarded time t_o in natural units for position $\boldsymbol{x}$ is given by solving (9.365),

$$t_o = t - |\boldsymbol{x} - \boldsymbol{x}_c|, \tag{9.372}$$

and the charge's proper velocity in natural units is simply

$$V^\mu(\tau_o) = (1, 0). \tag{9.373}$$

This gives

$$(x - x_c(\tau_o))_\mu V^\mu(\tau_o) = (x - x_c(\tau_o))_0 V^0(\tau_0) = (t - t_o) = |\boldsymbol{x} - \boldsymbol{x}_c|, \tag{9.374}$$

resulting in

$$A^0(x) = \frac{q}{4\pi|\boldsymbol{x} - \boldsymbol{x}_c|}, \quad \boldsymbol{A} = 0, \tag{9.375}$$

which is the previously derived Coulomb potential for a static charge.

For the case of a point charge moving with a uniform relative velocity of $\boldsymbol{v}(t)$, the proper velocity of the charge is given by (6.133) in natural units,

$$V^\mu(\tau) = \gamma_v(1, \boldsymbol{v}(t)). \tag{9.376}$$

The retarded time t_o is once again given by previous expression (9.372), but the solution is more complicated, since the trajectory $\boldsymbol{x}_c(\tau)$ is time-dependent. As a result, the retarded time t_o must satisfy the equation

$$t_o = t - |\boldsymbol{x} - \boldsymbol{x}_c(t_o)|. \tag{9.377}$$

It is assumed that (9.377) has a solution for t_o. Denoting the spatial vector $\boldsymbol{R}(t_o)$ by

$$\boldsymbol{R}(t_o) = \boldsymbol{x} - \boldsymbol{x}_c(t_o) \implies R(t_o) = |\boldsymbol{x} - \boldsymbol{x}_c(t_o)|, \tag{9.378}$$

it follows that

$$\begin{aligned}(x - x_c(\tau_o))_\mu V^\mu(\tau_o) &= \gamma_v(t - t_o) + \gamma_v \boldsymbol{R}(t_o) \cdot \boldsymbol{v}(t_o) \\ &= \gamma_v(R(t_o) + \boldsymbol{R}(t_o) \cdot \boldsymbol{v}(t_o)).\end{aligned} \tag{9.379}$$

This gives the Liénard–Wiechert potentials,

$$\begin{aligned}A^0(x) &= \frac{q}{4\pi|R(t_o) + \boldsymbol{R}(t_o) \cdot \boldsymbol{v}(t_o)|}, \\ \boldsymbol{A}(x) &= \frac{q\boldsymbol{v}(t_o)}{4\pi|R(t_o) + \boldsymbol{R}(t_o) \cdot \boldsymbol{v}(t_o)|},\end{aligned} \tag{9.380}$$

where the reader is reminded that the retarded time t_o is a function of both the space-time point x and the trajectory $\boldsymbol{x}_c(\tau)$. It can be shown that (9.380) reproduces the effect of boosting the Coulomb potential of (9.375) into a moving frame. In a moving frame, a nonzero $\boldsymbol{A}$ appears, resulting in a magnetic field. This is consistent with treating a current as an electric charge distribution viewed from a moving frame.

For the case of accelerated motion it is easiest to return to the general expression (9.364) and use it to find the field strength tensor through relation (9.65). In doing so it will be necessary to find the partial derivative of the delta function appearing in (9.364). The key step is to note that the chain rule gives

$$\begin{aligned}\partial^{\nu}\delta[f(x,\tau)] &= \frac{\partial}{\partial x_{\nu}}\delta[f(x,\tau)]\\ &= \frac{\partial f(x,\tau)}{\partial x_{\nu}}\,\frac{\mathrm{d}}{\mathrm{d}f(x,\tau)}\delta[f(x,\tau)]\\ &= \frac{\partial f(x,\tau)}{\partial x_{\nu}}\,\frac{\mathrm{d}\tau}{\mathrm{d}f(x,\tau)}\,\frac{\mathrm{d}}{\mathrm{d}\tau}\delta[f(x,\tau)],\end{aligned} \tag{9.381}$$

where

$$\frac{\mathrm{d}\tau}{\mathrm{d}f(x,\tau)} = \left(\frac{\mathrm{d}f(x,\tau)}{\mathrm{d}\tau}\right)^{-1} = \left(\frac{\mathrm{d}}{\mathrm{d}\tau}(x - x_c(\tau))^2\right)^{-1} = -\frac{1}{2\,V(\tau)\cdot(x - x_c(\tau))} \tag{9.382}$$

and

$$\frac{\partial}{\partial x_{\nu}}f(x,\tau) = 2\,(x - x_c)^{\nu}. \tag{9.383}$$

Using these results shows that

$$\partial^{\nu}\delta[(x - x_c(\tau))^2] = -\frac{(x - x_c(\tau))^{\nu}}{V(\tau)\cdot(x - x_c(\tau))}\,\frac{\mathrm{d}}{\mathrm{d}\tau}\delta[(x - x_c(\tau))^2]. \tag{9.384}$$

Returning this result to (9.364) gives

$$\begin{aligned}F^{\mu\nu} &= \partial^{\mu}A^{\nu} - \partial^{\nu}A^{\mu}\\ &= \frac{Q}{2\pi}\int \mathrm{d}\tau\;\theta(t - t_c)\\ &\quad\left[\frac{(x - x_c(\tau))^{\nu}V^{\mu}(\tau) - (x - x_c(\tau))^{\mu}V^{\nu}(\tau)}{V(\tau)\cdot(x - x_c(\tau))}\right]\\ &\quad\frac{\mathrm{d}}{\mathrm{d}\tau}\delta[(x - x_c(\tau))^2].\end{aligned} \tag{9.385}$$

In deriving (9.385), the independence of the proper velocity $\partial_{\mu}V^{\nu} = 0$ derived in (6.149) was used. In addition, the time derivative of the step function has been ignored, since

$$\begin{aligned}\delta[(x - x_c)^2]\partial_0\theta(x^0 - x_c^0) &= \delta[(x - x_c)^2]\delta(x^0 - x_c^0)\\ &= \delta[|\boldsymbol{x} - \boldsymbol{x}_c(\tau)|^2]\delta(x^0 - x_c^0),\end{aligned} \tag{9.386}$$

which is nonzero only for spatial points $\boldsymbol{x}$ on the classical trajectory where

$$|\boldsymbol{x} - \boldsymbol{x}_c(\tau)|^2 = 0. \tag{9.387}$$

Result (9.385) can now be integrated by parts, once again ignoring derivatives of the step function, and result (9.370) can be employed. This gives the final result

$$F^{\mu\nu}(x) = \frac{q}{4\pi V(\tau)\cdot(x - x_c(\tau))}\frac{\mathrm{d}}{\mathrm{d}\tau}\left[\frac{(x - x_c(\tau))^{\mu}V^{\nu}(\tau) - (x - x_c(\tau))^{\nu}V^{\mu}(\tau)}{V(\tau)\cdot(x - x_c(\tau))}\right]\Bigg|_{\tau=\tau_o}, \quad (9.388)$$

where, after taking the derivative, the entire expression is to be evaluated at the retarded proper time τ_o.

Exercise 9.13: Verify (9.388).

The derivative with respect to the proper time gives rise to both the proper velocity and the relativistic acceleration,

$$A^{\mu}(\tau) = \frac{\mathrm{d}V^{\mu}(\tau)}{\mathrm{d}\tau}. \quad (9.389)$$

Recalling that $V\cdot V = 1$ in natural units, it follows that

$$\begin{aligned}\frac{\mathrm{d}}{\mathrm{d}\tau}\left[\frac{(x - x_c(\tau))^{\mu}V^{\nu}(\tau)}{V(\tau)\cdot(x - x_c(\tau))}\right] = {} & \frac{(x - x_c(\tau))^{\mu}A^{\nu}(\tau)}{V(\tau)\cdot(x - x_c(\tau))} \\ & - \frac{(x - x_c(\tau))^{\mu}V^{\nu}(\tau)\, A(\tau)\cdot(x - x_c(\tau))}{(V(\tau)\cdot(x - x_c(\tau)))^2} \\ & + \frac{(x - x_c(\tau))^{\mu}V^{\nu}(\tau)}{(V(\tau)\cdot(x - x_c(\tau)))^2} \\ & - \frac{V^{\mu}(\tau)V^{\nu}(\tau)}{V(\tau)\cdot(x - x_c(\tau))},\end{aligned} \quad (9.390)$$

with a similar result for the second term in (9.388),

$$\begin{aligned}-\frac{\mathrm{d}}{\mathrm{d}\tau}\left[\frac{(x - x_c(\tau))^{\nu}V^{\mu}(\tau)}{V(\tau)\cdot(x - x_c(\tau))}\right] = {} & -\frac{(x - x_c(\tau))^{\nu}A^{\mu}(\tau)}{V(\tau)\cdot(x - x_c(\tau))} \\ & + \frac{(x - x_c(\tau))^{\nu}V^{\mu}(\tau)\, A(\tau)\cdot(x - x_c(\tau))}{(V(\tau)\cdot(x - x_c(\tau)))^2} \\ & - \frac{(x - x_c(\tau))^{\nu}V^{\mu}(\tau)}{(V(\tau)\cdot(x - x_c(\tau)))^2} \\ & + \frac{V^{\nu}(\tau)V^{\mu}(\tau)}{V(\tau)\cdot(x - x_c(\tau))}.\end{aligned} \quad (9.391)$$

These expressions can be evaluated using the forms for the proper velocity (6.152) and the relativistic acceleration (6.159) in terms of the relative velocity $\boldsymbol{v}$ and the relative acceleration $\boldsymbol{a}$ and returned to (9.388). This yields complicated expressions without providing a great deal of insight into the effects of acceleration.

However, these results simplify considerably in the nonrelativistic limit for the case that the instantaneous velocity of the charge is $\boldsymbol{v} = 0$ at the retarded time. For such a special case the proper velocity and the relativistic acceleration and the terms involving them become

$$\begin{aligned} (x^0 - x_c^0) &= |\boldsymbol{x} - \boldsymbol{x}_c|, \\ A^\mu &\approx (0, \boldsymbol{a}), \\ A \cdot (x - x_c) &\approx - \boldsymbol{a} \cdot (\boldsymbol{x} - \boldsymbol{x}_c), \\ V^\mu &\approx (1, 0), \\ V \cdot (x - x_c) &\approx |\boldsymbol{x} - \boldsymbol{x}_c|. \end{aligned} \tag{9.392}$$

It is straightforward to use the expressions of (9.392) in (9.390) and (9.391) to show that (9.388) yields

$$\begin{aligned} F^{0j}(x) = - E_j(x) &= \frac{q}{4\pi|\boldsymbol{x} - \boldsymbol{x}_c|} \\ &\left(a^j - \frac{(x - x_c)^j \boldsymbol{a} \cdot (\boldsymbol{x} - \boldsymbol{x}_c)}{|\boldsymbol{x} - \boldsymbol{x}_c|^2} - \frac{(x - x_c)^j}{|\boldsymbol{x} - \boldsymbol{x}_c|^2} \right) \Bigg|_{\tau=\tau_o}, \\ F^{jk}(x) = \varepsilon^{kj\ell} B_\ell(x) & \\ = \frac{q}{4\pi|\boldsymbol{x} - \boldsymbol{x}_c|} &\left(\frac{(x - x_c)^j a^k}{|\boldsymbol{x} - \boldsymbol{x}_c|} - \frac{(x - x_c)^k a^j}{|\boldsymbol{x} - \boldsymbol{x}_c|} \right) \Bigg|_{\tau=\tau_o}, \end{aligned} \tag{9.393}$$

where these expressions are to be evaluated at the retarded proper time τ_o. The expressions of (9.393) can be simplified further by noting that the vector $\hat{\boldsymbol{n}}$, defined by

$$\hat{\boldsymbol{n}} - \frac{(\boldsymbol{x} - \boldsymbol{x}_c)}{|\boldsymbol{x} - \boldsymbol{x}_c|}, \tag{9.394}$$

satisfies $\hat{\boldsymbol{n}} \cdot \hat{\boldsymbol{n}} = 1$ and points from the retarded position to the spatial point under consideration. This allows the expressions of (9.393) to be written as vectors,

$$\begin{aligned} \mathbf{E} &= \frac{q\,\hat{\boldsymbol{n}}}{4\pi|\boldsymbol{x} - \boldsymbol{x}_c|^2} - \frac{q}{4\pi|\boldsymbol{x} - \boldsymbol{x}_c|}(\boldsymbol{a} - (\hat{\boldsymbol{n}} \cdot \boldsymbol{a})\hat{\boldsymbol{n}}) \\ \mathbf{B} &= - \frac{q}{4\pi|\boldsymbol{x} - \boldsymbol{x}_c|}(\hat{\boldsymbol{n}} \times \boldsymbol{a}). \end{aligned} \tag{9.395}$$

Choosing $v(\tau_o) = 0$ has prevented the Liénard–Wiechert potentials of (9.380) from appearing. However, it reveals that the acceleration creates a contribution to both **E** and **B** that drops off as $1/r$ rather than the $1/r^2$ terms that occur as a result of including the Liénard–Wiechert potentials in $F^{\mu\nu}$. It is therefore the acceleration-dependent terms that describe the radiation of energy from the charge. In what follows only the acceleration-dependent terms will be considered.

Ignoring the first term in $\mathbf{E}$ as irrelevant to radiation, the *acceleration fields* are given by

$$\mathbf{E}_{ac} = -\frac{q}{4\pi|\boldsymbol{x} - \boldsymbol{x}_c|}(\boldsymbol{a} - (\hat{\boldsymbol{n}} \cdot \boldsymbol{a})\, \hat{\boldsymbol{n}}), \tag{9.396}$$

$$\mathbf{B}_{ac} = -\frac{q}{4\pi|\boldsymbol{x} - \boldsymbol{x}_c|}(\hat{\boldsymbol{n}} \times \boldsymbol{a}), \tag{9.397}$$

where $\boldsymbol{x}_c$ corresponds to the retarded position. Using the identities

$$\hat{\boldsymbol{n}} \cdot (\hat{\boldsymbol{n}} \times \boldsymbol{a}) = \boldsymbol{a} \cdot (\hat{\boldsymbol{n}} \times \boldsymbol{a}) = 0 \tag{9.398}$$

shows that the acceleration fields have the properties that

$$\hat{\boldsymbol{n}} \cdot \mathbf{E}_{ac} = \hat{\boldsymbol{n}} \cdot \mathbf{B}_{ac} = \mathbf{E}_{ac} \cdot \mathbf{B}_{ac} = 0. \tag{9.399}$$

Using $\hat{\boldsymbol{n}} \times \hat{\boldsymbol{n}} = 0$ shows that

$$\mathbf{B}_{ac} = \hat{\boldsymbol{n}} \times \mathbf{E}_{ac}. \tag{9.400}$$

Combining these results gives the Poynting vector for the acceleration fields,

$$\begin{aligned}\mathbf{S}_{ac} &= \mathbf{E}_{ac} \times \mathbf{B}_{ac} = \mathbf{E}_{ac} \times (\hat{\boldsymbol{n}} \times \mathbf{E}_{ac}) = \hat{\boldsymbol{n}}\, (\mathbf{E}_{ac} \cdot \mathbf{E}_{ac}) - \mathbf{E}_{ac}\, (\hat{\boldsymbol{n}} \cdot \mathbf{E}_{ac}) \\ &= \hat{\boldsymbol{n}}\, \mathbf{E}_{ac} \cdot \mathbf{E}_{ac}.\end{aligned} \tag{9.401}$$

Denoting the angle between $\hat{\boldsymbol{n}}$ and $\boldsymbol{a}$ by θ, it follows that

$$(\boldsymbol{a} - (\hat{\boldsymbol{n}} \cdot \boldsymbol{a})\hat{\boldsymbol{n}}) \cdot (\boldsymbol{a} - (\hat{\boldsymbol{n}} \cdot \boldsymbol{a})\hat{\boldsymbol{n}}) = a^2 - (\hat{\boldsymbol{n}} \cdot \boldsymbol{a})^2 = a^2(1 - \cos^2\theta) = a^2 \sin^2\theta. \tag{9.402}$$

The Poynting vector in natural units is therefore given by

$$\mathbf{S}_{ac} = \frac{q^2 a^2 \sin^2\theta}{16\pi^2|\boldsymbol{x} - \boldsymbol{x}_c|^2}\, \hat{\boldsymbol{n}}. \tag{9.403}$$

The Poynting vector is maximized at an angle perpendicular to the acceleration vector and corresponds to a transverse electromagnetic wave.

Result (9.181) and the discussion following it showed that the surface integral of the Poynting vector gives the energy flow rate out of the contained volume. Noting that a sphere centered on $\boldsymbol{x}_c$ has the radial vector $\hat{\boldsymbol{n}}$ and the radius $r = |\boldsymbol{x} - \boldsymbol{x}_c|$, while the surface element is written

$$\mathrm{d}\mathbf{S} = \hat{\boldsymbol{n}}\, r^2 \sin\theta\, \mathrm{d}\theta\, \mathrm{d}\phi = -\hat{\boldsymbol{n}}\, r^2\, \mathrm{d}(\cos\theta)\, \mathrm{d}\phi, \tag{9.404}$$

the energy per unit time flowing through an arbitrary sphere around the retarded position is

$$P = \oint_S \mathbf{S}_{ac} \cdot \mathrm{d}\boldsymbol{S} = \frac{q^2 a^2}{16\pi^2} \int_{-1}^{1} \sin^2\theta\, \mathrm{d}(\cos\theta) \int_0^{2\pi} \mathrm{d}\phi = \frac{q^2 a^2}{6\pi}. \tag{9.405}$$

Result (9.405) is known as the *Larmor formula.* If a is measured in length per unit time squared, (9.405) is converted to energy per unit time by a factor of $\hbar/c^2$, so that

$$P = \frac{\hbar q^2 a^2}{6\pi c^2}. \tag{9.406}$$

This expression is converted to MKSA units by using the scaling factor of (9.1), $q = 4\pi Q/e$, where Q is measured in Coulombs, and the expression for $\hbar$ given by (9.90), which yields the MKSA version of the Larmor radiation formula,

$$P = \frac{2}{3}\frac{KQ^2a^2}{c^3} = \frac{Q^2a^2}{6\pi\epsilon_0 c^3}. \tag{9.407}$$

Result (9.407) can be cast into a relativistically invariant form by noting that it can be written

$$P = \frac{Q^2}{6\pi\epsilon_0 m^2 c^3}\frac{\mathrm{d}\boldsymbol{p}}{\mathrm{d}t}\cdot\frac{\mathrm{d}\boldsymbol{p}}{\mathrm{d}t}. \tag{9.408}$$

This can be viewed as the nonrelativistic version of the expression

$$P = -\frac{Q^2}{6\pi\epsilon_0 m^2 c^3}\frac{\mathrm{d}p^\mu}{\mathrm{d}\tau}\frac{\mathrm{d}p_\mu}{\mathrm{d}\tau} = -\frac{Q^2}{6\pi\epsilon_0 c^3}A^\mu A_\mu. \tag{9.409}$$

The Larmor radiation formula was one of the key results that doomed the validity of classical physics at the atomic level. Early attempts to model the experimentally determined structure of the atom involved the Keplerian or planetary model, where the electron with a charge of $-Q$ moved in an orbit of radius r around the nucleus with a positive charge of Q at a speed of v under the influence of the attractive Coulomb force $Q^2/4\pi\epsilon_0 r^2$. This gave rise to a centripetal acceleration for an electron of mass m given by

$$\frac{Q^2}{4\pi\epsilon_0 r^2} = \frac{mv^2}{r} \implies a_c = \frac{v^2}{r} = \frac{Q^2}{4\pi\epsilon_0 m r^2}. \tag{9.410}$$

This centripetal acceleration therefore gave rise to a loss of energy by the electron. Using the nonrelativistic version of the Larmor formula (9.407), this loss occurs at the rate

$$P = \frac{Q^2 a_c^2}{6\pi\epsilon_0 c^3} = \frac{Q^6}{96\pi^3\epsilon_0^3 m^2 c^3 r^4}. \tag{9.411}$$

The total energy of the electron is given by

$$E = -\frac{Q^2}{8\pi\epsilon_0 r} \implies \frac{\mathrm{d}E}{\mathrm{d}t} = \frac{Q^2}{8\pi\epsilon_0 r^2}\frac{\mathrm{d}r}{\mathrm{d}t}. \tag{9.412}$$

Equating the two gives

$$\frac{\mathrm{d}E}{\mathrm{d}t} = -P \implies r^2\frac{\mathrm{d}r}{\mathrm{d}t} = -\frac{Q^4}{12\pi^2\epsilon_0^2 m^2 c^3}. \qquad (9.413)$$

Exercise 9.14: Solve (9.413) to show that the electron should spiral into the nucleus from the radius r orbit in the time

$$T = \frac{4\pi^2\epsilon_0^2 m^2 r^3 c^3}{Q^4}. \qquad (9.414)$$

Using atomic-level SI values of $r \approx 5 \times 10^{-11}$ m, $Q \approx 1.6 \times 10^{-19}$ C, $m \approx 9 \times 10^{-31}$ kg, and $\epsilon_0 \approx 8.9 \times 10^{-12}\ \mathrm{C}^2 \cdot \mathrm{s}^2\mathrm{kg}^{-1} \cdot \mathrm{m}^3$ gives $T \approx 10^{-11}$ s. The predicted collapse of the Keplerian atom was a serious factor in forcing physics to consider the possibility that classical physics breaks down in the subatomic realm, where it must be supplanted by quantum mechanics and quantum field theory.

9.14 The gauge field as a differential form

Lorentz covariant electrodynamics finds a natural expression in terms of differential forms. The starting point is the Lorentz invariant gauge field one-form,

$$A = A_\mu(x)\,\mathrm{d}x^\mu. \qquad (9.415)$$

The field strength two-form F is given by the exterior derivative of the one-form A. This follows from the definition (1.192) of exterior differentiation and the anti-symmetry of the wedge product,

$$F = \mathrm{d}A = \partial_\mu A_\nu\,\mathrm{d}x^\mu \wedge \mathrm{d}x^\nu = \frac{1}{2}(\partial_\mu A_\nu - \partial_\nu A_\mu)\,\mathrm{d}x^\mu \wedge \mathrm{d}x^\mu = \frac{1}{2}F_{\mu\nu}\,\mathrm{d}x^\mu \wedge \mathrm{d}x^\nu. \qquad (9.416)$$

The wedge product can be written

$$\mathrm{d}x^\mu \wedge \mathrm{d}x^\nu = \varepsilon^{\mu\nu\rho\sigma}\,\mathrm{d}S_{\rho\sigma}, \qquad (9.417)$$

where $\mathrm{d}S_{\rho\sigma}$ are the six two-dimensional oriented surfaces in Minkowski space. This allows the field strength two-form to be written as

$$F = \frac{1}{2}\varepsilon^{\mu\nu\rho\sigma}F_{\mu\nu}\,\mathrm{d}S_{\rho\sigma}. \qquad (9.418)$$

For example, the six components of this two-form include

$$\varepsilon^{jk0\ell}F_{jk}\,\mathrm{d}S_{0\ell} = -\varepsilon^{jk0\ell}\varepsilon_{jkr}B_r\,\mathrm{d}S_{0\ell} = B_\ell\,\mathrm{d}S_{0\ell} = \mathrm{d}t\,\mathbf{B}\cdot\mathrm{d}\boldsymbol{\ell}, \qquad (9.419)$$

which contains the magnetic line element. The electric field is also included,

$$\varepsilon^{0jk\ell}F_{0j}\mathrm{d}S_{k\ell} = E_j\varepsilon^{0jk\ell}\mathrm{d}S_{k\ell} = E_j\varepsilon^{0jk\ell}\varepsilon_{k\ell r}\mathrm{d}S_r = \mathbf{E}\cdot\mathrm{d}\boldsymbol{S}, \tag{9.420}$$

where $\mathrm{d}\boldsymbol{S}$ is a two-dimensional surface element.

The homogeneous Maxwell equations follow from the nilpotency of the exterior derivative, which gives

$$\mathrm{d}F = \mathrm{d}^2A = 0 \implies \mathrm{d}F = \frac{1}{2}\partial_\mu F_{\nu\rho}\,\mathrm{d}x^\mu \wedge \mathrm{d}x^\nu \wedge \mathrm{d}x^\rho = \frac{1}{2}\varepsilon^{\mu\nu\rho\sigma}\partial_\mu F_{\nu\rho}\,\mathrm{d}V_\sigma = 0, \tag{9.421}$$

where $\mathrm{d}V_\sigma$ are the four oriented three-dimensional volumes of Minkowski space. Since these are independent degrees of freedom, it follows that

$$\varepsilon^{\mu\nu\rho\sigma}\partial_\mu F_{\nu\rho} = \varepsilon^{\sigma\mu\nu\rho}\partial_\mu F_{\nu\rho} = 0, \tag{9.422}$$

which gives the covariant form of the homogeneous Maxwell equations found in (9.47).

A gauge transformation is given by

$$A \to A - \mathrm{d}\Lambda, \tag{9.423}$$

where Λ is the scalar gauge function. This follows from the definition of the exterior derivative of a scalar function,

$$\mathrm{d}\Lambda = \partial_\mu\Lambda(x)\,\mathrm{d}x^\mu. \tag{9.424}$$

As a result,

$$A - \mathrm{d}\Lambda = (A_\mu(x) - \partial_\mu\Lambda(x))\mathrm{d}x^\mu, \tag{9.425}$$

which reproduces the original definition of a gauge transformation made in (9.68). The gauge invariance of F follows from the nilpotency of d once again, which gives

$$F = \mathrm{d}A \to \mathrm{d}A - \mathrm{d}^2\Lambda = \mathrm{d}A = F. \tag{9.426}$$

However, the nilpotency of d depends on the differentiability of the functions on which it acts. To see this in the context of (9.426), it follows from the antisymmetry of the wedge product that

$$\begin{aligned}\mathrm{d}^2\Lambda &= \mathrm{d}(\partial_\mu\Lambda\,\mathrm{d}x^\mu) = \partial_\nu\partial_\mu\Lambda\,\mathrm{d}x^\nu \wedge \mathrm{d}x^\mu = \frac{1}{2}(\partial_\nu\partial_\mu - \partial_\mu\partial_\nu)\Lambda\,\mathrm{d}x^\mu \wedge \mathrm{d}x^\nu \\ &= \frac{1}{2}[\partial_\nu, \partial_\mu]\Lambda\,\mathrm{d}x^\nu \wedge \mathrm{d}x^\mu.\end{aligned} \tag{9.427}$$

The exterior derivative is nilpotent as long as

$$[\partial_\nu, \partial_\mu]\Lambda = 0. \tag{9.428}$$

The two-form that is dual to F is given by

$$^*F = -\frac{1}{4}\varepsilon_{\alpha\beta\mu\nu}F^{\mu\nu}\,\mathrm{d}x^\alpha \wedge \mathrm{d}x^\beta. \tag{9.429}$$

Similarly, the three-form representation of the electric current is given by

$$\begin{aligned} J &= \frac{1}{3!}\varepsilon_{\mu\nu\alpha\beta}J^{\mu}\,\mathrm{d}x^{\nu}\wedge\mathrm{d}x^{\alpha}\wedge\mathrm{d}x^{\beta} = \frac{1}{3!}\varepsilon^{\nu\alpha\beta\sigma}\varepsilon_{\mu\nu\alpha\beta}J^{\mu}\,\mathrm{d}V_{\sigma} \\ &= \delta^{\sigma}{}_{\mu}J^{\mu}\,\mathrm{d}V_{\sigma} = J^{\mu}\mathrm{d}V_{\mu}. \end{aligned} \tag{9.430}$$

The inhomogeneous Maxwell equations can then be written

$$\begin{aligned} \mathrm{d}^{*}F &= -\frac{1}{4}\varepsilon_{\alpha\beta\mu\nu}\partial_{\rho}F^{\mu\nu}\mathrm{d}x^{\rho}\wedge\mathrm{d}x^{\alpha}\wedge\mathrm{d}x^{\beta} = -\frac{1}{4}\varepsilon^{\rho\alpha\beta\sigma}\varepsilon_{\alpha\beta\mu\nu}\partial_{\rho}F^{\mu\nu}\mathrm{d}V_{\sigma} \\ &= -\frac{1}{2}(\delta^{\sigma}{}_{\mu}\delta^{\rho}{}_{\nu} - \delta^{\sigma}{}_{\nu}\delta^{\rho}{}_{\mu})\partial_{\rho}F^{\mu\nu}\mathrm{d}V_{\sigma} = \partial_{\mu}F^{\mu\sigma}\mathrm{d}V_{\sigma} = J^{\sigma}\mathrm{d}V_{\sigma} \\ &= J. \end{aligned} \tag{9.431}$$

Using $^{*}F$ the action (9.92) for the gauge field can be written as a four-form wedge product,

$$\begin{aligned} -\frac{1}{4}{}^{*}F\wedge F &= -\frac{1}{16}\varepsilon_{\alpha\beta\mu\nu}F^{\mu\nu}F_{\rho\sigma}\,\mathrm{d}x^{\alpha}\wedge\mathrm{d}x^{\beta}\wedge\mathrm{d}x^{\rho}\wedge\mathrm{d}x^{\sigma} \\ &= -\frac{1}{16}\varepsilon_{\alpha\beta\mu\nu}\varepsilon^{\alpha\beta\rho\sigma}F^{\mu\nu}F_{\rho\sigma}\,\mathrm{d}^{4}x \\ &= -\frac{1}{8}(\delta^{\rho}{}_{\mu}\,\delta^{\sigma}{}_{\nu} - \delta^{\rho}{}_{\nu}\,\delta^{\sigma}{}_{\mu})F^{\mu\nu}F_{\rho\sigma}\,\mathrm{d}^{4}x \\ &= -\frac{1}{4}F^{\mu\nu}F_{\mu\nu}\,\mathrm{d}^{4}x. \end{aligned} \tag{9.432}$$

As a final note, the four-form $F\wedge F$ is shown to be a perfect differential by using (1.196),

$$\mathrm{d}(A\wedge F) = \mathrm{d}A\wedge F - A\wedge\mathrm{d}F = \mathrm{d}A\wedge F = F\wedge F. \tag{9.433}$$

Using Stokes' theorem (1.204) gives

$$\int_{\mathcal{M}} F\wedge F = \int_{\mathcal{M}}\mathrm{d}(A\wedge F) = \int_{\partial\mathcal{M}} A\wedge F. \tag{9.434}$$

The integral of $F\wedge F$, where

$$F\wedge F = F_{\mu\nu}F_{\rho\sigma}\,\mathrm{d}x^{\mu}\wedge\mathrm{d}x^{\nu}\wedge\mathrm{d}x^{\rho}\wedge\mathrm{d}x^{\sigma} = \varepsilon^{\mu\nu\rho\sigma}F_{\mu\nu}F_{\rho\sigma}\,\mathrm{d}^{4}x, \tag{9.435}$$

is therefore determined by the behavior of F and A through the three-form $A\wedge F$ at the surface $\partial\mathcal{M}$ of Minkowski space $\mathcal{M}$. This results in no contribution to the variation of the action if such a term is included. However, the surface integral does not necessarily vanish and characterizes what are referred as the *topological properties* of the gauge field A. Aspects of this are discussed in the final chapter.

9.15 Magnetic monopoles

One of Maxwell's equations states that

$$\nabla \cdot \mathbf{B} = 0, \tag{9.436}$$

which is the magnetic counterpart of Gauss's law,

$$\nabla \cdot \mathbf{E} = J^0. \tag{9.437}$$

While pointlike sources of the electric field are readily available in the laboratory, there has never been a reproducible observation of a pointlike source of magnetic field, the so-called *magnetic monopole*, despite extensive searches. Magnetic materials, such as ferromagnets, are always dipolar in nature, presenting north and south poles that cannot be separated by breaking the magnet in half. Upon breaking a ferromagnet in half the result is two dipolar ferromagnets.

Nevertheless, (9.436) should be viewed as a law that is consistent with *current observations*. However, it plays a central role in defining the gauge field A_μ, since

$$\mathbf{B} = \nabla \times A \implies \nabla \cdot \mathbf{B} = \nabla \cdot (\nabla \times A) = 0, \tag{9.438}$$

and (9.438) is consistent with gauge invariance,

$$\mathbf{B}' = \nabla \times A' = \nabla \times A - \nabla \times \nabla\Lambda = \nabla \times A = \mathbf{B}. \tag{9.439}$$

The combination of the gauge field and the gauge invariance of its equations of motion has become a linchpin of modern physics and provides critical guidelines for the description of other naturally occurring forces. The question of why there are no magnetic monopoles remains an open area of conjecture. It is possible that magnetic monopoles are simply extremely rare, due to conditions present in the early stages of the Universe. This is one of the possible outgrowths of the *inflationary Big Bang* model of the early Universe. This model is discussed briefly in the next chapter.

However, Dirac pointed out that the presence of a magnetic monopole did not require relation (9.438) to be abandoned. He argued that there are singular vector potentials A that allow (9.436) to yield the field of a pointlike monopole. There are two relevant vector potentials in spherical coordinates,

$$\begin{aligned} A_+ &= \frac{g}{4\pi r}\frac{(1-\cos\theta)}{\sin\theta}\,\hat{\mathbf{e}}_\phi, \\ A_- &= -\frac{g}{4\pi r}\frac{(1+\cos\theta)}{\sin\theta}\,\hat{\mathbf{e}}_\phi. \end{aligned} \tag{9.440}$$

Exercise 9.15: Use the curl in spherical coordinates given by (1.159) to show that, for $\theta \neq 0$ for A_- and $\theta \neq \pi$ for A_+,

$$\nabla \times A_+ = \nabla \times A_- = \frac{g}{4\pi r^2}\,\hat{\mathbf{e}}_r. \tag{9.441}$$

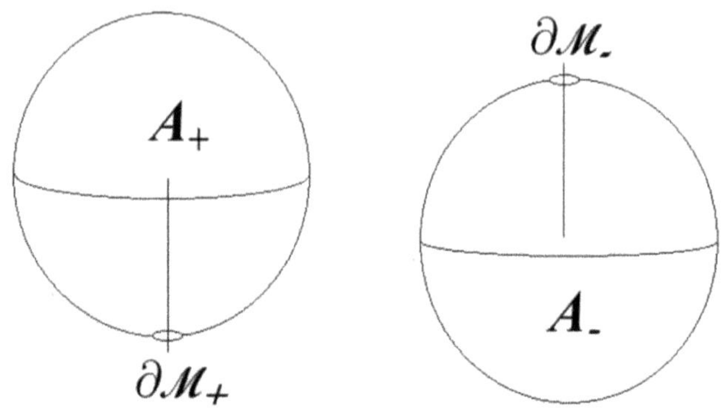

Figure 9.2. Punctured spheres $\mathcal{M}_+$ and $\mathcal{M}_-$ and their boundaries $\partial\mathcal{M}_+$ and $\partial\mathcal{M}_-$ used for finding the magnetic flux from the monopole potentials $\boldsymbol{A}_+$ and $\boldsymbol{A}_-$ via Stokes' theorem.

Result (9.441) is identical *almost everywhere* to the radial electric field of a pointlike charge at the origin originally found in (9.355). As result, the constant g is interpreted as the *magnetic charge* of a magnetic monopole. However, there is a subtlety that is exposed by applying Stokes' theorem to the integral of the magnetic flux through a *punctured sphere* of radius R with an infinitesimal boundary in the form of a circular hole. In the case of $\boldsymbol{A}_+$ the circular hole is at the south pole, while for $\boldsymbol{A}_-$ the circular hole is at the north pole. These are depicted in figure 9.2.

In the case of $\boldsymbol{A}_+$ there is a *string* of singularities that begins at the origin and extends to infinity along the z-axis at $\theta = \pi$. Likewise, in the case of $\boldsymbol{A}_-$ there is a *string* of singularities that begins at the origin and extends to infinity along the z-axis at $\theta = 0$. These singularities are referred to as the *Dirac string*. The potential $\boldsymbol{A}_-$ in the vicinity of the Dirac string at the north pole is given by the small θ form,

$$\boldsymbol{A}_- \approx -\frac{g}{2\pi R\theta}\,\hat{\mathbf{e}}_\phi, \tag{9.442}$$

where $\theta \to 0$ for an infinitesimally small hole. The radius of the hole is given by $r = R\theta$, so that Stokes' theorem gives

$$\int_{\mathcal{M}_-} \mathbf{B}_- \cdot \mathrm{d}\boldsymbol{S} = \int_{\mathcal{M}_-} \mathrm{d}A_- = \oint_{\partial\mathcal{M}_-} A_- \cdot \mathrm{d}\ell = \frac{g}{2\pi R\theta}\oint \mathrm{d}\ell = \frac{g}{2\pi R\theta}(2\pi r) = g, \tag{9.443}$$

where $\partial\mathcal{M}_-$ is clockwise around the north pole. The result is independent of the size of the punctured sphere $\mathcal{M}_-$, which is the same property as the field generated by an electric point charge. In effect, the potential creates an outward magnetic flux through the punctured sphere $\mathcal{M}_-$ while at the same time creating a singular inward flux through the infinitesimal puncture hole. As a result, the net magnetic flux through the total sphere is zero. This is consistent with **B** being given by (9.438) and the fact that the total sphere has *no boundary*, so that Stokes' theorem states that the *net magnetic flux* through the unpunctured sphere must be zero. The same result holds for $\boldsymbol{A}_+$ and the punctured sphere $\mathcal{M}_+$ with the hole at the north pole.

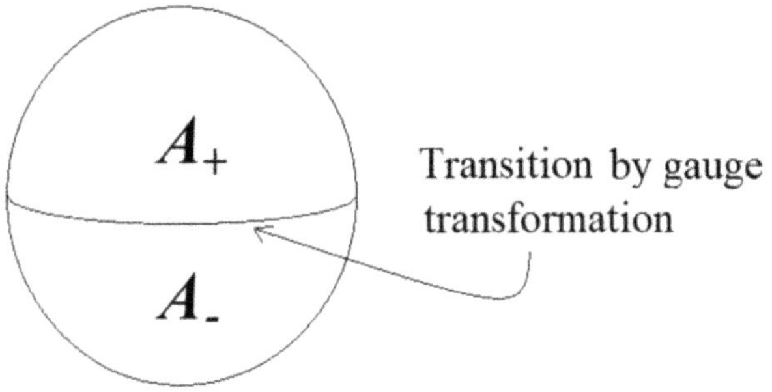

Figure 9.3. Combining A_+ in the northern hemisphere and A_- in the southern hemisphere using the gauge transformation (9.445) at the equator.

At first glance these singular fluxes would appear to render the Dirac potentials unphysical. However, there is a way to avoid this problem by moving the singularity. This begins by noting that the two potentials $\boldsymbol{A}_+$ and $\boldsymbol{A}_-$ are related by a gauge transformation,

$$\boldsymbol{A}_- = \boldsymbol{A}_+ - \frac{g\,\hat{\mathbf{e}}_\phi}{2\pi r \sin\theta} = \boldsymbol{A}_+ - \nabla\left(\frac{g\phi}{2\pi}\right), \tag{9.444}$$

so that the gauge function that relates the two potentials is given by

$$\Lambda = \frac{g\phi}{2\pi}. \tag{9.445}$$

The next step is cover the sphere around the magnetic monopole with two hemispheres, one for the northern hemisphere and one for the southern hemisphere, that are joined at the equator. The gauge potential $\boldsymbol{A}_+$ is used in the northern hemisphere and $\boldsymbol{A}_-$ in the southern hemisphere. Since the minimally coupled theory is gauge invariant the gauge transformation associated with (9.445) is performed at the equator and this transitions between the two potentials as depicted in figure 9.3. As a result, there is no Dirac string in either hemisphere and the net flux through the sphere is g. This is consistent with Stokes' theorem, since the boundaries of the two hemispheres at the equator yield line integrals in opposite directions, thereby canceling.

In order for this to be consistent, it is necessary that the field representing the magnetic charge is single valued at the equator. For the case of a spinor field ψ, the gauge transformation that transitions between the two potentials is given by (9.103), so that the solution for the magnetic matter field is transformed according to

$$\psi\left(R, \theta = \frac{\pi}{2}, \phi\right) = e^{ia\Lambda}\psi\left(R, \theta = \frac{\pi}{2}, \phi\right) = e^{iag\phi/2\pi}\psi\left(R, \theta = \frac{\pi}{2}, \phi\right). \tag{9.446}$$

Single valuedness gives the requirement that

$$\psi\left(R, \theta = \frac{\pi}{2}, \phi + 2\pi\right) = \psi\left(R, \theta = \frac{\pi}{2}, \phi\right) \implies e^{i\alpha g} = 1, \tag{9.447}$$

which has the solution

$$\alpha g = 2\pi n, \tag{9.448}$$

where n is an integer. Recalling that α is the dimensionless version of the fundamental unit of charge e, it follows that the definitions of (9.1) for ρ and $\mathbf{B}$ give the dimensional statement in SI units,

$$\alpha_d g_d = \left(\frac{e}{4\pi}\alpha\right)\left(\frac{Ke}{c}g\right) = (\alpha g)\left(\frac{Ke^2}{4\pi c}\right) = 2\pi n\hbar, \tag{9.449}$$

where the definition of $\hbar$ given by (9.90) was used. Result (9.449) is known as the *Dirac charge quantization condition.* The existence of magnetic monopoles, no matter how rare, requires that all electric charges must satisfy

$$\alpha_d = \frac{2\pi n}{g_d}\hbar, \tag{9.450}$$

so that all electric charges are integer multiples of the basic charge $2\pi\hbar/g_d$.

This method is an example of a general mathematical structure known as a *fiber bundle.* In this case the fiber bundle E is the local product of two *topological spaces,* referred to as the base space B and the fiber F. Loosely speaking, this means that the base space has a set of neighborhoods U_a that covers them and *locally* the fiber bundle can be written $B \times F$, where the product is understood in the same way that the product of two perpendicular line segments defines the two-dimensional Cartesian plane *locally*. In the case of the monopole, the base space B is the sphere S^2, the fiber F is the gauge potential A, and the set of neighborhoods $U_\pm$ consists of the two hemispheres. In addition, the bundle possesses a set of *invertible transition functions* f_{ab} that act on the fiber at the intersections of the covering $U_a \cap U_b$. In the case of the monopole the intersection $U_+ \cap U_-$ is the equator and the transition function on the fiber is the abelian gauge transformation. This very brief description does not do justice to the rich nature of fiber bundles nor their importance to gauge theory. The interested reader is directed to the references.

References and recommended further reading

Classical field theory monographs that cover electrodynamics include

- Landau L and Lifshitz E 1975 *The Classical Theory of Fields* 4th edn (Amsterdam: Elsevier)
- Kosyakov B 2007 *Introduction to the Classical Theory of Particles and Fields* (Berlin: Springer)
- Scheck F 2012 *Classical Field Theory* (Berlin: Springer)
- Nastase H 2019 *Classical Field Theory* (Cambridge: Cambridge University Press)

Monographs presenting the myriad applications of classical electrodynamics include

- Eyges W 1980 *The Classical Electromagnetic Field* (New York: Dover)
- Westgard J 1997 *Electrodynamics: A Concise Introduction* (Berlin: Springer)
- Greiner W 1998 *Classical Electrodynamics* (Berlin: Springer)
- Jackson J 1999 *Classical Electrodynamics* 3rd edn (Hoboken, NJ: Wiley)

The self-energy problem and classical attempts to solve it are presented in the excellent monograph

- Rohrlich F 2007 *Classical Charged Particles* 3rd edn (Singapore: World Scientific)

Classical electrodynamics is developed in terms of forms and exterior algebra in

- Darling R 1994 *Differential Forms and Connections* (Cambridge: Cambridge University Press)
- Scheck F 2012 *Classical Field Theory* (Berlin: Springer)

Monographs presenting the relationship of magnetic monopoles and fiber bundle theory include

- Naber G 1997 *Topology, Geometry, and Gauge Fields* (Berlin: Springer)
- Nakahara M 2003 *Geometry, Topology, and Physics* 2nd edn (New York: Taylor and Francis)

IOP Publishing

Classical Field Theory and the Stress-Energy Tensor
(Second Edition)

Mark S Swanson

Chapter 10

General relativity and gravitation

Special relativity consists of the set of Lorentz transformations that connects inertial frames, those moving with constant relative speed, in a manner that leaves the speed of light invariant. The universal laws of nature are required to take the same form in all such inertial frames. This is nothing more than Galileo's principle revised to respect the invariance of the speed of light. These laws must therefore be stated in terms of mathematical objects that transform as a representation of the Lorentz group, and these are the scalars, spinors, and vectors of earlier chapters. Result (6.64) showed that the Lorentz transformations are coordinate changes, $x'^{\mu} = x'^{\mu}(x)$, that leave the Minkowski metric tensor $\eta_{\mu\nu}$ *invariant*. This is also true of the Poincaré transformations defined by (8.232). This requires them to satisfy (6.90),

$$\frac{\partial^2 x'^{\mu}(x)}{\partial x^{\nu}\, \partial x^{\rho}} = 0. \tag{10.1}$$

Transformations to an accelerated frame of reference are not included in this group since

$$A^{\mu} = \frac{\mathrm{d}^2 x'^{\mu}}{\mathrm{d}\tau^2} \neq 0, \tag{10.2}$$

which violates this restriction. Accelerated frames are no longer inertial frames of reference. In particular, rotating frames are accompanied by the need to introduce *pseudoforces*, such as the Coriolis and centrifugal forces, in order to explain the motion observed from such a non-inertial frame.

It is an important observation that the laws of physics that are consistent with special relativity will not take *the same form* in an accelerated frame. The origin of

doi:10.1088/978-0-7503-3455-6ch10 © IOP Publishing Ltd 2022

this breakdown is the behavior of the partial derivative. As an example, for the gauge field A^μ, the Lorenz condition is given by

$$\partial_\mu A^\mu = 0. \tag{10.3}$$

In special relativity the Lorenz condition can be agreed upon by all inertial observers. However, this is true *only if consideration is limited to Lorentz transformations.* A transformation to a non-inertial frame gives

$$\begin{aligned}\frac{\partial A'^\mu(x')}{\partial x'^\mu} &= \frac{\partial x^\alpha}{\partial x'^\mu}\frac{\partial}{\partial x^\alpha}\left(\frac{\partial x'^\mu}{\partial x^\beta}A^\beta(x)\right)\\ &= \frac{\partial x^\alpha}{\partial x'^\mu}\frac{\partial x'^\mu}{\partial x^\beta}\frac{\partial A^\beta(x)}{\partial x^\alpha} + \frac{\partial x^\alpha}{\partial x'^\mu}\frac{\partial^2 x'^\mu}{\partial x^\alpha\,\partial x^\beta}A^\beta(x)\\ &= \frac{\partial A^\beta(x)}{\partial x^\beta} + \frac{\partial x^\alpha}{\partial x'^\mu}\frac{\partial^2 x'^\mu}{\partial x^\alpha\,\partial x^\beta}A^\beta(x),\end{aligned} \tag{10.4}$$

so that an additional term appears in the non-inertial frame.

Exercise 10.1: Show that a transformation to a non-inertial frame gives

$$\frac{\partial A'^\mu(x')}{\partial x'^\rho} = \frac{\partial x^\alpha}{\partial x'^\rho}\frac{\partial x'^\mu}{\partial x^\beta}\frac{\partial A^\beta(x)}{\partial x^\alpha} + \frac{\partial x^\alpha}{\partial x'^\rho}\frac{\partial^2 x'^\mu}{\partial x^\alpha\,\partial x^\beta}A^\beta(x). \tag{10.5}$$

The first term on the right-hand side of (10.5) is the usual second-rank mixed tensor Lorentz transformation rule, while the second term is new.

This is similar to the result obtained from Newtonian mechanics, where the non-Galilean transformation of (2.30) altered the nature of the force acting on an object. In that case a transformation to an appropriate accelerated frame was shown in (2.34) to cancel the effects of a local gravitational force *for all comoving inertial masses,* regardless of their mass, as long as the inertial mass and the gravitational mass are the same. This gave rise to the concept of a *freely falling frame of reference*, where the measurable effects of gravity *are absent*. Another way of saying this is that a freely falling object will follow a trajectory independent of its inertial mass. This has come to be known as the *weak principle of equivalence,* since it assumes that inertial mass and gravitational mass are the same.

Similarly, an accelerated frame of reference can give rise to effects identical to a local gravitational force. Einstein argued that a person in a closed elevator cannot tell if objects are falling in the elevator because of the elevator's acceleration or because of a nearby gravitating mass. Ignoring some subtleties, the *Einstein principle of equivalence* states that an accelerated frame of reference in the absence of *all forces* and the motion of an object in a gravitational field cannot be distinguished from each other by performing *local* experiments. This principle led Einstein to argue that including gravitation in field theories requires the equations of motion to remain *form invariant* under general coordinate transformations, which are referred to as *diffeomorphisms*. Loosely speaking, a diffeomorphism is a change of coordinates that is differentiable, single valued, and invertible, but otherwise *arbitrary*. Only a theory allowing this larger group of transformations would have the ability to transform away the effects of a gravitational field, at least locally, by an appropriate choice of accelerated frame.

Finally, the *strong principle of equivalence* includes both the weak and Einstein principles of equivalence. It includes the postulate that the gravitational motion of a small test body depends only on its initial position in space-time and its velocity, and not on its constitution, as well as the postulate that the outcome of any local experiment in a freely falling frame of reference is independent of the velocity of the laboratory and its location in space-time. These additions become significant in astronomical bodies such as neutron stars, where there is a significant level of binding energy due to gravitational attraction. Using mass and energy equivalence, such gravitational binding energy could affect the nature of gravitational interaction.

A theory including gravitation must also have a well-defined set of quantities that reflect the effects of gravitation on matter and energy in a manner that is also consistent with the invariance of the finite speed of light. The non-relativistic limit of these effects must be consistent with Newtonian gravitation. Finally, such a theory must reduce to the case of special relativity in the absence of a gravitational field or in a freely falling frame. A gravitational field theory must quantify the relationship between gravitation and general frames of reference, and so this version of gravitation has come to be known as *general relativity*. General relativity is an extremely rich subject, and the brief discussion presented in what follows can only sketch its foundations and some of its more famous ramifications.

10.1 The metric tensor and Einstein's principle of equivalence

It is not *a priori* obvious how to construct a theory that is invariant under diffeomorphisms. However, if the Minkowski metric $\eta_{\mu\nu}$ is assumed to transform as a second-rank tensor under diffeomorphisms, then the proper time element of special relativity remains invariant,

$$\begin{aligned} \eta'_{\mu\nu}(x')\,\mathrm{d}x'^{\mu}\,\mathrm{d}x'^{\nu} &= \frac{\partial x^{\alpha}}{\partial x'^{\mu}}\frac{\partial x^{\beta}}{\partial x'^{\mu}}\eta_{\alpha\beta}\frac{\partial x'^{\mu}}{\partial x^{\gamma}}\frac{\partial x'^{\nu}}{\partial x^{\rho}}\mathrm{d}x^{\gamma}\mathrm{d}x^{\rho} \\ &= \delta^{\alpha}{}_{\gamma}\delta^{\beta}{}_{\rho}\eta_{\alpha\beta}\,\mathrm{d}x^{\gamma}\mathrm{d}x^{\rho} = \eta_{\alpha\beta}\,\mathrm{d}x^{\alpha}\mathrm{d}x^{\beta}. \end{aligned} \tag{10.6}$$

As a result, proper time can still be used, *even when forces are present and the motion is no longer inertial*, as long as the Minkowski metric tensor is transformed using the associated diffeomorphism. In special relativity proper time allowed the analysis of accelerated motion using the definition of relativistic acceleration (6.145). However, in a non-inertial frame, the metric tensor will become dependent on space-time. Einstein's principle of equivalence then states that an observer in a gravitational field will see the same effect, i.e. a metric tensor that depends on space-time. The presence of gravitation is therefore represented by elevating the Minkowski metric $\eta_{\mu\nu}$ to a dynamical quantity, a second-rank symmetric invertible tensor, denoted by $g_{\mu\nu}(x)$, that is space-time dependent. In analogy with the metric tensor first introduced in (1.64), a space-time-dependent metric tensor corresponds to a non-Euclidean space-time manifold when gravitation is present.

This step has several immediate and important ramifications. The space-time-dependent metric $g_{\mu\nu}(x)$ is assumed to be invertible at every point and, like the Minkowski metric, its inverse is designated as $g^{\mu\nu}(x)$, so that

$$g^{\mu\nu}(x)\, g_{\nu\rho}(x) = \delta^{\mu}{}_{\rho}. \tag{10.7}$$

Since g is a symmetric 4×4 matrix, the existence of the inverse in (10.7) requires that

$$\det g < 0. \tag{10.8}$$

The negative sign is chosen for consistency with the *signature* of the Minkowski metric, $\det \eta = -1$. As with $\eta_{\mu\nu}$, the metric tensor still serves to map contravariant vectors into covariant vectors and vice versa, so that

$$\begin{aligned} dx_\mu = g_{\mu\nu}(x)\, dx^\nu &\implies dx^\mu = g^{\mu\nu}(x)\, dx_\nu, \\ \partial^\mu = g^{\mu\nu}(x)\, \partial_\nu &\implies \partial_\mu = g_{\mu\nu}(x)\, \partial^\nu. \end{aligned} \tag{10.9}$$

It is important to note that the statements of (10.9) are *local*, since that is where dx^μ, ∂_μ, and $g_{\mu\nu}(x)$ are defined. Because the manifold is now being treated as non-Euclidean, there are no well-defined position vectors of finite length. This is reflected in the element of proper time or length, now given by

$$c^2 d\tau^2 = ds^2 = g_{\mu\nu}(x)\, dx^\mu\, dx^\nu, \tag{10.10}$$

which has become a position-dependent quantity. It follows that identifying $g_{\mu\nu}$ with gravitation therefore has the result that gravitation affects proper time and contributes to length contraction and time dilation. The dynamic metric has introduced non-Euclidean geometry to space-time in order to model gravitation.

Riemann and Gauss showed that the metric tensor is the key to defining *differential geometry* for curved manifolds, where it generalizes the Pythagorean theorem of flat space. In special relativity the spatial part of the Minkowski metric yields, up to the minus sign, a distance identical to the Pythagorean theorem of Euclidean geometry,

$$dx_j\, dx^j = \eta_{jk} dx^j dx^k = -d\mathbf{x} \cdot d\mathbf{x}. \tag{10.11}$$

Since Euclidean geometry is defined for a flat surface, the Minkowski metric is commonly referred to as the metric of *flat space-time*. Riemannian geometry generalizes geometry to manifolds *with curvature*, to be defined in a later section, by using a position-dependent metric tensor $g_{\mu\nu}(x)$. The infinitesimal length element (10.10) generalizes the Pythagorean theorem to a manifold with curvature.

This allows general relativity to represent the physics of gravitation in terms of a deviation from the flat Minkowski metric, so that space-time itself becomes *dynamic*. It will be seen that the gravitational motion of point masses can be determined solely from the metric tensor in a Riemannian manifold. Since *all point masses* experience the *same metric tensor*, they will move identically in a gravitational field, just as they do in Newtonian gravity. While all other fields discussed so far are functions of space-time, the metric field $g_{\mu\nu}$ determines the properties of space-time itself.

Choosing $g_{\mu\nu}(x)$ as the field associated with gravitation allows a mathematical statement of Einstein's principle of equivalence. For an arbitrary metric tensor it will now be shown that it is always possible, *for a given point* x_o, to find a change of coordinates that transforms the metric tensor into the Minkowski metric $\eta_{\mu\nu}$ at that point in such a way that

$$\lim_{x\to x_o} \partial_\rho g_{\mu\nu}(x) = 0, \qquad \lim_{x\to x_o} g_{\mu\nu}(x) = \eta_{\mu\nu}. \tag{10.12}$$

In the neighborhood of x_o the metric tensor reduces to the Minkowski metric, so that the rules of special relativity hold in this reference frame. In Einstein's principle of equivalence this neighborhood therefore corresponds to a *local* inertial frame where the gravitational force is absent, identifying it as the freely falling or *local inertial frame* where the effects of gravitation have disappeared. At first glance, this seems a unique association of gravitation with the deviation of the metric tensor from the Minkowski metric. In the next section it will be seen that there are subtleties involved in this association. For the moment, (10.12) will be demonstrated and used in later sections.

Demonstrating the existence of the freely falling frame amounts to finding the local change of coordinates, $x^\mu = x^\mu(\xi)$, such that the metric tensor becomes the Minkowski metric in the neighborhood of x_o. The metric tensor in the freely falling ξ coordinate system is given by the tensor transformation rule,

$$g'_{ab}(\xi) = \frac{\partial x^\mu}{\partial \xi^a}\frac{\partial x^\nu}{\partial \xi^b} g_{\mu\nu}(x) = e^\mu{}_a(\xi)\, g_{\mu\nu}(x(\xi)) e^\nu{}_b(\xi), \tag{10.13}$$

where the *transformation matrix* is denoted by

$$e^\mu{}_a = \frac{\partial x^\mu}{\partial \xi^a}. \tag{10.14}$$

In what follows the Latin indices a, b, ... refer to the coordinates of the freely falling frame, while the Greek indices refer to an arbitrary coordinate system.

Exercise 10.2: Show that the symmetry of the metric, $g_{\mu\nu}(x) = g_{\nu\mu}(x)$, gives

$$g'_{ab}(\xi) = g'_{ba}(\xi). \tag{10.15}$$

The Latin characters a and b represent the indices of the metric in the new frame, which will be chosen so that the point $x_o = x_o(\xi_o)$ results in the Minkowski metric η *locally* at ξ_o. It is important to note that both the transformation matrix $e^\mu{}_a(\xi)$ and $g_{\mu\nu}(x(\xi))$ are dependent upon the new coordinates ξ in the presence of gravitation. As a result, both $g_{\mu\nu}$ and $e^\mu{}_a$ can be written in a Taylor series expansion around ξ_o. This allows the original metric tensor to be written in the neighborhood of ξ_o to $O(\xi)$ as

$$\begin{aligned} g_{\mu\nu}(x(\xi)) = g_{\mu\nu}(\xi) &= g_{\mu\nu}(\xi_o) + (\xi^b - \xi_o^b)\frac{\partial g_{\mu\nu}(\xi)}{\partial \xi^b}\bigg|_{\xi = \xi_o} \\ &\equiv g_{\mu\nu}(\xi_o) + (\xi^b - \xi_o^b)\,\partial_b g_{\mu\nu}(\xi_o). \end{aligned} \tag{10.16}$$

Similarly, the transformation matrix can be written to $O(\xi)$,

$$\begin{aligned} e^\mu{}_a(\xi) &= e^\mu{}_a(\xi_o) + (\xi^b - \xi_o^b)\frac{\partial e^\mu{}_a(\xi)}{\partial \xi^b}\bigg|_{\xi=\xi_o} \\ &\equiv e^\mu{}_a(\xi_o) + (\xi^b - \xi_o^b)\,\partial_b e^\mu{}_a(\xi_o). \end{aligned} \tag{10.17}$$

It is assumed that $g_{\mu\nu}(x)$ is known in the neighborhood of x_o, and this allows *all* the coefficients $\partial_b g_{\mu\nu}(\xi_o)$ of the metric tensor expansion (10.16) to be determined as functions of ξ_o. Because the transformation to the freely falling frame is unknown, the terms $e^\mu{}_a(\xi_o)$ and the coefficients $\partial_b e^\mu{}_a(\xi_o)$ of the transformation matrix expansion (10.17) are currently unknown functions of ξ_o. However, because the transformation is a diffeomorphism, the coefficients $\partial_b e^\mu{}_a(\xi_o)$ obey the following property:

$$\begin{aligned} \partial_b e^\mu{}_a(\xi_o) &= \frac{\partial e^\mu{}_a(\xi)}{\partial \xi^b}\bigg|_{\xi=\xi_o} = \frac{\partial^2 x^\mu(\xi)}{\partial \xi^b\,\partial \xi^a}\bigg|_{\xi = \xi_o} \\ &= \frac{\partial^2 x^\mu(\xi)}{\partial \xi^a\,\partial \xi^b}\bigg|_{\xi=\xi_o} = \frac{\partial e^\mu{}_b(\xi)}{\partial \xi^a}\bigg|_{\xi=\xi_o} = \partial_a e^\mu{}_b(\xi_o). \end{aligned} \tag{10.18}$$

Returning the two expressions (10.16) and (10.17) to (10.13) and retaining only terms to $O(\xi)$ gives the transformed metric to $O(\xi)$,

$$\begin{aligned} g'_{ab}(\xi) = e^\mu{}_a(\xi_o)g_{\mu\nu}(\xi_o)e^\nu{}_b(\xi_o) &+ \big(\partial_c e^\mu{}_a(\xi_o)g_{\mu\nu}(\xi_o)e^\nu{}_b(\xi_o) \\ &+ e^\mu{}_a(\xi_o)\partial_c g_{\mu\nu}(\xi_o)e^\nu{}_b(\xi_o) + e^\mu{}_a(\xi_o)g_{\mu\nu}(\xi_o)\partial_c e^\nu{}_b(\xi_o)\big)(\xi^c - \xi_o^c), \end{aligned} \tag{10.19}$$

where *all* functions and their derivatives on the right-hand side are evaluated at ξ_o. Relation (10.19) can now be solved for the transformation matrix functions and their derivatives, $e^\mu{}_b(\xi_o)$ and $\partial_c e^\mu{}_b(\xi_o)$, which will allow the principle of equivalence (10.12) to hold to $O(\xi)$.

The zeroth-order term in (10.19) can be set to η_{ab} in the following way. The metric tensor $g_{\mu\nu}(\xi_o)$ is a 4×4 symmetric matrix with a nonzero determinant, which allows $e^\mu{}_a(\xi_o)$ to be identified with the four contravariant eigenvectors of this matrix. This yields the eigenvalue equation

$$g_{\mu\nu}(\xi_o)e^{\nu}{}_{a}(\xi_o) = \lambda^{(a)}(\xi_o)e_{\mu a}(\xi_o). \tag{10.20}$$

It is important to note that the eigenvector $e_{\mu a}(\xi_o)$ transforms as a covariant second-rank tensor consistent with the action of the metric tensor in raising and lowering indices. The index a labels the four eigenvectors and their eigenvalues $\lambda^{(a)}$, so that there is no sum over (a) in (10.20). Because $\det g_{\mu\nu} < 0$, all four eigenvalues $\lambda^{(a)}$ will be nonzero. Since $g_{\mu\nu} = g_{\nu\mu}$ these eigenvectors will also satisfy

$$e^{\mu}{}_{a}(\xi_o)g_{\mu\nu}(\xi_o) = g_{\nu\mu}(\xi_o)e^{\mu}{}_{a}(\xi_o) = \lambda^{(a)}(\xi_o)e_{\nu a}(\xi_o). \tag{10.21}$$

These eigenvectors will be orthogonal, since the two eigenvalue equations (10.20) and (10.21) give

$$\lambda^{(a)}e_{\nu a}e^{\nu}{}_{b} = e^{\mu}{}_{a}g_{\mu\nu}e^{\nu}{}_{b} = \lambda^{(b)}e^{\mu}{}_{a}e_{\mu b} \Longrightarrow (\lambda^{(a)} - \lambda^{(b)})e_{\nu a}e^{\nu}{}_{b} = 0. \tag{10.22}$$

For the case in which the four eigenvalues are distinct, the four eigenvectors will be orthogonal. The eigenvectors can always be scaled by an arbitrary constant and still satisfy (10.20) and (10.21), and this allows the orthogonality condition to be written

$$e_{\nu a}(\xi_o)e^{\nu}{}_{b}(\xi_o) = c^{(a)}(\xi_o)\eta_{ab}, \tag{10.23}$$

where $c^{(a)}(\xi_o)$ is an arbitrary real factor. With this choice, the zeroth-order contribution to the transformed metric becomes

$$\begin{aligned} e^{\mu}{}_{a}(\xi_o)g_{\mu\nu}(\xi_o)e^{\nu}{}_{b}(\xi_o) &= \lambda^{(a)}(\xi_o)e^{\mu}{}_{a}(\xi_o)e^{\nu}{}_{b}(\xi_o) \\ &= \lambda^{(a)}(\xi_o)c^{(a)}(\xi_o)\eta_{ab} = \eta_{ab}, \end{aligned} \tag{10.24}$$

where $c^{(a)}(\xi_o) = 1/\lambda^{(a)}(\xi_o)$ was chosen. This yields the value of the transformation matrix $e^{\mu}{}_{a}(\xi_o)$ to the Minkowski metric at the point $x_o = x(\xi_o)$. This can be done at any chosen point in the manifold. However, it should be clear that the eigenvectors at one point will not necessarily be the eigenvectors at another point. In addition, more work is required to set the first-order term in (10.19) to zero.

The eigenvectors can be viewed as four contravariant vectors $\mathbf{e}_a$ labeled by the index a, so that $\mathbf{e}_a$ has the components $e^{\mu}{}_{a}$. These four vectors are referred to as the *tetrad basis*, and they have the zeroth-order property that at *every point* x it is possible to find a tetrad such that

$$g(\mathbf{e}_a, \mathbf{e}_b) = g_{\mu\nu}(x)\, e^{\mu}{}_{a}(x)\, e^{\nu}{}_{b}(x) = \eta_{ab}. \tag{10.25}$$

Before proceeding to remove the first-order term from (10.19), it is useful to note that the zeroth-order result (10.25) is always possible since $e^{\mu}{}_{a}(x)$ corresponds to 16 arbitrary parameters, while $g_{\mu\nu}(x)$ is a symmetric matrix and therefore corresponds to only ten free parameters. There are therefore $16 - 10 = 6$ free parameters left in the definition of $e^{\mu}{}_{a}(x)$. In order to understand the meaning of these six free parameters, four covariant four-vectors $\mathbf{e}^a(x)$ at the point x are defined using the four contravariant eigenvectors $\mathbf{e}_a(x)$, and their components are explicitly given by

$$e_{\alpha}{}^{a}(x) = g_{\alpha\beta}(x)\, e^{\beta}{}_{b}(x)\, \eta^{ba}. \tag{10.26}$$

Combining (10.26) with (10.21) and (10.23) gives the inner product

$$\begin{aligned}\mathbf{e}^a(x)\cdot\mathbf{e}_b(x) &= e_\alpha{}^a(x)\,e^\alpha{}_b(x) = g_{\alpha\beta}(x)\,e^\beta{}_c(x)\eta^{ca}e^\alpha{}_b(x)\\ &= \eta^{ca}e^\beta{}_c(x)e_{\beta b}(x)\lambda^{(b)}(x) = c^{(b)}(x)\lambda^{(b)}(x)\eta_{cb}\,\eta^{ca} = \eta_{cb}\eta^{ca}\\ &= \delta^a{}_b.\end{aligned} \tag{10.27}$$

The orthogonality property of (10.27) gives

$$\begin{aligned}(e_\alpha{}^a(x)\,e^\beta{}_a(x))e_\beta{}^b(x) &= e_\alpha{}^a(x)\big(e^\beta{}_a(x)e_\beta{}^b(x)\big)\\ &= e_\alpha{}^a(x)\,\delta^b{}_a = e_\alpha{}^b(x) = (\delta^\beta{}_\alpha)e_\beta{}^b(x),\end{aligned} \tag{10.28}$$

so that

$$e^\beta{}_a(x)\,e_\alpha{}^a(x) = \delta^\beta{}_\alpha. \tag{10.29}$$

Using these orthogonality relations allows the metric tensor relation (10.13) to be inverted, giving the very useful result

$$\begin{aligned}e_\mu{}^a(x)\,\eta_{ab}\,e_\nu{}^b(x) &= e_\mu{}^a(x)\,e^\rho{}_a(x)\,g_{\rho\sigma}(x)\,e^\sigma{}_b(x)\,e_\nu{}^b(x)\\ &= \delta^\rho{}_\mu\,g_{\rho\sigma}(x)\,\delta^\sigma{}_\nu = g_{\mu\nu}(x).\end{aligned} \tag{10.30}$$

This is identical to the earlier result (1.65) extended to a non-Euclidean space-time manifold and can be written

$$\mathbf{e}_\mu(x)\cdot\mathbf{e}_\nu(x) = g_{\mu\nu}(x), \tag{10.31}$$

where the inner product is given by the Minkowski metric acting on the Latin indices. Like the direction vectors of (1.30) the tetrads or *vierbeins* $\mathbf{e}^a(x)$ and $\mathbf{e}_a(x)$ provide a set of basis vectors at the space-time point x that serve as the four basis or direction vectors of the *freely falling frame*. These are the general relativistic version of the space-time basis vectors introduced in (6.58). From (6.64) the Minkowski metric appearing in (10.30) is invariant under a *local Lorentz transformation* $\boldsymbol{\Lambda}(x)$, one whose parameters $\lambda_{\mu\nu}$ are *space-time dependent*. The full transformation of the tetrad is given by

$$e'^{\,a}_\nu(x) = \frac{\partial x^\sigma}{\partial x'^\nu}\Lambda^a{}_c(x)e_\sigma{}^c(x). \tag{10.32}$$

Substituting this into (10.30) gives

$$\begin{aligned}e'^{\,a}_\mu(x')\,\eta_{ab}\,e'^{\,b}_\nu(x') &= \frac{\partial x^\rho}{\partial x'^\mu}\frac{\partial x^\sigma}{\partial x'^\nu}e_\sigma{}^c(x)\\ &\quad \Lambda^a{}_c(x)\,\eta_{ab}\,\Lambda^b{}_d(x)\,e_\rho{}^d(x)e_\mu{}^c(x)\,\eta_{cd}\,e_\nu{}^d(x)\\ &= \frac{\partial x^\rho}{\partial x'^\mu}\frac{\partial x^\sigma}{\partial x'^\nu}e_\rho{}^c(x)\,\eta_{cd}\,e_\sigma{}^d(x)\\ &= \frac{\partial x^\rho}{\partial x'^\mu}\frac{\partial x^\sigma}{\partial x'^\nu}g_{\rho\sigma}(x) = g'_{\mu\nu}(x').\end{aligned} \tag{10.33}$$

If the diffeomorphism is trivial, so that

$$\frac{\partial x^\rho}{\partial x'^\mu}\frac{\partial x^\sigma}{\partial x'^\nu} = \delta^\rho{}_\mu \delta^\sigma{}_\nu, \tag{10.34}$$

then $g_{\mu\nu}(x)$ is unchanged by the *local Lorentz transformation* of the tetrad indices. The tetrad basis vectors can therefore be transformed under the Lorentz group *differently at each point* without affecting the metric tensor $g_{\mu\nu}$. This is consistent with the Lorentz group having six free parameters, and these correspond to the six additional free parameters in the $e^\mu{}_a(\xi_o)$ transformation matrix used in (10.19).

Setting the first-order term in (10.19) to zero is critical to implementing the principle of equivalence, since it establishes a neighborhood in the freely falling frame where the effects of gravitation have vanished. It is more complicated since the expansion coefficient $\partial_c g_{\mu\nu}(\xi_o)$ corresponds to $4 \times 10 = 40$ degrees of freedom for a general metric. The first-order coefficient will vanish if $\partial_c e^\mu{}_a(\xi_o)$ is chosen to satisfy

$$\partial_c e^\mu{}_a(\xi_o) g_{\mu\nu}(\xi_o) e^\nu{}_b(\xi_o) = -\frac{1}{2} e^\mu{}_a(\xi_o) \partial_c g_{\mu\nu}(\xi_o) e^\nu{}_b(\xi_o). \tag{10.35}$$

This follows by substituting (10.35) into the first-order coefficient in (10.19) to obtain a second requirement,

$$\begin{aligned} &\partial_c e^\mu{}_a(\xi_o)\, g_{\mu\nu}(\xi_o)\, e^\nu{}_b(\xi_o) + e^\mu{}_a(\xi_o)\, \partial_c g_{\mu\nu}(\xi_o)\, e^\nu{}_b(\xi_o) \\ &+ e^\mu{}_a(\xi_o)\, g_{\mu\nu}(\xi_o) \partial_c\, e^\nu{}_b(\xi_o) \\ &= \frac{1}{2} e^\mu{}_a(\xi_o)\, \partial_c g_{\mu\nu}(\xi_o)\, e^\nu{}_b(\xi_o) + e^\mu{}_a(\xi_o)\, g_{\mu\nu}(\xi_o) \partial_c\, e^\nu{}_b(\xi_o) = 0, \end{aligned} \tag{10.36}$$

which requires

$$e^\mu{}_a(\xi_o)\, g_{\mu\nu}(\xi_o) \partial_c\, e^\nu{}_b(\xi_o) = -\frac{1}{2} e^\mu{}_a(\xi_o)\, \partial_c g_{\mu\nu}(\xi_o)\, e^\nu{}_b(\xi_o). \tag{10.37}$$

Exercise 10.3: Show that second requirement (10.37) is identical to (10.35) by exchanging $a \leftrightarrow b$ and $\mu \leftrightarrow \nu$ and using the symmetry of $g_{\mu\nu}$.

The requirement (10.35) is readily solved for $\partial_c e^\nu{}_a(\xi_o)$ by contracting both sides against $e^b{}_\rho(\xi_o) g^{\rho\sigma}(\xi_o)$ and using (10.7) and (10.29). The result is

$$\left.\frac{\partial}{\partial \xi^c} e^\sigma{}_a(\xi)\right|_{\xi=\xi_o} = -\frac{1}{2} e^\mu{}_a(\xi_o) \left.\frac{\partial}{\partial \xi^c} g_{\mu\nu}(\xi)\right|_{\xi=\xi_o} g^{\nu\sigma}(\xi_o). \tag{10.38}$$

Because all the factors on the right-hand side of (10.38) are known, the value of the derivatives of $e^\nu{}_a$ at ξ_o are now known. These derivatives are such that the Taylor series expansion (10.19) for g'_{ab} becomes η_{ab} to $O(\xi)$. The nonzero derivatives of $e^\nu{}_a$ correspond to nonzero second derivatives of the coordinate transformation, which distinguishes the transformation from a Lorentz transformation. As a result, the frame satisfying the Einstein principle of equivalence is an accelerated freely falling frame.

Removing the 40 degrees of freedom in $\partial_c g_{\mu\nu}$ is possible, since (10.18) shows that $\partial_c e^\mu{}_b$ can be viewed as four matrices indexed by μ that are symmetric in b and c. This provides $4 \times 10 = 40$ adjustable parameters that can be used to remove the forty degrees of freedom in the derivatives of the metric. As a result, the principle of equivalence can always be satisfied to $O(\xi)$. However, the quadratic terms in $(\xi - \xi_o)$ in the Taylor expansion of the metric *cannot be removed.* This starts by noting that $\partial_c\partial_d g_{\mu\nu}$ is both symmetric in c and d as well as μ and ν, giving $10 \times 10 = 100$ degrees of freedom. The second-order expansion coefficient $\partial_c\partial_b e^\mu{}_a$ is symmetric under interchanges of a, b, and c, since they represent partial derivatives. The formula for the number N of distinct partial derivatives of order n for a function of k variables is

$$N = \frac{(k + n - 1)!}{(k - 1)!\, n!}. \tag{10.39}$$

For the case that $k = 4$ this formula gives 20 distinct partial derivatives for $n = 3$, as well as the previous results of four and ten for $n = 1$ and $n = 2$, respectively. Therefore, there are only $4 \times 20 = 80$ free parameters in the second order of the Taylor expansion of $e^\mu{}_a$. As a result, quadratic and higher terms in $\xi^a - \xi^a_o$ will remain present in a general transformed metric, and there will be $100 - 80 = 20$ distinct parameters that quantify gravitation or curvature in a general manifold.

Contracting the definition (10.26) for $e_\mu{}^a(x)$ against $g^{\nu\mu}(x)$ and η_{ab} and using the definition (10.26) gives the local tetrad statement,

$$\begin{aligned} g^{\nu\mu}(x)\, e_\mu{}^a(x)\, \eta_{ab} &= g^{\nu\mu}(x)\, g_{\mu\beta}(x)\, e^\beta{}_c(x)\, \eta^{ca}\, \eta_{ab} \\ &= \delta^\nu{}_\beta\, e^\beta{}_c(x)\, \delta^c{}_b = e^\nu{}_b(x). \end{aligned} \tag{10.40}$$

Result (10.40) gives

$$g^{\nu\mu}(x)\, e_\mu{}^a(x)\, \eta_{ab} = e^\nu{}_b(x), \tag{10.41}$$

which shows that the Greek indices of the tetrad are raised and lowered using $g_{\mu\nu}$, while the Latin indices of the tetrad are raised and lowered using η_{ab}. Similarly, the Greek indices of the tetrad transform using a general coordinate diffeomorphism, while the Latin indices of the tetrad transform under the local Lorentz group. These are summarized for future reference,

$$\begin{aligned} e'_{\mu a} &= \frac{\partial x^\sigma}{\partial x'^\mu}\, \Lambda_a{}^c(x)\, e_{\sigma c}, \\ e'^{\rho a} &= \frac{\partial x'^\rho}{\partial x^\nu}\, \Lambda^a{}_b(x)\, e^{\nu b}, \end{aligned} \tag{10.42}$$

where the local Lorentz transformations of the tetrad indices have the property that

$$\Lambda^a{}_b(x)\, \Lambda_a{}^c(x) = \delta^c{}_b. \tag{10.43}$$

This completes the demonstration that the tetrad vectors give a local freely falling basis where the Minkowski metric of special relativity is used. It is useful to summarize the properties of the tetrads for future reference,

$$\begin{aligned} g_{\mu\nu}(x) &= \eta_{ab}e^{a}{}_{\mu}(x)e^{b}{}_{\nu}(x), \\ \eta_{ab} &= g^{\mu\nu}(x)e^{a}{}_{\mu}(x)e^{b}{}_{\nu}(x), \\ e_{\mu}{}^{a}(x)e^{\mu}{}_{b}(x) &= \delta^{a}{}_{b}, \\ e_{\mu}{}^{a}(x)e^{\nu}{}_{a}(x) &= \delta^{\nu}{}_{\mu}, \\ \partial_{c}e^{\mu}{}_{a}(x) &= -\frac{1}{2}e^{\sigma}{}_{a}(x)\partial_{c}g_{\sigma\nu}(x)g^{\nu\mu}(x); \end{aligned} \tag{10.44}$$

these hold at every point in the manifolds considered in the remainder of this book. Numerous additional properties will be derived as non-Euclidean geometry is developed. The last requirement can be rewritten using

$$\partial_{\rho} = e_{\rho}{}^{c}\partial_{c}, \tag{10.45}$$

which gives

$$\partial_{\rho}e^{\mu}{}_{a}(x) = -\frac{1}{2}e^{\sigma}{}_{a}(x)\partial_{\rho}g_{\sigma\nu}(x)g^{\nu\mu}(x), \tag{10.46}$$

which can be easier to verify.

A simple but illustrative example comes from considering the relationship of curvilinear coordinates to Cartesian coordinates in a Euclidean two-dimensional manifold. Using Cartesian coordinates $x^1 = x$ and $x^2 = y$, the metric tensor is given by examining the Pythagorean formula for the infinitesimal distance squared,

$$(\mathrm{d}s)^2 = \eta_{ab}\,\mathrm{d}x^a\mathrm{d}x^b = (\mathrm{d}x)^2 + (\mathrm{d}y)^2, \tag{10.47}$$

which gives the Euclidean metric

$$\eta = \begin{pmatrix} 1 & 0 \\ 0 & 1 \end{pmatrix}. \tag{10.48}$$

However, in polar coordinates $x^1 = r$ and $x^2 = \theta$, the infinitesimal distance squared is given by

$$(\mathrm{d}s)^2 = g_{\mu\nu}(x)\,\mathrm{d}x^\mu\mathrm{d}x^\nu = (\mathrm{d}r)^2 + r^2(\mathrm{d}\theta)^2, \tag{10.49}$$

which gives the polar coordinate metric

$$g(r) = \begin{pmatrix} 1 & 0 \\ 0 & r^2 \end{pmatrix}. \tag{10.50}$$

The zweibein representation of this is given by

$$e_r{}^x = \cos\theta, \quad e_r{}^y = \sin\theta, \quad e_\theta{}^x = -r\sin\theta, \quad e_\theta{}^y = r\cos\theta, \tag{10.51}$$

while the inverse zweibeins are given by

$$e^r{}_x = \cos\theta, \quad e^r{}_y = \sin\theta, \quad e^\theta{}_x = -\frac{1}{r}\sin\theta, \quad e^\theta{}_y = \frac{1}{r}\cos\theta. \tag{10.52}$$

Verifying that the identities of (10.44) hold is left as exercise. For example, the variant requirement (10.46) gives

$$\partial_r e^\theta{}_x = \partial_r\left(-\frac{1}{r}\sin\theta\right) = \frac{1}{r^2}\sin\theta, \tag{10.53}$$

while using the diagonal nature of the metric shows that the right-hand side of (10.46) gives

$$\begin{aligned} -\frac{1}{2}e^\sigma{}_x\partial_r g_{\sigma\nu}g^{\nu\theta} &= -\frac{1}{2}e^\theta{}_x\partial_r g_{22}(r)\, g^{22}(r) \\ &= -\frac{1}{2}\left(-\frac{1}{r}\sin\theta\right)\partial_r(r^2)\left(\frac{1}{r^2}\right) \\ &= \frac{1}{r^2}\sin\theta, \end{aligned} \tag{10.54}$$

which is consistent with (10.46).

> **Exercise 10.4:** Verify that the properties (10.44) hold for the zweibeins of (10.51) and (10.52).

The utility of the tetrad formalism is revealed by projecting an arbitrary vector at the point x onto the tetrad basis vectors at that location. The tetrad components of the vector A^a are defined as

$$A^a(x) = A \cdot \mathbf{e}^a = A^\mu(x)\, e_\mu{}^a(x), \tag{10.55}$$

where the covariant version is given by

$$A_a(x) = \eta_{ab}A^b(x). \tag{10.56}$$

It follows that

$$\begin{aligned} A_a(x)\, A^a(x) &= \eta_{ab}A^a(x)\, A^b(x) = \eta_{ab}\, e_\mu{}^a(x)\, e_\nu{}^b(x)\, A^\mu(x)\, A^\nu(x) \\ &= g_{\mu\nu}(x)\, A^\mu(x)\, A^\nu(x) = A_\mu(x)\, A^\mu(x), \end{aligned} \tag{10.57}$$

which shows that the inner product of a vector in the freely falling frame is consistent with its inner product in the presence of the metric $g_{\mu\nu}$.

In the previous two-dimensional Euclidean example, a vector in Cartesian components had the magnitude

$$A^2 = \eta_{ab}A^aA^b = (A^x)^2 + (A^y)^2, \tag{10.58}$$

while in polar coordinates it had the magnitude

$$A^2 = g_{\mu\nu}A^\mu A^\nu = (A^r)^2 + r^2(A^\theta)^2. \tag{10.59}$$

If a vector at (x, y) has the magnitude A and is directed at the angle α with respect to the x-axis, then its Cartesian components are given by

$$A^x = A\cos\alpha, \quad A^y = A\sin\alpha. \tag{10.60}$$

The vector components in polar coordinates can be found using the inverse zweibeins of (10.52),

$$\begin{aligned} A^r &= A^x e^r{}_x + A^y e^r{}_y = \cos\theta A^x + \sin\theta A^y \\ &= A\cos\alpha\cos\theta + A\sin\alpha\sin\theta = A\cos(\alpha - \theta), \\ A^\theta &= A^x e^\theta{}_x + A^y e^\theta{}_y = -A\cos\alpha\frac{\sin\theta}{r} + A\sin\alpha\frac{\cos\theta}{r} \\ &= \frac{1}{r}A\sin(\alpha - \theta). \end{aligned} \tag{10.61}$$

These components are consistent with the projections of the vector onto the polar coordinate unit vectors, as depicted in figure 10.1. Using (10.59) the results of (10.61) yield

$$A^2 = (A^r)^2 + r^2(A^\theta)^2 = A^2(\cos^2(\alpha - \theta) + \sin^2(\alpha - \theta)) = A^2, \tag{10.62}$$

so that it is identical to the Cartesian result.

The Cartan notation (6.62) for vectors and tensors in special relativity can be adapted to general relativity, so that in space-time component notation, the vector A is written

$$A(x) = A^\mu(x)\,\hat{\partial}_\mu = A^\mu(x)\,\hat{\mathbf{e}}_\mu(x), \tag{10.63}$$

where $\hat{\mathbf{e}}_\mu(x)$ are direction vectors which transform identically to the partial derivative ∂_μ. Expression (10.63) is manifestly invariant under diffeomorphisms. This can be adapted to the tetrad representation (10.55) to define the tetrad unit vectors $\hat{\mathbf{e}}_a$ required for the equality

$$A(x) = A^\mu(x)\,\hat{\mathbf{e}}_\mu(x) = A^a(x)\,\hat{\mathbf{e}}_a(x). \tag{10.64}$$

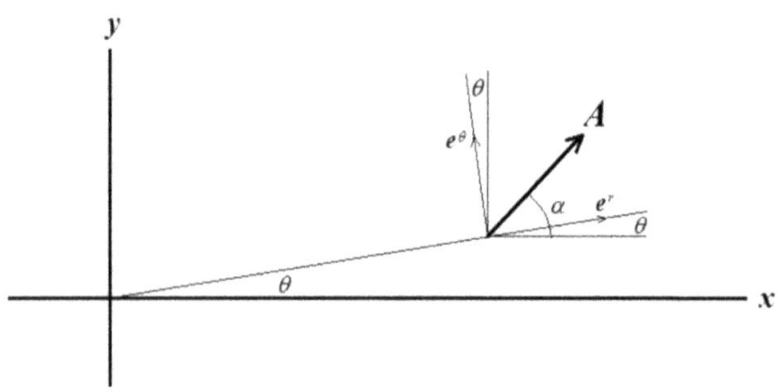

Figure 10.1. The zweibein in polar coordinates at an arbitrary point in a Euclidean plane.

Using (10.55) for $A^a(x)$ in (10.64) gives

$$A^\mu(x)\,\hat{\mathbf{e}}_\mu(x) = A^\mu(x) e_\mu{}^a(x)\hat{\mathbf{e}}_a(x) \implies \hat{\mathbf{e}}_\mu(x) = e_\mu{}^a(x)\,\hat{\mathbf{e}}_a(x). \tag{10.65}$$

Using the orthogonality relation (10.27) gives the relation of $\hat{\mathbf{e}}_a$ to $\hat{\mathbf{e}}_\mu$,

$$\hat{\mathbf{e}}_b(x) = \delta_b{}^a\,\hat{\mathbf{e}}_a(x) = e^\mu{}_b(x)\, e_\mu{}^a(x)\,\hat{\mathbf{e}}_a(x) = e^\mu{}_b(x)\,\hat{\mathbf{e}}_\mu(x). \tag{10.66}$$

For the simple two-dimensional example treated earlier, it follows that

$$\begin{aligned} \hat{\mathbf{e}}^r &= e^r{}_x\hat{\mathbf{e}}^x + e^r{}_y\hat{\mathbf{e}}^y = \cos\theta\,\hat{\mathbf{e}}^x + \sin\theta\,\hat{\mathbf{e}}^y, \\ \hat{\mathbf{e}}^\theta &= re^\theta{}_x\hat{\mathbf{e}}^x + re^\theta{}_y\hat{\mathbf{e}}^y = -\sin\theta\,\hat{\mathbf{e}}^x + \cos\theta\,\hat{\mathbf{e}}^y, \end{aligned} \tag{10.67}$$

where the factor of r gives $\hat{\mathbf{e}}^\theta$ the correct units. These are the same relationships found earlier in (1.38).

The results of this section show that it is consistent with the principle of equivalence to identify $g_{\mu\nu}(x)$ as the agent of gravitation. There is a set of *inertial or freely falling frames at every point* where $g_{\mu\nu}(x)$ becomes the Minkowski metric, $\eta_{\mu\nu}$, and its first derivative vanishes *in the neighborhood of* x. This does *not* mean that the metric becomes the Minkowski metric everywhere after the transformation. The eigenvalues and eigenvectors of the metric tensor will typically vary from point to point, and a different inertial frame must be chosen at each space-time point. This simply reflects the fact that the force of gravitation can vary from point to point. As a result, only freely falling observers *at the same point* will agree on the relative speed of two particles. This result is very different from special relativity, where inertial frames at *two different space-time points* are related by a Lorentz transformation that enables the two inertial observers to compensate for their relative motion. In the presence of gravitation the inertial frames at two different space-time points are no longer simply related if there is a difference in the metric between the two points.

10.2 The affine connection and the covariant derivative

The next step in quantifying a gravitational field theory is to define a *covariant derivative* D for tensors that is consistent with the generalized metric. A covariant derivative is defined to have the property that

$$D_\mu' A_\nu'(x') = \frac{\partial x^\rho}{\partial x'^\mu}\frac{\partial x^\sigma}{\partial x'^\nu} D_\rho A_\sigma(x), \tag{10.68}$$

when applied to a vector quantity. Its application to higher-rank tensor quantities follows the same tensor transformation rules given by (1.100). This behavior is required for a theory consistent with the principle of equivalence, since it will allow terms such as $D_\mu A^\nu D_\nu A^\mu$ to be *invariant* under *general* diffeomorphisms. The definition of the covariant derivative D starts by noting that a scalar quantity φ does not change form under a diffeomorphism, which gives

$$\begin{aligned}\partial'^{\mu}\varphi'(x')\,\partial'_{\mu}\varphi'(x') &= \frac{\partial x'^{\mu}}{\partial x^{\nu}}\partial^{\nu}\varphi(x)\frac{\partial x^{\rho}}{\partial x'^{\mu}}\partial_{\rho}\varphi(x) \\ &= \delta^{\rho}{}_{\nu}\partial^{\nu}\varphi(x)\partial_{\rho}\varphi(x) = \partial^{\nu}\varphi(x)\partial_{\nu}\varphi(x).\end{aligned} \tag{10.69}$$

As a result, for a scalar quantity φ the covariant derivative coincides with the partial derivative,

$$\mathrm{D}\varphi = \mathrm{d}\varphi \implies D_{\mu}\varphi\,\mathrm{d}x^{\mu} = \partial_{\mu}\varphi\,\mathrm{d}x^{\mu}. \tag{10.70}$$

10.2.1 Parallel transport of vectors and the covariant derivative

In the case of a vector, $D_{\mu}A^{\nu}$ must eliminate the second derivative term which appeared in (10.4) as a result of a general diffeomorphism. In addition, the principle of equivalence indicates that it should also reduce to $\partial_{\mu}A^{\nu}$ *in the local inertial frame* discussed in the previous section. The covariant derivative of a vector is therefore different from a scalar, since vectors are defined in the *tangent space* that exists at each point in a manifold. Obtaining the derivative of a vector requires finding the change in the vector between two infinitesimally separated points, and therefore the two vectors being compared are defined in *different tangent spaces.* In order to find the change between the two points, the vector must be *transported* from the tangent space at $x^{\mu} + \mathrm{d}x^{\mu}$ back to the tangent space at x^{μ}. This was briefly discussed following (1.56), where it was pointed out that this procedure is straightforward in a Euclidean space, since the two tangent spaces are *parallel.* In the case of curved or non-Euclidean manifolds, such as the circle and the sphere, the tangent spaces were depicted in figure 1.5, explicitly showing that tangent spaces at different points are no longer parallel. The vector in the tangent space at $x + \mathrm{d}x$ must be transported to x so that its components remain the same and the transported vector is parallel to the original vector, a process referred to as *parallel transport.* Parallel transport in a Euclidean space therefore corresponds to the *requirement* that the vector components A^{μ} are unchanged, and this has the simple mathematical statement

$$A^{\mu}(x + \mathrm{d}x) - A^{\mu}(x) = \mathrm{d}A^{\mu}(x) = 0. \tag{10.71}$$

However, in a non-Euclidean manifold, the tangent spaces at different points are no longer parallel and this complicates parallel transport.

The key step in understanding this is to examine the effect of a general diffeomorphism on $\mathrm{d}A^{\mu}$ *in a Euclidean space.* The result is

$$\begin{aligned}\mathrm{d}A'^{\mu}(x') &= \mathrm{d}\left(\frac{\partial x'^{\mu}}{\partial x^{\alpha}}A^{\alpha}(x)\right) \\ &= \frac{\partial^2 x'^{\mu}}{\partial x^{\alpha}\,\partial x^{\beta}}A^{\alpha}(x)\,\mathrm{d}x^{\beta} + \frac{\partial x'^{\mu}}{\partial x^{\alpha}}\,\mathrm{d}A^{\alpha}(x).\end{aligned} \tag{10.72}$$

Combining (10.72) with the following coordinate transformation relations,

$$\begin{aligned} \mathrm{d}x^\beta &= \frac{\partial x^\beta}{\partial x'^\rho}\,\mathrm{d}x'^\rho, \\ A^\alpha(x) &= \frac{\partial x^\alpha}{\partial x'^{\,\nu}} A'^\nu(x'), \end{aligned} \tag{10.73}$$

gives

$$\mathrm{d}A'^\mu(x') - \frac{\partial^2 x'^\mu}{\partial x^\alpha\,\partial x^\beta}\frac{\partial x^\alpha}{\partial x'^\nu}\frac{\partial x^\beta}{\partial x'^\rho}A'^{\nu}(x')\,\mathrm{d}x'^\rho = \frac{\partial x'^\mu}{\partial x^\alpha}\,\mathrm{d}A^\alpha(x). \tag{10.74}$$

Contracting both sides of (10.74) with $\partial x^\lambda/\partial x'^\mu$ and using the chain rule gives

$$\begin{aligned} \frac{\partial x^\lambda}{\partial x'^\mu}\frac{\partial x'^\mu}{\partial x^\alpha}\,\mathrm{d}A^\alpha(x) &= \delta^\lambda{}_\alpha \mathrm{d}A^\alpha(x) = \mathrm{d}A^\lambda(x) \\ &= \frac{\partial x^\lambda}{\partial x'^\mu}\Big(\mathrm{d}A'^\mu(x') + \Gamma'^\mu_{\nu\rho}(x')A'^{\nu}(x')\,\mathrm{d}x'^\rho\Big), \end{aligned} \tag{10.75}$$

where the *affine connection* $\Gamma'^\mu_{\nu\rho}(x')$ in the x'-coordinate system is defined as

$$\Gamma'^\mu_{\nu\rho}(x') \equiv -\frac{\partial^2 x'^\mu}{\partial x^\alpha\,\partial x^\beta}\frac{\partial x^\alpha}{\partial x'^\nu}\frac{\partial x^\beta}{\partial x'^\rho}. \tag{10.76}$$

A pair of extremely important points emerge from (10.75) and (10.76). The first is that the affine connection (10.76) vanishes if a Lorentz transformation is chosen, since the Lorentz transformation obeys (6.90), which requires its second derivative to vanish. The affine connection (10.76) has therefore emerged from a transformation to a non-inertial frame. The second is that the Einstein principle of equivalence then states that *a non-vanishing affine connection will be present in the case of a gravitational field.* In the absence of a gravitational field or in an inertial frame, the Einstein principle of equivalence requires the affine connection to vanish, which means $\Gamma^\mu_{\nu\rho}(x) = 0$ for all components in the absence of gravitation or in a freely falling frame.

At this stage of development it is not clear how the affine connection is related to the metric tensor. It will be seen that manifesting the Einstein principle of equivalence requires the affine connection to be a function of the metric $g_{\mu\nu}(x)$ which vanishes if $g_{\mu\nu}(x) = \eta_{\mu\nu}$. The second important point is that the right-hand side of (10.75) has the same form as the transformation rule for a contravariant vector under the diffeomorphism. Coupling this with the first result shows that the affine connection can be used to define the *differential of a vector*,

$$\begin{aligned} \mathrm{D}A^\mu(x) &\equiv \mathrm{d}A^\mu(x) + \Gamma^\mu_{\nu\rho}(x)A^\nu(x)\mathrm{d}x^\rho \\ &= \Big(\partial_\rho A^\mu(x) + \Gamma^\mu_{\nu\rho}(x)A^\nu(x)\Big)\mathrm{d}x^\rho \equiv (D_\rho A^\mu)\mathrm{d}x^\rho. \end{aligned} \tag{10.77}$$

This definition reduces to the usual differential $\mathrm{d}A^\mu(x)$ if $\Gamma^\mu_{\nu\rho}(x)$ vanishes, so that (10.77) is consistent with the Euclidean parallel transport statement (10.71), since $\Gamma^\mu_{\nu\rho}$ vanishes in a Euclidean space. For the special case of a transformation to an accelerated frame of reference in Minkowski space-time, result (10.75) shows that $\mathrm{D}A^\mu(x)$ transforms as a contravariant vector under a general diffeomorphism.

Another way to understand (10.77) is to use coordinate-free notation for the vector, so that

$$A(x) = A^{\mu}(x)\,\hat{\mathbf{e}}_{\mu}(x), \tag{10.78}$$

where $\hat{\mathbf{e}}_{\mu}(x)$ is a local set of basis vectors. As x changes, the change in $A(x)$ is a combination of the change in the components *and* the change in the basis vectors. Since $A(x)$ is a *scalar quantity* under diffeomorphisms, the covariant change in $A(x)$ coincides with the exact change, and is given by

$$\mathrm{D}A(x) = \mathrm{d}A(x) = \left(\frac{\partial A^{\mu}(x)}{\partial x^{\rho}} + A^{\mu}(x)\frac{\partial \hat{\mathbf{e}}_{\mu}(x)}{\partial x^{\rho}}\right)\mathrm{d}x^{\rho}. \tag{10.79}$$

The derivative of the basis vector is written as a superposition of the basis vectors,

$$\frac{\partial \hat{\mathbf{e}}_{\mu}(x)}{\partial x^{\rho}} = \partial_{\rho}\hat{\mathbf{e}}_{\mu}(x) = \Gamma^{\lambda}_{\mu\rho}(x)\hat{\mathbf{e}}_{\lambda}(x), \tag{10.80}$$

so that the affine connection is defined by (10.80). In order to justify this identification, (10.80) is returned to (10.79), which gives

$$\mathrm{D}A(x) = \left(\partial_{\rho}A^{\mu}(x) + \Gamma^{\mu}_{\lambda\rho}(x)\,A^{\lambda}(x)\right)\hat{\mathbf{e}}_{\mu}\mathrm{d}x^{\rho} = (D_{\rho}A^{\mu})\hat{\mathbf{e}}_{\mu}\mathrm{d}x^{\rho}. \tag{10.81}$$

This is the previous expression (10.77) for the covariant differential contracted with $\hat{\mathbf{e}}_{\mu}$, so that the identification (10.80) is justified.

For the general case of a non-Euclidean space-time it is assumed that the affine connection is present and results in $\mathrm{D}A^{\mu}$ transforming as a vector. This imposes the following requirement,

$$\begin{aligned} \mathrm{D}A'^{\mu} &= \mathrm{d}A'^{\mu}(x') + \Gamma'^{\mu}_{\sigma\delta}(x')A'^{\sigma}(x')\,\mathrm{d}x'^{\delta} \\ &= \frac{\partial x^{\mu}}{\partial x'^{\lambda}}\left(\mathrm{d}A^{\lambda}(x) + \Gamma^{\lambda}_{\alpha\beta}(x)A^{\alpha}(x)\,\mathrm{d}x^{\beta}\right). \end{aligned} \tag{10.82}$$

This requirement allows the transformation property of the general affine connection to be deduced by combining (10.82) with the following relations,

$$\begin{aligned} A'^{\sigma}(x') &= \frac{\partial x'^{\sigma}}{\partial x^{\alpha}}A^{\alpha}(x), \\ \mathrm{d}A'^{\mu}(x') &= \mathrm{d}\left(\frac{\partial x'^{\mu}}{\partial x^{\alpha}}A^{\alpha}(x)\right) \\ &= \frac{\partial^2 x'^{\mu}}{\partial x^{\alpha}\,\partial x^{\beta}}\,A^{\alpha}(x)\,\mathrm{d}x^{\beta} + \frac{\partial x'^{\mu}}{\partial x^{\lambda}}\,\mathrm{d}A^{\lambda}(x), \\ \mathrm{d}x'^{\lambda} &= \frac{\partial x'^{\lambda}}{\partial x^{\beta}}\,\mathrm{d}x^{\beta}. \end{aligned} \tag{10.83}$$

The requirement (10.82) becomes

$$\left(\frac{\partial x'^{\sigma}}{\partial x^{\alpha}}\frac{\partial x'^{\delta}}{\partial x^{\beta}}\Gamma'^{\mu}_{\sigma\delta}(x') - \frac{\partial x'^{\mu}}{\partial x^{\lambda}}\Gamma^{\lambda}_{\alpha\beta}(x) + \frac{\partial^2 x'^{\mu}}{\partial x^{\alpha}\,\partial x^{\beta}}\right)A^{\alpha}(x)\,\mathrm{d}x^{\beta} = 0. \tag{10.84}$$

Since A^{α} and $\mathrm{d}x^{\beta}$ are arbitrary, the requirement (10.84) is satisfied by

$$\frac{\partial x'^{\sigma}}{\partial x^{\alpha}}\frac{\partial x'^{\delta}}{\partial x^{\beta}}\Gamma'^{\mu}_{\sigma\delta}(x') = \frac{\partial x'^{\mu}}{\partial x^{\lambda}}\Gamma^{\lambda}_{\alpha\beta}(x) - \frac{\partial^2 x'^{\mu}}{\partial x^{\alpha}\,\partial x^{\beta}}. \tag{10.85}$$

Contracting both sides with

$$\frac{\partial x^{\alpha}}{\partial x'^{\nu}}\frac{\partial x^{\beta}}{\partial x'^{\rho}} \tag{10.86}$$

and using the chain rule gives

$$\Gamma'^{\mu}_{\nu\rho}(x') = \frac{\partial x'^{\mu}}{\partial x^{\lambda}}\frac{\partial x^{\alpha}}{\partial x'^{\nu}}\frac{\partial x^{\beta}}{\partial x'^{\rho}}\Gamma^{\lambda}_{\alpha\beta}(x) - \frac{\partial^2 x'^{\mu}}{\partial x^{\alpha}\,\partial x^{\beta}}\frac{\partial x^{\alpha}}{\partial x'^{\nu}}\frac{\partial x^{\beta}}{\partial x'^{\rho}}. \tag{10.87}$$

This shows that $\Gamma^{\mu}_{\nu\rho}$ does *not* transform as a third-rank tensor under a general diffeomorphism. However, for the case that $\Gamma^{\lambda}_{\alpha\beta}(x) = 0$ in the original coordinate system, the transformation rule (10.87) reproduces the transformation-induced affine connection of (10.76). For such a case, the two results (10.76) and (10.87) are consistent with each other. The appearance of the affine connection in a non-inertial frame is the counterpart to the Newtonian result of chapter 2, where an effective gravitational force appeared in an accelerated frame.

The expression for the affine connection given by (10.80) also yields the transformation rule (10.87). The derivation starts by noting that (10.80) gives

$$\begin{aligned}\frac{\partial}{\partial x'^{\rho}}\hat{\mathbf{e}}'_{\mu} &= \frac{\partial}{\partial x'^{\rho}}\left(\frac{\partial x^{\beta}}{\partial x'^{\mu}}\hat{\mathbf{e}}_{\beta}\right) = \frac{\partial^2 x^{\beta}}{\partial x'^{\rho}\,\partial x'^{\mu}}\hat{\mathbf{e}}_{\beta} + \frac{\partial x^{\alpha}}{\partial x'^{\rho}}\frac{\partial x^{\beta}}{\partial x'^{\mu}}\partial_{\alpha}\hat{\mathbf{e}}_{\beta} \\ &= \frac{\partial^2 x^{\sigma}}{\partial x'^{\rho}\,\partial x'^{\mu}}\hat{\mathbf{e}}_{\sigma} + \frac{\partial x^{\alpha}}{\partial x'^{\rho}}\frac{\partial x^{\beta}}{\partial x'^{\mu}}\Gamma^{\sigma}_{\alpha\beta}\hat{\mathbf{e}}_{\sigma}.\end{aligned} \tag{10.88}$$

Relation (10.80) then states that (10.88) is equal to

$$\Gamma'^{\lambda}_{\rho\mu}\hat{\mathbf{e}}'_{\lambda} = \Gamma'^{\lambda}_{\rho\mu}\frac{\partial x^{\sigma}}{\partial x'^{\lambda}}\hat{\mathbf{e}}_{\sigma}, \tag{10.89}$$

which gives

$$\Gamma'^{\lambda}_{\rho\mu}\frac{\partial x^{\sigma}}{\partial x'^{\lambda}} = \frac{\partial^2 x^{\sigma}}{\partial x'^{\rho}\,\partial x'^{\mu}} + \frac{\partial x^{\alpha}}{\partial x'^{\rho}}\frac{\partial x^{\beta}}{\partial x'^{\mu}}\Gamma^{\sigma}_{\alpha\beta}. \tag{10.90}$$

Contracting both sides of (10.90) with $\partial x'^{\nu}/\partial x^{\sigma}$ and using the chain rule gives

$$\Gamma'^{\nu}_{\rho\mu} = \frac{\partial^2 x^{\sigma}}{\partial x'^{\rho}\,\partial x'^{\mu}}\frac{\partial x'^{\nu}}{\partial x^{\sigma}} + \frac{\partial x^{\alpha}}{\partial x'^{\rho}}\frac{\partial x^{\beta}}{\partial x'^{\mu}}\frac{\partial x'^{\nu}}{\partial x^{\sigma}}\Gamma^{\sigma}_{\alpha\beta}. \tag{10.91}$$

It is left as an exercise to show that

$$\frac{\partial^2 x^\sigma}{\partial x'^\rho\, \partial x'^\mu}\frac{\partial x'^\nu}{\partial x^\sigma} = -\frac{\partial^2 x'^\nu}{\partial x^\alpha\, \partial x^\beta}\frac{\partial x^\alpha}{\partial x'^\rho}\frac{\partial x^\beta}{\partial x'^\mu}. \tag{10.92}$$

Exercise 10.5: Prove statement (10.92) by employing the identity

$$\frac{\partial^2 x^\sigma}{\partial x'^\rho\, \partial x'^\mu}\frac{\partial x'^\nu}{\partial x^\sigma} = \frac{\partial}{\partial x'^\rho}\left(\frac{\partial x^\sigma}{\partial x'^\mu}\frac{\partial x'^\nu}{\partial x^\sigma}\right) - \frac{\partial x^\sigma}{\partial x'^\mu}\frac{\partial}{\partial x'^\rho}\frac{\partial x'^\nu}{\partial x^\sigma}, \tag{10.93}$$

and then applying the chain rule.

Using (10.92) in (10.91) reproduces the transformation rule (10.87) for the affine connection.

Parallel transport of a vector can now be understood using relation (10.77). In a Euclidean space parallel transport of a vector corresponds to the requirement (10.71), so that

$$\mathrm{d}A^\mu(x) = \mathrm{d}x^\nu \partial_\nu A^\mu(x) = 0, \tag{10.94}$$

which must hold along each element $\mathrm{d}x^\nu$ of the path used to transport the vector. In a non-inertial frame *or in the presence of a gravitational field*, result (10.77) is used to generalize (10.94), giving the non-Euclidean parallel transport equation,

$$\mathrm{D}A^\mu = \mathrm{d}A^\mu(x) + \Gamma^\mu_{\nu\rho}(x) A^\nu(x)\, \mathrm{d}x^\rho = 0, \tag{10.95}$$

where the second term accounts for the non-parallelism of the infinitesimally separated tangent spaces.

Parallel transport in the general case picks a trajectory or path $x^\mu(\tau)$ for parallel transport that is parameterized by the element of invariant proper time, given in terms of the metric tensor by

$$\mathrm{d}\tau(x) = \frac{1}{c}\sqrt{g_{\mu\nu}(x)\, \mathrm{d}x^\mu\, \mathrm{d}x^\nu}. \tag{10.96}$$

A vector $A^\mu(x)$ becomes a function of τ along the trajectory. Once the affine connection is specified, the equation for *parallel transport* along the trajectory is

$$\begin{aligned} \mathrm{D}A^\mu(x(\tau)) &= \mathrm{d}A^\mu(x(\tau)) + \Gamma^\mu_{\rho\nu}(x(\tau))\, A^\rho(x(\tau))\, \mathrm{d}x^\nu(\tau) \\ &= \mathrm{d}\tau\left(\frac{\mathrm{d}A^\mu(\tau)}{\mathrm{d}\tau} + \Gamma^\mu_{\rho\nu}(\tau) A^\rho(\tau)\frac{\mathrm{d}x^\nu(\tau)}{\mathrm{d}\tau}\right) = 0. \end{aligned} \tag{10.97}$$

The result of parallel transporting a vector along the trajectory $x^\mu(\tau)$ is given by solving

$$\frac{\mathrm{D}A^\mu}{\mathrm{D}\tau} \equiv \frac{\mathrm{d}A^\mu(\tau)}{\mathrm{d}\tau} + \Gamma^\mu_{\rho\nu}(\tau)A^\rho(\tau)\frac{\mathrm{d}x^\nu(\tau)}{\mathrm{d}\tau} = 0. \tag{10.98}$$

An insight into this equation can be obtained by choosing the vector for parallel transport to be the proper velocity V^μ of a particle moving along the trajectory,

$$A^\rho \to \frac{\mathrm{d}x^\rho}{\mathrm{d}\tau} = V^\rho(\tau), \tag{10.99}$$

so that the parallel transport equation becomes

$$\frac{\mathrm{d}V^\mu(\tau)}{\mathrm{d}\tau} + \Gamma^\mu_{\rho\sigma}(\tau)V^\rho(\tau)V^\sigma(\tau) = 0. \tag{10.100}$$

If $\Gamma^\mu_{\rho\sigma}(\tau) = 0$, as it will be in the absence of gravitation and other forces or in an inertial frame, equation (10.100) reduces to

$$\frac{\mathrm{d}V^\mu(\tau)}{\mathrm{d}\tau} = \frac{1}{m}\frac{\mathrm{d}p^\mu}{\mathrm{d}\tau} = 0, \tag{10.101}$$

which is the Minkowski space statement (6.180) of free particle motion. Equation (10.100) is referred to as the *geodesic equation*, and a variational derivation of (10.100) is presented later in this chapter.

Using the coordinates of the trajectory gives

$$\frac{\mathrm{d}A^\mu(x(\tau))}{\mathrm{d}\tau}\,\mathrm{d}\tau = \frac{\partial A^\mu(x)}{\partial x^\nu}\frac{\partial x^\nu(\tau)}{\partial \tau}\,\mathrm{d}\tau = \partial_\nu A^\mu(x)\,\mathrm{d}x^\nu, \tag{10.102}$$

so that the covariant differential (10.77) can be written

$$\mathrm{D}A^\mu(x) = \big(\partial_\nu A^\mu(x) + \Gamma^\mu_{\rho\nu}(x)\,A^\rho(x)\big)\mathrm{d}x^\nu. \tag{10.103}$$

This enables the definition of the *covariant derivative* D_ν for a vector,

$$\mathrm{D}A^\mu(x) \equiv D_\nu A^\mu(x)\,\mathrm{d}x^\nu \implies D_\nu A^\mu(x) \equiv \partial_\nu A^\mu(x) + \Gamma^\mu_{\rho\nu}(x)A^\rho(x), \tag{10.104}$$

and the parallel transport equation can then be written as the covariant differential

$$\mathrm{D}A^\mu(x) = \mathrm{d}x^\nu D_\nu A^\mu(x) = 0, \tag{10.105}$$

which generalizes the Euclidean definition (10.94).

Result (10.75) indicates that D_ν transforms as the covariant components of a vector. The explicit transformation properties of $D_\nu A^\mu$ are found by contracting (10.87) with

$$A'^\nu(x') = \frac{\partial x'^\nu}{\partial x^\sigma}A^\sigma(x), \tag{10.106}$$

and then using the chain rule. The result is

$$\Gamma'^\mu_{\nu\rho}(x')A'^\nu(x') = \frac{\partial x'^\mu}{\partial x^\alpha}\frac{\partial x^\beta}{\partial x'^\rho}\Gamma^\alpha_{\sigma\beta}(x)A^\sigma(x) - \frac{\partial x^\alpha}{\partial x'^\rho}\frac{\partial^2 x'^\mu}{\partial x^\alpha\,\partial x^\beta}A^\beta(x). \tag{10.107}$$

Combining (10.107) with expression (10.5) gives

$$\begin{aligned}&\frac{\partial A'^{\mu}(x')}{\partial x'^{\rho}} + \Gamma'^{\mu}_{\nu\rho}(x')A'^{\nu}(x')\\ &= \frac{\partial x'^{\mu}}{\partial x^{\alpha}}\frac{\partial x^{\beta}}{\partial x'^{\rho}}\left(\frac{A^{\alpha}(x)}{\partial x^{\beta}} + \Gamma^{\alpha}_{\sigma\beta}(x)A^{\sigma}(x)\right),\end{aligned} \tag{10.108}$$

so that the terms involving the second derivative of the transformation are cancelled.

Result (10.108) shows that $D_\beta A^\alpha(x)$ does indeed transform under a general diffeomorphism as the components of a true second-rank mixed tensor. It also shows that $DA^\mu/D\tau$, as defined by (10.97), transforms as a vector under diffeomorphisms. It is important to remember that a parallel transported vector satisfies (10.105) along the path element $\mathrm{d}x^\nu$ joining the two points, but that does *not* mean that $D_\nu A^\mu$ vanishes. Instead, it means that $D_\nu A^\mu$ is orthogonal to $\mathrm{d}x^\nu$ at all points along the trajectory. In the case of a general metric tensor, the total change in the vector $A^\mu(x)$ between the two points x and $x + \mathrm{d}x$ is the combination of the change in the vector function, given by $\mathrm{d}A^\mu(x) = \mathrm{d}x^\nu\partial_\nu A^\mu(x)$, and the change in the orientation of the tangent space, given by $\Gamma^\mu_{\nu\rho}(x)A^\nu(x)\,\mathrm{d}x^\rho$. The sum of the two contributions is orthogonal to $\mathrm{d}x^\nu$.

Since it is a generalized derivative, the action of D_μ on products must obey the Leibniz property. This requires that

$$D_\nu(A_\mu A^\mu) = A^\mu D_\nu A_\mu + A_\mu D_\nu A^\mu = A^\mu D_\nu A_\mu + A_\mu(\partial_\nu A^\mu + \Gamma^\mu_{\alpha\nu}A^\alpha). \tag{10.109}$$

However, $A_\mu A^\mu$ is a scalar quantity, and so (10.70) gives

$$D_\nu(A_\mu A^\mu) = \partial_\nu(A_\mu A^\mu) = A^\mu\partial_\nu A_\mu + A_\mu\partial_\nu A^\mu. \tag{10.110}$$

Equating (10.109) and (10.110) and then relabeling dummy indices gives

$$A^\mu D_\nu A_\mu + A_\mu\Gamma^\mu_{\alpha\nu}A^\alpha - A^\mu\partial_\nu A_\mu = A^\mu(D_\nu A_\mu + \Gamma^\alpha_{\mu\nu}A_\alpha - \partial_\nu A_\mu) = 0. \tag{10.111}$$

This shows that the covariant derivative of a covariant vector is given by

$$D_\nu A_\mu(x) = \partial_\nu A_\mu(x) - \Gamma^\alpha_{\mu\nu}(x)A_\alpha(x). \tag{10.112}$$

This allows the definition of a covariant change ∇_X of a vector $Y^\mu(x)$ along the direction of the vector X^μ, which is given by

$$\nabla_X Y^\mu(x) = X^\nu(x)D_\nu Y^\mu(x). \tag{10.113}$$

Exercise 10.6: Using the second-rank mixed tensor $B^\mu{}_\rho(x) = A^\mu(x)A_\rho(x)$, the Leibniz property, (10.104), and (10.112), show that

$$D_\nu B^\mu{}_\rho(x) = \partial_\nu B^\mu{}_\rho(x) + \Gamma^\mu_{\alpha\nu}(x)B^\alpha{}_\rho(x) - \Gamma^\alpha_{\rho\nu}(x)B^\mu{}_\alpha(x). \tag{10.114}$$

10.2.2 Torsion

The geodesic equation (10.100) for the parallel transport of the proper velocity shows that only the symmetric part of the connection contributes,

$$\begin{aligned}&\frac{\mathrm{d}V^{\mu}(\tau)}{\mathrm{d}\tau}+\Gamma^{\mu}_{\rho\sigma}(\tau)V^{\rho}(\tau)V^{\sigma}(\tau)\\&=\frac{\mathrm{d}V^{\mu}(\tau)}{\mathrm{d}\tau}+\frac{1}{2}(\Gamma^{\mu}_{\rho\sigma}(\tau)+\Gamma^{\mu}_{\sigma\rho}(\tau))V^{\rho}(\tau)V^{\sigma}(\tau)=0,\end{aligned}\tag{10.115}$$

where

$$\tilde{\Gamma}^{\mu}_{\rho\sigma}(\tau)=\frac{1}{2}(\Gamma^{\mu}_{\rho\sigma}(\tau)+\Gamma^{\mu}_{\sigma\rho}(\tau))=\tilde{\Gamma}^{\mu}_{\sigma\rho}.\tag{10.116}$$

On the other hand, the antisymmetric part of the affine connection, denoted by

$$\mathcal{T}^{\mu}_{\nu\rho}=\frac{1}{2}(\Gamma^{\mu}_{\nu\rho}-\Gamma^{\mu}_{\rho\nu}),\tag{10.117}$$

and known as the *torsion tensor*, does not contribute, since relabeling dummy indices gives

$$\mathcal{T}^{\mu}_{\nu\rho}V^{\nu}V^{\rho}=\mathcal{T}^{\mu}_{\rho\nu}V^{\rho}V^{\nu}=-\mathcal{T}^{\mu}_{\nu\rho}V^{\nu}V^{\rho}\implies\mathcal{T}^{\mu}_{\nu\rho}V^{\nu}V^{\rho}=0.\tag{10.118}$$

However, the transformation rule (10.87) shows that the torsion transforms under a diffeomorphism as

$$\begin{aligned}\mathcal{T}'^{\mu}_{\nu\rho}(x')&=\frac{1}{2}(\Gamma'^{\mu}_{\nu\rho}(x')-\Gamma'^{\mu}_{\rho\nu}(x'))\\&=\frac{1}{2}\left(\frac{\partial x'^{\mu}}{\partial x^{\lambda}}\frac{\partial x^{\alpha}}{\partial x'^{\nu}}\frac{\partial x^{\beta}}{\partial x'^{\rho}}\Gamma^{\lambda}_{\alpha\beta}(x)\right.\\&\quad-\frac{\partial^2x'^{\mu}}{\partial x^{\alpha}\,\partial x^{\beta}}\frac{\partial x^{\alpha}}{\partial x'^{\nu}}\frac{\partial x^{\beta}}{\partial x'^{\rho}}-\frac{\partial x'^{\mu}}{\partial x^{\lambda}}\frac{\partial x^{\alpha}}{\partial x'^{\rho}}\frac{\partial x^{\beta}}{\partial x'^{\nu}}\Gamma^{\lambda}_{\alpha\beta}(x)\\&\quad\left.+\frac{\partial^2x'^{\mu}}{\partial x^{\alpha}\,\partial x^{\beta}}\frac{\partial x^{\alpha}}{\partial x'^{\rho}}\frac{\partial x^{\beta}}{\partial x'^{\nu}}\right)\\&=\frac{\partial x'^{\mu}}{\partial x^{\lambda}}\frac{\partial x^{\alpha}}{\partial x'^{\nu}}\frac{\partial x^{\beta}}{\partial x'^{\rho}}\left(\frac{1}{2}\Gamma^{\lambda}_{\alpha\beta}(x)-\frac{1}{2}\Gamma^{\lambda}_{\beta\alpha}(x)\right)\\&=\frac{\partial x'^{\mu}}{\partial x^{\lambda}}\frac{\partial x^{\alpha}}{\partial x'^{\nu}}\frac{\partial x^{\beta}}{\partial x'^{\rho}}\mathcal{T}^{\lambda}_{\alpha\beta}(x).\end{aligned}\tag{10.119}$$

This shows that the torsion is a true tensor under *arbitrary diffeomorphisms*.

If the torsion is nonzero then the quadrilateral defined in figure 10.2 will fail to close. To demonstrate this, we start by choosing two infinitesimal nonparallel but otherwise arbitrary tangent vectors at x, denoted by $\varepsilon X(x)$ and $\varepsilon Y(x)$ in coordinate-free notation, where ε is an infinitesimal parameter. The two vectors define two new positions in the manifold, $x+\varepsilon X(x)$ and $x+\varepsilon Y(x)$, respectively. The relevant

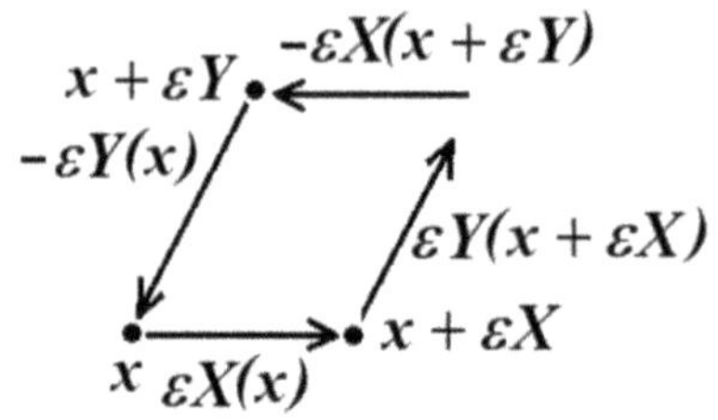

Figure 10.2. Torsion as the failure of the quadrilateral to close.

tangent vectors at those points are given by $\varepsilon Y(x + \varepsilon X(x))$ and $\varepsilon X(x + \varepsilon Y(x))$. These two tangent vectors can be found from the parallel transport of the respective tangent vectors to those points. The requirement for parallel transport using (10.113) gives

$$D_{\varepsilon X}(\varepsilon Y^\mu(x)) = 0 \implies \varepsilon X^\nu(x)\left(\varepsilon\partial_\nu Y^\mu(x) + \varepsilon\Gamma^\mu_{\rho\nu}(x)Y^\rho(x)\right) = 0. \tag{10.120}$$

Using the infinitesimal nature of ε and a Taylor series allows (10.120) to be written as

$$\varepsilon Y^\mu(x + \varepsilon X) - \varepsilon Y^\mu(x) = -\varepsilon^2\Gamma^\mu_{\rho\nu}(x)X^\nu(x)Y^\rho(x). \tag{10.121}$$

A similar treatment for the parallel transport of εX^μ gives

$$D_{\varepsilon Y}(\varepsilon X^\mu(x)) = 0 \implies \varepsilon X^\mu(x + \varepsilon Y) - \varepsilon X^\mu(x) = -\varepsilon^2\Gamma^\mu_{\rho\nu}(x)Y^\nu(x)X^\rho(x). \tag{10.122}$$

The infinitesimal quadrilateral depicted in figure 10.2 will close if

$$\varepsilon(X^\mu(x) + Y^\mu(x + \varepsilon X) - X^\mu(x + \varepsilon Y) - Y^\mu(x)) = 0. \tag{10.123}$$

Using (10.121) and (10.122) gives

$$\begin{aligned}&\varepsilon(X^\mu(x) + Y^\mu(x + \varepsilon X) - X^\mu(x + \varepsilon Y) - Y^\mu(x))\\&= -\varepsilon^2\Gamma^\mu_{\rho\nu}(x)(X^\nu(x)Y^\rho(x) - Y^\nu(x)X^\rho(x)).\end{aligned} \tag{10.124}$$

Switching dummy indices gives the criterion for closure of the quadrilateral,

$$\begin{aligned}-\varepsilon^2\Gamma^\mu_{\rho\nu}(x)\left(X^\nu(x)Y^\rho(x) - Y^\nu(x)X^\rho(x)\right) &= \varepsilon^2(\Gamma^\mu_{\rho\nu}(x) - \Gamma^\mu_{\nu\rho}(x))Y^\nu(x)X^\rho(x)\\&= 2\varepsilon^2\mathcal{T}^\mu_{\rho\nu}(x)Y^\nu(x)X^\rho(x) = 0.\end{aligned} \tag{10.125}$$

Since X^ρ and Y^ν are arbitrary, this shows that the quadrilateral defined by parallel transport closes only if the torsion tensor $\mathcal{T}^\mu_{\rho\nu}$ vanishes.

The Einstein principle of equivalence requires $\Gamma^\mu_{\nu\rho}$ to vanish in a freely falling frame. If the affine connection vanishes in the freely falling frame then so does the torsion tensor. Because torsion is a true tensor under diffeomorphisms, *if it is zero in any frame, it is zero in all other frames*, even non-inertial frames. However, the previous section showed that it is the metric tensor that reduces to the Minkowski metric in a freely falling frame and it is unclear at this stage of the discussion whether this reduces the affine connection to zero. The symmetric part of the affine connection will be denoted by

$$\tilde{\Gamma}^{\mu}_{\nu\rho}(x) = \frac{1}{2}(\Gamma^{\mu}_{\nu\rho}(x) + \Gamma^{\mu}_{\rho\nu}(x)). \tag{10.126}$$

However, if the Einstein principle of equivalence proves experimentally incorrect and $\Gamma^{\mu}_{\nu\rho}$ does not vanish in a freely falling frame, then the torsion tensor can be nonzero. As a result, the general affine connection will be written

$$\Gamma^{\mu}_{\nu\rho}(x) = \tilde{\Gamma}^{\mu}_{\nu\rho}(x) + \mathcal{T}^{\mu}_{\nu\rho}(x). \tag{10.127}$$

10.2.3 Metric compatibility and the Levi-Civita connection

The results for the affine connection can be related to the metric tensor by using $g_{\mu\nu}(x)$ to convert a contravariant vector into a covariant vector. For example, (10.114) shows that $D_\nu A^\mu(x)$ transforms as a mixed second-rank tensor, and therefore the tensor $g_{\rho\mu}(x)D_\nu A^\mu(x)$ will transform as a second-rank covariant tensor under a diffeomorphism. However, it follows from the Leibniz property that

$$\begin{aligned} D_\nu A_\rho(x) = D_\nu\big(g_{\rho\mu}(x)A^\mu(x)\big) &= A^\mu(x)D_\nu g_{\rho\mu}(x) \\ &+ g_{\rho\mu}(x)D_\nu A^\mu(x) \neq g_{\mu\rho}(x)D_\nu A^\mu(x). \end{aligned} \tag{10.128}$$

Instead, result (10.128) can be rewritten as

$$g_{\rho\mu}(x)D_\nu A^\mu(x) - D_\nu A_\rho(x) = A^\mu(x)D_\nu g_{\mu\rho}(x). \tag{10.129}$$

A physical insight into this is again obtained by invoking Einstein's principle of equivalence. In a freely falling frame ξ the affine connection is *assumed* to vanish, so that to $O(\xi - \xi_o)$, result (10.38) gives

$$D_\rho g_{\mu\nu}(\xi) = \frac{\partial}{\partial \xi^\rho} g'_{\mu\nu}(\xi) = 0, \tag{10.130}$$

in a sufficiently small neighborhood of the point ξ_o. Since $D_\rho g_{\mu\nu}$ transforms as a true tensor, consistency with this version of the principle of equivalence requires that (10.130) holds in all frames, so that

$$\lim_{x \to \xi_o} D_\rho g_{\mu\nu}(\xi) = 0. \tag{10.131}$$

Further insight into the relation of the covariant derivative to the metric follows if we consider the parallel transport of the magnitude of a vector along a path. If the magnitude of the vector remains constant along the path then

$$\mathrm{D}(A_\mu A^\mu) = \mathrm{D}(g_{\mu\nu}(x)A^\mu(x)A^\nu(x)) = dx^\rho D_\rho(g_{\mu\nu}(x)A^\mu(x)A^\nu(x)) = 0. \tag{10.132}$$

Combining the Leibniz property of the covariant derivative with the parallel transport equation (10.105) for A^μ gives

$$(dx^\rho D_\rho g_{\mu\nu}(x))A^\mu(x)A^\nu(x) + 2g_{\mu\nu}A^\nu(x)(dx^\rho D_\rho A^\mu(x)) \\ = A^\mu(x)A^\nu(x)dx^\rho D_\rho g_{\mu\nu}(x) = 0. \tag{10.133}$$

Since A^μ is arbitrary, the physical requirement that the magnitude of the vector remains constant at all points along the path gives

$$\mathrm{d}x^\rho D_\rho g_{\mu\nu}(x) = 0. \tag{10.134}$$

This must hold at all points in the manifold and for an arbitrary path. This requirement then gives

$$D_\rho g_{\mu\nu}(x) = 0. \tag{10.135}$$

For the case in which condition (10.135) holds, the affine connection is said to be *metric compatible*. In what follows, metric compatibility will be assumed to hold, although there are versions of general relativity for which this assumption is not made. It should be noted that (10.135) is consistent with the principle of equivalence as stated in (10.131). However, even if the affine connection does *not* vanish in the freely falling frame, (10.135) can still be enforced as a condition on the metric tensor.

Relation (10.135) can be used to express the *symmetric part* of a general affine connection in terms of $g_{\mu\nu}$ and the torsion tensor $\mathcal{T}^{\mu}_{\ \nu\rho}$. The cyclic permutations of (10.135) give

$$\begin{aligned} D_\rho g_{\mu\nu} &= \partial_\rho g_{\mu\nu} - \Gamma^{\lambda}_{\ \mu\nu} g_{\lambda\rho} - \Gamma^{\lambda}_{\ \mu\rho} g_{\nu\lambda} = 0, \\ D_\nu g_{\rho\mu} &= \partial_\mu g_{\nu\rho} - \Gamma^{\lambda}_{\ \mu\nu} g_{\lambda\rho} - \Gamma^{\lambda}_{\ \mu\rho} g_{\nu\lambda} = 0, \\ D_\mu g_{\nu\rho} &= \partial_\mu g_{\nu\rho} - \Gamma^{\lambda}_{\ \mu\nu} g_{\lambda\rho} - \Gamma^{\lambda}_{\ \mu\rho} g_{\nu\lambda} = 0. \end{aligned} \tag{10.136}$$

Using $g_{\mu\nu} = g_{\nu\mu}$ but differentiating $\Gamma^{\lambda}_{\ \mu\nu}$ and $\Gamma^{\lambda}_{\ \nu\mu}$ gives

$$D_\rho g_{\mu\nu} - D_\nu g_{\rho\mu} - D_\mu g_{\nu\rho} = \partial_\rho g_{\mu\nu} - \partial_\nu g_{\rho\mu} - \partial_\mu g_{\rho\nu} + (\Gamma^{\lambda}_{\ \mu\nu} \\ + \Gamma^{\lambda}_{\ \nu\mu}) g_{\lambda\rho} - 2\mathcal{T}^{\lambda}_{\ \mu\rho} g_{\nu\lambda} - 2\mathcal{T}^{\lambda}_{\ \nu\rho} g_{\mu\lambda} = 0. \tag{10.137}$$

Using $g_{\lambda\rho} g^{\sigma\rho} = \delta^\sigma_{\ \lambda}$ shows that the symmetric part of $\Gamma^{\sigma}_{\ \mu\nu}$ is given by

$$\begin{aligned} \Gamma_{S\ \mu\nu}^{\ \sigma} &= \frac{1}{2}(\Gamma^{\lambda}_{\ \mu\nu} + \Gamma^{\lambda}_{\ \nu\mu}) \\ &= \frac{1}{2} g^{\sigma\rho}\left(\partial_\mu g_{\nu\rho} + \partial_\nu g_{\mu\rho} - \partial_\rho g_{\mu\nu}\right) + g^{\sigma\rho}\left(\mathcal{T}^{\lambda}_{\ \mu\rho} g_{\nu\lambda} + \mathcal{T}^{\lambda}_{\ \nu\rho} g_{\mu\lambda}\right). \end{aligned} \tag{10.138}$$

Result (10.138) expresses the symmetric part of the affine connection in terms of the torsion tensor and derivatives of the metric tensor. It also shows that metric compatibility does *not* require the torsion tensor to vanish. Using (10.127) gives the affine connection for the case that the metric compatibility condition (10.135) holds,

$$\Gamma^{\sigma}_{\mu\nu} = \Gamma s^{\sigma}_{\mu\rho} + \mathcal{T}^{\sigma}_{\ \mu\nu} = \frac{1}{2} g^{\sigma\rho}\left(\partial_\mu g_{\nu\rho} + \partial_\nu g_{\mu\rho} - \partial_\rho g_{\mu\nu}\right) \\ + g^{\sigma\rho}\left(\mathcal{T}^{\lambda}_{\ \mu\rho}\, g_{\nu\lambda} + \mathcal{T}^{\lambda}_{\ \nu\rho}\, g_{\mu\lambda}\right) + \mathcal{T}^{\sigma}_{\ \mu\nu}. \tag{10.139}$$

The presence of the torsion tensor in the symmetric part of the affine connection adds a subtlety to the principle of equivalence. Result (10.139) shows that *if the torsion tensor vanishes* then the remaining affine connection is consistent with the Einstein equivalence principle. This is because a freely falling frame obeys (10.12) in the neighborhood of the point x_o where it is defined, so that

$$\lim_{x\to x_o} \partial_\rho g_{\mu\nu}(x) = \lim_{\xi\to\xi_o} \frac{\partial \xi^a}{\partial x_\rho}\frac{\partial}{\partial \xi^a} g_{\mu\nu}(\xi) = e_\rho{}^a \partial_a g_{\mu\nu}(\xi) = 0. \tag{10.140}$$

An observer in the freely falling frame at x will therefore claim $\Gamma^{\mu}_{\nu\rho}(x) = 0$ for the case in which the torsion vanishes. The resulting *torsion-free affine connection* is denoted by $\tilde{\Gamma}^{\sigma}_{\mu\nu}$ and is expressed *entirely* in terms of the metric tensor. It is explicitly given by

$$\tilde{\Gamma}^{\sigma}_{\mu\nu} = \frac{1}{2} g^{\sigma\rho}\left(\partial_\mu g_{\nu\rho} + \partial_\nu g_{\mu\rho} - \partial_\rho g_{\mu\nu}\right). \tag{10.141}$$

The specific form (10.141) is known as the *Levi-Civita connection* and occasionally as the *Christoffel symbol*. It is manifestly symmetric, $\tilde{\Gamma}^{\sigma}_{\mu\nu} = \tilde{\Gamma}^{\sigma}_{\nu\mu}$, and obeys the Einstein principle of equivalence by vanishing in the local inertial frame. It is the unique torsion-free and metric compatible affine connection.

It should be mentioned that nonzero torsion is postulated to be present in some variants of Einstein's version of general relativity, beginning with what is known as Einstein–Cartan general relativity. In these variants there are properties of matter, such as the spin of fermions or the rotation of a massive body, that act as the source of torsion even in freely falling frames. Since Einstein's version of gravitation assumes that torsion vanishes, the presence of torsion in these theories results in a deviation from the predictions of Einstein's gravitation for numerous physical phenomena, such as the precession rate of gyroscopes. However, the current measured deviations from Einstein's gravity all lie within the error limits of the experimental apparatus, so that there is no experimental indication at this time that Einstein's version of gravitation is incorrect. For completeness, torsion will be retained in further developments.

10.2.4 The affine connection and the tetrad formalism

In order to develop covariant derivatives for spinor fields, it is necessary to adapt the affine connection to the tetrad representation of (10.55). It was pointed out that an arbitrary vector $A^\mu(x)$ can be projected onto a tetrad basis at that point,

$$A^a(x) = e^a{}_\mu(x)A^\mu(x), \tag{10.142}$$

and that it transforms under the *local Lorentz group* rather than diffeomorphisms. As the vector of (10.142) undergoes parallel transport, the tetrad basis will change, since the metric tensor appearing in (10.30) will change. Because of the product nature of the tetrad version, the affine connection for the tetrad version will differ from the Levi-Civita connection. It is therefore denoted by $\Omega^a_{\mu b}$ and is referred to as the *spin connection*, since it will be adapted to parallel transport spinor fields. By analogy to previous results, the covariant derivative of A^a is written in terms of the differential,

$$\mathrm{D}A^a(x) = \tilde{D}_\mu A^a(x)\mathrm{d}x^\mu = (\partial_\mu A^a(x) + \Omega^a_{\mu b}(x)A^b(x))\,\mathrm{d}x^\mu. \tag{10.143}$$

The covariant derivative of a mixed-representation tensor can be deduced using (10.142) in (10.143), with the result that

$$D_\mu A^a = \partial_\mu(e_\nu{}^a A^\nu) + \Omega^a_{\mu b}e_\nu{}^b A^\nu = (\partial_\mu e_\nu{}^a)A^\nu + e_\nu{}^a(\partial_\mu A^\nu) + \Omega^a_{\mu b}e_\nu{}^b A^\nu. \tag{10.144}$$

The Leibniz nature of D_μ gives

$$\begin{aligned} D_\mu A^a &= D_\mu(e_\nu{}^a A^\nu) = (D_\mu e_\nu{}^a)A^\nu + e_\nu{}^a D_\mu A^\nu \\ &= (D_\mu e_\nu{}^a)A^\nu + e_\nu{}^a\left(\partial_\mu A^\nu + \Gamma^\nu{}_{\rho\mu}A^\rho\right). \end{aligned} \tag{10.145}$$

Equating (10.144) and (10.145) and canceling the common derivative term gives

$$(D_\mu e_\nu{}^a)A^\nu = \left(\partial_\mu e_\nu{}^a + \Omega^a_{\mu b}e_\nu{}^b - e_\rho{}^a\Gamma^\rho_{\nu\mu}\right)A^\nu. \tag{10.146}$$

Since A^ν is arbitrary, it follows that

$$D_\mu e_\nu{}^a = \partial_\mu e_\nu{}^a + \Omega^a_{\mu b}e_\nu{}^b - \Gamma^\rho_{\nu\mu}e_\rho{}^a. \tag{10.147}$$

This gives the expected result that the Latin tetrad indices transport using the spin connection and the Greek vector indices transport using the affine connection.

So far, there has been no relationship between the spin connection and the affine connection. The first step in finding this relationship is to apply metric compatibility to the freely falling frame where the Minkowski metric is present. This gives

$$D_\lambda\eta^{ab} = \partial_\lambda\eta^{ab} + \Omega^a_{\lambda c}\eta^{cb} + \Omega^b_{\lambda c}\eta^{ac} = \Omega_\lambda{}^{ab} + \Omega_\lambda{}^{ba} = 0, \tag{10.148}$$

where property (10.41) was used. This requires the spin connection to be antisymmetric in the tetrad indices,

$$\Omega_\lambda{}^{ab} = -\Omega_\lambda{}^{ba} \implies \Omega_{\lambda a}{}^b = -\Omega^b_{\lambda a}. \tag{10.149}$$

Metric compatibility for $g_{\mu\nu}$ is evaluated using the tetrad identity (10.30) and result (10.148),

$$\begin{aligned} D_\rho g_{\mu\nu} &= D_\rho(\eta_{ab}e_\mu{}^a e_\nu{}^b) \\ &= (D_\rho\eta_{ab})e_\mu{}^a e_\nu{}^b + \eta_{ab}(D_\rho e_\mu{}^a)e_\nu{}^b + \eta_{ab}e_\mu{}^a(D_\rho e_\nu{}^b) \\ &= 2\eta_{ab}e_\mu{}^a(D_\rho e_\nu{}^b) = 0, \end{aligned} \tag{10.150}$$

where $g_{\mu\nu} = g_{\nu\mu}$ was used. The last expression in (10.150) is contracted with $e^{\mu}{}_{c}\eta^{cd}$ to give

$$D_{\mu}e_{\nu}{}^{d} = \partial_{\mu}e_{\nu}{}^{d} + \Omega^{d}_{\mu b}e_{\nu}{}^{b} - \Gamma^{\rho}_{\nu\mu}e_{\rho}{}^{d} = 0. \tag{10.151}$$

Result (10.151) is often referred to as the *tetrad postulate*, although here it has been derived by combining the covariant derivative (10.147) with metric compatibility and the tetrad identity (10.30). It is important to note that (10.151) immediately gives

$$\partial_{\mu}e_{\nu}{}^{a} = \Gamma^{\rho}_{\nu\mu}e_{\rho}{}^{a} - \Omega^{a}_{\mu b}e_{\nu}{}^{b}, \tag{10.152}$$

so that the individual tetrad basis vectors obey (10.80) *only if the spin connection vanishes.*

Contracting (10.151) with $e^{\sigma}{}_{d}$ and using (10.44) gives

$$\Gamma^{\sigma}_{\nu\mu} = \partial_{\mu}e_{\nu}{}^{a}e^{\sigma}{}_{a} - \Omega^{a}_{\mu b}e_{\nu}{}^{b}e^{\sigma}{}_{a}. \tag{10.153}$$

Expression (10.153) is not symmetric in μ and ν, instead yielding the following expression for the torsion,

$$\frac{1}{2}(\Gamma^{\sigma}_{\mu\nu} - \Gamma^{\sigma}_{\nu\mu})e_{\sigma}{}^{a} = \mathcal{T}^{\sigma}_{\mu\nu}e_{\sigma}{}^{a} = \mathcal{T}^{a}_{\mu\nu} = \frac{1}{2}\left(\partial_{\mu}e_{\nu}{}^{a} - \partial_{\nu}e_{\mu}{}^{a} + \Omega^{a}_{\mu b}e_{\nu}{}^{b} - \Omega^{a}_{\nu b}e_{\mu}{}^{b}\right). \tag{10.154}$$

The spin connection for the case that torsion vanishes is denoted by $\omega_{\mu}{}^{ab}$ and must satisfy

$$\partial_{\mu}e_{\nu}{}^{a} - \partial_{\nu}e_{\mu}{}^{a} + \omega_{\mu}{}^{ab}e_{\nu b} - \omega_{\nu}{}^{ab}e_{\mu b} = 0, \tag{10.155}$$

where

$$e^{\rho a} = \eta^{ab}e^{\rho}{}_{b}, \quad e_{\rho a} = \eta_{ab}e_{\rho}{}^{b}. \tag{10.156}$$

Using the tetrad properties of (10.44) gives a means to find the *torsion-free* spin connection components. The solution to (10.155) is given by the manifestly antisymmetric expression,

$$\begin{aligned}\omega_{\mu}{}^{ab} = &-\frac{1}{2}e^{\lambda a}\left(\partial_{\lambda}e_{\mu}{}^{b} - \partial_{\mu}e_{\lambda}{}^{b}\right) + \frac{1}{2}e^{\lambda b}\left(\partial_{\lambda}e_{\mu}{}^{a} - \partial_{\mu}e_{\lambda}{}^{a}\right)\\ &-\frac{1}{2}e_{\mu}{}^{c}e^{\lambda a}e^{\rho b}\left(\partial_{\lambda}e_{\rho c} - \partial_{\rho}e_{\lambda c}\right).\end{aligned} \tag{10.157}$$

Exercise 10.7: Verify that the third term in (10.157) is antisymmetric in *a* and *b*. Show by direct substitution that (10.157) satisfies (10.155).

Using (10.151) shows that a vector expressed in the tetrad basis obeys the parallel transport condition,

$$\begin{aligned} \mathrm{D}A^a(x) &= \mathrm{D}(e_\mu{}^a(x)A^\mu(x)) = e_\mu{}^a \mathrm{d}x^\rho D_\rho A^\mu + A^\mu \mathrm{d}x^\rho D_\rho e_\mu{}^a \\ &= e_\mu{}^a \mathrm{d}x^\rho D_\rho A^\mu = 0. \end{aligned} \tag{10.158}$$

Contracting (10.151) with $e_c{}^\mu$ gives

$$\Omega_\mu{}^{ab} = \Gamma^\rho_{\nu\mu} e_\rho{}^a e^{\nu b} - e^{\nu a}\partial_\mu e_\nu{}^b, \tag{10.159}$$

which is not manifestly antisymmetric. Using the antisymmetric nature of the spin connection gives

$$\Omega_\mu{}^{ab} = -\Omega_\mu{}^{ba} = -\Gamma^\rho_{\nu\mu} e_\rho{}^b e^{\nu a} + e^{\nu b}\partial_\mu e_\nu{}^a, \tag{10.160}$$

so that the manifestly antisymmetric version is given by

$$\begin{aligned} \Omega_\mu{}^{ab} &= \frac{1}{2}(\Omega_\mu{}^{ab} - \Omega_\mu{}^{ba}) = \frac{1}{2}\Gamma^\rho_{\nu\mu}\left(e_\rho{}^a e^{\nu b} - e_\rho{}^b e^{\nu a}\right) \\ &\quad + \frac{1}{2}e^{\nu b}\partial_\mu e_\nu{}^a - \frac{1}{2}e^{\nu a}\partial_\mu e_\nu{}^b. \end{aligned} \tag{10.161}$$

The transformation property of the spin connection can now be found by combining the transformation properties of the tetrad given by (10.32) and the transformation properties of the affine connection given by (10.87). The result is

$$\begin{aligned} \Omega_\mu^{\prime ab}(x') &= \frac{\partial x^\nu}{\partial x'^\mu}\Big(\Lambda^a{}_c(x)\,\Lambda^b{}_d(x)\,\Omega_\nu{}^{cd}(x) \\ &\quad - \frac{1}{2}\Lambda^a{}_c(x)\,\partial_\nu\Lambda^{bc}(x) + \frac{1}{2}\Lambda^b{}_c(x)\,\partial_\nu\Lambda^{ac}(x)\Big). \end{aligned} \tag{10.162}$$

The term (10.76) is canceled and replaced with the derivatives of the local Lorentz transformation. Result (10.162) is critical to formulating the Dirac equation for spinor fields in general relativity. This is discussed in detail in the next chapter.

Exercise 10.8: Verify (10.162).

It is useful to note that the torsion-free Levi-Civita connection $\tilde{\Gamma}^\sigma_{\mu\nu}$, given by (10.141), can be used to derive the torsion-free spin connection $\omega_\mu{}^{ab}$ of (10.157). This starts with the tetrad postulate, which gives

$$\begin{aligned} e^{\nu a}D_\mu e_\nu{}^b &= e^{\nu a}\partial_\mu e_\nu{}^b + \Omega^b_{\mu c} e_\nu{}^c e^{\nu a} - \Gamma^\rho_{\mu\nu} e_\rho{}^b e^{\nu a} \\ &= e^{\nu a}\partial_\mu e_\nu{}^b + \Omega_\mu{}^{ba} - \Gamma^\rho_{\mu\nu} e_\rho{}^b e^{\nu a} = 0. \end{aligned} \tag{10.163}$$

Using the antisymmetry of the spin connection in (10.163) gives

$$\Omega_\mu{}^{ab} = -\Gamma^\rho_{\nu\mu} e_\rho{}^b e^{\nu a} + e^{\nu a} \partial_\mu e_\nu{}^b. \tag{10.164}$$

Using the torsion-free Levi-Civita connection $\tilde{\Gamma}^\rho_{\nu\mu}$, given by (10.141), in (10.164) gives the torsion-free spin connection $\omega_\mu{}^{ab}$,

$$\begin{aligned}\omega_\mu{}^{ab} &= -\tilde{\Gamma}^\rho_{\nu\mu} e_\rho{}^b e^{\nu a} + e^{\nu a} \partial_\mu e_\nu{}^b \\ &= -\frac{1}{2} e_\rho{}^b e^{\nu a} g^{\rho\sigma} (\partial_\nu g_{\mu\sigma} + \partial_\mu g_{\nu\sigma} - \partial_\sigma g_{\nu\mu}) + e^{\nu a} \partial_\mu e_\nu{}^b.\end{aligned} \tag{10.165}$$

The equivalence is demonstrated using the tetrad identity (10.30) in (10.165), so that the following replacements are made,

$$g_{\mu\sigma} = e_\mu{}^c e_{\sigma c}, \quad g_{\nu\sigma} = e_\nu{}^c e_{\sigma c}, \quad g_{\nu\mu} = e_\nu{}^c e_{\mu c}, \quad g^{\rho\sigma} = e^{\rho d} e^\sigma{}_d. \tag{10.166}$$

After using the properties (10.44) of the inverse vielbeins, expression (10.165) becomes

$$\begin{aligned}\omega_\mu{}^{ab} = &-\frac{1}{2} e^{\lambda a} \left(\partial_\lambda e_\mu{}^b - \partial_\mu e_\lambda{}^b\right) + \frac{1}{2} e^{\lambda b} \left(\partial_\lambda e_\mu{}^a - \partial_\mu e_\lambda{}^a\right) \\ &-\frac{1}{2} e^{\lambda a} e^{\rho b} \left(\partial_\lambda e_{\rho c} - \partial_\rho e_{\lambda c}\right) e_\mu{}^c,\end{aligned} \tag{10.167}$$

which is identical to the torsion-free spin connection found in (10.157).

> **Exercise 10.9:** Verify that (10.165) becomes (10.167) after the substitutions of (10.166).

10.2.5 Examples: the Euclidean plane and the sphere

At first glance, the two-dimensional Euclidean plane would appear to have no need for the geometric tools assembled so far. However, if the polar coordinates $x^1 = r$ and $x^2 = \theta$ are used, then the metric tensor is given by (10.50). While remaining diagonal, it is explicitly dependent on r, since $g_{22} = r^2$. The Levi-Civita connection components are calculated using (10.141). The nonzero components are given by

$$\begin{aligned}\Gamma^1_{22} &= \frac{1}{2} g^{1\rho} \left(\partial_\theta g_{2\rho} + \partial_\theta g_{2\rho} - \partial_\rho g_{22}\right) = -\frac{1}{2} g^{11} \partial_r g_{22} = -\frac{1}{2} \partial_r (r^2) = -r, \\ \Gamma^2_{12} &= \frac{1}{2} g^{2\rho} \left(\partial_r g_{2\rho} + \partial_\theta g_{1\rho} - \partial_\rho g_{12}\right) = \frac{1}{2} g^{22} \partial_r g_{22} = \frac{1}{2} \left(\frac{1}{r^2}\right) \partial_r (r^2) = \frac{1}{r}.\end{aligned} \tag{10.168}$$

Exercise 10.10: Verify that all others components of the Levi-Civita connection vanish.

The Levi-Civita connection found in (10.168) can be used to perform parallel transport of a vector in the Euclidean plane using polar coordinates. For simplicity, the path will be a circle, so that $\mathrm{d}r = 0$. The vector will have the boundary conditions that $A^r(0) = 0$ and $A^\theta(0) = A_0^\theta/r$ at $\theta = 0$, where A_0^θ is an arbitrary constant. The components of the parallel transported vector must satisfy the parallel transport equation (10.105), so that for the path of a circle and the connection (10.168) it follows that

$$\begin{aligned} \mathrm{d}x^\mu D_\mu A^r(\theta) &= \mathrm{d}\theta D_\theta A^r(\theta) = \mathrm{d}\theta\big(\partial_\theta A^r(\theta) + \Gamma^1_{22}A^\theta(\theta)\big) = 0, \\ \mathrm{d}x^\mu D_\mu A^\theta(\theta) &= \mathrm{d}\theta D_\theta A^\theta(\theta) = \mathrm{d}\theta\big(\partial_\theta A^\theta(\theta) + \Gamma^2_{12}A^r(\theta)\big) = 0. \end{aligned} \tag{10.169}$$

This yields two partial differential equations,

$$\begin{aligned} \partial_\theta A^r(\theta) - rA^\theta(\theta) &= 0, \\ \partial_\theta A^\theta(\theta) + \frac{1}{r}A^r(\theta) &= 0. \end{aligned} \tag{10.170}$$

These coupled equations are solved by differentiating them both to give

$$\begin{aligned} \frac{\partial^2}{\partial\theta^2}A^r(\theta) - r\frac{\partial}{\partial\theta}A^\theta(\theta) &= 0, \\ \frac{\partial^2}{\partial\theta^2}A^\theta(\theta) + \frac{1}{r}\frac{\partial}{\partial\theta}A^r(\theta) &= 0. \end{aligned} \tag{10.171}$$

Using the equations of (10.170) in (10.171) gives

$$\begin{aligned} \frac{\partial^2}{\partial\theta^2}A^r(\theta) + A^r(\theta) &= 0, \\ \frac{\partial^2}{\partial\theta^2}A^\theta(\theta) + A^\theta(\theta) &= 0. \end{aligned} \tag{10.172}$$

The differential equations of both (10.171) and (10.172) are easily solved in a way that is consistent with the boundary conditions chosen earlier, so that

$$\begin{aligned} A^r(\theta) &= A_0^\theta \sin(\theta), \\ A^\theta(\theta) &= \frac{1}{r}A_0^\theta \cos(\theta). \end{aligned} \tag{10.173}$$

This shows that the magnitude of the vector is given by the constant A_0^θ as it is parallel transported. This is easily visualized, showing that the parallel transported vector is radial at $\theta = \pi/2$ and $\theta = 3\pi/2$. It also shows that the parallel transported vector returns to its initial value after $\theta = 2\pi$, when the transport around the circle is completed.

Another simple but illustrative example of truly non-Euclidean geometry and the effects of parallel transport is that of a sphere with radius R. Because the sphere can be embedded in familiar three-dimensional space, many of the abstract properties of Riemannian geometry can be visualized. The tangent spaces of the sphere are two-dimensional planes at each point and are depicted in figure 1.5. A tangent vector defined at a point lies in the tangent plane at that point, but these tangent planes change orientation at different points on the sphere. Choosing polar coordinates for the sphere gives the distance $\mathrm{d}s^2$ along the surface associated with a change in polar angle $\mathrm{d}\theta$ and azimuthal angle $\mathrm{d}\phi$ as

$$\mathrm{d}s^2 = \mathcal{R}^2\,\mathrm{d}\theta^2 + \mathcal{R}^2\sin^2\theta\,\mathrm{d}\phi^2, \tag{10.174}$$

where $\mathcal{R}$ is the constant radius of the sphere. Treating θ and ϕ as the two spatial variables x^1 and x^2, expression (10.174) gives the components of the metric tensor for a sphere in polar coordinates,

$$\begin{aligned} g_{11}(\theta,\phi) &= \mathcal{R}^2, \\ g_{22}(\theta,\phi) &= \mathcal{R}^2\sin^2\theta, \\ g_{12}(\theta,\phi) &= 0. \end{aligned} \tag{10.175}$$

The metric tensor can therefore be visualized as a diagonal 2×2 matrix. The inverse metric tensor is a diagonal matrix as well, with the elements

$$\begin{aligned} g^{12}(\theta,\phi) &= 0, \\ g^{11}(\theta,\phi) &= \frac{1}{\mathcal{R}^2}, \\ g^{22}(\theta,\phi) &= \frac{1}{(\mathcal{R}\sin\theta)^2}. \end{aligned} \tag{10.176}$$

The partial derivatives in the vector basis are given by

$$\partial_\mu = \left(\frac{\partial}{\partial\theta}, \frac{\partial}{\partial\phi}\right). \tag{10.177}$$

Using expression (10.138) for the Levi-Civita connection and the diagonal nature of the metric gives

$$\Gamma^1_{\mu\nu} = \frac{1}{2}g^{11}(\partial_\nu g_{1\mu} + \partial_\mu g_{1\nu} - \partial_1 g_{\mu\nu}). \tag{10.178}$$

Replacing index one with two yields a similar expression,

$$\Gamma^2_{\mu\nu} = \frac{1}{2}g^{22}(\partial_\nu g_{2\mu} + \partial_\mu g_{2\nu} - \partial_2 g_{\mu\nu}). \tag{10.179}$$

Using the form for the metric and its inverse shows that the components of the Levi-Civita connection are

$$\begin{aligned}\Gamma^1_{22} &= \frac{1}{2}\frac{1}{\mathcal{R}^2}(\partial_\phi g_{12} + \partial_\phi g_{12} - \partial_\theta g_{22}) = -\sin\theta\cos\theta,\\ \Gamma^2_{12} = \Gamma^2_{21} &= \frac{1}{2}\frac{1}{(\mathcal{R}\sin\theta)^2}(\partial_\theta g_{22} + \partial_\phi g_{21} - \partial_\phi g_{12}) = \cot\theta.\end{aligned} \tag{10.180}$$

All other components of the Levi-Civita connection vanish.

These can now be used to parallel transport a vector in the sphere's tangent space with the polar components denoted by $A^\theta = A^1$ and $A^\phi = A^2$. For simplicity, the starting point of the path for transport will be chosen to be $\phi = 0$. The polar angle θ will be arbitrary but held constant along the path, so that the path corresponds to a latitude of the sphere. As a result, the path is characterized by $\mathrm{d}\theta = \mathrm{d}x^1 = 0$. The parallel transport equation becomes

$$\mathrm{d}x^\mu(\partial_\mu A^\nu + \Gamma^\nu_{\ \rho\mu}A^\rho) = \mathrm{d}\phi\,(\partial_2 A^\nu + \Gamma^\nu_{\ \rho 2}A^\rho) = 0. \tag{10.181}$$

Expression (10.181) yields two distinct equations,

$$\begin{aligned}\partial_2 A^1 + \Gamma^1_{12}A^1 + \Gamma^1_{22}A^2 &= \frac{\partial A^\theta}{\partial\phi} + \Gamma^1_{22}A^\phi = \frac{\partial A^\theta}{\partial\phi} - \sin\theta\cos\theta\, A^\phi = 0,\\ \partial_2 A^2 + \Gamma^2_{12}A^1 + \Gamma^2_{22}A^2 &= \frac{\partial A^\phi}{\partial\phi} + \Gamma^2_{21}A^\theta = \frac{\partial A^\phi}{\partial\phi} + \cot\theta\, A^\theta = 0.\end{aligned} \tag{10.182}$$

Like the equations of (10.171) the coupled differential equations of (10.182) can be solved by taking the derivative of the first equation with respect to ϕ, which gives

$$\frac{\partial^2 A^\theta}{\partial\phi^2} - \sin\theta\cos\theta\,\frac{\partial A^\phi}{\partial\phi} = 0. \tag{10.183}$$

This can be used in the second equation of (10.182) to eliminate $\partial_\phi A^\phi$. This yields a differential equation for the parallel transported component A^θ,

$$\frac{\partial A^\phi}{\partial\phi} = \frac{1}{\sin\theta\cos\theta}\frac{\partial^2 A^\theta}{\partial\phi^2} \implies \frac{\partial^2 A^\theta}{\partial\phi^2} + \cos^2\theta\, A^\theta = 0. \tag{10.184}$$

Because $\cos\theta$ is being held constant the solution to (10.184) is straightforward. If the initial value of A^θ at $\phi = 0$ is A^θ_0, then the solution consistent with the boundary conditions for transport around the latitude is

$$A^\theta(\phi) = A^\theta_0\cos(\phi\cos\theta). \tag{10.185}$$

Exercise 10.11: Show that the azimuthal component A^ϕ of the vector undergoes parallel transport identically, so that

$$A^\phi(\phi) = A^\phi_0\cos(\phi\cos\theta), \tag{10.186}$$

where $A^\phi_0 = A^\phi(0)$.

Result (10.186) shows that parallel transport along the equator, where $\theta = \pi/2$, leaves the vector unchanged. This will be shown to be a general property of a *geodesic*, which is the path of minimal length on a non-Euclidean surface. It is also important to note that after a full trip around the latitude back to the starting point, the parallel transported components are given by

$$A^{\mu}(2\pi) = A_0^{\mu} \cos(2\pi \cos\theta). \tag{10.187}$$

This is an outgrowth of the underlying curvature of the sphere. The basis vectors undergo a rotation as the tangent plane is transported along the latitude, and this causes the parallel transport to rotate the vector in the tangent plane. The difference between the initial vector and the final vector after a closed path is completed is the hallmark of curvature in a manifold and is referred as *holonomy*. In that regard, result (10.187) shows that

$$\theta \approx 0 \implies A_{\mu}(\phi) \approx A_0^{\mu} \cos(\phi), \tag{10.188}$$

which is identical to the result (10.173) for parallel transport around a circle in the Euclidean plane. This shows that there is a neighborhood of the north pole that is Euclidean, which is the counterpart of the local inertial frame that exists in general relativity.

It is also instructive to apply the vielbein formalism to the sphere. The Euclidean metric for the two-dimensional surface of the sphere is given by the equivalent of the Minkowski metric,

$$\eta = \begin{pmatrix} 1 & 0 \\ 0 & 1 \end{pmatrix}. \tag{10.189}$$

The two zweibeins that reproduce the properties of (10.44) have the components given by

$$e_{\mu}{}^{1}(\theta) = (\mathcal{R}, 0), \quad e_{\mu}{}^{2}(\theta) = (0, \mathcal{R}\sin\theta). \tag{10.190}$$

The inverse zweibeins $e^{\mu}{}_{b}$ are given by

$$e^{\mu}{}_{1}(\theta) = \left(\frac{1}{\mathcal{R}}, 0\right), \quad e^{\mu}{}_{2}(\theta) = \left(0, \frac{1}{\mathcal{R}\sin\theta}\right). \tag{10.191}$$

These definitions are easily seen to satisfy the first four of the required properties listed in (10.44),

$$\begin{aligned} e_{\mu}{}^{a}(\theta)e^{\mu}{}_{b}(\theta) &= \delta^{a}{}_{b}, \\ e_{\mu}{}^{a}(\theta)e^{\nu}{}_{a}(\theta) &= \delta^{\nu}{}_{\mu}, \\ \eta_{ab}e_{\mu}{}^{a}(\theta)e_{\nu}{}^{b}(\theta) &= g_{\mu\nu}(\theta), \\ g^{\mu\nu}e_{\mu}{}^{a}(\theta)e_{\nu}{}^{b}(\theta) &= \eta^{ab}. \end{aligned} \tag{10.192}$$

Using the Euclidean metric (10.189) shows that

$$e_{\mu a} = \eta_{ab}e_{\mu}{}^{b} = e_{\mu}{}^{a}, \tag{10.193}$$

which will be useful in what follows.

The partial derivatives in the zweibein basis are given by

$$\tilde{\partial}_a = e^{\mu}{}_a \partial_\mu = \left(\frac{1}{\mathcal{R}}\frac{\partial}{\partial\theta}, \frac{1}{\mathcal{R}\sin\theta}\frac{\partial}{\partial\phi}\right), \tag{10.194}$$

where $\tilde{\partial}$ is being used to distinguish the partial derivatives of (10.194) in the zweibein basis from the partial derivatives (10.177) in the vector basis. The last of the requirements listed in (10.44),

$$\tilde{\partial}_c e^{\sigma}{}_a = -\frac{1}{2} e^{\mu}{}_a \tilde{\partial}_c g_{\mu\nu} g^{\nu\sigma}, \tag{10.195}$$

is demonstrated using the derivatives of (10.194). For the case that $c = 2$ in (10.195), both sides vanish, since $e^{\sigma}{}_a$ and $g_{\mu\nu}$ are independent of ϕ.

> **Exercise 10.12:** For the case that $c = 1$ in (10.195), show that all cases except for $(\sigma, a) = (2, 2)$ are zero on both sides.

For the case that $c = 1$ and both σ and a are 2, then the left-hand side of (10.195) becomes

$$\tilde{\partial}_1 e^2{}_2 = \frac{1}{\mathcal{R}}\frac{\partial}{\partial\theta}\left(\frac{1}{\mathcal{R}\sin\theta}\right) = -\frac{1}{\mathcal{R}^2}\frac{\cot\theta}{\sin\theta}. \tag{10.196}$$

Using the diagonal nature of the metric the right-hand side of (10.195) is given by

$$\begin{aligned} -\frac{1}{2} e^{\mu}{}_2 \tilde{\partial}_1 g_{\mu\nu} g^{\nu 2} &= -\frac{1}{2} e^2{}_2 (\tilde{\partial}_1 g_{22}) g^{22} \\ &= -\frac{1}{2}\frac{1}{\mathcal{R}\sin\theta}\left(\frac{1}{\mathcal{R}}\frac{\partial}{\partial\theta}(\mathcal{R}^2\sin^2\theta)\right)\frac{1}{\mathcal{R}^2\sin^2\theta} \\ &= -\frac{1}{\mathcal{R}^2}\frac{\cot\theta}{\sin\theta}, \end{aligned} \tag{10.197}$$

which is identical to (10.196). As a result, the last zweibein requirement (10.195) is satisfied. This shows there is a neighborhood of every point on the sphere where the metric becomes Euclidean, which was found earlier in (10.188) using parallel transport around an infinitesimal latitude.

It is tedious but straightforward to find the spin connection $\Omega_\mu{}^{ab}$ using (10.159) and the Levi-Civita connection (10.180). The component $\Omega_2{}^{12}$ is given by

$$\begin{aligned} \Omega_2{}^{12} &= \Gamma^{\rho}_{\nu 2} e_\rho{}^1 e^{\nu 2} - e^{\nu 1}\partial_\phi e_\nu{}^2 = \Gamma^{\rho}_{\nu 2} e_\rho{}^1 e^{\nu 2} \\ &= \Gamma^1_{22} e_1{}^1 e^{22} = (-\sin\theta\cos\theta)(\mathcal{R})\left(\frac{1}{\mathcal{R}\sin\theta}\right) = -\cos\theta. \end{aligned} \tag{10.198}$$

Similarly, the component $\Omega_2{}^{21}$ is given by

$$\begin{aligned}\Omega_2{}^{21} &= \Gamma^{\rho}_{\nu 2} e_\rho{}^2 e^{\nu 1} - e^{\nu 2}\partial_\phi e_\nu{}^1 = \Gamma^{\rho}_{\nu 2} e_\rho{}^2 e^{\nu 1} \\ &= \Gamma^2_{12} e_2{}^2 e^{11} = (\cot\theta)(\mathcal{R}\sin\theta)\left(\frac{1}{\mathcal{R}}\right) = \cos\theta.\end{aligned} \tag{10.199}$$

Despite the use of formula (10.159), which is not manifestly antisymmetric, the result satisfies $\Omega_2{}^{12} = -\Omega_2{}^{21}$.

Exercise 10.13: Show that the spin connection for the sphere obeys

$$\Omega_1{}^{ab} = 0. \tag{10.200}$$

Using the zweibein basis (10.190) and the spin connection results of (10.199) and (10.200) allows the torsion tensor for the sphere to be found from (10.154). For example, $\mathcal{T}^1_{12}$ is given by

$$\mathcal{T}^1_{12} = \frac{1}{2}\left(\partial_\theta e_2{}^1 - \partial_\phi e_1{}^1 + \Omega_1{}^{1b} e_{2b} - \Omega_2{}^{1b} e_{1b}\right) = -\frac{1}{2}\Omega_2{}^{12} e_{12} = 0. \tag{10.201}$$

Similarly, $\mathcal{T}^2_{12}$ is given by

$$\begin{aligned}\mathcal{T}^2_{12} &= \frac{1}{2}\left(\partial_\theta e_2{}^2 - \partial_\phi e_1{}^2 + \Omega_1{}^{2b} e_{2b} - \Omega_2{}^{2b} e_{1b}\right) \\ &= -\frac{1}{2}(\partial_\theta e_2{}^2 - \Omega_2{}^{21} e_{11}) = -\frac{1}{2}(\partial_\theta(\mathcal{R}\sin\theta) - (\cos\theta)(\mathcal{R})) = 0.\end{aligned} \tag{10.202}$$

This shows that the sphere as defined by the metric (10.175) and the Levi-Civita connection has zero torsion. This is to be expected from using the torsion-free Levi-Civita connection.

10.3 The curvature tensor

The holonomy for the parallel transport of a vector around a closed loop demonstrated for the case of the sphere in (10.187) provides the basis used to quantify curvature.

10.3.1 Defining the curvature tensor

In order to define the curvature at a point a vector will be parallel transported to a second point along two infinitesimal path elements, dx^μ and dy^μ. Performing the parallel transport using two different orders, either dx^μ or dy^μ first, transports the vector to the same point by two different paths that form an infinitesimal closed loop. The infinitesimal nature of the path elements allows a Taylor series expansion, yielding an expression for the holonomy of the closed loop. Using the same method as (10.121) and (10.122) shows that parallel transport of the vector $A^\mu(x)$ along the path element $\mathrm{d}x^\mu$ gives

$$\begin{aligned} D_{\mathrm{d}x}A^\mu(x) = \mathrm{d}x^\nu D_\nu A^\mu(x) = 0 \implies \\ A^\mu(x+\mathrm{d}x) = A^\mu(x) - \Gamma^\mu_{\alpha\nu}(x)\mathrm{d}x^\alpha A^\nu(x). \end{aligned} \tag{10.203}$$

Subsequently transporting the vector obtained from (10.203) along $\mathrm{d}y^\mu$ gives the first version of the final resulting vector, denoted by $A_1{}^\mu$,

$$A_1{}^\mu(x+\mathrm{d}x+\mathrm{d}y) = A^\mu(x+\mathrm{d}x) - \Gamma^\mu_{\alpha\nu}(x+\mathrm{d}x)\mathrm{d}y^\alpha A^\nu(x+\mathrm{d}x). \tag{10.204}$$

A Taylor series expansion to the first order gives

$$\Gamma^\mu_{\alpha\nu}(x+\mathrm{d}x) \approx \Gamma^\mu_{\nu\rho}(x) + \mathrm{d}x^\beta\partial_\beta\Gamma^\mu_{\alpha\nu}(x). \tag{10.205}$$

Combining (10.205) with (10.203) and discarding terms to the third order in the differentials shows that the first form (10.204) for the parallel transport can be written

$$\begin{aligned} A_1{}^\mu(x+\mathrm{d}x+\mathrm{d}y) &= A^\mu(x) - \Gamma^\mu_{\alpha\nu}(x)\,\mathrm{d}x^\alpha A^\nu(x) \\ &\quad - \left(\Gamma^\mu_{\nu\rho}(x) + \mathrm{d}x^\beta\partial_\beta\Gamma^\mu_{\alpha\nu}(x)\right)\mathrm{d}y^\alpha \\ &\quad \left(A^\nu(x) - \Gamma^\nu_{\sigma\rho}(x)\mathrm{d}x^\sigma A^\rho(x)\right) \\ &= A^\mu(x) - \Gamma^\mu_{\alpha\nu}(x)(\mathrm{d}x^\alpha + \mathrm{d}y^\alpha)A^\nu(x) \\ &\quad - \left(\partial_\beta\Gamma^\mu_{\alpha\nu}(x) - \Gamma^\mu_{\alpha\lambda}(x)\Gamma^\lambda_{\beta\nu}(x)\right)\mathrm{d}x^\alpha\mathrm{d}y^\beta A^\nu(x). \end{aligned} \tag{10.206}$$

The result of parallel transport in the reverse order is found from (10.206) by exchanging $\mathrm{d}x^\alpha$ and $\mathrm{d}y^\alpha$, with the result that the second form for the parallel transport, denoted by A_2^μ, is given by

$$\begin{aligned} A_2^\mu(x+\mathrm{d}y+\mathrm{d}x) &= A^\mu(x) - \Gamma^\mu_{\alpha\nu}(x)(\mathrm{d}x^\alpha + \mathrm{d}y^\alpha)A^\nu(x) \\ &\quad - \left(\partial_\beta\Gamma^\mu_{\alpha\nu}(x) - \Gamma^\mu_{\alpha\lambda}(x)\Gamma^\lambda_{\beta\nu}(x)\right)\mathrm{d}y^\alpha\mathrm{d}x^\beta A^\nu(x). \end{aligned} \tag{10.207}$$

The holonomy H^μ associated with the infinitesimal loop is given by

$$\begin{aligned} H^\mu \equiv A_1^\mu - A_2^\mu &= (\partial_\alpha\Gamma^\mu_{\beta\nu} - \partial_\beta\Gamma^\mu_{\alpha\nu} + \Gamma^\mu_{\alpha\lambda}\Gamma^\lambda_{\beta\nu} - \Gamma^\mu_{\beta\lambda}\Gamma^\lambda_{\alpha\nu}) \\ &\quad \mathrm{d}x^\alpha\mathrm{d}y^\beta A^\nu(x) \\ &\equiv R^\mu{}_{\nu\alpha\beta}(x)\,\mathrm{d}x^\alpha\mathrm{d}y^\beta A^\nu(x), \end{aligned} \tag{10.208}$$

where the *Riemann curvature tensor* $R^\mu{}_{\nu\alpha\beta}$ is defined as

$$R^\mu{}_{\nu\alpha\beta} = \partial_\alpha\Gamma^\mu_{\beta\nu} - \partial_\beta\Gamma^\mu_{\alpha\nu} + \Gamma^\mu_{\alpha\lambda}\Gamma^\lambda_{\beta\nu} - \Gamma^\mu_{\beta\lambda}\Gamma^\lambda_{\alpha\nu}. \tag{10.209}$$

Because it is written in terms of the affine connection, which is not a tensor, it is not apparent that the Riemann curvature tensor actually transforms as a tensor.

Using the transformation properties (10.87) for the affine connection in (10.209), it can be shown that the Riemann curvature is a tensor. However, a simpler method is to note that the curvature can also be expressed in terms of the covariant derivative by rewriting the parallel transport along the two possible paths as

$$\begin{aligned} A_1^{\mu}(x + \mathrm{d}x + \mathrm{d}y) &= A^{\mu}(x + \mathrm{d}x) + \mathrm{d}y^{\beta} D_{\beta} A^{\mu}(x + \mathrm{d}x) \\ &= A^{\mu}(x) + (\mathrm{d}x^{\beta} + \mathrm{d}y^{\beta}) D_{\beta} A^{\mu}(x) + \mathrm{d}x^{\alpha} \mathrm{d}y^{\beta} D_{\alpha} D_{\beta} A^{\mu}(x) \\ A_2^{\mu}(x + \mathrm{d}x + \mathrm{d}y) &= A^{\mu}(x + \mathrm{d}y) + \mathrm{d}x^{\alpha} D_{\alpha} A^{\mu}(x + \mathrm{d}y) \\ &= A^{\mu}(x) + (\mathrm{d}x^{\alpha} + \mathrm{d}y^{\alpha}) D_{\alpha} A^{\mu}(x) + \mathrm{d}x^{\alpha} \mathrm{d}y^{\beta} D_{\beta} D_{\alpha} A^{\rho}(x). \end{aligned} \tag{10.210}$$

Taking the difference of the two results and comparing it to (10.208) shows that

$$\begin{aligned} &H^{\mu} = A_1^{\mu} - A_2^{\mu} = \mathrm{d}x^{\alpha} \mathrm{d}y^{\beta} [D_{\alpha}, D_{\beta}] A^{\mu}(x) = R^{\mu}{}_{\nu\alpha\beta}(x)\, \mathrm{d}x^{\alpha} \mathrm{d}y^{\beta} A^{\nu}(x) \\ &\Longrightarrow R^{\mu}{}_{\nu\alpha\beta} A^{\nu} = [D_{\alpha}, D_{\beta}] A^{\mu}. \end{aligned} \tag{10.211}$$

Result (10.211) shows that the curvature transforms as a tensor, since all the quantities involved on the right-hand side of the equation transform as tensors under a diffeomorphism. It is also possible to use (10.87) in (10.209) to explicitly demonstrate the curvature's tensor nature. Result (10.211) shows that the commutator of the covariant derivative gives the curvature tensor. This result will be quite useful later.

10.3.2 Properties of the curvature tensor

Until the affine connection is specified, the specific properties of the Riemann tensor will be undefined. However, by construction, it has the property that

$$R^{\mu}{}_{\nu\alpha\beta} = -R^{\mu}{}_{\nu\beta\alpha}. \tag{10.212}$$

The Riemann curvature tensor in Einstein's version of gravitation is found using the *torsion-free* Levi-Civita connection (10.141) for $\Gamma^{\mu}_{\nu\rho}$. Since $R^{\mu}{}_{\nu\alpha\beta}$ is a true tensor, its properties are most easily examined by writing it in its totally covariant form,

$$R_{\mu\nu\alpha\beta} = g_{\mu\sigma} R^{\sigma}{}_{\nu\alpha\beta}, \tag{10.213}$$

and evaluating it in the local inertial frame. It follows from the discussion subsequent to (10.38) that the only contributions to the curvature tensor in the inertial frame come from the second derivatives of the metric tensor present in the terms of the form $\partial_{\alpha}\Gamma^{\mu}_{\nu\rho}$. Denoting the metric tensor in the inertial frame as $g'_{\mu\nu}$ and using only the Levi-Civita connection (10.141), it follows that $R'_{\mu\nu\alpha\beta}$ in the inertial frame for such a choice is given by

$$\begin{aligned} R'_{\mu\nu\alpha\beta} &\approx g'_{\mu\sigma} \partial_{\alpha} \Gamma^{\sigma}_{\beta\nu} - g'_{\mu\sigma} \partial_{\beta} \Gamma^{\sigma}_{\alpha\nu} \\ &= \frac{1}{2}\left(\partial_{\alpha}\partial_{\beta} g'_{\nu\mu} + \partial_{\alpha}\partial_{\nu} g'_{\beta\mu} - \partial_{\mu}\partial_{\alpha} g'_{\beta\nu}\right) - \frac{1}{2}\left(\partial_{\alpha}\partial_{\beta} g'_{\nu\mu} + \partial_{\beta}\partial_{\nu} g'_{\beta\mu} - \partial_{\mu}\partial_{\beta} g'_{\alpha\nu}\right) \\ &= \frac{1}{2}\left(\partial_{\alpha}\partial_{\nu} g'_{\beta\mu} - \partial_{\beta}\partial_{\nu} g'_{\alpha\mu} + \partial_{\mu}\partial_{\beta} g'_{\alpha\nu} - \partial_{\mu}\partial_{\alpha} g'_{\beta\nu}\right). \end{aligned} \tag{10.214}$$

This result reveals that the Riemann tensor obtained from the Levi-Civita connection possesses an additional set of properties. Using $g_{\mu\nu} = g_{\nu\mu}$ and assuming $\partial_\mu\partial_\nu = \partial_\nu\partial_\mu$ shows that (10.214) gives

$$\begin{aligned} R'_{\alpha\beta\mu\nu} &= \frac{1}{2}(\partial_\mu\partial_\beta g'_{\nu\alpha} - \partial_\nu\partial_\beta g'_{\mu\alpha} + \partial_\alpha\partial_\nu g'_{\mu\beta} - \partial_\alpha\partial_\mu g'_{\nu\beta}) \\ &= \frac{1}{2}(\partial_\mu\partial_\beta g'_{\alpha\nu} - \partial_\beta\partial_\nu g'_{\alpha\mu} + \partial_\alpha\partial_\nu g'_{\beta\mu} - \partial_\mu\partial_\alpha g'_{\beta\nu}) \\ &= R'_{\mu\nu\alpha\beta}. \end{aligned} \tag{10.215}$$

This property is easily seen in the inertial frame, but it remains true in all frames because the curvature is a tensor. As a result, the prime will be dropped, so that

$$R_{\alpha\beta\mu\nu} = R_{\mu\nu\alpha\beta}. \tag{10.216}$$

Result (10.214) can be used to show that the curvature tensor has the additional properties

$$R_{\mu\nu\alpha\beta} = -R_{\nu\mu\alpha\beta}, \tag{10.217}$$

$$R_{\mu\nu\alpha\beta} + R_{\mu\beta\nu\alpha} + R_{\mu\alpha\beta\nu} = 0, \tag{10.218}$$

where (10.218) is known as the *first Bianchi identity*.

> **Exercise 10.14:** Use (10.214) to verify the two identities (10.217) and (10.218).

It should be stressed that these properties of the curvature tensor assumed a torsion-free connection. Although the Riemann tensor has $4^4 = 256$ components in four dimensions, the additional identities of (10.212), (10.216), (10.217), and (10.218) reduce the number of independent components in the curvature to 20, which is precisely the number predicted earlier by (10.39).

An important identity emerges from the local inertial frame version of the curvature (10.214) found from the Levi-Civita connection. In the local inertial frame the connection vanishes, so that the covariant derivative becomes the partial derivative, $D_\mu \to \partial_\mu$. Using the inertial frame representation (10.214) of the torsion-free Riemann curvature tensor shows that the cyclic permutation of the indices ρ, α, and β yields

$$\begin{aligned} \partial_\rho R'_{\mu\nu\alpha\beta} &= \partial_\rho\partial_\mu\partial_\beta g'_{\alpha\nu} - \partial_\rho\partial_\beta\partial_\nu g'_{\alpha\mu} + \partial_\rho\partial_\alpha\partial_\nu g'_{\beta\mu} - \partial_\rho\partial_\mu\partial_\alpha g'_{\beta\nu}, \\ \partial_\beta R'_{\mu\nu\rho\alpha} &= \partial_\beta\partial_\mu\partial_\alpha g'_{\rho\nu} - \partial_\beta\partial_\alpha\partial_\nu g'_{\rho\mu} + \partial_\beta\partial_\rho\partial_\nu g'_{\alpha\mu} - \partial_\beta\partial_\mu\partial_\rho g'_{\alpha\nu}, \\ \partial_\alpha R'_{\mu\nu\beta\rho} &= \partial_\alpha\partial_\mu\partial_\rho g'_{\beta\nu} - \partial_\alpha\partial_\rho\partial_\nu g'_{\beta\mu} + \partial_\alpha\partial_\beta\partial_\nu g'_{\rho\mu} - \partial_\alpha\partial_\mu\partial_\beta g'_{\rho\nu}. \end{aligned} \tag{10.219}$$

Assuming that the order of the partial derivatives is irrelevant, this gives

$$\partial_\rho R'_{\mu\nu\alpha\beta} + \partial_\beta R'_{\mu\nu\rho\alpha} + \partial_\alpha R'_{\mu\nu\beta\rho} = 0. \tag{10.220}$$

For a general frame *this process is reversed*, with partial derivatives replaced by covariant derivatives, $\partial_\mu \to D_\mu$. The identity in a general frame is then written

$$D_\rho R'_{\mu\nu\alpha\beta} + D_\beta R'_{\mu\nu\rho\alpha} + D_\alpha R'_{\mu\nu\beta\rho} = 0, \tag{10.221}$$

which is known as the *second Bianchi identity*. It bears a formal similarity to the electromagnetic Bianchi identity (9.57). It is important to note that (10.221) implicitly assumes that the torsion vanishes, since it was derived by using only the Levi-Civita connection to compute the curvature tensor. A version of (10.221) in which the torsion does not vanish will be derived using differential forms later in this section. Due to the symmetries of the curvature tensor, there are many variants of (10.221).

10.3.3 Two-dimensional examples of curvature

Computing the curvature tensor for a two-dimensional surface is comparatively easy. For example, the affine connection for the sphere was found in the previous section, and using those results in (10.209) shows that there is one distinct component to the curvature tensor,

$$\begin{aligned} R^1{}_{212} &= \partial_1\Gamma^1_{22} - \partial_2\Gamma^1_{12} + \Gamma^1_{11}\Gamma^1_{22} + \Gamma^1_{12}\Gamma^2_{22} - \Gamma^1_{21}\Gamma^1_{12} - \Gamma^1_{22}\Gamma^2_{12} \\ &= \partial_\theta(-\sin\theta\cos\theta) - 0 + 0 + 0 - 0 - (-\sin\theta\cos\theta)\cdot\cot\theta = \sin^2\theta. \end{aligned} \tag{10.222}$$

This gives the one unique curvature component of the sphere,

$$R_{1212} = g_{11}R^1{}_{212} = R^2\sin^2\theta. \tag{10.223}$$

Because of the numerous identities that obeyed the Riemann curvature tensor, there are other non-vanishing components. For example, identity (10.217) shows that

$$R^2{}_{112} = -R^1{}_{212} = -\sin^2\theta. \tag{10.224}$$

The Riemann tensor has the important property that it vanishes in a flat manifold, *even if the coordinate system being used yields a nonzero affine connection*. For example, it was found earlier in (10.168) that using polar coordinates (r, θ) for the Euclidean plane gives rise to nonzero affine connection components. Using these affine connection components,

$$\Gamma^1_{22} = -r, \quad \Gamma^2_{12} = \frac{1}{r}, \tag{10.225}$$

shows that the Riemann tensor component for the Euclidean plane component is given by

$$
\begin{aligned}
R^{1}{}_{212} &= \partial_1\Gamma^1_{22} - \partial_2\Gamma^1_{12} + \Gamma^1_{11}\Gamma^1_{22} + \Gamma^1_{12}\Gamma^2_{22} - \Gamma^1_{21}\Gamma^1_{12} - \Gamma^1_{22}\Gamma^2_{12} \\
&= \partial_r(-r) - 0 + 0 + 0 - 0 - (-r)\cdot\frac{1}{r} = 0,
\end{aligned}
\tag{10.226}
$$

which is quite different from result (10.224) for the sphere. Despite a non-vanishing connection in polar coordinates, the Riemann tensor vanishes everywhere, which is consistent with the flat nature of the manifold.

If the metric tensor is constant everywhere then the Levi-Civita connection vanishes, rendering the Riemann curvature tensor computed from it zero everywhere. Solely using the Levi-Civita connection yields a flat Riemannian manifold if the metric tensor is constant. Similarly, if the Riemann curvature tensor vanishes everywhere, regardless of the connection used, then (10.208) shows that the components of the tetrad basis vectors $e_\mu{}^a$ for the local inertial frame at one point can be parallel transported to every point in the manifold with *no holonomy*. For such a case, the metric tensor at any given point can be parallel transported to all other points in the manifold with no ambiguity depending on path. This gives a *unique* metric tensor at each point in the manifold,

$$
g_{\mu\nu} = \eta_{ab}e^a{}_\mu e^b{}_\nu. \tag{10.227}
$$

The Riemann curvature tensor is therefore the true indicator of the presence of a gravitational field in space-time. However, the equivalence principle in Einstein's gravitation still allows an inertial frame where there are no observable gravitational effects on motion, *even if curvature induced by the gravitational field is present.*

10.3.4 Tensors derived from the curvature

Several important tensors can be formed from the Riemann tensor by contracting its indices. The *Ricci tensor* is given by

$$
R_{\mu\nu} = R^\lambda{}_{\mu\lambda\nu} = g^{\lambda\beta}R_{\beta\mu\lambda\nu}, \tag{10.228}
$$

and from it, the *curvature scalar* is defined as

$$
R = R^\mu{}_\mu = g^{\mu\nu}R_{\mu\nu} = g^{\mu\nu}g^{\lambda\beta}R_{\beta\mu\lambda\nu}. \tag{10.229}
$$

The Riemann curvature property (10.216) gives

$$
g^{\mu\alpha}R_{\mu\nu\alpha\beta} = g^{\mu\alpha}R_{\alpha\beta\mu\nu} \implies R_{\nu\beta} = R^\alpha{}_{\nu\alpha\beta} = R^\mu{}_{\beta\mu\nu} = R_{\beta\nu}, \tag{10.230}
$$

showing that the Ricci tensor is symmetric. However, the identity (10.216) was determined by assuming the absence of torsion, and so the symmetric nature of the Ricci tensor is unclear in the presence of torsion.

For example, the nonzero components of the Ricci tensor for the sphere are

$$
\begin{aligned}
R_{11} &= g^{11}R_{1212} = \sin^2\theta, \\
R_{22} &= g^{22}R_{2121} = 1.
\end{aligned}
\tag{10.231}
$$

The curvature scalar for the sphere is then given by

$$R = g^{11}R_{11} + g^{22}R_{22} = \frac{2}{R^2}, \tag{10.232}$$

so that the curvature scalar is determined entirely by the radius of the sphere.

In the absence of torsion the curvature scalar R and the Ricci tensor $R_{\mu\nu}$ can be related by first contracting the second Bianchi identity (10.221) with $g^{\alpha\lambda}g^{\beta\nu}$, which gives

$$g^{\alpha\lambda}g^{\beta\nu}D_\lambda R_{\mu\nu\alpha\beta} + g^{\alpha\lambda}g^{\beta\nu}D_\nu R_{\lambda\mu\alpha\beta} + g^{\alpha\lambda}g^{\beta\nu}D_\mu R_{\nu\lambda\alpha\beta} = 0. \tag{10.233}$$

Combining metric compatibility and the curvature properties (10.212) and (10.217) gives

$$\begin{aligned} g^{\beta\nu}D_\lambda R_{\mu\nu\alpha\beta} &= D_\lambda(g^{\beta\nu}R_{\mu\nu\alpha\beta}) - R_{\mu\nu\alpha\beta}D_\lambda g^{\beta\nu} = D_\lambda(g^{\beta\nu}R_{\mu\nu\alpha\beta}) \\ &= D_\lambda(g^{\beta\nu}R_{\nu\mu\beta\alpha}) = D_\lambda R_{\mu\alpha}. \end{aligned} \tag{10.234}$$

Similarly, it follows that the second term in (10.233) is given by

$$g^{\alpha\lambda}D_\nu R_{\lambda\mu\alpha\beta} = D_\nu(g^{\alpha\lambda}R_{\lambda\mu\alpha\beta}) = D_\nu R_{\mu\beta}. \tag{10.235}$$

The final term in (10.233) can be written

$$\begin{aligned} g^{\alpha\lambda}g^{\beta\nu}D_\mu R_{\nu\lambda\alpha\beta} &= D_\mu(g^{\alpha\lambda}g^{\beta\nu}R_{\nu\lambda\alpha\beta}) = -D_\mu(g^{\alpha\lambda}g^{\beta\nu}R_{\nu\lambda\beta\alpha}) \\ &= -D_\mu(g^{\alpha\lambda}R_{\lambda\alpha}) = -D_\mu R. \end{aligned} \tag{10.236}$$

Returning (10.234), (10.235), and (10.236) to (10.233) gives

$$\begin{aligned} g^{\alpha\lambda}D_\lambda R_{\mu\alpha} + g^{\beta\nu}D_\nu R_{\mu\beta} - D_\mu R &= D^\alpha R_{\mu\alpha} + D^\beta R_{\mu\beta} - D_\mu R \\ &= 2D^\alpha R_{\mu\alpha} - D_\mu R = 0. \end{aligned} \tag{10.237}$$

This result allows the definition of the *Einstein tensor*,

$$G_{\mu\nu} = R_{\mu\nu} - \frac{1}{2}g_{\mu\nu}R. \tag{10.238}$$

Because the metric tensor $g_{\mu\nu}$ and the Ricci tensor $R_{\mu\nu}$ are symmetric, the Einstein tensor is also symmetric. Metric compatibility gives

$$D^\nu G_{\mu\nu} = D^\nu\left(R_{\mu\nu} - \frac{1}{2}g_{\mu\nu}R\right) = D^\nu R_{\mu\nu} - \frac{1}{2}g_{\mu\nu}D^\nu R = D^\nu R_{\mu\nu} - \frac{1}{2}D_\mu R = 0, \tag{10.239}$$

so that the Einstein tensor is *covariantly conserved.* Finally, the trace of the Einstein tensor is given by

$$\mathrm{Tr}\, G = G^\mu{}_\mu = g^{\mu\nu}R_{\mu\nu} - \frac{1}{2}g^{\mu\nu}g_{\mu\nu}R = R - \frac{1}{2}\delta^\mu{}_\mu R = -R. \tag{10.240}$$

The Einstein tensor plays a central role in general relativity.

10.4 The connection and curvature in differential geometry

The definition and use of the affine connection and the curvature in coordinate notation is accompanied by a proliferation of indices which can be cumbersome and mistake prone. Modern differential geometry allows a much more streamlined version of these definitions and relationships by adapting them to forms and coordinate-free notation. Since these techniques are often overlooked or introduced without explanation in physics texts, this section will present a reasonably concise introduction, drawing on the concepts of tensors and forms defined in chapter 1.

The relation of torsion to the tetrad formalism, defined by (10.154), can be written in terms of forms. The tetrad or vielbein one-form is defined as

$$\boldsymbol{e}^a = e_\mu{}^a \mathrm{d}x^\mu, \tag{10.241}$$

the spin connection one-form is defined as

$$\boldsymbol{\Omega}^{ab} = \Omega_\mu{}^{ab} \mathrm{d}x^\mu, \tag{10.242}$$

and the torsion two-form is defined as

$$\mathbf{T}^a = e^a{}_\lambda T^\lambda_{\ \mu\nu}\, \mathrm{d}x^\mu \wedge \mathrm{d}x^\nu. \tag{10.243}$$

The next step is to introduce the covariant exterior derivative d_Ω, whose action on the tetrad one-form is defined as

$$\mathrm{d}_\Omega \boldsymbol{e}^a = \mathrm{d}\boldsymbol{e}^a + \boldsymbol{\Omega}^a{}_b \wedge \boldsymbol{e}^b, \tag{10.244}$$

where d is the exterior derivative as defined in (1.192).

Using the covariant exterior derivative (10.244) allows the torsion relationship (10.154) to be written

$$\mathbf{T}^a = \mathrm{d}_\Omega \boldsymbol{e}^a = \mathrm{d}\boldsymbol{e}^a + \boldsymbol{\Omega}^a{}_b \wedge \boldsymbol{e}^b, \tag{10.245}$$

which is known as the *first Maurer–Cartan structure equation*. The proof of (10.245) begins by using the definition of the wedge product (1.178) and its antisymmetry, which gives the exterior derivative

$$\mathrm{d}\boldsymbol{e}^a = \partial_\mu e_\nu{}^a \mathrm{d}x^\mu \wedge \mathrm{d}x^\nu = \frac{1}{2}(\partial_\mu e_\nu{}^a - \partial_\nu e_\mu{}^a)\mathrm{d}x^\mu \wedge \mathrm{d}x^\nu. \tag{10.246}$$

The second step is to apply the same treatment to the second term in (10.245), which gives

$$\boldsymbol{\Omega}^a{}_b \wedge \boldsymbol{e}^b = \Omega^a_{\mu b} e_\nu{}^b \mathrm{d}x^\mu \wedge \mathrm{d}x^\nu = \frac{1}{2}(\Omega^a_{\mu b} e_\nu{}^b - \Omega^a_{\nu b} e_\mu{}^b)\mathrm{d}x^\mu \wedge \mathrm{d}x^\nu. \tag{10.247}$$

Comparing the sum of (10.246) and (10.247) with form (10.154) for the torsion tensor written in terms of the spin connection shows that

$$\begin{aligned} \mathrm{d}_\Omega e^a &= \mathrm{d}e^a + \boldsymbol{\Omega}^a{}_b \wedge e^b = \frac{1}{2}(\partial_\mu e_\nu{}^a - \partial_\nu e_\mu{}^a + \Omega^a_{\mu b} e_\nu{}^b - \Omega^a_{\nu b} e_\mu{}^b)\mathrm{d}x^\mu \wedge \mathrm{d}x^\nu \\ &= T^a_{\mu\nu}\mathrm{d}x^\mu \wedge \mathrm{d}x^\nu = \mathbf{T}^a, \end{aligned} \tag{10.248}$$

which completes the demonstration of (10.245).

By construction the curvature tensor obeys

$$R^\mu{}_{\nu\alpha\beta} = -R^\mu{}_{\nu\beta\alpha}, \tag{10.249}$$

and so it can be contracted with the wedge product $\mathrm{d}x^\alpha \wedge \mathrm{d}y^\beta$ to create the *curvature two-form*,

$$\boldsymbol{R}^\mu{}_\nu = \frac{1}{2} R^\mu{}_{\nu\alpha\beta}\, \mathrm{d}x^\alpha \wedge \mathrm{d}y^\beta. \tag{10.250}$$

The wedge product is proportional to the infinitesimal oriented area $\mathrm{d}S_{\rho\sigma}$,

$$\mathrm{d}x^\alpha \wedge \mathrm{d}y^\beta = \varepsilon^{\alpha\beta\rho\sigma}\mathrm{d}S_{\rho\sigma}, \tag{10.251}$$

which shows that the infinitesimal parallel transport holonomy (10.208) can be written

$$H^\mu = A_1^\mu - A_2^\mu = \varepsilon^{\alpha\beta\rho\sigma} R^\mu{}_{\nu\alpha\beta}\, A^\nu\, \mathrm{d}S_{\rho\sigma}. \tag{10.252}$$

The coordinate version of the *second Maurer–Cartan structure equation* can be found by relating the curvature two-form $\boldsymbol{R}^\mu{}_\nu$ defined in (10.250) to the affine connection written as a one-form. The connection one-form is denoted by $\boldsymbol{\gamma}^\mu{}_\nu$ and is given by

$$\boldsymbol{\gamma}^\mu{}_\nu = \Gamma^\mu_{\beta\nu}\mathrm{d}x^\beta. \tag{10.253}$$

It will be useful for future developments to examine the transformation property of (10.253). This is given by combining the affine connection transformation (10.87) with the transformation of the differential to obtain

$$\begin{aligned} \gamma'^\mu{}_\nu &= \Gamma'^\mu_{\beta\nu}\mathrm{d}x'^\beta = \frac{\partial x'^\mu}{\partial x^\rho}\frac{\partial x^\sigma}{\partial x'^\nu}\Gamma^\rho_{\beta\sigma}\mathrm{d}x^\beta - \frac{\partial^2 x'^\mu}{\partial x^\alpha\,\partial x^\beta}\mathrm{d}x^\alpha\frac{\partial x^\beta}{\partial x'^\nu} \\ &= \frac{\partial x'^\mu}{\partial x^\rho}\frac{\partial x^\sigma}{\partial x'^\nu}\gamma^\rho{}_\sigma - \mathrm{d}\left(\frac{\partial x'^\mu}{\partial x^\beta}\right)\frac{\partial x^\beta}{\partial x'^\nu}. \end{aligned} \tag{10.254}$$

Applying the exterior derivative and using the antisymmetry of the wedge product gives

$$\mathrm{d}\boldsymbol{\gamma}^\mu{}_\nu = \partial_\alpha\Gamma^\mu_{\beta\nu}\,\mathrm{d}x^\alpha \wedge \mathrm{d}x^\beta = \frac{1}{2}(\partial_\alpha\Gamma^\mu_{\beta\nu} - \partial_\beta\Gamma^\mu_{\alpha\nu})\,\mathrm{d}x^\alpha \wedge \mathrm{d}x^\beta. \tag{10.255}$$

Similarly, contracting the wedge product of (10.253) with itself gives

$$\boldsymbol{\gamma}^{\mu}{}_{\lambda} \wedge \boldsymbol{\gamma}^{\lambda}{}_{\nu} = \Gamma^{\mu}_{\alpha\lambda}\Gamma^{\lambda}_{\beta\nu}\mathrm{d}x^{\alpha} \wedge \mathrm{d}x^{\beta} = \frac{1}{2}(\Gamma^{\mu}_{\alpha\lambda}\Gamma^{\lambda}_{\beta\nu} - \Gamma^{\mu}_{\beta\lambda}\Gamma^{\lambda}_{\alpha\nu})\,\mathrm{d}x^{\alpha} \wedge \mathrm{d}x^{\beta}. \tag{10.256}$$

Adding (10.255) and (10.256) gives

$$\begin{aligned}\mathrm{d}\boldsymbol{\gamma}^{\mu}{}_{\nu} + \boldsymbol{\gamma}^{\mu}{}_{\lambda} \wedge \boldsymbol{\gamma}^{\lambda}{}_{\nu} &= \frac{1}{2}(\partial_{\alpha}\Gamma^{\mu}_{\beta\nu} - \partial_{\beta}\Gamma^{\mu}_{\alpha\nu} + \Gamma^{\mu}_{\alpha\lambda}\Gamma^{\lambda}_{\beta\nu} - \Gamma^{\mu}_{\beta\lambda}\Gamma^{\lambda}_{\alpha\nu})\,\mathrm{d}x^{\alpha} \wedge \mathrm{d}x^{\beta} \\ &= \frac{1}{2}R^{\mu}{}_{\nu\alpha\beta}\,\mathrm{d}x^{\alpha} \wedge \mathrm{d}x^{\beta} = \boldsymbol{R}^{\mu}{}_{\nu}.\end{aligned} \tag{10.257}$$

This yields the coordinate basis version of the second Maurer–Cartan structure equation,

$$\boldsymbol{R}^{\mu}{}_{\nu} = \mathrm{d}\boldsymbol{\gamma}^{\mu}{}_{\nu} + \boldsymbol{\gamma}^{\mu}{}_{\lambda} \wedge \boldsymbol{\gamma}^{\lambda}{}_{\nu}. \tag{10.258}$$

The second Maurer–Cartan structure equation is often written in the tetrad basis, where the spin connection one-form replaces the affine connection one-form,

$$\mathbf{R}^{a}{}_{b} = \mathrm{d}_{\Omega}\boldsymbol{\Omega}^{a}{}_{b} = \mathrm{d}\boldsymbol{\Omega}^{a}{}_{b} + \boldsymbol{\Omega}^{a}{}_{c} \wedge \boldsymbol{\Omega}^{c}{}_{b}. \tag{10.259}$$

The second Maurer–Cartan structure equation (10.259) in the tetrad basis is understood to be the *definition* of the curvature in the tetrad basis.

The exterior derivative of $\boldsymbol{R}^{\mu}{}_{\nu}$ is the three-form given by

$$\mathrm{d}\boldsymbol{R}^{\mu}{}_{\nu} = \frac{1}{2}\partial_{\rho}R^{\mu}{}_{\nu\alpha\beta}\,\mathrm{d}x^{\rho} \wedge \mathrm{d}x^{\alpha} \wedge \mathrm{d}x^{\beta}. \tag{10.260}$$

The wedge product has the cyclic permutation property

$$\mathrm{d}x^{\rho} \wedge \mathrm{d}x^{\alpha} \wedge \mathrm{d}x^{\beta} = \mathrm{d}x^{\beta} \wedge \mathrm{d}x^{\rho} \wedge \mathrm{d}x^{\alpha} = \mathrm{d}x^{\alpha} \wedge \mathrm{d}x^{\beta} \wedge \mathrm{d}x^{\rho}, \tag{10.261}$$

so that the exterior derivative of the curvature two-form can be written

$$\mathrm{d}\boldsymbol{R}^{\mu}{}_{\nu} = \frac{1}{6}\left(\partial_{\rho}R^{\mu}{}_{\nu\alpha\beta} + \partial_{\beta}R^{\mu}{}_{\nu\rho\alpha} + \partial_{\alpha}R^{\mu}{}_{\nu\beta\rho}\right)\mathrm{d}x^{\rho} \wedge \mathrm{d}x^{\alpha} \wedge \mathrm{d}x^{\beta}. \tag{10.262}$$

Using the coordinate version (10.258) of the second Maurer–Cartan structure equation and property (1.196) of the exterior derivative for wedge products gives

$$\begin{aligned}\mathrm{d}\boldsymbol{R}^{\mu}{}_{\nu} &= \mathrm{d}^2\boldsymbol{\gamma}^{\mu}{}_{\nu} + \mathrm{d}\boldsymbol{\gamma}^{\mu}{}_{\lambda} \wedge \boldsymbol{\gamma}^{\lambda}{}_{\nu} - \boldsymbol{\gamma}^{\mu}{}_{\lambda} \wedge \mathrm{d}\boldsymbol{\gamma}^{\lambda}{}_{\nu} \\ &= \mathrm{d}\boldsymbol{\gamma}^{\mu}{}_{\lambda} \wedge \boldsymbol{\gamma}^{\lambda}{}_{\nu} - \boldsymbol{\gamma}^{\mu}{}_{\lambda} \wedge \mathrm{d}\boldsymbol{\gamma}^{\lambda}{}_{\nu} \\ &= (\boldsymbol{R}^{\mu}{}_{\lambda} - \boldsymbol{\gamma}^{\mu}{}_{\sigma} \wedge \boldsymbol{\gamma}^{\sigma}{}_{\lambda}) \wedge \boldsymbol{\gamma}^{\lambda}{}_{\nu} - \boldsymbol{\gamma}^{\mu}{}_{\lambda} \wedge (\boldsymbol{R}^{\lambda}{}_{\nu} - \boldsymbol{\gamma}^{\lambda}{}_{\sigma} \wedge \boldsymbol{\gamma}^{\sigma}{}_{\nu}) \\ &= \boldsymbol{R}^{\mu}{}_{\lambda} \wedge \boldsymbol{\gamma}^{\lambda}{}_{\nu} - \boldsymbol{\gamma}^{\mu}{}_{\lambda} \wedge \boldsymbol{R}^{\lambda}{}_{\nu} - \boldsymbol{\gamma}^{\mu}{}_{\sigma} \wedge \boldsymbol{\gamma}^{\sigma}{}_{\lambda} \wedge \boldsymbol{\gamma}^{\lambda}{}_{\nu} + \boldsymbol{\gamma}^{\mu}{}_{\lambda} \wedge \boldsymbol{\gamma}^{\lambda}{}_{\sigma} \wedge \boldsymbol{\gamma}^{\sigma}{}_{\nu} \\ &= \boldsymbol{R}^{\mu}{}_{\lambda} \wedge \boldsymbol{\gamma}^{\lambda}{}_{\nu} - \boldsymbol{\gamma}^{\mu}{}_{\lambda} \wedge \boldsymbol{R}^{\lambda}{}_{\nu} \\ &= \frac{1}{2}R^{\mu}{}_{\lambda\alpha\beta}\Gamma^{\lambda}_{\rho\nu}\mathrm{d}x^{\alpha} \wedge \mathrm{d}x^{\beta} \wedge \mathrm{d}x^{\rho} - \frac{1}{2}\Gamma^{\mu}_{\rho\lambda}R^{\lambda}{}_{\nu\alpha\beta}\mathrm{d}x^{\rho} \wedge \mathrm{d}x^{\alpha} \wedge \mathrm{d}x^{\beta} \\ &= \frac{1}{2}(\Gamma^{\lambda}_{\rho\nu}R^{\mu}{}_{\lambda\alpha\beta} - \Gamma^{\mu}_{\rho\lambda}R^{\lambda}{}_{\nu\alpha\beta})\mathrm{d}x^{\rho} \wedge \mathrm{d}x^{\alpha} \wedge \mathrm{d}x^{\beta}.\end{aligned} \tag{10.263}$$

Applying the same cyclic permutations as were used in (10.261) shows that the final expression in (10.263) can be written

$$\begin{aligned} \mathrm{d}\boldsymbol{R}^{\mu}{}_{\nu} = \frac{1}{6}\Big(&\Gamma^{\lambda}_{\rho\nu}R^{\mu}{}_{\lambda\alpha\beta} - \Gamma^{\mu}_{\rho\lambda}R^{\mu}{}_{\nu\alpha\beta} + \Gamma^{\lambda}_{\beta\nu}R^{\mu}{}_{\lambda\rho\alpha} - \Gamma^{\mu}_{\beta\lambda}R^{\lambda}{}_{\nu\rho\alpha} \\ &+ \Gamma^{\lambda}_{\alpha\nu}R^{\mu}{}_{\lambda\beta\rho} - \Gamma^{\mu}_{\alpha\lambda}R^{\lambda}{}_{\nu\beta\rho}\Big)\, \mathrm{d}x^{\rho}\wedge \mathrm{d}x^{\alpha}\wedge \mathrm{d}x^{\beta}. \end{aligned} \tag{10.264}$$

Setting the two results (10.262) and (10.264) equal to each other and dropping the common factor of 1/6 gives

$$\begin{aligned} 0 = \Big(&\partial_{\rho}R^{\mu}{}_{\nu\alpha\beta} - \Gamma^{\lambda}_{\rho\nu}R^{\mu}{}_{\lambda\alpha\beta} + \Gamma^{\mu}_{\rho\lambda}R^{\mu}{}_{\nu\alpha\beta} \\ &+ \partial_{\beta}R^{\mu}{}_{\nu\rho\alpha} - \Gamma^{\lambda}_{\beta\nu}R^{\mu}{}_{\lambda\rho\alpha} + \Gamma^{\mu}_{\beta\lambda}R^{\lambda}{}_{\nu\rho\alpha} \\ &+ \partial_{\alpha}R^{\mu}{}_{\nu\beta\rho} - \Gamma^{\lambda}_{\alpha\nu}R^{\mu}{}_{\lambda\beta\rho} + \Gamma^{\mu}_{\alpha\lambda}R^{\lambda}{}_{\nu\beta\rho}\Big)\mathrm{d}x^{\rho}\wedge \mathrm{d}x^{\alpha}\wedge \mathrm{d}x^{\beta}. \end{aligned} \tag{10.265}$$

Using the definition of the covariant derivative of $R^{\mu}{}_{\nu\alpha\beta}$ gives

$$\begin{aligned} \partial_{\rho}R^{\mu}{}_{\nu\alpha\beta} - \Gamma^{\lambda}_{\rho\nu}R^{\mu}{}_{\lambda\alpha\beta} + \Gamma^{\mu}_{\rho\lambda}R^{\mu}{}_{\nu\alpha\beta} &= D_{\rho}R^{\mu}{}_{\nu\alpha\beta} + \Gamma^{\lambda}_{\rho\alpha}R^{\mu}{}_{\nu\lambda\beta} + \Gamma^{\lambda}_{\rho\beta}R^{\mu}{}_{\nu\alpha\lambda}, \\ \partial_{\beta}R^{\mu}{}_{\nu\rho\alpha} - \Gamma^{\lambda}_{\beta\nu}R^{\mu}{}_{\lambda\rho\alpha} + \Gamma^{\mu}_{\beta\lambda}R^{\lambda}{}_{\nu\rho\alpha} &= D_{\beta}R^{\mu}{}_{\nu\rho\alpha} + \Gamma^{\lambda}_{\beta\rho}R^{\mu}{}_{\nu\lambda\alpha} + \Gamma^{\lambda}_{\beta\alpha}R^{\mu}{}_{\nu\rho\lambda}, \\ \partial_{\alpha}R^{\mu}{}_{\nu\beta\rho} - \Gamma^{\lambda}_{\alpha\nu}R^{\mu}{}_{\lambda\beta\rho} + \Gamma^{\mu}_{\alpha\lambda}R^{\lambda}{}_{\nu\beta\rho} &= D_{\alpha}R^{\mu}{}_{\nu\beta\rho} + \Gamma^{\lambda}_{\alpha\beta}R^{\mu}{}_{\nu\lambda\rho} + \Gamma^{\lambda}_{\alpha\rho}R^{\mu}{}_{\nu\beta\lambda}. \end{aligned} \tag{10.266}$$

Using the results of (10.266) in (10.265) gives

$$\begin{aligned} 0 = \Big(&D_{\rho}R^{\mu}{}_{\nu\alpha\beta} + \Gamma^{\lambda}_{\rho\alpha}R^{\mu}{}_{\nu\lambda\beta} + \Gamma^{\lambda}_{\rho\beta}R^{\mu}{}_{\nu\alpha\lambda} \\ &+ D_{\beta}R^{\mu}{}_{\nu\rho\alpha} + \Gamma^{\lambda}_{\beta\rho}R^{\mu}{}_{\nu\lambda\alpha} + \Gamma^{\lambda}_{\beta\alpha}R^{\mu}{}_{\nu\rho\lambda} \\ &+ D_{\alpha}R^{\mu}{}_{\nu\beta\rho} + \Gamma^{\lambda}_{\alpha\beta}R^{\mu}{}_{\nu\lambda\rho} + \Gamma^{\lambda}_{\alpha\rho}R^{\mu}{}_{\nu\beta\lambda}\Big)\mathrm{d}x^{\rho}\wedge \mathrm{d}x^{\alpha}\wedge \mathrm{d}x^{\beta}. \end{aligned} \tag{10.267}$$

Using the curvature tensor property (10.212) gives

$$R^{\mu}{}_{\nu\beta\lambda} = -R^{\mu}{}_{\nu\lambda\beta},\quad R^{\mu}{}_{\nu\alpha\lambda} = -R^{\mu}{}_{\nu\lambda\alpha},\quad R^{\mu}{}_{\nu\rho\lambda} = -R^{\mu}{}_{\nu\rho\lambda}. \tag{10.268}$$

Recalling the definition of the torsion tensor $\mathcal{T}^{\lambda}_{\alpha\beta}$ given by (10.117) shows that (10.267) gives the second Bianchi identity in the presence of torsion,

$$D_{\rho}R^{\mu}{}_{\nu\alpha\beta} + 2\,\mathcal{T}^{\lambda}_{\alpha\beta}R^{\mu}{}_{\nu\lambda\rho} + D_{\beta}R^{\mu}{}_{\nu\rho\alpha} + 2\,\mathcal{T}^{\lambda}_{\rho\alpha}R^{\mu}{}_{\nu\lambda\beta} + D_{\alpha}R^{\mu}{}_{\nu\beta\rho} + 2\,\mathcal{T}^{\lambda}_{\beta\rho}R^{\mu}{}_{\nu\lambda\alpha} = 0. \tag{10.269}$$

In the event that the torsion tensor vanishes, (10.269) becomes the second Bianchi identity (10.221) derived using the torsion-free Levi-Civita connection.

Expressions such as (10.266) benefit greatly from a more streamlined approach. Modern differential geometry employs coordinate-free notation for tensors as well as the covariant scalar gradient ∇_X introduced in (10.113),

$$\nabla_X = X^{\mu}D_{\mu}, \tag{10.270}$$

where D_μ is the covariant derivative of (10.68). The covariant derivative is particularly simple when applied to a scalar φ, where (10.70) shows that

$$\nabla_X \varphi = X^\mu \partial_\mu \varphi. \tag{10.271}$$

Applying ∇_X to a vector Y in coordinate-free notation therefore gives

$$\begin{aligned}\nabla_X Y &= X^\mu \partial_\mu (Y^\nu \hat{\mathbf{e}}_\nu) = (X^\mu \partial_\mu Y^\nu)\hat{\mathbf{e}}_\nu + X^\mu Y^\nu \partial_\mu \hat{\mathbf{e}}_\nu \\ &= (X^\mu \partial_\mu Y^\nu)\hat{\mathbf{e}}_\nu + X^\mu Y^\nu \Gamma^\sigma_{\mu\nu} \hat{\mathbf{e}}_\sigma,\end{aligned} \tag{10.272}$$

where relation (10.80) was used to identify the affine connection in terms of the basis vector derivative.

Expression (10.272) gives

$$\nabla_Y X = (Y^\mu \partial_\mu Y^\nu)\hat{\mathbf{e}}_\nu + Y^\mu X^\nu \Gamma^\sigma_{\mu\nu} \hat{\mathbf{e}}_\sigma. \tag{10.273}$$

Switching dummy indices shows that the two expressions (10.272) and (10.273) combine to give

$$\nabla_X Y - \nabla_Y X = (X^\mu \partial_\mu Y^\nu - Y^\mu \partial_\mu X^\nu)\hat{\mathbf{e}}_\nu + (\Gamma^\sigma_{\mu\nu} - \Gamma^\sigma_{\nu\mu}) X^\mu Y^\nu \hat{\mathbf{e}}_\sigma. \tag{10.274}$$

The first expression on the right-hand side of (10.274) is recognized as the Lie bracket (1.219) of the two vectors,

$$[X, Y] = (X^\mu \partial_\mu Y^\nu - Y^\mu \partial_\mu X^\nu)\hat{\mathbf{e}}_\nu \equiv L^\nu(X, Y)\, \hat{\mathbf{e}}_\nu, \tag{10.275}$$

where the Cartan identification $\hat{\mathbf{e}}_\nu = \partial_\nu$ has been made. The second expression involves the torsion tensor $\mathcal{T}^\sigma_{\mu\nu}$ defined by (10.117). Treating the torsion tensor as a function of vectors, as in (1.104), allows the second term to be written as coordinate-free vector,

$$\frac{1}{2}(\Gamma^\sigma_{\mu\nu} - \Gamma^\sigma_{\nu\mu}) X^\mu Y^\nu \hat{\mathbf{e}}_\sigma = \mathcal{T}^\sigma_{\mu\nu} X^\mu Y^\nu \hat{\mathbf{e}}_\sigma = \mathcal{T}^\sigma(X, Y)\hat{\mathbf{e}}_\sigma \equiv \mathcal{T}(X, Y). \tag{10.276}$$

The coordinate-free expressions (10.274), (10.275), and (10.276) combine to define the action of the torsion tensor as a function of vectors,

$$\mathcal{T}(X, Y) = \frac{1}{2}\nabla_X Y - \frac{1}{2}\nabla_Y X - \frac{1}{2}[X, Y]. \tag{10.277}$$

The factor of $\frac{1}{2}$ arises from the definition of torsion given in (10.117) and is absent in some presentations.

A related expression for the curvature tensor is anticipated by (10.211), which states that

$$R^\mu{}_{\nu\alpha\beta} Z^\nu X^\alpha Y^\beta = X^\alpha Y^\beta [D_\alpha, D_\beta] Z^\mu. \tag{10.278}$$

It is left as an exercise to show that

$$\begin{aligned}[\nabla_X, \nabla_Y]Z &= L^\beta(X, Y)(\partial_\beta Z^\nu \hat{\mathbf{e}}_\nu + Z^\nu \Gamma^\lambda_{\beta\nu} \hat{\mathbf{e}}_\lambda) \\ &\quad + (X^\alpha Y^\beta - X^\beta Y^\alpha) Z^\nu (\partial_\alpha \Gamma^\mu_{\beta\nu} + \Gamma^\mu_{\alpha\lambda}\Gamma^\lambda_{\beta\nu})\hat{\mathbf{e}}_\mu,\end{aligned} \tag{10.279}$$

where Z is an arbitrary vector and $L^{\beta}(X, Y) = [X, Y]^{\beta}$ was defined in (10.275). It is also left as part of the same exercise to show that

$$\nabla_{[X,Y]}Z = L^{\beta}(X, Y)\partial_{\beta}(Z^{\nu}\hat{\mathbf{e}}_{\nu}) = L^{\beta}(X, Y)(\partial_{\beta}Z^{\nu}\hat{\mathbf{e}}_{\nu} + Z^{\nu}\Gamma^{\lambda}_{\beta\nu}\hat{\mathbf{e}}_{\lambda}). \tag{10.280}$$

Exercise 10.15: Verify (10.279) and (10.280).

Combining (10.279) and (10.280) and relabeling dummy indices gives the coordinate-free definition of the curvature tensor,

$$\begin{aligned}[\nabla_X, \nabla_Y]Z - \nabla_{[X,Y]}Z &= R^{\mu}{}_{\nu\alpha\beta}Z^{\nu}X^{\alpha}Y^{\beta}\hat{\mathbf{e}}_{\mu} = R^{\mu}(Z, X, Y)\hat{\mathbf{e}}_{\mu} \\ &\equiv R(Z, X, Y).\end{aligned} \tag{10.281}$$

10.5 Variational techniques in general relativity

Apart from the principle of equivalence very little of this chapter has considered the physics embodied in general relativity. In that regard the goal of modern field theory is to derive its equations of motion from the variation of an action that reflects the symmetries of the physical system under consideration. The principle of equivalence has led inexorably to the conclusion that a theory of gravitation requires its associated action to be invariant under general coordinate diffeomorphisms and that $g_{\mu\nu}$ is the field of gravitation. In order to define the action it is necessary to have a relevant definition of $\hbar$, similar to the definition (9.88) made for electrodynamics. Using Newton's universal gravitational constant G, it follows that $\hbar$ can be defined in terms of an inertial mass M as

$$\hbar = \frac{GM^2}{c}. \tag{10.282}$$

However, unlike the case of electrodynamics and the fundamental charge e, there is no obvious fundamental mass M that occurs in nature. In other words, the particle masses occurring in nature are not integer multiples of some fundamental mass, in contrast to the electric charges that occur, which are integer multiples of a fundamental charge. Instead, G, c, and Planck's constant $\hbar$ are combined to form what is called the *Planck length*,

$$\ell_P = \sqrt{\frac{\hbar G}{c^3}}, \tag{10.283}$$

which has the value $\ell_P \approx 1.6 \times 10^{-35}$ m in SI units. This length is generally believed to be the scale of *quantum gravity*, a topic of considerable research and far outside the scope of this book. As a result, Planck's constant will be used to create a dimensionless action.

10.5.1 General variational properties

The curvature scalar R will provide the means to construct an action, but first some general results must be established. The diffeomorphism matrix $\mathbf{M}$ and its inverse $\mathbf{M}^{-1}$ have the elements

$$M^{\mu}{}_{\nu}(x) = \frac{\partial x'^{\mu}}{\partial x^{\nu}}, \quad (M^{-1})^{\rho}_{\mu}(x) = \frac{\partial x^{\rho}}{\partial x'^{\mu}}. \tag{10.284}$$

The matrix $\mathbf{M}$ is no longer the orthogonal matrix of special relativity, and so $\det \mathbf{M} = 1$ *no longer holds.* This has the immediate consequence that a general diffeomorphism causes the space-time volume element d^4x to pick up a nontrivial Jacobian,

$$\mathrm{d}^4x' = \mathrm{d}^4x \, |\det \mathbf{M}|. \tag{10.285}$$

The metric tensor can be treated as the matrix $\boldsymbol{g}$. In so doing it is important understand that the elements of the metric matrix are the *covariant components* of the metric tensor. Using the covariant components shows that the metric matrix transforms according to

$$g'_{\mu\nu}(x') = \frac{\partial x^{\alpha}}{\partial x^{\mu'}} \frac{\partial x^{\beta}}{\partial x^{\nu'}} g_{\alpha\beta}(x) = (M^{-1})^{\alpha}{}_{\mu}(x) g_{\alpha\beta}(x) (M^{-1})^{\beta}{}_{\nu}(x). \tag{10.286}$$

This shows that the determinant of $\boldsymbol{g}$ transforms as

$$\det \boldsymbol{g}' = \det(\mathbf{M}^{-1} \cdot \boldsymbol{g} \cdot \mathbf{M}^{-1}) = (\det \mathbf{M})^{-2} \det \boldsymbol{g}. \tag{10.287}$$

For simplicity the determinant of the metric tensor is denoted by g, so that

$$g(x) = \det \boldsymbol{g}(x). \tag{10.288}$$

There is a subtlety that arises from the choice of the signature of the metric used in this text, which is that g is negative, just as $\det \eta$ is negative. This becomes important when the square root of the determinant is considered. In that regard it follows from (10.287) that a general diffeomorphism gives

$$\sqrt{-g'} = \frac{\sqrt{-g}}{|\det \mathbf{M}|}, \tag{10.289}$$

where the absolute value preserves the signature of the metric. The *diffeomorphism invariant volume element* can therefore be written $\mathrm{d}^4x\sqrt{|g|}$, since this gives

$$\mathrm{d}^4x \sqrt{|g|} \implies \mathrm{d}^4x' \sqrt{|g'|} = \mathrm{d}^4x \sqrt{|g|}. \tag{10.290}$$

Applying variational techniques to an action in which the invariant volume element is present requires an understanding of the behavior of the determinant g under the infinitesimal variation of $g_{\mu\nu}$,

$$g_{\mu\nu} \rightarrow g_{\mu\nu} + \delta g_{\mu\nu}. \tag{10.291}$$

Deriving a simple form for the variation of the determinant is expedited by using the matrix identities (7.77) and (7.79),

$$\ln \det \boldsymbol{g} = \operatorname{Tr} \ln \boldsymbol{g}, \tag{10.292}$$

$$\operatorname{Tr}(\ln \mathbf{A} \cdot \mathbf{B}) = \operatorname{Tr}(\ln \mathbf{A} + \ln \mathbf{B}). \tag{10.293}$$

The analysis begins by noting that $g \rightarrow g + \delta g$ gives

$$\begin{aligned} \delta \ln(-g) &= \ln(-g - \delta g) - \ln(-g) = \ln\left(-g\left(1 + \frac{\delta g}{g}\right)\right) - \ln(-g) \\ &= \ln\left(1 + \frac{\delta g}{g}\right) = \frac{\delta g}{g}. \end{aligned} \tag{10.294}$$

Treating the metric tensor as the matrix $\boldsymbol{g}$ with the elements $g_{\mu\nu}$ shows that

$$\begin{aligned} \delta \operatorname{Tr}(\ln \boldsymbol{g}) &= \operatorname{Tr}(\ln(\boldsymbol{g} + \delta\boldsymbol{g}) - \ln \boldsymbol{g}) \\ &= \operatorname{Tr}(\ln(\boldsymbol{g} \cdot (1 + \boldsymbol{g}^{-1} \cdot \delta\boldsymbol{g})) - \ln \boldsymbol{g}) \\ &= \operatorname{Tr}(\ln(1 + \boldsymbol{g}^{-1} \cdot \delta\boldsymbol{g})) = \operatorname{Tr}(\boldsymbol{g}^{-1} \cdot \delta\boldsymbol{g}) = g^{\mu\nu}\delta g_{\mu\nu}, \end{aligned} \tag{10.295}$$

where the elements of $\boldsymbol{g}^{-1}$ are $g^{\mu\nu}$. The same result follows from the cyclic property (7.68) of the trace operation, which gives

$$\begin{aligned} \delta \operatorname{Tr}(\ln \boldsymbol{g}) &= \delta \operatorname{Tr}\left(\sum_{n=1}^{\infty} \frac{1}{n}(\boldsymbol{g} - \mathbf{I})^n\right) \\ &= \operatorname{Tr}\left(\sum_{n=0}^{\infty}(\boldsymbol{g} - \mathbf{I})^n \cdot \delta\boldsymbol{g}\right) = \operatorname{Tr}\left\{\left(\sum_{n=0}^{\infty}(\boldsymbol{g} - \mathbf{I})^n\right) \cdot \delta\boldsymbol{g}\right\} \\ &= \operatorname{Tr}(\boldsymbol{g}^{-1} \cdot \delta\boldsymbol{g}) = g^{\mu\nu}\delta g_{\mu\nu}. \end{aligned} \tag{10.296}$$

This gives

$$\delta \ln(|g|) = g^{\mu\nu}\delta g_{\mu\nu}. \tag{10.297}$$

Combining this with (10.294) gives *Jacobi's formula*,

$$\delta g = g\, g^{\mu\nu}\delta g_{\mu\nu}. \tag{10.298}$$

By noting that

$$\delta(g^{\mu\rho}g_{\rho\nu}) = \delta(\delta^{\mu}{}_{\nu}) = 0 \implies \delta g^{\mu\rho}g_{\rho\nu} + g^{\mu\rho}\delta g_{\rho\nu} = 0, \tag{10.299}$$

it follows that

$$\delta g^{\mu\rho} g_{\rho\nu} = -g^{\mu\rho} \delta g_{\rho\nu}. \tag{10.300}$$

Contracting (10.300) with $g^{\nu\sigma}$ gives

$$\delta g^{\mu\sigma} = -g^{\mu\rho} g^{\nu\sigma} \delta g_{\rho\nu}. \tag{10.301}$$

Using (10.300) in (10.298) gives a useful variant of Jacobi's formula,

$$\delta g = -g \; g_{\mu\nu} \; \delta g^{\mu\nu}. \tag{10.302}$$

Exercise 10.16: Using the approach of (10.296) show that

$$\partial_\rho \operatorname{Tr}(\ln \boldsymbol{g}) = g^{\mu\nu} \partial_\rho g_{\mu\nu}. \tag{10.303}$$

Using (10.292) shows that

$$\partial_\rho \operatorname{Tr}(\ln \boldsymbol{g}) = \partial_\rho \ln \det \boldsymbol{g} = \frac{1}{\det \boldsymbol{g}} \partial_\rho \det \boldsymbol{g} = \frac{1}{g} \partial_\rho g. \tag{10.304}$$

Equating (10.303) and (10.304) gives

$$\partial_\rho g = g \; g^{\mu\nu} \partial_\rho g_{\mu\nu}. \tag{10.305}$$

For the choice that the affine connection $\Gamma^{\nu}_{\mu\rho}$ is given by the Levi-Civita connection (10.141), the symmetry of the metric tensor $g_{\mu\nu} = g_{\nu\mu}$ shows that the contraction of the affine connection indices gives

$$\begin{aligned} \Gamma^{\mu}_{\mu\rho} &= \frac{1}{2} g^{\mu\nu}\left(\partial_\mu g_{\nu\rho} + \partial_\rho g_{\nu\mu} - \partial_\nu g_{\mu\rho}\right) \\ &= \frac{1}{2} g^{\mu\nu}\left(\partial_\mu g_{\nu\rho} - \partial_\nu g_{\mu\rho}\right) + \frac{1}{2} g^{\mu\nu} \partial_\rho g_{\nu\mu} = \frac{1}{2} g^{\mu\nu} \partial_\rho g_{\mu\nu} = \frac{\partial_\rho g}{2g}, \end{aligned} \tag{10.306}$$

where relation (10.305) was used. This gives

$$\partial_\rho g = 2g \; \Gamma^{\mu}_{\mu\rho}. \tag{10.307}$$

It is important to note that (10.307) is independent of the sign of g, so it can be written

$$-\partial_\rho g = -2g \; \Gamma^{\mu}_{\mu\rho} \implies \partial_\rho |g| = 2|g| \Gamma^{\mu}_{\mu\rho}. \tag{10.308}$$

Using (10.308) shows that

$$\begin{aligned} \sqrt{|g|} \; \partial_\mu X^\mu &= \partial_\mu(\sqrt{|g|} \; X^\mu) - X^\mu \partial_\mu \sqrt{|g|} = \partial_\mu(\sqrt{|g|} \; X^\mu) - \frac{1}{2} X^\mu \frac{\partial_\mu |g|}{\sqrt{|g|}} \\ &= \partial_\mu(\sqrt{|g|} \; X^\mu) - \sqrt{|g|} \; \Gamma^{\nu}_{\nu\mu} X^\mu. \end{aligned} \tag{10.309}$$

Rearranging (10.309) gives

$$\sqrt{|g|}\left(\partial_\mu X^\mu + \Gamma^\nu_{\nu\mu} A^\mu\right) = \sqrt{|g|}\, D_\mu X^\mu = \partial_\mu(\sqrt{|g|}\ X^\mu). \tag{10.310}$$

Result (10.310) allows the diffeomorphic extension of the divergence theorem,

$$\int \mathrm{d}^4x\ \sqrt{|g|}\ D_\mu X^\mu = \int \mathrm{d}^4x\ \partial_\mu(\sqrt{|g|}\ X^\mu) = \oint_S \mathrm{d}S_\mu\ \sqrt{|g|}\ X^\mu, \tag{10.311}$$

where $\mathrm{d}S_\mu$ are the four three-volume elements that form the surface of four-dimensional space. Result (10.311) allows integration by parts in a general manifold *as long as covariant derivatives are present.*

10.5.2 Point-mass motion in curved space-time

An important application of variational techniques is to derive the condition for *geodesic motion.* The goal is to extend the relativistic motion of a free particle derived from the Minkowski space action (6.179) to the case that the proper time element is given by the diffeomorphic invariant expression

$$c\ \mathrm{d}\tau = \sqrt{g_{\mu\nu}(x)\ \mathrm{d}x^\mu\ \mathrm{d}x^\nu}. \tag{10.312}$$

The geometric analogy is that of finding a parameterized path that *minimizes the distance between two points,* which is defined as a *geodesic.*

The procedure is quite similar to (6.179), so that the action to be extremized is given by

$$S = \int_{\tau_0}^{\tau_f} \mathrm{d}\tau\ \mathcal{L}_G = \int_{\tau_0}^{\tau_f} \mathrm{d}\tau\ \frac{1}{2} m\ g_{\mu\nu}(x)\ \frac{\mathrm{d}x^\mu}{\mathrm{d}\tau}\frac{\mathrm{d}x^\nu}{\mathrm{d}\tau}, \tag{10.313}$$

where an arbitrary symmetric and invertible metric tensor characterizes the space-time. Varying the action according to $x^\mu \to x^\mu + \delta x^\mu$ gives, to the first order

$$\delta S = \int_{\tau_0}^{\tau_f} \mathrm{d}\tau\ m\left(g_{\mu\rho}\frac{\mathrm{d}x^\mu}{\mathrm{d}\tau}\frac{\mathrm{d}\delta x^\rho}{\mathrm{d}\tau} + \frac{1}{2}\,\partial_\rho g_{\mu\nu}\frac{\mathrm{d}x^\mu}{\mathrm{d}\tau}\frac{\mathrm{d}x^\nu}{\mathrm{d}\tau}\delta x^\rho\right). \tag{10.314}$$

Expression (10.314) can be integrated by parts and the endpoints dropped, which yields

$$\delta S = \int_{\tau_0}^{\tau_f} \mathrm{d}\tau\ m\left(-\frac{\mathrm{d}}{\mathrm{d}\tau}\left(g_{\mu\rho}\frac{\mathrm{d}x^\mu}{\mathrm{d}\tau}\right) + \frac{1}{2}\partial_\rho g_{\mu\nu}\frac{\mathrm{d}x^\mu}{\mathrm{d}\tau}\frac{\mathrm{d}x^\nu}{\mathrm{d}\tau}\right)\delta x^\rho. \tag{10.315}$$

Demanding that the variation vanishes gives a useful property of the motion along an extremal path,

$$\frac{1}{2}\partial_\rho g_{\mu\nu}\frac{\mathrm{d}x^\mu}{\mathrm{d}\tau}\frac{\mathrm{d}x^\nu}{\mathrm{d}\tau} = \frac{\mathrm{d}}{\mathrm{d}\tau}\left(g_{\mu\rho}\frac{\mathrm{d}x^\mu}{\mathrm{d}\tau}\right) = \frac{\mathrm{d}g_{\mu\rho}}{\mathrm{d}\tau}\frac{\mathrm{d}x^\mu}{\mathrm{d}\tau} + g_{\mu\rho}\frac{\mathrm{d}^2x^\mu}{\mathrm{d}\tau^2}. \tag{10.316}$$

It is important to note that (10.316) is independent of mass. This will be valuable later in finding constants of gravitational motion. Using the chain rule, the symmetry of the metric tensor, and relabeling dummy indices gives

$$\frac{\mathrm{d}g_{\mu\rho}}{\mathrm{d}\tau}\frac{\mathrm{d}x^\mu}{\mathrm{d}\tau} = \partial_\nu g_{\mu\rho}\frac{\mathrm{d}x^\nu}{\mathrm{d}\tau}\frac{\mathrm{d}x^\mu}{\mathrm{d}\tau} = \frac{1}{2}\left(\partial_\nu g_{\rho\mu} + \partial_\mu g_{\rho\nu}\right)\frac{\mathrm{d}x^\mu}{\mathrm{d}\tau}\frac{\mathrm{d}x^\nu}{\mathrm{d}\tau}. \tag{10.317}$$

Using this result in (10.316) shows that the trajectory that extremizes the action satisfies

$$g_{\mu\rho}\frac{\mathrm{d}^2x^\mu}{\mathrm{d}\tau^2} + \frac{1}{2}\left(\partial_\nu g_{\rho\mu} + \partial_\mu g_{\rho\nu} - \partial_\rho g_{\mu\nu}\right)\frac{\mathrm{d}x^\mu}{\mathrm{d}\tau}\frac{\mathrm{d}x^\nu}{\mathrm{d}\tau} = 0. \tag{10.318}$$

Contracting this against $g^{\alpha\rho}$ and using the definition (10.141) of the Levi-Civita connection, the geodesic path that extremizes the action satisfies

$$\frac{\mathrm{d}^2x^\alpha}{\mathrm{d}\tau^2} + \Gamma^\alpha_{\mu\nu}\frac{\mathrm{d}x^\mu}{\mathrm{d}\tau}\frac{\mathrm{d}x^\nu}{\mathrm{d}\tau} = \frac{\mathrm{d}V^\alpha}{\mathrm{d}\tau} + \Gamma^\alpha_{\mu\nu}V^\mu V^\nu = 0, \tag{10.319}$$

which is identical to the parallel transport equation for the proper velocity V^α found in (10.100). Result (10.319) is independent of the test mass m for the case that gravity alone is acting on the mass and so, as Galileo observed long ago, all point masses will fall identically. This is the means by which gravitation is included in the general relativistic version of Newton's law of motion. Any additional forces f^μ appear on the right-hand side of (10.319) as f^μ/m, since the left-hand side of (10.319) has been written solely in terms of acceleration. In the presence of only gravitation, *all freely falling objects move along a geodesic determined by the metric.*

Result (10.319) can also be adapted to find the minimal-length path between two points in a manifold using the action

$$L = \int_{x_o}^{x_f} \mathrm{d}s = \int_{x_o}^{x_f} \sqrt{g_{\mu\nu}(x)\,\mathrm{d}x^\mu\,\mathrm{d}x^\nu}. \tag{10.320}$$

By parameterizing the minimal-length path as $x^\mu = x^\mu(\lambda)$, so that

$$\mathrm{d}x^\mu = \frac{\mathrm{d}x^\mu}{\mathrm{d}\lambda}\mathrm{d}\lambda, \tag{10.321}$$

it is straightforward to repeat the previous steps (10.314) through (10.319) to show that the minimal-length path is given by solving

$$\frac{\mathrm{d}^2x^\alpha}{\mathrm{d}\lambda^2} + \Gamma^\alpha_{\mu\nu}\frac{\mathrm{d}x^\mu}{\mathrm{d}\lambda}\frac{\mathrm{d}x^\nu}{\mathrm{d}\lambda} = 0. \tag{10.322}$$

For example, the minimal-length path on a sphere in angular coordinates is found using the Levi-Civita connection (10.180), which gives two simultaneous equations,

$$\frac{\mathrm{d}^2x^1}{\mathrm{d}\lambda^2} + \Gamma^1_{22}\frac{\mathrm{d}x^2}{\mathrm{d}\lambda}\frac{\mathrm{d}x^2}{\mathrm{d}\lambda} = \frac{\mathrm{d}^2\theta}{\mathrm{d}\lambda^2} - \sin\theta\cos\theta\left(\frac{\mathrm{d}\phi}{\mathrm{d}\lambda}\right)^2 = 0, \tag{10.323}$$

$$\frac{d^2x^2}{d\lambda^2} + \Gamma^2_{12}\frac{dx^1}{d\lambda}\frac{dx^2}{d\lambda} + \Gamma^2_{21}\frac{dx^2}{d\lambda}\frac{dx^1}{d\lambda} = \frac{d^2\phi}{d\lambda^2} + 2\cot\theta\frac{d\theta}{d\lambda}\frac{d\phi}{d\lambda} = 0. \qquad (10.324)$$

The second equation can be written

$$\frac{d}{d\lambda}\left(\sin^2\theta\frac{d\phi}{d\lambda}\right) = 0 \implies \sin^2\theta\frac{d\phi}{d\lambda} = \kappa, \qquad (10.325)$$

where κ is a constant.

Exercise 10.17: Using (10.325) in (10.323) show that the minimal-length path corresponds to a circle with its center at the center of the sphere, known as a *great circle*.

10.5.3 Diffeomorphic invariant field theory

In general, a diffeomorphic version of scalar and vector field theory can obtained from the action approach by using the space-time volume element (10.290), replacing ∂_μ with D_μ, and performing tensor contractions using $g_{\mu\nu}$ rather than the Minkowski metric. However, spinor fields will require further developments, since there are no spinor representations of a diffeomorphism.

As an example, a diffeomorphic invariant action for a scalar field φ is given by

$$S = \int d^4x\,\sqrt{|g|}\left(\frac{1}{2}g^{\mu\nu}D_\mu\varphi\, D_\nu\varphi - \frac{1}{2}m^2\varphi^2\right) \equiv \int d^4x\,\sqrt{|g|}\,\mathcal{L}(\varphi, D\varphi), \qquad (10.326)$$

where $D_\mu\varphi = \partial_\mu\varphi$ for a scalar field. Clearly, (10.326) reduces to the previous scalar field action (8.20) in the absence of gravitation or in an inertial frame. The equation of motion for φ is determined by the demand that the action is stationary under the variation $\varphi \to \varphi + \delta\varphi$. In order to integrate by parts and discard surface terms using (10.311), the Euler–Lagrange equation must be cast in terms of covariant derivatives. The result is the manifestly diffeomorphic invariant scalar Klein–Gordon equation,

$$\frac{\partial\mathcal{L}}{\partial\varphi} - D_\mu\frac{\partial\mathcal{L}}{\partial(D_\mu\varphi)} = -m^2\varphi - D_\mu(g^{\mu\nu}D_\nu\varphi) = -(g^{\mu\nu}D_\mu D_\nu + m^2)\varphi = 0, \qquad (10.327)$$

where metric compatibility, $D_\mu g^{\mu\nu} = 0$, was assumed to hold. The differential operator $g^{\mu\nu}D_\mu D_\nu$ is the generalization of the d'Alembertian to a curved space and reduces to the usual d'Alembertian in the absence of gravitation or in an inertial frame. For the case of a scalar field the d'Alembertian is given by

$$\Box_g = g^{\mu\nu}\partial_\mu\partial_\nu. \qquad (10.328)$$

Solving (10.327) obviously depends on the form for $g^{\mu\nu}$, which in turn requires determining the equation of motion for $g_{\mu\nu}$.

In the case of the electromagnetic vector field A_μ, the field strength tensor $F_{\mu\nu}$ of (9.53) is now found using the covariant derivative (10.112), so that

$$\begin{aligned} F_{\mu\nu} &= D_\mu A_\nu - D_\nu A_\nu = \partial_\mu A_\nu - \partial_\nu A_\mu + (\Gamma^\lambda_{\mu\nu} - \Gamma^\lambda_{\nu\mu})A_\lambda \\ &= \partial_\mu A_\nu - \partial_\nu A_\mu + 2\,\mathcal{T}^\lambda_{\mu\nu} A_\lambda, \end{aligned} \tag{10.329}$$

where the torsion tensor is present in the general case. If the torsion-free Levi-Civita connection is used, then the field strength tensor is unchanged. However, if torsion is present then the gauge invariance of the field strength tensor is *broken*. It was shown earlier in (9.311) that a mass term for the electromagnetic field also breaks gauge invariance. Note that the electromagnetic Lagrangian in the presence of torsion is given by

$$\mathcal{L}_{EM} = -\frac{1}{2}F_{\mu\nu}F^{\mu\nu} = -\frac{1}{2}\bar{F}_{\mu\nu}\bar{F}^{\mu\nu} + \bar{F}^{\mu\nu}\mathcal{T}^\lambda_{\mu\nu}A_\lambda + g^{\mu\alpha}g^{\nu\beta}\mathcal{T}^\lambda_{\alpha\beta}\mathcal{T}^\rho_{\mu\nu}A_\lambda A_\rho, \tag{10.330}$$

where

$$\bar{F}_{\mu\nu} = \partial_\mu A_\nu - \partial_\nu A_\mu. \tag{10.331}$$

The torsion term of (10.329) gives rise to a term that is quadratic in A_λ, so that torsion can result in a massive electromagnetic field and a massive photon. It also shows that torsion, if present, is directly coupled to the electric and magnetic fields that form the elements of $\bar{F}_{\mu\nu}$. It is useful for future purposes to note that the free electromagnetic Lagrangian is given by

$$\mathcal{L}_{EM} = -\frac{1}{2}g^{\mu\rho}g^{\nu\sigma}F_{\mu\nu}F_{\rho\sigma}, \tag{10.332}$$

which shows that the electromagnetic field is coupled to the gravitational field through the metric tensor.

10.6 The generalized stress–energy tensor

Like Lorentz invariance in the case of special relativistic fields, diffeomorphism invariance is a symmetry of the action in general relativity. Therefore, Noether's theorem provides the associated conserved current (8.223). This requires identifying the variation of the fields and coordinates that constitute the infinitesimal version of the symmetry operation. An infinitesimal diffeomorphism can be written

$$x'^\mu = x^\mu + \xi^\mu(x), \tag{10.333}$$

where the terms of $O(\xi^2)$ are ignorable. This gives the infinitesimal transformation matrix,

$$\frac{\partial x'^\mu}{\partial x^\nu} = \delta^\mu{}_\nu + \frac{\partial \xi^\mu}{\partial x^\nu}. \tag{10.334}$$

Using (10.334) shows that to $O(\xi)$, the metric tensor becomes

$$g'^{\mu\nu}(x') = \frac{\partial x'^{\mu}}{\partial x^{\rho}}\frac{\partial x'^{\nu}}{\partial x^{\sigma}} g^{\rho\sigma}(x) = g^{\mu\nu} + \frac{\partial \xi^{\mu}}{\partial x^{\rho}} g^{\rho\nu} + \frac{\partial \xi^{\nu}}{\partial x^{\rho}} g^{\mu\rho}. \qquad (10.335)$$

The symmetry variation of the metric $\bar{\delta} g^{\mu\nu}$ in Noether's theorem is found from its definition (8.210) by using $\delta x^{\mu} = \xi^{\mu}$ and (10.335), so that

$$\bar{\delta} g^{\mu\nu} = g'^{\mu\nu}(x') - \delta x^{\rho}\partial_{\rho} g^{\mu\nu}(x) - g^{\mu\nu}(x) = \frac{\partial \xi^{\mu}}{\partial x^{\rho}} g^{\rho\nu} + \frac{\partial \xi^{\nu}}{\partial x^{\rho}} g^{\mu\rho} - \xi^{\rho}\partial_{\rho} g^{\mu\nu}. \qquad (10.336)$$

Metric compatibility (10.135) requires

$$\partial_{\rho} g^{\mu\nu} = -\Gamma^{\mu}_{\ \rho\sigma} g^{\sigma\nu} - \Gamma^{\nu}_{\ \rho\sigma} g^{\mu\sigma}, \qquad (10.337)$$

so that (10.336) becomes

$$\bar{\delta} g^{\mu\nu} = g^{\nu\sigma}(\partial_{\sigma}\xi^{\mu} + \Gamma^{\mu}_{\ \sigma\rho}\xi^{\rho}) + g^{\mu\sigma}(\partial_{\sigma}\xi^{\nu} + \Gamma^{\nu}_{\ \sigma\rho}\xi^{\rho}) = D^{\nu}\xi^{\mu} + D^{\mu}\xi^{\nu}. \qquad (10.338)$$

Using the inverse of result (10.301) in conjunction with (10.338) gives

$$\begin{aligned} &\bar{\delta} g^{\mu\nu} = -g^{\mu\alpha} g^{\beta\nu} \bar{\delta} g_{\alpha\beta} \\ \Longrightarrow \quad &\bar{\delta} g_{\alpha\beta} = -g_{\alpha\mu} g_{\beta\nu} \bar{\delta} g^{\mu\nu} = -D_{\beta}\xi_{\alpha} - D_{\alpha}\xi_{\beta}. \end{aligned} \qquad (10.339)$$

The Lagrangian that gives the equation of motion for $g_{\mu\nu}$ will be derived in the next section. However, the remaining part of the Lagrangian with the *matter–energy fields* φ^{α}, denoted by $\mathcal{L}_M(\varphi_{\alpha}, D\varphi_{\alpha})$, will also involve the metric tensor as well as its derivatives if $\Gamma^{\mu}_{\ \rho\sigma}$ is present. Since the action is invariant under a diffeomorphism, its variation under the diffeomorphism will vanish. As in Noether's theorem, the matter action can be expanded to the first order using a symmetry variation, $\bar{\delta}\varphi_{\alpha}$, around the solutions to the equations of motion for φ_{α}, so that

$$\varphi_{\alpha} \rightarrow \varphi_{\alpha} + \bar{\delta}\varphi_{\alpha}, \qquad (10.340)$$

while the metric tensor is varied according to

$$g_{\mu\nu} \rightarrow g_{\mu\nu} + \bar{\delta} g_{\mu\nu}, \qquad (10.341)$$

and the inverse metric tensor undergoes a similar variation,

$$g^{\mu\nu} \rightarrow g^{\mu\nu} + \bar{\delta} g^{\mu\nu}. \qquad (10.342)$$

Since the matter action is assumed to be diffeomorphic invariant, the variation of the action created by (10.342) must vanish, giving rise to a conservation law. The variation of the matter action $\bar{\delta} S_M$ yields two integrals, one proportional to $\bar{\delta} g^{\mu\nu}$ and denoted by $\bar{\delta} S_g$, the other proportional to $\bar{\delta}\varphi_{\alpha}$ and denoted by $\bar{\delta} S_{\varphi}$, so that

$$0 = \bar{\delta} S_M = \bar{\delta} S_g + \bar{\delta} S_{\varphi}. \qquad (10.343)$$

The integral proportional to $\bar{\delta} g^{\mu\nu}$ is given by the Taylor series terms of $O(\bar{\delta} g^{\mu\nu})$,

$$\bar{\delta} S_g = \int \mathrm{d}^4x \left(\bar{\delta} g^{\mu\nu} \frac{\partial}{\partial g^{\mu\nu}} \left(\sqrt{|g|}\, \mathcal{L}_M \right) + \bar{\delta}(\partial_\rho g^{\mu\nu}) \frac{\partial}{\partial(\partial_\rho g^{\mu\nu})} \left(\sqrt{|g|}\, \mathcal{L}_M \right) \right). \tag{10.344}$$

Using property (8.209) of the symmetry variation,

$$\bar{\delta}(\partial_\rho g^{\mu\nu}) = \partial_\rho \bar{\delta} g^{\mu\nu}, \tag{10.345}$$

enables an integration by parts in which the surface terms are discarded, since $\bar{\delta} g^{\mu\nu}$ vanishes there, so that

$$\begin{aligned} \bar{\delta} S_g &= \int \mathrm{d}^4x\, \bar{\delta} g^{\mu\nu} \left(\frac{\partial}{\partial g^{\mu\nu}} \left(\sqrt{|g|}\, \mathcal{L}_M \right) - \partial_\rho \frac{\partial}{\partial(\partial_\rho g^{\mu\nu})} \left(\sqrt{|g|}\, \mathcal{L}_M \right) \right) \\ &\equiv \int \mathrm{d}^4x \sqrt{|g|}\, \bar{\delta} g^{\mu\nu} \frac{1}{2} T_{\mu\nu}. \end{aligned} \tag{10.346}$$

The definition of $T_{\mu\nu}$ shows that it is a manifestly symmetric tensor given by

$$T_{\mu\nu} = \frac{2}{\sqrt{|g|}} \left(\frac{\partial(\sqrt{|g|}\, \mathcal{L}_M)}{\partial g^{\mu\nu}} - \partial_\rho \frac{\partial(\sqrt{|g|}\, \mathcal{L}_M)}{\partial(\partial_\rho g^{\mu\nu})} \right). \tag{10.347}$$

The variation of the action for the field is given by

$$\bar{\delta} S_\varphi = \int \mathrm{d}^4x \sqrt{|g|}\, \bar{\delta} \varphi_\alpha \left(\frac{\partial \mathcal{L}_M}{\partial \varphi_\alpha} - D_\mu \frac{\partial \mathcal{L}_M}{\partial(D_\mu \varphi_\alpha)} \right). \tag{10.348}$$

The variation $\bar{\delta} S_\varphi$ given by (10.348) vanishes, since the integrand is evaluated using the solutions to the Euler–Lagrange equations for φ_α.

Because the variation of the action is a symmetry, it follows that

$$0 = \bar{\delta} S_M = \bar{\delta} S_g + \bar{\delta} S_\varphi = \bar{\delta} S_g, \tag{10.349}$$

so that (10.346) must also vanish. Substituting (10.338) into (10.346), using (10.311) to integrate by parts and drop surface terms, as well as using the property of (10.347) that $T_{\mu\nu} = T_{\nu\mu}$ finally gives

$$\begin{aligned} \bar{\delta} S_M &= \int \mathrm{d}^4x \sqrt{|g|}\, \frac{1}{2} T_{\mu\nu} (D^\mu \xi^\nu + D^\nu \xi^\mu) \\ &= - \int \mathrm{d}^4x \sqrt{|g|}\, \xi^\nu D^\mu T_{\mu\nu} = 0. \end{aligned} \tag{10.350}$$

Since ξ^ν is arbitrary, it follows that

$$D^\mu T_{\mu\nu} = 0, \tag{10.351}$$

so that $T_{\mu\nu}$ is a *covariantly conserved symmetric tensor*. The contravariant version of result (10.351) is obtained if the variation of the metric given by (10.341) is used.

Exercise 10.18: Show that the metric variation of (10.341) gives

$$D_\mu T^{\mu\nu} = 0, \tag{10.352}$$

where

$$T^{\mu\nu} = -\frac{2}{\sqrt{|g|}}\left(\frac{\partial(\sqrt{|g|}\ \mathcal{L}_M)}{\partial g_{\mu\nu}} - \partial_\rho \frac{\partial(\sqrt{|g|}\ \mathcal{L}_M)}{\partial(\partial_\rho g_{\mu\nu})}\right). \tag{10.353}$$

The origin of the minus sign is relation (10.339) for the infinitesimal variation.

A specific example of the tensor $T_{\mu\nu}$ and its relation to $T^{\mu\nu}$ is obtained by using the diffeomorphic invariant scalar field Lagrangian (10.326) in (10.347), so that

$$\sqrt{|g|}\,\mathcal{L}_M = \frac{1}{2}\sqrt{|g|}\left(g^{\mu\nu}D_\mu\varphi\ D_\nu\varphi - \frac{1}{2}m^2\varphi^2\right). \tag{10.354}$$

Its evaluation starts by noting that the Jacobi formula of (10.302) gives

$$\frac{\partial g}{\partial g^{\mu\nu}} = -g\ g_{\mu\nu} \implies \frac{\partial(-g)}{\partial g^{\mu\nu}} = g\ g_{\mu\nu}, \tag{10.355}$$

which in turn gives

$$\begin{aligned}\frac{\partial}{\partial g^{\mu\nu}}\sqrt{|g|} &= \frac{\partial}{\partial g^{\mu\nu}}\sqrt{-g} = \frac{1}{2\sqrt{-g}}\frac{\partial(-g)}{\partial g^{\mu\nu}} = \frac{g}{2\sqrt{-g}}\ g_{\mu\nu}\\ &= -\frac{1}{2}\sqrt{-g}\ g_{\mu\nu} = -\frac{1}{2}\sqrt{|g|}\ g_{\mu\nu}.\end{aligned} \tag{10.356}$$

The resulting expression for $T_{\mu\nu}$ is

$$T_{\mu\nu} = 2\frac{\partial \mathcal{L}_M}{\partial g^{\mu\nu}} + \frac{2}{\sqrt{|g|}}\frac{\partial\sqrt{|g|}}{\partial g^{\mu\nu}}\mathcal{L} = D_\mu\varphi D_\nu\varphi - \frac{1}{2}g_{\mu\nu}(D_\rho\varphi D^\rho\varphi - m^2\varphi^2). \tag{10.357}$$

Exercise 10.19: Use the equation of motion (10.327) to show that (10.357) obeys the covariant conservation law (10.351).

In an inertial frame or in the absence of gravitation, result (10.357) is identical to the canonical stress–energy tensor (8.239) for a free scalar field. As a result, the tensor $T_{\mu\nu}$ defined by (10.347) is identified as the *stress–energy tensor*. The new definition is manifestly symmetric and a tensor under diffeomorphisms.

Similarly, the tensor $T^{\mu\nu}$ is found by using the Jacobi formula

$$\frac{\partial g}{\partial g_{\mu\nu}} = g\, g^{\mu\nu} \implies \frac{\partial(-g)}{\partial g_{\mu\nu}} = -g\, g^{\mu\nu}, \tag{10.358}$$

which in turn gives

$$\begin{aligned}\frac{\partial}{\partial g_{\mu\nu}}\sqrt{|g|} &= \frac{\partial}{\partial g_{\mu\nu}}\sqrt{-g} = \frac{1}{2\sqrt{-g}}\frac{\partial(-g)}{\partial g^{\mu\nu}} = -\frac{g}{2\sqrt{-g}}\, g^{\mu\nu} \\ &= \frac{1}{2}\sqrt{-g}\, g^{\mu\nu} = \frac{1}{2}\sqrt{|g|}\, g^{\mu\nu}.\end{aligned} \tag{10.359}$$

The metric variation of the Lagrangian gives

$$\frac{1}{2}\bar{\delta}g^{\rho\sigma}D_\rho\varphi D_\sigma\varphi = -\frac{1}{2}g^{\rho\mu}g^{\nu\sigma}\bar{\delta}g_{\mu\nu}D_\rho\varphi D_\sigma\varphi = -\frac{1}{2}\bar{\delta}g_{\mu\nu}D^\mu\varphi D^\nu\varphi. \tag{10.360}$$

The resulting expression for $T^{\mu\nu}$ is

$$T^{\mu\nu} = -2\frac{\partial\mathcal{L}_M}{\partial g_{\mu\nu}} - \frac{2}{\sqrt{|g|}}\frac{\partial\sqrt{|g|}}{\partial g_{\mu\nu}}\mathcal{L} = D^\mu\varphi D^\nu\varphi - \frac{1}{2}g^{\mu\nu}(D_\rho\varphi D^\rho\varphi - m^2\varphi^2), \tag{10.361}$$

which is the contravariant version of (10.357).

The definition of the stress–energy tensor can be summarized using functional derivatives,

$$T_{\mu\nu}(x) = \frac{\delta S_M}{\delta g^{\mu\nu}(x)}, \quad T^{\mu\nu}(x) = -\frac{\delta S_M}{\delta g_{\mu\nu}(x)}, \tag{10.362}$$

where S_M is the diffeomorphic invariant matter–energy action.

Exercise 10.20: Verify that (10.362) reproduces the results of (10.357) and (10.361) for the specific scalar action (10.326).

Unlike the canonical stress–energy tensor of the Noether theorem, the definition of $T_{\mu\nu}$ given by (10.362) is unambiguous as long as the action for the associated matter fields is unambiguous. In the case of electrodynamics it was necessary to add the terms of (9.156) to the canonical stress–energy tensor using the Belinfante method to achieve gauge invariance for the stress–energy tensor. This problem is alleviated by using the definition (10.357) for the stress–energy tensor. As an example, the covariant extension of the free electromagnetic Lagrangian $\mathcal{L}_{EM}$ given by (10.332) has $g^{\mu\nu}$ present, but no derivatives of $g^{\mu\nu}$ appear *if torsion is absent*, so that the torsion-free version of $T^{EM}_{\mu\nu}$ is given by

$$
\begin{aligned}
T^{EM}_{\mu\nu} &= \frac{2}{\sqrt{|g|}}\frac{\partial(\sqrt{|g|}\ \mathcal{L}_{EM})}{\partial g^{\mu\nu}} = 2\frac{\partial \mathcal{L}_{EM}}{\partial g^{\mu\nu}} - g_{\mu\nu}\mathcal{L}_{EM} \\
&= -\frac{1}{2}F_{\mu}{}^{\sigma}F_{\nu\sigma} - \frac{1}{2}F_{\sigma\mu}F^{\sigma}{}_{\nu} + \frac{1}{2}g_{\mu\nu}F_{\rho\sigma}F^{\rho\sigma} \\
&= g^{\sigma\rho}F_{\mu\rho}F_{\sigma\nu} + \frac{1}{2}g_{\mu\nu}F_{\rho\sigma}F^{\rho\sigma} = F_{\mu\rho}F^{\rho}{}_{\nu} + \frac{1}{2}g_{\mu\nu}F_{\rho\sigma}F^{\rho\sigma},
\end{aligned}
\tag{10.363}
$$

where the antisymmetry $F_{\mu\nu} = -F_{\nu\mu}$ was used to combine the two terms. This is identical to the Belinfante version of the electromagnetic stress–energy tensor given by (9.158). If torsion is absent then (10.329) shows that the field strength tensor $F_{\mu\nu}$ is manifestly gauge invariant. It then follows that the electromagnetic stress–energy tensor (10.363) is also gauge invariant in the absence of torsion.

10.7 Einstein's field equation

All the necessary tools for working with gravitation have been assembled, except for the equations that determine the metric tensor. To qualify, general relativity's equations must yield Newtonian gravitation in the non-relativistic limit. In Newtonian physics it is the presence of inertial mass that is responsible for gravitation. Since special relativity shows the equivalence of mass and energy, it is logical to expect that in a relativistic version of gravity, both mass and energy act as the source of gravitation. This indicates a central role for the stress–energy tensor, which has already emerged in (10.347) by varying $g_{\mu\nu}$ in the action for the matter fields. Therefore, unlike Einstein's original derivation of the field equations, a variational approach will be used to derive the equations of motion.

10.7.1 The variational approach

The action for gravitation must be a scalar under diffeomorphisms. The goal is to generate an equation of motion that is a second-order partial differential equation that couples $g_{\mu\nu}$ to matter and energy and reduces to Newtonian gravitation in the non-relativistic limit. Such a theory must involve the Riemann curvature tensor, since it alone determines the true presence of curvature and therefore gravitation. The simplest possible choice for the Lagrangian $\mathcal{L}_G$ of gravity that meets these requirements is proportional to the Riemann curvature scalar,

$$
\mathcal{L}_G = -\kappa R, \tag{10.364}
$$

where κ is a constant that is determined by matching the results to Newtonian gravity in the weak gravitational limit. This choice for the Lagrangian is referred to as the *Hilbert action* for gravitation and *is not unique*. Numerous modifications of the Hilbert action have been made over the years in an effort to determine whether they better fit observational data. The choice of the Hilbert action sets the units of κ by the requirement that κR have the units of inverse length to the fourth power. Since $g_{\mu\nu}$ is a dimensionless quantity, the definitions of the Riemann curvature tensor (10.209) and the Riemann curvature scalar (10.229) show that R has the units of inverse length squared due to the presence of second derivatives. Therefore, κ must have the units of inverse length squared or an inverse area.

The Lagrangian density can be written in terms of the Ricci tensor (10.228) as

$$\mathcal{L}_G = -\kappa R = -\kappa\, g^{\mu\nu} R_{\mu\nu}. \tag{10.365}$$

Performing the inverse metric variation (10.342) gives

$$\begin{aligned}\delta S_G &= -\kappa \int \mathrm{d}^4x \left(R\,\delta\sqrt{|g|} + \sqrt{|g|}\,\delta R \right) \\ &= -\kappa \int \mathrm{d}^4x\, \sqrt{|g|} \left(R\,\frac{\delta\sqrt{|g|}}{\sqrt{|g|}} + R_{\mu\nu}\,\delta g^{\mu\nu} + g^{\mu\nu}\delta R_{\mu\nu} \right).\end{aligned} \tag{10.366}$$

Using (10.302) the first term in the parentheses of (10.366) becomes

$$R\,\frac{\delta\sqrt{|g|}}{\sqrt{|g|}} = -\frac{1}{2} R\, g_{\mu\nu}\,\delta g^{\mu\nu}. \tag{10.367}$$

Since the affine connection is a function of the metric, $\Gamma^{\mu}_{\nu\rho}(g)$, the third term in the parentheses of (10.366) can use the definition (10.209) to cast the variation of the Ricci tensor in terms of the variation $\delta\Gamma^{\mu}_{\nu\rho}$ of the affine connection. The variation of the Riemann tensor is therefore given by

$$\delta R^{\rho}{}_{\sigma\mu\nu} = \partial_\mu \delta\Gamma^{\rho}_{\nu\sigma} - \partial_\nu \delta\Gamma^{\rho}_{\mu\sigma} + \Gamma^{\lambda}_{\nu\sigma}\delta\Gamma^{\rho}_{\mu\lambda} + \Gamma^{\rho}_{\mu\lambda}\delta\Gamma^{\lambda}_{\nu\sigma} - \Gamma^{\lambda}_{\mu\sigma}\delta\Gamma^{\rho}_{\nu\lambda} - \Gamma^{\rho}_{\nu\lambda}\delta\Gamma^{\lambda}_{\mu\sigma}. \tag{10.368}$$

The key step is to note that the variation of the affine connection,

$$\delta\Gamma^{\rho}_{\mu\nu} = \Gamma^{\mu}_{\nu\rho}(g+\delta g) - \Gamma^{\mu}_{\nu\rho}(g), \tag{10.369}$$

transforms as a true tensor, since the non-tensorial term (10.76) in the transformation rule (10.87) is canceled by taking the difference in (10.369). As a result, the partial derivatives in (10.368) can be rewritten using the covariant derivative,

$$\partial_\mu \delta\Gamma^{\rho}_{\nu\sigma} = D_\mu \delta\Gamma^{\rho}_{\nu\sigma} - \Gamma^{\rho}_{\lambda\mu}\delta\Gamma^{\lambda}_{\nu\sigma} + \Gamma^{\lambda}_{\nu\mu}\delta\Gamma^{\rho}_{\lambda\sigma} + \Gamma^{\lambda}_{\sigma\mu}\delta\Gamma^{\rho}_{\nu\lambda}. \tag{10.370}$$

Combining (10.370) with the similar expression,

$$\partial_\nu \delta\Gamma^{\rho}_{\mu\sigma} = D_\nu \delta\Gamma^{\rho}_{\mu\sigma} - \Gamma^{\rho}_{\lambda\nu}\delta\Gamma^{\lambda}_{\mu\sigma} + \Gamma^{\lambda}_{\mu\nu}\delta\Gamma^{\rho}_{\lambda\sigma} + \Gamma^{\lambda}_{\sigma\nu}\delta\Gamma^{\rho}_{\mu\lambda}, \tag{10.371}$$

allows (10.368) to be written

$$\begin{aligned}\delta R^{\rho}{}_{\sigma\mu\nu} &= D_\mu \delta\Gamma^{\rho}_{\nu\sigma} - D_\nu \delta\Gamma^{\rho}_{\mu\sigma} \\ &\quad + 2\left(T^{\lambda}_{\ \nu\mu}\delta\Gamma^{\rho}_{\lambda\sigma} + T^{\rho}_{\ \mu\lambda}\delta\Gamma^{\lambda}_{\nu\sigma} + T^{\lambda}_{\ \sigma\mu}\delta\Gamma^{\rho}_{\nu\lambda} + T^{\rho}_{\ \lambda\nu}\delta\Gamma^{\lambda}_{\mu\sigma} + T^{\lambda}_{\ \nu\sigma}\delta\Gamma^{\rho}_{\mu\lambda} \right),\end{aligned} \tag{10.372}$$

where the torsion tensor has appeared. This gives the variation of the Ricci tensor as

$$\begin{aligned}\delta R_{\mu\nu} = \delta R^{\rho}{}_{\mu\rho\nu} &= D_\rho \delta\Gamma^{\rho}_{\nu\mu} - D_\nu \delta\Gamma^{\rho}_{\rho\mu} \\ &\quad + 2T^{\rho}_{\ \rho\lambda}\delta\Gamma^{\lambda}_{\nu\mu} + 2T^{\lambda}_{\ \mu\rho}\delta\Gamma^{\rho}_{\nu\lambda} + 2T^{\lambda}_{\ \mu\nu}\delta\Gamma^{\rho}_{\rho\mu}.\end{aligned} \tag{10.373}$$

The expression (10.373) is not symmetric, even in the absence of torsion. From this point forward the torsion tensor will be set to zero for consistency with Einstein's version of gravitation.

The symmetric part of (10.373) is retained by contracting it with the symmetric metric tensor. Using metric compatibility and relabeling indices gives the final result,

$$g^{\mu\nu}\delta R_{\mu\nu} = D_\rho(g^{\mu\nu}\delta\Gamma^{\rho}_{\nu\mu}) - D_\nu(g^{\mu\nu}\delta\Gamma^{\rho}_{\rho\mu}) = D_\rho\left(g^{\mu\nu}\delta\Gamma^{\rho}_{\nu\mu} - g^{\mu\rho}\delta\Gamma^{\sigma}_{\sigma\mu}\right). \tag{10.374}$$

The third term in the parentheses of (10.366) is therefore a total covariant divergence and can be discarded by using the divergence theorem (10.311) and the boundary conditions imposed on the variation $\delta g_{\mu\nu}$. The variation (10.366) becomes

$$\delta S_G = -\kappa\int \mathrm{d}^4x\sqrt{|g|}\ (R_{\mu\nu} - \frac{1}{2}g_{\mu\nu}R)\ \delta g^{\mu\nu} = -\kappa\int \mathrm{d}^4x\sqrt{|g|}\ G_{\mu\nu}\ \delta g^{\mu\nu}, \tag{10.375}$$

where $G_{\mu\nu}$ is the Einstein tensor of (10.239).

Writing the total action of gravitation and matter as the sum $S_M + S_G$, the variational principle gives the equation of motion

$$\delta S_G + \delta S_M = \int \mathrm{d}^4x\ \sqrt{g}\left(-\kappa G_{\mu\nu} + \frac{1}{2}T_{\mu\nu}\right)\delta g^{\mu\nu} = 0 \tag{10.376}$$

from which follows the field equation for $g_{\mu\nu}$ proposed by Einstein in 1916,

$$G_{\mu\nu} = R_{\mu\nu} - \frac{1}{2}g_{\mu\nu}R = \frac{1}{2\kappa}T_{\mu\nu}. \tag{10.377}$$

Both sides are covariantly conserved tensors and (10.377) shows that *the metric of space-time is determined by the energy present.* Einstein did not use the action approach to derive (10.377). Instead, he made the argument that mass and energy should both create gravitation and that both sides of (10.377) are covariantly conserved. The motion of matter as well as light is then determined by (10.319) and the associated affine connection. Contracting both sides of (10.377) against $g^{\mu\nu}$ and using $\delta^\mu{}_\mu = 4$ gives

$$g^{\mu\nu}R_{\mu\nu} - \frac{1}{2}R\delta^\mu{}_\mu = R - 2R = -R = \frac{1}{2\kappa}T^\rho{}_\rho. \tag{10.378}$$

Substituting the R determined by (10.378) into the Einstein equation yields a second equivalent form,

$$R_{\mu\nu} = \frac{1}{2\kappa}\left(T_{\mu\nu} - \frac{1}{2}g_{\mu\nu}T^\rho{}_\rho\right), \tag{10.379}$$

so that the Ricci tensor determined by matter and energy is both symmetric and traceless. Switching from natural units is achieved by simply replacing κ with $\hbar c\kappa$ in (10.379). Since the units of $T_{\mu\nu}$ and κ are energy per spatial volume and inverse area, respectively, $T_{\mu\nu}/\hbar c\kappa$ has the required units of inverse area.

It is important to note that the Hilbert action for gravitation is not unique, and many modifications and additional terms have been considered. For example, Einstein effectively modified the Lagrangian to

$$\mathcal{L}_G = R - 2\Lambda, \tag{10.380}$$

where Λ is a scalar quantity with units of inverse length squared and is known as the *cosmological constant*. Adding this term gives rise to an additional term in (10.377) when the metric is varied so that the stress–energy tensor is effectively modified to

$$T'_{\mu\nu} = T_{\mu\nu} + \rho_v c^2 g_{\mu\nu}, \tag{10.381}$$

where the quantity ρ_v, given by

$$\rho_v = -\frac{c^2}{8\pi G}\Lambda, \tag{10.382}$$

acts as an effective vacuum mass density. The quantity $\rho_v c^2$ is referred to as the *vacuum energy density*, since it can be viewed as a constant energy density term present in the Lagrangian describing the matter–energy fields. This highlights a very stark difference between (10.377) and Newtonian mechanics. In Newtonian mechanics only energy *differences* create force and motion, whereas the metric determined from (10.377) is altered by different *absolute values* of the vacuum energy. The cosmological constant is discussed further in the later section on cosmology.

10.7.2 The Newtonian limit

In order to connect Einstein's equation (10.377) to standard non-relativistic Newtonian gravitation, the attractive force $\boldsymbol{F}(\boldsymbol{x})$ on a point mass m in Newtonian gravitation is written in terms of the gravitating object's mass density $\rho(\boldsymbol{x})$ as

$$\begin{aligned}\boldsymbol{F}(\boldsymbol{x}) &= -mG\int_V \mathrm{d}^3x' \frac{(\boldsymbol{x}-\boldsymbol{x}')\rho(\boldsymbol{x}')}{|\boldsymbol{x}-\boldsymbol{x}'|^3} \\ &= m\nabla\int_V \mathrm{d}^3x' \frac{G\,\rho(\boldsymbol{x}')}{|\boldsymbol{x}-\boldsymbol{x}'|} \equiv -m\nabla\Phi(x),\end{aligned} \tag{10.383}$$

where V is the volume of the gravitating object and Φ is the Newtonian gravitational potential. Combining the definition (10.383) with result (9.13),

$$\nabla^2\left(\frac{1}{|\boldsymbol{x}-\boldsymbol{x}'|}\right) = -4\pi\delta^3(\boldsymbol{x}-\boldsymbol{x}'), \tag{10.384}$$

shows that Φ obeys Poisson's equation,

$$\nabla^2\Phi(\boldsymbol{x}) = 4\pi G\,\rho(\boldsymbol{x}), \tag{10.385}$$

and that the acceleration of the test mass is given by

$$\boldsymbol{a}(\boldsymbol{x}) = -\nabla\Phi(\boldsymbol{x}), \tag{10.386}$$

where Φ has the units of velocity squared.

Connecting Einstein's equation to these Newtonian results is made in the *weak field limit*, where the metric tensor $g_{\mu\nu}$ is written in terms of the Minkowski metric $\eta_{\mu\nu}$ and a symmetric matrix $h_{\mu\nu}$,

$$g_{\mu\nu} = \eta_{\mu\nu} + h_{\mu\nu}. \tag{10.387}$$

In the weak field limit it is assumed that $h_{\mu\nu} \ll 1$, which allows terms $O(h^2)$ to be discarded. To $O(h)$ the inverse metric tensor is given by

$$g^{\mu\nu} = \eta^{\mu\nu} - h^{\mu\nu}, \tag{10.388}$$

where

$$h^{\mu\nu} = \eta^{\mu\rho}\eta^{\nu\sigma}h_{\rho\sigma}. \tag{10.389}$$

Exercise 10.21: Show to $O(h)$ that the forms (10.387) and (10.388) satisfy the required relationship

$$g_{\mu\nu} = g_{\mu\rho}g_{\nu\sigma}g^{\rho\sigma}. \tag{10.390}$$

In the weak field approximation the Levi-Civita connection (10.141) becomes

$$\Gamma^{\rho}_{\mu\nu} = \frac{1}{2}\eta^{\rho\alpha}(\partial_\mu h_{\alpha\nu} + \partial_\nu h_{\alpha\mu} - \partial_\alpha h_{\mu\nu}), \tag{10.391}$$

where $\partial_\mu g_{\alpha\nu} = \partial_\mu h_{\alpha\nu}$ was used and $O(h^2)$ terms were discarded. Next, the geodesic equation (10.319) for the motion of a point mass in general relativity is treated in the non-relativistic low velocity limit, so that the spatial components of the proper velocity become

$$V^j = \frac{\mathrm{d}x^j}{\mathrm{d}\tau} \approx \frac{\mathrm{d}x^j}{\mathrm{d}t} = 0. \tag{10.392}$$

The four-vector proper velocity is then given by

$$V^\mu \approx (c, 0). \tag{10.393}$$

The acceleration is given by

$$\frac{\mathrm{d}^2x^j}{\mathrm{d}\tau^2} \approx \frac{\mathrm{d}^2x^j}{\mathrm{d}t^2} = a^j, \tag{10.394}$$

which is the Newtonian acceleration. For this case, the geodesic equation (10.319) becomes

$$\frac{d^2x^j}{d\tau^2} + \Gamma^j_{\mu\nu}\frac{dx^j}{d\tau}\frac{dx^j}{d\tau} = a^j + c^2\Gamma^j_{00} = 0. \tag{10.395}$$

It is also assumed that the field is *static* throughout space, so that $\partial_0 h_{\mu\nu} = 0$. Using the weak field affine connection of (10.391) as well as $\partial_0 h_{\mu\nu} = 0$ gives

$$\Gamma^j_{00} = \frac{1}{2}\eta^{j\alpha}(\partial_0 h_{\alpha 0} + \partial_0 h_{\alpha 0} - \partial_\alpha h_{00}) = -\frac{1}{2}\partial^j h_{00} = \frac{1}{2}\partial_j h_{00}, \tag{10.396}$$

so that the geodesic equation (10.395) becomes

$$a^j = -\frac{1}{2}c^2\partial_j h_{00}. \tag{10.397}$$

This is identical to the equation of motion (10.386) for a point mass in static Newtonian gravity with the identification

$$h_{00} = \frac{2}{c^2}\Phi, \tag{10.398}$$

which is consistent with the assumption that $h_{00} \ll 1$ and the requirement that Φ has the units of velocity squared.

The next step is to examine the Einstein equation (10.379) in the weak field limit. Result (10.391) shows that the quadratic terms in $\Gamma^\rho_{\mu\nu}$ in the Riemann tensor (10.209) can be discarded as $O(h^2)$, and this gives the weak field Riemann curvature tensor,

$$R^\rho{}_{\mu\alpha\nu} = \partial_\alpha\Gamma^\rho_{\nu\mu} - \partial_\nu\Gamma^\rho_{\alpha\mu}. \tag{10.399}$$

The Ricci tensor is then given by

$$R_{\mu\nu} = R^\rho{}_{\mu\rho\nu} = \partial_\rho\Gamma^\rho_{\mu\nu} - \partial_\nu\Gamma^\rho_{\rho\mu}. \tag{10.400}$$

Using (10.391) shows that the weak field Ricci tensor is

$$\begin{aligned} R_{\mu\nu} = \partial_\rho\Gamma^\rho_{\mu\nu} - \partial_\nu\Gamma^\rho_{\mu\rho} = -\frac{1}{2}\,\eta^{\rho\alpha}\partial_\alpha\partial_\rho h_{\nu\mu} + \frac{1}{2} \\ \eta^{\rho\alpha}(\partial_\mu\partial_\rho h_{\alpha\nu} - \partial_\mu\partial_\nu h_{\alpha\rho} + \partial_\alpha\partial_\nu h_{\rho\mu}). \end{aligned} \tag{10.401}$$

The vector A_μ is defined as

$$A_\mu = \partial_\rho h^\rho{}_\mu - \frac{1}{2}\partial_\mu h^\rho{}_\rho, \tag{10.402}$$

and because $\eta^{\rho\alpha}$ is constant, this gives

$$\eta^{\rho\alpha}(\partial_\mu\partial_\rho h_{\alpha\nu} - \frac{1}{2}\partial_\mu\partial_\nu h_{\alpha\rho}) = \partial_\mu\partial_\rho h^\rho{}_\nu - \frac{1}{2}\partial_\mu\partial_\nu h^\rho{}_\rho = \partial_\mu A_\nu. \tag{10.403}$$

The weak field Ricci tensor (10.401) can therefore be written as

$$R_{\mu\nu} = -\frac{1}{2}\eta^{\rho\alpha}\partial_\alpha\partial_\rho h_{\nu\mu} + \frac{1}{2}(\partial_\mu A_\nu + \partial_\nu A_\mu) = -\frac{1}{2}\Box h_{\nu\mu} + \frac{1}{2}(\partial_\mu A_\nu + \partial_\nu A_\mu). \quad (10.404)$$

The Minkowski space version of the d'Alembertian has appeared in the first term of (10.404).

Expression (10.404) is greatly simplified by applying the infinitesimal change of coordinates (10.335) that was used to derive the stress–energy tensor,

$$x'^\mu = x^\mu - \xi^\mu \implies \frac{\partial x^\mu}{\partial x'^\nu} = \delta^\mu{}_\nu + \frac{\partial \xi^\mu}{\partial x'^\nu} \approx \delta^\mu{}_\nu + \frac{\partial \xi^\mu}{\partial x^\nu}, \quad (10.405)$$

where only terms of $O(\xi)$ were retained. For such a change of coordinates the metric tensor transforms as

$$\begin{aligned} g'_{\mu\nu} &= \frac{\partial x^\alpha}{\partial x'^\mu}\frac{\partial x^\beta}{\partial x'^\nu} g_{\alpha\beta} = (\delta^\alpha{}_\mu + \partial_\mu\xi^\alpha)(\delta^\beta{}_\nu + \partial_\nu\xi^\beta)(\eta_{\alpha\beta} + h_{\alpha\beta}) \\ &= \eta_{\mu\nu} + h_{\mu\nu} + \partial_\mu\xi_\nu + \partial_\nu\xi_\mu, \end{aligned} \quad (10.406)$$

where terms of $O(h\xi)$ were dropped. This infinitesimal coordinate transformation has the effect that $h_{\mu\nu}$ is transformed according to

$$h'_{\mu\nu} = h_{\mu\nu} + \partial_\mu\xi_\nu + \partial_\nu\xi_\mu, \quad (10.407)$$

which gives

$$\begin{aligned} \partial_\rho h'^\rho{}_\mu &= \partial_\rho h^\rho{}_\mu + \partial_\rho\partial^\rho\xi_\mu + \partial_\mu\partial_\rho\xi^\rho \\ h'^\rho{}_\rho &= h^\rho{}_\rho + 2\partial_\rho\xi^\rho. \end{aligned} \quad (10.408)$$

Using these results shows that the vector A_μ of (10.402) becomes

$$A'_\mu = \partial_\rho h'^\rho{}_\mu - \frac{1}{2}\partial_\mu h'^\rho{}_\rho = \partial_\rho h^\rho{}_\mu - \frac{1}{2}\partial_\mu h^\rho{}_\rho + \partial_\rho\partial^\rho\xi_\mu = A_\mu + \Box\xi_\mu. \quad (10.409)$$

This coordinate change is structurally similar to a gauge transformation in electrodynamics. Using an infinitesimal diffeomorphism, the infinitesimal vector A^μ appearing in the weak field Ricci tensor (10.404) can be set to zero by choosing

$$\Box\xi_\mu = -A_\mu. \quad (10.410)$$

This process is identical in form to enforcing the Lorentz condition in electrodynamics, and so this is often referred to as *gauge fixing* in general relativity. For this choice of gauge the weak field Ricci tensor (10.404) becomes

$$R_{\mu\nu} = -\frac{1}{2}\Box h_{\mu\nu}. \quad (10.411)$$

Using (10.411) in (10.379) gives the Einstein field equation for the weak field case,

$$R_{\mu\nu} = -\frac{1}{2}\Box h_{\mu\nu} = \frac{1}{2\hbar\kappa c}\left(T_{\mu\nu} - \frac{1}{2}\eta_{\mu\nu}T^{\rho}{}_{\rho}\right). \tag{10.412}$$

The remaining step is to choose a form for $T_{\mu\nu}$. In general relativity it is common practice to approximate matter as a fluid governed by a thermodynamic equation of state. In the case of a static fluid, the stress–energy tensor (6.256) can be used. In the static case the fluid stress–energy tensor has

$$\begin{aligned} T_{00} &= \rho c^2, \\ T_{jk} &= -\eta_{jk}P, \end{aligned} \tag{10.413}$$

where ρ is the mass density and P is the pressure. For typical Newtonian conditions the pressure P is negligible compared to the mass–energy density ρc^2, so that $T_{00} = \rho c^2$, $T_{jk} \approx 0$, and $T^{\rho}{}_{\rho} = \rho c^2$. For the case of a static $h_{\mu\nu}$, equation (10.412) becomes

$$\nabla^2 h_{00} = \frac{c}{2\hbar\kappa}\rho, \quad \nabla^2 h_{jk} = -\frac{c}{2\hbar\kappa}\rho\, \eta_{jk}. \tag{10.414}$$

The choice for $\hbar\kappa$ given by

$$\hbar\kappa = \frac{c^3}{16\pi G}, \tag{10.415}$$

coupled with the previous identification $h_{00} = 2\Phi/c^2$, results in the first equation of (10.414) becoming the desired Newtonian gravitation equation (10.385), $\nabla^2\Phi = 4\pi G\rho$. Using (10.415) in the second equation shows that the weak field Einstein equation is solved by

$$h_{jk} = -\frac{2\Phi}{c^2}\eta_{jk}. \tag{10.416}$$

Using (10.415) yields the final form for the Einstein equation in terms of the gravitational constant G,

$$R_{\mu\nu} - \frac{1}{2}g_{\mu\nu}R = \frac{8\pi G}{c^4}T_{\mu\nu}. \tag{10.417}$$

10.7.3 Gravitational time dilation and Doppler shift

The weak field results can be used show that gravitation, like relative motion, affects length and time measurements. A stationary observer in a gravitational field undergoes the same effects as an accelerated observer in special relativity due to the presence of a nontrivial metric. A *stationary clock* is a clock that has a fixed spatial position, so that $\mathrm{d}x^j = 0$. Proper time intervals for the stationary clock are given from the metric by

$$\mathrm{d}\tau = \mathrm{d}t\sqrt{g_{00}}, \tag{10.418}$$

so that the time interval for the clock's period is related to the proper time interval of each tick by

$$\Delta t = \frac{\Delta\tau}{\sqrt{g_{00}}}. \tag{10.419}$$

An inertial observer will claim that the stationary clock is ticking with an apparent frequency f that is inversely proportional to the period Δt of the clock, so that

$$f = f(x) = \frac{1}{\Delta t} = \frac{1}{\Delta\tau}\sqrt{g_{00}(x)}. \tag{10.420}$$

When the clock is moved to a new stationary location the apparent frequency changes to

$$f' = f(x') = \frac{1}{\Delta\tau}\sqrt{g_{00}(x')}. \tag{10.421}$$

Since the proper time is the same for observers at the two locations, the ratio gives

$$f' = f\sqrt{\frac{g_{00}(x')}{g_{00}(x)}}. \tag{10.422}$$

For the weak field case (10.387) and (10.398) give to $O(h_{00})$,

$$g_{00} = 1 + h_{00} = 1 + \frac{2}{c^2}\Phi, \tag{10.423}$$

so that

$$\begin{aligned}\sqrt{\frac{g_{00}(x')}{g_{00}(x)}} &= \sqrt{\frac{1 + h_{00}(x')}{1 + h_{00}(x)}} \approx \left(1 + \frac{1}{2}h_{00}(x')\right)\left(1 - \frac{1}{2}h_{00}(x)\right)\\ &= 1 + \frac{1}{2}(h_{00}(x') - h_{00}(x))\\ &= 1 + \frac{1}{c^2}\Phi(x') - \frac{1}{c^2}\Phi(x).\end{aligned} \tag{10.424}$$

The frequency change (10.422) is therefore

$$\begin{aligned} f' = f + \frac{f}{c^2}(\Phi(x') - \Phi(x)) \implies \frac{\Delta f}{f} &= \frac{f' - f}{f}\\ &= \frac{1}{c^2}(\Phi(x') - \Phi(x)).\end{aligned} \tag{10.425}$$

The Newtonian gravitational potential at a distance r outside of a spherical mass M is $\Phi(r) = -GM/r$. For the case that $r' < r$ this gives

$$\Delta f = -\frac{GM}{c^2}\left(\frac{1}{r'} - \frac{1}{r}\right)f = -\frac{GM}{c^2}\left(\frac{r - r'}{rr'}\right)f < 0. \tag{10.426}$$

According to a fixed observer, a clock or oscillator will have a frequency that is lower as it is brought closer to a gravitating body.

This causes the frequency of light waves emitted by an oscillator closer to the gravitating body to arrive with a *lower* frequency than those emitted by an identical source further from the body. This is known as the *gravitational redshift* and can be understood in terms of energy conservation. It was shown in the Klein–Gordon case (8.58) and the electromagnetic case (9.259) that the modes of an associated wave have an associated energy of $E_f = \hbar\omega = 2\pi\hbar f$. The wave mode can be assigned an effective inertial mass

$$m_{eff} = \frac{E_f}{c^2} = \frac{2\pi\hbar f}{c^2}. \tag{10.427}$$

Treating the emerging wave mode as having an effective mass enables the change in gravitational potential energy to be found in terms of a frequency change as the mode moves away from a gravitating object,

$$\begin{aligned} \Delta E = E_{f'} - E_f &= 2\pi\hbar(f' - f) \\ &= m_{eff}(\Phi(x') - \Phi(x)) = 2\pi\hbar\frac{f}{c^2}(\Phi(x') - \Phi(x)), \\ \Longrightarrow f' - f &= \frac{f}{c^2}(\Phi(x') - \Phi(x)), \end{aligned} \tag{10.428}$$

which is precisely result (10.425). Light, despite being a wave and having no inertial mass associated with it, is affected by gravity since it is a form of energy, and inertial mass and energy are interchangeable. In turn, this means that light is deflected by gravitating bodies, and this constituted one of the early experimental confirmations of general relativity. This is discussed later in this chapter.

10.8 Vacuum solutions to Einstein's equation

A vacuum in general relativity simply means that the stress–energy tensor $T_{\mu\nu}$ is set to zero, indicating the absence of any matter or energy. For a vacuum Einstein's equation becomes

$$R_{\mu\nu} = 0. \tag{10.429}$$

10.8.1 Gravitational waves

In the weak field limit, (10.412) shows that this becomes

$$\Box h_{\mu\nu} = (\partial_0{}^2 - \nabla^2)h_{\mu\nu} = 0, \tag{10.430}$$

which is the wave equation. Just as Maxwell's equations predict light waves, Einstein's equation, at least in the weak field case, predicts the existence of *gravitational waves* which travel at the speed of light.

Extracting the gravitational wave equation from the Einstein equation required using the vector gauge fixing condition (10.410), which (10.409) shows results in

$$A^\mu = \partial_\rho h^{\rho\mu} - \partial^\mu h^\rho{}_\rho = 0. \tag{10.431}$$

Like the Lorenz condition in electrodynamics, the gauge condition (10.410) is invariant under infinitesimal diffeomorphisms that satisfy $\Box\zeta^\mu = 0$. However, this is precisely the same equation that $h_{\mu\nu}$ obeys, so that such a transformation can be used to remove four additional degrees of freedom from the weak field metric by choosing ζ_μ to satisfy

$$h_{\mu\nu} - \partial_\mu\zeta_\nu - \partial_\nu\zeta_\mu = 0. \tag{10.432}$$

Since $h_{\mu\nu}$ is a symmetric 4×4 tensor, it begins with ten degrees of freedom. The gauge condition $A^\mu = 0$ removes four, while four more are extraneous and removed using the residual diffeomorphism invariance represented by ζ_μ. The final result shows that there are only two independent degrees of freedom, a result identical to the electromagnetic wave.

The standard choice used to enforce this reduction in degrees of freedom is to remove four with

$$h_{0j} = h^\rho{}_\rho = 0. \tag{10.433}$$

For this choice the gauge condition (10.431) becomes

$$A^\mu = \partial_\rho h^{\rho\mu} = 0, \tag{10.434}$$

which removes an additional four degrees of freedom by enforcing

$$\partial_0 h^{00} = \partial_j h^{jk} = 0. \tag{10.435}$$

This is referred to as the *transverse–traceless or TT gauge.*

A simple solution to the equation of motion (10.430) consistent with the TT gauge conditions can be written

$$h_{\mu\nu} = \alpha_{\mu\nu}\cos(k \cdot x), \tag{10.436}$$

where the symmetric matrix constant $\alpha_{\mu\nu}$ satisfies

$$\begin{aligned} \alpha_{j0} &= \alpha_{j0} = 0 \\ \alpha^\rho{}_\rho &= 0, \\ k^\mu\alpha_{\mu\nu} &= 0, \end{aligned} \tag{10.437}$$

along with the usual massless dispersion relation $k_\mu k^\mu = 0$. These conditions are consistent with (10.435), since they give

$$\begin{aligned} k^\mu\alpha_{\mu\ell} &= k^0\alpha_{0\ell} + k^j\alpha_{j\ell} = k^j\alpha_{j\ell} = 0, \\ k^\mu\alpha_{\mu 0} &= k^0\alpha_{00} + k^j\alpha_{j0} = k^0\alpha_{00} = 0, \end{aligned} \tag{10.438}$$

which are the relations of (10.435). For the case that a weak field gravitational wave is traveling in the z-direction, so that $\boldsymbol{k} = (0, 0, k)$, the gauge conditions give the eight constraints,

$$\alpha_{0j} = \alpha_{3j} = \alpha_{00} = 0, \quad \alpha_{11} + \alpha_{22} = 0. \tag{10.439}$$

In turn, this shows that

$$\begin{aligned} h_{11}(x) = -h_{22}(x) &\equiv h_{+}\cos(k \cdot x), \\ h_{12}(x) = h_{21}(x) &\equiv h_{\times}\cos(k \cdot x), \end{aligned} \tag{10.440}$$

which are the two modes of a gravitational wave, with all others vanishing.

As a wave moving in the z-direction with the h_+ polarization passes two adjacent masses, the distances in the x and y directions between the two alternately increase and decrease due to the fluctuation of the metric. This allows an interferometer similar to the Michelson–Morley experiment discussed to detect the passage of a gravitational wave. As depicted in figure 10.3, a laser light source emits a beam of wavelength λ. A beam splitter passes the beam in perpendicular directions to suspended mirrors, where the two beams are directed back to be rejoined on the screen. The distances between the two mirrors and the beam splitter are L_x and L_y. For the idealized case, as the h_+ polarized wave with angular frequency ω passes, the two distances change according to the changing metric, which gives

$$\begin{aligned} L_x' &= g_{xx}L_x = L_x(1 + h_{+}\cos(\omega t)), \\ L_y' &= g_{yy}L_y = L_y(1 - h_{+}\cos(\omega t)). \end{aligned} \tag{10.441}$$

Assuming $L_x = L_y = L$ for simplicity, the net path difference between the two rejoining beams is given by

$$\Delta L = |2L_x' - 2L_y'| = 2L|h_{+}\cos\omega t|, \tag{10.442}$$

which oscillates between zero and $2L|h_+|$ with a frequency of ω. For the case that $2L|h_+|$ is an integer multiple of the wavelength λ, the screen will display an oscillation of constructive interference to destructive interference and back at the frequency of the gravitational wave. Clearly, the larger the value of L, the more sensitive the interferometer. This is the

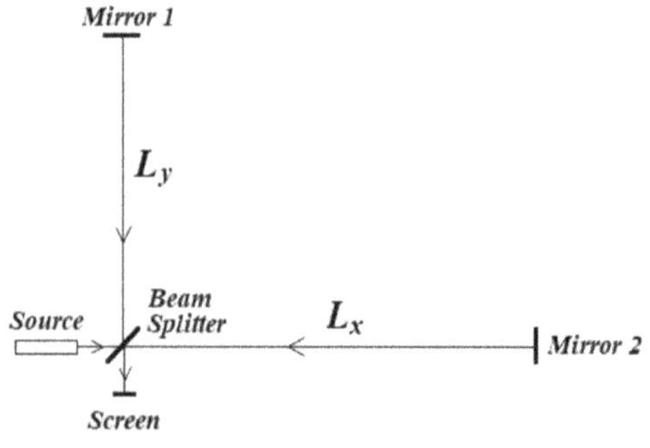

Figure 10.3. An interferometer for observing gravitational waves.

basis of the LIGO (Laser Interferometer Gravitational-wave Observatory) laboratories, for which $L \approx 4$ km. Noise in the form of seismic disturbances is screened by having several LIGO laboratories separated by great distances, so that a true observation of a gravitational wave will occur at both almost simultaneously. The first successful observation of gravitational waves occurred in 2015. The origins of gravitational waves are astronomical-scale events, such as collisions of black holes. These are discussed in the next section.

It is widely believed that the gravitational waves of general relativity can be viewed as composed of *gravitons*, which would be the counterpart to the photon of electromagnetic waves. Like electromagnetic waves, the gravitational wave travels at the speed of light and therefore the constituent gravitons would need to be massless. At the time of writing, unlike the case of electromagnetism, a consistent quantized version of general relativity does not exist, despite strenuous efforts to develop it. The primary problem in quantizing gravitation is the extreme nonlinearity of the Einstein equation of motion. In the case of electromagnetism the interaction term is linear in the gauge field which, by comparison, is much simpler than the nonlinear terms involving the metric tensor in the Ricci tensor. The problem is compounded by the weakness of the gravitational force compared to hadronic and electromagnetic interactions, which makes the experimental detection of a single graviton virtually impossible. For example, in SI units the gravitational force between two protons separated by the distance r is given by

$$F_G = \frac{Gm^2}{r^2} = \frac{1.7 \times 10^{-64}}{r^2}, \tag{10.443}$$

while the electrostatic force between the same two protons is given by

$$F_e = \frac{Ke^2}{r^2} = \frac{2.3 \times 10^{-28}}{r^2}, \tag{10.444}$$

which gives

$$\frac{F_e}{F_G} \approx 1.4 \times 10^{36}. \tag{10.445}$$

The electric force completely dominates the interaction of charged inertial masses. In the case of protons there is also a hadronic force even stronger than the electric force for close separations.

In chapters 8 and 9 it was shown that fields may possess an intrinsic angular momentum, referred to as spin or helicity. In natural units the spinor field representing electrons has a helicity of $\pm\frac{1}{2}$, while the helicity of the gauge field representing photons is ± 1. The identification of these quantities as the spin projection of *particles* comes from the successful quantization of the associated field theories and the experimental measurement of the intrinsic angular momentum of these particles. However, the helicity results of the spinor and vector fields can be anticipated using classical field theory by applying Noether's theorem to the *normalized classical waves* associated with these fields. In the case of electromagnetism the spin or helicity of the photon was identified classically by analyzing circularly polarized waves written in terms of the gauge field. Since the weak field metric tensor is the quantity that is propagating in a gravitational wave, it is straightforward to analyze the rotational content of the graviton, since the rotated metric tensor transforms according to

$$\delta g_{jk} = \theta^a T^a_{j\ell} g^{\ell k} + \theta^a T^a_{kr} g^{jr}. \tag{10.446}$$

The gravitational wave receives two contributions from the same rotation matrix as the electromagnetic field, and so it has a helicity of ± 2.

10.8.2 The Schwarzschild metric

One of the most studied vacuum solutions to the vacuum Einstein equation is that of the *Schwarzschild metric*, derived from spherical coordinates and given by

$$c^2\, \mathrm{d}\tau^2 = \left(1 - \frac{2GM}{rc^2}\right)c^2 \mathrm{d}t^2 - \left(1 - \frac{2GM}{rc^2}\right)^{-1} \mathrm{d}r^2 - r^2\, \mathrm{d}\theta^2 - r^2 \sin^2\theta\, \mathrm{d}\phi^2. \tag{10.447}$$

The metric of (10.447) is spherically symmetric and reduces to the Minkowski metric in the event that $M \to 0$. Identifying the *Schwarzschild radius* r_s as

$$r_s = \frac{2GM}{c^2}, \tag{10.448}$$

allows the metric to be written

$$c^2\, \mathrm{d}\tau^2 = \left(1 - \frac{r_s}{r}\right)c^2 \mathrm{d}t^2 - \left(1 - \frac{r_s}{r}\right)^{-1} \mathrm{d}r^2 - r^2\, \mathrm{d}\theta^2 - r^2 \sin^2\theta\, \mathrm{d}\phi^2. \tag{10.449}$$

Proving that the metric satisfies the Einstein equation begins by finding the Levi-Civita connection components. The nonzero components are given by

$$\begin{gathered} \Gamma^t_{rt} = -\Gamma^r_{rr} = \frac{r_s}{2r(r - r_s)}, \quad \Gamma^r_{tt} = \frac{r_s(r - r_s)}{2r^3}, \quad \Gamma^r_{\phi\phi} = -(r - r_s)\sin^2\theta, \\ \Gamma^r_{\theta\theta} = r_s - r, \quad \Gamma^\theta_{r\theta} = \Gamma^\phi_{r\phi} = \frac{1}{r}, \quad \Gamma^\theta_{\phi\phi} = -\sin\theta\cos\theta, \quad \Gamma^\phi_{\theta\phi} = \cot\theta. \end{gathered} \tag{10.450}$$

Exercise 10.22: Verify the results of (10.450) using the metric components of (10.449).

Using the expression for the Riemann curvature tensor (10.209) allows the Ricci tensor to be directly calculated from the affine connection,

$$R_{\mu\nu} = R^\lambda{}_{\mu\lambda\nu} = \partial_\nu \Gamma^\lambda_{\mu\lambda} - \partial_\lambda \Gamma^\lambda_{\mu\nu} + \Gamma^\alpha_{\mu\lambda}\Gamma^\lambda_{\nu\alpha} - \Gamma^\alpha_{\mu\nu}\Gamma^\lambda_{\lambda\alpha}. \tag{10.451}$$

This is expedited by using the results

$$
\begin{aligned}
\Gamma^{\lambda}_{\lambda t} &= \Gamma^{t}_{tt} + \Gamma^{r}_{rt} + \Gamma^{\theta}_{\theta t} + \Gamma^{\phi}_{\phi t} = 0,\\
\Gamma^{\lambda}_{\lambda r} &= \Gamma^{t}_{tr} + \Gamma^{r}_{rr} + \Gamma^{\theta}_{\theta r} + \Gamma^{\phi}_{\phi r} = \frac{2}{r},\\
\Gamma^{\lambda}_{\lambda \theta} &= \Gamma^{t}_{t\theta} + \Gamma^{r}_{r\theta} + \Gamma^{\theta}_{\theta\theta} + \Gamma^{\phi}_{\phi\theta} = \cot\theta,\\
\Gamma^{\lambda}_{\lambda \phi} &= \Gamma^{t}_{t\phi} + \Gamma^{r}_{r\phi} + \Gamma^{\theta}_{\theta\phi} + \Gamma^{\phi}_{\phi\phi} = \cot\theta.
\end{aligned}
\tag{10.452}
$$

Showing that $R_{\mu\nu}$ vanishes is tedious and mistake prone. For example, the component R_{rr} is given by

$$
R_{rr} = \partial_r \Gamma^{\lambda}_{r\lambda} - \partial_\lambda \Gamma^{\lambda}_{rr} + \Gamma^{\alpha}_{r\lambda}\Gamma^{\lambda}_{r\alpha} - \Gamma^{\alpha}_{rr}\Gamma^{\lambda}_{\lambda\alpha}. \tag{10.453}
$$

Using the affine connection (10.450) and results (10.452) gives

$$
\begin{aligned}
\partial_r \Gamma^{\lambda}_{r\lambda} &= -\frac{2}{r^2},\\
-\partial_\lambda \Gamma^{\lambda}_{rr} &= -\partial_r \Gamma^{r}_{rr} = -\frac{r_s(2r - r_s)}{2r^2(r - r_s)^2},\\
\Gamma^{\alpha}_{r\lambda}\Gamma^{\lambda}_{r\alpha} &= (\Gamma^{t}_{rt})^2 + (\Gamma^{r}_{rr})^2 + (\Gamma^{\theta}_{r\theta})^2 + (\Gamma^{\phi}_{r\phi})^2 = \frac{r_s^2}{2r^2(r - r_s)^2} + \frac{2}{r^2},\\
-\Gamma^{\alpha}_{rr}\Gamma^{\lambda}_{\lambda\alpha} &= -\Gamma^{r}_{rr}\Gamma^{\lambda}_{\lambda r} = \frac{r_s}{r^2(r - r_s)}.
\end{aligned}
\tag{10.454}
$$

The sum of the four terms in (10.454) vanishes. The reader is recommended to use the general-relativity-oriented calculational tools available in *Mathematica* in order to verify that the remaining components of the Ricci tensor also vanish for the Schwarzschild metric.

Exercise 10.23: Show that the Schwarzschild metric gives

$$
R_{\mu\nu} = 0 \tag{10.455}
$$

Because it is a vacuum solution, the Schwarzschild metric is interpreted as the space-time *outside* of a spherical mass M. The Schwarzschild metric reduces to the previous weak field analysis (10.398) for the case that the Newtonian potential is weak,

$$
\Phi = -\frac{GM}{r} \approx 0 \implies g_{00} = 1 + h_{00} = 1 - \frac{2GM}{rc^2} = 1 + \frac{2}{c^2}\Phi, \tag{10.456}
$$

with a result identical to (10.416) for h_{jk}. However, because it is a solution of the Einstein equation, (10.447) holds even for the strong field case. If the *entire mass M* lies *inside* the radial distance

$$r_s = \frac{2GM}{c^2}, \tag{10.457}$$

called the *Schwarzschild radius*, then the object is referred to as a *black hole* and the Schwarzschild metric is valid for all $r > R_s$. However, if that is not the case then the Schwarzschild metric will break down for a radial value greater than r_s. Newtonian physics predicts that the escape velocity of a spherical mass with the Schwarzschild radius is the speed of light. Since $g_{00} \to 0$ as $r \to r_s$, the gravitational redshift for light emitted from the Schwarzschild radius becomes infinite. This is the relativistic counterpart of light never escaping the black hole, since the energy of the light wave becomes zero. The spherical surface around the black hole at the Schwarzschild radius is therefore known as *the event horizon*.

10.8.3 Killing vectors and motion

Since the components of the affine connection for the Schwarzschild metric are known, it is possible to examine motion by solving the geodesic equation (10.319) to find the relativistic proper velocity V^μ of a point mass in a gravitational field. However, there are simpler methods similar to the conservation methods used in Newtonian physics. These are provided by *Killing vectors* K_μ. Each Killing vector gives a quantity, $K_\mu V^\mu$, that is constant or *conserved* along the geodesic of motion *in the absence of other forces*. This is the general relativistic version of conservation of energy along the classical trajectory. An immediate example is the four-velocity $K^\mu = V^\mu$, since the definition of the metric gives

$$c^2\,\mathrm{d}\tau^2 = g_{\mu\nu}\mathrm{d}x^\mu\,\mathrm{d}x^\nu = g_{\mu\nu}\frac{\mathrm{d}x^\mu}{\mathrm{d}\tau}\frac{\mathrm{d}x^\nu}{\mathrm{d}\tau}\mathrm{d}\tau^2 = g_{\mu\nu}V^\mu V^\nu\,\mathrm{d}\tau^2. \tag{10.458}$$

This shows that

$$g_{\mu\nu}V^\mu V^\nu = V_\mu V^\mu = c^2, \tag{10.459}$$

so that $V^\mu V_\mu$ is constant, just as it is in special relativity.

Other Killing vectors exist if there is a *symmetry* of the metric tensor. This can be understood using the Euler–Lagrange equation for the geodesic Lagrangian (10.313) *in the absence of other forces*. If the metric tensor is independent of a specific coordinate x^α *in some coordinate system*, then it is invariant under the symmetry operation $x^\alpha \to x^\alpha + a$, which follows from the infinitesimal Taylor expansion

$$g_{\mu\nu}(x^\alpha + a) = g_{\mu\nu}(x^\alpha) + a\,\delta^\rho{}_\alpha\partial_\rho g_{\mu\nu}(x) = g_{\mu\nu}(x) \Longrightarrow \delta^\rho{}_\alpha\partial_\rho g_{\mu\nu}(x) = 0. \tag{10.460}$$

Using the general relativistic point particle action defined by (10.313) and given by

$$\mathcal{L}_G = \frac{1}{2}mg_{\mu\nu}\frac{\mathrm{d}x^\mu}{\mathrm{d}\tau}\frac{\mathrm{d}x^\nu}{\mathrm{d}\tau}, \tag{10.461}$$

shows that the Euler–Lagrange equation obtained from (10.313) for the specific coordinate x^α gives

$$\frac{\partial \mathcal{L}_G}{\partial x^\alpha} - \frac{\mathrm{d}}{\mathrm{d}\tau}\frac{\partial \mathcal{L}_G}{\partial \dot{x}^\alpha} = -\frac{\mathrm{d}}{\mathrm{d}\tau}\left(\frac{\partial \mathcal{L}_G}{\partial \dot{x}^\alpha}\right) = -m\frac{\mathrm{d}}{\mathrm{d}\tau}\left(g_{\alpha\nu}\frac{\mathrm{d}x^\nu}{\mathrm{d}\tau}\right) = -m\frac{\mathrm{d}}{\mathrm{d}\tau}(g_{\alpha\nu}V^\nu) = 0. \quad (10.462)$$

For this simple case, the Killing vector is constant and given by

$$K^\mu = \delta^\mu{}_\alpha, \quad (10.463)$$

since (10.462) shows that quantity $K_\nu V^\nu$ obeys

$$\frac{\mathrm{d}}{\mathrm{d}\tau}(K_\nu V^\nu) = \frac{\mathrm{d}}{\mathrm{d}\tau}(K^\mu g_{\mu\nu} V^\nu) = \frac{\mathrm{d}}{\mathrm{d}\tau}(\delta^\mu{}_\alpha g_{\mu\nu} V^\nu) = \frac{\mathrm{d}}{\mathrm{d}\tau}(g_{\alpha\nu} V^\nu) = 0. \quad (10.464)$$

Since it is true in one coordinate system, it is true in all.

A general Killing vector K may be space-time dependent, and in that case it is defined by demanding that the action of its Lie derivative on the metric tensor, given by (1.226), satisfies

$$(\mathcal{L}_K g)_{\alpha\beta} = K^\mu \partial_\mu g_{\alpha\beta} + g_{\mu\beta}\partial_\alpha K^\mu + g_{\alpha\mu}\partial_\beta K^\mu = 0. \quad (10.465)$$

For such a case the vector K is said to be an *isometry* of the metric. Clearly, in the case of a constant Killing vector such as (10.463), expressions (10.464) and (10.465) are both satisfied. Proving the general case begins by noting that the trajectory x^μ satisfies the variational result (10.316). Combining that result with the chain rule gives

$$\begin{aligned}\frac{\mathrm{d}}{\mathrm{d}\tau}\left(K_\beta V^\beta\right) &= \frac{\mathrm{d}}{\mathrm{d}\tau}\left(K^\mu g_{\mu\beta} V^\beta\right) = \frac{\mathrm{d}K^\mu}{\mathrm{d}\tau}\left(g_{\mu\beta}V^\beta\right) + K^\mu \frac{\mathrm{d}}{\mathrm{d}\tau}\left(g_{\mu\beta}V^\beta\right)\\ &= \left(\partial_\alpha K^\mu g_{\mu\beta} + \frac{1}{2}K^\mu \partial_\mu g_{\alpha\beta}\right)V^\alpha V^\beta.\end{aligned} \quad (10.466)$$

Switching the dummy indices α and β and using the symmetry of the metric tensor allows the first term in the parentheses to be rewritten as two terms, yielding

$$\begin{aligned}\frac{\mathrm{d}}{\mathrm{d}\tau}\left(K_\beta V^\beta\right) &= \frac{1}{2}\left(K^\rho \partial_\rho g_{\alpha\beta} + g_{\rho\beta}\partial_\alpha K^\rho + g_{\alpha\rho}\partial_\beta K^\rho\right)V^\alpha V^\beta\\ &= \frac{1}{2}(\mathcal{L}_K g)_{\alpha\beta} V^\alpha V^\beta = 0,\end{aligned} \quad (10.467)$$

which is the desired result.

The motion of an inertial point mass in the Schwarzschild metric can be analyzed relatively easily using its Killing vectors. In the spherical coordinates of (10.447), the Schwarzschild metric is *static*, $\partial_0 g_{\alpha\beta} = 0$, and so it possesses the Killing vector $K^\mu = \delta^\mu{}_0$. In addition, it is rotationally symmetric, so that $\partial_\phi g_{\alpha\beta} = 0$, and therefore $R^\mu = \delta^\mu{}_3$ is a second Killing vector. These two Killing vectors correspond to two conserved quantities which can serve as the initial conditions for the motion of a point mass m in the Schwarzschild gravitational field. The first Killing vector gives

$$\frac{\mathrm{d}}{\mathrm{d}\tau}\left(g_{0\mu}\frac{\mathrm{d}x^\mu}{\mathrm{d}\tau}\right) = \frac{\mathrm{d}}{\mathrm{d}\tau}\left\{\left(1 - \frac{r_s}{r}\right)\frac{\mathrm{d}t}{\mathrm{d}\tau}\right\} = 0 \implies \left(1 - \frac{r_s}{r}\right)\frac{\mathrm{d}t}{\mathrm{d}\tau} = \frac{\mathcal{E}}{mc^2}, \quad (10.468)$$

where the constant $\mathcal{E}$ is the relativistic kinetic energy. This identification is made by using the limit $r \to \infty$, which gives

$$\lim_{r\to\infty}\left(1 - \frac{r_s}{r}\right)\frac{\mathrm{d}t}{\mathrm{d}\tau} = \lim_{r\to\infty}\left(1 - \frac{r_s}{r}\right)\frac{K}{mc^2} = \frac{\mathcal{E}}{mc^2}. \tag{10.469}$$

Assuming that planar motion occurs, it follows that

$$\frac{\mathrm{d}\theta}{\mathrm{d}\tau} = 0, \tag{10.470}$$

and so the polar angle can be set to $\theta = \pi/2$ for simplicity. The second Killing vector then gives

$$\frac{\mathrm{d}}{\mathrm{d}\tau}\left(g_{3\mu}\frac{\mathrm{d}x^\mu}{\mathrm{d}\tau}\right) = \frac{\mathrm{d}}{\mathrm{d}\tau}\left\{r^2\frac{\mathrm{d}\phi}{\mathrm{d}\tau}\right\} = 0 \implies r^2\frac{\mathrm{d}\phi}{\mathrm{d}\tau} = \frac{L}{m}, \tag{10.471}$$

where the constant L is the total angular momentum.

The necessary tools have been assembled to find the equation for radial motion in the Schwarzschild metric. This follows from the proper velocity, which gives

$$\begin{aligned} c^2 = g_{\mu\nu}V^\mu V^\nu &= c^2 g_{00}\left(\frac{\mathrm{d}t}{\mathrm{d}\tau}\right)^2 - g_{rr}\left(\frac{\mathrm{d}r}{\mathrm{d}\tau}\right)^2 - g_{\phi\phi}\left(\frac{\mathrm{d}\phi}{\mathrm{d}\tau}\right)^2 \\ &= c^2\left(1 - \frac{r_s}{r}\right)\left(\frac{\mathrm{d}t}{\mathrm{d}\tau}\right)^2 - \left(1 - \frac{r_s}{r}\right)^{-1}\left(\frac{\mathrm{d}r}{\mathrm{d}\tau}\right)^2 - r^2\left(\frac{\mathrm{d}\phi}{\mathrm{d}\tau}\right)^2. \end{aligned} \tag{10.472}$$

Multiplying both sides by $1 - (r_s/r)$ and using (10.468) and (10.471) gives

$$\frac{1}{2}\left(\frac{\mathrm{d}r}{\mathrm{d}\tau}\right)^2 + \left(\frac{L^2}{2m^2r^2} - \frac{GM}{r} - \frac{L^2 r_s}{2m^2r^3}\right) = \frac{1}{2}c^2\left(\frac{\mathcal{E}^2}{m^2c^4} - 1\right), \tag{10.473}$$

where the static Newtonian gravitational potential is obtained from

$$-\frac{r_s c^2}{2r} = -\frac{GM}{r}. \tag{10.474}$$

The effective radial gravitational potential, $\Phi_{\mathrm{eff}}(r)$, is enclosed by the second pair of parentheses. It is the familiar Newtonian gravitational potential along with the centrifugal barrier due to angular momentum, but it is modified by the third term containing the Schwarzschild radius. This term causes a deviation from the Newtonian Kepler orbits by altering the effective radial force, which is now given by

$$f(r) = -m\frac{\mathrm{d}}{\mathrm{d}r}\Phi_{\mathrm{eff}}(r) = \frac{L^2}{mr^3}\left(1 - \frac{3r_s}{2r}\right) - \frac{GMm}{r^2}. \tag{10.475}$$

In the case of a black hole, the modification serves to destabilize low-lying circular orbits, since $f(r)$ does not have a zero if $r < \frac{3}{2}r_s$, whereas the original Newtonian potential does. In the case of the Sun, the modification of the radial potential is most pronounced for inner planetary orbits and explains the small *anomalous precession of the perihelion of Mercury's orbit*, an early successful application of general relativity.

While this instability of low-lying orbits around a black hole has been demonstrated for the motion of a point mass, the orbital restriction $r = 1.5r_s$ can also be demonstrated by finding the orbital radius of light around a black hole. Light follows a null or lightlike geodesic, so that for $\mathrm{d}r = 0$ and $\theta = \pi/2$, its proper time is given by

$$c^2\mathrm{d}\tau^2 = 0 \implies \left(1 - \frac{r_s}{r}\right)c^2\mathrm{d}t^2 - r^2\mathrm{d}\phi^2 = 0. \tag{10.476}$$

This immediately gives

$$\left(\frac{\mathrm{d}\phi}{\mathrm{d}t}\right)^2 = \frac{c^2}{r^2}\left(1 - \frac{r_s}{r}\right). \tag{10.477}$$

A second expression for the orbital motion can be found from the geodesic equation for r. Assuming that a circular orbit is stable, it follows that the radius is constant, so that

$$\frac{\mathrm{d}^2r}{\mathrm{d}\tau^2} = \frac{\mathrm{d}r}{\mathrm{d}\tau} = 0. \tag{10.478}$$

The geodesic equation for r then gives

$$\frac{\mathrm{d}^2r}{\mathrm{d}\tau^2} + \Gamma^r_{\mu\nu}\frac{\mathrm{d}x^\mu}{\mathrm{d}\tau}\frac{\mathrm{d}x^\nu}{\mathrm{d}\tau} = \Gamma^r_{tt}\left(c\frac{\mathrm{d}t}{\mathrm{d}\tau}\right)^2 + \Gamma^r_{\phi\phi}\left(\frac{\mathrm{d}\phi}{\mathrm{d}\tau}\right)^2 = 0. \tag{10.479}$$

Dividing by γ_v^2 and using the affine connection components of (10.450) gives

$$\left(\frac{\mathrm{d}\phi}{\mathrm{d}t}\right)^2 = -c^2\frac{\Gamma^r_{tt}}{\Gamma^r_{\phi\phi}} = \frac{c^2r_s}{2r^3}. \tag{10.480}$$

Equating (10.477) and (10.480) and solving for r gives the previous result for the lowest-lying stable orbit,

$$r = \frac{3}{2}r_s. \tag{10.481}$$

This is referred to as the radius of the *photon ring* around a black hole. Remarkable visual evidence for the existence of the photon ring around the black hole at the center of the Galaxy M87 was released by the global collaboration EHT (Event Horizon Telescope) in 2019.

The Schwarzschild metric (10.447) becomes singular at $r = r_s$. However, the gravitational field in terms of the Riemann curvature tensor is well-behaved. Its components at $r = r_s$ can be computed and are found to be proportional to $1/r^3$. This indicates the existence of a *singularity* at $r = 0$ in the interior of a black hole, but that the apparent singularity at the event horizon is a result of the choice of coordinates for the metric. It is recommended that the reader study *Kruskal–Szekeres coordinates*, which better reveal the properties of a black hole.

10.8.4 Gravitational infall for the Schwarzschild metric

It is interesting to examine the time required for a point mass to fall from rest to the event horizon using the Schwarzschild metric. The velocity of infall was computed in (6.227) for the case of the Newtonian gravitational potential. However, radial infall can now be analyzed using the equation of motion for the Schwarzschild metric given by (10.473).

There are two ways to proceed. The first is to compute the *proper time* that is required for the infalling mass to reach the event horizon at $r = r_S$. For the case of radial motion there is no orbital angular momentum and $L = 0$. Choosing $\mathcal{E} = mc^2$ for simplicity, relation (10.473) gives

$$\frac{\mathrm{d}r}{\mathrm{d}\tau} = -\sqrt{\frac{2GM}{r}}, \tag{10.482}$$

where the minus sign is chosen for the square root to reflect an infalling situation. Integrating equation (10.482) yields the time T of infall for the case that the initial radius is R and the final radius is r,

$$-\sqrt{2GM}\int_0^{\mathcal{T}} \mathrm{d}\tau = -\sqrt{2GM}\,\mathcal{T} = \int_R^r \mathrm{d}r\,\sqrt{r} = \frac{2}{3}(r^{3/2} - R^{3/2}). \tag{10.483}$$

Using the definition of the Schwarzschild radius allows the proper infall time to be written

$$\mathcal{T} = \frac{2}{3c}\left(\sqrt{\frac{R^3}{r_S}} - \sqrt{\frac{r^3}{r_S}}\right). \tag{10.484}$$

For the case that R is finite, this result shows that the infalling observer will record a finite amount of time in order to reach the event horizon at $r = r_S$.

The second way to proceed is to compute the time T that an observer at a fixed position records until the infalling mass reaches r. Using the Schwarzschild metric (10.449) gives

$$\frac{\mathrm{d}r}{\mathrm{d}\tau} = \frac{\mathrm{d}t}{\mathrm{d}\tau}\frac{\mathrm{d}r}{\mathrm{d}t} = \left(1 - \frac{r_S}{r}\right)^{-1}\frac{\mathrm{d}r}{\mathrm{d}t}, \tag{10.485}$$

so that equation (10.482) becomes

$$\frac{\mathrm{d}r}{\mathrm{d}t} = -\left(1 - \frac{r_S}{r}\right)\sqrt{\frac{2GM}{r}} = -c\left(1 - \frac{r_S}{r}\right)\sqrt{\frac{r_S}{r}}. \tag{10.486}$$

Equation (10.486) gives

$$T = \int_0^T dt = -\frac{1}{c\sqrt{r_s}} \int_R^r dr \frac{r^{3/2}}{r - r_s} = \frac{2r_s}{3c}\left\{\left(\frac{R}{r_s}\right)^{3/2} - \left(\frac{r}{r_s}\right)^{3/2} + 3\sqrt{\frac{R}{r_s}} - 3\sqrt{\frac{r}{r_s}} + \frac{3}{2} ln\left(\frac{\left(\sqrt{R/r_s} + 1\right)\left(\sqrt{r/r_s} + 1\right)}{\left(\sqrt{R/r_s} - 1\right)\left(\sqrt{r/r_s} - 1\right)}\right)\right\}. \tag{10.487}$$

The first two terms are the proper time found in (10.484), while the remaining terms constitute the general relativistic time dilation effect recorded by the fixed observer.

Result (10.487) shows that the time T recorded by the fixed observer diverges logarithmically as $r \rightarrow r_S$. This gives the paradoxical result that a fixed observer will never see the infalling mass reach the event horizon. This raises the question of how a black hole can ever increase its mass and grow in size according to a fixed observer, even though the infalling mass reaches the event horizon in a finite amount of proper time. The resolution of this apparent paradox is found by noting that the Schwarzschild radius of the mass M black hole *plus the infalling mass m* changes to

$$r_s' = \frac{G(M + m)}{c^2} = \frac{GM}{c^2}\left(1 + \frac{m}{M}\right) = r_s\left(1 + \frac{m}{M}\right). \tag{10.488}$$

According to the fixed observer, the infalling mass m will be inside the *modified event horizon* when it reaches the radius

$$r = r_s' = r_s + \frac{mr_s}{M}. \tag{10.489}$$

For such a case, it follows that the black hole plus mass will have become a single black hole when

$$\sqrt{\frac{r}{r_s}} = \sqrt{1 + \frac{m}{M}} \approx 1 + \frac{m}{2M}, \tag{10.490}$$

where it was assumed that $m/M \ll 1$. Using this value for r in (10.487) shows that the fixed observer will claim that the additional time beyond the proper time of (10.484) required for the mass to be absorbed by the black hole is given by

$$\Delta T = 2\sqrt{\frac{Rr_s}{c^2}} - \frac{2r_s}{c} - \frac{mr_s}{Mc} + \left(\frac{r_s}{c}\right) \ln\left(\frac{\sqrt{R/r_s} + 1}{\sqrt{R/r_s} - 1}\right) + \left(\frac{r_s}{c}\right)\ln\left(1 + \frac{2M}{m}\right). \tag{10.491}$$

As an example, the black hole at the center of the Galaxy Cygnus A has a mass of $M \approx 5 \times 10^{39}$ kg or 2.5×10^9 solar masses, which yields a Schwarzschild radius of $r_s \approx 7.5 \times 10^{12}$ m, or roughly 50 times the radius of the solar system. For the case of an electron, $m \approx 9.1 \times 10^{-31}$ kg, falling into the Cygnus A black hole from the distance $R = 2r_s$, the additional time taken to reach the new event horizon and modify the size of the black hole comes primarily from the last logarithm, and is given by

$$\Delta T \approx 4.1 \times 10^6 \text{ s}, \tag{10.492}$$

or about 47.5 days. This is the extreme case of gravitational time dilation for radial infall recorded by the stationary observer, since all other massive particles that can be absorbed are larger. Visible light has an effective mass that is smaller, $m_{\text{eff}} \approx 3 \times 10^{-36}$ kg, which does not alter (10.492) significantly. This resolves the paradox of matter and energy absorption by a black hole.

10.9 Kaluza–Klein theory

One of the earliest attempts at unifying forces originated with the work of Kaluza and Klein, who focused on unifying gravitation and electromagnetism in a geometric framework similar to general relativity. They used a five-dimensional variant of Einstein's theory of gravitation to include electromagnetism as a geometric effect.

The basic idea is to expand the number of space-time dimensions from four to five by adding a fourth spatial dimension that is compact on a very small scale. The Kaluza–Klein mechanism is easily understood in the case of a cylindrical manifold where there are two spatial dimensions, the x-coordinate, with an infinite range, and the y-coordinate, a circular dimension characterized by the radius R. The y-coordinate can be replaced by the angular coordinate θ, so that $y = R\theta$. For the case of a massless scalar field $\varphi(x, y, t)$, the equation of motion is given by

$$\begin{aligned}\Box\varphi(x, y, t) &= \left(\frac{1}{c^2}\frac{\partial^2}{\partial t^2} - \frac{\partial^2}{\partial x^2} - \frac{\partial^2}{\partial y^2}\right)\varphi(x, y, t) \\ &= \left(\frac{1}{c^2}\frac{\partial^2}{\partial t^2} - \frac{\partial^2}{\partial x^2} - \frac{1}{R^2}\frac{\partial^2}{\partial \theta^2}\right)\varphi(x, \theta, t) = 0.\end{aligned} \tag{10.493}$$

Because the field must be periodic in the angle θ, the general solution to (10.493) is given by

$$\varphi(x, \theta, t) = \sum_{n=-\infty}^{\infty} \int_{-\infty}^{\infty} \frac{\mathrm{d}k}{2\pi}\, \tilde{\varphi}_n(k) \exp i(\omega_n(k)t - kx - n\theta), \tag{10.494}$$

where $\tilde{\varphi}_n^*(k) = \tilde{\varphi}_{-n}(-k)$ and

$$\frac{1}{c^2}\omega_n^2(k) = k^2 + \frac{n^2}{R^2}. \tag{10.495}$$

The energy of the individual mode $\tilde{\varphi}_n(k)$ is given by adapting the three-dimensional result (8.58), so that

$$E_n(k) = \hbar c\, \omega_n(k) = \hbar c\sqrt{k^2 + n^2/R^2}\,. \tag{10.496}$$

If $R \to 0$ the energy diverges, unless $n = 0$. As a result, for $R \approx 0$ the physically accessible modes of the solution are dominated by the $n = 0$ mode, so that $\omega_0(k) = ck$, and the field becomes

$$\varphi(x, y, t) \approx \int_{-\infty}^{\infty} \frac{\mathrm{d}k}{2\pi} \tilde{\varphi}_0(k) e^{ik(ct-x)}. \tag{10.497}$$

In effect, for $R \approx 0$ the cylindrical manifold behaves as a one-dimensional line for low energies. This allows the compact dimension to be experimentally invisible at low energies, yielding a field theory that is effectively independent of the compact dimension. This is sometimes referred to as the *cylinder condition* in analogy to result (10.497).

Kaluza and later Klein adapted this concept to five-dimensional space-time by adding a fourth spatial dimension x^4 that is compact, with $R \approx 10^{-32}$ m. This corresponds to an $n = 1$ energy level of $\approx 3 \times 10^6$ J or $\approx 1.8 \times 10^{24}$ eV, which is $\approx 10^{13}$ times greater than currently available experimental energies. For such a small radius the fields, including the metric tensor, become effectively independent of x^4. The five-dimensional metric tensor is denoted by g_{AB}, where A and B range from zero to four. Restricting this to Greek indices gives $g_{\alpha\beta}$, which stands for the four-dimensional metric tensor. The motion of a particle of mass m in the five-dimensional space-time is given by simply extending the four-dimensional action (10.313),

$$\mathcal{L} = \frac{1}{2} m g_{AB} \frac{\mathrm{d}x^A}{\mathrm{d}\tau} \frac{\mathrm{d}x^B}{\mathrm{d}\tau}. \tag{10.498}$$

The metric tensor g_{AB} is assumed to obey the cylinder condition, so that it depends only on the four-dimensional space-time coordinates x^μ and is independent of x^4.

In what follows a simplified version of the Kaluza–Klein mechanism will be presented. Rather than treating the five-dimensional metric tensor as arbitrary and solving the Einstein equation for a choice of stress–energy tensor, the five-dimensional metric tensor will be approximated as the Minkowski metric for the four-dimensional part, with space-time dependent components only for g_{A4}. The metric tensor is written in terms of the four-dimensional electromagnetic gauge field $A_\mu(x)$ and is assumed to take the form

$$\boldsymbol{g} = \begin{pmatrix} 1 & 0 & 0 & 0 & \kappa A_0(x) \\ 0 & -1 & 0 & 0 & \kappa A_1(x) \\ 0 & 0 & -1 & 0 & \kappa A_2(x) \\ 0 & 0 & 0 & -1 & \kappa A_3(x) \\ \kappa A_0(x) & \kappa A_1(x) & \kappa A_2(x) & \kappa A_3(x) & \psi(x) \end{pmatrix}, \tag{10.499}$$

where κ is a constant with units of length, so that the metric tensor is dimensionless. Restricting Greek letters to zero through three, the components of the metric tensor can be written

$$g_{\mu\nu} = \eta_{\mu\nu}, \quad g_{4\mu} = g_{\mu 4} = \kappa A_\mu, \quad g_{44} = \psi(x) = -1 + \kappa^2 A_\mu A^\mu. \tag{10.500}$$

This choice for $\psi(x)$ allows the metric tensor to be easily inverted. The inverse metric tensor is given by

$$g^{-1} = \begin{pmatrix} 1-\kappa^2A_0^2 & \kappa^2A_0A_1 & \kappa^2A_0A_2 & \kappa^2A_0A_3 & \kappa A_0 \\ \kappa^2A_1A_0 & -1-\kappa^2A_1^2 & -\kappa^2A_1A_2 & -\kappa^2A_1A_3 & -\kappa A_1 \\ \kappa^2A_2A_0 & -\kappa^2A_2A_1 & -1-\kappa^2A_2^2 & -\kappa^2A_2A_3 & -\kappa A_2 \\ \kappa^2A_3A_0 & -\kappa^2A_3A_1 & -\kappa^2A_3A_2 & -1-\kappa^2A_3^2 & -\kappa A_3 \\ \kappa A_0 & -\kappa A_1 & -\kappa A_2 & -\kappa A_3 & -1 \end{pmatrix}, \tag{10.501}$$

which matrix multiplication shows is the inverse of $\boldsymbol{g}$. Restricting Greek letters to zero through three and using $A^0 = A_0$ and $A^j = -A_j$ allows the components of the inverse metric tensor to be written

$$g^{\mu\nu} = \eta^{\mu\nu} - \kappa^2A^\mu A^\nu, \quad g^{4\mu} = g^{\mu 4} = \kappa A^\mu, \quad g^{44} = -1. \tag{10.502}$$

These satisfy the inverse relationship,

$$g^{AB}g_{BC} = \delta^A{}_C. \tag{10.503}$$

As examples,

$$\begin{aligned} g^{\mu B}g_{B\nu} &= g^{\mu\rho}g_{\rho\nu} + g^{\mu 4}g_{4\nu} = (\eta^{\mu\rho} - \kappa^2A^\mu A^\rho)\eta_{\rho\nu} + \kappa^2A^\mu A_\nu = \delta^\mu{}_\nu, \\ g^{\mu B}g_{B4} &= g^{\mu\nu}g_{\nu 4} + g^{\mu 4}g_{44} \\ &= (\eta^{\mu\nu} - \kappa^2A^\mu A^\nu)\kappa A_\nu + \kappa A^\mu(-1 + \kappa^2A_\nu A^\nu) = 0, \\ g^{4B}g_{B\mu} &= g^{4\nu}g_{\nu\mu} + g^{44}g_{4\mu} = \kappa A^\nu\eta_{\nu\mu} - \kappa A_\mu = 0. \end{aligned} \tag{10.504}$$

Exercise 10.24: Complete the proof by showing that

$$g^{4B}g_{B4} = 1. \tag{10.505}$$

There are several ways to compute the equation of motion. Rather than finding the Levi-Civita connection and using it in the geodesic equation (10.100), it is entirely equivalent and far simpler to use the Lagrangian (10.498), which gives the relativistic Euler–Lagrange equation of motion

$$\begin{aligned} \frac{\mathrm{d}}{\mathrm{d}\tau}\frac{\partial\mathcal{L}}{\partial\dot{x}_A} - \frac{\partial\mathcal{L}}{\partial x_A} = \frac{\mathrm{d}}{\mathrm{d}\tau}\left(mg^{AB}(x)\frac{\mathrm{d}x_B}{\mathrm{d}\tau}\right) \\ - \frac{1}{2}m\frac{\partial g^{BC}(x)}{\partial x_A}\frac{\mathrm{d}x_B}{\mathrm{d}\tau}\frac{\mathrm{d}x_C}{\mathrm{d}\tau} = 0. \end{aligned} \tag{10.506}$$

It is useful to consider the $A = 4$ case first. Using $\partial g^{BC}/\partial x_4 = 0$ results in the simple version of the Killing equation that was derived in (10.462),

$$\begin{aligned}\frac{\mathrm{d}}{\mathrm{d}\tau}\left(mg^{4B}(x)\frac{\mathrm{d}x_B}{\mathrm{d}\tau}\right) &= \frac{\mathrm{d}}{\mathrm{d}\tau}\left(mg^{4\mu}\frac{\mathrm{d}x_\mu}{\mathrm{d}\tau} + mg^{44}\frac{\mathrm{d}x_4}{\mathrm{d}\tau}\right)\\ &= \frac{\mathrm{d}}{\mathrm{d}\tau}\left(m\kappa A^\mu\frac{\mathrm{d}x_\mu}{\mathrm{d}\tau} - m\frac{\mathrm{d}x_4}{\mathrm{d}\tau}\right) = 0.\end{aligned} \tag{10.507}$$

This gives the result that

$$\frac{\mathrm{d}x_4}{\mathrm{d}\tau} = \alpha + \kappa A^\nu\frac{\mathrm{d}x_\nu}{\mathrm{d}\tau}, \tag{10.508}$$

where α is a constant to be determined.

Setting the index $A = \mu$ in (10.506) and dropping the factor of m gives

$$\begin{aligned}&\frac{\mathrm{d}}{\mathrm{d}\tau}\left(g^{\mu B}\frac{\mathrm{d}x_B}{\mathrm{d}\tau}\right) - \frac{1}{2}(\partial^\mu g^{BC})\frac{\mathrm{d}x_B}{\mathrm{d}\tau}\frac{\mathrm{d}x_C}{\mathrm{d}\tau}\\ &= \frac{\mathrm{d}}{\mathrm{d}\tau}\left(g^{\mu\nu}\frac{\mathrm{d}x_\nu}{\mathrm{d}\tau} + g^{\mu 4}\frac{\mathrm{d}x_4}{\mathrm{d}\tau}\right) - \frac{1}{2}(\partial^\mu g^{\rho\sigma})\frac{\mathrm{d}x_\rho}{\mathrm{d}\tau}\frac{\mathrm{d}x_\sigma}{\mathrm{d}\tau} - (\partial^\mu g^{4\rho})\frac{\mathrm{d}x_4}{\mathrm{d}\tau}\frac{\mathrm{d}x_\rho}{\mathrm{d}\tau} = 0,\end{aligned} \tag{10.509}$$

where $\partial g^{44}/\partial x^\mu = 0$ and the symmetry of g^{AB} were used. Using the expressions for $g^{\mu\nu}$, $g^{\mu 4}$, and $\mathrm{d}x_4/\mathrm{d}\tau$ gives

$$\begin{aligned}g^{\mu\nu}\frac{\mathrm{d}x_\nu}{\mathrm{d}\tau} + g^{\mu 4}\frac{\mathrm{d}x_4}{\mathrm{d}\tau} &= (\eta^{\mu\nu} - \kappa^2 A^\mu A^\nu)\frac{\mathrm{d}x_\nu}{\mathrm{d}\tau}\\ &\quad + \kappa A^\mu\left(\alpha + \kappa A^\nu\frac{\mathrm{d}x_\nu}{\mathrm{d}\tau}\right) = \frac{\mathrm{d}x^\mu}{\mathrm{d}\tau} + \alpha\kappa A^\mu.\end{aligned} \tag{10.510}$$

Similarly, it follows that

$$\begin{aligned}&-\frac{1}{2}\,\partial_\mu g^{\rho\sigma}\frac{\mathrm{d}x_\rho}{\mathrm{d}\tau}\frac{\mathrm{d}x_\sigma}{\mathrm{d}\tau} - \partial_\mu g^{4\rho}\frac{\mathrm{d}x_4}{\mathrm{d}\tau}\frac{\mathrm{d}x_\rho}{\mathrm{d}\tau}\\ &= -\frac{1}{2}\partial_\mu(\eta^{\rho\sigma} - \kappa^2 A^\rho A^\sigma)\frac{\mathrm{d}x_\rho}{\mathrm{d}\tau}\frac{\mathrm{d}x_\sigma}{\mathrm{d}\tau} - \partial_\mu(\kappa A^\rho)\left(\alpha + \kappa A^\sigma\frac{\mathrm{d}x_\sigma}{\mathrm{d}\tau}\right)\frac{\mathrm{d}x_\rho}{\mathrm{d}\tau}\\ &= -\alpha\kappa\,\partial^\mu A^\rho\frac{\mathrm{d}x_\rho}{\mathrm{d}\tau},\end{aligned} \tag{10.511}$$

which used the identity

$$\begin{aligned}\frac{1}{2}\partial^\mu(A^\rho A^\sigma)\frac{\mathrm{d}x_\rho}{\mathrm{d}\tau}\frac{\mathrm{d}x_\sigma}{\mathrm{d}\tau} &= \frac{1}{2}(A^\sigma\partial^\mu A^\rho + A^\rho\partial^\mu A^\sigma)\frac{\mathrm{d}x_\rho}{\mathrm{d}\tau}\frac{\mathrm{d}x_\sigma}{\mathrm{d}\tau}\\ &= A^\sigma(\partial^\mu A^\rho)\frac{\mathrm{d}x_\rho}{\mathrm{d}\tau}\frac{\mathrm{d}x_\sigma}{\mathrm{d}\tau}.\end{aligned} \tag{10.512}$$

This reduces (10.509) to

$$\begin{aligned}\frac{\mathrm{d}}{\mathrm{d}\tau}\left(\frac{\mathrm{d}x^\mu}{\mathrm{d}\tau} + \alpha\kappa A^\mu\right) - \alpha\kappa\,(\partial^\mu A^\rho)\frac{\mathrm{d}x_\rho}{\mathrm{d}\tau} &= \frac{\mathrm{d}^2x^\mu}{\mathrm{d}\tau^2} + \alpha\kappa(\partial^\rho A^\mu - \partial^\mu A^\rho)\frac{\mathrm{d}x_\rho}{\mathrm{d}\tau} \\ &= \frac{\mathrm{d}^2x^\mu}{\mathrm{d}\tau^2} - \alpha\kappa F^{\mu\rho}\frac{\mathrm{d}x_\rho}{\mathrm{d}\tau} = 0.\end{aligned} \tag{10.513}$$

Setting $\alpha\kappa$ equal to the charge q of the particle reproduces the contravariant version of the covariant electromagnetic Lorentz force law derived in (9.128).

The Kaluza–Klein approach has also been adapted to Yang–Mills field theories, which are discussed in the next chapter, by increasing the number of compact dimensions. It is also used in string theory, where the extra dimensions needed to make string theory consistent quantum mechanically can be hidden through the compactification process.

10.10 Basic cosmology

Despite being the weakest of the natural forces, gravitation is the dominant force at the largest scales of the Universe. Not only is it long ranged, but unlike the only other long ranged force, electromagnetism, it is solely attractive and therefore cannot be screened. Soon after its inception Einstein's version of gravity was applied to the large-scale behavior of the Universe, ushering in the era of *scientific cosmology*.

10.10.1 The cosmological principle

The application of general relativity to cosmology depends critically on observational astronomy, which has determined that, at very large scales, the visible part of the Universe is virtually *isotropic* and *homogeneous* in its mass distribution. This has been elevated to the *cosmological principle*, which states that all observers, regardless of location, will agree on the large-scale isotropic and homogeneous nature of the Universe, and that there is no preferred observer in the grand scheme of things. This restricts the spatial metric of the Universe to one of three possibilities, uniform and infinite in volume, uniform and finite in volume, or uniform and neither infinite nor finite in volume. These three options are referred to as open, closed, and flat. Each of these three possibilities can be modeled using a reasonably simple metric.

Before finding the form of the three possible metrics, it is important to characterize the stress–energy tensor for the uniformly distributed mass and energy of the Universe. Typically, this is done by using the stress–energy tensor for a *non-interacting electrically neutral isotropic and homogeneous fluid*, referred to earlier as a *perfect fluid*. The perfect fluid stress–energy tensor of special relativity, given by (6.260), is adapted to general relativity by replacing the Minkowski metric $\eta^{\mu\nu}$ with the metric tensor $g^{\mu\nu}$, so that

$$T^{\mu\nu} = \left(\rho + \frac{P}{c^2}\right)V^\mu V^\nu - g^{\mu\nu}P, \tag{10.514}$$

where ρ is the mass density and P is the pressure. The pressure P is assumed to be obtained from an equation of state for the matter, so that it is determined by the mass density ρ and the temperature T. If the matter is dispersed or cold, and therefore not interacting through collisions, it is commonly referred to as *dust* and its contribution to the pressure P vanishes. This is the current state of the visible matter in the Universe at the very large scale. However, radiant energy also contributes to the stress–energy tensor, supplementing the mass density with the effective mass of the thermal blackbody energy density ϵ_r,

$$\rho_r = \frac{\epsilon_r}{c^2} = \frac{\sigma}{c^2}T^4, \tag{10.515}$$

as well as contributing a pressure P_r to the total pressure due to radiation given by

$$P_r = \frac{1}{3}\epsilon_r. \tag{10.516}$$

These are quantum mechanical effects, which this text must borrow. Including both the polarizations of light and neutrinos gives the meter, kilogram, second, and ampere (MKSA) value

$$\sigma \approx 4.1 \times 10^{-16}\ \mathrm{J/m^3 \cdot K^4}. \tag{10.517}$$

The net energy density ϵ is given by the sum of the mass energy density and the radiation energy density,

$$\epsilon = \rho c^2 + \epsilon_r, \tag{10.518}$$

while P designates the pressure from the matter equation of state combined with the radiation pressure P_r. As discussed earlier in (10.382), it is also possible that a vacuum energy is present, which is proportional to the *cosmological constant* Λ. For such a case the stress–energy tensor is modified to read

$$\theta_{\mu\nu} = T_{\mu\nu} - \frac{c^4\Lambda}{8\pi G}g_{\mu\nu}, \tag{10.519}$$

where $T_{\mu\nu}$ is the stress–energy tensor for matter and energy.

General relativity allows a frame of reference to be selected at each point for which the mass–energy fluid of the Universe appears stationary. Such a *comoving frame* is characterized by the proper velocity

$$V^\mu = (c, 0). \tag{10.520}$$

This coordinate system has the advantage that the stress–energy tensor (10.514) reduces to a simple pair of expressions,

$$\begin{aligned} T_{00} &= \rho c^2 + \epsilon_r + (1 - g_{00})P = \epsilon + (1 - g_{00})P, \\ T_{jk} &= -g_{jk}P. \end{aligned} \tag{10.521}$$

If a cosmological constant is included in the energy, these expressions are modified to read

$$\theta_{00} = \epsilon + (1 - g_{00})P - \frac{c^4\Lambda}{8\pi G} g_{00},$$
$$\theta_{jk} = -g_{jk}\left(P + \frac{c^4\Lambda}{8\pi G}\right). \tag{10.522}$$

This shows that the vacuum energy density ρ_v and the vacuum pressure P_v are equal and opposite, so that

$$P_v = -\rho_v. \tag{10.523}$$

As outlined earlier, the metric for a spatially isotropic Universe can have only three distinct geometries: *three-sphere*, *three-hyperboloid*, or *flat*. These are the three-dimensional extensions of the spatially isotropic surfaces given by the two-dimensional sphere, the two-dimensional hyperboloid, and the Euclidean plane. These two-dimensional versions are depicted in figure 10.4. In each of the two-dimensional cases, the surfaces are isotropic in the radial direction. The three depicted surfaces exhaust the possibilities. The corresponding three-dimensional geometries can be defined in a manner identical to their two-dimensional versions. The metric of a two-sphere with radius R is found by beginning with the flat three-dimensional Euclidean metric

$$\mathrm{d}s^2 = \mathrm{d}x^2 + \mathrm{d}y^2 + \mathrm{d}z^2, \tag{10.524}$$

and imposing the constraint

$$x^2 + y^2 + z^2 = R^2. \tag{10.525}$$

This constraint is automatically satisfied by employing the standard spherical coordinates,

$$x = R\sin\theta\cos\phi, \quad y = R\sin\theta\sin\phi, \quad z = R\cos\theta. \tag{10.526}$$

Substituting the differentials $\mathrm{d}x$, $\mathrm{d}y$, and $\mathrm{d}z$ in terms of $\mathrm{d}\theta$ and $\mathrm{d}\phi$ into the line element $\mathrm{d}s^2$ reproduces the metric of a sphere,

$$\mathrm{d}s^2 = R^2\,\mathrm{d}\theta^2 + R^2\sin^2\theta\,\mathrm{d}\phi^2. \tag{10.527}$$

Similarly, the three-dimensional spherical and hyperbolic spaces are defined by introducing a Euclidean four-dimensional volume that satisfies

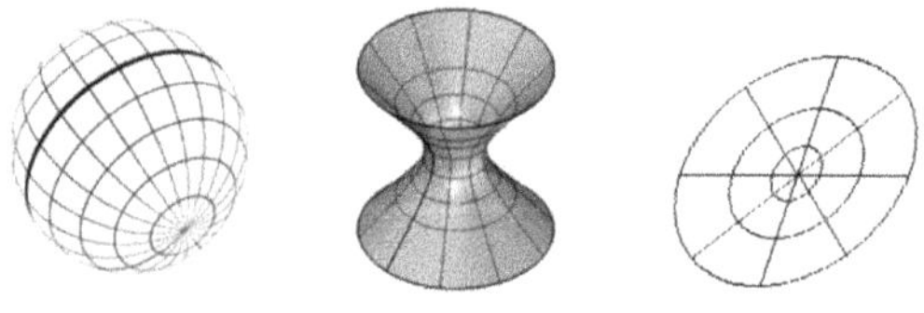

Figure 10.4. The two-dimensional counterparts to the closed, open, or flat isotropic Universe.

$$ds^2 = dx^2 + dy^2 + dz^2 + dw^2, \tag{10.528}$$

and imposing one of the following two constraints,

$$\begin{aligned} \text{three–sphere} &\Longrightarrow x^2 + y^2 + z^2 + w^2 = R^2, \\ \text{three–hyperboloid} &\Longrightarrow x^2 + y^2 + z^2 - w^2 = R^2. \end{aligned} \tag{10.529}$$

For the three-sphere the appropriate extension of spherical coordinates is given by introducing three angles ψ, θ, and ϕ, which have the ranges

$$0 \leqslant \psi \leqslant \pi, \quad 0 \leqslant \theta \leqslant \pi, \quad 0 \leqslant \phi \leqslant 2\pi. \tag{10.530}$$

These can be used to characterize a point on the three-sphere as

$$\begin{aligned} x &= R \sin\psi \sin\theta \cos\phi, \\ y &= R \sin\psi \sin\theta \sin\phi, \\ z &= R \sin\psi \cos\theta, \\ w &= R \cos\psi. \end{aligned} \tag{10.531}$$

Inserting these expressions into the three-sphere constraint of (10.529) shows that it is satisfied. Expressing the differentials in the flat metric in terms of the three angles induces the three-sphere metric

$$ds^2 = R^2 d\psi^2 + R^2 \sin^2\psi \, d\theta^2 + R^2 \sin^2\psi \sin^2\theta \, d\phi^2. \tag{10.532}$$

Exercise 10.25: Verify (10.532).

Using the dimensionless variable a allows a radial coordinate r to be defined,

$$r = \left(\frac{R}{a}\right) \sin\psi \Longrightarrow d\psi = \frac{a}{R\sqrt{1 - (a^2 r^2/R^2)}} \, dr. \tag{10.533}$$

This allows the three-sphere metric (10.532) to be expressed in spherical coordinates as

$$ds^2 = a^2 \left(\frac{dr^2}{1 - (a^2 r^2/R^2)} + r^2 d\theta^2 + r^2 \sin^2\theta \, d\phi^2 \right), \tag{10.534}$$

where a is an arbitrary scale factor for the volume that can be chosen to be R, fixing the range of r to be zero to one. The volume of the three-sphere can be computed by using the invariant three-volume,

$$d^3x \sqrt{g} = d^3x \sqrt{g_{11} g_{22} g_{33}} = d\psi \, d\theta \, d\phi \, R^3 \sin^2\psi \sin\theta, \tag{10.535}$$

which gives the volume V of the three-sphere,

$$V = R^3 \int_0^{\pi} \mathrm{d}\psi \, \sin^2\psi \int_0^{\pi} \mathrm{d}\theta \, \sin\theta \int_0^{2\pi} \mathrm{d}\phi = 2\pi^2 R^3, \tag{10.536}$$

so that the scale factor R determines the finite volume of the three-sphere. Because the volume is finite, the three-sphere is *closed.*

An identical process is followed for the three-hyperboloid by introducing the hyperbolic variable,

$$-\infty \leqslant \chi \leqslant \infty, \tag{10.537}$$

and satisfying the three-hyperboloid constraint of (10.529) by writing

$$\begin{aligned} x &= R \sinh\chi \sin\theta \cos\phi, \\ y &= R \sinh\chi \sin\theta \sin\phi, \\ z &= R \sinh\chi \cos\theta, \\ w &= R \cosh\chi. \end{aligned} \tag{10.538}$$

This gives the metric of the three-hyperboloid as

$$\mathrm{d}s^2 = R^2 \, \mathrm{d}\chi^2 + R^2 \sinh^2\chi \, \mathrm{d}\theta^2 + R^2 \sinh^2\chi \sin^2\theta \, \mathrm{d}\phi^2. \tag{10.539}$$

Defining the radial variable

$$r = \left(\frac{R}{a}\right) \sinh\chi \implies \mathrm{d}\chi = \frac{a}{R\sqrt{1 + (a^2 r^2/R^2)}} \, \mathrm{d}r, \tag{10.540}$$

gives the three-hyperboloid metric as

$$\mathrm{d}s^2 = a^2 \left(\frac{\mathrm{d}r^2}{1 + a^2 r^2/R^2} + r^2 \mathrm{d}\theta^2 + r^2 \sin^2\theta \, \mathrm{d}\phi^2 \right), \tag{10.541}$$

where a can again be set to R. The invariant volume element of the three-hyperboloid is given by

$$\mathrm{d}^3x \, \sqrt{g} = \mathrm{d}\chi \, \mathrm{d}\theta \, \mathrm{d}\phi \, R^3 \sinh^2\chi \, \sin\theta, \tag{10.542}$$

which shows that the volume of the three-hyperboloid is infinite. Because the volume is infinite, the three-hyperboloid is *open.* The *flat case* corresponds to a Euclidean three-volume with a variable volume scale, so that it has the metric

$$\mathrm{d}s^2 = a^2(\mathrm{d}r^2 + r^2 \mathrm{d}\theta^2 + r^2 \sin^2\theta \, \mathrm{d}\phi^2), \tag{10.543}$$

where a is the scale factor of a manifold that has no spatial curvature. These three possibilities, closed, open, and flat, exhaust the possible types of a spatially isotropic Universe.

10.10.2 The Friedmann–Robertson–Walker metric and Hubble's law

Solving the Einstein equation for a homogeneous and isotropic Universe uses the *Friedmann–Robertson–Walker (FRW) metric*, which allows all three possible cases for the geometry of space by writing

$$c^2\,\mathrm{d}\tau^2 = c^2\,\mathrm{d}t^2 - a^2(t)\left(\frac{\mathrm{d}r^2}{1-\kappa r^2} + r^2\,\mathrm{d}\theta^2 + r^2\sin^2\theta\,\mathrm{d}\phi^2\right). \tag{10.544}$$

The constant κ has units of inverse length squared, while $a(t)$ is a dimensionless time-dependent scale factor. The choices $k < 0$, $k = 0$, and $k > 0$ correspond to open (hyperbolic), flat, and closed (spherical) spatial volumes, respectively. Because the metric is dependent on both time and radius, the components of the Levi-Civita connection are greater in number, given by

$$\begin{aligned}
&\Gamma^0_{rr} = \frac{a\dot{a}}{1-\kappa r^2}, \quad \Gamma^0_{\theta\theta} = a\dot{a}r^2, \quad \Gamma^0_{rr} = a\dot{a}r^2\sin^2\theta,\\
&\Gamma^r_{0r} = \Gamma^\theta_{0\theta} = \Gamma^\phi_{0\phi} = \frac{\dot{a}}{a}, \quad \Gamma^\theta_{\phi\phi} = -\sin\theta\cos\theta,\\
&\Gamma^r_{\theta\theta} = -r(1-\kappa r^2), \quad \Gamma^r_{\phi\phi} = -r(1-\kappa r^2),\\
&\Gamma^\theta_{r\theta} = \Gamma^\phi_{r\phi} = \frac{1}{r}, \quad \Gamma^\phi_{\theta\phi} = \cot\theta,
\end{aligned} \tag{10.545}$$

where $\dot{a} = \partial_0 a$ and $\ddot{a} = \partial_0^2 a$. The Ricci tensor components are computed using the connections of (10.545) in the formula (10.451). The result is diagonal,

$$\begin{aligned}
R_{00} &= -3\frac{\ddot{a}}{a},\\
R_{rr} &= \frac{a\ddot{a} + 2\dot{a}^2 + 2\kappa}{1-\kappa r^2},\\
R_{\theta\theta} &= r^2(a\ddot{a} + 2\dot{a}^2 + 2\kappa),\\
R_{\phi\phi} &= r^2\sin^2\theta(a\ddot{a} + 2\dot{a}^2 + 2\kappa).
\end{aligned} \tag{10.546}$$

Finally, the curvature scalar is given by

$$\begin{aligned}
R = R^\mu{}_\mu = g^{\mu\nu}R_{\nu\mu} &= R_{00} - \frac{1-\kappa r^2}{a^2}R_{rr} - \frac{1}{a^2r^2}R_{\theta\theta}\\
&\quad - \frac{1}{a^2r^2\sin^2\theta}R_{\phi\phi}\\
&= -\frac{6}{a^2}(a\ddot{a} + \dot{a}^2 + \kappa).
\end{aligned} \tag{10.547}$$

Exercise 10.26: Verify results (10.545), (10.546), and (10.547).

The Einstein equations of (10.417) can now be found by combining the results (10.546) and (10.547) with the comoving stress–energy tensor of (10.522) that is consistent with the FRW metric. Using $g_{00} = 1$ in the stress–energy tensor of (10.522), the first equation is given by

$$R_{00} - \frac{1}{2}g_{00}R = \frac{8\pi G}{c^4}T_{00} \implies \frac{3}{a^2}(\kappa + \dot{a}^2) = \frac{8\pi G\epsilon}{c^4}. \tag{10.548}$$

Similarly, using $g_{rr} = a^2(1 - \kappa r^2)$ gives the second equation

$$R_{rr} - \frac{1}{2}g_{rr}R = -\left(\frac{2a\ddot{a} + \dot{a}^2 + \kappa}{1 - \kappa r^2}\right) = \frac{8\pi G}{c^4}T_{rr} = -\frac{8\pi G}{c^4}g_{rr}P = \frac{8\pi G a^2 P}{c^4(1 - \kappa r^2)}, \tag{10.549}$$

which simplifies to

$$-\left(\frac{2\ddot{a}}{a} + \frac{\dot{a}^2 + \kappa}{a^2}\right) = \frac{8\pi GP}{c^4}. \tag{10.550}$$

The two equations (10.548) and (10.550) are known as the *Friedmann equations.*

Exercise 10.27: Show that the two remaining Einstein equations involving $R_{\theta\theta}$ and $R_{\phi\phi}$ yield equations identical to (10.550).

The *Hubble factor* $H(t)$ is defined as

$$H(t) = c\frac{\dot{a}}{a} = \frac{1}{a}\frac{\mathrm{d}a}{\mathrm{d}t}, \tag{10.551}$$

and its value is an inverse time that gives the rate of change in the distance scale factor $a(t)$. For the case that the energy density and pressure are *uniform*, the FRW metric gives the distance separating two comoving infinitesimally separated radial positions as

$$R = a(t)\frac{\mathrm{d}r}{\sqrt{1 - \kappa r^2}} = a(t)\,\Delta r. \tag{10.552}$$

This shows that the separation R is changing according to

$$v_r = \frac{\mathrm{d}R}{\mathrm{d}t} = \frac{\mathrm{d}a}{\mathrm{d}t}\Delta r = \frac{1}{a}\frac{\mathrm{d}a}{\mathrm{d}t}(a(t)\Delta r) = HR. \tag{10.553}$$

Result (10.553) is known as *Hubble's Law* and v_r gives the apparent *recession rate* between the two points. The astronomer Edwin Hubble observed that the redshift of distant galaxies was proportional to their distance and was isotropic at great distances. This phenomenon can be interpreted as the Doppler shift due to a recession rate that is proportional to distance and isotropic. In the context of general relativity it is understood as a lengthening of the wavelength of the light as it propagates through a space that is increasing in scale. This follows by considering a light wave with wavelength λ traveling through the FRW space for one period, given by $T = \lambda/c$. The new wavelength λ' is given by

$$\lambda' = \lambda + v_r T = \lambda\left(1 + \frac{v_r}{c}\right) = \lambda\left(1 + \frac{H}{c}R\right) \implies \Delta\lambda = \lambda' - \lambda = \lambda\frac{H}{c}R, \quad (10.554)$$

which shows that the redshift of the light is proportional to the distance from the source. It also shows that the *cosmological redshift* (10.554) is somewhat different from the standard relativistic redshift found earlier in (6.44). Hubble's groundbreaking observations supported both the cosmological principle and an FRW metric where $a(t)$ is increasing.

10.10.3 The Big Bang model

Combining the Friedmann equations with the known behavior of matter and radiation has led to a picture of the evolution of the Universe known as the *Big Bang model.* It is rich in content and full of uncertainties, and this section will only present a few of the key results.

The first step in understanding the nature and evolution of the Universe comes from the first Friedmann equation (10.548). If the Universe is flat then $\kappa = 0$, and this immediately relates the current value of the Hubble factor, denoted H_0, to the *energy density*, denoted ϵ_c,

$$\epsilon_c = \frac{3{H_0}^2 c^2}{8\pi G}. \quad (10.555)$$

If the energy density deviates from ϵ_c it can be written

$$\epsilon = \epsilon_c + \Delta\epsilon, \quad (10.556)$$

where $\Delta\epsilon$ can be positive, zero, or negative. Returning (10.556) to the first Friedmann equation (10.548) gives

$$\frac{3\kappa}{a^2} = \frac{8\pi G}{c^2}\Delta\epsilon. \quad (10.557)$$

If $\Delta\epsilon$ is negative, so that $\epsilon < \epsilon_c$, then $\kappa < 0$ and the Universe is an open hyperboloid. If $\Delta\epsilon$ is positive, so that $\epsilon > \epsilon_c$, then $\kappa > 0$ and the Universe is a closed hypersphere.

In a matter-dominated Universe the energy density is found from the density of matter ρ, and this gives a critical matter density that determines the sign of κ,

$$\epsilon_c = \rho_c c^2 \implies \rho_c = \frac{3{H_0}^2}{8\pi G}. \quad (10.558)$$

The current measured value of the Hubble factor is $H_0 \approx 2.3 \times 10^{-18}$ per second, and this corresponds to a critical density of 9.2×10^{-27} kg/m^3. By terrestrial standards this is a high vacuum, corresponding to roughly six protons per cubic meter. However, intergalactic space does not necessarily attain this value, and this remains one of the numerous experimental uncertainties in applying the model. This result has a Newtonian counterpart. If the mass density ρ of the Universe is uniform, then a sphere with radius R contains the mass

$$M = \frac{4}{3}\pi R^3 \rho. \tag{10.559}$$

The Newtonian escape velocity for matter on the surface of the sphere is

$$v_e = \sqrt{\frac{2GM}{R}} = \sqrt{\frac{8}{3}\pi G\rho R^2} = R\sqrt{\frac{8}{3}\pi G\rho}. \tag{10.560}$$

If the recession velocity v_r of that matter given by Hubble's law (10.553) is greater than the escape velocity v_e, then the matter in the Universe will continue to move apart and the Universe is open. If v_r is less than the escape velocity, then the matter will eventually collapse and the Universe is closed. The critical density occurs when the recession velocity v_r and the escape velocity v_e coincide, and this gives

$$H_0 R = R\sqrt{\frac{8}{3}\pi G\rho_c} \implies \rho_c = \frac{3H_0{}^2}{8\pi G}, \tag{10.561}$$

which identifies the same critical density as (10.558).

A volume V in the FRW metric evolves as the scale factor changes according to

$$V = a^3 V_0 \implies \partial_0 V = 3a^2 \dot{a} V_0, \tag{10.562}$$

where V_0 is a fixed volume element. Combining (10.562) with the second Friedmann equation (10.550) gives

$$\begin{aligned} -\frac{P}{V_0}\partial_0 V = -\,3Pa^2\dot{a} &= \frac{3c^4}{8\pi G}\left(\frac{2\ddot{a}}{a} + \frac{\dot{a}^2 + \kappa}{a^2}\right)(a^2\dot{a}) \\ &= \frac{3c^4}{8\pi G}(2a\ddot{a} + \dot{a}^3 + \kappa\dot{a}) \\ &= \frac{3c^4}{8\pi G}\partial_0(a\dot{a}^2 + \kappa a) \\ &= \frac{3c^4}{8\pi G}\partial_0((\dot{a}^2 + \kappa)a). \end{aligned} \tag{10.563}$$

Using the first Friedmann equation (10.548) in the last term of (10.564) gives

$$-\frac{P}{V_0}\partial_0 V = \partial_0(\epsilon a^3) = \frac{1}{V_0}\partial_0(\epsilon a^3 V_0) = \frac{1}{V_0}\partial_0 U, \tag{10.564}$$

where U is the energy present in the volume V. Canceling the common factor of $1/V_0$, the right-hand side of (10.564) is the rate of change in energy in the volume, while the left-hand side is the rate of work done by pressure,

$$\partial_0 W = -P\,\partial_0 V. \tag{10.565}$$

Result (10.564) shows that the first law of thermodynamics, $\mathrm{d}U = \mathrm{d}W$, emerges from (10.550).

If the matter can be treated as non-interacting dust, then the pressure can be ignored. This shows that the matter mass energy behaves according to

$$dU = d(\epsilon a^3 V_0) = 0 \implies \epsilon = \frac{k}{a^3}, \tag{10.566}$$

where k is a constant. If the dominant energy density is electromagnetic radiation created by blackbody radiation, then the radiation pressure (10.516) must be used. The behavior of the radiation pressure in the Friedmann metric can be deduced by assuming that

$$\epsilon_r = ka^n, \tag{10.567}$$

where k is a constant and n is a constant to be determined. The rate of change in the radiation energy is then given by

$$dU_r = d(\epsilon_r a^3 V_o) = kV_o\, d(a^{n+3}) = (3+n)kV_0\, a^{n+2}\, da. \tag{10.568}$$

The first law gives

$$\begin{aligned} dU_r &= -P_r d(a^3 V_0) = -\frac{1}{3}\epsilon_r d(a^3 V_0) \\ &= -\frac{1}{3}ka^n(3a^2 V_0 da) = -kV_0\, a^{n+2}\, da. \end{aligned} \tag{10.569}$$

The required equality of (10.568) and (10.569) identifies $n = -4$. The radiation energy density therefore behaves according to

$$\epsilon_r \propto \frac{1}{a^4}. \tag{10.570}$$

Since the radiation energy density is given by Stefan's law, $\epsilon_r = 4\sigma T^4/c$, this shows that the temperature behaves according to

$$T \propto \frac{1}{a}. \tag{10.571}$$

This is consistent with Wien's law that states the predominant wavelength of a blackbody is inversely proportional to the temperature, since (10.553) and (10.554) show that the wavelength grows as a increases.

It was pointed out in (10.523) that the vacuum energy density ρ_v and the vacuum pressure P_v are equal and opposite. Assuming that the vacuum energy density behaves as

$$\rho_v = \alpha a^n, \tag{10.572}$$

where α is a constant and n is to be determined, it follows that the change in the vacuum energy is once again given by

$$dU_v = d(\rho_v V) = d(\alpha a^{n+3} V_0) = \alpha V_0(n+3)a^{n+2}. \tag{10.573}$$

However, since ρ_v and P_v are equal and opposite, the work done by vacuum pressure is given by

$$-P_v \mathrm{d}V = \alpha a^n \mathrm{d}(a^3 V_0) = 3\alpha V_0 a^{n+2}. \tag{10.574}$$

Equating the two results requires $n = 0$, showing that the vacuum energy density and pressure are unaffected by the changing scale factor a.

These results can be used in the Friedmann equation to determine the behavior of the scale factor for the various forms of energy in a *flat Universe* where $\kappa = 0$. If the constant vacuum energy density ρ_v is dominant, then the first Friedmann equation (10.548) shows that the flat Universe scale factor evolves according to

$$\frac{\dot{a}^2}{a^2} = \frac{8\pi G\rho_v}{3} \implies a = a_0 e^{H_v(t-t_o)}, \tag{10.575}$$

where a_0 is the scale factor at t_o and the Hubble factor in units of inverse time is given by

$$H_v = c\frac{\dot{a}}{a} = \sqrt{\frac{8\pi G\rho_v}{3c^2}}. \tag{10.576}$$

Exponential growth of the scale factor is referred to as a *de Sitter phase*, and any such period of exponential growth of the scale factor is referred to as *inflation*. For the case of a vacuum-energy-dominated open Universe, where $\kappa < 0$, the first Friedmann equation becomes

$$\frac{\dot{a}^2}{a^2} = \frac{8\pi G\rho_v}{3c^4} + \frac{|\kappa|}{a^2}. \tag{10.577}$$

It is assumed that the vacuum energy density is positive, and for that case, the right-hand side of (10.577) is always positive. This shows that a is an increasing function of time and that the second term on the right-hand side will therefore tend to zero, so that an open Universe dominated by vacuum energy will also enter a de Sitter phase and quickly become identical to a flat Universe dominated by vacuum energy. The deviation from the flat-space Universe is then quickly driven to zero. A period of inflation early in the history of the Universe, lasting roughly 10^{-34} seconds, would explain why the current density of the Universe appears to be so close to the critical value. It would also explain the absence of magnetic monopoles, which would have become separated before standard matter coalesced. The details of this are in the references. On the other hand, if the Universe is closed, so that $\kappa > 0$, and is dominated by vacuum energy, then the second term on the right-hand side of (10.577) will be negative. This can result in the impossible situation that $\dot{a}^2$ becomes negative.

A possible source of inflation is the potential energy of a scalar field $U(\varphi)$, sometimes referred to as the *inflaton field*. Extremely large vacuum energy density terms naturally occur in the physics of elementary particles, and so quantum field theory has direct implications for early cosmology. It is argued that in the early Universe the temperature was high enough to force the inflaton field into a metastable state, and the energy density of the field is estimated to have been on the order of 5×10^{93} Jm^{-3}.

Exercise 10.28: Show that the energy density 5×10^{93} Jm^{-3} is sufficient to cause the scale factor a to increase by a factor 10^{25} in 10^{-32} seconds.

The Universe is conjectured to have inflated from 10^{-29} meters in diameter to 10^{-4} meters in diameter in a time interval of 10^{-32} seconds.

However, it is important to note that some physicists have been critical of the inflationary model. One of the more pressing difficulties in modeling inflation is to find a mechanism that allows inflation both to occur early in the evolution of the Universe and then to terminate, leading to the current state of the Universe, where inflation is not present. Such a mechanism is tied to the quantum behavior of fields under extreme conditions, and there are many computational and conceptual difficulties. It also often requires fine tuning of the initial conditions and the dynamics, a situation that has led some to reject inflation as an extremely improbable occurrence in the early Universe. Others remain confident in the viability of inflation. The interested reader is referred to the literature.

Setting aside the questions surrounding an early period of inflation, if dust dominates the energy landscape, as it does in the current epoch of the Universe, then the flat Universe scale factor evolves according to

$$\frac{\dot{a}^2}{a^2} \propto \frac{1}{a^3} \Longrightarrow a = a_0\left(\frac{t}{t_o}\right)^{2/3} \Longrightarrow H_m = \frac{1}{a}\frac{\mathrm{d}a}{\mathrm{d}t} = \frac{2}{3t}, \tag{10.578}$$

where a factor of c has been canceled in the Friedmann equation so that H_m is measured in inverse time. Similarly, in a radiation-dominated flat Universe, the scale factor evolves according to

$$\frac{\dot{a}^2}{a^2} \propto \frac{1}{a^4} \Longrightarrow a = a_0\left(\frac{t}{t_o}\right)^{1/2} \Longrightarrow H_r = \frac{\dot{a}}{a} = \frac{1}{2t}. \tag{10.579}$$

The common feature of both these results is the growth of the scale factor and the concurrent decrease in the matter and radiation energy densities as the temperature drops. It is possible to model a static Universe by introducing a negative vacuum energy density that is exactly opposite to the matter and radiation energy density of the Universe at some point in its evolution. When the scale factor a reaches a value where the right-hand side of the Friedmann equation vanishes, then $\dot{a} = 0$ and a flat Universe becomes static, although it is unstable against fluctuations in the matter density. This was Einstein's reason for introducing the cosmological constant, since, at that time, Hubble had not yet provided the observational evidence for an expanding Universe.

Tracing backward in time therefore presents a Universe that is ever smaller in volume, since a is decreasing, and ever hotter, since $T \propto 1/a$. Even if inflation did not occur, this eventually leads back to a moment where the volume of the Universe is so small and its contents so hot that the known behaviors of matter and energy break down. This is the moment that early critics of the idea derisively referred to as the *Big Bang*, a name which has stuck. In order to trace back any further requires an understanding of quantum gravity, and such a theory is currently not available in a consistent form.

Because the blackbody radiation energy density scales as $1/a^4$ while matter energy scales as $1/a^3$, at very early times, radiation energy would have been dominant and the Universe would have been too hot for normal atomic matter to exist. Electromagnetic blackbody radiation has an energy density given by

$$\epsilon = \frac{4\sigma}{c}T^4, \tag{10.580}$$

where σ is Stefan's constant. Including the contributions of neutrinos raises the radiation energy density to

$$\epsilon \approx \frac{6.7\sigma}{c}T^4. \tag{10.581}$$

Using this energy density in the first Friedmann equation along with the Hubble factor (10.579) for a radiation-dominated flat Universe gives

$$\frac{3}{4t^2} \approx \frac{8\pi G}{c^2}\left(\frac{6.7\sigma}{c}T^4\right) \implies t \approx \sqrt{\frac{c^3}{71\pi G\sigma}}\,\frac{1}{T^2}, \tag{10.582}$$

which relates the age of the Universe to its temperature. In SI units the constant in (10.582) is given by

$$\sqrt{\frac{c^3}{71\pi G\sigma}} \approx 1.8 \times 10^{20}\ \mathrm{s \cdot K^2}. \tag{10.583}$$

Deuterium nuclei can form at temperatures below $\approx 1 \times 10^9$ K, and this occurred at $t \approx 180$ seconds. As the Universe continued to cool, greater percentages of electrons and nuclei formed into atoms. The temperature of $T = 4000$ K marks the point at which the atomic form of matter would become prevalent, and this occurred at $t \approx 1.1 \times 10^{13}$ seconds or about 360 000 years. At that point the blackbody radiation would cease interacting with individual protons and electrons. In effect, the blackbody radiation that was present would decouple from matter, since only those wavelengths corresponding to atomic spectra would be absorbed and scattered. The details of this depend on the quantum mechanical absorption and emission spectrums of the atomic matter that had formed.

As it continued to expand, the Universe became dominated by matter, while the decoupled blackbody radiation underwent an increase in wavelength as the scale of the Universe increased. This ambient blackbody radiation has been detected and is remarkably uniform, coming from all directions in the sky in concordance with the cosmological principle. It now lies in the microwave spectrum, corresponding to blackbody radiation at $T = 2.7$ K, revealing that the Universe has increased in scale by over a factor of 10^3 since atomic matter first formed. Initially predicted by Alpher and Herman in 1948 and later observed in 1965 by Penzias and Wilson, it is known as the *cosmic microwave background* (CMB) and carries information about the earliest stages of the Universe after the Big Bang. Its extreme uniformity provides experimental support for the occurrence of an inflationary period in the early Universe. The subsequent evolution of the Universe was the result of a combination of vacuum energy and matter. Using a combined energy model, the current age of the Universe is estimated at

$$t \approx \frac{1}{H_0} \approx 4.4 \times 10^{17}\ \text{s} \approx 1.4 \times 10^{10}\ \text{years}. \tag{10.584}$$

In the time since the Big Bang, the greatest distance light could have traveled is therefore 14 billion light-years or about 4000 megaparsecs. This constitutes the observable size of the Universe, and is referred to as the *cosmic horizon.*

Many theoretical questions and experimental uncertainties remain; in particular, the observationally inferred existence of *dark matter* and *dark vacuum energy*. Dark matter does not interact with light as the visible matter of galaxies and interstellar space does, but its presence and its inertial mass have been inferred from the anomalous rotation rates for galaxies. Similarly, dark vacuum energy has been inferred from the redshifts of type I supernovas, which indicate the expansion rate of the Universe is increasing. These two sources of gravitation will play a key role in determining the past as well as the future of the Universe. As the time of writing, these forms of energy and matter have yet to be experimentally identified and are an open theoretical question.

References and recommended further reading

The foundational papers of general relativity are available in

- Einstein A *et al* 1952 *The Principle of Relativity* (New York: Dover)

Classical field theory texts that cover general relativity as a topic include

- Landau L and Lifshitz E 1975 *The Classical Theory of Fields* 4th edn (Amsterdam: Elsevier)
- Scheck F 2012 *Classical Field Theory* (Berlin: Springer)
- Nastase H 2019 *Classical Field Theory* (Cambridge: Cambridge University Press)

Monographs written specifically as introductory texts for general relativity include

- Schutz B 1990 *A First Course in General Relativity* (Cambridge: Cambridge University Press)
- Hartle J 2003 *Gravity: An Introduction to Einstein's General Relativity* (San Francisco, CA: Addison-Wesley)
- Carroll S 2004 *Spacetime and Geometry: An Introduction to General Relativity* (San Francisco, CA: Addison-Wesley)
- Walecka J 2007 *Introduction to General Relativity* (Cambridge: Cambridge University Press)

Monographs written as reference texts for general relativity include

- Weinberg S 1972 *Gravitation and Cosmology: Principles and Applications of the General Theory of Relativity* (New York: Wiley)
- Wald R 1984 *General Relativity* (Chicago: Chicago University Press)
- Misner C, Thorne K and Wheeler J 2017 *Gravitation* (Princeton, NJ: Princeton University Press)

Monographs developing general relativity in terms of differential forms and connections include

- Stewart J 1991 *Advanced General Relativity* (Cambridge: Cambridge University Press)
- Ludvigsen M 1999 *General Relativity: A Geometric Approach* (Cambridge: Cambridge University Press)

Monographs developing Kaluza–Klein theory in various contexts include

- Kaku M 1993 *Quantum Field Theory* (Oxford: Oxford University Press)
- Wesson P 2006 *Five-Dimensional Physics* (Singapore: World Scientific)

Monographs concentrating solely on cosmology and the inflationary model include

- Narlikar J 1983 *Introduction to Cosmology* (Cambridge: Cambridge University Press)
- Islam J 1992 *An Introduction to Mathematical Cosmology* (Cambridge: Cambridge University Press)
- Peebles P 1993 *Principles of Physical Cosmology* (Princeton, NJ: Princeton University Press)
- Bernstein J 1995 *An Introduction to Cosmology* (Englewood Cliffs, NJ: Prentice-Hall)
- Kolb E and Turner M 2018 *The Early Universe* (New York: Taylor and Francis)

IOP Publishing

Classical Field Theory and the Stress–Energy Tensor
(Second Edition)

Mark S Swanson

Chapter 11

Yang–Mills fields and connections

A major theme in physics has been the identification of similarities or symmetries in the patterns of naturally occurring forces. In the subatomic realm this classification scheme goes by the general name of *isospin*. In its earliest form the similarities between the neutron and proton allowed them to be treated as two manifestations of the *nucleon*. The neutron bispinor n and proton bispinor p were grouped together into an *isospin doublet* $(n\ p)$, in effect a matrix vector built from the two bispinors. This led to the generalization of electric charge, as well as the exploration of how the nucleon doublet entered into interactions with itself and other particles. In the process, physicists understood how the gauge symmetry and gauge field of electrodynamics could be generalized to isospin, and this gave rise to the extremely important concept of the *Yang–Mills field*. Subsequently, it was found that electrodynamics, general relativity, and Yang–Mills fields share many similar structures, in particular, the presence of a *connection*, and can be expressed in terms of *forms* and *curvature*. In what follows, *natural units* are employed, where all powers of $\hbar$ and c are suppressed, and it is assumed that all fields, masses, and energies are expressed in terms of powers of inverse length.

11.1 Unitary symmetry and isospin

Unitary symmetry generalizes the gauge invariance of electrodynamics to isospin groupings of matter fields. To make this concrete, an *isospin vector* of *complex scalar matter fields* Φ is used,

$$\Phi = \begin{pmatrix} \varphi_1 \\ \varphi_2 \\ \vdots \\ \varphi_n \end{pmatrix}, \tag{11.1}$$

doi:10.1088/978-0-7503-3455-6ch11 © IOP Publishing Ltd 2022

where the φ_j fields form the n components of Φ and $\varphi_j^* \neq \varphi_j$. This grouping reflects the idea that these fields share similar properties of interaction and energy. It may be the case that these fields share an identical mass and, if that is the case, a simple real-valued Lorentz invariant bilinear Lagrangian that reflects this is given by

$$\mathcal{L}_0 = \frac{1}{2}\partial_\mu \mathbf{\Phi}^\dagger \cdot \partial^\mu \mathbf{\Phi} - \frac{1}{2}m^2 \mathbf{\Phi}^\dagger \cdot \mathbf{\Phi}, \tag{11.2}$$

where the inner product in complex isospin space is given by

$$\mathbf{\Phi}^\dagger \cdot \mathbf{\Phi} = \varphi_j^* \varphi_j. \tag{11.3}$$

The Lagrangian (11.2) gives the equation of motion,

$$\left(\Box + m^2\right)\mathbf{\Phi} = 0 \implies \left(\Box + m^2\right)\varphi_j = 0 \tag{11.4}$$

so that the equation of motion is identical for all n components of $\mathbf{\Phi}$.

The Lagrangian (11.2) also possesses two symmetries. The first is the *discrete symmetry*,

$$\mathbf{\Phi}' = -\mathbf{\Phi}. \tag{11.5}$$

It also possesses a *global unitary symmetry*, since it is invariant under the action

$$\mathbf{\Phi}' = \mathbf{U} \cdot \mathbf{\Phi}, \tag{11.6}$$

where the $n \times n$ matrix U obeys

$$\begin{aligned} \partial_\mu \mathbf{U} &= 0, \\ \mathbf{U}^\dagger \cdot \mathbf{U} &= \mathbf{I}. \end{aligned} \tag{11.7}$$

This symmetry is referred to as *global*, since the same transformation $\mathbf{U}$ is applied at *all* space-time points. The matrix $\mathbf{U}$ therefore forms a representation of the unitary group of dimension n, and $\mathbf{\Phi}$ is the representation space of the group. Restricting our attention to the *fundamental representation of the special unitary group* SU(n) introduced in (7.124), the matrix $\mathbf{U}$ can be written using the exponential map as

$$\mathbf{U} = \exp(i\lambda_a \boldsymbol{\tau}^a), \tag{11.8}$$

where $\boldsymbol{\tau}^a$ are the $n^2 - 1$ Hermitian generators of SU(n) and λ_a are $n^2 - 1$ real parameters. The generators obey the Lie algebra (7.118),

$$[\boldsymbol{\tau}^a, \boldsymbol{\tau}^b] = i f^{abc} \boldsymbol{\tau}^c, \tag{11.9}$$

where the f^{abc} are the structure constants of the group. It is useful to remember that the Hermitian generators can be redefined as $\boldsymbol{\tau}^a = -i\mathbf{T}^a$, and these non-Hermitian generators satisfy

$$[\boldsymbol{T}^a, \boldsymbol{T}^b] = -f^{abc} \boldsymbol{T}^c. \tag{11.10}$$

By definition, the structure constants are antisymmetric, so that $f^{abc} = -f^{acb}$. It is possible to view $\boldsymbol{\tau}^a$ as the basis elements of the group space. This was developed in detail for SU(2) subsequent to definition (7.130).

Noether's theorem can be used to identify the conserved quantities associated with this symmetry. The infinitesimal form of the symmetry is given by

$$\mathbf{U} \approx \mathbf{I} + i\lambda_a\boldsymbol{\tau}^a, \tag{11.11}$$

so that the infinitesimal symmetry transformation of the isospin field is

$$\delta\boldsymbol{\Phi} = i\lambda_a\boldsymbol{\tau}^a \cdot \boldsymbol{\Phi}. \tag{11.12}$$

Noether's theorem (8.223) gives the conserved current associated with this symmetry

$$\boldsymbol{J}_\mu = i\lambda_a\left(\partial_\mu\boldsymbol{\Phi}^\dagger\boldsymbol{\tau}^a\boldsymbol{\Phi} - \boldsymbol{\Phi}^\dagger\boldsymbol{\tau}^a\partial_\mu\boldsymbol{\Phi}\right), \tag{11.13}$$

and it follows from the equation of motion (11.4) that $\partial_\mu\boldsymbol{J}^\mu = 0$.

Since there are $n^2 - 1$ arbitrary parameters λ_a, result (11.13) appears to give $n^2 - 1$ distinct conserved charge densities,

$$\boldsymbol{J}_0^a = i\partial_0\boldsymbol{\Phi}^\dagger\boldsymbol{\tau}^a\boldsymbol{\Phi} - i\boldsymbol{\Phi}^\dagger\boldsymbol{\tau}^a\partial_0\boldsymbol{\Phi}. \tag{11.14}$$

While this is true, there is a subtlety that must be considered. For an arbitrary choice of parameters, the matrix $\delta\mathbf{U} = i\lambda_a\boldsymbol{\tau}^a$ can be diagonalized by a unitary transformation. Since the components of $\boldsymbol{\Phi}$ are indistinguishable, this unitary transformation can be absorbed into the definition of $\boldsymbol{\Phi}$. The result is that a subset of the charges obtained in this way will have the same eigenvalues and therefore can be redefined *as the same charge*.

The *unique* conserved charges associated with the unitary symmetry are therefore given by the *diagonal generators* of SU(n), since these reflect all the possibilities of the diagonalization process. These generators are referred to as the *Cartan subalgebra* of the group. A simple and illustrative example is given by SU(2), where the transformation matrix is given by

$$\delta\mathbf{U} = \lambda_a\boldsymbol{\tau}^a = \frac{1}{2}\lambda_a\boldsymbol{\sigma}^a = \frac{1}{2}\begin{pmatrix} \lambda_3 & \lambda_1 - i\lambda_2 \\ \lambda_1 + i\lambda_2 & -\lambda_3 \end{pmatrix}. \tag{11.15}$$

Matrix (11.15) has two eigenvalues,

$$\lambda_\pm = \pm\sqrt{\lambda_1^2 + \lambda_2^2 + \lambda_3^2} = \pm\lambda, \tag{11.16}$$

so that the diagonalized version $\delta\mathbf{U}_d$ of (11.15) is given by

$$\delta\mathbf{U}_d = \frac{1}{2}\begin{pmatrix} \lambda_+ & 0 \\ 0 & \lambda_- \end{pmatrix} = \frac{1}{2}\begin{pmatrix} \lambda & 0 \\ 0 & -\lambda \end{pmatrix} = \lambda\frac{1}{2}\boldsymbol{\sigma}^3 = \lambda\boldsymbol{\tau}^3. \tag{11.17}$$

The generator $\boldsymbol{\tau}^3$ is the single diagonal generator of SU(2) and so it is the single member of the Cartan subalgebra for SU(2). It can be shown that SU(n) has $n - 1$ diagonal generators, and therefore there are only $n - 1$ *distinct charges*, often referred to as *hypercharges*, associated with the Lagrangian (11.2). This gives the generalization of electric charge to the SU(n) quantity

$$Q = Q^a \boldsymbol{\tau}_c^a, \tag{11.18}$$

where it is understood that only the diagonal Cartan subgroup generators $\boldsymbol{\tau}_c$ enter the sum. The generalized current can then be written

$$J_\mu = i\partial_\mu \boldsymbol{\Phi}^\dagger Q \boldsymbol{\Phi} - i\boldsymbol{\Phi} Q \partial_\mu \boldsymbol{\Phi}. \tag{11.19}$$

11.2 Nonabelian gauge fields

The next step is to elevate **U** to a *local symmetry* by allowing the group parameters to be space-time dependent. As with electrodynamics and gravitation, the key step is note that $\partial_\mu \boldsymbol{\Phi}$ is no longer invariant under the transformation and requires the definition of a *gauge covariant derivative* D_μ such that

$$D_\mu' \boldsymbol{\Phi}' = \mathbf{U} D_\mu \mathbf{U}^{-1} \mathbf{U} \boldsymbol{\Phi} = \mathbf{U} D_\mu \boldsymbol{\Phi}, \tag{11.20}$$

along with a similar expression for the Hermitian adjoint,

$$\left(D_\mu' \boldsymbol{\Phi}'\right)^\dagger = \left(D_\mu \boldsymbol{\Phi}\right)^\dagger \mathbf{U}^\dagger. \tag{11.21}$$

Finding the gauge covariant derivative starts by noting that

$$\partial_\mu\left(\mathbf{U}\boldsymbol{\Phi}\right) = \left(\partial_\mu \mathbf{U}\right)\boldsymbol{\Phi} + \mathbf{U}\partial_\mu \boldsymbol{\Phi} = \left(\partial_\mu \mathbf{U}\mathbf{U}^{-1}\right)\mathbf{U}\boldsymbol{\Phi} + \mathbf{U}\partial_\mu \boldsymbol{\Phi}. \tag{11.22}$$

The next step is to introduce an $n \times n$ matrix A_μ, which is defined to transform under SU(n) as

$$A_\mu' = \mathbf{U} A_\mu \mathbf{U}^{-1} + \frac{i}{g}\partial_\mu \mathbf{U}\mathbf{U}^{-1}, \tag{11.23}$$

where g will act as a coupling constant. The gauge covariant derivative can then be written as

$$D_\mu = \mathbf{I}\,\partial_\mu + igA_\mu, \tag{11.24}$$

since this gives

$$\begin{aligned} D_\mu' \boldsymbol{\Phi}' &= \left(\mathbf{I}\,\partial_\mu + igA_\mu'\right)\boldsymbol{\Phi}' = \left(\mathbf{I}\,\partial_\mu + ig\mathbf{U}A_\mu \mathbf{U}^{-1} - \partial_\mu \mathbf{U}\mathbf{U}^{-1}\right)\mathbf{U}\boldsymbol{\Phi} \\ &= \mathbf{U}\left(\mathbf{I}\,\partial_\mu + igA_\mu\right)\boldsymbol{\Phi} = \mathbf{U}D_\mu \boldsymbol{\Phi}. \end{aligned} \tag{11.25}$$

Since $\mathbf{U}^{-1} = \mathbf{U}^\dagger$ for a unitary symmetry, the base Lagrangian (11.2) can be given a generalized form of gauge invariance by replacing ∂_μ with D_μ,

$$\mathcal{L} = \frac{1}{2} D_\mu^\dagger \boldsymbol{\Phi}^\dagger \cdot D^\mu \boldsymbol{\Phi} - \frac{1}{2} m^2 \boldsymbol{\Phi}^\dagger \cdot \boldsymbol{\Phi};\,. \tag{11.26}$$

For many groups, including SU(n), the generators can be normalized to obey the condition

$$\mathrm{Tr}(\boldsymbol{\tau}^a\boldsymbol{\tau}^b) = \frac{1}{2}\delta^{ab}. \tag{11.27}$$

For such a case, the final step is to introduce the Yang–Mills gauge fields A_μ^a by using the matrix A_μ of (11.23) to define

$$A_\mu^a = 2\,\mathrm{Tr}\left(A_\mu\boldsymbol{\tau}^a\right). \tag{11.28}$$

This allows the matrix introduced in (11.23) to be written as a superposition of the Yang–Mills gauge fields A_μ^a,

$$A_\mu = A_\mu^a\boldsymbol{\tau}^a, \tag{11.29}$$

since using (11.27) shows that this is consistent with (11.28),

$$2\,\mathrm{Tr}\left(A_\mu\boldsymbol{\tau}^a\right) = 2A_\mu^b\,\mathrm{Tr}\left(\boldsymbol{\tau}^a\boldsymbol{\tau}^b\right) = A_\mu^b\delta^{ab} = A_\mu^a. \tag{11.30}$$

Transformation (11.23) reproduces the gauge field transformation (9.68) of electrodynamics if $\mathbf{U}$ is chosen to be

$$\mathbf{U} = \exp(ig\Lambda), \quad \mathbf{U}^{-1} = \exp(-ig\Lambda), \tag{11.31}$$

since this gives

$$A_\mu' = \mathbf{U}A_\mu\mathbf{U}^{-1} + \frac{i}{g}\partial_\mu\mathbf{U}\mathbf{U}^{-1} = A_\mu - \partial_\mu\Lambda. \tag{11.32}$$

Because the gauge transformations $\mathbf{U}$ of electrodynamics commute,

$$\mathbf{U}\mathbf{U}' = \exp ig(\Lambda + \Lambda') = \exp ig(\Lambda' + \Lambda) = \mathbf{U}'\mathbf{U}, \tag{11.33}$$

electrodynamics is often referred to as an *abelian gauge theory* and the one-parameter abelian gauge group is denoted by $\mathbf{U}(1)$.

In the *nonabelian* case, substituting the infinitesimal forms

$$\begin{aligned}\mathbf{U} &\approx \mathbf{I} + ig\Lambda_b\boldsymbol{\tau}^b,\\ \mathbf{U}^{-1} &\approx \mathbf{I} - ig\Lambda_c\boldsymbol{\tau}^c,\end{aligned} \tag{11.34}$$

into (11.23), using the Lie algebra (11.9), and retaining only terms of $O(\Lambda)$ gives the *infinitesimal form* of the nonabelian gauge transformation,

$$\begin{aligned}A_\mu^{a\,\prime}\boldsymbol{\tau}^a &= A_\mu^a\boldsymbol{\tau}^a - \partial_\mu\Lambda^a\boldsymbol{\tau}^a + ig\Lambda^b A_\mu^c\left[\boldsymbol{\tau}^b, \boldsymbol{\tau}^c\right]\\ &= \left(A_\mu^a - \partial_\mu\Lambda^a - gf^{abc}\Lambda^b A_\mu^c\right)\boldsymbol{\tau}^a.\end{aligned} \tag{11.35}$$

The infinitesimal nonabelian gauge transformation,

$$A_\mu^{a\,\prime} = A_\mu^a - \partial\Lambda^a - gf^{abc}\Lambda^b A_\mu^c, \tag{11.36}$$

shows that the nonabelian gauge fields transform into each other under the action of the Lie algebra. This is very different than multiple copies of the gauge field of electrodynamics. The gauge invariant version of the charged scalar matter field can now be obtained by the replacement $\partial_\mu \mathbf{\Phi} \to D_\mu \mathbf{\Phi}$. Like electrodynamics, this is referred to as *minimal coupling* for the scalar fields and yields a set of nonlinear equations of motion. Additional gauge invariant terms are typically added to the scalar field Lagrangian in the form of a potential function $V(\mathbf{\Phi}^\dagger \cdot \mathbf{\Phi})$, where the argument $\mathbf{\Phi}^\dagger \cdot \mathbf{\Phi}$ is a manifestly gauge invariant scalar. The final form for the minimally coupled scalar matter field is

$$\mathcal{L}_\mathrm{M} = \frac{1}{2}\left(D_\mu \mathbf{\Phi}\right)^\dagger \cdot D^\mu \mathbf{\Phi} - V(\mathbf{\Phi}^\dagger \cdot \mathbf{\Phi}). \tag{11.37}$$

In the case that there is only one complex scalar field ϕ there will be a single gauge field, identified as the electromagnetic field. For the case of a single scalar field the associated electric current is

$$J_\mu = ie\ \partial_\mu \phi^* \phi - ie\ \phi^* \partial_\mu \phi, \tag{11.38}$$

and the theory is invariant under the abelian gauge transformation $\phi \to e^{i\Lambda}\phi$. This gives a complex scalar field that is minimally coupled to the electromagnetic field. It is the starting point for understanding the Higgs mechanism and spontaneously broken symmetry, which is presented later in this chapter. Similarly, a multiplet of spinor fields $\mathbf{\Psi}$ with the components ψ^a can be minimally coupled to Yang–Mills fields by the replacement $\gamma^\mu \partial_\mu \to \gamma^\mu D_\mu$. The γ^μ matrices act on the spinor components of ψ^a for each spinor field, while the Yang–Mills fields are coupled to the component index.

Constructing the action for the gauge field is made simpler by noting that the gauge covariant derivative can be used to define the gauge field equivalent of *curvature*. In the case of abelian electrodynamics the commutator of the covariant derivative gives

$$\left[D_\mu, D_\nu\right] = ie\left(\partial_\mu A_\nu - \partial_\nu A_\mu\right) = ieF_{\mu\nu}, \tag{11.39}$$

yielding the gauge invariant electrodynamic field tensor. In the case of general relativity it was shown in (10.211) that the action of the commutator of the covariant derivative on a vector gave

$$\left[D_\mu, D_\nu\right]V^\rho = R^\rho{}_{\mu\nu\sigma}V^\sigma, \tag{11.40}$$

yielding the Riemann tensor. In the case of Yang–Mills fields the commutator of the gauge covariant derivative (11.24) gives

$$\begin{aligned}
\left[D_\mu, D_\nu\right] &= ig\partial_\mu A_\nu - ig\partial_\nu A_\mu - g^2\left[A_\mu, A_\nu\right] \\
&= ig\partial_\mu A_\nu^a \boldsymbol{\tau}^a - ig\partial_\nu A_\mu^a \boldsymbol{\tau}^a - g^2 A_\mu^b A_\nu^c\left[\boldsymbol{\tau}^b, \boldsymbol{\tau}^c\right] \\
&= ig\left(\partial_\mu A_\nu^a - \partial_\nu A_\mu^a - gf^{abc}A_\mu^b A_\nu^c\right)\boldsymbol{\tau}^a \\
&\equiv igF_{\mu\nu}^{\ a}\boldsymbol{\tau}^a \equiv igF_{\mu\nu},
\end{aligned} \tag{11.41}$$

where the nonabelian field strength components are given by

$$F^a_{\mu\nu} = \partial_\mu A^a_\nu - \partial_\nu A^a_\mu - g f^{abc} A^b_\mu A^c_\nu. \tag{11.42}$$

These are the isospin components of the antisymmetric nonabelian field strength tensor $F_{\mu\nu}$. The coupling constant g, as expected from its role in the gauge transformation $\mathbf{U}$, has no dimensions. If the *gauge group* is abelian, then $f^{abc} = 0$ and the theory becomes a set of copies of abelian electrodynamics.

Unlike the abelian case, the field strength tensor is *not* gauge invariant. Result (11.25) shows that

$$D'_\mu \Phi' = \mathbf{U} D_\mu \Phi = \mathbf{U} D_\mu \mathbf{U}^{-1} \mathbf{U} \Phi = \mathbf{U} D_\mu \mathbf{U}^{-1} \Phi' \implies D'_\mu = \mathbf{U} D_\mu \mathbf{U}^{-1}. \tag{11.43}$$

As a result,

$$\left[D'_\mu, D'_\nu\right] = \left[\mathbf{U} D_\mu \mathbf{U}^{-1}, \mathbf{U} D_\nu \mathbf{U}^{-1}\right] = \mathbf{U}\left[D_\mu, D_\nu\right]\mathbf{U}^{-1}, \tag{11.44}$$

which shows that the nonabelian field strength has the transformation property

$$F'_{\mu\nu} = \mathbf{U} F_{\mu\nu} \mathbf{U}^{-1}. \tag{11.45}$$

The nonabelian field strength $F_{\mu\nu}$ is *gauge covariant* rather than *gauge invariant*. This has an important physical consequence. In abelian electrodynamics the field strength tensor is gauge invariant, and so its components, the magnetic and electric field, are independent of any gauge function and are therefore *observable*. However, in the nonabelian case, the field strength is not invariant under gauge transformations, and this means that its individual components *are not directly observable*. Any physical phenomena described by nonabelian gauge fields have more complicated properties in terms of the gauge fields.

The covariant derivative of the field tensor $F^{\mu\nu}$ is defined as

$$D_\rho F^{\mu\nu} = \partial_\rho F^{\mu\nu} + ig\left[A_\rho, F^{\mu\nu}\right]. \tag{11.46}$$

This definition has the required Leibniz property since its action on a product of nonabelian quantities B and C is given by

$$\begin{aligned} D_\mu(B \cdot C) &= \partial_\mu(B \cdot C) + ig\left[A_\mu, (B \cdot C)\right] \\ &= \partial_\mu B \cdot C + B \cdot \partial_\mu C + igA_\mu \cdot B \cdot C - igB \cdot C \cdot A_\mu \\ &= \partial_\mu B \cdot C + ig\left[A_\mu, B\right] \cdot C + B \cdot \partial_\mu C + igB \cdot \left[A_\mu, C\right] \\ &= \left(D_\mu B\right) \cdot C + B \cdot \left(D_\mu C\right). \end{aligned} \tag{11.47}$$

The covariant derivative (11.46) can be broken into its individual components using (11.9),

$$D_\rho F^{a\mu\nu} \boldsymbol{\tau}^a = \partial_\rho F^{a\mu\nu} \boldsymbol{\tau}^a + igA^b_\rho F^{c\mu\nu}[\boldsymbol{\tau}^b, \boldsymbol{\tau}^c] = \left(\partial_\rho F^{a\mu\nu} - g f^{abc} A^b_\rho F^{c\mu\nu}\right)\boldsymbol{\tau}^a, \tag{11.48}$$

so that

$$D_\rho F^{a\mu\nu} = \partial_\rho F^{a\mu\nu} - gf^{abc}A_\rho^b F^{c\mu\nu}. \tag{11.49}$$

The Lagrangian for Yang–Mills fields must be *gauge invariant*. Assuming the generator normalization of (11.27) is valid, the gauge invariant Yang–Mills Lagrangian can be defined as

$$\mathcal{L}_{\mathrm{YM}} = -\frac{1}{2}\,\mathrm{Tr}(F_{\mu\nu}F^{\mu\nu}) = -\frac{1}{2}F_{\mu\nu}^a F^{b\mu\nu}\,\mathrm{Tr}(\boldsymbol{\tau}^a\boldsymbol{\tau}^b) = -\frac{1}{4}F_{\mu\nu}^a F^{a\mu\nu}. \tag{11.50}$$

The Lagrangian (11.50) is invariant under a nonabelian gauge transformation, since the trace of a product of matrices is invariant under cyclic permutations of the product, from which it follows that

$$\begin{aligned}\mathcal{L}'_{\mathrm{YM}} &= -\frac{1}{2}\,\mathrm{Tr}(\mathbf{U}F_{\mu\nu}\mathbf{U}^{-1}\mathbf{U}F^{\mu\nu}\mathbf{U}^{-1}) \quad = -\frac{1}{2}\,\mathrm{Tr}(\mathbf{U}^{-1}\mathbf{U}F_{\mu\nu}F^{\mu\nu})\\ &= -\frac{1}{2}\,\mathrm{Tr}(F_{\mu\nu}F^{\mu\nu}) = \mathcal{L}_{\mathrm{YM}}.\end{aligned} \tag{11.51}$$

The Yang–Mills action is fundamentally different from that of quadratic abelian electrodynamics, since the Lagrangian density contains cubic and quartic terms in the gauge fields.

The combination of the Yang–Mills Lagrangian (11.50) and a matter-field Lagrangian $\mathcal{L}_M$ gives the Euler–Lagrange equation for the Yang–Mills fields,

$$\frac{\partial\mathcal{L}}{\partial A_\nu^a} - \partial_\mu\frac{\partial\mathcal{L}}{\partial\left(\partial_\mu A_\nu^a\right)} = \partial_\mu F^{a\mu\nu} - gf^{abc}A_\mu^b F^{c\mu\nu} - j^{a\nu} = D_\mu F^{a\mu\nu} - j^{a\nu} = 0, \tag{11.52}$$

where the nonabelian matter current is given by

$$j^{a\mu} = -\frac{\partial\mathcal{L}_M}{\partial A_\mu^a}, \tag{11.53}$$

the details of which depend on the choice for the matter Lagrangian $\mathcal{L}_M$. The profound differences between Yang–Mills fields and abelian electrodynamics are apparent from (11.52). The Yang–Mills field equations are *nonlinear* even in the absence of matter fields, whereas abelian electrodynamics is linear. This has the important result that classical Yang–Mills fields *do not* obey the principle of linear superposition as classical electromagnetic fields do. In particular, the presence of quadratic and cubic terms in the gauge fields of the equation of motion precludes simple wave solutions even if $j^{a\mu} = 0$.

The equation of motion can be written in matrix form by defining $j^\mu = j^{a\mu}\boldsymbol{\tau}^a$, since

$$\begin{aligned}D_\mu F^{\mu\nu} &= \partial_\mu F^{a\mu\nu}\boldsymbol{\tau}^a + igA_\mu^b F^{c\mu\nu}[\boldsymbol{\tau}^b, \boldsymbol{\tau}^c]\\ &= \partial_\mu F^{a\mu\nu}\boldsymbol{\tau}^a - gf^{abc}A_\mu^b F^{c\mu\nu}\boldsymbol{\tau}^a = j^{a\nu}\boldsymbol{\tau}^a,\end{aligned} \tag{11.54}$$

so that (11.52) can be written in matrix form as

$$D_\mu F^{\mu\nu} = j^\nu . \tag{11.55}$$

Applying a gauge transformation to this equation results in

$$D'_\mu F'^{\mu\nu} = \mathbf{U}D_\mu\mathbf{U}^{-1}\mathbf{U}F^{\mu\nu}\mathbf{U}^{-1} = \mathbf{U}D_\mu F^{\mu\nu}\mathbf{U}^{-1} = \mathbf{U}j^\nu\mathbf{U}^{-1}. \tag{11.56}$$

In order to be consistent this equation requires that the nonabelian matter current j^ν transforms as

$$j'^\mu = \mathbf{U}j^\mu\mathbf{U}^{-1}. \tag{11.57}$$

If this is the case, regardless of its detailed form, the matter current j^μ is *covariantly* conserved. This is demonstrated by applying D_ν to (11.54) and using (11.41) and the antisymmetry of $F^{\mu\nu}$ to obtain

$$\begin{aligned} D_\nu j^\nu = D_\nu D_\mu F^{\mu\nu} &= \frac{1}{2}[D_\mu, D_\nu]F^{\mu\nu} = \frac{1}{2}ig[F_{\mu\nu}, F^{\mu\nu}] \\ &= -gf^{abc}F^b_{\mu\nu}F^{c\mu\nu}\boldsymbol{\tau}^a = 0, \end{aligned} \tag{11.58}$$

which vanishes due to the antisymmetry of the structure constants.

Exercise 11.1: Show that (11.58) gives

$$\partial_\mu j^{a\mu} - gf^{abc}A^b_\mu j^{c\mu} = 0. \tag{11.59}$$

Similarly, the nonabelian version of the homogeneous Maxwell equations, (9.57), is obtained by using the covariant derivative and the Jacobi identity (7.9), which gives

$$\left[D_\nu, \left[D_\rho, D_\sigma\right]\right] + \left[D_\sigma, \left[D_\nu, D_\rho\right]\right] + \left[D_\rho, \left[D_\sigma, D_\nu\right]\right] = 0. \tag{11.60}$$

The individual double commutators are evaluated using (11.41), so that

$$\left[D_\nu, \left[D_\rho, D_\sigma\right]\right] = ig\left[D_\nu, F_{\rho\sigma}\right] = igD_\nu F_{\rho\sigma}. \tag{11.61}$$

The result is the *nonabelian Bianchi identity*,

$$D_\nu F_{\rho\sigma} + D_\sigma F_{\nu\rho} + D_\rho F_{\sigma\nu} = 0 \implies D_\nu F^a_{\rho\sigma} + D_\sigma F^a_{\nu\rho} + D_\rho F^a_{\sigma\nu} = 0. \tag{11.62}$$

Result (11.62) reduces to multiple copies of the abelian Bianchi identity (9.57) for the case that the structure constants vanish.

There are two currents in a Yang–Mills theory that are conserved in the usual way. The first is the nonabelian extension of charge current,

$$J^{a\mu} = \partial_\nu F^{a\nu\mu}. \tag{11.63}$$

This is conserved, since $[\partial_\mu, \partial_\nu] = 0$ and

$$\partial_\mu J^{a\mu} = \partial_\mu \partial_\nu F^{a\nu\mu} = \frac{1}{2}[\partial_\mu, \partial_\nu]F^{a\nu\mu} = 0. \tag{11.64}$$

Using the equation of motion for the Yang–Mills fields shows that $J^{a\mu}$ has a second form,

$$D_\nu F^{a\nu\mu} = j^{a\mu} = 0 \implies J^{a\mu} = \partial_\nu F^{a\nu\mu} = f^{abc} A_\nu^b F^{c\nu\mu} + j^{a\mu}. \tag{11.65}$$

This current demonstrates another significant difference from the abelian case, which is that the conserved nonabelian charge current receives a contribution from *both* the nonabelian gauge fields and the matter current, whereas the abelian conserved current depends only on the matter current.

The second conserved current, ${}^*J^{a\mu}$, is given by

$${}^*J^{a\nu} \equiv \frac{1}{2}\varepsilon^{\nu\alpha\beta\gamma}\partial_\alpha F^a_{\beta\gamma}, \tag{11.66}$$

which is conserved since

$$\partial_\nu {}^*J^{a\nu} = \frac{1}{2}\varepsilon^{\nu\alpha\beta\gamma}\partial_\nu\partial_\alpha F^a_{\beta\gamma} = \frac{1}{4}\varepsilon^{\nu\alpha\beta\gamma}[\partial_\nu, \partial_\alpha]F^a_{\beta\gamma} = 0. \tag{11.67}$$

Its magnetic nature is revealed by the charge density, given by

$${}^*J^{a0} = \varepsilon^{ijk}\partial_i F^a_{jk} = -\nabla \cdot \mathbf{B}^a, \tag{11.68}$$

where the nonabelian magnetic field is identified analogously to the abelian magnetic field (9.32),

$$B^a_\ell = -\frac{1}{2}\varepsilon^{jk\ell}F^a_{jk}. \tag{11.69}$$

If there is a material source of nonabelian magnetic field, then $\nabla \cdot \mathbf{B}^a \neq 0$, and this current is nonzero. A similar current would also be present in abelian electrodynamics if matter that behaved as a magnetic monopole were discovered and needed to be included in the matter Lagrangian. However, unlike abelian electrodynamics, where a charged current is required to create a magnetic field, there are *static* nonzero magnetic field solutions to the Yang–Mills gauge field equations of motion *in the absence of a matter field.* These will be discussed in a following section.

It was pointed out earlier that, unlike electrodynamics, $F_{\mu\nu}$ cannot be experimentally observed, since it is *not* independent of the arbitrary gauge functions Λ^a and is therefore *not unique*. In that regard, Yang–Mills theory was not developed from a body of experimental observations of reproducible phenomena as electrodynamics was. Instead, it is a theoretical extension of abelian gauge theory and its connection to experiment has arisen from observations of particle processes modeled by using a Yang–Mills theory to define a quantum field theory. This is the principal reason that most attention has been focused on the quantum properties of Yang–Mills field theories rather than on their classical behavior.

A very successful use of Yang–Mills fields is that of the *electroweak (Glashow–Weinberg–Salam) theory*, in which both the electromagnetic and the weak force are united as manifestations of an SU(2) × U(1) gauge theory. The three gauge fields associated with SU(2) and the one gauge field associated with U(1) combine into the massless photon and three massive vector particles through an effect known as the *Higgs mechanism*. This is presented in a later section. Another application is the *quark model of hadrons*, which postulates that hadronic matter is composed of spinor particles, known as *quarks*, interacting through Yang–Mills gauge fields, known as *gluons*. The theory is known as *quantum chromodynamics* and derives its name from the assumption that there are three *colors* of quarks—the name for the index of the isospin triplet of spinor fields—that form the representation space for the gauge group SU(3). Single quarks are never observed, and there are strong indications that the nonlinear nature of the gluonic fields creates an effect known as *confinement*, with the result that all hadronic matter consists of pairs and triplets of quarks. In particular, the *quantum mechanical* result known as *asymptotic freedom* shows that the effective coupling constant g in the quantum field theory becomes negligible at high momentum or, equivalently, short distances. Much effort has gone into extracting the hadronic spectrum from the associated Yang–Mills quantum field theory, primarily through numerical simulation of quantum processes via *lattice gauge theory*. Of the two theories, the electroweak model is arguably better understood, since its particle processes are both theoretically well-defined and accessible through experiment. Some aspects of the SU(2) version of classical quantum chromodynamics are discussed in a later section.

11.3 The Yang–Mills stress–energy tensor and force equation

The stress–energy tensor for Yang–Mills theory follows from (11.50) and the generalized definition (10.347). In Minkowski space the result is

$$\Theta^{\mu}{}_{\nu} = F^{a\mu\sigma}F^{a}_{\sigma\nu} + \frac{1}{4}\delta^{\mu}{}_{\nu}F^{a\alpha\beta}F^{a}_{\alpha\beta} + T^{\mu}{}_{\nu} \equiv Y^{\mu}{}_{\nu} + T^{\mu}{}_{\nu}, \tag{11.70}$$

where T^{μ}_{ν} is the contribution from the matter-field action. The exact nature of $T^{\mu}{}_{\nu}$ depends on the types of matter field and the details of their interactions. However, by construction, the sum of the gauge field and matter contributions, $\Theta^{\mu}{}_{\nu}$, is covariantly conserved, and if the affine connection vanishes, as it does in the absence of a gravitational field or in a freely falling frame, then $\partial_{\mu}\Theta^{\mu}{}_{\nu} = 0$. The gauge field contribution is manifestly gauge invariant, since it can be written

$$Y^{\mu}{}_{\nu} = 2\,\mathrm{Tr}\left(F^{\mu\sigma}F_{\sigma\nu} + \frac{1}{4}\delta^{\mu}{}_{\nu}F^{\alpha\beta}F_{\alpha\beta}\right). \tag{11.71}$$

The contribution of the Yang–Mills fields to the stress–energy tensor is highly nonlinear due to the presence of cubic and quartic terms in the gauge fields.

Restricting attention to the gauge field contribution in flat space and employing the equation of motion (11.52) shows that the divergence of the Yang–Mills contribution is given by

$$
\begin{aligned}
\partial_\mu Y^\mu{}_\nu &= \partial_\mu F^{a\mu\sigma} F^a_{\sigma\nu} + F^{a\mu\sigma}\partial_\mu F^a_{\sigma\nu} + \frac{1}{2}F^{a\alpha\beta}\partial_\nu F^a_{\alpha\beta} \\
&= (gf^{abc}A^b_\mu F^{c\mu\sigma} + j^{a\sigma})F^a_{\sigma\nu} + \frac{1}{2}F^{a\alpha\beta}\left(\partial_\alpha F^a_{\beta\nu} + \partial_\nu F^a_{\alpha\beta} + \partial_\beta F^a_{\nu\alpha}\right),
\end{aligned}
\tag{11.72}
$$

which employed the following identity derived from the antisymmetry of $F^a_{\mu\nu}$,

$$
F^{a\mu\sigma}\partial_\mu F^a_{\sigma\nu} = \frac{1}{2}F^{a\alpha\beta}\partial_\alpha F^a_{\beta\nu} + \frac{1}{2}F^{a\alpha\beta}\partial_\beta F^a_{\nu\alpha}. \tag{11.73}
$$

In electrodynamics the second term in (11.72) is zero in the absence of magnetic monopoles, since it corresponds to the homogeneous Maxwell equations of (9.57). When the field tensor was written in terms of the gauge field, the sum of the three terms vanished *unless the gauge field was singular*, $[\partial_\alpha, \partial_\beta]A_\mu \neq 0$. However, even the requirement for nonsingular gauge fields *is insufficient* to make the second term vanish in Yang–Mills theory, because of the quadratic term in the gauge fields that is present in $F^a_{\mu\nu}$. Using the Bianchi identity (11.62), the second term can be written

$$
\begin{aligned}
\frac{1}{2}F^{a\alpha\beta}\left(\partial_\alpha F^a_{\beta\nu} + \partial_\nu F^a_{\alpha\beta} + \partial_\beta F^a_{\nu\alpha}\right) &= \frac{1}{2}F^{a\alpha\beta}\left(D_\alpha F^a_{\beta\nu} + D_\nu F^a_{\alpha\beta} + D_\beta F^a_{\nu\alpha}\right) \\
&\quad + \frac{1}{2}gF^{a\alpha\beta}f^{abc}\left(A^b_\alpha F^c_{\beta\nu} + A^b_\nu F^c_{\alpha\beta}\right. \\
&\quad \left. + A^b_\beta F^c_{\nu\alpha}\right) \\
&= gF^{a\alpha\beta}f^{abc}A^b_\alpha F^c_{\beta\nu} = -gf^{abc}A^b_\mu F^{c\mu\sigma}F^a_{\sigma\nu},
\end{aligned}
\tag{11.74}
$$

where $f^{abc}F^{a\alpha\beta}F^c_{\alpha\beta} = 0$ and $f^{cba} = -f^{abc}$ were used. This term cancels the identical contribution in the first term of (11.72), giving the final result

$$
\partial_\mu Y^\mu{}_\nu = j^{a\mu}F^a_{\mu\nu}. \tag{11.75}
$$

Therefore, any energy or momentum change in the gauge field configuration can occur solely through the direct interaction of the matter current with the gauge fields, not through the self-interaction of the gauge fields, as it might appear at first glance from (11.72). This is the same as the abelian gauge field theory of electrodynamics. This must be the case in order for the total energy and momentum of the matter plus gauge fields to be conserved, since the mechanism of exchange must involve *both* the matter and gauge fields.

11.4 Spontaneous breakdown of symmetry

The Yang–Mills gauge field Lagrangian does not allow a quadratic mass term of the form $m^2A^{a\mu}A^a_\mu$ for the simple reason that such a term is not gauge invariant. This was discussed in an earlier section on the massive Proca field. However, through gauge invariant interactions with a scalar matter field, the gauge fields can behave as if they are massive. The resulting theory no longer directly exhibits the gauge invariance of the original Lagrangian, and so this is known as *spontaneous breakdown of symmetry*.

This analysis begins with an action that is the combination of (11.37) and (11.50) involving a multiplet of n complex scalar fields $\mathbf{\Phi}$ and the associated gauge fields A_μ^a for the choice of symmetry group,

$$\mathcal{L} = -\frac{1}{4}F_{\mu\nu}^a F^{a\mu\nu} + \frac{1}{2}(D_\mu \mathbf{\Phi})^\dagger \cdot D^\mu \mathbf{\Phi} - V(\mathbf{\Phi}^\dagger \cdot \mathbf{\Phi}). \tag{11.76}$$

Spontaneous breakdown of symmetry can occur if the scalar field components φ^a possess a *stable* nonzero solution, denoted by φ_0^a, to its equation of motion. For the case that the solution is a *constant*, this requires that the potential function satisfies

$$\left.\frac{\partial V(\mathbf{\Phi})}{\partial \varphi^a}\right|_{\varphi^a = \varphi_0^a} = 0, \tag{11.77}$$

where it is assumed that $\varphi_0^a \neq 0$ and is a constant. The fields in (11.76) will then fluctuate around this solution in the same manner that the density fluctuated around the stationary fluid solution in (4.188).

The Lagrangian (11.76) is expanded around the solution, so that

$$\varphi^a \rightarrow \varphi^a + \varphi_0^a. \tag{11.78}$$

Since the gauge covariant derivative term in (11.76) contains the subterm $\frac{1}{2}g^2\mathbf{\Phi}^\dagger A_\mu A^\mu \mathbf{\Phi}$, expanding this around the constant solution φ_0^a gives rise to a fluctuation term that is quadratic in the gauge field,

$$V\left(A_\mu\right) = \frac{1}{2}g^2\left(\varphi_0^j \boldsymbol{\tau}_{jk}^a\right)\left(\boldsymbol{\tau}_{k\ell}^b \varphi_0^\ell\right)A_\mu^a A^{b\mu} \equiv \frac{1}{2}M_{ab}^2(\varphi_0)A_\mu^a A^{b\mu}. \tag{11.79}$$

The matrix $M_{ab}^2(\varphi_0)$ corresponds to an effective mass for the gauge fields, since it is a constant matrix whose real elements have the units of inverse length squared inherited from the field solution φ_0^a. The eigenvalues of M_{ab}^2 give the masses of the gauge fields, and the eigenvectors associated with each eigenvalue give the combinations of gauge fields that correspond to the mass eigenvalue.

For simplicity of notation, the potential V in the action (11.76) will be expressed in terms of real scalar quantities by writing

$$\varphi^j = \varphi_R^j + i\varphi_I^j \equiv \varphi^j + i\varphi^{j+n}, \tag{11.80}$$

where n is the dimension of the group generators. In this way, V can be written as $V(\varphi)$, where the range of the indices j used for φ^j has been doubled to include both the real and the imaginary parts of the original complex field φ. Expanding the potential in a Taylor series and using $\partial V(\varphi_0)/\partial\varphi_0^j = 0$ gives

$$\begin{aligned} V(\varphi + \varphi_0) &= V(\varphi_0) + \frac{1}{2}\frac{\partial^2 V(\varphi_0)}{\partial\varphi_0^j\,\partial\varphi_0^k}\varphi^j\varphi^k + \dots \\ &\equiv V(\varphi_0) + \frac{1}{2}H_{jk}^{(2)}(\varphi_0)\,\varphi^j\varphi^k + \dots \end{aligned} \tag{11.81}$$

where the ellipsis refers to all the higher-order terms in the Taylor series. Like the gauge field mass matrix (11.79), the second derivative in (11.81) generates a matrix, $H^{(2)}_{jk}(\varphi_0)$, which determines the scalar field masses through its eigenvalues. The result of this process apparently gives a *minimally coupled* version of massive gauge fields interacting with a set of scalar fields.

However, there are several important problems with this procedure that must be dealt with. The spin of the massless abelian gauge field discussed in chapter 9 was the result of circular polarization for the electromagnetic wave, and it was shown in (9.310) that there are only two helicities available, corresponding to $\pm\hbar$. Since the electromagnetic wave is massless it is impossible to boost into a frame where the wave's momentum is zero and the helicity is zero. Since a massive gauge field no longer travels at the speed of light, it is possible to boost into the frame of a circularly polarized wave, so that there is no longitudinal momentum, resulting in a spin projection or helicity that is zero. In analyzing the massive Proca field it was shown that it does indeed have three possible polarizations or degrees of freedom, one of which results in the nonzero longitudinal contribution (9.326). A massive gauge field therefore requires a nonzero longitudinal component, $\nabla \cdot \boldsymbol{A} \neq 0$. This component is absent for the massless version of the gauge field. If spontaneous symmetry breakdown does create a massive gauge field, it must provide a longitudinal component for the gauge field.

The second problem arises from the residual gauge invariance of the scalar potential $V(\boldsymbol{\Phi}^\dagger \cdot \boldsymbol{\Phi})$ appearing in the action (11.76). If this potential is expressed in terms of $\boldsymbol{\Phi}_0$, where $\boldsymbol{\Phi}_0$ has the components φ_0^j, it is invariant under the gauge transformation $\mathbf{U} \cdot \boldsymbol{\Phi}_0$. For an infinitesimal transformation, the action on the real and imaginary parts is obtained from

$$\begin{aligned}\delta\varphi_0^j &= \delta\varphi_{R0}^j + i\delta\varphi_{I0}^j = \delta\varphi_0^j + i\delta\varphi_0^{j+n} = ig\Lambda^a\boldsymbol{\tau}^a_{jk}\varphi_0^k \\ &= ig\Lambda^a\boldsymbol{\tau}^a_{jk}\varphi_0^k - g\Lambda^a\boldsymbol{\tau}^a_{jk}\varphi_0^{k+n},\end{aligned} \tag{11.82}$$

so that the real parts transform as

$$\delta\varphi_0^j = -g\Lambda^a\boldsymbol{\tau}^a_{jk}\varphi_0^{k+n}, \tag{11.83}$$

while the imaginary parts transform as

$$\delta\varphi_0^{j+n} = g\Lambda^a\boldsymbol{\tau}^a_{jk}\varphi_0^k. \tag{11.84}$$

The generators of group $\boldsymbol{\tau}^a$ break into two distinct classes, those that leave the solution *invariant*,

$$\delta\varphi_0 = 0, \tag{11.85}$$

and those that do not,

$$\delta\varphi_0 \neq 0. \tag{11.86}$$

The latter generators are no longer a symmetry of the theory, since φ_0, which defines the fluctuations of the theory, is changed by a gauge transformation that involves them. These generators are said to correspond to a *broken symmetry*. The details of this depend upon both the generators and the fluctuation solution chosen. Because the potential is still invariant under the gauge transformation, it follows that

$$V(\varphi_0 + \delta\varphi_0) = V(\varphi_0). \tag{11.87}$$

Using $\partial V(\varphi_0)/\partial\varphi_0^{\,j} = 0$, a Taylor series expansion to $O(\delta\varphi_0^{\,2})$ gives

$$V(\varphi_0 + \delta\varphi_0) - V(\varphi_0) = \frac{1}{2}\frac{\partial^2 V(\varphi_0)}{\partial\varphi_0^{\,j}\,\partial\varphi_0^{\,k}}\delta\varphi_0^{\,j}\,\delta\varphi_0^{\,k} = 0. \tag{11.88}$$

The same scalar mass matrix $H_{ij}^{(2)}(\varphi_0)$ found in (11.81) has appeared in (11.88). For the generators that correspond to a broken symmetry and are therefore nonzero, $\delta\varphi_0 \neq 0$, (11.88) shows that

$$H_{jk}^{(2)}\delta\varphi_0^{\,k} = 0, \tag{11.89}$$

so that $\delta\varphi_0$ is an eigenvalue of the scalar mass matrix with an eigenvalue of zero. Since the Λ^a appearing in $\delta\varphi_0$ are arbitrary, this means that for every generator that corresponds to a broken symmetry, there is a zero-mass eigenvalue and therefore a scalar field combination that has zero mass. This is referred to as the *Goldstone theorem*, and the zero-mass scalar fields that result from the broken symmetry are referred to as Goldstone modes, or, in the terminology of quantum field theory, *Goldstone bosons*. Because there are no zero-mass scalar fields in nature, this appears to render spontaneous breakdown of symmetry physically irrelevant. Fortunately, there is a loophole in the Goldstone theorem, known as the *Higgs mechanism*, that enables the Goldstone bosons to become the longitudinal component of the massive gauge fields, thereby solving both problems.

Although this chapter addresses Yang–Mills theories, the Higgs mechanism is most easily demonstrated in the abelian context, where there is only one gauge field A_μ coupled to a single complex scalar field φ. The Lagrangian is given by the abelian version of (11.76),

$$\mathcal{L} = -\frac{1}{4}F_{\mu\nu}F^{\mu\nu} + (\partial_\mu - igA_\mu)\varphi^*(\partial^\mu + igA^\mu)\varphi - V(\varphi^*, \varphi), \tag{11.90}$$

which is invariant under the usual abelian gauge transformations,

$$\varphi \to e^{ig\Lambda}\varphi, \quad A_\mu \to A_\mu - \partial_\mu\Lambda. \tag{11.91}$$

The potential is chosen to have the explicitly gauge invariant form,

$$V(\varphi^*, \varphi) = -\alpha^2\varphi^*\varphi + \frac{1}{2}\lambda(\varphi^*\varphi)^2. \tag{11.92}$$

This is the complex version of the Higgs potential analyzed in (8.112). The complex Higgs potential has a minimum at

$$\frac{\partial V(\varphi^*, \varphi)}{\partial \varphi^*} = -\alpha^2\varphi + \lambda(\varphi^*\varphi)\varphi = 0 \implies \varphi_0 = \gamma e^{i\theta(x)/\gamma}, \tag{11.93}$$

where $\gamma = \sqrt{\alpha^2/\lambda}$ and $\theta(x)$ is, for the moment, an *arbitrary* function with units of inverse length. Rather than simply translating the complex field φ by the solution, the complex scalar field is given a polar decomposition,

$$\varphi = (\phi + \gamma)e^{i\theta/\gamma}, \tag{11.94}$$

where ϕ is a *real field* with the units of inverse length. Substituting this form into (11.90) gives the effective Lagrangian,

$$\begin{aligned}\mathcal{L} = &-\frac{1}{4}F_{\mu\nu}F^{\mu\nu} + \partial_\mu\phi\,\partial^\mu\phi - 2\alpha^2\phi^2 - 2\lambda\gamma\phi^3 - \frac{1}{2}\lambda\phi^4\\ &+ g^2(\phi+\gamma)^2\left(A_\mu A^\mu + \frac{2}{g\gamma}A^\mu\partial_\mu\theta + \frac{1}{g^2\gamma^2}\partial_\mu\theta\partial^\mu\theta\right).\end{aligned} \tag{11.95}$$

Exercise 11.2: Verify (11.95) is the result of expanding (11.90) using (11.94).

Ignoring, for the moment, cubic and higher terms in the three fields, the Lagrangian (11.95) contains a quadratic Lagrangian for θ,

$$\mathcal{L}_\theta = \partial_\mu\theta\,\partial^\mu\theta + 2g\gamma\partial_\mu\theta A^\mu. \tag{11.96}$$

This gives the equation of motion for θ,

$$2\partial_\mu\left(\partial^\mu\theta + g\gamma A^\mu\right) = 2\Box\theta + 2g\gamma\partial_\mu A^\mu = 2\Box\theta = 0, \tag{11.97}$$

where the Lorenz condition for A^μ was used. This identifies θ as the Goldstone field, since there is no mass term in (11.97).

The Lagrangian (11.95) can be rewritten in a very useful form as

$$\begin{aligned}\mathcal{L} = &-\frac{1}{4}F_{\mu\nu}F^{\mu\nu} + \partial_\mu\phi\,\partial^\mu\phi - 2\alpha^2\phi^2 - 2\lambda\gamma\phi^3 - \frac{1}{2}\lambda\phi^4\\ &+ g^2(\phi+\gamma)^2\left(A_\mu + \frac{\partial_\mu\theta}{g\gamma}\right)\left(A^\mu + \frac{\partial^\mu\theta}{g\gamma}\right).\end{aligned} \tag{11.98}$$

The Lagrangian (11.98) is invariant under the simultaneous operations $A_\mu \to A_\mu - \partial_\mu\Lambda$ and $\theta \to \theta + g\gamma\Lambda$, corresponding to the original gauge invariance

(11.91). The Higgs mechanism can now be expressed by introducing the vector field Z_μ through the definition

$$A_\mu = Z_\mu - \frac{1}{g\gamma}\partial_\mu\theta. \tag{11.99}$$

Since definition (11.99) is identical to a gauge transformation, this transmutes $F_{\mu\nu}$ into

$$F_{\mu\nu} \to \bar{F}_{\mu\nu} = \partial_\mu Z_\nu - \partial_\nu Z_\mu. \tag{11.100}$$

After this redefinition the Lagrangian (11.98) contains the quadratic terms

$$\mathcal{L}_0 = -\frac{1}{4}\bar{F}_{\mu\nu}\bar{F}^{\mu\nu} + g^2\gamma^2 Z_\mu Z^\mu + \partial_\mu\phi\,\partial^\mu\phi - 2\alpha^2\phi^2. \tag{11.101}$$

This is a combination of the Klein–Gordon Lagrangian (8.25) for a real scalar field with mass $\sqrt{2}\,\alpha$ and the Proca Lagrangian (9.313) for a vector field with mass $\sqrt{2}\,g\gamma$. The remaining terms are cubic and higher in ϕ and Z^μ,

$$\mathcal{L}_I = -2\lambda\gamma\phi^3 - \frac{1}{2}\lambda\phi^4 + g^2(\phi^2 + \gamma\phi)Z_\mu Z^\mu, \tag{11.102}$$

and these represent the nonlinear interaction of the field ϕ with itself and the Proca field.

The field A_μ is transverse, since it satisfies $\nabla \cdot \boldsymbol{A} \neq 0$. However, the field Z_μ is not transverse, since (11.99) gives

$$\nabla \cdot \boldsymbol{Z} = \nabla \cdot \boldsymbol{A} + \frac{1}{g\gamma}\nabla^2\theta = \frac{1}{g\gamma}\nabla^2\theta \neq 0. \tag{11.103}$$

The massive Proca field has obtained a longitudinal component by absorbing the Goldstone mode into its definition. The nonabelian version is more complicated, but it is essentially the same mechanism, i.e. all would-be Goldstone modes disappear by a gauge transformation to become the longitudinal components of massive vector fields.

11.5 Aspects of classical solutions for Yang–Mills fields

In order to make the classical aspects of Yang–Mills theories simpler and more explicit, the case of SU(2) will be considered in the remainder of this chapter. The structure constants are given by the Levi-Civita symbol,

$$f^{abc} = \varepsilon^{abc}, \tag{11.104}$$

and the generators are the Pauli spin matrices,

$$\tau^a = \frac{1}{2}\sigma^a, \tag{11.105}$$

which clearly satisfy (11.27).

For the case that the matter current is zero, the Yang–Mills equation (11.55) becomes

$$D_\mu F^{\mu\nu} = \partial_\mu F^{\mu\nu} + ig\left[A_\mu, F^{\mu\nu}\right] = 0. \tag{11.106}$$

It is obvious that $A_\mu = 0$ is a solution of (11.106). If $\mathbf{U}(x)$ is an element of the local SU(2) gauge group, equation (11.106) is covariant under the gauge transformation generated by $\mathbf{U}(x)$. As a result, equation (11.106) is also solved by the result of gauge transforming $A_\mu = 0$. Using the form of the gauge transformation shown in (11.23), the simplest classical solutions are therefore given by

$$A_\mu = \frac{i}{g}\partial_\mu \mathbf{U}\mathbf{U}^{-1}. \tag{11.107}$$

Using the identity

$$\partial_\mu(\mathbf{U}\mathbf{U}^{-1}) = 0 \implies \partial_\mu \mathbf{U}^{-1} = -\mathbf{U}^{-1}\partial_\mu \mathbf{U}\mathbf{U}^{-1}, \tag{11.108}$$

it follows that

$$\partial_\mu A_\nu = \frac{i}{g}\partial_\mu\partial_\nu \mathbf{U}\mathbf{U}^{-1} + \frac{i}{g}\partial_\mu \mathbf{U}\partial_\nu \mathbf{U}^{-1} = \frac{i}{g}\partial_\mu\partial_\nu \mathbf{U}\mathbf{U}^{-1} - \frac{i}{g}\partial_\nu \mathbf{U}\mathbf{U}^{-1}\partial_\mu \mathbf{U}\mathbf{U}^{-1}. \tag{11.109}$$

Using this result, it follows that the field strength tensor associated with (11.107) is given by

$$\begin{aligned} F_{\mu\nu} &= \partial_\mu A_\nu - \partial_\nu A_\mu + ig\left[A_\mu, A_\nu\right] \\ &= \frac{i}{g}\left(\partial_\mu \mathbf{U}\mathbf{U}^{-1}\partial_\nu \mathbf{U}\mathbf{U}^{-1} - \partial_\nu \mathbf{U}\mathbf{U}^{-1}\partial_\mu \mathbf{U}\mathbf{U}^{-1}\right) \\ &\quad - \frac{i}{g}\left[\partial_\mu \mathbf{U}\mathbf{U}^{-1}, \partial_\nu \mathbf{U}\mathbf{U}^{-1}\right] = 0. \end{aligned} \tag{11.110}$$

Since $F_{\mu\nu} = 0$ this is a solution of (11.52) for $j_\mu = 0$, and it is referred to as a *pure gauge solution*. It is the nonabelian counterpart to setting $A_\mu = \partial_\mu\Lambda$ in electrodynamics, although it is not as trivial.

Like electrodynamics, the gauge covariance of the equation of motion represents a mismatch between the observable degrees of freedom and the gauge field degrees of freedom. As in electrodynamics, this mismatch is remedied by imposing a gauge condition on each of the Yang–Mills fields. The Lorenz gauge, $\partial_\mu A^{a\mu} = 0$, the Coulomb gauge, $\nabla \cdot \boldsymbol{A}^a = 0$, or the temporal gauge, $A_0^a = 0$, are typically chosen. For static problems the Coulomb and Lorenz gauges are equivalent. In abelian gauge theory a gauge transformation of the Coulomb condition gives

$$\nabla \cdot \boldsymbol{A}' = \nabla \cdot \boldsymbol{A} - \nabla^2\Lambda = -\nabla^2\Lambda = 0. \tag{11.111}$$

In an infinite spatial volume there is no *finite* solution to this equation for Λ such that $\nabla\Lambda \neq 0$. Therefore, in the abelian case, enforcing the Coulomb condition results in a *unique solution* for $\boldsymbol{A}$.

11.5.1 The Gribov ambiguity

However, in the Yang–Mills case, these gauge conditions *do not uniquely specify the fields*. This can be seen by considering the gauge field configuration $A_j = 0$, which trivially satisfies $\nabla \cdot \boldsymbol{A} = 0$. A gauge transformation on A_j gives

$$A'_j = \mathbf{U}A_j\mathbf{U}^{-1} + \frac{i}{g}\partial_j\mathbf{U}\mathbf{U}^{-1} = \frac{i}{g}\partial_j\mathbf{U}\mathbf{U}^{-1}, \tag{11.112}$$

so that the zero field is replaced by the pure gauge solution found earlier. This solution must also satisfy $\nabla \cdot \boldsymbol{A}' = 0$, and this requirement becomes the equation

$$\nabla \cdot (\nabla\mathbf{U}\mathbf{U}^{-1}) = \nabla^2\mathbf{U}\mathbf{U}^{-1} + \nabla\mathbf{U} \cdot \nabla\mathbf{U}^{-1} = 0. \tag{11.113}$$

In the case of SU(2) it is possible to find a nontrivial static pure gauge solution to this equation. Using the generators of SU(2) it takes the form

$$\mathbf{U} = \mathbf{I}\cos\omega(r) + i\frac{1}{r}\boldsymbol{\sigma}^j x^j \sin\omega(r), \quad r^2 = \boldsymbol{x} \cdot \boldsymbol{x}, \tag{11.114}$$

where $\omega(r)$ is a function that is determined so that the associated pure gauge solution is transverse. It follows that

$$\mathbf{U}\mathbf{U}^\dagger = \mathbf{I}\cos^2\omega + \frac{1}{r^2}x^j x^k \boldsymbol{\sigma}^j\boldsymbol{\sigma}^k \sin^2\omega = \mathbf{I}(\cos^2\omega + \sin^2\omega) = \mathbf{I}, \tag{11.115}$$

so that it is unitary. Its determinant is given by

$$\begin{aligned} \det\mathbf{U} = \det&\begin{pmatrix} \cos\omega + i(x^3/r)\sin\omega & ((ix^1 + x^2)/r)\sin\omega \\ ((ix^1 - x^2)/r)\sin\omega & \cos\omega - i(x^3/r)\sin\omega \end{pmatrix} \\ &= \cos^2\omega + \sin^2\omega = 1, \end{aligned} \tag{11.116}$$

so that the matrix (11.114) is a member of SU(2). For the elements of $\mathbf{U}$ to be well-behaved, the boundary condition

$$\lim_{r \to 0} \omega(r) = 0 \tag{11.117}$$

is required, since this enforces

$$\lim_{r \to 0} \mathbf{U}(r) = \mathbf{I}. \tag{11.118}$$

Demonstrating that (11.114) provides a nontrivial solution to (11.113) begins by using spherical coordinates. Replacing the radial coordinate with the variable $t = \ln r$ gives

$$\begin{aligned} \nabla &= \frac{1}{r}\left(\frac{\partial}{\partial t}, \frac{\partial}{\partial\theta}, \frac{1}{\sin\theta}\frac{\partial}{\partial\phi}\right), \\ \nabla^2 &= \frac{1}{r^2}\left(\frac{\partial^2}{\partial t^2} + \frac{\partial}{\partial t} + \cot\theta\frac{\partial}{\partial\theta} + \frac{\partial^2}{\partial\theta^2} + \frac{1}{\sin\theta}\frac{\partial^2}{\partial\phi^2}\right), \end{aligned} \tag{11.119}$$

while $\mathbf{U}(x)$ becomes

$$\mathbf{U} = \begin{pmatrix} \cos\omega + i\cos\theta\sin\omega & ie^{-i\phi}\sin\theta\sin\omega \\ ie^{i\phi}\sin\theta\sin\omega & \cos\omega - i\cos\theta\sin\omega \end{pmatrix}, \tag{11.120}$$

where it is understood that ω is now a function of t. Dropping the common factor of $1/r^2$, equation (11.113) becomes

$$\begin{aligned} &\left(\frac{\partial^2}{\partial t^2} + \frac{\partial}{\partial t} + \cot\theta\frac{\partial}{\partial\theta} + \frac{\partial^2}{\partial\theta^2} + \frac{1}{\sin\theta}\frac{\partial^2}{\partial\phi^2}\right)\mathbf{U}\mathbf{U}^{-1} \\ &+ \frac{\partial\mathbf{U}}{\partial t}\frac{\partial\mathbf{U}^{-1}}{\partial t} + \frac{\partial\mathbf{U}}{\partial\theta}\frac{\partial\mathbf{U}^{-1}}{\partial\theta} + \frac{1}{\sin^2\theta}\frac{\partial\mathbf{U}}{\partial\phi}\frac{\partial\mathbf{U}^{-1}}{\partial\phi} = 0. \end{aligned} \tag{11.121}$$

Evaluating the eight terms of (11.121) is facilitated by noting that, for example,

$$\frac{\partial^2\mathbf{U}}{\partial t^2}\mathbf{U}^{-1} + \frac{\partial\mathbf{U}}{\partial t}\frac{\partial\mathbf{U}^{-1}}{\partial t} = \frac{\partial}{\partial t}\left(\frac{\partial\mathbf{U}}{\partial t}\mathbf{U}^{-1}\right). \tag{11.122}$$

To carry the step further, it follows that

$$\begin{aligned} \frac{\partial\mathbf{U}}{\partial t}\mathbf{U}^{-1} &= \begin{pmatrix} -\sin\omega + i\cos\theta\cos\omega & ie^{-i\phi}\sin\theta\cos\omega \\ ie^{i\phi}\sin\theta\cos\omega & -\sin\omega - i\cos\theta\cos\omega \end{pmatrix} \times \\ &\times \begin{pmatrix} \cos\omega + i\cos\theta\sin\omega & -ie^{-i\phi}\sin\theta\sin\omega \\ -ie^{i\phi}\sin\theta\sin\omega & \cos\omega - i\cos\theta\sin\omega \end{pmatrix}\frac{\partial\omega}{\partial t} \\ &= \begin{pmatrix} i\cos\theta & ie^{-i\phi}\sin\theta \\ -ie^{i\phi}\sin\theta & -i\cos\theta \end{pmatrix}\frac{\mathrm{d}\omega}{\mathrm{d}t}, \end{aligned} \tag{11.123}$$

so that

$$\frac{\partial}{\partial t}\left(\frac{\partial\mathbf{U}}{\partial t}\mathbf{U}^{-1}\right) = \begin{pmatrix} i\cos\theta & ie^{-i\phi}\sin\theta \\ -ie^{i\phi}\sin\theta & -i\cos\theta \end{pmatrix}\frac{\mathrm{d}^2\omega}{\mathrm{d}t^2}. \tag{11.124}$$

The result of the three terms in (11.121) that involve t is

$$\left(\frac{\partial^2\mathbf{U}}{\partial t^2} + \frac{\partial\mathbf{U}}{\partial t}\right)\mathbf{U}^{-1} + \frac{\partial\mathbf{U}}{\partial t}\frac{\partial\mathbf{U}^{-1}}{\partial t} = \begin{pmatrix} i\cos\theta & ie^{-i\phi}\sin\theta \\ -ie^{i\phi}\sin\theta & -i\cos\theta \end{pmatrix}\left(\frac{\mathrm{d}^2\omega}{\mathrm{d}t^2} + \frac{\mathrm{d}\omega}{\mathrm{d}t}\right). \tag{11.125}$$

The results for the other five terms are given by

$$\begin{aligned} &\frac{\partial\mathbf{U}}{\partial\theta}\frac{\partial\mathbf{U}^{-1}}{\partial\theta} + \frac{\partial^2\mathbf{U}}{\partial\theta^2}\mathbf{U}^{-1} = \frac{\partial}{\partial\theta}\left(\frac{\partial\mathbf{U}}{\partial\theta}\mathbf{U}^{-1}\right) \\ &= -\begin{pmatrix} i\cos\theta & ie^{-i\phi}\sin\theta \\ -ie^{i\phi}\sin\theta & -i\cos\theta \end{pmatrix}\sin\omega\cos\omega, \end{aligned} \tag{11.126}$$

and

$$\frac{1}{\sin^2\theta}\left(\frac{\partial \mathbf{U}}{\partial\phi}\frac{\partial \mathbf{U}^{-1}}{\partial\phi} + \frac{\partial^2\mathbf{U}}{\partial\phi^2}\mathbf{U}^{-1}\right) + \cot\theta\frac{\partial \mathbf{U}}{\partial\theta}\mathbf{U}^{-1} = -\begin{pmatrix} i\cos\theta & ie^{-i\phi}\sin\theta \\ -ie^{i\phi}\sin\theta & -i\cos\theta \end{pmatrix}\sin\omega\cos\omega. \tag{11.127}$$

Exercise 11.3: Verify results (11.126) and (11.127).

Combining results (11.125), (11.126), and (11.127) in (11.121) yields the following equation,

$$\begin{pmatrix} i\cos\theta & ie^{-i\phi}\sin\theta \\ -ie^{i\phi}\sin\theta & -i\cos\theta \end{pmatrix}\left(\frac{\mathrm{d}^2\omega}{\mathrm{d}t^2} + \frac{\mathrm{d}\omega}{\mathrm{d}t} - 2\sin\omega\cos\omega\right) = 0. \tag{11.128}$$

Using the trigonometric identity $2\sin\omega\cos\omega = \sin 2\omega$ and defining $\bar{\omega} = 2\omega$ gives the equation

$$\frac{\mathrm{d}^2\bar{\omega}}{\mathrm{d}t^2} + \frac{\mathrm{d}\bar{\omega}}{\mathrm{d}t} - 2\sin\bar{\omega} = 0. \tag{11.129}$$

The boundary condition on $\bar{\omega}$ becomes

$$\lim_{r\to 0}\bar{\omega} = \lim_{t\to-\infty}\bar{\omega} = 0. \tag{11.130}$$

This equation is known as the *Gribov pendulum equation,* since it is the same equation as that for a vertical damped pendulum, where $\bar{\omega}$ is interpreted as the angular coordinate of a mass on a fixed length swinging in the presence of a uniform gravitational acceleration. Using this simple physical analogy, (11.129) has an infinite number of solutions consistent with the initial condition (11.130). These solutions are indexed by an integer n, corresponding to the number of full rotations the pendulum makes during its motion. This is reflected in the symmetry $\bar{\omega} \to \bar{\omega} + 2\pi n$ that equation (11.129) possesses. The existence of nontrivial transverse pure gauge solutions is referred to as the *Gribov ambiguity* and the individual solutions are referred to as *Gribov copies.* This shows that the Yang–Mills field is not uniquely fixed by the Coulomb gauge condition. The problem arises in this Yang–Mills field theory because the SU(2) gauge group is isomorphic to the three-sphere S^3. This results in pure gauge configurations being characterized by the number of times they wrap around the three-sphere. This problem persists when the theory is treated quantum mechanically, where the different solutions give rise to different sectors of the quantized theory.

The individual nonabelian gauge fields are found from the matrix $A_\mu = i\partial_\mu \mathbf{U}\mathbf{U}^{-1}/g$ using the definition (11.28). As an example, for the SU(2) case of (11.114), it follows that

$$A_\theta^3 = 2\,\mathrm{Tr}\left(\frac{i}{g}\frac{\partial \mathbf{U}}{\partial \theta}\mathbf{U}^{-1}\frac{1}{2}\boldsymbol{\sigma}^3\right) = -\frac{2}{g}\sin^2\theta\,\sin^2\omega. \tag{11.131}$$

This demonstrates that the pure gauge solution (11.114) is *nonperturbative*, since the limit $g \to 0$ is undefined. This is yet another aspect of nonabelian gauge theory that is starkly different from abelian electrodynamics.

11.5.2 The Wu–Yang monopole

There are gauge field solutions to (11.52) that yield a magnetic field in the absence of matter fields. In the case of an SU(2) invariant theory, the *static* gauge field configuration known as the *Wu–Yang monopole* is defined using the Levi-Civita symbol as

$$A_j^a = \varepsilon^{jab}\frac{x^b}{gr^2}, \quad r^2 = \boldsymbol{x}\cdot\boldsymbol{x}, \quad A_0^a = 0. \tag{11.132}$$

This particular mixture of gauge indices and spatial indices can occur because there are three gauge parameters in SU(2), matching the number of spatial dimensions. At a deeper level, this occurs because SU(2) and the three-sphere S^3 are isomorphic. Configuration (11.132) satisfies the Coulomb condition,

$$\nabla\cdot\boldsymbol{A}^a = \frac{\varepsilon^{jab}\delta^{bj}}{gr^2} - 2\frac{\varepsilon^{jab}x^b x^j}{gr^4} = 0, \tag{11.133}$$

since both terms vanish individually.

Using (11.132) shows that

$$\begin{aligned}\partial_j A_k^a - \partial_k A_j^a &= \frac{\varepsilon^{kaj}}{gr^2} - 2\varepsilon^{kab}\frac{x^j x^b}{gr^4} - \frac{\varepsilon^{jak}}{gr^2} + 2\varepsilon^{jab}\frac{x^k x^\ell}{gr^4} \\ &= 2\frac{\varepsilon^{ajk}}{gr^2} - 2\varepsilon^{kab}\frac{x^j x^b}{gr^4} + 2\varepsilon^{jab}\frac{x^k x^b}{gr^4}.\end{aligned} \tag{11.134}$$

The familiar Levi-Civita identity,

$$\varepsilon^{abc}\varepsilon^{ade} = \delta^{bd}\delta^{ce} - \delta^{be}\delta^{dc}, \tag{11.135}$$

results in

$$gf^{abc}A_j^b A_k^c = g\varepsilon^{abc}\varepsilon^{bjd}\varepsilon^{cke}\frac{x^d x^e}{g^2 r^4} = \varepsilon^{jkb}\frac{x^a x^b}{gr^4}. \tag{11.136}$$

Combining (11.134) and (11.136) gives

$$\begin{aligned} F^a_{jk} &= \partial_j A^a_k - \partial_k A^a_j - g f^{abc} A^b_j A^c_k \\ &= 2\frac{\varepsilon^{ajk}}{gr^2} - 2\varepsilon^{kab}\frac{x^j x^b}{gr^4} + 2\varepsilon^{jab}\frac{x^k x^b}{gr^4} - \varepsilon^{jkb}\frac{x^a x^b}{gr^4}. \end{aligned} \tag{11.137}$$

Using (11.132) and (11.137) gives

$$\partial_j F^a_{jk} = g f^{abc} A^b_j F^c_{jk} = -\frac{\varepsilon^{kab} x^b}{gr^4}, \tag{11.138}$$

which shows that the Wu–Yang monopole satisfies the matter-free equation of motion,

$$\partial_j F^a_{jk} - g f^{abc} A^b_j F^c_{jk} = 0. \tag{11.139}$$

Using (11.137) and (11.135) shows that the associated magnetic field (11.69) is given by

$$B^a_\ell = -\frac{1}{2} F^a_{jk} \varepsilon^{jk\ell} = -\frac{x^a x^\ell}{gr^4}. \tag{11.140}$$

Exercise 11.4: Verify (11.140).

Result (11.140) is well-defined everywhere, except at the origin. It shows that $\mathbf{B}^a \cdot \hat{\boldsymbol{x}}$ is given by

$$\mathbf{B}^a_j \hat{x}^j = \mathbf{B}^a \cdot \frac{\boldsymbol{x}}{r} = -\frac{x^a}{gr^3}, \tag{11.141}$$

so that it is radial in isospin space. Result (11.140) also shows that

$$\nabla \cdot \mathbf{B}^a = 0. \tag{11.142}$$

Although (11.140) is singular at $r = 0$, the magnetic flux is given by

$$\mathrm{d}\Phi^a = \mathbf{B}^a \cdot \mathrm{d}\boldsymbol{S} = \frac{x^a}{gr^2}\,\mathrm{d}\Omega. \tag{11.143}$$

The flux element has the same sign as x^a, so that it is positive over half of the surface and negative over the other half, resulting in a vanishing surface integral. This is consistent with the assumption of the absence of material magnetic monopoles, but demonstrates that even matter-free space may have nonabelian magnetic fields present.

11.6 Yang–Mills fields, forms, and connections

In abelian electrodynamics the field strength two-form,

$$F = F_{\mu\nu}\, \mathrm{d}x^\mu \wedge \mathrm{d}x^\nu, \tag{11.144}$$

was derived from the action of the exterior derivative d on the gauge field one-form, $A = A_\mu \mathrm{d}x^\mu$, where

$$\frac{1}{2}F = \mathrm{d}A. \tag{11.145}$$

This definition is invariant under the gauge transformation $A \to A - \mathrm{d}\Lambda$, where Λ is a zero-form, since the exterior derivative is nilpotent, so that $\mathrm{d}^2\Lambda = 0$.

Like the abelian gauge field, the Yang–Mills gauge fields have a natural expression in terms of forms, but the definition of the exterior derivative must be generalized to the *nonabelian case*. This is done using the matrix formulation of the gauge fields introduced earlier, where $A_{\mu jk} = A_\mu^a \boldsymbol{\tau}_{jk}^a$ was used to define the covariant derivative D_μ by its action,

$$D_\mu F_{\nu\rho} = \partial_\mu F_{\nu\rho} + ig\left[A_\mu, F_{\nu\rho}\right]. \tag{11.146}$$

The nonabelian one-form $A = A_\mu \mathrm{d}x^\mu$ is referred to as the *gauge connection* and its transformation property is given by (11.23) as

$$A' = A'_\mu \mathrm{d}x^\mu = \mathbf{U}A_\mu \mathbf{U}^{-1}\mathrm{d}x^\mu + \frac{i}{g}\partial_\mu \mathbf{U}\mathbf{U}^{-1}\mathrm{d}x^\mu \equiv \mathbf{U}A\mathbf{U}^{-1} + \frac{i}{g}\mathrm{d}\mathbf{U}\mathbf{U}^{-1}, \tag{11.147}$$

where the exterior derivative of a group element is given by

$$\mathrm{d}\mathbf{U} = \partial_\mu \mathbf{U}\mathrm{d}x^\mu. \tag{11.148}$$

The gauge connection one-form is playing the same role as the affine connection one-form $\boldsymbol{\gamma}^\mu{}_\nu$, defined by (10.253), did in general relativity for the second Maurer–Cartan structure equation (10.258). For general relativity, the transformation matrix $\mathbf{U}$ has the matrix elements given by the tensor matrix

$$\mathbf{U}^\mu{}_\nu = \frac{\partial x'^\mu}{\partial x^\nu}, \tag{11.149}$$

so that its inverse $\mathbf{U}^{-1}$ has the elements

$$(\mathbf{U}^{-1})^\alpha_\beta = \frac{\partial x^\alpha}{\partial x'^\beta}. \tag{11.150}$$

The two space-time indices of the affine connection one-form $\boldsymbol{\gamma}^\mu{}_\nu$ play a similar role to the two group indices of the gauge connection A_{jk}. This follows from the transformation property of the affine connection one-form $\boldsymbol{\gamma}^\mu{}_\nu$ given by (10.254), which can be written in terms of the transformation matrices (11.149) and (11.150) as

$$\boldsymbol{\gamma}'^\mu{}_\nu = \mathbf{U}^\mu{}_\alpha \boldsymbol{\gamma}^\alpha{}_\beta (\mathbf{U}^{-1})^\beta{}_\nu - \mathrm{d}\mathbf{U}^\mu{}_\alpha (\mathbf{U}^{-1})^\alpha{}_\nu \tag{11.151}$$

where

$$\mathrm{d}\mathbf{U}^\mu{}_\alpha = \partial_\rho \mathbf{U}^\mu{}_\alpha \mathrm{d}x^\rho. \tag{11.152}$$

Apart from the factor of i/g, the transformation (11.151) of the affine connection one-form is identical in form to the Yang–Mills transformation (11.147).

The next step is to identify a *covariant exterior derivative*, denoted d_A, which extends the exterior derivative d to the case where there is a *Lie valued* gauge connection one-form present. Lie valued simply means that the connection has the form $A_\mu^a \boldsymbol{\tau}_{ij}^a$, where the $\boldsymbol{\tau}^a$ are the generators for a Lie group. The covariant exterior derivative is linear in the connection A and written

$$\mathrm{d}_A = \mathbf{I}\,\mathrm{d} + igA, \tag{11.153}$$

where $\mathbf{I}$ is the unit matrix with the same dimensions as A. The action on another Lie valued form B is defined to be identical to the covariant derivative of general relativity written as a form,

$$\mathrm{d}_A B = \mathrm{d}B + igA \wedge B. \tag{11.154}$$

If B is a one-form then

$$A \wedge B = A_\mu \mathrm{d}x^\mu \wedge B_\nu \mathrm{d}x^\nu = A_\mu B_\nu\, \mathrm{d}x^\mu \wedge \mathrm{d}x^\nu, \tag{11.155}$$

so the wedge product is a Lie valued two-form determined by the antisymmetric part of the Lie valued product $A_\mu B_\nu$,

$$\begin{aligned} A_\mu B_\nu\, \mathrm{d}x^\mu \wedge \mathrm{d}x^\nu &= A_\mu^a B_\nu^b \left(\boldsymbol{\tau}^a \cdot \boldsymbol{\tau}^b\right) \mathrm{d}x^\mu \wedge \mathrm{d}x^\nu \\ &= \frac{1}{2}(A_\mu^a B_\nu^b - A_\nu^a B_\mu^b)(\boldsymbol{\tau}^a \cdot \boldsymbol{\tau}^b)\mathrm{d}x^\mu \wedge \mathrm{d}x^\nu. \end{aligned} \tag{11.156}$$

This wedge product has the property that

$$\begin{aligned} A \wedge A &= A_\mu^a A_\nu^b \boldsymbol{\tau}^a \cdot \boldsymbol{\tau}^b \mathrm{d}x^\mu \wedge \mathrm{d}x^\nu \\ &= \frac{1}{2}\left(A_\mu^a A_\nu^b \boldsymbol{\tau}^a \cdot \boldsymbol{\tau}^b + A_\mu^b A_\nu^a \boldsymbol{\tau}^b \cdot \boldsymbol{\tau}^a\right)\mathrm{d}x^\mu \wedge \mathrm{d}x^\nu \\ &= \frac{1}{2}A_\mu^a A_\nu^b [\boldsymbol{\tau}^a, \boldsymbol{\tau}^b]\mathrm{d}x^\mu \wedge \mathrm{d}x^\nu \\ &= \frac{1}{2}if^{abc}A_\mu^a A_\nu^b \boldsymbol{\tau}^c \mathrm{d}x^\mu \wedge \mathrm{d}x^\nu. \end{aligned} \tag{11.157}$$

It follows that the gauge covariant exterior derivative gives

$$\begin{aligned} \mathrm{d}_A A &= \mathrm{d}A + igA \wedge A \\ &= \left(\partial_\mu A_\nu^a \boldsymbol{\tau}^a - \frac{1}{2}gf^{abc}A_\mu^b A_\nu^c\, \boldsymbol{\tau}^a\right)\mathrm{d}x^\mu \wedge \mathrm{d}x^\nu \\ &= \frac{1}{2}F_{\mu\nu}^a \boldsymbol{\tau}^a \mathrm{d}x^\mu \wedge \mathrm{d}x^\nu = \frac{1}{2}F, \end{aligned} \tag{11.158}$$

resulting in the Lie valued *curvature two-form* $\frac{1}{2}F$, a result identical to abelian electrodynamics and the second Maurer–Cartan structure equation (10.258) of general relativity. Although d is nilpotent, the covariant exterior derivative is *not*. Using the nilpotency of d, result (11.158), and rule (1.196) for exterior differentiation of wedge products gives

$$\frac{1}{2}\mathrm{d}F = \mathrm{d}^2 A + ig\,\mathrm{d}A \wedge A - igA \wedge \mathrm{d}A = ig\left(\frac{1}{2}F - igA \wedge A\right) \wedge A - igA \wedge \left(\frac{1}{2}F - igA \wedge A\right) = \frac{1}{2}ig(F \wedge A - A \wedge F). \tag{11.159}$$

This is combined with

$$\mathrm{d}F = \mathrm{d}_A F - igA \wedge F = \mathrm{d}_A^2 A - igA \wedge F, \tag{11.160}$$

which gives

$$\mathrm{d}_A^2 A = igF \wedge A, \tag{11.161}$$

showing that d_A is not nilpotent. The antisymmetry of the wedge product and the properties of the Levi-Civita symbol give

$$\begin{aligned} igA \wedge F - igF \wedge A &= ig\left(A_\mu^b F_{\nu\rho}^c - F_{\mu\nu}^b A_\rho^c\right) \boldsymbol{\tau}^b \boldsymbol{\tau}^c \,\mathrm{d}x^\mu \wedge \mathrm{d}x^\nu \wedge \mathrm{d}x^\rho \\ &= igA_\mu^b F_{\nu\rho}^c \left[\boldsymbol{\tau}^b, \boldsymbol{\tau}^c\right] \varepsilon^{\mu\nu\rho\sigma} \mathrm{d}V_\sigma \\ &= -gf^{abc} A_\rho^b F_{\mu\nu}^c \boldsymbol{\tau}^a \varepsilon^{\mu\nu\rho\sigma} \mathrm{d}V_\sigma, \end{aligned} \tag{11.162}$$

while the exterior derivative gives

$$\mathrm{d}F = \partial_\rho F_{\mu\nu}^a \boldsymbol{\tau}^a \,\mathrm{d}x^\rho \wedge \mathrm{d}x^\mu \wedge \mathrm{d}x^\nu = \partial_\rho F_{\mu\nu}^a \boldsymbol{\tau}^a \varepsilon^{\mu\nu\rho\sigma} \mathrm{d}V_\sigma. \tag{11.163}$$

The Bianchi identity (11.62) follows by combining (11.162) and (11.163) in (11.159), which gives

$$\mathrm{d}F + igA \wedge F - igF \wedge A = \varepsilon^{\mu\nu\rho\sigma} \mathrm{d}V_\sigma D_\rho F_{\mu\nu}^a \boldsymbol{\tau}^a = 0, \tag{11.164}$$

where the covariant derivative D_ρ has appeared, reproducing (11.62),

$$\varepsilon^{\mu\nu\rho\sigma} D_\rho F_{\mu\nu}^a = 0. \tag{11.165}$$

An important property of Yang–Mills fields is revealed by including of a term proportional to $\mathrm{Tr}(F \wedge F)$ in the action. This takes the form $\varepsilon^{\mu\nu\rho\sigma} F_{\mu\nu}^a F_{\rho\sigma}^a$, which is formally similar to the term considered in (9.434) for the abelian case. Using the relation $F = \mathrm{d}A + igA \wedge A$ gives

$$\mathrm{Tr}(F \wedge F) = \mathrm{Tr}(\mathrm{d}A \wedge \mathrm{d}A + ig\, A \wedge A \wedge \mathrm{d}A + ig\,\mathrm{d}A \wedge A \wedge A - g^2 A \wedge A \wedge A \wedge A). \tag{11.166}$$

Performing a cyclic permutation of the trace, which leaves it invariant, and relabeling dummy indices shows that the last term in (11.166) vanishes,

$$\begin{aligned}\mathrm{Tr}\big(A_\mu A_\nu A_\rho A_\sigma\big)\,\varepsilon^{\mu\nu\rho\sigma} &= \mathrm{Tr}\big(A_\sigma A_\mu A_\nu A_\rho\big)\,\varepsilon^{\mu\nu\rho\sigma}\\ &= \mathrm{Tr}\big(A_\mu A_\nu A_\rho A_\sigma\big)\,\varepsilon^{\nu\rho\sigma\mu}\\ &= -\,\mathrm{Tr}\big(A_\mu A_\nu A_\rho A_\sigma\big)\,\varepsilon^{\mu\nu\rho\sigma}.\end{aligned} \tag{11.167}$$

Performing a cyclic permutation of the second and third terms in (11.166) shows that

$$\mathrm{Tr}(ig\,A\wedge A\wedge \mathrm{d}A + ig\,\mathrm{d}A\wedge A\wedge A) = 2ig\,\mathrm{Tr}(\mathrm{d}A\wedge A\wedge A). \tag{11.168}$$

Cyclic permutations of the trace, the antisymmetry of the wedge product, and property (1.196) of the exterior derivative give

$$\begin{aligned}\mathrm{d}\mathrm{Tr}(A\wedge A\wedge A) &= \mathrm{Tr}(\mathrm{d}A\wedge A\wedge A - A\wedge \mathrm{d}A\wedge A + A\wedge A\wedge \mathrm{d}A)\\ &= 3\,\mathrm{Tr}(\mathrm{d}A\wedge A\wedge A)\end{aligned} \tag{11.169}$$

Using this result shows that (11.168) can be written

$$2ig\,\mathrm{Tr}(\mathrm{d}A\wedge A\wedge A) = \mathrm{d}\mathrm{Tr}\left(\frac{2}{3}igA\wedge A\wedge A\right). \tag{11.170}$$

Similarly, using the nilpotency of d the first term in (11.166) can be written

$$\mathrm{Tr}(\mathrm{d}A\wedge \mathrm{d}A) = \mathrm{d}\,\mathrm{Tr}(A\wedge \mathrm{d}A). \tag{11.171}$$

Combining (11.167), (11.170), and (11.171) shows that

$$\mathrm{Tr}(F\wedge F) = \mathrm{d}K, \tag{11.172}$$

where K is given by

$$K = \mathrm{Tr}\left(A\wedge \mathrm{d}A + \frac{2}{3}igA\wedge A\wedge A\right). \tag{11.173}$$

The three-form K given by (11.173) is known as the *Chern–Simons form*.

Because it is gauge invariant, the Yang–Mills action can include a term proportional to

$$\mathrm{Tr}\big(F\wedge F\big) = \varepsilon^{\mu\nu\rho\sigma}F^a_{\mu\nu}F^a_{\rho\sigma}. \tag{11.174}$$

However, result (11.172) and Stokes' theorem shows that it becomes a surface term when integrated. In the derivation of the equations of motion using the calculus of variations, it was assumed that the variations of the fields vanished at the boundary or surface of the space-time manifold. Nevertheless, there are pure gauge solutions which yield a non-vanishing result even as a surface term. Of particular interest in the quantized version of Yang–Mills theory are pure gauge solutions that match a *Euclidean metric*. These are obtained from the Minkowski metric by using the Wick rotation (8.100), $x_0 \to -ix_0$, which results in

$$x_\mu x^\mu = -x_0^2 - x^i x^i \equiv -r^2, \tag{11.175}$$

so that Minkowski space-time becomes a four-dimensional Euclidean manifold. The pure gauge solutions in Euclidean space are often called *instantons*. In an SU(2) theory they are obtained from the SU(2) group element

$$\mathbf{U} = \frac{1}{r^n}(x_0\mathbf{I} + ix^j\boldsymbol{\sigma}^j)^n, \tag{11.176}$$

where n is an integer. Demonstrating that $\mathbf{U}$ is an member of SU(2) follows from $\mathbf{U}^\dagger\mathbf{U} = \mathbf{I}$, which holds since there are n factors of the form

$$\frac{1}{r^2}(x_0\mathbf{I} + ix^j\boldsymbol{\sigma}^j)(x_0\mathbf{I} - ix^k\boldsymbol{\sigma}^k) = \frac{1}{r^2}(x_0^2\mathbf{I} + \frac{1}{2}x^j x^k\{\boldsymbol{\sigma}^j, \boldsymbol{\sigma}^k\}) = \frac{r^2}{r^2}\mathbf{I} = \mathbf{I}, \tag{11.177}$$

and $\det \mathbf{U} = 1$, which holds since there n factors of the form

$$\det\left(\frac{x_0}{r}\mathbf{I} + i\frac{x^j}{r}\boldsymbol{\sigma}^j\right) = \det\begin{pmatrix} (x_0 + ix^3)/r & i(x^1 - ix^2)/r \\ i(x^1 + ix^2)/r & (x_0 - ix^3)/r \end{pmatrix} = \frac{r^2}{r^2} = 1. \tag{11.178}$$

Demonstrating the nature of instanton solutions requires a significant amount of algebraic manipulation. Limiting attention to the case $n = 1$ gives

$$\begin{aligned} A_0 &= -\frac{i}{g}\partial_0\mathbf{U}\mathbf{U}^{-1} = -\frac{1}{gr^2}\begin{pmatrix} x^3 & x^1 - ix^2 \\ x^1 + ix^2 & -x^3 \end{pmatrix}, \\ A_1 &= -\frac{i}{g}\partial_1\mathbf{U}\mathbf{U}^{-1} = \frac{1}{gr^2}\begin{pmatrix} x^2 & x^0 + ix^3 \\ x^0 - ix^3 & -x^2 \end{pmatrix}, \\ A_2 &= -\frac{i}{g}\partial_2\mathbf{U}\mathbf{U}^{-1} = -\frac{1}{gr^2}\begin{pmatrix} x^1 & ix^0 - x^3 \\ -ix^0 - x^3 & -x^1 \end{pmatrix}, \\ A_3 &= -\frac{i}{g}\partial_3\mathbf{U}\mathbf{U}^{-1} = \frac{1}{gr^2}\begin{pmatrix} x^0 & -ix^1 - x^2 \\ ix^1 - x^2 & -x^0 \end{pmatrix}. \end{aligned} \tag{11.179}$$

Using the pure gauge configurations of (11.179) it follows that

$$\mathrm{Tr}(A \wedge A \wedge A) = \mathrm{d}S_\mu \varepsilon^{\mu\nu\rho\sigma}\,\mathrm{Tr}(A_\nu A_\rho A_\sigma) = \mathrm{d}S_\mu \frac{12x^\mu}{g^3 r^4}. \tag{11.180}$$

For example, it is reasonably straightforward to show that

$$\mathrm{Tr}(A_1 A_2 A_3) = \frac{2x^0}{g^3 r^4}, \quad \mathrm{Tr}(A_1 A_2 A_0) = -\frac{2x^3}{g^3 r^4}, \tag{11.181}$$

with the factor of twelve in (11.180) resulting from the six permutations of the Levi-Civita symbol.

Because (11.179) are pure gauge solutions, result (11.110) shows that the field strength two-form F is zero. Relation (11.158) then gives

$$\frac{1}{2}F = \mathrm{d}A + igA \wedge A = 0 \implies \mathrm{d}A = -igA \wedge A, \qquad (11.182)$$

which shows that the Chern–Simons form (11.173) is given by

$$K = A \wedge \mathrm{d}A + \frac{2ig}{3}A \wedge A \wedge A = -\frac{ig}{3}A \wedge A \wedge A. \qquad (11.183)$$

The boundary of a four-dimensional Euclidean space is the three-sphere S^3, which has the volume found in (10.536),

$$V = 2\pi^2 R^3 \implies \int_{S^3} \mathrm{d}S^\mu = V\frac{x^\mu}{R}. \qquad (11.184)$$

The net contribution of the instanton to the Euclidean action is given by

$$\begin{aligned} i\int \mathrm{d}^4x\, \varepsilon^{\mu\nu\rho\sigma}F_{\mu\nu}F_{\rho\sigma} &= \frac{g}{3}\int \mathrm{d}^4x\, \mathrm{d}\,\mathrm{Tr}\Big(A \wedge A \wedge A\Big) \\ &= \frac{g}{3}\int_{S^3} \mathrm{d}S_\mu \frac{12x^\mu}{g^3R^4} = \frac{4x^\mu}{g^2R^4}V\frac{x_\mu}{R} = \frac{8\pi^2}{g^2}. \end{aligned} \qquad (11.185)$$

For the case that $n \neq 1$, the result is $8\pi^2 n/g^2$. The details of this calculation, as well as subtleties of group theory, can be found in texts on quantum field theory, where the ground state of the quantum theory and the instanton contributions to the action are related.

11.7 Spinor fields in general relativity

Since there is no spinor representation of general diffeomorphisms, the Dirac equation can be defined in the presence of gravitation only by making it invariant under the local Lorentz group, in effect making SO(3, 1) into a gauge symmetry. This reflects the fact that there is a local freely falling frame where the Minkowski-space version of the Dirac equation is valid. This process begins by using the tetrad basis $e^\mu{}_a$ and relation (10.30) to define the space-time-dependent version of the Dirac matrices,

$$\gamma^\mu(x) = \gamma^a e_a{}^\mu(x), \qquad (11.186)$$

where the Latin index Dirac matrices γ^a are the Minkowski-space versions given by (8.168) for the Weyl representation. The γ^μ matrices have the critical property that

$$\frac{1}{2}\{\gamma^\mu, \gamma^\nu\} = \frac{1}{2}\{\gamma^a, \gamma^b\}e_a{}^\mu e_b{}^\nu = \mathbf{I}\,\eta^{ab}e_a{}^\mu e_b{}^\nu = \mathbf{I}\,g^{\mu\nu}, \qquad (11.187)$$

where (10.30) was used. This enables a Dirac algebra that is consistent with the presence of a space-time-dependent metric.

In what follows, a single massless Dirac spinor field ψ will be considered for simplicity. The bispinor Lorentz transformation is elevated from the original spinor Lorentz transformation (7.264) to a local transformation, given by

$$\mathbf{S} = \exp\left(-\frac{1}{2}\lambda^{ab}(x)\,\mathbf{J}_{ab}\right), \tag{11.188}$$

where

$$\mathbf{J}_{ab} = \frac{1}{4}[\gamma_a, \gamma_b], \tag{11.189}$$

and the Lorentz transformation parameters $\lambda_{ab}(x)$ are now space-time dependent. The next step is to define the gauge connection $\Gamma_\mu(x)$,

$$\Gamma_\mu(x) = \frac{1}{2}\Omega_\mu{}^{ab}(x)\,\mathbf{J}_{ab}, \tag{11.190}$$

where $\Omega_\mu{}^{ab}$ is the spin connection introduced in (10.143), which is found in terms of the tetrad basis and the torsion-free affine connection in (10.161) and (10.167). Its transformation properties are given by (10.162),

$$\begin{aligned}\Omega_\mu^{\prime ab}(x') = {} & \frac{\partial x^\nu}{\partial x'^\mu}(\Lambda^a{}_c(x)\,\Lambda^b{}_d(x)\,\Omega_\nu{}^{cd}(x) \\ & -\frac{1}{2}\Lambda^a{}_c(x)\,\partial_\nu\Lambda^{bc}(x) + \frac{1}{2}\Lambda^b{}_c(x)\,\partial_\nu\Lambda^{ac}(x).\end{aligned} \tag{11.191}$$

The Dirac Lagrangian consistent with diffeomorphisms is given by

$$\mathcal{L}_\psi = i\bar{\psi}(x)\,\gamma^\mu(x)\big(\partial_\mu + \Gamma_\mu(x)\big)\psi(x). \tag{11.192}$$

Under the local Lorentz transformation (11.188) the action becomes

$$\mathcal{L}'_{\psi'} = i\bar{\psi}\mathbf{S}^{-1}\gamma'^\mu(x)\left(\partial'_\mu + \Gamma'_\mu(x')\right)\mathbf{S}\psi. \tag{11.193}$$

The first step in demonstrating that this is invariant is to use the transformation properties of the tetrad, which gives

$$\gamma'^\mu(x') = \gamma^a e'^\mu{}_a(x') = \gamma^a\Lambda_a{}^b(x)\,\frac{\partial x'^\mu}{\partial x^\nu}\,e^\nu{}_b(x). \tag{11.194}$$

Because γ^a is the Minkowski-space version of the Dirac matrices, relation (7.274) gives

$$\begin{aligned}\mathbf{S}^{-1}(x)\gamma'^\mu(x')\mathbf{S}(x) &= \mathbf{S}^{-1}(x)\,\gamma^a\,\mathbf{S}(x)\frac{\partial x'^\mu}{\partial x^\nu}\Lambda_a{}^b(x)e^\nu{}_b(x) \\ &= \Lambda^a{}_c(x)\,\gamma^c\,\frac{\partial x'^\mu}{\partial x^\nu}\Lambda_a{}^b(x)e^\nu{}_b(x) \\ &= \delta^b{}_c\frac{\partial x'^\mu}{\partial x^\nu}\gamma^c e^\nu{}_b(x) = \frac{\partial x'^\mu}{\partial x^\nu}\gamma^b e^\nu{}_b(x) \\ &= \frac{\partial x'^\mu}{\partial x^\nu}\gamma^\nu(x).\end{aligned} \tag{11.195}$$

Using (11.195) shows that the first term in (11.193) becomes

$$\bar{\psi}\mathbf{S}^{-1}\left(\gamma'^{\mu}(x)\partial'_{\mu}\right)\mathbf{S}\psi = \bar{\psi}\mathbf{S}^{-1}\gamma'^{\mu}(x)\,\mathbf{S}\,\frac{\partial x^{\rho}}{\partial x'^{\mu}}\partial_{\rho}\psi$$
$$+\,\bar{\psi}\mathbf{S}^{-1}\gamma'^{\mu}(x)\,\mathbf{S}\,\frac{\partial x^{\rho}}{\partial x'^{\mu}}\,\mathbf{S}^{-1}(\partial_{\rho}\mathbf{S})\psi \qquad (11.196)$$
$$= \bar{\psi}\gamma^{\nu}(x)\partial_{\nu}\psi + \bar{\psi}\gamma^{\nu}(x)\,\mathbf{S}^{-1}(\partial_{\nu}\mathbf{S})\psi.$$

Using (11.195) once again shows that the second term in (11.193) is given by

$$\bar{\psi}\mathbf{S}^{-1}\gamma'^{\mu}(x)\Gamma'_{\mu}\left(x'\right)\mathbf{S}\psi = \bar{\psi}\mathbf{S}^{-1}\gamma'^{\mu}(x)\,\mathbf{S}\,\mathbf{S}^{-1}\Gamma'_{\mu}\left(x'\right)\mathbf{S}\psi$$
$$= \bar{\psi}\gamma^{\nu}(x)\,\frac{\partial x'^{\mu}}{\partial x^{\nu}}\,\mathbf{S}^{-1}\Gamma'_{\mu}(x')\,\mathbf{S}\psi. \qquad (11.197)$$

Using the spin connection transformation (11.191) gives

$$\mathbf{S}^{-1}(x)\,\Gamma'_{\mu}(x')\,\mathbf{S}(x) = \frac{1}{2}\Omega'^{ab}_{\mu}(x')\,\mathbf{S}^{-1}(x)\mathbf{J}_{ab}\,\mathbf{S}(x)$$
$$= \frac{\partial x^{\nu}}{\partial x'^{\mu}}\left(\frac{1}{2}\Lambda^{a}{}_{c}(x\,\Lambda^{b}{}_{d}(x)\,\Omega_{\nu}{}^{cd}(x)\right. \qquad (11.198)$$
$$\left.-\,\frac{1}{4}\Lambda^{a}{}_{c}(x)\,\partial_{\nu}\Lambda^{bc}(x) + \frac{1}{4}\Lambda^{b}{}_{c}(x)\,\partial_{\nu}\Lambda^{ac}(x)\right)\mathbf{S}^{-1}\mathbf{J}_{ab}\mathbf{S}.$$

The Minkowski-space result (7.275) can be extended to the local case, which gives

$$\mathbf{S}^{-1}(x)\,\mathbf{J}_{ab}\,\mathbf{S}(x) = \Lambda_{a}{}^{e}(x)\Lambda_{b}{}^{f}(x)\,\mathbf{J}_{ef}. \qquad (11.199)$$

Returning this to (11.198) and using (6.68), which remains valid at each point x, gives

$$\gamma'^{\mu}(x)\mathbf{S}^{-1}\Gamma'_{\mu}\left(x'\right)\mathbf{S} = \gamma^{\rho}(x)\,\frac{\partial x'^{\mu}}{\partial x^{\rho}}\frac{\partial x^{\nu}}{\partial x'^{\mu}}$$
$$\left(\frac{1}{2}\Omega_{\nu}{}^{cd}(x)\,\mathbf{J}_{cd} - \frac{1}{4}\partial_{\nu}\Lambda^{ac}(x)\Lambda_{a}{}^{b}(x)(\mathbf{J}_{cb} - \mathbf{J}_{bc})\right) \qquad (11.200)$$
$$= \gamma^{\nu}(x)\left(\Gamma_{\nu}(x) + \frac{1}{2}\partial_{\nu}\Lambda^{ac}(x)\,\Lambda_{a}{}^{b}(x)\,\mathbf{J}_{bc}\right).$$

Combining (11.196) and (11.200) gives

$$\mathcal{L}'_{\psi'} = i\bar{\psi}\mathbf{S}^{-1}\gamma'^{\mu}\left(\partial'_{\mu} + \Gamma'_{\mu}\left(x'\right)\right)\mathbf{S}\psi$$
$$= i\bar{\psi}\gamma^{\nu}(x)(\partial_{\nu} + \Gamma_{\nu}(x))\psi \qquad (11.201)$$
$$+\, i\bar{\psi}\gamma^{\nu}(x)\left(\frac{1}{2}\partial_{\nu}\Lambda^{ac}(x)\,\Lambda_{a}{}^{b}(x)\,\mathbf{J}_{bc} + \mathbf{S}^{-1}\partial_{\nu}\mathbf{S}\right)\psi.$$

This is identical to the mechanism of gauge invariance, where the gauge field transforms according to (11.23). In this case Γ_{μ} is playing the role of the gauge field connection. The terms involving the local Lorentz transformation can be evaluated for the case of infinitesimal parameters, which gives to $O(\lambda)$,

$$\begin{aligned} \Lambda_a{}^b &\approx \delta_a{}^b + \lambda_a{}^b, \\ \frac{1}{2}\partial_\nu \Lambda^{ac}\Lambda_a{}^b \mathbf{J}_{bc} &\approx \frac{1}{2}\partial_\nu \lambda^{bc}\mathbf{J}_{bc}, \\ \mathbf{S} &\approx \mathbf{I} - \frac{1}{2}\lambda^{ab}\mathbf{J}_{ab}, \\ \mathbf{S}^{-1}\partial_\nu \mathbf{S} &\approx -\frac{1}{2}\partial_\nu \lambda^{ab}\mathbf{J}_{ab}. \end{aligned} \tag{11.202}$$

This shows that the two remaining transformation-dependent terms in (11.201) cancel for the infinitesimal case, which means they cancel for the arbitrary case, with the result that $\mathcal{L}'_{\psi'} = \mathcal{L}_\psi$. The spin connection allows the effects of the gravitational field on the spinor field to be modeled. This process can be followed for any type of field by simply using the appropriate form for the Lorentz generators $\mathbf{J}_{ab}$.

This allows the spinor action to be written in the manifestly symmetric manner as

$$S_\psi = \int \mathrm{d}^4x \sqrt{|g|}\, \frac{1}{2} i g^{\mu\nu}\bar{\psi}\gamma_\mu(\partial_\nu + \Gamma_\nu)\psi. \tag{11.203}$$

Using the functional derivative version of the stress–energy tensor given by (10.362) gives the manifestly symmetric stress–energy tensor for a massless spinor field,

$$\begin{aligned} T^{\mu\nu} &= \frac{\delta S_\psi}{\delta g_{\mu\nu}} = \frac{1}{2}\frac{\delta S_\psi}{\delta g_{\mu\nu}} + \frac{1}{2}\frac{\delta S_\psi}{\delta g_{\nu\mu}} \\ &= \frac{1}{2} i\bar{\psi}(\gamma^\mu(\partial^\nu + \Gamma^\nu) + \gamma^\nu(\partial^\mu + \Gamma^\mu))\psi - i g^{\mu\nu}\bar{\psi}\gamma^\rho(\partial_\rho + \Gamma_\rho)\psi. \end{aligned} \tag{11.204}$$

Evaluating (11.204) using a solution to the equation of motion gives

$$T^{\mu\nu} = \frac{1}{2} i\bar{\psi}(\gamma^\mu(\partial^\nu + \Gamma^\nu) + \gamma^\nu(\partial^\mu + \Gamma^\mu))\psi. \tag{11.205}$$

For the case of Minkowski space the torsion-free spin connection (10.157) vanishes, since (10.46) shows that the tetrads are constant. As a result, the Minkowski-space massless spinor stress–energy tensor is given by

$$T^{\mu\nu} = \frac{1}{2} i\bar{\psi}(\gamma^\mu\partial^\nu + \gamma^\nu\partial^\mu)\psi. \tag{11.206}$$

Up to a factor of one-half this is identical to the first term in the symmetrized version (9.216) used to evaluate the angular momentum of the spinor field. The symmetrized version (9.216) is found by adding the complex conjugate and dividing the sum by two. This is equivalent to starting with a manifestly real and symmetric action.

11.8 Yang–Mills fields and the Gribov instability

The theory known as quantum chromodynamics is an SU(3) gauge theory. In its fundamental representation it describes three spinor *quarks* interacting through the eight gauge fields referred to as *gluons* that are associated with the eight generators of SU(3). It is generally believed that this theory underlies hadronic physics, explaining the extensive variety of strongly interacting massive particles, ranging from nucleons to the scalar and vector mesons detected in interactions. However, single quarks have never been detected in any experiments, and there are indications from the quantum version of the theory that the gluons create an interaction that grows with separation, so that the quarks are *confined* into pairs and triplets. Lattice gauge analysis of the quantum theory indicates that nonabelian magnetic effects may be responsible for creating such a confining potential, but the details of this remain unclear at this time.

Analysis of this problem in the research literature has focused on the role of quantum mechanical effects in this process. Since this is a monograph on classical field theory, it is meaningful to ask if there is any insight into this problem at the classical level. In regard to that question, this monograph will focus on the classical version of what is called the *Gribov instability*. In the context of the quantum theory, Gribov demonstrated that the Coulomb phase of quantum chromodynamics is unstable for strong coupling. This can be interpreted as a need to replace the Coulomb potential by a confining potential, often modeled as a *flux tube* joining the quarks into a bound state. The analysis to be presented here supports the existence of the Gribov instability, even at the classical level. For simplicity, this will be demonstrated for the case that the gauge group is reduced to SU(2) with three gauge fields interacting with a spinor doublet.

In earlier analysis of abelian gauge theory the Coulomb potential emerged by solving the classical equation of motion in the presence of a static point charge, while the Coulomb force was understood by considering the field (9.359) around a static charge pair. A similar approach will now be taken in the nonabelian case. Before proceeding it is important to note that the stress–energy tensor analyzed earlier shows that the force density equations generated by *static* charges are *formally* similar in both the abelian and nonabelian cases. This follows from the expression for the spatial force density,

$$f_j = \partial_\mu Y^\mu{}_j = \partial_0 Y^0{}_j = j_0^a F^{a0}{}_j = j_0^a E_j^a, \tag{11.207}$$

where the spatial divergence $\partial_k Y^k{}_j$ has been dropped, since it becomes a surface term in a spatial integral. At the classical level this shows that the force on static charges comes from the nonabelian electric field, and does not indicate an obvious role for the nonabelian magnetic field in the case of *static* charges.

In the simpler case of SU(2) the nonabelian point charges can be associated with the charge density j_0^3, since $\boldsymbol{\tau}^3$ is usually chosen to be diagonal. The ansatz is made that only A_0^3 is nonzero, while all the other fields and components are *initially* zero. This is consistent with the previous stress–energy tensor analysis, since the electric field $E_j^3 = F_{0j}^3$ is the primary agent of force. The two situations to be analyzed, which are similar to the abelian case, are given by the nonabelian charge density of a *static* particle at the origin,

$$j_0^3 = g\,\delta^3(\boldsymbol{x}), \tag{11.208}$$

and the case where there are two oppositely charged static point particles at separate locations,

$$j_0^3 = g\,\delta^3(\boldsymbol{x} - \boldsymbol{a}) - g\,\delta^3(\boldsymbol{x} + \boldsymbol{a}). \tag{11.209}$$

Since $j^{1\mu} = j^{2\mu} = 0$, the equations of motion are given by

$$\partial_\mu F^{3\mu\nu} - gf^{3bc}A_\mu^b F^{c\mu\nu} = j^{3\nu}, \quad \partial_\mu F^{(1,2)\mu\nu} - gf^{(1,2)bc}A_\mu^b F^{c\mu\nu} = 0. \tag{11.210}$$

Solving (11.210) uses the Lorentz gauge. For the assumption that only A_0^3 is nonzero, the field strength reduces to

$$F_{j0}^3 = \partial_j A_0^3 - \partial_0 A_j^3 - gf^{3bc}A_j^b A_0^c = \partial_j A_0^3, \quad gf^{3bc}A_j^b A_0^c = 0, \tag{11.211}$$

so that the equations of motion reduce to a form *identical* to the abelian case,

$$\partial^j F_{j0}^3 = -\nabla^2 A_0^3 = j_0^3. \tag{11.212}$$

The solution to (11.212) for the case of point charges gave rise to the Coulomb potential in (9.359). Therefore, it would appear that the Coulomb potential and force are also the solution for the Yang–Mills case. Since the Coulomb potential definitely does *not* confine electric charge, it is natural to conclude that reducing the Yang–Mills equations of motion to (11.212) must be an error, and that solutions involving the magnetic fields must be sought.

Unfortunately, there are no known analytical solutions to the equations of motion (11.52) that involve both the magnetic and electric fields in the presence of point charges. However, there is a reasonably simple way to investigate the effects of the nonabelian magnetic fields in shaping the nature of the force, and that is to consider magnetic fluctuations around the solution to (11.212). If these fluctuations are unstable, growing arbitrarily large or *becoming complex*, then the Coulombic solution (11.212) is not physically viable. This was done with fluids in chapter 4, where freely propagating density fluctuations or sound waves in a static fluid were found. In that case it meant that a static fluid was stable, since the sound waves, which carry energy, did not grow arbitrarily large spatially or temporally.

In the case of Yang–Mills fields, this corresponds to writing the variables appearing in the equations of motion as

$$A_\mu^a = C_\mu^a + \xi_\mu^a, \tag{11.213}$$

where C_μ^a is the solution to (11.212) and ξ_μ^a is an infinitesimally small *time-dependent fluctuation* around the static classical solution. Inserting this into the Yang–Mills field tensor generates a *fluctuation term*, denoted by $\delta F_{\mu\nu}^a$, linear in ξ_μ^a and given by

$$\delta F^a_{\mu\nu} = \partial_\mu \xi^a_\nu - \partial_\nu \xi^a_\mu - gf^{abc}\left(\xi^b_\mu C^c_\nu + C^b_\mu \xi^c_\nu\right). \tag{11.214}$$

Letting $F^a_{\mu\nu}$ denote the field tensor evaluated using C^a_μ, the equation of motion to $O(\xi)$ becomes

$$\partial_\mu F^{a\mu\nu} + gf^{abc}A^b_\mu F^{c\mu\nu} + \partial_\mu \delta F^{a\mu\nu} - gf^{abc}\xi^b_\mu F^{c\mu\nu} - gf^{abc}C^b_\mu\, \delta F^{c\mu\nu} = j^{a\nu}. \tag{11.215}$$

Assuming that the solution satisfies (11.210), the linearized equation of motion (11.215) becomes

$$\partial_\mu \delta F^{a\mu\nu} - gf^{abc}\xi^b_\mu F^{c\mu\nu} - gf^{abc}C^b_\mu\, \delta F^{c\mu\nu} = 0. \tag{11.216}$$

Using (11.214), the Lorenz condition,

$$\partial_\mu \xi^{a\mu} = \partial_\mu C^{a\mu} = 0, \tag{11.217}$$

and the SU(2) relation (11.135), equation (11.216) becomes

$$\begin{aligned} &\Box \xi^{a\nu} - g\varepsilon^{abc}\left[\xi^{b\mu}\partial_\mu C^{c\nu} + C^{b\mu}\partial_\mu \xi^{c\nu} + \xi^b_\mu\left(\partial^\mu C^{c\nu} - \partial^\nu C^{c\mu}\right) + C^b_\mu\left(\partial^\mu \xi^{c\nu} - \partial^\nu \xi^{c\mu}\right)\right] \\ &+ g^2\Big(C^{a\mu}\xi^b_\mu C^{b\nu} - C^{a\nu}C^{b\mu}\xi^b_\mu + \xi^{a\mu}C^b_\mu C^{b\nu} - C^{a\nu}C^b_\mu \xi^{b\mu} + C^{a\mu}C^b_\mu \xi^{b\nu} \\ &\qquad - \xi^{a\nu}C^b_\mu C^{b\mu}\Big) = 0. \end{aligned} \tag{11.218}$$

Result (11.218) describes the infinitesimal fluctuations around $C^{a\mu}$ for the case that $C^{a\mu}$ solves the Yang–Mills equation.

In order to demonstrate the Gribov instability, the case of (11.212) is chosen, where only C^3_0 is nonzero and satisfies $\partial_0 C^3_0 = 0$. For this special case the various components of the fluctuation equation (11.218) reduce to

$$\begin{aligned} \Box \xi^{3\mu} &= 0, \\ \Box \xi^{10} - 2g\xi^{2j}\partial_j C^3_0 + gC^{30}\partial_0 \xi^{20} &= 0, \\ \Box \xi^{20} + 2g\xi^{1j}\partial_j C^3_0 - gC^{30}\partial_0 \xi^{10} &= 0, \\ \Box \xi^1_j + 2gC^3_0\partial_0 \xi^2_j + g\xi^2_0\partial_j C^3_0 - gC^3_0\partial_j \xi^2_0 - g^2 C^3_0 C^3_0 \xi^1_j &= 0, \\ \Box \xi^2_j - 2gC^3_0\partial_0 \xi^1_j - g\xi^1_0\partial_j C^3_0 + gC^3_0\partial_j \xi^1_0 - g^2 C^3_0 C^3_0 \xi^2_j &= 0. \end{aligned} \tag{11.219}$$

Exercise 11.5: Demonstrate that (11.218) yields the fluctuation equations of (11.219).

The ξ^3_μ fluctuations are completely decoupled from the C^3_0 field and if they are initially infinitesimal and real they will remain so. However, the $\xi^{1,\,2}_\mu$ fluctuations are coupled to the solution C^3_0 for the A^3_μ field. As mentioned earlier, the gauge field carries nonabelian charge and therefore the gauge fields created by the charges affect the other gauge fields. Solving the four coupled equations of (11.219) begins by noting that they support standing waves. Substituting the standing wave forms,

$$\begin{aligned} \xi_0^1 &= \varphi_0(\boldsymbol{x})\sin \omega t, \\ \xi_0^2 &= \varphi_0(\boldsymbol{x})\cos \omega t, \\ \xi_j^1 &= \varphi_j(\boldsymbol{x})\cos \omega t, \\ \xi_j^2 &= -\varphi_j(\boldsymbol{x})\sin \omega t, \end{aligned} \tag{11.220}$$

shows that the last four equations of (11.219) reduce to two equations independent of t,

$$-\nabla^2\varphi_0 + 2g\,\varphi^j\,\partial_j C_0^3 - g\omega\,C_0^3\,\varphi_0 = \omega^2\varphi_0, \tag{11.221}$$

$$-\nabla^2\varphi_j - 2g\omega\,C_0^3\,\varphi_j + g\varphi_0\,\partial_j C_0^3 - g\,C_0^3\,\partial_j\varphi_0 - g^2 C_0^3 C_0^3\,\varphi_j = \omega^2\varphi_j. \tag{11.222}$$

In the derivation of (11.218) it was assumed that the two Lorenz conditions hold, which requires

$$\partial_\mu \xi^{1\mu} = \partial_\mu \xi^{2\mu} = 0. \tag{11.223}$$

Using the standing waves (11.220) in the Lorenz conditions (11.223) reduces both of them to

$$\omega\varphi_0 + \nabla\cdot\boldsymbol{\phi} = 0. \tag{11.224}$$

The Lorenz condition is satisfied by choosing

$$\varphi_0 = 0, \quad \nabla\cdot\boldsymbol{\phi} = 0, \tag{11.225}$$

which is identical to result (9.248) for the transverse electromagnetic wave. For the conditions of (11.225), equation (11.221) reduces to the single condition

$$\varphi^j\partial_j C_0^3 = 0, \tag{11.226}$$

while (11.222) can be written in the useful form

$$-\nabla^2\varphi_j - \left(\omega + gC_0^3\right)^2\varphi_j = 0. \tag{11.227}$$

Further progress in understanding the fluctuation fields requires a specific form for C_0^3. For the case of a *single static* nonabelian point charge located at the origin, the only solution to (11.212) for an infinite volume is the Coulomb potential,

$$C_0^3 = \frac{g}{r}, \tag{11.228}$$

where g is the coupling constant governing the interaction of the nonabelian vector fields with the matter current. For this case equation (11.222) becomes

$$-\nabla^2\varphi_j - \frac{2g^2\omega}{r}\varphi_j - \frac{g^4}{r^2}\varphi_j = \omega^2\varphi_j, \tag{11.229}$$

along with the transversality requirement $\nabla \cdot \boldsymbol{\phi} = 0$. Because of the spherical symmetry of C_0^3 it is most convenient to work in spherical coordinates. Since C_0^3 is a function solely of r equation (11.226) is satisfied if $\varphi_r = 0$. In order to maintain azimuthal symmetry for the fluctuations, the fluctuations will be assumed to be functions of r and θ and independent of ϕ. The condition $\nabla \cdot \boldsymbol{\phi} = 0$ is then satisfied if $\varphi_\theta = 0$ and the single remaining component φ_ϕ is a function of r and θ, so that $\partial\varphi_\phi/\partial\phi = 0$. Using the expression for the Laplacian in spherical coordinates, equation (11.229) becomes

$$\left(\frac{\partial^2}{\partial r^2} + \frac{2}{r}\frac{\partial}{\partial r} + \frac{1}{r^2}\frac{\partial^2}{\partial\theta^2} + \frac{\cot\theta}{r^2}\frac{\partial}{\partial\theta} + \frac{2g^2\omega}{r} + \frac{g^4}{r^2} + \omega^2\right)\varphi_\phi(r, \theta) = 0. \tag{11.230}$$

The next step is to write $\varphi_\phi = M(r, \theta)/r$ and then scale the radial coordinate by $|\omega|$, so that $r = x/|\omega|$. The absolute value is required so that x and r have the same range. Equation (11.230) becomes

$$\left(\frac{\partial^2}{\partial x^2} + \frac{1}{x^2}\frac{\partial^2}{\partial\theta^2} + \frac{\cot\theta}{x^2}\frac{\partial}{\partial\theta} \mp \frac{2g^2}{x} + \frac{g^4}{x^2} + 1\right)M(x, \theta) - 0, \tag{11.231}$$

where the sign of the fourth term is given by $|\omega|/\omega$. Equation (11.231) can be solved by separation of variables, where the function $M(x, \theta)$ is written

$$M(x, \theta) = R_l(x)P_l(\theta). \tag{11.232}$$

Substituting this form into equation (11.231) and dividing by $M(x, \theta)$ gives

$$\begin{aligned}&\frac{1}{R_l(x)}\left(\frac{\partial^2}{\partial x^2} \mp \frac{2g^2}{x} + \frac{g^4}{x^2} + 1\right)R_l(x)\\&+ \frac{1}{x^2}\frac{1}{P_l(\theta)}\left(\frac{\partial^2}{\partial\theta^2} + \cot\theta\,\frac{\partial}{\partial\theta}\right)P_l(\theta) = 0.\end{aligned} \tag{11.233}$$

The function $P_l(\theta)$ must be single valued and is chosen to render the angular part of (11.233) into a constant, so that the $1/x^2$ factor can be absorbed into the terms involving $R_l(x)$. The solutions are the *Legendre polynomials* in $\cos\theta$, $P_l(\theta)$, which satisfy

$$\left(-\frac{\partial^2}{\partial\theta^2} - \cot\theta\,\frac{\partial}{\partial\theta}\right)P_l(\theta) = l(l+1)P_l(\theta), \tag{11.234}$$

where l must be a *positive integer*, including the value zero. In their normalized form the first few are given by $P_0 = 1/\sqrt{4\pi}$, $P_1 = \cos\theta\sqrt{3/4\pi}$, and $P_2 = (3\cos^2\theta - 1)\sqrt{5/16\pi}$. Like the Fourier transform, the Legendre polynomials form a complete and orthonormal set of angular functions that can be used to expand an arbitrary single-valued function of θ.

Returning result (11.234) to (11.233) gives the equation for the radial behavior of the fluctuations $\xi_j^{(1,\,2)}$,

$$\left(\frac{\mathrm{d}^2}{\mathrm{d}x^2} + 1 \mp \frac{2g^2}{x} + \frac{g^4 - l(l+1)}{x^2}\right)R_l(x) = 0. \tag{11.235}$$

Without the g^4/x^2 term *and* without restricting R_l to being real, (11.235) is similar to the quantum mechanical Schrödinger equation for the wave function of a charge moving in the Coulomb potential $\pm 2g^2/x$. In the Schrödinger equation the term proportional to $l(l+1)$ is referred to as the *centrifugal barrier*, where it serves as a repulsive potential affecting particles with orbital angular momentum. Equation (11.235) can be directly related to a much-studied differential equation known as the *Coulomb wave equation.* The final step in this process is to define the parameter s so that

$$\begin{aligned} s(s+1) &= l(l+1) - g^4, \\ \Longrightarrow\ s &= -\frac{1}{2} \pm \frac{1}{2}\sqrt{1 + 4l(l+1) - 4g^4}. \end{aligned} \tag{11.236}$$

Equation (11.235) can then be written

$$\left(\frac{\mathrm{d}^2}{\mathrm{d}x^2} + 1 \mp \frac{2g^2}{x} - \frac{s(s+1)}{x^2}\right)R_l(x) = 0, \tag{11.237}$$

which is the Coulomb wave equation.

In order to understand the solutions to (11.237) it is useful to examine the two extremes for x. For large x, (11.237) becomes

$$\frac{\mathrm{d}^2 R_l}{\mathrm{d}x^2} \approx -R_l, \tag{11.238}$$

which has the solution $R_l \approx \sin x$. Since $x = |\omega| r$ this describes an oscillatory radial behavior at great distances from the charge governed by the frequency ω. At the other extreme, for small x (11.237) becomes

$$\frac{\mathrm{d}^2 R_l}{\mathrm{d}x^2} \approx \frac{s(s+1)}{x^2} R_l, \tag{11.239}$$

which gives the solution $R_l \approx x^{-s}$ in the proximity of the charge. Depending on the sign of s, R_l can be singular or well-behaved as $r \to 0$. The solutions to (11.237) are therefore real and well-behaved over the entire range of x, *as long as the value of s is real.*

However, (11.236) shows that if the coupling constant g is too large, s becomes complex and a *complex fluctuation mode* appears in a theory where the fields are required to be real functions. Such a result represents an *unstable fluctuation mode*, since it corresponds to a complex energy. This is the Gribov instability, which is derived in the context of quantum chromodynamics using the eigenvalues of the quark differential operator. For $l = 0$ result (11.236) shows that the classical SU(2) Gribov instability occurs if

$$g^2 > \frac{1}{2}, \tag{11.240}$$

and this condition is met in SU(3) QCD where $g \approx 1$. For such a case magnetic fluctuations $\xi_j^{(1,\,2)}$ around the *negative centrifugal barrier*, $-g^4/x^2$, destabilize the case of a single nonabelian point charge. There is no stable single-particle *classical solution* for large coupling constant values.

The origin of the destabilizing negative potential energy for the nonabelian magnetic fields is the quartic coupling term present in the Lagrangian. After using the SU(2) structure constants, the Lagrangian contains a quartic term given by

$$\mathcal{L}_Q = -\frac{1}{4}g^2\left(A_\mu^a A^{a\mu} A_\nu^b A^{b\nu} - A_\mu^a A^{b\mu} A_\nu^a A^{b\nu}\right). \tag{11.241}$$

Using the fluctuation translation (11.213) in (11.241) for the special case of (11.228) and retaining only quadratic terms in the fluctuations ξ_μ^a gives

$$\mathcal{L}_Q = -\frac{1}{2}g^2 C_0^3 C^{30}\left(A_j^1 A^{1j} + A_j^2 A^{2j}\right). \tag{11.242}$$

It is this term that gives rise to the negative centrifugal barrier in (11.237), ultimately destabilizing the Coulomb phase of classical SU(2) gauge theory for $g^2 > 1/2$.

The next case to consider is two oppositely charged point particles, so that the nonabelian dipole charge is given by

$$j_0^3 = g\,\delta^3(\boldsymbol{x} - \boldsymbol{a}) - g\,\delta^3(\boldsymbol{x} + \boldsymbol{a}). \tag{11.243}$$

For simplicity, the two charges can be chosen to lie on the z-axis at $z = a$ and $z = -a$, so that the charge density is axially symmetric around the z-axis. Once again, the assumption is that A_0^3 is the only nonzero component; it then satisfies

$$-\nabla^2 A_0^3 = j_0^3 = g\,\delta^3(\boldsymbol{x} - \boldsymbol{a}) - g\,\delta^3(\boldsymbol{x} + \boldsymbol{a}). \tag{11.244}$$

In the abelian case this gave rise to the dipole field of (9.359), which is simply the superposition of two Coulomb fields. Again, such a result is not a confining potential. However, the two Coulomb fields of such a solution will both suffer from the Gribov instability for fluctuations with strong coupling, indicating that (11.244) must have another solution that is very different from the abelian dipole Coulomb potential.

The two point charges can be assumed to lie along the z-axis, so that the charge density is given by

$$j_0^3 = g\,\delta^2(\rho)(\delta(z - a) - \delta(z + a)) = \frac{g}{2\pi\rho}\delta(\rho)(\delta(z - a) - \delta(z + a)). \tag{11.245}$$

The azimuthal symmetry of the charge distribution results in A_0^3 depending solely on the two cylindrical coordinates ρ and z. Using the Laplacian in cylindrical coordinates (1.156) along with the independence from ϕ gives

$$-\left(\frac{\partial^2}{\partial\rho^2} + \frac{1}{\rho}\frac{\partial}{\partial\rho} + \frac{\partial^2}{\partial z^2}\right)A_0^3(\rho, z) = \frac{g}{2\pi\rho}\delta(\rho)(\delta(z - a) - \delta(z + a)). \tag{11.246}$$

In order to make progress in solving (11.246) the radial delta function will be replaced by a radial function $F(\rho)$, so that

$$\delta^2(\rho) \rightarrow \frac{1}{\rho_o}F(\rho), \tag{11.247}$$

where ρ_o is a length scale for the solution and $F(\rho)$ is assumed to be sharply peaked at around $\rho = 0$ with units of inverse length. For such an easing of the singular nature of the charge distribution, equation (11.246) becomes

$$-\left(\frac{\partial^2}{\partial\rho^2} + \frac{1}{\rho}\frac{\partial}{\partial\rho} + \frac{\partial^2}{\partial z^2}\right)A_0^3(\rho, z) = gF(\rho)(\delta(z - a) - \delta(z + a)). \tag{11.248}$$

The solution to (11.248) is given by the ansatz

$$\begin{aligned} A_0^3 &= F(\rho)\sin(2\pi nz/a)\,\theta(a - z)\,\theta(z + a) \equiv F(\rho)G(z), \\ G(z) &= \sin(2\pi nz/a)\theta(a - z)\theta(z + a), \end{aligned} \tag{11.249}$$

where $\theta(z)$ is the Heaviside step function and n is an *arbitrary* integer. Since (11.249) vanishes for $z > a$ and $z < -a$, the solution begins at one charge and terminates at the other. If a solution for $F(\rho)$ can be found that is sharply peaked at around $\rho = 0$, then (11.249) describes a *flux tube* between the two charges. Using the results

$$\begin{aligned} \frac{\partial}{\partial z}\theta(a \pm z) &= \pm\,\delta(z \pm a), \\ G(z)\,\delta(z \pm a) &= 0, \\ \cos(2\pi nz/a)\,\delta(z \pm a) &= \delta(z \pm a), \end{aligned} \tag{11.250}$$

shows that the function $G(z)$ satisfies

$$-\frac{\partial^2}{\partial z^2}G(z) = \frac{4n^2\pi^2}{a^2}G(z) + \left(\frac{2n\pi}{a}\delta(z - a) - \frac{2n\pi}{a}\delta(z + a)\right). \tag{11.251}$$

Using (11.251) in equation (11.244) gives

$$\begin{aligned} -\nabla^2 A_0^3 = G(z)&\left(-\frac{\partial^2}{\partial\rho^2} - \frac{1}{\rho}\frac{\partial}{\partial\rho} + \frac{4\pi^2 n^2}{a^2}\right)F(\rho) \\ &+ \frac{2\pi n}{a}F(\rho)[\delta(z - a) - \delta(z + a)]. \end{aligned} \tag{11.252}$$

For a given charge separation a it is possible to choose an integer n that defines a *constant characteristic length* ρ_o given by

$$\rho_o = \frac{ga}{2\pi n}, \tag{11.253}$$

where g is the coupling constant. In this approach, a length scale emerges from the separation a and the dimensionless integer n. This definition treats the separation of the two point charges as

$$a = n\left(\frac{2\pi\rho_o}{g}\right), \tag{11.254}$$

and if ρ_o is to be a constant length, the separation of the point charges is no longer arbitrary. Instead, like the wavelengths of a standing wave, the separation of the two charges is an integer multiple of $2\pi\rho_o/g$. The function $G(z)$ becomes

$$G(z) = \sin\left(\frac{gz}{\rho_o}\right)\theta(a - z)\,\theta(z + a). \tag{11.255}$$

Using this definition and (11.252) shows that the nonabelian dipole equation (11.244) is satisfied by (11.249) if both of the following hold,

$$\left(-\frac{\partial^2}{\partial\rho^2} - \frac{1}{\rho}\frac{\partial}{\partial\rho} + \frac{g^2}{\rho_o^2}\right)F(\rho) = 0, \quad \frac{1}{\rho_o}F(\rho) \approx \delta^2(\rho). \tag{11.256}$$

The first equation in (11.256) is a variant of *Bessel's equation* and the solution to it that best matches the second required boundary condition is the singular *modified Bessel function of order zero*, denoted by $K_0(g\rho/\rho_o)$. This Bessel function is logarithmically divergent as $\rho \to 0$ and exponentially damped for arguments greater than one, and so its radial behavior matches that of a flux tube between the two charges. The solution is written

$$F(\rho) = \alpha K_0(g\rho/\rho_o), \tag{11.257}$$

and the coefficient α is fixed by replacing the second requirement of (11.256) with

$$\begin{aligned}\int \mathrm{d}^2S\,\delta^2(\rho) &\approx \frac{\alpha}{\rho_o}\int_0^{2\pi}\mathrm{d}\phi\int_0^{\rho_0/g}\mathrm{d}\rho\,\rho K_0\left(\frac{g\rho}{\rho_o}\right)\\ &= \frac{2\pi\alpha\rho_o}{g^2}\int_0^1 \mathrm{d}x\,xK_0(x) = 1.\end{aligned} \tag{11.258}$$

This matches the flux inside the radius ρ_o/g to that of a delta function. This is possible since the characteristic length ρ_o can be made arbitrarily small in the limit $n \to \infty$. The length ρ_o/g can then be identified as the classical nonabelian charge radius R, which then functions as the flux tube radius. In this limit the flux tube has a radius of zero, becoming a string joining the two charges. The integral over x can be exactly evaluated to obtain

$$\frac{2\pi\alpha\rho_o}{g^2}(1 - K_1(1)) = 1, \tag{11.259}$$

where K_1 is the second modified Bessel function. This gives

$$1 - K_1(1) \approx 0.5, \tag{11.260}$$

which fixes the coefficient α to be

$$\alpha \approx \frac{g^2}{\pi\rho_o} = \frac{g}{\pi R}. \tag{11.261}$$

The z component of the nonabelian electric field can now be found by using these forms for α and R, and it is given by

$$E_z^3 = -\partial_z A_0^3 \approx -\frac{g}{\pi R^2}K_0\left(\frac{\rho}{R}\right)\cos\left(\frac{z}{R}\right)\theta(a - z)\,\theta(z + a). \tag{11.262}$$

Lattice gauge analysis of SU(2) quantum chromodynamics produces numerical results that qualitatively match quite well with (11.262) for the radial dependence of the chromoelectric field of a dipole. Result (11.262) gives rise to a constant force of attraction between the two pointlike nonabelian charges, regardless of their separation. This follows from the force on the charge at $z = a$, which is given by

$$F_z(\rho, a) = gE_z^3(\rho, a) = -\frac{g^2}{\pi R^2}K_0\left(\frac{\rho}{R}\right)\cos\left(\frac{a}{R}\right) = -\frac{g^2}{\pi R^2}K_0\left(\frac{\rho}{R}\right), \tag{11.263}$$

where $R = \rho_o/g = a/2\pi n$ was used. Because the extended charge distribution was used, it is necessary to average the force over the cross-sectional area πR^2, which gives

$$\begin{aligned} F_z(a) &= \frac{1}{\pi R^2}\int_0^{2\pi} \mathrm{d}\phi \int_0^R \mathrm{d}\rho\, \rho F_z(\rho, a) \\ &= -\frac{2g^2}{\pi R^2}\int_0^1 \mathrm{d}x\, xK_0(x) \approx -\frac{g^2}{\pi R^2}. \end{aligned} \tag{11.264}$$

This is referred to as the *string tension*, and the mechanical work required to separate the two charges by the distance a is therefore

$$E = -F_z(a)a \approx \left(\frac{g^2}{\pi R^2}\right)a, \tag{11.265}$$

which is clearly a confining potential. Such a linear potential is commonly used to find the excitation energies of quark bound states in the quantum approach to hadronic dynamics by combining it with the Coulomb potential g/a in the interior of the flux tube.

The question remains as to why the flux tube solution to (11.212) is the physically correct choice, rather than the Coulomb solution (9.359). As mentioned earlier, the answer can be found by examining the stability of fluctuations around the Coulomb choice (9.359) versus the confining choice (11.249). These fluctuations are once again governed by the same equations (11.221) and (11.222), with the form for C_0^3 given by the respective choice of Coulombic or confining vector potential. Because both forms for C_0^3 are axially symmetric, the only nonzero magnetic fluctuation is once again φ_ϕ. In the vicinity of each charge the Coulombic dipole solution once again gives rise to a negative centrifugal barrier, and therefore it will suffer from the same instability as the single-particle case for strong coupling. The radial part of the flux tube solution,

$$C_0^3(\rho, z) = \alpha \sin(z/R)K_0(\rho/R)\theta(a - z)\theta(z + a), \tag{11.266}$$

creates a potential in (11.222) that is exponentially suppressed for radial distances $\rho > R$. As a result, $C_0^3 \to 0$ outside the flux tube and the magnetic fluctuation equation is

$$\left(\nabla^2 + \omega^2\right)\varphi_\phi \approx 0. \tag{11.267}$$

In cylindrical coordinates this has the simple and well-behaved oscillatory solutions $J_0(\omega\rho)$ and $Y_0(\omega\rho)$, where J_0 and Y_0 are zeroth-order Bessel functions of the first and second kind.

In the interior of the flux tube the Bessel function is logarithmically divergent, so that C_0^3 is given by

$$C_0^3(\rho, z) = \lim_{\rho \to 0} \alpha K_0\left(\frac{\rho}{R}\right)\sin\left(\frac{z}{R}\right) \to -\alpha \ln\left(\frac{\rho}{R}\right)\cos\left(\frac{z}{R}\right). \tag{11.268}$$

An exact solution to the fluctuation equation as $\rho \to 0$ is not available. However, it is possible to show that the solution to (11.222) is well-behaved as $\rho \to 0$ using the following simple scaling argument. For the case that φ_ϕ is a solution to (11.222) for small ρ, it satisfies

$$\left(\frac{\partial^2}{\partial\rho^2} + \frac{1}{\rho}\frac{\partial}{\partial\rho} + \frac{\partial^2}{\partial z^2} + (\omega - \alpha\cos(z/R)\ln(\rho/R))^2\right)\varphi_\phi(\rho, z) = 0. \tag{11.269}$$

The radial coordinate is scaled according to $\rho \to \lambda\rho$, which results in (11.269) becoming

$$\left(\frac{\partial^2}{\partial\rho^2} + \frac{1}{\rho}\frac{\partial}{\partial\rho} + \lambda^2\frac{\partial^2}{\partial z^2} + (\lambda\omega - \alpha\cos(z/R)\lambda\ln(\lambda\rho/R))^2\right)\varphi_\phi(\lambda\rho, z) = 0. \tag{11.270}$$

In the limit $\lambda \to 0$, the solution to (11.269) solves the equation

$$\left(\frac{\partial^2}{\partial\rho^2} + \frac{1}{\rho}\frac{\partial}{\partial\rho}\right)\varphi_\phi(\lambda\rho, z) = 0, \tag{11.271}$$

which has the simple solution

$$\lim_{\lambda \to 0} \varphi_\phi(\lambda\rho, z) = a - b \ln\left(\frac{\rho}{R}\right). \tag{11.272}$$

In the string-like limit that $R \to 0$, the flux tube solution exhibits stability, since $C_0^3 \to 0$ everywhere in the space outside the flux tube.

11.9 Classical string theory

String theory has its origin in the hadronic physics discussed in the previous section, where pairs of quarks are modeled as being confined through a string-like flux tube. As the confined pair moves, the string joining them traces out a surface. It was initially thought that the myriad of light mesons that occur in hadronic processes could be understood as excitation states of the quark pair plus the confining string. Such a string would have endpoints corresponding to the positions of the two point particle quarks, corresponding to an *open string*.

Whereas a point particle traces out a *worldline* as it moves, such an open string would trace out a surface or *worldsheet* as it moved, as depicted in figure 11.1. The action for a point particle can be viewed as the length of its worldline. This is most easily seen from the relativistic free point particle action (6.100), where

$$S = -mc^2 \int \mathrm{d}\tau = -mc \int \sqrt{\mathrm{d}x^\mu \mathrm{d}x_\mu} = -mc \int \mathrm{d}s. \tag{11.273}$$

It is a property of the particle's worldline that it is *parameterization invariant* when the metric tensor is included in its definition, so that

$$\mathrm{d}s^2 = g_{\mu\nu}\mathrm{d}x^\mu \mathrm{d}x^\nu. \tag{11.274}$$

As an extension of the particle's one-dimensional worldline, the action for the string is the parameterization invariant area of the worldsheet.

11.9.1 The string action

Developing the expression for the invariant area requires finding the metric for the worldsheet. In the case of the particle, the position of the particle was given by $X^\mu(\tau)$. For the case of the string worldsheet, the positions of the points on the worldsheet

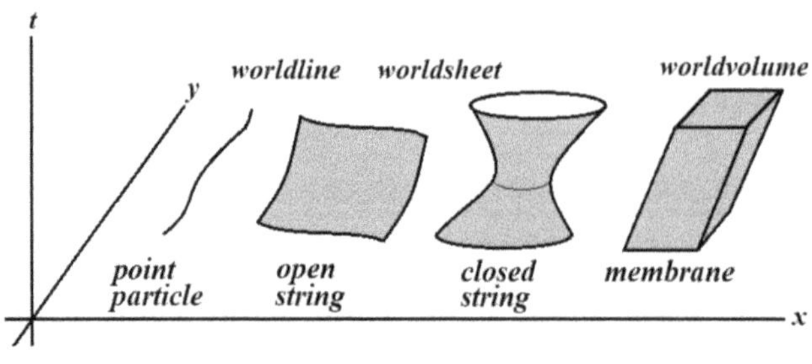

Figure 11.1. The actions for the particle, the open and closed strings, and the membrane.

are given by $X^\mu(\xi_1, \xi_2)$, where ξ_1 and ξ_2 are the two parameters required to describe a surface. Because strings and string theory are formulated in space-time dimensions higher than four, the position vector components X^μ are distinguished by capitalization and are referred to as the coordinates of the *target space*. For a point on the worldsheet, the direction vectors are given by the expression first introduced in (1.32),

$$e_a^\mu = \frac{\partial X^\mu}{\partial \xi^a}, \tag{11.275}$$

where μ gives the components of the two direction vectors in the target space. The components of the metric tensor for the worldsheet are then derived from expression (1.65), which gives

$$\gamma_{ab}(\xi_1, \xi_2) = \mathbf{e}_a \cdot \mathbf{e}_b = g_{\mu\nu} e_a^\mu e_b^\nu = \frac{\partial X^\mu}{\partial \xi^a} \frac{\partial X_\mu}{\partial \xi^b}, \tag{11.276}$$

where γ is referred to as the *induced metric*. The induced metric was previously encountered in (10.175) for the case of a sphere. Like the string worldsheet the sphere is a surface parameterized by θ and ϕ, and its induced metric (10.175) can be found from formula (11.276) by using the definition of spherical coordinates (1.42) and fixing the radius r to be constant.

> **Exercise 11.6:** Show that parametrizing the sphere using the spherical coordinates θ and ϕ in (11.276) gives the induced metric (10.175) for a sphere.

Here, $\mathbf{e}_a$ provide the basis vectors for the tangent space at each point P described by (ξ_1, ξ_2) on the worldsheet. An arbitrary vector Z in the worldsheet tangent space at P has components given by

$$Z^\mu = Z^a e_a^\mu = Z_\mu^a \frac{\partial X^\mu}{\partial \xi^a} = Z^1 \frac{\partial X^\mu}{\partial \xi^1} + Z^2 \frac{\partial X^\mu}{\partial \xi^2}, \tag{11.277}$$

so that

$$\frac{Z^\mu}{Z^1} = \frac{\partial X^\mu}{\partial \xi^1} + \lambda \frac{\partial X^\mu}{\partial \xi^2}, \tag{11.278}$$

where $\lambda = Z^2/Z^1$. This gives the relation

$$\frac{Z^\mu Z_\mu}{(Z_1)^2} = \frac{\partial X^\mu}{\partial \xi^1} \frac{\partial X_\mu}{\partial \xi^1} + 2\lambda \frac{\partial X^\mu}{\partial \xi^2} \frac{\partial X_\mu}{\partial \xi^1} + \lambda^2 \frac{\partial X^\mu}{\partial \xi^2} \frac{\partial X_\mu}{\partial \xi^2}. \tag{11.279}$$

The product $Z^\mu Z_\mu$ is real and can be positive for a space-like vector or negative for a time-like vector. The parameter λ is also real and has a range $(-\infty, \infty)$. This means that the quadratic equation (11.279) for λ must have real solutions, which are given by

$$\lambda = \left(-\frac{\partial X^\mu}{\partial \xi^2}\frac{\partial X_\mu}{\partial \xi^1} \pm \sqrt{\left(\frac{\partial X^\mu}{\partial \xi^2}\frac{\partial X_\mu}{\partial \xi^1}\right)^2 - \frac{\partial X^\mu}{\partial \xi^2}\frac{\partial X_\mu}{\partial \xi^2}\left(\frac{\partial X^\mu}{\partial \xi^1}\frac{\partial X_\mu}{\partial \xi^1} - \frac{Z^\mu Z_\mu}{(Z_1)^2}\right)}\right)\left(\frac{\partial X^\mu}{\partial \xi^2}\frac{\partial X_\mu}{\partial \xi^2}\right)^{-1}. \tag{11.280}$$

The square root must therefore have an argument that is positive. For a lightlike vector, $Z^\mu Z_\mu = 0$, this gives the inequality

$$\left(\frac{\partial X^\mu}{\partial \xi^2}\frac{\partial X_\mu}{\partial \xi^1}\right)^2 - \left(\frac{\partial X^\mu}{\partial \xi^2}\frac{\partial X_\mu}{\partial \xi^2}\right)\left(\frac{\partial X^\mu}{\partial \xi^1}\frac{\partial X_\mu}{\partial \xi^1}\right) \geqslant 0. \tag{11.281}$$

Expression (11.281) can be written in terms of the induced metric γ as

$$\gamma_{12}\gamma_{21} - \gamma_{11}\gamma_{22} = -\det\gamma \geqslant 0. \tag{11.282}$$

This shows that the two parameters can be identified as a time-like parameter τ and a space-like parameter σ, so that

$$\gamma_{11} = \frac{\partial X^\mu}{\partial \tau}\frac{\partial X_\mu}{\partial \tau}, \quad \gamma_{22} = \frac{\partial X^\mu}{\partial \sigma}\frac{\partial X_\mu}{\partial \sigma}, \quad \gamma_{12} = \gamma_{21} = \frac{\partial X^\mu}{\partial \tau}\frac{\partial X_\mu}{\partial \sigma}. \tag{11.283}$$

Now that the metric tensor of the string worldsheet is known, the reparameterization invariant area A of the open string worldsheet can be written as the two-dimensional version of the diffeomorphic volume element (10.290). This gives the open string action

$$\begin{aligned} S = \alpha A = \alpha \int \mathrm{d}A &= \alpha \int_0^{\tau_f}\int_0^{\sigma_f} \mathrm{d}\tau\, \mathrm{d}\sigma \sqrt{|\det\gamma|} \\ &= \alpha \int_0^{\tau_f}\int_0^{\sigma_f} \mathrm{d}\tau\, \mathrm{d}\sigma \sqrt{-\det\gamma}, \end{aligned} \tag{11.284}$$

where the string worldsheet is parameterized so that the open ends have the coordinates $X^\mu(\tau, 0)$ and $X^\mu(\tau, \sigma_f)$ for the value of the parameter τ. The fact that expression (11.284) gives the area can be demonstrated by considering two infinitesimal vectors $\mathrm{d}v_1$ and $\mathrm{d}v_2$ in the tangent space of the worldsheet at point P. The infinitesimal area of the parallelogram formed by two infinitesimal *Euclidean* vectors oriented along the two direction vectors is given in terms of their magnitudes $\mathrm{d}v_1$ and $\mathrm{d}v_2$ and the angle θ between them by

$$\begin{aligned} \mathrm{d}A = \mathrm{d}v_1\mathrm{d}v_2\, |\sin\theta| &= \mathrm{d}v_1\mathrm{d}v_2 \left| \sqrt{1 - \cos^2\theta} \right| \\ &= \left| \sqrt{(\mathrm{d}v_1)^2(\mathrm{d}v_2)^2 - (\mathrm{d}v_1\mathrm{d}v_2\cos\theta)^2} \right| \\ &= \left| \sqrt{(\mathrm{d}v_1 \cdot \mathrm{d}v_1)(\mathrm{d}v_2 \cdot \mathrm{d}v_2) - (\mathrm{d}v_1 \cdot \mathrm{d}v_2)^2} \right| \\ &= \mathrm{d}\xi^1\, \mathrm{d}\xi^2 \sqrt{\left| \left(\frac{\partial r}{\partial \xi^1} \cdot \frac{\partial r}{\partial \xi^1}\right)\left(\frac{\partial r}{\partial \xi^2} \cdot \frac{\partial r}{\partial \xi^2}\right) - \left(\frac{\partial r}{\partial \xi^1} \cdot \frac{\partial r}{\partial \xi^2}\right)^2 \right|}. \end{aligned} \tag{11.285}$$

Up to the minus sign generated by the Minkowski nature of space-time, (11.285) is identical to (11.284). The constant α must have units of mass per time in order for the action to have units of $\hbar$. This is given by setting $\alpha = -T/c$, where T is the string tension. Expression (11.284) is known as the *Nambu–Goto action.*

11.9.2 The string equation of motion

The variational approach treats the Nambu–Goto action as a functional of two quantities,

$$\begin{aligned} \dot{X}^\mu &= \frac{\partial X^\mu}{\partial \tau}, \\ \bar{X}^\mu &= \frac{\partial X^\mu}{\partial \sigma}, \end{aligned} \tag{11.286}$$

so that

$$\begin{aligned} S &= -\frac{T}{c}\int_0^{\tau_f}\int_0^{\sigma_f} \mathrm{d}\tau\, \mathrm{d}\sigma \sqrt{\left(\dot{X}^\mu \bar{X}_\mu\right)^2 - \left(\dot{X}^\mu \dot{X}_\mu\right)\left(\bar{X}^\mu \bar{X}_\mu\right)} \\ &= \int_0^{\tau_f}\int_0^{\sigma_f} \mathrm{d}\tau\, \mathrm{d}\sigma\, \mathcal{L}(\dot{X}, \bar{X}), \end{aligned} \tag{11.287}$$

where the Lagrangian density is given by

$$\mathcal{L}(\dot{X}, \bar{X}) = -\frac{T}{c}\sqrt{\left(\dot{X}^\mu \bar{X}_\mu\right)^2 - \left(\dot{X}^\mu \dot{X}_\mu\right)\left(\bar{X}^\mu \bar{X}_\mu\right)}. \tag{11.288}$$

The variation of the action is given by varying the string worldsheet according to

$$\begin{aligned} \dot{X}^\mu &\to \dot{X}^\mu + \delta\dot{X}^\mu = \dot{X}^\mu + \frac{\partial \delta X^\mu}{\partial \tau}, \\ \bar{X}^\mu &\to \bar{X}^\mu + \delta\bar{X}^\mu = \bar{X}^\mu + \frac{\partial \delta X^\mu}{\partial \sigma}. \end{aligned} \tag{11.289}$$

The resulting variation is very similar to that of the nonrelativistic string of chapter 3,

$$\delta S = \int_0^{\tau_f} \mathrm{d}\tau \int_0^{\sigma_f} \mathrm{d}\sigma \left(\frac{\partial \mathcal{L}(\dot{X}, \bar{X})}{\partial \dot{X}^\mu} \frac{\partial \delta X^\mu}{\partial \tau} + \frac{\partial \mathcal{L}(\dot{X}, \bar{X})}{\partial \bar{X}^\mu} \frac{\partial \delta X^\mu}{\partial \sigma} \right). \tag{11.290}$$

Dirichlet boundary conditions for the open string still specify the shape at $\tau = 0$ and $\tau = \tau_f$, which requires that the initial and final variations vanish,

$$\delta X^\mu\left(0, \sigma\right) = \delta X^\mu\left(\tau_f, \sigma\right) = 0. \tag{11.291}$$

However, in the case of the open string, the endpoints are *no longer fixed* as were those of the nonrelativistic string of chapter 3. As a result, integrating (11.290) by parts gives

$$\delta S = \int_0^{\tau_f} d\tau \left(\delta X^\mu \frac{\partial \mathcal{L}(\dot{X}, \bar{X})}{\partial \bar{X}^\mu}\right)\Bigg|_{\sigma=0}^{\sigma=\sigma_f} - \int_0^{\tau_f} d\tau \int_0^{\sigma_f} d\sigma\, \delta X^\mu \left(\frac{\partial}{\partial \tau}\frac{\partial \mathcal{L}(\dot{X}, \bar{X})}{\partial \dot{X}^\mu} + \frac{\partial}{\partial \sigma}\frac{\partial \mathcal{L}(\dot{X}, \bar{X})}{\partial \bar{X}^\mu}\right), \tag{11.292}$$

where the boundary conditions of (11.291) result in the endpoint contributions for the integration by parts of $\partial\delta X^\mu/\partial\tau$ vanishing. In order for the variation of the action to vanish, two conditions must hold:

$$\delta X^\mu(\tau, \sigma)\left(\frac{\partial}{\partial \tau}\frac{\partial \mathcal{L}(\dot{X}, \bar{X})}{\partial \dot{X}^\mu(\tau, \sigma)} + \frac{\partial}{\partial \sigma}\frac{\partial \mathcal{L}(\dot{X}, \bar{X})}{\partial \bar{X}^\mu(\tau, \sigma)}\right) = 0, \tag{11.293}$$

where τ and σ are arbitrary members of the two intervals $(0, \tau_f)$ and $(0, \sigma_f)$, and the endpoint conditions

$$\left(\delta X^\mu \frac{\partial \mathcal{L}(\dot{X}, \bar{X})}{\partial \bar{X}^\mu}\right)\Bigg|_{\sigma=0}^{\sigma=\sigma_f} = \delta X^\mu(\tau, \sigma_f)\frac{\partial \mathcal{L}(\dot{X}, \bar{X})}{\partial \bar{X}^\mu(\tau, \sigma_f)} - \delta X^\mu(\tau, 0)\frac{\partial \mathcal{L}(\dot{X}, \bar{X})}{\partial \bar{X}^\mu(\tau, 0)} = 0. \tag{11.294}$$

Since $\delta X^\mu(\tau, \sigma)$ is arbitrary in (11.293), it follows that the equation of motion is given by

$$\frac{\partial}{\partial \tau}\frac{\partial \mathcal{L}(\dot{X}, \bar{X})}{\partial \dot{X}^\mu(\tau, \sigma)} + \frac{\partial}{\partial \sigma}\frac{\partial \mathcal{L}(\dot{X}, \bar{X})}{\partial \bar{X}^\mu(\tau, \sigma)} = 0, \tag{11.295}$$

which holds for all but the endpoints of the string.

The endpoints of the string are governed by requirement (11.294), which can be implemented in one of two ways. The first is to set the variation δX^μ to zero at one or both of the ends for a given coordinate X^μ. This choice corresponds to fixing the endpoint of the string. For example, for the case of the endpoint $X^\mu(\tau, \sigma_f)$, it follows that setting $\delta X^\mu(\tau, \sigma_f) = 0$ causes the motion along X^μ to be zero at $\sigma = \sigma_f$, since

$$\delta X^\mu(\tau, \sigma_f) = 0 \implies \frac{\partial X^\mu(\tau, \sigma_f)}{\partial \tau} = 0. \tag{11.296}$$

These are referred to as the *Dirichlet conditions* for the endpoint, since they are consistent with fixing the shape of the string initially and finally. Dirichlet conditions are not possible for X^0, since $\partial X^0/\partial\tau \neq 0$. The second possible choice is to set

$$\frac{\partial \mathcal{L}(\dot{X}, \bar{X})}{\partial \dot{X}^\mu(\tau, \sigma_f)} = -\frac{T}{c}\frac{(\dot{X}\cdot\bar{X})\bar{X}_\mu - (\bar{X}\cdot\bar{X})\dot{X}_\mu}{\sqrt{(\dot{X}\cdot\bar{X})^2 - (\dot{X}\cdot\dot{X})(\bar{X}\cdot\bar{X})}}\Bigg|_{\sigma=\sigma_f} = 0, \tag{11.297}$$

which is referred to as the *free endpoint condition* since it allows $\delta X^\mu(\tau, \sigma_f)$ to be nonzero, which is required for motion in the X^μ direction at $\sigma = \sigma_f$. Free endpoint conditions must be chosen for X^0. Clearly, the same choices are available for the endpoint given by $\sigma = 0$. For the case of free endpoint conditions the open string is chosen to obey Neumann boundary conditions, which gives

$$\left.\frac{\partial X^\mu(\tau, \sigma}{\partial \sigma}\right)\Bigg|_{\sigma=0,\sigma_f} = 0, \tag{11.298}$$

which allows the endpoints to be located anywhere. This results in a choice of $2d$ endpoint boundary conditions for the open string, where d is the dimension of the target space-time.

Treating the string as an object in and of itself shows that it need not be open. A much-studied object is the *closed string*, whose worldsheet is depicted in figure 11.1. In the study of its quantum behavior it was discovered that it possesses a spin-2 excitation. This raised the hope that gravitation, both classical and quantum, could be modeled using the closed-string model in an appropriate target space compactified to four-dimensional space-time using the Kaluza–Klein mechanism. Such considerations lie far outside of the scope of this monograph. However, the closed string must obey the boundary condition

$$X^\mu\big(\tau, \sigma = 0\big) = X^\mu\big(\tau, \sigma_f\big), \tag{11.299}$$

which does not allow the free endpoint condition (11.297).

Result (11.297) demonstrates the highly nonlinear nature of the equation of motion (11.293). Solving the equations of motion is much easier when the *Polyakov action* for the string is used, which treats the induced metric γ_{ab} as a functional variable. The Polyakov action is given by

$$S = -\frac{T}{2c}\int \mathrm{d}\tau\, \mathrm{d}\sigma\, \sqrt{-\gamma}\, \gamma^{ab}\partial_a X^\mu \partial_b X_\mu, \tag{11.300}$$

where $\gamma = \det \boldsymbol{\gamma}$. Demonstrating that this form of the action is identical to the Nambu–Goto action begins with the variational result first derived in (10.298), which gives

$$\delta\gamma = \gamma\, \gamma^{ab}\delta\gamma_{ab} \implies \frac{\delta\sqrt{-\gamma}}{\delta\gamma_{ab}} = \frac{1}{2}\sqrt{-\gamma}\, \gamma^{ab}. \tag{11.301}$$

The equation of motion for γ^{ab} derived from the Polyakov action is given by

$$\frac{\delta S}{\delta\gamma_{ab}} = -\frac{T}{2c}\left(\frac{1}{2}\sqrt{-\gamma}\, \gamma^{ab}\gamma^{cd}\partial_c X^\mu \partial_d X_\mu + \sqrt{-\gamma}\, \partial^a X^\mu \partial^b X_\mu\right) = 0. \tag{11.302}$$

Solving this for γ_{ab} and using the properties of the determinant for a 2×2 matrix gives

$$\gamma_{ab} = -2\frac{\partial_a X^\mu \partial_b X_\mu}{\gamma^{cd}\partial_c X^\nu \partial_d X_\nu} \implies (\gamma^{cd}\partial_c X^\nu \partial_d X_\nu)^2 = \frac{4\det\big(\partial_a X^\mu \partial_b X_\mu\big)}{(\det \boldsymbol{\gamma})}, \tag{11.303}$$

where the determinant in the numerator uses the a, b indices. The second result gives

$$\gamma^{cd}\partial_c X^\nu \partial_d X_\nu = 2\frac{\sqrt{-\det(\partial_a X^\mu \partial_b X_\mu)}}{\sqrt{-\gamma}}. \tag{11.304}$$

Substituting (11.304) into the Polyakov action (11.300) gives the Nambu–Goto action (11.284).

Solving the equation of motion for the Polyakov action is most easily done by using its parameterization invariance to diagonalize the induced metric γ at a chosen point on the worldsheet. This is a parameterization where the time-like and space-like direction vectors are orthogonal in all tangent spaces of the string worldsheet. This is identical to the tetrad basis vectors of (10.25) developed for general relativity, only in the case of the string these correspond to zweibeins and are referred to as the *conformal gauge*. The Polyakov action becomes

$$S = -\frac{T}{c}\int \mathrm{d}\tau\, \mathrm{d}\sigma\, \frac{1}{2}\eta^{ab}\partial_a X^\mu \partial_b X_\mu, \tag{11.305}$$

where η is the two-dimensional version of the Minkowski metric. The equation of motion in the conformal gauge is given by

$$\eta^{ab}\partial_a\partial_b X^\mu(\tau, \sigma) = \left(\frac{\partial^2}{\partial\tau^2} - \frac{\partial^2}{\partial\sigma^2}\right)X^\mu(\tau, \sigma) = 0. \tag{11.306}$$

Not surprisingly, the two-dimensional d'Alembertian in natural units has appeared.

The general solution to (11.306) depends on the boundary conditions of the string. For simplicity, the dimensionless variables $\bar{\tau}$ and $\bar{\sigma}$ are defined as

$$\bar{\tau} = \frac{c\tau}{\sigma_f}, \quad \bar{\sigma} = \frac{\sigma}{\sigma_f}, \tag{11.307}$$

so that the conformal gauge equation of motion (11.306) becomes

$$\left(\frac{\partial^2}{\partial\bar{\tau}^2} - \frac{\partial^2}{\partial\bar{\sigma}^2}\right)X^\mu(\bar{\tau}, \bar{\sigma}) = 0. \tag{11.308}$$

For a closed string the solution to (11.308) can be written as a superposition of left- and right-moving waves that are periodic in $\bar{\sigma}$ over the interval [0, 1]. This gives

$$\begin{aligned} X_L^\mu(\bar{\tau}, \bar{\sigma}) &= \frac{1}{2}X_0^\mu + V_0^\mu(\bar{\tau} + \bar{\sigma}) + \frac{i}{2\pi}\sum_{n\neq 0}\frac{1}{n}\alpha_n^\mu \exp(-i2\pi n(\bar{\tau} + \bar{\sigma})), \\ X_R^\mu(\bar{\tau}, \bar{\sigma}) &= \frac{1}{2}X_0^\mu + V_0^\mu(\bar{\tau} - \bar{\sigma}) + \frac{i}{2\pi}\sum_{n\neq 0}\frac{1}{n}\tilde{\alpha}_n^\mu \exp(-i2\pi n(\bar{\tau} - \bar{\sigma})), \\ X^\mu(\bar{\tau}, \bar{\sigma}) &= X_L^\mu(\bar{\tau}, \bar{\sigma}) + X_R^\mu(\bar{\tau}, \bar{\sigma}), \end{aligned} \tag{11.309}$$

where the sum is over all n except $n = 0$. In order for the two expansions to be real, the Fourier coefficients must obey

$$(\alpha_n^\mu)^* = \alpha_{-n}^\mu, \quad (\tilde{\alpha}_n^\mu)^* = \tilde{\alpha}_{-n}^\mu, \tag{11.310}$$

while the two constants X_0^μ and V_0^μ must be real. The sum of the two expansions satisfies the closed-string boundary condition (11.299). The open string must satisfy the Neumann boundary conditions (11.298). This is consistent with the solution being given by a superposition of standing waves,

$$X^\mu(\bar{\tau}, \bar{\sigma}) = X_0^\mu + V_0^\mu \bar{\tau} + \frac{i}{2\pi} \sum_{n \neq 0} \frac{1}{n} \alpha_n^\mu \exp(-i\pi n \bar{\tau}) \cos(n\pi\bar{\sigma}), \tag{11.311}$$

which clearly satisfies (11.298). In addition, the coefficients must satisfy (11.310).

The definition of the stress–energy tensor for the string coincides with the equation of motion (11.302) for γ^{ab} obtained from the Polyakov action, so that the string stress–energy tensor must vanish. In turn, this places an infinite set of constraints on the Fourier coefficients.

Exercise 11.7: Show that for the stress–energy tensor of the closed string to vanish,

$$\sum_{n=-\infty}^{\infty} L_n \exp\left(i2\pi n(\bar{\tau} + \bar{\sigma})\right) = \sum_{n=-\infty}^{\infty} \tilde{L}_n \exp\left(i2\pi n(\bar{\tau} - \bar{\sigma})\right) = 0 \tag{11.312}$$

are required, where

$$\begin{aligned} L_n &= \frac{1}{2} \sum_{m=-\infty}^{\infty} = \alpha_{n-m}^\mu \alpha_{\mu m}, \\ \tilde{L}_n &= \frac{1}{2} \sum_{m=-\infty}^{\infty} = \tilde{\alpha}_{n-m}^\mu \tilde{\alpha}_{\mu m}, \end{aligned} \tag{11.313}$$

and $\alpha_0^\mu = \tilde{\alpha}_0^\mu = V_0^\mu$.

11.9.3 D-branes

Higher-dimensional objects are available in this approach to dynamics. For example, a two-dimensional surface that propagates is referred to as a *membrane*, and as it propagates it traces out a three-dimensional *worldvolume*, as depicted in figure 11.1. The Nambu–Goto action can be generalized to such p-dimensional objects by using the parameterization invariant worldvolume defined using the induced metric. For example, the membrane can be parameterized as

$$X^{\mu} = X^{\mu}(\tau, \sigma_1, \sigma_2), \tag{11.314}$$

where the two parameters σ_1 and σ_2 describe the space-like position $\boldsymbol{\sigma}$ on the membrane. The induced metric for the membrane is given by the same approach as that of the string,

$$\gamma_{ab} = \partial_a X^{\mu}(\tau, \boldsymbol{\sigma})\, \partial_b X_{\mu}(\tau, \boldsymbol{\sigma}). \tag{11.315}$$

The action is given by

$$S = -\alpha \int \mathrm{d}\tau\, \mathrm{d}\sigma_1\, \mathrm{d}\sigma_2\, \sqrt{-\gamma}, \tag{11.316}$$

where α is chosen to give units of $\hbar$. A simple version of a membrane is described by a plane that is orthogonal to all but two of the spatial directions. For example, in a four-dimensional target space-time, the coordinate system can be chosen so that the membrane is perpendicular to the z-axis, in effect parallel to the x–y plane.

Like the open string the open membrane is finite in extension in both spatial directions. This requires a set of boundary conditions for the equation of motion generated by the action. Choosing Dirichlet boundary conditions led to the membrane being referred to as a D2-brane. Similarly, higher-dimensional objects are referred to as Dp-branes. One of the more intriguing aspects of Dp-branes is their interaction with open strings. Polchinski showed that open strings could bind to Dp-branes by choosing free endpoint boundary conditions for the brane coordinates and Dirichlet conditions for the coordinates orthogonal to the brane. In effect, the string endpoints are free to move in the volume of the brane. A simple example of this is given by an open string bound to a D1-brane, as depicted in figure 11.2. The D1-brane consists of a line in the y direction at a specific value of x. The string endpoint is free to slide in the y direction, but it cannot move in the x direction due to the Dirichlet condition (11.296). This has led to the interesting possibility that the particles that are present in the observed four-dimensional space-time are open strings bound to a $D3$-brane in a higher-dimensional Universe.

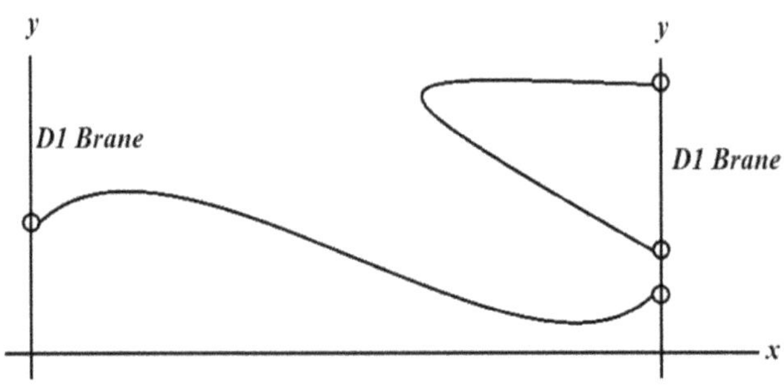

Figure 11.2. The open string attached to D1-branes.

References and recommended further reading

The application of unitary symmetry to particle physics originated with

- Heisenberg W 1932 *Z. Phys.* **77 1**

The development of local nonabelian gauge fields originated with

- Yang C N and Mills R 1954 *Phys. Rev.* **96 379**

Classical field theory texts that cover unitary symmetry include

- Kosyakov B 2007 *Introduction to the Classical Theory of Particles and Fields* (Berlin: Springer)
- Scheck F 2012 *Classical Field Theory* (Berlin: Springer)
- Nastase H 2019 *Classical Field Theory* (Cambridge: Cambridge University Press)

Most of the physical applications of unitary symmetry have been in the context of particle physics and its associated quantum field theory. As a result, the content of chapter 11 includes many concepts that are also available in quantum field theory textbooks, where they are often initially stated using a classical framework. This includes the development of nonabelian symmetry, Yang–Mills fields, spontaneously broken symmetry, and pure gauge solutions to the equations of motion. The reader interested in this perspective is referred to the following quantum field theory textbooks for useful insights into unitary symmetry in field theory and its pertinence to particle physics.

- Peskin M and Schroeder D 1995 *An Introduction to Quantum Field Theory* (Reading, MA: Addison-Wesley)
- Weinberg S 1996 *The Quantum Theory of Fields* vol II (Cambridge: Cambridge University Press)
- Srednicki M 2007 *Quantum Field Theory* (Cambridge: Cambridge University Press)
- Tully C 2011 *Elementary Particle Physics in a Nutshell* (Princeton, NJ: Princeton University Press)

Classical solutions to nonlinear field theory equations of motion are presented in detail in

- Actor A 1979 *Rev. Mod. Phys.* **51** 461
- Rajaraman R 1987 *Solitons and Instantons* (Amsterdam: North-Holland)

Nonabelian gauge theory is presented in terms of differential forms in

- Naber G 1997 *Topology, Geometry, and Gauge Fields* (Berlin: Springer)
- Frankel T 1997 *The Geometry of Physics* (Cambridge: Cambridge University Press)
- Nakahara M 2003 *Geometry, Topology, and Physics* 2nd edn (New York: Taylor and Francis)

The Gribov instability was first noted in

- Gribov V 1977 *Lecture at 12th Winter School of the Leningrad Nuclear Physics Institute* (https://http://www.slac.stanford.edu/pubs/slactrans/trans01/slac-trans-0176.pdf)
- Magg M 1978 *Phys. Lett.* **B74** 246

Classical string theory is presented in

- Zweibach B 2004 *A First Course in String Theory* (Cambridge: Cambridge University Press)
- Szabo R 2004 *An Introduction to String Theory and D-brane Dynamics* (London: Imperial College Press)
- Schomerus V 2017 *A Primer on String Theory* (Cambridge: Cambridge University Press)

www.ingramcontent.com/pod-product-compliance
Ingram Content Group UK Ltd.
Pitfield, Milton Keynes, MK11 3LW, UK
UKHW051911100926
8556IPUK00005B/843